Fungi of Temperate Europe

VOLUME 2

Fungi of Temperate Europe

VOLUME 2

Thomas Læssøe
& Jens H. Petersen

Princeton University Press
Princeton and Oxford

Contents

Volume 2

Form groups and fungal wheels

The fungi that are included in the two volumes of this publication have been organized into groups of morphologically similar fungi, which we have called 'form groups'. With some experience, these groups can be recognized with the naked eye or with a hand lens. It is important to remember that form groups do not necessarily reflect natural relationships, since morphology only reflects such relationships to a certain degree. For example, fungi with polypore morphology have evolved multiple times and, as a result, many of the taxa concerned are not closely related – even though they may look similar. The section on polypores in this book therefore covers all the species with polypore morphology, regardless of their current classification.

There is one major exception to this general rule. We have divided the fungi, irrespective of their macroscopic appearance, into the two major phylogenetic groups: the *Basidiomycota* and the *Ascomycota*. At the end of Volume 2, we have also included sections on asexual fungi, and organisms that belong in other kingdoms but resemble true fungi, *e.g.* the slime moulds.

The fungal wheels are the key feature of this publication. We have adopted a highly pragmatic approach when preparing these wheels, so that, for example, even if you find yourself looking at the wheel that covers cyphelloids (*Basidiomycota*) (p. 1076), you will find cross-references to similar discoid fungi in the *Ascomycota* (p. 1372). For this reason, prior knowledge of fungal systematics is not a prerequisite.

The wheels on this and the following page spread are organized based on how the sexual spores are produced. In order

Fruitbodies with external spore production on an exposed hymenium, see also page 31.

clavarioids
page 1092
cyphelloids
page 1076
Dacrymycetales
page 1154
jelly fungi
page1162
clavate or
coralloid
with hanging, tiny
discs or tubes
± rubbery; basidia
like tuning-forks;
spores often
septate
often gelatinous;
spores repeating
rusts and smuts
page 1198
basidiomycotes
basidia
one-celled,
'normal'
with
deviating
basidia
parasitize plants
without
fruitbodies
fruitbodies
(when present) with
external spore
production
and active
spore release
ascomycotes
discoid to clavate
apothecia;
asci with lids
operculate
discomycetes
page 1278
discoid, slit-
like to clavate
apothecia;
asci without
lids
fruitbodies with
internal spore
production,
with or without
active spore release
without
fruitbodies
thalli with
enclosed
algae and/or
cyanobacteria
inoperculate
discomycetes
page 1348
without
fruitbodies
asci on living
host tissue; on
leaves and fruits
asexual
lichens
page 1484
see next page spread
asexual
fungi
page 1626
Taphrina
page 1622

to decide which wheel to use, you will first have to determine whether your chosen fungus produces spores from an external hymenium (see p. 31 and the illustration p. 42 bottom left). The hymenium consists of a palisade of basidia or asci, and typically forms a smooth surface on the fruitbody. Fruitbodies with internal spore production do not display such a surface. Instead, you may find small openings (ostioles), through which the spores can be ejected. Form groups with external production and active dispersal of the sexual spores (meiospores) from hymenia are found on the previous page spread, and groups with internal spore production are covered on this page spread.

On this page spread, the groups with spores that are actively released are shown at the upper/left side; the groups with fruitbodies that have passive spore dispersal, or have a powdery or slimy inner spore mass, are represented in the lower/right side of the wheel. For passive spore dispersal to take place an external force is often required, *e.g.* rain drops or animal activity.

The wheels here and throughout this publication can be used in various ways. By checking the images in the outer 'rim', a quick overview can be obtained of the variation in the fungal groups covered. Images in the pale blue sections (see, *e.g.*, the wheel on page 46) represent fungi that in some way recall species from the form group covered by the wheel, and which are treated in other wheels in this publication; the relevant page number is given for ease of cross-referencing. The inner part of the wheel shows other features that could be considered, particularly microscopic characters such as spore morphology (see, *e.g.*, the wheel on page 60 and 176).

We recognize that the wheel approach to fungal identification will never reach the same level of accuracy as a traditional analytical key. The distinguishing features of some fungi are not always clear-cut, and several wheels and species accounts may need to be consulted in order to reach a satisfactory conclusion. However, we hope that the approach taken in this publication will be easier to use than trying to follow the often highly technical keys found in other identification guides.

Laboulbeniales
page 1620
externally on living insects and other arthropods
powdery mildews (Erysiphales)
page 1500
on leaves and stems; parasitic
Eurotiomycetes
page 1500
on plant litter, feathers, hair etc.
asexual fungi
page 1626
conidia dry or in slimy masses
calicioids
page 1494
very small with dusty heads; mainly on wood and bark
cleistothecia-like (chasmothecia)
cleistothecia
without fruitbodies
with active spore release
ascomycotes
fruitbodies with internal spore production with or without active spore release
without active spore (meiospore) release
basidiomycotes
mix
puffballs and the like
page1228
larger, with powdery contents
initially ± onion-shaped, then splitting star-like
earthstars and the like
page 1228
spores in small egg-like structures (peridioles)
bird's nest fungi
page 1216
foetid spore mass; insect-dispersed
stinkhorns and the like
page 1222
± subterranean; animal-dispersed
truffles
page 1256

Polypores

The polypores is a form group that is characterized by a tubular hymenophore where the spores are formed on a hymenium inside the downwards-pointing tubes. The tubes are normally inseparable from each other and from the cap flesh. The pores (tube mouths) may be spherical, angular, stretched, labyrinthine or gill-like. The number of pores per millimetre is an important character, best measured with a translucent ruler using a lens. The pore edges are sterile. Fruitbodies can be completely appresed to the substrate or have ± well-developed caps, and rarely also stems; they may be soft and annual or tough to hard, and in some cases perennial. Perennial fruitbodies may be recognized by typically having several tube layers and broad growth zones on the caps.

All polypores belong to *Basidiomycota* and have one-celled basidia. The only exception is *Aporpium* (page 1186), which belongs to the jelly fungi and has longitudinally divided basidia. Most species with tough or hard fruitbodies have a complicated hyphal structure, which, besides the normal generative hyphae, consists of branched binding hyphae and/or skeletal hyphae (page 35). Fruitbodies with two hyphal types are termed dimitic, and those with three are referred to as trimitic.

Almost all polypores are decomposers and typically occur on wood. They produce a brown rot or a white rot (page 14). A few species are ectomycorrhizal, e.g. *Albatrellus* & *Scutiger* (page 828) and *Coltricia* (page 843).

Other similar fungi:
– the boletes are always soft-fleshed, and their tubes are not well-joined and can be loosened from the cap flesh; most are mycorrhizal (page 760).
– merulioid corticioids have a ± folded hymenophore where also the edges are covered with hymenium (page 974).
– cyphelloids are small, cup-shaped *Basidiomycota*. They may occur in dense groups and mimic a polypore (page 1076).

Further reading: 17, 37, 161, 163, 193, 219, 220, 235, 282, 283, 308, 309, 339, 356, 366, 367.

perennial polypores
page 854
annual, capped and adnate polypores
page 886
fruitbodies perennial; hard or tough
polypores
form the spores in tubes on the underside of the fruitbody. Mostly wood decayers.
fruitbodies annual, with caps but possibly partly resupinate (reflexed)
fruitbodies annual, tough or soft
annual, resupinate (without caps but sometimes with small projections)
annual, completely resupinate (flat) polypores
page 932

Polypores with a stem

This form group includes somewhat boletoid fruitbodies and more tongue-shaped fruitbodies, that may have a short lateral stem. All form annual fruitbodies. In one genus, *Fistulina*, the tubes are not joined. Most species produce hyaline, smooth spores, but a few have brown, ornamented spores. Prominent brown setae are found in one genus.

The group includes both ectomycorrhizals, *e.g. Albatrellus*, *Scutiger* and *Coltricia*, decomposers on soil, *e.g. Xanthoporus*, and wood decayers that produce a white or brown rot.

OTHER SIMILAR FUNGI:
– boletes with decurrent tubes are soft-fleshed, and usually have tubes that are easy to loosen from the cap flesh. *Gyrodon* has firmly attached tubes, but these stain blue when bruised (pages 760 & 795).
– in clustered polypores the individual caps may also have a stem (page 846).
– broadly adnate polypores on top of stumps and trunks may develop a contracted base and thus recall polypores with a stem, but usually there will be more normal fruitbodies in the same fructification (pages 854 & 886).

See also the overview and the references to further reading on page 824.

Micro-drawings: spores, cystidia and setae.

Approximate species number applies to temperate Europe.

Cerioporus
2 species &
Picipes
4 species
page 836
Lentinus
4+ species
page 838
Neofavolus
1 species
page 839
Jahnoporus
1 species
page 833
Ganoderma
7 species, 3 with a stem
page 840
Podofomes
1 species
page 842
Pelloporus
3 species, 1 with a stem
page 842
Coltricia
5 species
page 843
Phaeolus
(see page 853)
Fistulina
1 species
page 845
Favolaschia
1 species
page 845
unusually formed, normally broadly adnate fruitbodies
pages 854 & 886
fasciculate polypores
page 846
mostly with 4–9 pores/mm; cap smooth or hairy
1–7 pores/mm; stem black with age; cap ± smooth
with rudimentary stem and 1–2 pores/mm; on deciduous wood
felty-hairy; from buried conifer wood
with lacquer-like cap crust that can be melted
flesh ± pale brown; with a black zone below the cap felt; on *Abies*
cap and stem brown felty; flesh with duplex structure
with concentric zones and thin felty covering
brown-felty with labyrinthine pores
with red juice; tubes in mature fruitbodies not joined; mostly on *Quercus*
small and thin-stemmed; with large pores
wood decayers, white rot
seta
flesh colour
polypores with a stem
mycorrhizal
wood decayers, brown rot
white rot
other similar fungi
soil-inhabiting decomposers

Xanthoporus syringae is a soil-inhabiting, pale yellow-fleshed annual polypore with a stem and a vivid yellow pore surface. The smooth, sometimes zonate cap surface is more dull apricot-orange. It has 3–5 ± angular pores per mm. Spores smooth, hyaline, inamyloid and measure 4–5 × 3–4 µm. Occurs on ± disturbed soil, often nitrogen enriched; a decomposer.

Could recall *Gyrodon lividus* (page 795), but in that species the pores stain markedly blue.

Widespread, mainly in the hemiboreal–boreal zones, very rare and mostly absent from the nemoral zones and the western Atlantic parts of temperate Europe; mainly May–September.

Thomas Læssøe/Emilio Pini

Scutiger pes-caprae is a dark-scaled, reddish-brown, annual polypore with an excentric stem and white, rather coarse, 0.5–2 mm large pores. The stem is rather thick and sometimes yellowish. The inamyloid, smooth spores measure 8–11 × 6–8 µm. Mycorrhizal with *Fagus* and conifers.

Polyporus tuberaster ▷▷ has larger scales, a thinner and more fibrillose cap margin and even larger pores. *Albatrellus cristatus* ▷ is a more yellow, irregularly shaped polypore. *Jahnoporus hirtus* ▷▷ has much larger spores and smaller cap scales.

Found mostly in central Europe, but reaches as far north as Denmark; very rare; August–November.

Jens Maarbjerg/Jens H. Petersen

Albatrellus is the largest genus among the mycorrhizal, fleshy polypores. The species generally have small, white pores and often ± amyloid, smooth spores.

•• ***Albatrellus ovinus*** is a large to very large, pale, mild-tasting,annual polypore with small, whitish pores, 3–5 per mm, that stain yellow when bruised. The cap cracks with age and becomes pale brown. The pale flesh stains black with Melzer's reagent. The hyphae lack clamps. The somewhat thick-walled, inamyloid, hyaline spores measure 4–5 × 3–3.5 µm. Mostly mycorrhizal with *Picea* in rich, older forests, mostly on calcareous soils.

Albatrellus confluens ▷▷ has clamps, a somewhat bitter taste, non-staining pores and more vivid cap colours. *Albatrellus subrubescens* ▷▷ is small and slender, and acquires reddish spots with age.

Fairly common in the hemiboreal–boreal zones including upland areas in central Europe, but absent or nearly so from the nemoral zones; July–October.

Albatrellus cristatus is a rather small, irregular-shaped, annual polypore with a yellow-brown cap, an excentric stem, decurrent tubes and 2–4 white pores per mm. The somewhat amyloid, smooth, hyaline spores measure 4–7 × 4–5 µm. Mycorrhizal with *Fagus* and *Quercus*, mostly in heath forests, which may be calcareous.

Xanthoporus syringae ◁ has vivid yellow pores. *Albatrellus confluens* ▷▷ is more brownish-rose and tastes bitter.

Occurs in southern Norway, southern Sweden and central Europe; July October.

Claes Ingvert

Jens H. Petersen/Jens Maarbjerg

Albatrellus confluens is a large, thick-fleshed, annual polypore with a red- brown-tinged, quickly cracking cap surface and white, non-staining pores, 3–5 per mm. The taste is bitter. The hyphae have clamps. The somewhat amyloid, hyaline, smooth spores measure 4–5 × 2.5–3.5 µm. Mycorrhizal with both *Pinus* and *Picea*, and perhaps also with deciduous trees.

Albatrellus ovinus ◁◁ lacks clamps, stains ± yellow, is completely mild-tasting and generally somewhat paler.

Widespread in more continental parts of the hemiboreal–boreal zones, occasional to locally common, rare in central Europe; July–October.

Thomas Læssøe

Albatrellus subrubescens is a small, slender, thin-fleshed, annual polypore that quickly develops red-brown spots. The pores are pale and do not stain yellow but may have red spots, 2–4 per mm. The amyloid spores measure 3.5–4.5 × 2.5–3.5 µm. Mainly mycorrhizal with *Pinus*, more rarely with *Picea*, and apparently with deciduous trees in central Europe, typically on calcareous soils.

Albatrellus citrinus ⋉ has amyloid spores and thin-fleshed, very pale fruitbodies. It occurs only with *Picea* on calcareous soils.

Has an eastern distribution in hemiboreal–boreal Scandinavia and also in central Europe; July–October.

Jan Vesterholt

Species of **Boletopsis** are soil-inhabiting, mycorrhizal polypores with ± central stems and coarsely ornamented, pale brown spores. They are related to *Thelephora* and some of the hydnoids (pages 956 & 1054).

Boletopsis grisea is a fleshy, firm and rather tough-fleshed, annual polypore with relatively pale colours. The pores are white, 3–4 per mm. The ± brown-tinged spores are large, 5–6(–6.5) × 4–5 µm, and have bifurcated warts. Mainly mycorrhizal with *Pinus* on sandy soils.

Boletopsis leucomelaena ▽ is darker on the cap and mostly, if not exclusively, associated with *Picea*.

Boletopsis perplexa ⋉, described from Scotland with *Pinus*, macroscopically mostly recalls *B. leucomelaena*.

Widespread in the boreal and to a lesser extent the hemiboreal parts of Scandinavia; July–October.

Boletopsis leucomelaena is a rather compact, soil-inhabiting, annual polypore with a central, fairly thick stem. The flesh is rather fragile. The cap is very dark and the pores are whitish, 1–3 per mm. The somewhat brown-tinged spores have bifurcated warts and measure 5–6.5 × 4–5 µm. Mainly mycorrhizal with *Picea*, but apparently also with deciduous trees.

Boletopsis grisea △ has somewhat paler colours, tougher flesh and a preference for dry *Picea* forests. It is possible that specimens that occur with deciduous trees constitute a separate species.

Mainly in the hemiboreal zone of Scandinavia, very rare farther south; July–October.

Jan Vesterholt

Thomas Læssøe

Thomas Læssøe

Thomas Læssøe

Buglossoporus quercinus is a thick- and pale-fleshed, tongue- or hoof-shaped annual polypore. The cap lacks a crust and has a yellow-brown surface. The pores are ± white and rounded, 2–4 per mm. In very fresh fruitbodies the tubes, pores and flesh become vinaceous when bruised. The flesh has a perfumed smell. The smooth, thin-walled, hyaline, inamyloid spores measure 6–8 × 2.5–3.5 µm. Produces a brown rot in veteran *Quercus*. Typically found in ancient woodlands and deer parks with many old trees.

Very young and very old specimens could be confused with *Laetiporus sulphureus* (page 848), but fresh, mature fruitbodies are unlikely to be confused with any other polypore on old *Quercus*. The fruitbodies are usually eaten very quickly by insects and slugs. *Odoria alborubescens* (page 904) has a somewhat similar smell and stain in the flesh, but is broadly attached, less firm and occurs on *Fagus*.

Widespread in the nemoral–hemiboreal zones, but very rare to rare; May–October. A typical summer species that is found predominantly July–August.

Fomitopsis betulina is a rather large, annual polypore with a very short, lateral stem, white, tough but soft flesh, and a thin, paper-like cap crust. The pores are rounded to angular, 3–5 per mm. Has generative hyphae with clamps. The smooth, hyaline spores measure 5–6 × 1.5–1.7 µm. Produces a brown rot in old, living *Betula* and continues to degrade the wood after the death of the host tree.

Buglossoporus quercinus ◁ is somewhat similar but rare, and only occurs on old *Quercus*. *Favolus pseudobetulinus* ⋉ has an eastern (continental) distribution and occurs on *Populus*. See also *Fomitopsis*, page 867.

Widespread and very common; all year, but in best condition in autumn.

The flesh may be used for mounting insects. *Fomitopsis betulina* is often seen with stromata of *Trichoderma pulvinata* (page 1529) and sometimes with *Melanospora lagenaria* (page 1589).

Jahnoporus hirtus has a felty-hairy brown cap, an excentric, rather short and thick, brown stem and whitish flesh. The pores are whitish and angular; 1–2 per mm. The spores are hyaline, smooth, fusiform and measure 12.5–17 × 4.5–5.5 µm. Occurs on soil (on roots or possibly buried wood) under conifers.

Somewhat recalls a *Scutiger* ◁◁ or *Albatrellus* ◁◁ but those genera have much smaller spores and do not have brown hairs on the cap. *Pelloporus tomentosus* ▷▷, *Podofomes trogii* ▷▷ and *Coltricia* ▷▷ all have brown flesh.

Known from southern Germany and other countries in central Europe; autumn.

Michael Wood

Thomas Læssøe

Polyporus, ***Cerioporus***, ***Picipes*** ***Lentinus*** and ***Neofavolus*** were earlier combined in the genus *Polyporus*. Molecular studies have shown that genus to be polyphyletic, including both gilled, *e.g. Lentinus*, *Panus* (pages 81 & 75), and broadly adnate polypores, *e.g. Dichomitus* and *Datronia* (pages 859 & 911).

Polyporus tuberaster is a small to rather large, fleshy, pointed-scaly, yellow-brown, annual polypore with a fimbriate cap margin and 1-2 mm large, angular, elongated pores. Has clamps. The spores are hyaline, smooth and measure 10–16 × 4.5–6 µm. Small fruitbodies may sit directly on attached branches, but larger fruitbodies typically connect to large pseudosclerotia that partly consist of soil (lower picture). Mostly in rich, deciduous forests.

Cerioporus squamosus ▷ has coarser scales, the margin is not finely fimbriate, and it does not form pseudosclerotia.

Widespread in the nemoral zones, less so in the hemiboreal zone, occasional to rather common; May–October.

The genus ***Cerioporus*** includes polypores with stems. It is closely related to the broadly attached species in *Datronia* (page 911).

• ***Cerioporus squamosus*** is a large to huge annual polypore with ± adpressed, coarse, yellow to dark brown scales, a dark, thick, excentric stem and thick white flesh. The pores are large, 1–2 mm wide, angular and elongated. Smells strongly farinaceous or of very ripe water melon. It may grow singly or in large, tiered groups. The smooth, hyaline spores measure 14–17 × 5–6 µm. Occurs in gardens, parks and deciduous forests on various deciduous hosts, *e.g. Fagus, Ulmus* and *Fraxinus*.

Polyporus tuberaster ◁ is typically smaller, has a more central stem and more fimbriate-upright scales that form a fringe at the cap margin.

Widespread and common; April–November.

The middle image depicts very young, emerging fruitbodies.

Cerioporus varius is a buff to yellow-brown, tough-fleshed, annual polypore with an excentric to central stem that gradually blackens from the base. The fruitbodies are very variable in size. The pores are pale grey and rounded, 6–9 per mm. Has clamps. The cylindrical spores measure 8.5–11 × 3–3.5 µm. Occurs on various kinds of deciduous wood; very frequent in *Salix* carrs but also common in commercial forests.

Polyporus hygrocybe ⋉ (known from the French Alps) differs by its ellipsoid spores. Species in the genus *Picipes* ▽ also acquire black stems with age but generally have more red-brown or grey-brown caps and are mostly somewhat more robust.

Widespread and very common; all year, but at its best in the autumn.

The genus ***Picipes*** includes small-pored (4–7 pores per mm), medium-sized polypores with a tendency to acquire a black stem with age. Young fruitbodies are mostly pruinose but become smoother and darker with age.

Picipes badius is a large, thin- fleshed, somewhat shiny, typically chestnut-brown, excentrically, short- and dark-stemmed, annual polypore with 4–7 pores per mm. Hyphae lack clamps. The spores measure 6–9 × 3–4µm. Occurs on the trunks of deciduous trees and on stumps in deciduous forests with plenty of dead wood.

Picipes tubaeformis ▷ is typically more funnel-shaped and smaller and occurs in very wet, large willow carrs. *Picipes melanopus* ▷ is typically more greyish and occurs on buried wood. Both species have generative hyphae with clamps.

Widespread, common to occasional; May–January.

Thomas Læssøe

Picipes melanopus is a somewhat compact, typically greyish, annual polypore that, with age, acquires a dark, rather excentric stem. Pores whitish, round, 5–7 per mm. Hyphae have clamps. Spores 7–9(–10) × 3–4 µm. Typically on buried wood from deciduous trees, mostly on calcareous soils in forests with long continuity.

Picipes badius ◁ and *P. tubaeformis* ▽ become more shiny with orange- to red-brown caps. The former lacks clamps; the latter is more funnel-shaped and normally occurs in very wet habitats.

Widespread, rare; May–December.

Picipes tubaeformis is a medium-sized, annual polypore with a deeply funnel- shaped, grey-brown to red-brown, gradually more shiny cap and black stem. Pores whitish, round, 4–7 per mm. Has clamps. The spores are hyaline, smooth and measure 7–9 × 3–3.5 µm. Typically occurs above soil level on *Salix* trunks and branches, often in very wet *Salix* carrs, rarely on other woody hosts.

May be confused with *Cerioporus varius* ◁, which never has funnel-shaped fruitbodies, and with *Picipes badius* ◁, which is not usually so distinctly funnel-shaped, lacks clamps and mostly occurs on larger trunks of deciduous trees on rich soil. Young specimens may be confused with *P. melanopus* △, which usually occurs on buried branches on ± calcareous soils in drier habitats.

Widespread, but apparently local and seemingly absent from central Europe; April–December.

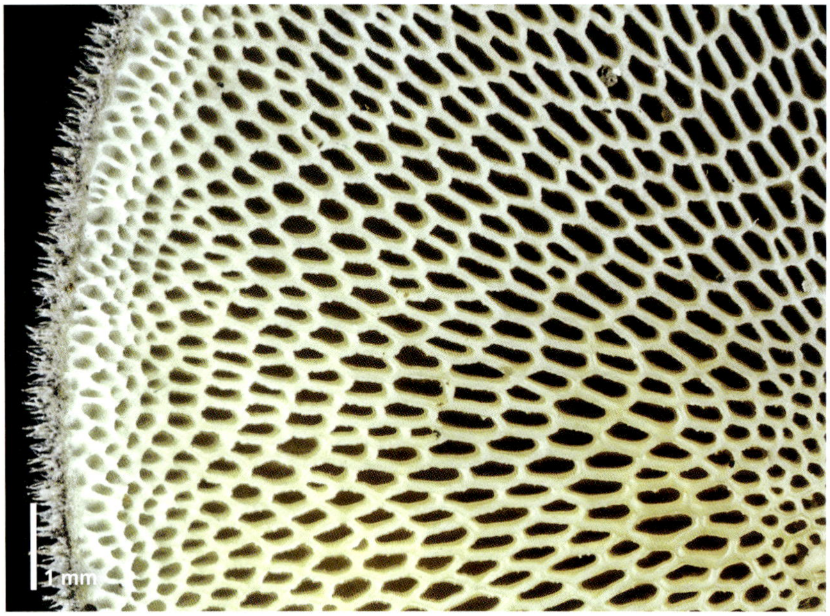

The genus ***Lentinus*** includes polypores with medium-sized to small pores (1–7 pores per mm). The fruitbodies are generally rather small and have fine hairs or small scales, and a brown or grey (not black) stem. Other species in the genus have gills (page 80).

Lentinus brumalis is a grey-brown, matt, felty, rather small, annual polypore with a central, thin, brown-scaled stem and relatively coarse, somewhat elongated pores, 2–4 per mm (easily seen without a hand lens). Has clamps. The cylindrical to somewhat allantoid, hyaline, smooth spores measure 6–7 × 2–2.5 µm. Occurs on deciduous wood, typically of small dimensions.

Lentinus substrictus ▷ looks similar but has much smaller pores that are only easily seen with a hand lens.

Widespread and very common to common; mostly seen in good condition during the winter.

Jaroslav Malý

Lentinus arcularius is a very coarse-pored, rather small polypore with a central, brown, scaly stem. Pores angular, 1–2 per mm. Has clamps. The spores measure 7–9 × 2.5–3 µm. Occurs almost exclusively on deciduous wood.

Neofavolus alveolaris ▷ is more yellow-brown and has a ± excentric, only rudimentary stem and larger spores.

Occurs in central Europe including the Netherlands and possibly farther north, uncommon; mainly October–March.

Lentinus substrictus is a greyish, rather small, finely scaly, annual polypore with a central, greyish, fine-scaled stem, a fimbriate cap margin and very fine, pale pores, 5–7 per mm. Has clamps. The hyaline, smooth, allantoid to cylindrical spores measure 5–7 × 2 µm. Produces a white rot in wood from various deciduous trees, mostly in damp forests and marshes.

Lentinus brumalis ◁ has coarser pores and is typically found during the winter.

Widespread and rather common to common; March–October.

Thomas Læssøe

Neofavolus alveolaris is a rather small, yellow-brown, radially fibrillose to appressed-scaly, annual polypore with large, hexagonal pores, 1–2 per mm. The fruitbody has a narrow attachment, but no proper stem. The spores measure 10–13(–14.5) × 4–5 µm. Occurs on branches of various deciduous trees.

Polyporus tuberaster ◁◁ is typically larger with a ± evident stem and darker scales. *Lentinus arcularius* ◁ has a more central stem, less vivid colours and smaller spores.

Known mainly from central Europe, uncommon; mainly April-September.

Irina Ukhanova

Thomas Læssøe

Thomas Læssøe

Hermann Jahn

Ganoderma is a large genus of brown-fleshed polypores with a distinct lacquered cap crust that is meltable. Some species are annual, relatively pale-fleshed, soft and have stems (this page spread); others are perennial, dark-fleshed, hard, and broadly attached (page 871). All have unique, double-walled, brown, ornamented spores, and are white-rotters.

Ganoderma lucidum is a rather large, annual, brilliant orange-red polypore with a short to very long, smooth and shiny, reddish stem. The crust, which can be melted with a flame, varies from reddish-yellow, orange-red to almost black. The flesh is rather soft and pale brown. The cap crust consists of a palisade of thick-walled, clavate cells. The brown, double-walled spores measure 7–11 × 6–8 µm. Occurs on both deciduous and coniferous wood, mostly in forests with long continuity.

Dark forms are commonly misidentified as *G. carnosum* ▽, but that species occurs only on *Abies* in central Europe, is always very dark and is umbonate where the stem meets the cap; the spore ornament should also be coarser. *Ganoderma valesiacum* ⋉ occurs on *Larix* and develops a cracking cap crust. The Asian *G. sichuanense* ⋉ is used in folk medicine as 'lingzhi' or 'reishi'.

Widespread, occasional, ± absent from the boreal zone; all year.

Ganoderma carnosum is a rather large, very dark, annual polypore with an evident, ± cylindrical stem that typically forms an umbo where it joins the cap. The crust can be melted with a flame. The flesh is pale brown. The brown spores measure 10–13 × 7–8(–8.5) µm. Occurs only on the wood of *Abies*.

Ganoderma lucidum △ is mostly paler and redder, but may also be very dark. It normally does not have an umbo where the stem joins the cap and the spore ornament should be somewhat finer.

Widespread in central Europe, uncommon; all year.

Ganoderma resinaceum is a large, soft- and pale-fleshed, annual polypore that is ± constricted at the base but, rarely, can have a proper stem. On relatively young specimens the cap is almost three-coloured: innermost dark red-brown, then orange, and ± cream at the margin. The crust can be melted with a flame. The double-walled, brown spores measure 9–11.5 4.5–7 µm (on the top image the cap surfaces are coloured by freshly deposited spores). Forms a pale central rot, usually in living *Quercus*, but also known from a number of other deciduous hosts.

Ganoderma lucidum ◁ is smaller with an evident, cylindrical stem, and usually has more orange-red to deep wine-red colours. *Ganoderma pfeifferi* (page 871) is perennial, hard and has dark red-brown flesh.

Widespread and occasional to rare in the nemoral zones, very rare in the hemiboreal zone; mostly August–January.

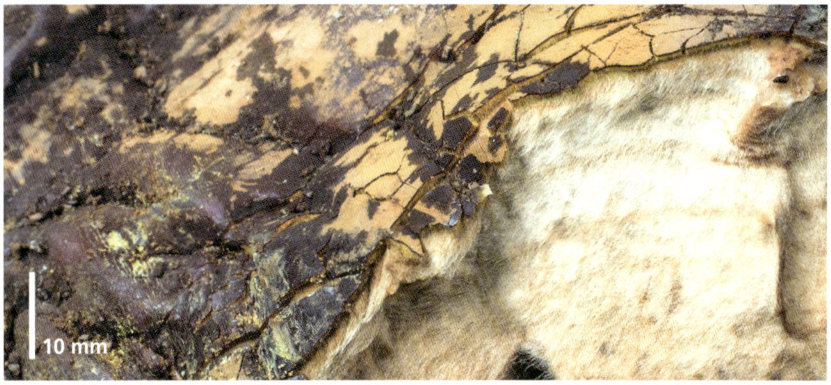

Thomas Læssøe

Jean Després

Pelloporus tomentosus is a ± centrally stemmed, annual polypore with a red-brown, felty-hairy cap with yellow-brown flesh. The cap surface is either azonate or slightly zonate. The pores are pale brown to grey-brown, reflecting and angular, 2–4 per mm. The flesh has a duplex structure with an upper, felty layer of ± erect hyphae several millimetre thick, and a lower layer with ± horizontal hyphae. Has large, red-brown setae in the hymenium. The hyaline, non-dextrinoid spores measure 5–6 × 3–4 µm. Occurs on or around conifer trunks, mostly associated with *Picea*.

Also known as *Onnia tomentosa*. May recall the larger *Phaeolus schweinitzii* (page 853) but that species has olive-tinged, ± labyrinthine pores and no setae. *Pelloporus leporinus* (page 926) and *P. triqueter* ⋉ are broadly adnate and have hooked setae. Species of *Coltricia* ▷ are thinner-fleshed and do not have setae. *Jahnoporus hirtus* ◁◁ is hairy but has whitish flesh.

Mostly in eastern hemiboreal areas, where it can be locally common; mainly August-October.

Jean-Marc Moingeon

Podofomes trogii is a dark brown, ± felty and radially wrinkled, annual polypore with an evident stem. The pores are pale grey, 2–4 per mm. The flesh is brown, ± trimitic and rather complex in structure. The thin-walled, inamyloid, non-dextrinoid spores measure 4.5–6 × 2.5–3.5 µm. Occurs on the roots and stumps of *Abies*.

Recalls *Ischnoderma* (page 925) but has an evident stem. *Pelloporus tomentosus* △ has a thicker felt on the cap surface. *Jahnoporus hirtus* ◁◁ is also more hairy and has whitish flesh.

Found in central Europe, rare; mostly autumn.

Coltricia is a genus of soil-inhabiting, brown- and thin-fleshed, tough polypores with a central stem. The flesh and pores do not have setae. At least partly mycorrhizal.

Jan Vesterholt

Coltricia perennis is a regularly shaped, mostly distinctly zonate, dull yellow-brown to red-brown polypore with 2–4 angular pores per mm. The ± cylindrical, ± thick-walled, slightly pigmented, ± dextrinoid spores measure 6–9(–10) × 3.5–5(–5.5) µm. Occurs on sandy soils, mostly in conifer plantations, but also on poor soils in deciduous forests.

Coltricia confluens ▷▷ forms more composite, irregular fruitbodies on richer soils. *Coltricia cinnamomea* ▷▷ has a more shiny cap and relatively wider spores, 6.5–8 × 5–6 µm. *Coltricia focicola* ⋉ has somewhat coarser pores, more cylindrical spores and typically occurs on burnt ground. *Coltricia montagnei* ⋉ is a thicker-fleshed species with concentrically stretched pores and larger spores.

Widespread, common to occasional; mostly July–October.

Coltricia confluens is a dull, yellow-brown, somewhat rugose-irregular shaped, almost non-zonate polypore with 2–4 pores per mm. The ellipsoid to cylindrical-ellipsoid spores measure 6–9(–10) × 3.5–5(–5.5) µm. Occurs on ± disturbed, rich soils in parks and open forests, including on burnt ground.

Coltricia perennis ◁◁ is more regularly shaped, has very obvious zones on the cap and normally occurs in poorer habitats.

Widespread in the nemoral–hemiboreal zones but rarely recorded; all year, but in best condition July–October.

Coltricia cinnamomea is a small, shiny, smooth, annual polypore with red-brown colours. Has 2–4 pores per mm. The broadly ellipsoid spores measure 6.5–8 × 5–6 µm. Occurs on soil, including on burnt ground, in deciduous forests, more rarely with conifers. Probably mycorrhizal.

Coltricia montagnei ⋉ is normally thicker-fleshed with concentrically stretched pores and has larger spores measuring 9–12(–14) × 5–6(–7) µm. *Coltricia perennis* ◁◁ has a dull cap surface and a different spore shape.

Widespread in central Europe and also recorded in Scandinavia, but not in Denmark and very rare in the Netherlands; all year.

Fistulina hepatica is a very distinctive, flesh-pink to red, juicy, ± tongue-shaped, short-stemmed annual polypore in which the yellowish to pinkish tube layer is formed of free tubes – best seen in fully mature specimens. The flesh exudes a red juice, is somewhat zonate-marbled, and has a weak smell, but a strong, sour taste. The spores are smooth, hyaline and measure 4–6 × 3–4 µm. It may form brain-like structures, upon which asexual spores are formed. Produces a brown rot in living *Quercus* trees, usually fruiting low down, more rarely higher up or on stumps. It may also occur on *Castanea* and very rarely on other deciduous trees.

A unique fungus in Europe, having a 'fruitbody' that may be interpreted as a composite of numerous, single, one-tubed fruitbodies (it is relatively closely related to *Schizophyllum* (page 62), another composite fungus, and both reside inside the *Agaricales*). Outside Europe, several other species occur.

Widespread in the nemoral–hemiboreal zones, occasional to locally common; July–November.

Favolaschia calocera is a vivid orange-yellow to reddish apricot-orange, soft-fleshed, poroid fungus with a lateral stem of variable length. The angular, orange pores are large and have a fimbriate margin. It has warty and clavate cystidia (acanthocysts), found both on the cap surface and on the pore edges. The faintly amyloid, smooth spores measure 9–12.5 × 6.5–8.5 µm. Occurs on deciduous wood.

Very distinctive; no similar species in Europe but several in the tropics.

Currently spreading from southern Europe (Spain and Italy) to parts of temperate Europe (*e.g.* south-western UK); most of the year.

Clustered polypores

This group includes species with numerous caps from a narrow, common base or from a thick, short stem. All species in the group form annual, rather soft-fleshed fruitbodies and are typically decomposing parasites on trees.

A number of these species are prized edibles, *e.g. Grifola*, *Cladomeris umbellatus* and *Laetiporus*, and less so *Meripilus* and *Fistulina*. *Phaeolus* is used for dyeing wool. *Laetiporus* is known to cause mild poisoning when cooked insufficiently.

OTHER SIMILAR FUNGI:
– polypores with stems are similar, but only have one or a few caps (page 826).
– perennial, broadly adnate polypores may form numerous, layered caps, but do not have a common stem or constricted base (pages 854 & 886).
– *Sparassis* and similar fungi lack pores (page 952).

See also the overview and the references to further reading on page 824.

Micro-drawings:
spores outwards
then cystidia.
cy. = cystidia.
cl. = clamps

Approximate
species number
applies to
temperate
Europe.

Thomas Læssøe

•• ***Laetiporus sulphureus*** is a very large, fleshy, mostly imbricate, ± orange-yellow to yellow or rarely almost white annual polypore. Pores vivid yellow, rounded to angular, 3–4 per mm. Dead fruitbodies are white and crumbly. The yellowish flesh is up to 5 cm thick and has binding hyphae (dimitic). The generative hyphae lack clamps. The spores are hyaline, smooth and measure 5–7 × 3.5–5 µm. An asexual state produces chlamydospores. Forms a brown rot, typically in rather old trees, with a preference for *Quercus*, *Prunus* and *Salix*, but also other deciduous trees and, in England, also rather common on *Taxus*. It is the main cause of hollowed *Quercus*.

Fresh, living fruitbodies may only be confused with other closely related *Laetiporus* species, *e.g. L. montanus* × described from *Picea* in central Europe. There are other closely related species on conifers in North America, and it is possible that European material may be split further.

Widespread and rather common, ± absent from the boreal zone; mainly May–September.

●● ***Cladomeris umbellata*** is a soft-fleshed annual polypore with numerous circular, ± centrally stemmed caps from a common thick, rooting stem. The pores are whitish and ± stretched, 1–3 per mm. The spores are smooth and hyaline and measure 7.5–10 × 3–4 µm. Arises from large underground pseudosclerotia (partly made of hyphae, partly of soil); typically in older *Fagus* forests, but may also occur with other deciduous trees and perhaps conifers. Produces a white rot, but also parasitizes *Armillaria* (page 232) during sclerotial growth.

Grifola frondosa ▽ is similar but has tongue-shaped caps.

Widespread in the nemoral–hemiboreal zones, but rather rare to very rare; June–October (can appear even in the middle of dry summers).

An important medicinal fungus in Asia, but cultivation is hampered by the dependence on *Armillaria*.

Michal Mikšík/Jens H. Petersen

●● ***Grifola frondosa*** is a large, fleshy, grey-brown, clustered or imbricate annual polypore with tongue-shaped caps and whitish to cream, angular and floccose pores, 2–4 per mm. The flesh is white and firm. Has skeletal hyphae and clamps. The spores are smooth and measure 6–7 × 4–4.5 µm. Produces a white rot in living deciduous trees, mostly *Quercus*, but has a wide range of hosts. It mainly fruits at the very base of trunks, rarely farther up.

Meripilus giganteus ▷▷ is larger, a more yellow-brown and has blackening flesh and pores. *Cladomeris umbellata* △ has small caps with ± central stems.

Widespread in the nemoral–hemiboreal zones, but absent from the boreal zone, rare to occasional; mainly August–December.

Grifola frondosa is an important medicinal fungus in Asia and is commercially cultivated, often under the name mai-take.

• ***Meripilus giganteus*** is a large to very large, ± clustered, pale brown, ± zonate, soft-fleshed, annual polypore. The flesh, and particularly the whitish pores, stain grey-black when bruised. Has 3–5 rounded pores per mm. The flesh is monomitic and the smooth, hyaline, inamyloid spores measure 6–7 × 4.5–6 µm. Occurs at the base of living deciduous trees, almost always *Fagus*, or around stumps. It produces a white rot.

Grifola frondosa ◁◁ does not stain black, is smaller, more grey than brown and has coarser pores. The conifer-associated *Bondarzewia mesenterica* ▷ has somewhat smaller fruitbodies with larger pores that do not blacken when bruised, and amyloid, ornamented spores.

Widespread and common in the nemoral zones, less so in the hemiboreal zone and absent from the boreal zone; mainly July–November.

Old decaying fruitbodies are often invaded by various fungi, such as *Hypomyces aurantius* (page 1544, *Melanospora lagenaria* (page 1589) and *Collybia cookei* (page 279).

Bondarzewia mesenterica is a soft-fleshed, pale brown, somewhat zonate, smooth, shelf-shaped, sometimes imbricate annual polypore. The pores are pale greyish and ± angular, 1–3 per mm. The flesh is dimitic and has skeletal hyphae. The coarsely ornamented, amyloid spores measure 6–8 × 5–7 µm. Occurs on living roots or on dead conifers, mainly, or perhaps exclusively, on *Abies*.

May recall *Meripilus giganteus* ◁ but that species blackens when bruised, has completely different spores and, normally, larger fruitbodies with finer pores.

Distributed in central Europe in areas where the host, *Abies*, occurs; apparently not known from plantations farther north; mainly July October.

Lucie Zíbarová

Jens Maarbjerg

Abortiporus biennis is an irregularly shaped, imbricate, sometimes short-stemmed, annual polypore with rather large, irregular and ± jagged, pale, red-staining pores, 1–3 per mm. Growing fruitbodies sometimes exude red drops. The spores are hyaline, smooth and measure 4–7 × 3.5 µm; usually also has chlamydospores. Occurs mainly on buried deciduous wood, mostly in man-made habitats, *e.g.* on remnants of stumps in lawns, rarely on trunks in more natural settings.

Daedaleopsis confragosa (page 910) also reddens but is tougher, far more regularly shaped and occurs on branches and trunks. *Abortiporus biennis* may recall a small version of *Grifola frondosa* ◁◁ or *Meripilus giganteus* ◁◁, but these species do not stain red (the latter stains grey to black). *Loweomyces fractipes* ⋉ is very similar, with single caps and a lateral stem, but does not redden and the pores are well-defined and smaller (4–5 per mm).

Widespread in the nemoral–hemiboreal zones, fairly common to rather rare; June–January.

Phaeolus schweinitzii is a circular or shelf-shaped, often imbricate, rust-brown, rather soft-fleshed annual polypore with a vivid yellow margin when growing actively. The pores are ± grey-green, ± labyrinthine and rather coarse, 1–2 per mm. Has large gloeocystidia and monomitic flesh. The smooth, inamyloid, hyaline spores measure 6–9 × 4.5–5 µm. Occurs on roots at the base of living conifers or on the lower part of trunks of living or dead conifers, rarely on deciduous wood (*e.g. Prunus avium* and *Sorbus*); it produces a brown rot.

Pelloporus tomentosus (page 842) has large, brown setae in the hymenium. Species of *Gloeophyllum* (page 868) are more tough-fleshed and thinner, and usually have elongated pores.

Widespread and rather common; all year, but in best condition July–October.

Perennial polypores

All polypores where the fruitbodies continue to produce new tube layers in successive years belong here. The growth periods can often be seen as raised concentric zones on the caps. The group includes most polypores with hard fruitbodies. In general, the fruitbodies have one growth period per year, but *Fomes fomentarius* has two growth periods: a strong one in the spring and a weaker one the autumn; this creates two zones of growth per year.

Some species are borderline perennial. For example, species of *Trametes*, *Cerrena* and *Antrodia* may, during a mild spring, continue to grow and thus produce two-layered fruitbodies. However, these fruitbodies typically rot away during the summer.

Most perennial polypores have a complicated hyphal system with both generative hyphae with septa, and branched binding hyphae and/or thick-walled skeletal hyphae without septa (page 35).

The species can be divided into white- and brown-rotters; although the two types may look surprisingly similar, they are not closely related.

OTHER SIMILAR FUNGI:

– similar but annual species of polypore are divided into two groups: those with caps (pileate) and those without (resupinate) (pages 886 & 932).

See also the overview and the references to further reading on page 824.

Micro-drawings: spores outwards, followed on the inside by cystidia and setae. cl. = clamps.

Approximate species number applies to temperate Europe.

aricifomes
1 species
page 866
high,
hoof-shaped;
flesh very
bitter
Fomitopsis
1 species
page 867
Rhodofomes
1 species,
page 866
pores
pink
flesh pale
yellow in
young parts
Osmoporus
1 species
page 868
with strong,
sweetish smell
Gloeophyllum
4 species
page 868
pores mostly
gill-like,
stretched
Fomes
1 species
page 870
spore-deposit white;
fruitbody
with mycelial core
Ganoderma
7 species
page 871
BROWN ROT
WHITE ROT
flesh colour
spore-deposit red-
brown; mainly on
deciduous trees
Phellinus
10+ species
page 874
mostly with hoof-
shaped, usually cracked
caps; on deciduous
trees
perennial
polypores
– pileate or
resupinate; on
wood.
- cl.
mostly with hoof-
shaped, cracked caps;
on deciduous trees or
conifers
Fomitiporia
4 species
page 879
mostly resupinate or
with sharp-edged
caps; mostly on
deciduous trees
other
similar
fungi
setae
cl.
-/l+
with caps or
resupinate;
with black lines in the
flesh; on conifers
Fuscoporia
5 species
page 880
± hairy and
with large,
± labyrinthine pores;
on conifers
with small caps;
mostly on Salix
Phellopilus
1 species
page 884
annual, resupinate polypores
page 932
with dark line in
the flesh; mostly
on Ribes or
Euonymus
Porodaedalea
3 species
page 883
Phellinopsis
1 species
page 885
Phylloporia
1 species, page 884
annual, pileate polypores
page 886

Heterobasidion annosum fruits close to ground level and has irregular, highly variable, shelf-shaped or completely resupinate, pale-fleshed perennial fruitbodies with (2–)3–4(–5) ± cream pores per mm. The cap crust is mostly smooth and cannot be melted with a flame. The flesh is dimitic with ± dextrinoid skeletal hyphae. The very finely echinulate spores may be ± amyloid and measure 5.5–6.2 × 4–5.2 µm. Produces a pinkish-buff white rot inside the base of conifers (lower image), mostly *Pinus*, and more rarely in deciduous trees. Especially common in uniform, dense conifer plantations, where it may cause extensive damage to the crop.

There are two further, very similar, European species of *Heterobasidion*: *H. abietinum* ▷ is only known from *Abies* and has 2–3(–4) pores per mm; *H. parviporum* ▷ has (3–)4–6(–7) pores per mm and, when young, the cap surface has a finely felty covering. The latter usually prefers *Picea*. There are two further species in the complex in North America. The brown-rotting *Neoantrodia serialis* (page 896) lacks a cap crust.

Widespread and common; all year.

Sphaerostilbella broomeana (page 1541) can occur on old fruitbodies.

Species of ***Heterobasidion*** are troublesome plant pathogens. They mainly attack conifers through the roots and thereafter produce a pale brown rot from below (actually a white rot, despite the colour). Ultimately, the trunks are completely hollowed. It is difficult to identify the species by morphology. The brown-rotting *Fomitopsis* ▷▷ is similar but has a cap crust that can be melted with a flame. See also *Neoantrodia* (page 896)

Heterobasidion parviporum is a perennial polypore with a finely felty cap margin and surface when young (hand lens required!) and (3–)4–6(–7) pores per mm. The cap is ± orange-brown and the pores ± cream. The finely echinulate spores measure 4.3–5.4 × 3.7–4.3 µm. Occurs mainly on conifers, particularly *Picea*, but also *Abies*, *Larix*, *etc.*; it attacks the roots and eventually kills the tree. Produces a white rot.

Very hard to recognize by morphological characters alone. *Heterobasidion annosum* ◁ should lack felt at the cap margin and on the cap surface, have fewer pores per mm and apparently occurs mainly on *Pinus*. *Heterobasidion abietinum* ▽ should be exclusively associated with *Abies*.

Apparently widespread and common; all year.

Heterobasidion abietinum is, with age, a rather thick perennial polypore with 2–3(–4) pores per mm and a ± smooth cap with a ± rounded margin. The spores measure 4.7–5.8 × 3.8–4.7 µm. Should apparently be restricted to *Abies* where it produces a white rot up the trunk via the roots.

The number of pores per mm corresponds to *H. annosum* ◁ but that species should typically be thinner and have a sharper and thinner cap margin. *Heterobasidion parviporum* △ has smaller pores.

The exact distribution is not understood, but it is probably not rare; all year.

Jerzy Opioła/Wikipedia

Pekka Helo

Haploporus has ornamented spores and fruits high up in deciduous trees.

Haploporus odorus is a ± hoof-shaped, rather pale, mostly high-fruiting, white-rotting, perennial polypore with a very strong aniseed-like smell; it has 4–5 whitish pores per mm. The aniseed smell is so strong that the fungus is typically detected at a distance. The flesh is trimitic and tough. The finely ornamented, ± dextrinoid spores measure 5–6.5 × 3–4.5 µm. Occurs almost exclusively on living *Salix caprea*, where it produces a white rot.

Trametes suaveolens (page 921) has larger pores, is more southern and typically occurs on other *Salix* species; it has smooth spores.

Boreal, rather uncommon and northern (a taiga-species); all year. In northern Scandinavia this species is used as an indicator species for damp, boreal mixed forest of high conservation value. It is included on IUCN's international Red List of Threatened Species. It has been considered a sacred fungus amongst North American indigenous tribes, probably due to the intense smell.

Jacob Heilmann-Clausen

Haploporus tuberculosus is a resupinate, but mostly irregularly shaped, fairly pale perennial polypore with rather coarse, irregular pores, 2–3 per mm. The flesh is trimitic. The large, hyaline spores have a longitudinal striation and measure 13–15(–18) × 6–8.5 µm. Typically on attached dead *Quercus* branches, rarely on other deciduous hosts; it is a white-rotter.

The very characteristic spores help if identification is in doubt. May superficially recall a species of *Antrodia* (page 896), *Aporpium* (page 1186) or *Dichomitus campestris* ▷.

Widespread and occasional in the hemiboreal zone, but rare in the nemoral zones and absent from the boreal zone; all year.

Dichomitus campestris is a resupinate to almost pileate, pale-fleshed, annual to perennial polypore that acquires a black margin along the entire base. The pores are rather coarse, 1–2(–3) per mm. The flesh is dimitic with clamps and branching, thick-walled hyphae; the generative hyphae have clamps. The large, hyaline, cylindrical spores are smooth and measure 13–19 × 4–5.5 µm. Occurs on deciduous wood, usually *Quercus* and *Corylus*, mostly on living hosts, typically on the trunks of *Corylus* or on branches in the canopy of *Quercus*; it is a white-rotter.

Dichomitus effibulatus ⋉ occurs in Atlantic parts of the UK, and probably elsewhere, on *Prunus*, *Corylus*, *Ulex*, *Rosa*, *Rubus* and *Carpinus* and is best told by the lack of clamps. *Dichomitus squalens* ⋉ is a northern, conifer-associated species. Large-spored species of *Antrodia* (page 896) do not have branching thick-walled hyphae in the flesh.

Widespread and fairly common in the hemiboreal zone, rare in the nemoral zones and to the west; all year, but in best condition in the autumn.

Thomas Læssøe

Jan Vesterholt

Cerrena unicolor is a rather thin, shelf-shaped to partly resupinate, pale fleshed and white spored, mostly perennial polypore with labyrinthine to almost spiny, ± greyish pores, 1–3 per mm, and a strigose-hairy, somewhat zonate cap surface. There is a dark line between the felty layer and the flesh. The flesh is trimitic. Has thin-walled, rounded cystidia that arise in the flesh. The hyaline, smooth spores measure 5–7 × 2.5–4 µm. Known from numerous, deciduous hosts, but *Betula* is the dominant host – mostly found in damp forest types; it is a white-rotter.

Small, first-year specimens are very similar to *Irpex lacteus* (page 1048), but that species has large cystidia with crystals. The annual polypores *Trametes hirsuta* (page 920) and *T. pubescens* (page 920) have paler, rounded to angular pores and lack the dark line in the flesh.

Widespread and common in the hemiboreal–boreal zones, becoming scarcer towards the south; all year.

Physisporinus crocatus is a rather thick- and pale-fleshed, but resupinate, annual to perennial polypore with vivid brownish to brownish-rose colours on the margin and pores. Stains dark when bruised. Has 5–7 pores per mm. The spores are ± spherical and measure 3.5–5.5 × 3.5–5 µm. Occurs on the trunks of both deciduous and coniferous trees; a white-rotter.

Also known as *Rigidoporus crocatus*. *Physisporinus undatus* ⋉ is less vivid and somewhat more gelatinous in texture; it has tramal cystidia with crystals on the tips. See also the annual species of *Physisporinus* (page 944).

Rare and eastern; all year.

Jacob Heilmann-Clausen

Rigidoporus is a genus of polypores that includes species with mainly annual, resupinate fruitbodies (page 945) as well as species with perennial, capped fruitbodies (covered here). All produce a white rot. Most have crystal-bearing cystidia, and such species were formerly placed in *Oxyporus*.

Rigidoporus populinus is a relatively pale- and thick-fleshed, mostly imbricate and confluent perennial polypore that, with age, is overgrown with mosses. Has 5–7 round, chalk-white pores per mm. The cystidia are thick-walled and crystal-bearing, and the somewhat thick-walled spores measure 3.5–4.5 × 2.5–4 µm. Occurs on deciduous trees, particularly in scrub, hedgerows and parks; common hosts include *Salix*, *Acer*, *Populus* and *Malus*.

Commonly reported as *Oxyporus populinus*. *Bjerkandera fumosa* (page 913) is thinner, annual and has a dark line between the flesh and the tube layer. *Rigidoporus obducens* is very similar, but completely resupinate, and mainly annual (page 945).

Widespread and common; all year.

Thomas Læssøe

Thomas Læssøe

Rigidoporus ulmarius is a large to extremely large, hoof-shaped to shelf-shaped, pale-fleshed perennial polypore. The surface is pale, and the margin initially rounded and pale brown to pale brownish-orange. Over time it becomes covered in algae and mosses. The pores are pale brown to pale brownish-orange, rounded to angular, 5–6 per mm. The hyaline spores are ± thick-walled, inamyloid and measure 6–8 × 5–6.5 µm. Occurs on living deciduous trees or on stumps, mostly on *Ulmus*, but also on a number of other deciduous hosts; it is a white-rotter.

Rigidoporus liniatus ⋉ is smaller, thinner and produces smaller spores, and in Europe occurs only in greenhouses, mine shafts, *etc.* *Perenniporia fraxinea* ▷ has less vividly coloured pores and dextrinoid spores that have a germ pore.

Widespread but mainly western and rare in the nemoral zones, not present in other zones; all year.

A specimen at Kew Gardens, UK, for years held the world record for the largest fungal fruitbody, measuring 150 cm by 132 cm in diameter, and having a circumference of 242 cm (lower image).

Perenniporia is a genus of perennial, resupinate or capped, white-rotting polypores with pale brown tube layers and flesh. The species have thick-walled, pale, sometimes dextrinoid, spores.

Perenniporia fraxinea is a rather large to large, shelf- to hoof-shaped, perennial polypore with dull yellowish to greyish colours and pale, dimitic, tough flesh. The margin is ± rounded and cream. The pale pores are rounded, 4–6 per mm. The hyaline spores are thick-walled, have a germ pore, are ± dextrinoid and measure 6–8 × 5–6.5 µm. Typically fruits at the base of old, living, deciduous trees, *e.g. Quercus* and *Fagus*; produces a white rot.

Perenniporia ochroleuca ⋉, which occurs in the UK, but is mostly found farther south in Europe, has smaller hoof-shaped fruitbodies and much larger spores. *Rigidoporus ulmarius* ◁ has more brightly coloured pores and inamyloid spores that lack a germ pore.

Widespread, but rare, in the nemoral zones, very rare in the hemiboreal zone; all year.

Perenniporia medulla-panis is a reflexed to resupinate, corky, whitish, perennial polypore with a dark brown cap and 4–7 pores per mm. The skeletal hyphae are ± dextrinoid. Spores ± dextrinoid, 4.5–6 × 3.5–4.5 µm, mostly truncate. Produces a white rot in a number of deciduous trees but usually *Quercus*.

The similar *P. meridionalis* ⋉ has larger spores measuring 6.3–8.4 × 4.5–6.7 µm; *P. tenuis* ⋉ is yellower and has ± non-dextrinoid spores. *Yuchengia narymica* ⋉ is annual and has amyloid skeletal hyphae.

Widespread in the nemoral–hemiboreal zones, but rather rare; all year.

Perenniporia subacida is a resupinate, ivory-white to buff, perennial polypore with a whitish margin and 4–5(–6) rounded pores per mm. It has very distinctly dextrinoid skeletal hyphae. The ± thick-walled, somewhat dextrinoid, ovoid spores measure 4.5–6 × 3.5–4.5 µm. Occurs mostly on *Picea*, very rarely on deciduous wood.

Other similar species may have amyloid or iodine-inert skeletal hyphae, or spores with a truncate end.

An eastern, hemiboreal–boreal species, occasional; all year.

Daedalea quercina is a thick-fleshed, very tough, pale brown, ± zoned perennial polypore with coarse, thick-walled, gill-like to labyrinthine pores. The fruitbodies are mostly ± confluent and may stain greyish-purple with age. The flesh is pale pinkish-buff and trimitic. The hyaline, smooth spores measure 5–6 × 2.5–3.5 µm. It produces a brown rot in *Quercus* and *Castanea*, but, farther south in Europe, also in other hosts. Found mostly on old stumps, but may also occur on thick, dead, attached branches on living trees.

The colours may recall *Daedaleopsis confragosa* (page 910) but that species is mostly annual, reddens, is thinner and has much thinner, narrower labyrinths; it is mainly found on *Salix*. *Trametes betulina* (page 919) may also look similar but is thinner, the hymenophore is even more gill-like and the caps are felty-hairy.

Widespread and common in the nemoral–hemiboreal zones; all year.

Jean-Marc Moingeon

Jean-Marc Moingeon

Laricifomes officinalis is a large hoof-shaped to almost cylindrical, perennial (often becoming very old), pale-fleshed polypore with rather pale colours and a strange, crumbly flesh that tastes extremely bitter. Has (1–)4–5 pores per mm. The spores are hyaline, smooth and measure 6–9 × 3–4 µm. Occurs on old, living *Larix*, but outside Europe also on other conifers; it produces a brown rot.

The distinctive consistency of the flesh, the taste, and occurrence on old, living *Larix* should make identification staightforward.

Widespread in Central Europe, but nowadays very rare; all year.

Laricifomes has, for millennia, been used to treat infections. It goes under the vernacular names agarikon (gharikon) and quinine conk.

Thomas Læssøe

Rhodofomes roseus is a rather small, very dark *Fomitopsis*-like perennial polypore with striking pink pores and, during active growth, a similarly coloured cap margin. The fruitbodies are distinctly hoof-shaped, and the flesh is brown with a rose tinge. Has 3–5 pores per mm. Produces a brown rot in *Picea* trunks, typically in sites with high biodiversity. May occur rarely on other conifers and even on deciduous wood, and has also been reported from construction timber within buildings

Distinctive and unlikely to be misidentified in temperate Europe.

Widespread in the boreal, continental zone in Scandinavia and parts of central Europe; all year.

Fomitopsis pinicola is a rather small to very large, hoof- or shelf-shaped, perennial polypore with greyish-black colours in the inner, oldest parts, and zones of brownish-red, orange and yellow closer to the growing margin. The surface is lacquered and melts when heated with a flame. The flesh is ± cream, yellowish to pale brown. Has a distinctly sour-aromatic odour. The rounded pores are tinged yellow, 3–5 per mm. The hyaline spores measure 6–9 × 3.5–4.5 µm. Produces a distinct, cubic, brown rot in standing and fallen trunks of both coniferous and deciduous trees.

Fomes fomentarius ▷▷ smells less strongly, has no yellowish tinges on the tubes and pores (more grey) and also has a darker flesh; it also produces a white rather than a brown rot. Species of *Heterobasidion* ◁◁ do not have meltable cap crusts and mostly have larger pores. *Ganoderma pfeifferi* ▷▷ has a meltable crust, but much browner flesh and is generally powdered by its brown spores.

Widespread and common in most of the area, but rare in the British Isles; all year.

On older specimens, two or three species of *Trichoderma* (page 1528) can be found, and also *Nemania reticulata* (page 1564).

Osmoporus & **Gloeophyllum** are brown-rotting, brown-fleshed polypores with round, oblong or almost gill-like pores. They form tough, annual to perennial fruitbodies and occur predominantly on conifer wood, often sun-exposed.

Osmoporus odoratus is an orange to inwards brownish-black, thick-fleshed, felty perennial polypore with a very strong, aromatic, orange- to aniseed-like smell and 1–2 rounded pores per mm. The flesh is tough and brown. Lacks setae and cystidia. The spores are smooth, hyaline and measure 7.5–10.5(–12) × 3–4.5 µm. Mostly found at the top of conifer stumps but may also form regular caps from the sides of dead wood; it produces a brown rot.

No similar species has ± orange colours and a strong aniseed-like smell. It was formerly accepted in the genus *Gloeophyllum* ▽ ▷.

Widespread, common in most of the hemiboreal–boreal zones, but becoming scarcer in the western part of the nemoral–hemiboreal zones; all year.

Gloeophyllum abietinum is a dark, felty perennial polypore with an almost gill-like hymenophore of very thick 'pores', 8–12 pores/gills per cm at the margin of the cap. The hyaline spores measure 9–12 × 3–4 µm. Produces a brown rot, mostly in conifers, mainly *Pinus*, but occasionally also in deciduous wood. It prefers dry, sun-exposed wood, *e.g.* driftwood on beaches.

The very similar *G. sepiarium* ▷ has more crowded pores/gills (around 15–20 per cm).

Widespread, but rather rare; all year.

Gloeophyllum sepiarium is a vivid rust-brown to blackish-brown, felty, zonate perennial polypore with a labyrinthine or gill-like hymenophore of very elongated pores, 15–20 pores/gills per cm at the margin of the cap. The hyaline spores measure 8–11(–12) × 3–5 µm. Produces a brown rot, mostly on sun-exposed wood, usually coniferous (mainly *Picea*), rarely on deciduous wood; a common cause of rot in wooden houses.

Gloeophyllum abietinum ◁ has fewer pores/gills at the margin (8–12 per cm). *Gloeophyllum trabeum* ▽ has more regular, oblong or labyrinthine pores (about 4 pores per mm at the margin) and a smoother cap surface.

Widespread and common; all year.

Gloeophyllum trabeum is a rather thick-fleshed, grey-brown, almost smooth perennial polypore with a ± labyrinthine hymenophore, 3–4 pores per mm. The hyaline spores measure 7–10 × 3–4.5 µm. Occurs on both deciduous and coniferous wood, mostly on construction timber, but in southern parts of temperate Europe also outdoors on unworked wood.

Gloeophyllum sepiarium △ is more felty-hairy, more vividly coloured and has fewer 'pores' per mm.

Widespread in the nemoral zones but mostly rare, absent from the hemiboreal–boreal zones; all year.

Jan Vesterholt

Thomas Læssøe

Fomes fomentarius is a large, hoof-shaped, hard, grey to almost black, mostly distinctly zonate perennial polypore. The cap crust mostly has a reddish-brown zone close to the finely felty margin and it is not meltable with a flame. Pores rounded, pale grey, 4–5 per mm. The flesh is yellow-brown with an aberrant, marbled area (a mycelial core) where the fruitbody started (see arrow). Individual pore layers are mostly more than 6 mm thick. Spore-deposit white (middle image). The smooth spores measure 12–20 × 4–7 µm. A white-rotter. Found on a number of deciduous trees but *Fagus* and *Betula* are the dominant hosts; occurs both on standing dying trees and fallen trunks.

Fomitopsis pinicola ◁◁ has paler flesh and yellowish tinges on the pores, a strong smell and a meltable crust. *Ganoderma* ▷ has no mycelial core and a more obvious 'lacquered' crust that is usually powdered by brown spores. *Phellinus* ▷▷ typically has a cracking surface, narrower pore layers and much smaller spores.

Widespread and common in most of the area, although less so in southern UK; all year.

Ganoderma includes both, pale-fleshed annual polypores with ± stipitate fruitbodies (page 840) and broadly adnate, dark-fleshed perennial polypores. The double-walled, brown, ornamented spores are mostly easily detectable by rubbing a wet finger on the cap surface. White rotters.

Ganoderma pfeifferi is a rather large to very large, ± hoof-shaped, hard, mostly zonate perennial polypore with an apricot-orange, resinous, finely wrinkled-veined surface that cracks and can be melted with a flame. Has 5–6 pores per mm. During winter the tube layer is covered by a yellow wax (lower image). The odour is pleasantly sweetish. Spores are double-walled, cinnamon-brown (see below the upper fruitbody) and measure 9–12 × 6–9 µm. Occurs at the base of large living trees, mostly *Fagus*, or on stumps; rarely on smaller trees, *e.g. Prunus*.

No other dark-fleshed *Ganoderma* species has a meltable cap crust. *Fomitopsis pinicola* ⊲⊲ has a meltable cap crust, but pale flesh and white spores.

Widespread and occasional in the nemoral zones, very rare in the hemiboreal zone; all year.

Ganoderma applanatum is a large, mostly rather thin, perennial polypore with a rather thin, grey-brown, non-meltable crust that, when fresh, can be cracked with a nail. The brown flesh and tube layers often gradually become filled with white hyphae, and there is usually a thin flesh layer between the individual, more than 6 mm thick pore layers. Pores white but dark at touch (inset image), 5–6 pores per mm. The brown spores that powder the cap measure 6–8.5 × 5–6 µm. Occurs on deciduous wood, mostly stumps and fallen trunks, rarely on conifers.

Ganoderma adspersum ▷ has a 1–2 mm thick cap crust that cannot be cracked with a nail, and lacks flesh layers between the tube layers. *Ganoderma pfeifferi* ◁◁ has a resinous, meltable cap crust and often yellowish tinges on the pores. *Fomes fomentarius* ◁◁ and *Fomitopsis pinicola* ◁◁ have white spore-deposits.

Widespread and common, less so in the boreal zone; all year.

The pore layer often has chimney-like galls caused by the fly *Agathomyia wankowiczii* (inset image), which is specific to *G. applanatum*.

Ganoderma adspersum is a large, perennial polypore with a 1–2 mm thick crust that cannot be cracked with a nail. The pore layers are more than 6 mm thick. The flesh lacks distinct, white, secondary hyphal growth and sterile layers between the pore layers. Pores white but dark at touch, 5–6 pores per mm. This species never has galls on the pore side. The double-walled, brown spores that powder the cap measure 8.5–10 (–12) × 5–7.5 µm. Typically fruits low down on older, living deciduous trees, mostly sun-baked, *e.g.* in avenues or in park-like forests; *Tilia*, *Fagus*, *Quercus* and *Aesculus* are common hosts.

Also known as *G. australe* and *G. europaeum*. *Ganoderma applanatum* ◁ has a cap crust that can be cracked with a nail and usually has thin flesh layers between the tube layers. Species of *Phellinus* ▷▷ are white-spored, mostly have caps with deep cracks and the tube layers are thinner than 6 mm. *Fomes fomentarius* ◁◁ is also white-spored and has a mycelium core at the base.

Widespread in the nemoral zones, common to rather rare, very rare in the hemiboreal zone; all year.

Thomas Læssøe

Thomas Læssøe

Thomas Læssøe

The genus ***Phellinus*** includes perennial, brown-fleshed, white-rotting polypores with pale spores and brown setae. The caps are ± hoof-shaped, dark and usually have deep cracks. In comparison to similar, brown-fleshed genera, such as *Fomes* ◁◁ and *Ganoderma* ◁◁, the annual tube layers are narrow, less than 6 mm.

Phellinus igniarius is a large, hoof-shaped, rather dark, ± cracking, perennial polypore with wide concentric zones. Has 4–6 pores per mm. Has setae in the hymenium that measure 14–17 × 4–6 µm, and thick-walled, hyaline, non-dextrinoid spores measuring 5–6.5 × 4.5–6 µm. Occurs mainly on living *Salix* but also found on other hosts (proven via DNA analysis).

Phellinus alni ▷ is a sister species with a partly different host range; it probably does not occur on *Salix*. *Phellinus populicola* ▷▷ and *P. tremulae* ▷▷ both occur on *Populus*, including *P. tremula*. *Phellinus nigricans* ▷ occurs on *Betula* and has sharp-edged caps with crowded zones and particularly deep cracks.

Widespread, and rather common; all year.

Phellinus alni is a hoof-shaped, dark, perennial polypore. The spores measure 5–6.5 × 4.5–6 µm. Occurs on various deciduous trees, but probably not on *Salix* or *Populus*.

Often treated as a variety or a synonym of *P. igniarius* ◁, since no clear morphological characters seem to separate species in the *P. igniarius* complex. It has, at least partly, a different range of hosts compared to *P. igniarius* but can probably only be identified with certainty via DNA-sequencing.

Widespread, common to occasional, very rare in the nemoral zones; all year.

Phellinus nigricans is a rather large, hoof-shaped, but very sharp-edged, dark and deeply cracked-furrowed perennial polypore. Has 5–6 pores per mm and rather short hymenial setae. The hyaline, thick-walled, non-dextrinoid spores measure 5–6.5 × 4.5–6 µm. Known only from ± living or recently dead, mostly standing, *Betula* trunks.

Phellinus lundellii ▷▷, which also prefers *Betula*, has a more rounded cap margin and is often reflexed; it is almost smooth on the cap surface.

Widespread and common in the boreal zone, becoming scarcer towards the south and absent from the nemoral zones; all year.

Phellinus populicola is a large to very large, grey-black, flattened, hoof-shaped perennial polypore with a broad, rounded margin. Has 4–6 pores per mm and short setae in the hymenium. The hyaline, thick-walled spores measure 5–6 × 4–5 µm. Occurs on the trunks of large, mostly standing and living *Populus* species, including *P. tremula*. The fungus and the host tree may coexist for decades, sometimes resulting in the host partly overgrowing the fungus fruitbodies.

Phellinus tremulae ▷ is more steeply hoof-shaped and sharp-edged, and produces smaller fruitbodies that are usually situated either under a branch, or where a branch used to be.

Widespread, common to occasional, becoming scarcer in the nemoral zones and very rare in the UK; all year.

Phellinus tremulae is a rather small, steeply hoof-shaped and somewhat sharp-edged, deeply furrowed perennial polypore. It has short hymenial setae. The thick-walled, hyaline spores measure 4.5–5 × 4–4.5 µm. Occurs on *Populus tremula*, mostly just below a side branch.

Phellinus populicola ◁ forms larger fruitbodies with a more rounded margin.

Widespread and common in the hemiboreal–boreal zones, rare in the nemoral zones, very rare in the UK; all year.

Thomas Læssøe

Phellinus lundellii is a rather smooth, medium-sized, greyish, ± reflexed, but thick-fleshed perennial polypore. It has 5–6 pores per mm and short hymenial setae. The thick-walled, hyaline spores measure 4.5–6 × 4–5 µm. Occurs mostly on standing or fallen *Betula* trunks, but also known from other deciduous trees, including *Alnus*.

Phellinus nigricans ◁◁ occurs on *Betula* but is sharp-edged, hoof-shaped, very deeply furrowed and zonate; *P. alni* ◁◁ is distinctly hoof-shaped with a rounded margin.

Widespread and rather common in the hemiboreal–boreal zones, very rare in the nemoral zones; all year.

Phellinus laevigatus is a resupinate greyish, perennial polypore with 8–10 pores per mm. It typically produces very long fruitbodies. The hymenial setae measure 13–20(–30) × 6–9 µm. The spore-deposit is white (seen in image). The spores are hyaline and measure 4–5 × 3–4 µm. Almost exclusively associated with fallen, sun-baked *Betula* trunks.

Phellinus lundellii ◁◁ and *P. nigricans* ◁◁ also occur on *Betula* but form caps and have only 5–6 pores per mm.

Widespread and fairly common in the hemiboreal–boreal zones, rare to absent from many areas in the nemoral zones; all year.

Phellinus pomaceus is a medium-sized, rather pale, hoof-shaped to reflexed, perennial polypore that rarely shows significant cracking on the cap. The margin is typically rather rounded. Has 4–6 pores per mm and short hymenial setae. The thick-walled, hyaline, non-dextrinoid spores measure 4–5 × 3–4.5 µm. Almost exclusively associated with species of *Prunus* (e.g. *P. spinosa*, *P. domesticus* and *P. cerasifera*) but is also known from other deciduous trees, particularly *Crataegus*. The fungus attacks mostly old and weakened trees, where it causes a white rot.

Phellinus lundellii ◁◁ looks rather similar but, like other capped *Phellinus* species, occurs on different hosts.

Widespread and very common in the nemoral zones, but becoming scarcer towards the north and absent from the boreal zone; all year.

Thomas Læssøe

Fomitiporia forms resupinate or rounded, hoof-shaped, ± pale grey fruitbodies on deciduous and coniferous trees. It generally has no, or at most only a few, hymenial setae but, in terms of other characters, resembles *Phellinus* ◁◁.

Fomitiporia robusta is a large, ± hoof-shaped, perennial, coarsely cracking polypore with 5–6 pores per mm and shiny, yellow-brown flesh. Has no, or only a few, hymenial setae. The spores are dextrinoid, hyaline, thick-walled and measure 6–8.5 × 5.5–7 µm. Occurs on living *Quercus*.

Other species of capped *Phellinus* and *Fomitoporia* very rarely occur on *Quercus*. *Fomes fomentarius* ◁◁ does not crack, has a mycelial core and completely different spores.

Widespread in the nemoral–hemiboreal zones, occasional to very rare; all year.

Fomitiporia hartigii is a large, hoof-shaped to reflexed perennial polypore that is finely felty close to the margin. Has 5–7 pores per mm. Lacks hymenial setae. The dextrinoid spores measure 6–8 × 5–7 µm. Associated with conifer wood, mostly on *Abies* trunks.

The host and dextrinoid spores should make identification straightforward.

Widespread within the natural distribution of *Abies* in southern Central Europe, occasional; all year.

Zygmunt Augustowski

Fomitiporia punctata is a perennial, resupinate, with age several centimetre thick, dull-coloured polypore with small pores, 5–7 per mm. Lacks hymenial setae. The strongly dextrinoid spores measure 6.5–8.5 × 5.5–7 µm. Known from a large number of deciduous hosts, but most commonly found on *Salix* and *Corylus*.

Fuscoporia ferrea ▽ may also become thick, but is more reddish in colour, has hymenial setae and the cylindrical spores are non-dextrinoid.

Widespread and fairly common in the hemiboreal zone, rarer in the nemoral and boreal zones; all year.

Jens Maarbjeerg

The genus ***Fuscoporia*** forms fruitbodies that are resupinate or reflexed with rather thin caps and predominantly rust-brown colours. Has setae in the hymenium. Most species occur on deciduous wood.

Fuscoporia ferrea is a ± resupinate, ± rust-brown perennial polypore that sometimes has small false caps, with 5–7 pores per mm. Has hymenial setae, but lacks setae in the woolly margin (a × 20 hand lens and good light required). The hyaline, non-dextrinoid spores measure 5–7.5 × 2–2.5 µm. Occurs on various deciduous trees and rarely on conifers; very frequent on the trunks and attached branches of *Quercus*.

Fuscoporia ferruginosa ▷ has dark brown, pointed setae in the margin. *Fomitiporia punctata* △, which prefers *Salix* and *Corylus*, has no setae in the margin or hymenium, and has ± spherical, dextrinoid spores.

Widespread and very common in the nemoral and western hemiboreal zones, rare or absent from parts of the hemiboreal zone and from the boreal zone; all year.

Fuscoporia ferruginosa is a resupinate, rust-brown to grey-brown perennial polypore with (3–)4–6 pores per mm. Has prominent rust-coloured setae in the woolly margin of the fruitbody (a × 20 hand lens and good light required – see the lower picture). The spores are ± cylindrical, hyaline, non-dextrinoid and measure 5–7 × 3–3.5 µm. Occurs on various types of deciduous wood, but particularly *Fagus* and *Corylus*.

Fuscoporia ferrea ◁ has no setae in the margin of the fruitbody.

Widespread and common to rather common in the nemoral and hemiboreal zones; all year.

Fuscoporia contigua is a resupinate, ± red-brown to grey-brown perennial polypore with rather coarse pores, typically 1–4 per mm. It has very long setae in the flesh and hymenium that measure 40–120 µm. The hyaline, non-dextrinoid spores measure 5–7 × 3–3.5 µm. Occurs on a range of deciduous wood, but particularly *Hippophaë*; also known from construction timber in houses.

Fuscoporea ferrea ◁ and *F. ferruginosa* △ have finer pores.

Widespread in the nemoral–hemiboreal zones, occasional to rare; all year.

Thomas Læssøe/Jens H. Petersen

Thomas Læssøe

Fuscoporia torulosa is a rather large, capped, somewhat felty perennial polypore, with 4–6 pores per mm. Has rather few, somewhat projecting, straight, large setae that measure 20–50 × 6–11 µm. The hyaline, non-dextrinoid spores measure 4–6 × 3–4 µm. Occurs on a wide range of deciduous and coniferous hosts; it fruits mostly at soil level on living trees and is therefore often partly covered by mosses and lichens.

Similar species of *Fuscoporia* have a smooth cap surface or hooked hymenial setae.

Very rare in the northern part of the nemoral zones and absent from Scandinavia, but increasing towards the south – a common species in the Mediterranean; all year.

Jan Vesterholt

Fuscoporia viticola is a capped to completely resupinate, rust-brown perennial polypore with 3–5 pores per mm. It has hymenial setae up to 75 µm long. The cylindrical, non-dextrinoid spores measure 7–9 × 1.5–2 µm. Almost exclusively associated with *Pinus* and *Picea*.

The long, narrow spores are a good character for separating this species from similar resupinate *Fuscoporia*.

Widespread and common in continental parts of the hemiboreal–boreal zones and in parts of central Europe, but absent from the UK, the Netherlands, Denmark, *etc*; all year.

The genus **Porodaedalea** forms medium-sized fruitbodies with ± hairy-felty caps and large, with age, labyrinthine pores. Has setae.

Porodaedalea chrysoloma is a medium-sized, sharp-edged perennial polypore with large, rounded to labyrinthine pores, 1–3 per mm. The fruitbodies are ± reflexed and the cap surface is deeply felty. Has a dark line where the felt and flesh meet. The setae measure 30–60 × 7–11 µm. The spores are hyaline to brownish-yellow, non-dextrinoid and measure 4–5 × 3–4.5 µm; Qav. 1.1–1.2. Associated with conifer wood, mostly *Picea*, *Abies* and *Larix*.

Porodaedalea pini ▽ occurs almost exclusively on *Pinus* and has slightly more ellipsoid spores (Qav. 1.2).

A rather continental species in the hemiboreal–boreal zones, absent or almost so from the nemoral zones, including the UK; all year.

Porodaedalea pini is a hoof-shaped, sharp-edged, perennial polypore with large, rounded, stretched to somewhat labyrinthine pores, 1–3 per mm. The cap margin is hairy and delimited towards the flesh by a dark line. The setae measure 50–65 × 9–12 µm. The hyaline to yellowish, non-dextrinoid spores measure 4.5–6(–6.5) × 3.5–5 µm; Qav. 1.2. Associated with living, older *Pinus* trees.

Porodaedalea chrysoloma △ is rather similar but does not occur on *Pinus*.

Widespread and common in the hemiboreal–boreal zones, but rare to very rare in the nemoral zones; all year.

Michael Krikorev/Lucie Zíbarová

Phellopilus nigrolimitatus is a resupinate to reflexed perennial polypore with a finely felty to smooth cap, and 5–6 pores per mm. The yellow-brown flesh has thin black lines. Has numerous dark setae. Spores are carrot-shaped, hyaline, non-dextrinoid, 7–10 × 2–2.5 µm. Produces a poroid rot in fallen *Picea* and *Pinus* trunks (right image), mostly in forests with long continuity, but also found in buildings.

The distinctive spores and dark lines in the flesh are the best characters for separating this species from other similar polypores. Species of *Porodaedalea* ⊲⊲ also have dark lines, but have larger, labyrinthine pores.

Widespread in the hemiboreal–boreal zones, farther south only in houses, fairly common; all year.

Lucie Zíbarová

Phylloporia ribis is a rather small, capped perennial polypore that generally forms composite fruitbodies surrounding the substrate. The cap has a dense felt above that is separated by a dark line from the flesh below (a duplex structure). Lacks setae. The spores are yellowish, somewhat thick-walled and measure 3–4.5 × 2.5–3 µm. Fruits at soil level, usually on old *Ribes*, but also on *Euonymus* and numerous other hosts, mostly in dry, hot places such as south-facing coastal slopes.

Phellinopsis conchata ⊳ has setae and no separating line between the cap felt and the main flesh.

Widespread in the nemoral–hemiboreal zones, occasional to rare, possibly in decline; all year.

Phellinopsis conchata is a rather thin, capped, perennial polypore that typically forms many-layered caps with a rough, non-hairy surface that is generally covered by mosses and algae. The pores are cinnamon-brown and very fine, 5–7 per mm. The hymenium has numerous, ± straight to deformed setae measuring 20–50 × 7–9 µm. The hyaline, non-dextrinoid spores measure 5–6.5 × 4–4.5 µm. Most commonly occurs on *Salix* in calcareous sites, but is known from a number of other deciduous hosts and from different soils.

Phylloporia ribis ◁ has a dark line between the upper felt and the flesh (a duplex structure), lacks setae and occurs typically on other hosts, such as *Ribes* and *Euonymus*.

Widespread and rather common, less so towards the west; all year.

The smallest beetle in Europe, *Baranowskiella ehnstromi,* lives in the pores of *Phellinopsis conchata*.

Annual, capped polypores

This group includes the reflexed and capped polypores that have fruitbodies which rot away after one season. The species form either soft or tough, but never hard, fruitbodies. All capped and stemless species with soft fruitbodies belong here.

Continued on next page spread . . .

Micro-drawings: spores on the outside followed by cystidia.
cl. = clamps.

Approximate species number applies to temperate Europe.

Tyromyces & *Loweomyces*
6 species
page 908
Datronia
2 species
page 911
Abortiporus
(see page 852)
Daedaleopsis
2 species
page 910
Bjerkandera
2 species
page 912
Trichaptum
4 species
page 914
Cerrena
(see page 860)
Dichomitus
(see page 859)
Funalia
2 species
page 916
Frantisekia
1 species
page 917
Trametopsis
1 species
page 917
Trametes
9 species
page 918
medium-wide; mostly on deciduous wood
narrow; with dark back side; on deciduous wood
medium-wide; pores labyrinthine and reddening; on deciduous wood
wide; pores oblong and reddening; on deciduous wood
medium-wide; with grey pores or brown line; on deciduous wood
narrow; with violet colours and irregular pores
medium-wide; with pale grey, labyrinthine pores
narrow; with darker margin; mostly high on *Quercus* or *Corylus*
medium-wide; very felty and with large pores
medium-wide; with yellowish colours and small pores
medium-wide; with large pores that split into teeth
narrow to wide; upper side densely zonate, mostly hairy; pores small to large, rounded to stretched
flesh white, cream, pale pinkish-buff or pale blue
annual, capped polypores
with broadly adnate, soft- to tough-fleshed caps; on wood
similar groups
capped corticioids
page 962
capped, perennial polypores
page 854
flesh orange to dark brown
see next page spread

Annual, capped polypores

continued from previous page spread . . .

Some annual polypores may consist solely of generative hyphae and are termed monomitic; such fruitbodies are typically soft and easy to break. When either skeletal hyphae or binding hyphae are also present the fruitbodies are termed dimitic; if all three hyphal types are present they are termed trimitic. Dimitic and trimitic annual fruitbodies are typically tough and leathery. (See page 35 for more on hyphal systems)

Other similar fungi:

– capped, broadly adnate, perennial polypores over time form several layers of tubes. They also often have raised zones on the surface (page 854).

– clustered polypores are annual and form numerous caps from a common stem (page 846).

– capped corticioids do not have pores; a hand lens is required (page 962).

See also the overview and the references to further reading on page 824.

Micro-drawings: spores outermost, followed by cystidia or setae.
cl. = clamps.

Approximate species number applies to temperate Europe.

annual, capped polypores
with broadly adnate, soft- to tough-fleshed caps; on wood
flesh dark brown
Daedaleopsis
2 species
page 910
pores oblong and reddening; on deciduous wood
Hapalopilus
2 species
page 924
base stains violet; on deciduous wood
Ischnoderma
2 species
page 925
with dark crust below the cap felt; pores ± light-reflecting; on deciduous and conifer wood
Pelloporus
3 species
page 926
flesh with duplex structure; pores ± light-reflecting; on conifer wood
Mensularia
3 species
page 926
pores ± light-reflecting; caps rather narrow, ± smooth; on deciduous wood
Inonotus
5 species
page 928
pores ± light-reflecting; caps wide, deeply tomentose; on deciduous wood
pores ± light-reflecting; caps very large, ± smooth, flesh without mycelial core; on Quercus
Pseudoinonotus
1 species
page 930
pores ± light-reflecting; flesh with mycelial core; on deciduous wood
Inocutis
2 species
page 931
pores ± light-reflecting, labyrinthine
Phaeolus
(see page 853)
not light-reflecting; pores gill-like
Gloeophyllum
(see page 869)
setae
- cl.
- cl.
- cl.
- cl.
- cl.
- cl.
WHITE ROT
BROWN ROT
similar fungi
capped corticioids
page 962
capped, perennial polypores
page 854

Pekka Helo

Amylocystis lapponicus is an juicy, somewhat reddening or browning, with age, brown and felty annual polypore. The pores are angular, and, with age, somewhat labyrinthine, 2–3 per mm. The flesh is monomitic and has thick-walled, ± amyloid hyphae. The cystidia are amyloid and thick-walled, either with or without crystals at the top. The inamyloid, smooth, thin-walled spores measure 8–11 × 2.5–3.5 µm. A brown-rotter, typically on *Picea abies* trunks in forests of high conservation value, rarely on other conifers. It is used as an indicator species for boreal conifer forest with long continuity.

The browning *Postia fragilis* ▷▷ is somewhat similar, but old caps are not as felty and it does not have amyloid cystidia.

Occasional in mainly eastern parts of the boreal zone and to a much lesser extent the hemiboreal zone, but also present as a rarity in the uplands of central Europe; mostly in the autumn.

Jan Vesterholt

Lucie Zíbarová

Leptoporus mollis is a capped, strikingly pink to wine-red, soft-fleshed annual polypore with paler, ± rounded pores, 2–3 per mm. Lacks cystidia and clamps. The spores measure 5–6 × 1.5–2 µm. Occurs on *Picea* and *Pinus*; a brown-rotter.

The unclamped hyphae combined with the vivid colour make this species easy to recognize. *Rhodonia placenta* (page 936) has ± similar colours but is resupinate and clamped. *Rhodofomes roseus* (page 866) is woody, hard and perennial.

Widespread and occasional in the hemiboreal–boreal zones, ± absent from the nemoral zones and rare in central European uplands; autumn.

The genera ***Amaropostia, Calcipostia, Fuscopostia, Oligoporus*, *Postia*** & ***Spongioporus*** constitutes a complex of closely related, brown-rotting, soft-fleshed (monomitic) resupinate (page 936) to capped polypores that tend to have very small, hyaline spores.

Amaropostia stiptica is a ± hoof-shaped, whitish, soft, very bitter-tasting annual polypore. The cap has a rugose-felty surface. Has 4–6 ± rounded, white pores per mm. When fresh usually exudes whitish drops from the pores. Lacks cystidia but has clamps. The spores are ± cylindrical and measure 3.5–5 × 1.5–2 µm. A brown-rotter in conifer wood, mostly *Picea*, but may also occur on deciduous wood.

Calcipostia guttulata ▷▷, also typically a *Picea*-associated fungus, is only slightly bitter-tasting, has greenish and pink tinges and usually a very strong guttulation.

Widespread and very common, but becoming scarcer towards the north; mainly July–November.

Postia wakefieldiae is a thin- and soft-fleshed annual polypore, mostly growing imbricate or as a rosette, at first white, later tinged red-brown. Has 4–5 rounded to angular, white pores per mm. The spores measure 3.5–4.5 × 2–2.5 µm. A brown-rotter of coniferous and deciduous wood.

Often treated as *Oligoporus wakefieldiae*. *Spongiporus floriformis* ⋉ has almost pure white fruitbodies and perhaps somewhat finer pores; *S. balsameus* ⋉ has somewhat larger spores and thick-walled cystidia; and *Fuscopostia leucomallella* ▷▷ has somewhat larger spores and gloeocystidia.

Only nemoral; mainly August–November.

Fuscopostia fragilis is a ± hoof-shaped, solitary or imbricate, smooth to somewhat felty annual polypore that stains red-brown when bruised and with age. The pores are white, 2–4 per mm. Lacks cystidia. The cylindrical to allantoid spores measure 4–6 × 1.7–2.1 µm. A brown-rotter on conifer wood, mostly *Picea*.

Fuscopostia lateritia ⋉ prefers *Pinus* and has narrower spores. *Amylocystis lapponicus* ◁◁ has thick-walled, amyloid cystidia.

Widespread, common to rather rare; mainly August–November.

Calcipostia guttulata is a rather large, mostly shelf-shaped annual polypore that, when fresh, has pinkish tinges. The pores are whitish, ± green tinged and rounded, 4–6 per mm. The taste slowly becomes bitter in the back of the mouth. Lacks cystidia. The spores measure 3.5–5 × 2–2.5 µm. A brown-rotter on conifer wood, mostly *Picea*; very rarely on deciduous wood.

Amaropostia stiptica ◁◁ is more pure white and is immediately distinctly bitter-tasting. Other species in the group are either smaller, have different colours or are completely mild-tasting.

Widespread and expanding in the nemoral zones, *e.g.* in the Netherlands and Denmark, but very rare in the UK, and occasional in the hemiboreal–boreal zones; mainly July–November.

In southern parts of Sweden, *Calcipostia guttulata* is considered to be a useful indicator species for conifer woodlands of high conservation value. However, in Denmark it occurs in plantations with stumps and has extended its range across most of the country over a short time period of time.

Oligoporus ptychogaster is an annual polypore that typically occurs in an asexual, yellow-brown, cushion-shaped state, which dissolves into brown conidia that measure approx. 7.5 × 4 µm. Occasionally the poroid sexual state is present at the base of the cushions (see arrow). This has rounded, whitish pores, measuring 2–4 per mm and lacks cystidia. The spores measure 4.5–5.5 × 2–3 µm. Occurs almost exclusively on *Picea*, rarely on other conifers and very rarely on deciduous wood.

Old myzetozoa (slime moulds), mostly in the genus *Reticularia* (page 1649), may look superficially like old specimens of the asexual state. *Oligoporus rennyi* (page 937) has a white–yellowish margin with asexual spores, and is completely resupinate.

Widespread and very common to common, rare in the boreal zone; mainly August–November.

Postia caesia is a hoof- to shelf-shaped, occasionally imbricate, dark blue, hairy annual polypore with 4–5 bluish pores per mm. Lacks cystidia. The hyphae in the flesh are thin- to slightly thick-walled. The spores are, when fresh, somewhat amyloid, and measure 4.5–5 × 1.4–1.6 µm; Q 2.95–3.36; the spore-deposit is bluish. A brown-rotter on conifer wood, mostly *Picea* and *Abies*, rarely on deciduous wood, including *Fagus*.

Postia simulans ⋉ also occurs on conifers but has finer pores, tends to be less pubescent and has slightly larger spores with a higher Q; *P. cyanescens* ⋉, on *Picea*, has thin fruitbodies and long, narrow basidiospores; Qav. 3.92; *P. alni* ▽ is less strongly blue, less hairy, has finer pores, narrower spores and occurs on deciduous wood; *P. subcaesius* ⋉, also on deciduous wood, has larger, more pubescent fruitbodies compared to *P. alni* and a spore Q of 3.6–3.9. *Postia populi* ⋉ has a spore Q of 3.8–4.4, is almost exclusively confined to wood of *Populus*, and usually does not show much blue on the cap.

Widespread and common; mainly August–November.

The fruitbodies of e.g. *Postia caesia* can be parasitized by *Tremella polyporina* ⋉ but the fruitbodies of that species are not visible externally.

Postia alni is an almost smooth, pale blue annual polypore. It has 5–6 fine pores per mm. The spores measure 4.9–5.3 × 1.2 µm; Q 4–4.3. Produces a brown rot in various deciduous hosts.

Postia subcaesia ⋉, also on deciduous wood, has larger, wider, more hairy fruitbodies with slightly broader spores; Q 3.6–3.9; it is apparently also a more southern species. *Postia populi* ⋉ only rarely has blue tinges and is apparently confined to *Populus* wood. *Postia caesia* △ is mostly a conifer species, is more hairy and has broader spores.

Widespread and common; mostly August–November.

Postia lactea is a rather large, ± hoof-shaped, white, pale brown to greyish, mild-tasting annual polypore with pale pores, 3–5 per mm. Lacks cystidia. The ± cylindrical spores measure 4.5–6 × 1–1.5 µm. Typically on deciduous wood, but occasionally also on conifers. A brown-rotter.

The name *Postia tephroleuca* is sometimes applied to very grey forms (lower image). *Amaropostia stiptica* ◁◁ is whiter, more rugose on the surface and very bitter-tasting; it mostly grows on conifers. *Tyromyces chioneus* ▷▷ produces a white rot.

Widespread and common; mostly July–January.

Postia lactea is often parasitized by *Protocrea pallida* (page 1533), which colours the hymenium orange with its perithecia.

Fuscopostia leucomallella is a reflexed annual polypore with ± brownish, ± zonate caps. The pores are pale, 3–4 per mm. Has gloeocystidia and allantoid spores that measure 4.5–6 × 1–1.7 µm. Occurs on conifer wood, predominantly *Pinus*.

Fuscopostia fragilis ◁◁ stains red-brown when bruised and lacks gloeocystidia. *Postia wakefieldiae* ◁◁ is thinner and also lacks gloeocystidia.

Widespread and fairly common in the hemiboreal–boreal zones, more scattered in the nemoral zones; mostly August–October.

Amyloporia (page 935), ***Antrodia*** ▷, ***Cartilosoma*** ▷, ***Flavidoporia*** (page 934), ***Fibroporia*** (page 936) and ***Neoantrodia*** ▽ are brown-rotters with a dimitic hyphal system (skeletal hyphae) and predominantly tough flesh. They are resupinate, reflexed or capped and occur mainly on conifers. All species were formerly included in *Antrodia*.

Neoantrodia serialis is an annual to rarely biennial, very tough, reflexed to almost completely resupinate, to imbricate polypore with a somewhat brownish cap. The pores are rounded, white, 2–4 per mm. The pale flesh is dominated by skeletal hyphae. The spores measure 6.3–8.2 × 2–3.5 µm. Almost always on conifer wood, mostly *Picea*; a brown-rotter.

Resupinate, annual forms of *Heterobasidion* (page 856) have cream pores, an evident cap crust (a section plus hand lens required!), dextrinoid skeletal hyphae, much smaller spores and form a pinkish-buff to white white rot. Species of *Antrodia* ▷ (and the like) are not as tough-fleshed. *Neoantrodia leucaena* ⋉ occurs on *Populus* and has wider spores.

Widespread and common; all year, but at its best in late autumn.

The discomycete *Hyphodiscus hymeniophilus* (page 1390) provokes a red stain on *Neoantrodia serialis*. The tiny, greyish apothecia are rather uncommon, but also easily overlooked.

Jerzy Opioła/Wikipedia

Antrodia heteromorpha is a rather coarse-pored, reflexed, fairly tough annual polypore with 1–2 pores per mm and up to 3 cm thick flesh, which is dimitic. The spores are ± cylindrical and measure 10–13 × 5–7 µm. Produces a brown rot in conifer wood, mostly *Picea*.

Antrodia albida ⋉ is sometimes recognized as a sibling species on deciduous wood, but is treated here as a synonym. *Cartilosoma ramentacea* ▽ is very soft-fleshed and has somewhat smaller spores; it prefers *Pinus*.

Widespread but uncommon in the hemiboreal-boreal zones of Scandinavia, rare in central Europe; mostly during winter.

Lucie Zíbarová

Cartilosoma ramentaceum is a rather soft-fleshed and fragile annual polypore with relatively small, reflexed, almost discoid, pale fruitbodies, 0.5–2 pores per mm. When drying the margin typically detaches. The spores measure (5–)6–8 × 2.5–3.2 µm. Occurs on *Pinus*, mostly in sandy plantations; a brown-rotter.

Antrodia heteromorpha △ has tougher, more expanded fruitbodies and larger spores.

Widespread, but rather uncommon; all year, but in best condition during autumn and early winter.

Species of ***Antrodiella*** are very tough, dimitic–trimitic and capped or ± resupinate (page 942). The majority are associated with other polypore fruitbodies or with wood decomposed by other polypores; they are white-rotters. Most of the 'hosts' are also white-rotters, except for the brown-rotting *Fomitopsis pinicola* (page 867), which is associated with *A. citrinella* (page 942).

Antrodiella serpula is a vivid yellow, capped, reflexed, very tough annual polypore with 4–6 pores per mm. The flesh is dimitic–trimitic and the spores measure 3.5–4.5 × 1.5–2 µm. Always found on or near old fruitbodies of *Mensularia* ▷▷ on various deciduous trees.

Also known as *A. hoehnelii*. Other *Antrodiella* species are not as yellow and have a different ecology.

Widespread and rather common, becoming scarcer towards the north; mostly June–November.

Antrodiella pallescens is generally a capped, somewhat zonate, pale brown, very tough, annual polypore. Usually has 5–7 pores per mm, but somewhat fewer on fruitbodies on vertical substrates. The spores measure 3–4 × 1.7–2 µm. Occurs on or close to fruitbodies of *Fomes fomentarius* and on wood decomposed by *Fomes*; it is a white-rotter.

This species has frequently been misnamed as *A. semisupina* ⋉, an American species. *Antrodiella faginea* ⋉ is similar but is associated with species of *Phellinus* (page 874). *Inonotus* ▷▷ and closely related genera have somewhat smaller, ellipsoid spores, whereas the spores in *A. pallescens* are ± flat-sided or concave.

Probably widespread and common; mostly in the autumn.

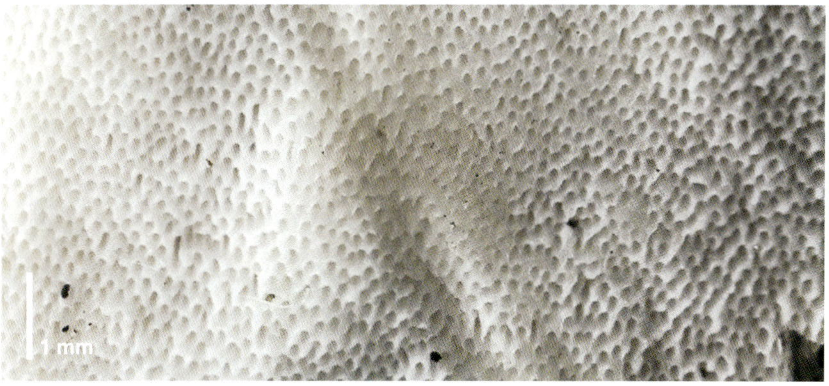

Gloeoporus dichrous is a reflexed-capped, whitish, non-zonate annual polypore with short, apricot, rounded pores, 4–6 per mm, that stain dark when bruised. Has woolly mycelium at the base or at the margin of resupinate parts. The tube layer is rubbery, may be loosened from the flesh and has a hymenium in both the tubes and on the pore margins. Hyphae have clamps, no cystidia. The somewhat allantoid, hyaline spores measure 3.5–5.5 × 0.7–1.5 µm. On deciduous wood, typically *Betula* or *Quercus*, often in open, older *Betula* forests. May also overgrow other polypores.

Skeletocutis amorpha ▷ is rather similar, but occurs on conifers. *Meruliopsis taxicola* (page 977) is resupinate, a similar colour and has a hymenium at the pore margins, but occurs on *Pinus*. *Byssomerulius corium* (page 977) has a wrinkled-veined hymenium.

Widespread and common in the hemiboreal–boreal zones, but occasional to rare in the nemoral zones, especially towards the west, apparently increasing (on *Quercus*); all year, but at its best in the autumn.

Diplomitoporus flavescens is a ± reflexed, cream, tough annual polypore with a non-zonate, finely felty surface. The pores are rounded to angular, 2–3 to per mm. The flesh is dimitic; the hyphae have clamps, and cystidia are absent. The hyaline, allantoid spores measure 5–8 × 2–3 µm. A specialist on old *Pinus*, where it forms a white rot.

Neoantrodia primaeva ⋉, a brown-rotter, has larger spores. *Dichomitus squalens* ⋉, a white-rotter, has branching binding hyphae and larger spores.

Widespread, but local and rare; all year.

Species of ***Skeletocutis*** form resupinate, reflexed to capped, relatively soft and pliable but, at the same time, mostly tough fruitbodies. They are dimitic with skeletal hyphae and the generative hyphae typically have prominent crystals on the surface, particularly in the pores. The spores are very small (typically 0.5–1.5 µm wide) and allantoid. See also the resupinate species page 941.

Skeletocutis amorpha is a reflexed, annual polypore with a white, downy cap and at first white, but with age and when bruised, ± vivid orange rounded to angular pores, 3–4 per mm. The fruitbody has a cartilaginous layer between the tube layer and the flesh. It has hyaline hyphae with incrusted material in the pores, but no proper cystidia. The hyaline, allantoid spores measure 3–4.5 × 1.2–1.5 µm. Forms a white rot, mostly in *Pinus*, but also in other conifers.

Skeletocutis carneogrisea ▽ is also common on *Pinus* but does not redden as much and has even narrower, more allantoid spores.

Widespread and common; all year, but in best condition during the autumn.

Skeletocutis carneogrisea is a reflexed, ± pale grey, rather soft, but tough annual polypore with 3–4 greyish to somewhat pinkish-tinted pores per mm. It has a cartilaginous layer between the tube layer and the flesh. Spores are allantoid and measure 3–4(–4.5) × 1–1.3 µm. On and around old fruitbodies of *Trichaptum* ▷▷, mostly on *Pinus* and *Picea*.

The paler, completely resupinate *S. kuehneri* ⋉ and *Antrodiella parasitica* (page 942) also occur with *Trichaptum* ▷▷; the first has narrower and the second wider spores. *Skeletocutis amorpha* △ is more vividly coloured, and spores are less curved.

Widespread and common, has apparently increased; all year.

Tuomo Niemelä

Skeletocutis lilacina is a reflexed, whitish annual polypore with 5–7 strikingly lilac, angular pores per mm. The flesh is dimitic and the hyphae in the pores are covered in crystals. The hyaline, allantoid spores measure 3–4 × 0.7–1 µm. Only known from trunks of *Picea*.

Species of *Trichaptum* ▷▷ are the same colour but far larger and usually have more irregular pores.

Very rare, with scattered records from central Europe and Scandinavia; mainly in the autumn.

Jens H. Petersen/Thomas Læssøe

Skeletocutis nemoralis is a ± capped, very fine pored, annual to, rarely, perennial *Skeletocutis* that usually has a blue-green tinge to the pores. The cap is at first pale but becomes pinkish-buff and then dark brown from the base. Has 7–10 pores per mm, which are hardly visible without a hand lens. Lacks cystidia, has clamps. The allantoid spores measure 3–5 × 0.5–1 µm. Produces a white rot in deciduous wood, especially *Fraxinus*, but also in other deciduous trees; mostly in mull-rich places.

Forms a species complex with several rather similar species in Europe and elsewhere. Whilst *S. nemoralis* is common on *Fraxinus*, *S. pallescens* ◁◁ prefers other deciduous hosts. Formerly named *S. nivea* ⋉ but that is an Austral–Asian species. Normally a rather easy complex to identify due to the relatively fleshy fruitbodies with extremely small pores, but the exact width of the very narrow spores may be hard to measure.

Apparently widespread and very common in the nemoral zones and possibly farther north; all year, but in best condition during the autumn.

Skeletocutis nemoralis is very frequently parasitized by *Protocrea farinosa* (see page 1533).

Climacocystis borealis is a rather large, very juicy, ± hoof-shaped, mostly imbricate, pale yellow-brown annual polypore with a bitter taste. The cap surface is strigose-hairy. Has cream, angular pores, 1–2 per mm. The cystidia are thick-walled and crystal-bearing. Has clamps and monomitic flesh that does not have a particularly evident duplex structure. The smooth, inamyloid, egg-shaped spores measure 4.5–6.5 × 3–4.5 µm. Almost always on larger trunks of conifers, but is also reported from deciduous wood; a white-rotter.

The combination of large, pale fruitbodies with a covering of dense, long hairs and occurrence on conifers should be sufficient to ensure an accurate identification. If there is any doubt, the disitinctive cystidia are a useful character.

Widespread and rather common in the hemiboreal–boreal zones (including central Europe), but rare to absent from the nemoral zones; mostly August–November.

Jan Vesterholt

Thomas Læssøe

Thomas Læssøe

Odoria alborubescens is a large, ± hoof-shaped, coarsely felty, pinkish-tinted, soft and juicy, annual polypore. The flesh becomes somewhat wax-like when dry. The pores are pale, 2–4 per mm. The flesh is thick, monomitic, smells sweetish-aromatic, sometimes almost nauseating, and becomes mostly wine-red with KOH. The smooth, inamyloid spores measure 5–7.5 × 4–5 µm. Forms a white rot, typically in large, mostly fallen trunks of *Fagus* in forests with long continuity and a good supply of sufficiently old trees. May also fruit in the cavities of standing veteran trees.

Aurantioporus fissilis ▷ is very similar but does not stain wine-red with KOH, smells sour and has narrower spores. *Buglossoporus quercinus* (page 832) has a somewhat similar smell and discolours pinkish, but it is narrowly attached and occurs on *Quercus*, where it forms a brown rot.

Widespread in the nemoral zones, rather rare to very rare; mostly July–November.

Thomas Læssøe

Aurantiporus fissilis is a large, soft, greasy, somewhat felty, pale, ± pinkish and yellowish annual polypore. Has 2–3 pores per mm. The flesh is thick, pale, finely zonate and becomes somewhat wax-like when dry. Smells sour and either hardly changes colour with KOH or becomes blackish in old and dried out parts (inset picture). The spores measure 4–5 × 3–4 µm. There are also spherical asexual spores (chlamydospores) in the flesh. Forms a white rot in old, living or dead deciduous trees, *e.g.* *Malus* and *Fagus*.

Odoria alborubescens ◁ normally stains red with KOH, has wider spores and smells aromatic. The other species in the genus, *Aurantiporus croceus* ▷▷, is bright orange-yellow.

Widespread in the nemoral–hemiboreal zones, but rare to occasonal; June–November.

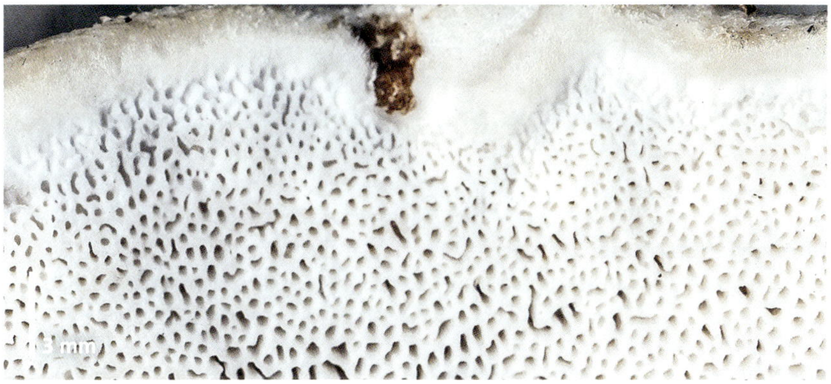

Jens H. Petersen/Thomas Kehlet

Marco Facchini

Spongipellis is a genus of large, whitish and white-rotting annual polypores with strikingly thick-walled, broadly ellipsoid spores and a monomitic hyphal system with clamps.

Spongipellis spumeus is a large, ± hoof-shaped, whitish to pale clay-buff annual polypore with a hairy-felty cap surface and 1–3 rounded pores per mm. The flesh is monomitic and lacks cystidia; has large clamps. The smooth, cyanophilous, almost spherical to broadly ellipsoid, thick-walled spores measure 6–7 × 5–6 µm. Forms a white rot in living and dead deciduous trees, particularly *Acer*.

The southern *S. litschaueri* ⋉ has coarser, disintegrating pores, 0.5–1 per mm. *Aurantiporus fissilis* ◁◁ is more felty and has narrower, thinner-walled spores.

Widespread, rare in western, nemoral parts of temperate Europe, but more common to the east; all year, but in best condition from late summer to October.

Thomas Læssøe

Spongipellis pachyodon is a capped to ± reflexed annual polypore with pores that quickly transform into long, flat spines. The thick-walled, cyanophilous spores measure 6–7 × 5–5.5 µm. Forms a white rot in various deciduous trees, but usually *Quercus* and *Fagus*, and mostly in places with long continuity.

The combination of rather thick flesh, coarse 'teeth' and thick-walled spores should make identification straightforward. *Climacodon septentrionalis* (page 1056) is larger, has round spines and different spores.

Widespread, but very rare, and absent from the hemiboreal–boreal zones; all year, but best in the autumn.

Spongipellis delectans is a very pale, shelf-shaped annual polypore with a finely felty cap surface and labyrinthine, cream pores, (0.5–)1–2 per mm. The thick-walled, cyanophilous spores measure 6–7 × 5–6 µm. Almost exclusively associated with dead or dying *Fagus* in places with long continuity and a good supply of suitable trees; a white-rotter.

Spongipellis litschaueri ⋉ is more thick-fleshed, 2–7 cm when dry as opposed to 1–2 cm in *S. delectans*. May look like *Climacocystis borealis* ◁◁ but that species has a strigose-hairy cap and occurs on conifers.

A rare, scattered, nemoral species; mostly October–January.

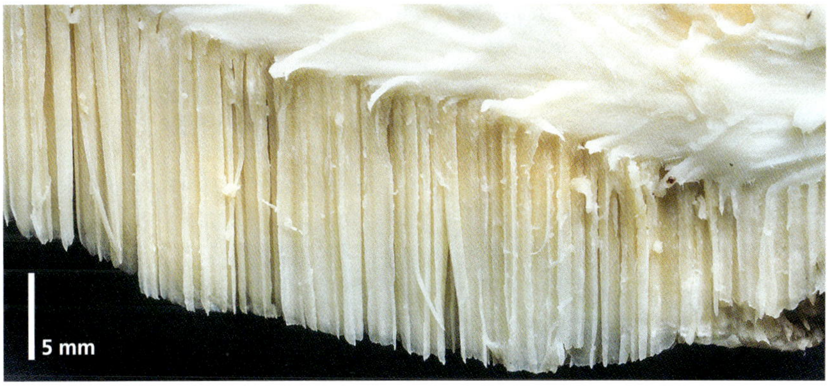

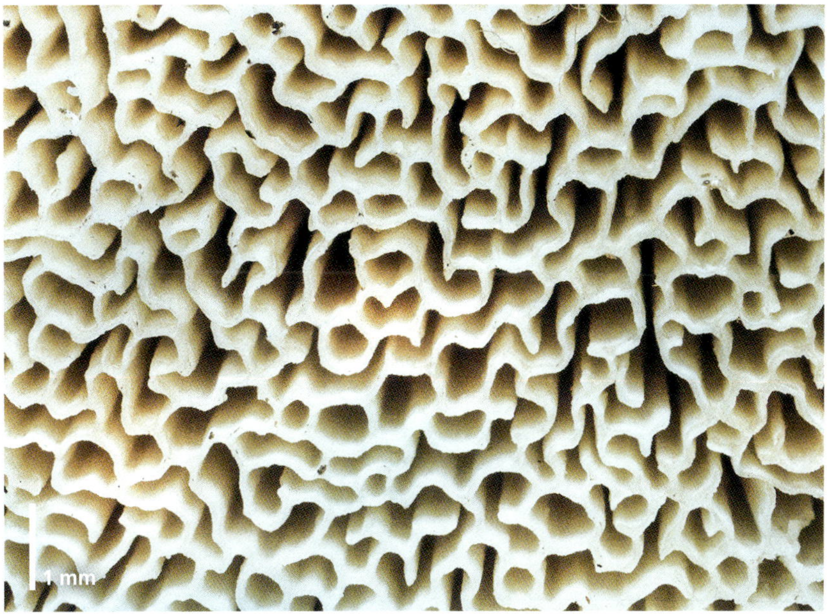

Thomas Læssøe

Tyromyces includes soft-fleshed, white-rotting polypores that are monomitic or dimitic with skeletal hyphae.

Tyromyces chioneus is a pale, somewhat greyish, almost non-zonate, hoof- to shelf-shaped, soft, annual polypore. The pores are white and small, 5–7 per mm. The flesh has strongly branching hyphae. The hyaline, cylindrical spores measure 4–5 × 1.5–2 µm. Forms a white rot in various deciduous trees, mostly in *Fagus* in the nemoral zones, but has a wide host range.

Postia lactea ◁◁ is very similar but produces a brown rot, lacks the branching generative hyphae and has somewhat narrower spores.

Widespread and fairly common; mostly August–November.

Jacob Heilmann-Clausen

Tyromyces kmetii is a hoof-shaped, yellowish to apricot-orange, soft, annual polypore. The cap surface stains pink with KOH. Has 3–5 pores per mm. The hyaline, broadly ellipsoid spores measure 4–4.5 × 2–2.5 µm. Forms a white rot in various deciduous trees, *e.g. Alnus*, *Betula* and *Sorbus*.

The golden, hoof-shaped fruit-bodies separates this species from other coloured species of *Tyromyces*. Brown species of *Postia* ◁◁ occur mainly on conifers and produce a brown rot.

Widespread in the hemiboreal–boreal zones, including the uplands in central Europe, but rare; mainly autumn.

Loweomyces wynneae is a very irregularly shaped, mostly partly sterile, yellow-brown, rather thin- and soft-fleshed annual polypore. Pores white, angular, 3–4 per mm. Has a strong, aromatic smell and orange-brown hyphal cords. The hyaline spores measure 3.5–4 × 2–2.5 µm. Creeps from twig to twig on the forest floor, on damp mull soils, *e.g.* at the border of boggy *Fraxinus* stands; a white-rotter.

Abortiporus biennis (page 852) reddens on the pores and is usually larger and much tougher. It also has no prominent hyphal cords and a different smell.

Widespread in the nemoral–hemiboreal zones, but rather rare; mostly August–October.

Daedaleopsis form medium-sized, leathery fruitbodies with pale brown flesh and pores that, with age, become radially stretched and may end in a gill-like pattern. The pores stain red on fresh fruitbodies. The hyphal structure is trimitic, and all are white-rotters.

Daedaleopsis confragosa is a rather flat, ± zonate, yellow-brown to brownish-red annual polypore that is somewhat constricted at the base but still broadly attached. It has 1–2 red-staining pores per mm (measured across) that are at first ± rounded, later elongated. Neither the margin nor the surface have hairs. The very tough trimitic flesh is pale brown. The spores are hyaline, smooth and measure 9–11 × 2–2.5 µm. Occurs on deciduous wood, mostly on attached branches in damp or wet forests, particularly in older Salix carrs.

Both *D. confragosa* var. *tricolor* ⋉ and *D. septentrionalis* ▷ quickly develop thick 'gills' (not pores) and are darker brown. *Trametes betulina* ▷▷ has a felty cap and smaller spores.

Widespread and common; all year, but in best condition during the autumn.

Daedaleopsis septentrionalis is a red-brown annual polypore with radiating, almost unbranched, gill-like, reddening pores and a zoned cap. The spores measure (5.4–)6–8.4(–10.1) × (1.7–)1.8–2.4(–2.7) µm. Occurs on *Betula* or, very rarely, on *Alnus* or *Sorbus*.

Daedaleopsis confragosa var. *tricolor* ◁ also has ± gill-like pores but grows on various deciduous trees farther south. *Trametes betulina* ▷▷ is very similar, but is zonate, felty-hairy on the cap and has smaller spores.

Rather common in the Scandinavian taiga; all year.

Tuomo Niemelä

Datronia mollis is a very variable, reflexed, rather irregular annual polypore with a very dark brown, rugose, hairy to smooth cap surface and irregular, rather large, greyish pores, 1–2 per mm. The back side of the fruitbodies is black, like the cap, and loosens easily. The flesh is trimitic and is two-layered with a very dark upper part and a pale brown lower part. The smooth, thin-walled, hyaline, inamyloid spores measure 10–12 × 3–4 µm. Forms a white rot on a number of deciduous hosts, rarely on conifers.

The similar *Datroniella scutellata* ⋉ forms small caps on *Alnus* in the alpine zones. *Datronia stereoides* ⋉ is completely resupinate, and has 4–5 hexagonal pores per mm. Species of *Datronia* are closely related to species of *Cerioporus* (page 834) and accepted in that genus by some authorities.

Widespread and common; all year, but best in the autumn.

Bjerkandera is a small genus with rather small, imbricate caps that may have resupinate parts or even completely resupinate fruitbodies. The pores are mostly ± greyish. The hyphal system is monomitic. All species are white-rotters.

Bjerkandera adusta is a small-capped, partly to rarely completely resupinate (lower image), semi-tough, annual polypore with an indistinctly zonate, finely felty to hairy, brownish cap. The grey pores darken when bruised, 6–7 per mm. The flesh is thin, pale brown and rather tough. Lacks cystidia. The hyaline spores measure 4.5–6 × 2.5–3.5 µm. Occurs on various deciduous trees, rarely on conifer wood.

Bjerkandera fumosa ▷ is thicker, has paler pores and a darker zone just above the tubes. Resupinate forms may recall *Cinereomyces lindbladii* (page 943) but that species has 3–4 pores per mm that do not change colour, and skeletal hyphae that dissolve in KOH.

Widespread and very common; all year, but in best condition during the autumn.

Bjerkandera fumosa is a sordid brown, capped, somewhat felty, annual polypore with very pale grey-brown, rounded pores, 4–5 per mm. The flesh is rather tough and ± pinkish-buff, but has a distinct dark line above the tube layer. May have a somewhat liquorice-like odour. The hyaline spores measure 5.5–7 × 2.5–3.5 µm. Occurs on deciduous wood, mostly in damp forests, boggy woodlands, *etc.*

Bjerkandera adusta ◁ is thinner and has darker grey pores that darken further when bruised; it also lacks the dark zone above the tubes but the tubes themselves are darker grey. *Trametes suaveolens* ▷▷ is much tougher, has larger pores (1–2 per mm), no dark line in the flesh and a much stronger, aniseed-like, smell. Grey forms of *Postia lactea* ◁◁ are softer and more fragile.

Widespread and common, becoming scarcer towards the north; mostly September–January.

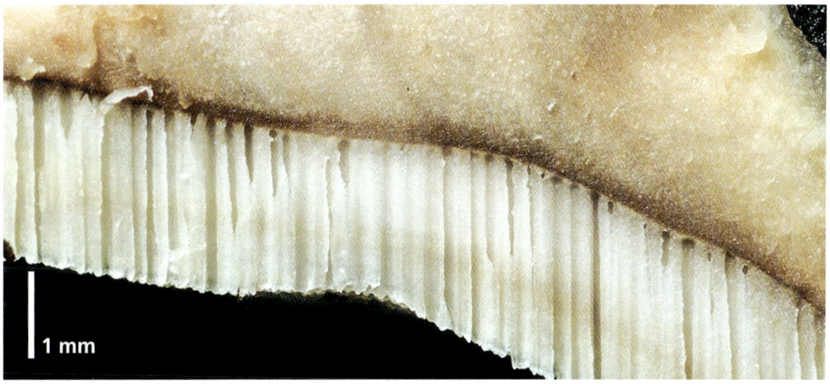

Trichaptum is a genus of thin-capped, appressed-hairy annual polypores with lilac to violet colours on the pores, which usually split irregularly with age. The flesh is ± tough, dimitic with skeletal hyphae, and has a ± distinct duplex structure. Closely related to *Phellinus*, *etc.* in the *Hymenochaetales*. All species are white-rotters.

Trichaptum fuscoviolaceum is a reflexed annual polypore with a hymenophore that is at first ± pore-like (2–3 pores per mm) but which quickly disintegrates into coarse teeth. Has thick-walled, crystal-bearing cystidia. The hyaline spores measure 6–7 × 2.5–3 µm. Occurs on conifer wood, especially *Pinus*.

Also known as *T. hollii*. *Trichaptum abietinum* ▽ has pores that never split into teeth; it prefers *Picea*. The more northern and eastern *T. laricinum* ⋉ has pores that split to form gills, ± as in *Gloeophyllum sepiarium* (page 869); it occurs on conifers.

Widespread and common in the hemiboreal–boreal zones, becoming scarcer towards the south but common in dune plantations in Denmark; all year.

Trichaptum abietinum is a reflexed annual polypore with ± violet, rounded to angular pores, 3–4 per mm. Has thick-walled, crystal-bearing cystidia. The hyaline spores measure 5.5–7.5 × 2–3 µm. Occurs on conifer wood, mostly *Picea*, rarely on deciduous wood.

Trichaptum fuscoviolaceum △ occurs mostly on *Pinus* and has splitting, dentate pores. *Trichaptum laricinum* ⋉, also on conifers, has pores that split to form gills. *Trichaptum biforme* ▷ is larger and occurs on deciduous wood.

Widespread and very common; all year-round, but best during the winter.

Trichaptum biforme is a rather large, capped to reflexed, imbricate annual polypore with a rather pale cap that may be violet at the margin. The hymenophore is at first ± pore-like with 2–4 pores per mm, but with age the pores disintegrate to form 'teeth'. Has thick-walled, crystal-bearing cystidia. The hyaline spores measure 5–7 × 2–2.5 µm. Occurs on deciduous wood, particularly *Betula*.

Also known as *T. pargamenum*. *Trichaptum fuscoviolaceum* ◁ also has pores that soon split into 'teeth' but produces smaller caps and is more prone to forming resupinate parts; it occurs on conifers, mostly *Pinus*.

An eastern species that is quite frequent in eastern parts of the hemiboreal-boreal zones (the taiga zone) of the Baltic area, but rare or absent from most of temperate Europe; all year.

Thomas Læssøe

Jacob Heilmann-Clausen

Funalia form rather thick, ± triangular, large-pored, leathery fruitbodies with a thick felt on the cap surface and pale brown colours on the cap, the tube layer and in the flesh. The hyphal structure is trimitic. White-rotters on sun-exposed deciduous wood.

Funalia trogii is a rather coarse-pored, annual polypore with a thick, felty layer on the cap. The flesh is pale brown and the pores angular to labyrinthine, 1–2 per mm. The hyaline spores measure 8–11 × 3–3.5 µm. Produces a white rot in various deciduous trees but mostly *Salix* and *Populus*, both in standing and fallen, mostly sun-baked, trunks.

Funalia gallica ▽ has larger spores (9–13 × 3–4 µm) and darker flesh.

Widespread, rather frequent in southern nemoral areas but very rare in hemiboreal Scandinavia; all year, but in best condition during the autumn.

Thomas Læssøe

Funalia gallica is an annual polypore with long, rigid, brownish hairs, brown flesh and 1–2 irregular pores per mm. The hyaline spores measure 9–13 × 3–4 µm. Occurs on various deciduous trees in warm places.

Funalia trogii △ has smaller spores. *Trametopsis cervina* ▷ has paler flesh, finer pores and a less hairy surface.

Very rare in the northern part of temperate Europe, increasing in abundance farther south; all year, but best in the autumn/early winter.

Frantisekia mentschulensis is a reflexed, cream-yellow to apricot-orange, annual polypore that usually starts as a resupinate fruitbody from which rather narrow, imbricate caps are formed. Has 5–7 angular to somewhat labyrinthine pores per mm, which, on vertical parts, break up into larger pores/'teeth'. The flesh is moderately tough, somewhat bitter-tasting and monomitic with both thin- and thick-walled, clamped generative hyphae. Lacks cystidia. The ± cylindrical spores are hyaline, thin-walled and measure 3.5–5 × 1.7–2 µm. Occurs on the trunks of deciduous trees, *e.g. Fagus*; a white-rotter.

Antrodiella serpula ◁◁ is also yellowish but it is very tough and grows on or next to old *Mensularia* ▷▷ fruitbodies.

Apparently spreading in parts of southern Scandinavia (nemoral zone), although it is still rare, but also known from central Europe; in best condition from July through into autumn.

Ole Martin/Thomas Læssøe

Ole Martin

Trametopsis cervina is typically a very imbricate *Trametes*-like annual polypore with a ± hairy, brownish, almost non-zonate cap and ± brownish pores that almost break into 'teeth', 1–2 per mm. The flesh is tough, dimitic with skeletal hyphae, and pale brown. The hyaline spores measure 7–9(–10) × 2.5–3 µm. Occurs on a number of deciduous hosts, especially *Fagus*; a white-rotter.

May recall species of *Funalia* ◁ but these are even darker fleshed and have larger spores.

Widespread in central Europe, but apparently extending its range northwards, *e.g.* to the Netherlands; all year, but mainly autumn.

Morten Christensen

Trametes form predominantly annual, thin but tough, leathery caps that mostly appear in crowded, imbricate formations. The flesh is whitish and trimitic. All are white-rotters, mostly on deciduous trees.

Trametes versicolor is a thin- and pale-fleshed, felty, densely zonate, tough annual polypore that mostly has almost blackish, narrow, ± smooth zones with a silky sheen. Has 4–5 rounded pores per mm. The hyaline spores measure 5–6.5 × 2–2.5 µm. Occurs on all kinds of deciduous wood, usually on stumps, very rarely on conifers. In the nemoral zones *Fagus* is by far the most dominant host.

Trametes ochracea ▷ normally has a paler brown, fine-felty to almost smooth cap surface. It is somewhat thicker and less flexible, coarser-pored (3–4 pores per mm) and has somewhat larger spores (6–7 × 2–2.5 µm); it prefers *Betula* and *Populus*.

Widespread and very common; all year, but usually deformed by the spring.

Trametes ochracea is a fine-felty to almost smooth, zonate, pale brown, tough, annual polypore with 3–4 rounded pores per mm. The hyaline spores measure 6–7 × 2–2.5 (–3) µm. Occurs on deciduous wood, mostly on *Betula* and *Populus* (including *P. tremula*); it forms a white rot.

The cap of *Trametes versicolor* ◁ is thinner and more contrasty, with alternating felty and almost smooth zones and dark, narrow zones; it may have slightly more pores per mm and somewhat shorter spores.

Widespread, occasional to common; all year, but ± deformed by the spring.

Trametes betulina is a rather pale, felty-hairy, zonate, annual polypore with at first somewhat labyrinthine, later more gill-like, pale cream 'pores', 10–15 per cm at the margin. The flesh is trimitic and very tough and pale. The hyaline spores measure 5–6 × 2–3 µm. Produces a white rot in various deciduous trees, mostly in rather sun-exposed wood; *Fagus* and *Betula* are the most common hosts in the nemoral zones.

Also known as *Lenzites betulina*. *Daedalea quercina* (page 865) has far coarser gills on much thicker fruitbodies. *Daedaleopsis confragosa* and *D. septentrionalis* ◁◁ stain red on the pores/gills when fresh and have a smooth cap surface.

Widespread and fairly common to common; all year, in best condition in the autumn.

Andrew Khitsun

Trametes pubescens is a gregarious and mostly imbricate, rather small and pale, downy to finely-felty, ± zonate, annual polypore. The pores are whitish to cream and 3–5-angled, 2–3 per mm. There is limited contrast between the zones at the surface. The cap typically has radial wrinkles/furrows. The flesh is tough and trimitic. The hyaline spores measure 5–7 × 2–2.5 µm. Occurs on trunks of deciduous trees, particularly *Betula*.

Trametes hirsuta ▽ is more strigose-hairy and the pores are more greyish; it prefers *Fagus* in the nemoral zones.

Widespread in hemiboreal-boreal Scandinavia, mainly in subalpine *Betula* forests, but rare or very rare and often misidentified in the nemoral zones; all year, but quickly devoured by insects and best seen in mid-autumn.

Trametes hirsuta is an annual polypore that is covered by long, rigid hairs in indistinct zones. The angular pores are greyish, 2–4 per mm. May have a sweetish, ± somewhat aniseed-like smell. The hyaline spores measure 6–9 × 2–2.5 µm. Occurs on sun-exposed wood of deciduous trees, rarely on conifers; a white-rotter.

Trametes pubescens △ is more downy, with whitish to cream pores; it prefers *Betula* in the boreal zone. *Trametes suaveolens* ▷ smells strongly of aniseed, has a smooth cap surface and larger pores; it prefers *Salix*.

Widespread, common to very common; all year, but best in the autumn.

Trametes gibbosa is an annual to biennial, thick, rigid, almost smooth, indistinctly zonate and basally umbonate polypore with very tough, trimitic flesh. With age it typically acquires a green layer of algae. The pores are radially stretched, up to 5 mm long, 1–2 per mm measured across. The hyaline spores measure 4–5 × 2–2.5 µm. Occurs on deciduous wood, predominantly on *Fagus* stumps.

Trametes hirsuta ◁ has similar colours and habitat but is thinner, is covered by long, rigid hairs and has angular, greyer pores. *Trametes suaveolens* ▽ has rounded to angular pores and smells strongly of aniseed.

Widespread and very common in the nemoral zones and extending into the hemiboreal zone, absent from the boreal zone; all year.

Trametes suaveolens is an almost hoof-shaped, pale, mostly non-zonate, pale-fleshed, smooth annual polypore with a strong odour of aniseed. The pores are rounded to angular and large, 1–2 per mm. The hyaline spores measure 8–12 × 4–4.5 µm. Occurs on deciduous wood, predominantly *Salix*.

The large-pored, aniseed-scented fruitbodies are characteristic. *Bjerkandera fumosa* ◁◁ occurs in the same habitats, but is thinner, smells less strongly, has smaller spores and a dark zone in the flesh. *Trametes hirsuta* ◁ may have a weak, sweetish smell but has long hairs on the cap surface.

Widespread in the nemoral–hemiboreal zones, but occasional to rather rare; all year, but best in autumn.

Václav Hanzlík

Thomas Læssøe

Jens Maarbjerg

Pycnoporellus fulgens is a distinctive, annual, bright orange, capped, only slightly tough polypore. When young the caps are distinctly strigose-hairy. Has 1–3 pores per mm. The pores are somewhat paler than the upper side and disintegrate into 'teeth' with age. Has cystidia-like cells in the hymenium and monomitic flesh. The hyaline spores measure 4–10 × 3–4 µm. Occurs with the brown-rotting *Fomitopsis pinicola* (page 867), mostly on *Picea*, but also on other coniferous and deciduous trees.

Pycnoporus cinnabarinus ▽ is a very tough and more fine-pored polypore that is not associated with brown rot and *Fomitopsis pinicola*. See also *Pycnoporellus alboluteus* (page 934).

Local to rather common in the hemiboreal–boreal zones, has recently spread into southern nemoral Scandinavia and is also present in central Europe. It is absent from western Atlantic parts of temperate Europe; mostly July–November.

The fruitbodies disappear very quickly, mainly due to attacks from beetles.

Pycnoporus cinnabarinus is an orange-red to scarlet, tough-fleshed *Trametes*-like annual polypore with ± scarlet, round to irregular pores, 2–4 per mm. The flesh is red to orange and trimitic. The hyaline spores measure 6–7 × 2.5–3 µm. Occurs on sun-exposed wood of deciduous trees; produces a white rot.

Pycnoporellus fulgens △ has larger pores, a softer texture and occurs with *Fomitopsis pinicola* on both coniferous and deciduous wood (page 867).

Widespread (all zones), common in more continental parts but rare in western, Atlantic areas; all year, but in best condition during the autumn.

Thomas Læssøe

Aurantiporus croceus is a rather large, annual, bright yellow-orange, capped or partly resupinate polypore. The tubes are orange with somewhat angular pores, 2–4 per mm. The surfaces and flesh stain vivid purplish red with a drop of KOH (inset picture). The flesh is soft and juicy, monomitic, orange and dries to a greasy substance. The spores are thin-walled, hyaline and measure 4–5 × 2.8–3.5 µm. Forms fruitbodies and a white rot in very old, living or dead *Quercus*, but farther south also on *Castanea*.

Hapalopilus rutilans ▷▷ is a smaller, matt, brownish-red polypore that stains purplish-red with KOH. The orange *Pycnoporellus fulgens* ◁ and *Pycnoporus cinnabarinus* ◁ are tougher and occur on other hosts. *Aurantiporus fissilis* ◁◁, the other species in the genus, is ± whitish.

A southeastern nemoral–hemiboreal species, very rare to rare, not found in the western parts of temperate Europe, *e.g.* the UK; mostly August–December.

†† ***Hapalopilus rutilans*** is a rather small, typically ± triangular, smooth, completely cinnamon-brown, annual polypore. A violet pigment appears when it is treated with ammonia or KOH. Has 2–3-edged pores per mm. There are no setae or cystidia in the flesh or hymenium. The spores are hyaline, smooth, thin-walled and measure 3–4 × 2–2.5 µm. The hyphal system is monomitic. Produces a white rot in various deciduous trees, mostly on attached branches, often on *Sorbus*.

The very similar, small but somewhat larger-spored *H. ribicola* ⋉ occurs on old *Ribes* shrubs. The rare, resupinate *H. eupatorii* ⋉ occurs on coarse herbaceous stems.

Widespread and rather common; year-round.

Hapalopilus rutilans is a very popular fungus for dyeing as it produces vivid violet colours. It contains very high levels of polyporic acid and is extremely poisonous if eaten, causing kidney and liver failure, and also resulting in violet urine.

Species of ***Ischnoderma*** form wide caps with small to large resupinate parts. The annual fruitbodies may also be constricted at the base. The flesh and tube layers are at first pale and soft but, with age, become cinnamon-brown and harder; thicker-walled hyphae are also produced. The flesh is dimitic with skeletal hyphae. White-rotters.

Ischnoderma benzoinum is a dark brown, zonate, tarred-felty, annual polypore with rather pale brown, dimitic flesh. Has 4–6 rounded to ± angular pores per mm. The ellipsoid, smooth, hyaline, thin-walled spores measure 5–6 × 2–2.5 µm. Occurs on coniferous wood, mostly *Picea*, but rarely also on deciduous wood.

Ischnoderma resinosum ▽ is somewhat thicker, usually more rounded at the margin and has slightly paler flesh; it occurs mostly on *Fagus*. *Podofomes trogii* (page 842) has a stem. Species of *Inonotus* and the like ▷▷ have darker flesh and setae in the hymenium.

Rather common and widespread and apparently increasing, at least in part of the area; all year, but in best condition September–December.

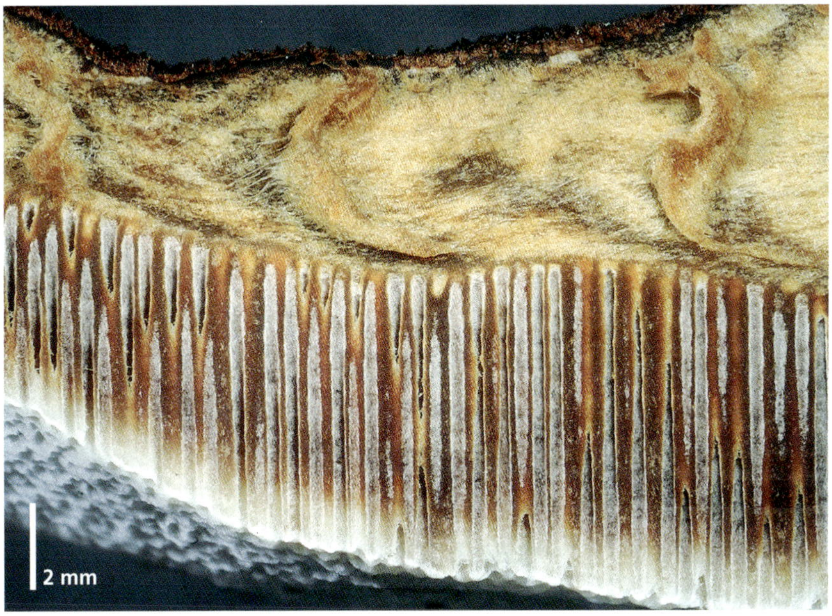

Ischnoderma resinosum is a rather thick-fleshed, annual, indistinctly zonate polypore with a felt-like cap surface and 4–6 fine, rounded pores per mm. The margin is rounded and may exude coloured drops, and the flesh smells somewhat like oranges. The ellipsoid, hyaline, inamyloid spores measure 5–6 × 2–2.5 µm. Grows on larger trunks of deciduous trees, mainly *Fagus*.

Resembles the perhaps somewhat darker *I. benzoinum* △, which mostly grows on conifer wood.

Widespread in the nemoral zones, rare to occasional and on the increase at least in southern Scandinavia; may be found year-round, but in best condition September–December.

Thomas Læssøe

Pekka Helo

Pelloporus leporinus is an imbricate, orange-brown, felty annual polypore with 3–4 rounded to hexagonal, light-reflecting pores per mm. The felt becomes denser with age and the cap then appears smoother. The fruitbodies are very short-lived. The flesh is red-brown, monomitic and has a duplex structure: an upper, several millimetre thick felty layer of ± erect hyphae and a lower flesh-layer with ± horizontal hyphae. Has curved, long setae in the hymenium. The spores measure 5.5–7 × 3–4.5 µm. Occurs on living or dead *Picea abies*, towards the east also on *Larix*.

Also known as *Inonotus leporinus* and *Onnia leporina*. *Pelloporus triqueter* ⋉ is a southern, *Pinus*-associated species. *Pelloporus tomentosus* (page 842) differs by having a ± central stem.

A rare species both in hemiboreal–boreal Scandinavia and in central Europe; most of the year.

Mensularia nodulosa is a nodulose-rugose, ± imbricate, reflexed, yellow-brown annual polypore with 1–4, somewhat irregular, light-reflecting pores per mm. Has straight setae in the hymenium. The spores are pale and measure 4.5–5 × 3.5–4 µm. Associated with *Fagus*, mostly on larger fallen trunks and thick branches.

Mensularia radiata ▷ is very rare on *Fagus*, has ± curved setae and is more capped. The larger *Inonotus cuticularis* ▷▷ occurs on living *Fagus* and has branching setae in the cap felt.

Widespread and common in the nemoral zones; August–November, but present in a dead state year-round.

Cosmospora coccinea (page 1536) is found on dead *M. nodulosa* fruitbodies.

Species of ***Mensularia*** typically form rather large clusters of narrow, imbricate annual fruitbodies, mostly with decurrent resupinate parts. The flesh is at first rather soft (monomitic), orange-brown to cinnamon-brown and, in fresh specimens, the pores are pale and light-reflecting. Has pale yellow-brown spores, and setae in the hymenium. White-rotters.

Mensularia radiata is an imbricate, at first apricot-orange, later red-brown, somewhat nodulose uneven annual polypore with 2–5 angular, light-reflecting pores per mm. Has ± hooked setae in the hymenium. The spores are pale yellow-brown and measure 5–6.5 × 3–4.5 µm. Typically produces a white rot in standing dead or dying deciduous trees, mostly *Alnus*, *Betula* and *Corylus*.

Mensularia nodulosa ◁ has smaller caps, straight setae and is associated with *Fagus*. Old fruitbodies of *Inocutis rheades* ▷▷ have an evident mycelial core in the inner part of the flesh.

Widespread and common; mostly August–October, but may be found in a dead state year-round.

Species in the genus ***Inonotus*** form large, flat to hoof-shaped, often imbricate or completely resupinate (page 951) annual fruitbodies. They have moderately tough flesh (monomitic to dimitic with skeletal hyphae), light-reflecting pores, few to numerous setae and yellow spores. All are white-rotters.

Inonotus cuticularis is a large, imbricate, yellow-brown to red-brown, felty-hairy annual polypore with 3–4 light-reflecting, rounded pores per mm. The thick flesh is yellow-brown, later red-brown and does not have a mycelial core. Has straight and hooked setae in the hymenium and branching setae in the cap felt. The yellow spores measure 6–8 × 4.5–5.5 µm. Almost exclusively in wounds on living *Fagus*.

Inonotus hispidus ▷ is even larger and prefers other hosts, particularly *Fraxinus*; it has larger spores and lacks branching setae in the cap felt. *Inocutis rheades* ▷▷ has flesh with a mycelial core and occurs mainly on *Populus tremula*.

Widespread and occasional in the nemoral zones, rare in the hemiboreal zone; mostly July–October, but may be found in a dead state year-round.

The light-reflecting pores are demonstrated in the images below.

Inonotus hispidus is a very large, flat to hoof-shaped and often imbricate, orange-brown, strongly strigose-hairy annual polypore with 2–3 rounded pores per mm. Has a few to numerous curved setae in the hymenium, but no branching setae in the cap felt. The yellow, ± thick-walled spores measure 8–11 × 6–8 µm. Forms a central rot in standing deciduous trees, particularly *Fraxinus*, but also *Malus*, *Pyrus*, *Crataegus* and *Sorbus*, mostly in open places.

May look similar to *I. cuticularis* ◁ but that species prefers *Fagus*, is somewhat smaller, has smaller spores and has branching setae in the cap felt.

Widespread and occasional in the nemoral zones, but rare and southern in the hemiboreal zone; July–October, but may be found in a dead state year-round.

Pseudoinonotus dryadeus is an annual, large, hoof- to shelf-shaped, pale yellow-brown, felty polypore, which, when actively growing, exudes conspicuous golden drops at the cap margin. Has yellow-brown, monomitic, rather tough flesh and 3–5 rounded to angular, light-reflecting pores per mm. Has curved, brown setae in the hymenium. The hyaline, somewhat thick-walled, dextrinoid spores measure 6–8 × 5–7 µm. Occurs almost exclusively at the base of older, living *Quercus*, where it forms a white rot.

Inocutis dryophila ⋉ also occurs on old *Quercus* but is usually positioned somewhat farther up the trunk, has a mycelial core, brownish spores and lacks setae.

Widespread in the nemoral–hemiboreal zones, occasional; mostly July–October, but may be found in a dead, decaying state year-round.

Inocutis rheades is an annual, typically imbricate, orange-brown, coarsely felty polypore with 2–3 rounded to angular, light-reflecting pores per mm. The inner part of the flesh has an evident mycelial core (see arrow on cut fruitbody). Lacks setae in the flesh and hymenium. The rusty brown, inamyloid, non-dextrinoid spores measure 5–6 × 3.5–4 µm. Found mostly on *Populus tremula*, only rarely on other *Populus* species or other deciduous trees; it is often positioned rather high up in standing trees.

Inocutis dryophila ⋉ and *Fomes fomentarius* (see page 870) also have mycelial cores but the former only occurs on *Quercus* boles, while the latter is hard and perennial. Large specimens of *Mensularia radiata* ◁◁ may look similar but have no mycelial core and usually occur on *Alnus* and *Betula*.

Widespread and fairly common in the hemiboreal–boreal zones, but rather rare in the nemoral zones and apparently absent from the UK; mostly July–October, but may be found in a dead state year-round.

Annual, resupinate polypores

This group includes polypores with annual, flat fruitbodies that completely lack caps. Some species may, however, form small protuberances (false caps) without an evident cap surface. Although some species can be recognized in the field, it is a very difficult group to identify, and in most cases correct identification requires microscopy. Many species have tough, long-lasting fruitbodies that persist long after spore production has ceased.

Sterile material is easily sampled but cannot be identified. It may be a good idea to check if the specimen is able to deposit spores on a coverslip or microscope slide. Keep the specimen moist in order to enhance the chance of obtaining a good spore-deposit.

All species are wood-degrading and most produce a white rot.

OTHER SIMILAR FUNGI:

– annual reflexed (polypores with a resupinate decurrent lower part) and capped polypores (page 886).

– perennial, resupinate polypores (page 854).

– merulioid corticioids (corticioids with wrinkled-veined hymenophore) may look similar but the ‘pores’ are rounded and have fertile pore mouths – in true polypores the pore mouths are sterile (page 974).

– gregarious cyphelloids may recall resupinate polypores (page 1076).

See also the overview and the references to further reading on page 824.

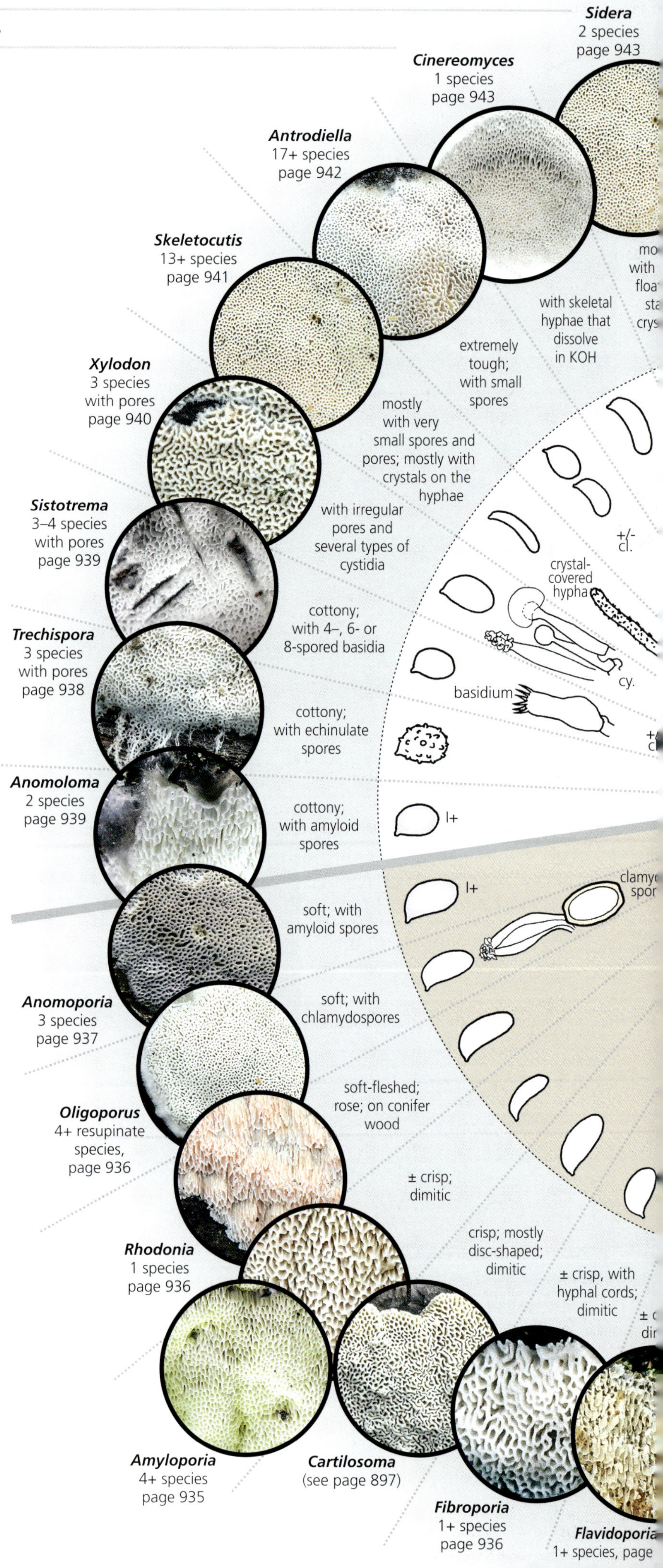

Micro-drawings: spores first, followed on the inside by cystidia.
cl. = clamps.
cy. = cystidia.

Approximate species number applies to temperate Europe.

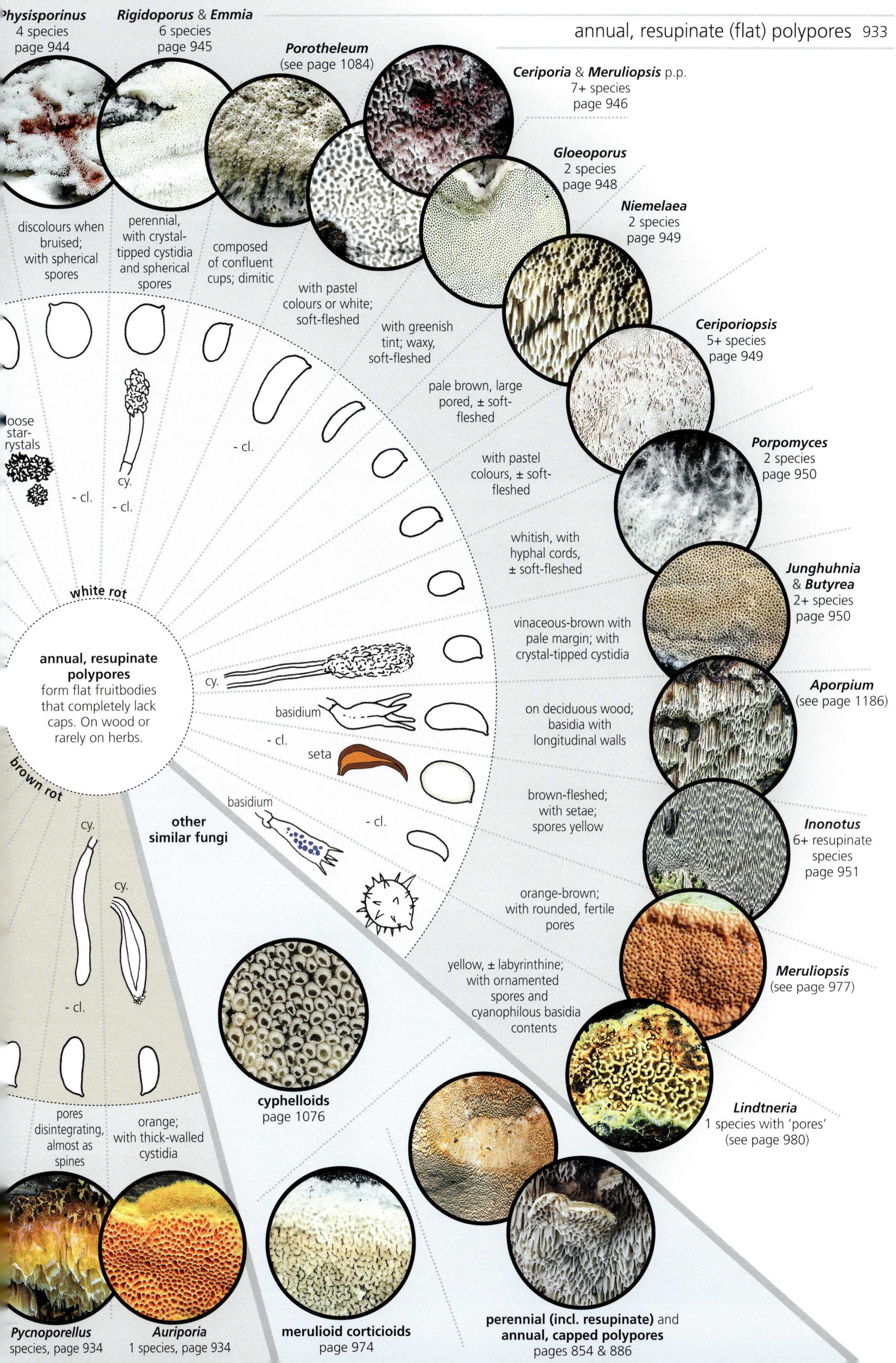
Physisporinus
4 species
page 944
Rigidoporus & Emmia
6 species
page 945
Porotheleum
(see page 1084)
Ceriporia & Meruliopsis p.p.
7+ species
page 946
Gloeoporus
2 species
page 948
Niemelaea
2 species
page 949
Ceriporiopsis
5+ species
page 949
Porpomyces
2 species
page 950
Junghuhnia
& Butyrea
2+ species
page 950
Aporpium
(see page 1186)
Inonotus
6+ resupinate
species
page 951
Meruliopsis
(see page 977)
Lindtneria
1 species with 'pores'
(see page 980)
discolours when bruised; with spherical spores
perennial, with crystal-tipped cystidia and spherical spores
composed of confluent cups; dimitic
with pastel colours or white; soft-fleshed
with greenish tint; waxy, soft-fleshed
pale brown, large pored, ± soft-fleshed
with pastel colours, ± soft-fleshed
whitish, with hyphal cords, ± soft-fleshed
vinaceous-brown with pale margin; with crystal-tipped cystidia
on deciduous wood; basidia with longitudinal walls
brown-fleshed; with setae; spores yellow
orange-brown; with rounded, fertile pores
yellow, ± labyrinthine; with ornamented spores and cyanophilous basidia contents
loose star-crystals
- cl.
cy.
- cl.
- cl.
white rot
annual, resupinate polypores
form flat fruitbodies that completely lack caps. On wood or rarely on herbs.
cy.
basidium
- cl.
seta
basidium
- cl.
brown rot
other similar fungi
cy.
cy.
- cl.
pores disintegrating, almost as spines
orange; with thick-walled cystidia
cyphelloids
page 1076
Pycnoporellus
species, page 934
Auriporia
1 species, page 934
merulioid corticioids
page 974
perennial (incl. resupinate) and
annual, capped polypores
pages 854 & 886

Teppo Helo

Pycnoporellus alboluteus is a ± resupinate annual polypore with vivid orange-yellow colours and coarse, disintegrating pores, 0.3–1 per mm. The hyphal system is monomitic. Spores measure 9–12 × 3–3.5 µm. Occurs on brown rot formed by *Fomitopsis pinicola* (page 867) on large, dead conifer trunks, mostly *Picea*. A good indicator species of forests with high consevation value.

Pycnoporellus fulgens (page 922) is capped and has 1–3 pores per mm. Other similar orange species do not have dentate disintegrating pores.

A rarity in hemiboreal–boreal Scandinavia and eastern central Europe; mostly summer–autumn.

Jean-Pierre Dechaume

Auriporia aurulenta is a striking, orange, yellow-rimmed, rather thick, resupinate annual polypore with 2–3 pores per mm. Has thick-walled, fusiform cystidia with crystals at the top. The smooth, inamyloid spores measure 4.5–6 × 2–3 µm. Forms a brown rot in conifer wood, very rarely deciduous wood.

The rare *Hapalopilus aurantiacus* ⋉ stains blood-red with KOH and produces a white rot. The equally rare, pinker *Aurantiporus priscus* ⋉ (*Erastia salmonicolor* ss. auct.) also produces a white rot.

Known mainly from the uplands in central Europe; all year.

Thomas Læssøe

Flavidoporia mellita is a pale, coarse-pored, resupinate, brown-rotting annual polypore. The pores are ± buff, large, angular but splitting and thin-walled, 1–1.5 per mm. Lacks cystidia and does not react with iodine. The spores measure (4.5–)5–9 × 3–4.5 µm and have a flat to concave side. Mostly found on *Salix* and *Populus*.

Also known as *Antrodia mellita*. *Antrodia heteromorpha* (page 897) has larger spores, is more whitish and has a tendency to form caps; *A. macra* ⋉ has smaller pores and longer spores. *Fibroporia vaillantii* ▷▷ is white and smaller-pored.

Seems to be confined to hemi-boreal–boreal Scandinavia; all year.

Amyloporia ▽, ***Antrodia*** (page 896), ***Cartilosoma*** (page 896), ***Flavidoporia*** ◁, ***Fibroporia*** ▷▷ and ***Neoantrodia*** (page 896)) are brown-rotters with a dimitic hyphal structure with skeletal hyphae and usually tough flesh. Most species are resupinate or reflexed and occur on conifers.

Amyloporia xantha is a ± resupinate annual polypore, which, on vertical substrates, forms small, cap-like protuberances. Dry fruitbodies are fragile and break into cubes. The flesh is dimitic with amyloid skeletal hyphae. The pores are pale lemon-yellow and somewhat irregular, 4–6 per mm. They react blue with Melzer's reagent. Fresh fruitbodies smell like lemon-scented soap and taste bitter. Lacks cystidia but has clamps. The hyaline, allantoid spores measure 4–5 × 1.2–1.5 µm. Forms a brown rot, almost always on conifer wood, particularly *Larix*.

Amyloporia carbonica ⋉ also occurs on conifer wood but has larger spores (4.5–7 × 3 µm) and pores, which also stain blue with iodine; it is also paler. *Amyloporia alpina* ⋉ has the same colours, but only 2–5 pores per mm; the flesh stains purple-red with KOH.

Widespread and common in many areas, occasional in others; all year.

Thomas Læssøe

Amyloporia sinuosa is a resupinate, pale brown annual polypore with 1–3, finely labyrinthine pores per mm. It becomes darker brown when dry. Lacks cystidia. The ± allantoid spores measure 4–6(–7) × 1–2 µm. Forms a brown rot in conifer wood, also in construction timber in houses.

The combination of labyrinthine pores, fruitbodies that become brown when dried and small spores should make identification straightforward. *Flavidoporia mellita* ◁ is more yellowish and has wider spores. *Fibroporia vaillantii* ▷▷ also has wider spores.

Rather rare outside buildings in the nemoral zones, but more common toward the north; all year.

Thomas Læssøe

Fibroporia vaillantii is a white, resupinate, rather soft-fleshed annual polypore with 2–3 whitish, irregular pores per mm, and a fringed margin with hyphal cords. When dry and old it is brown-tinged. Lacks cystidia. The broadly ellipsoid, hyaline spores measure 4.5–7 × 3–3.6 µm. Forms a brown rot in conifer wood, also in timber, including wood work in houses, mines, *etc.*, rare on deciduous wood.

The rather soft, white fruitbodies, fimbriate margin and rather large spores should make identification relatively straightforward.

Rare in the wild but frequent in houses; all year.

Rhodonia placenta is a soft-fleshed, resupinate, ± flesh-pink annual polypore, which, on vertical substrates, produces small, cap-like protuberances. Has 3–4 pores per mm. The margin is ± fimbriate. Lacks cystidia but has clamps. The ± cylindrical, hyaline spores measure 5–7 × 2.5–3 µm. Forms a brown rot in conifer wood, mostly *Picea*.

Aurantiporus priscus ⋉ (*Erastia salmonicolor* ss. auct.) has shorter spores and produces a white rot. *Leptoporus mollis* (page 890) has similar rose colours but forms caps.

Rare in the nemoral zones, but apparently spreading, occasional farther north; mostly in the autumn.

Jan Vesterholt

Species in the genus ***Oligoporus*** are brown-rotting, soft-fleshed (monomitic), polypores (see also page 891).

Oligoporus sericeomollis is a bitter-tasting, white resupinate annual polypore with easily loosening fruitbodies that have 3–6 ± rounded pores per mm. Has thick-walled, crystal-bearing cystidia. The hyaline spores measure 4–5 × 2–2.5 µm. Produces a brown rot in conifer wood, mostly *Picea*, very rarely on deciduous wood.

The thick-walled cystidia combined with the taste make this species easy to recognize.

Widespread, but uncommon; autumn.

Oligoporus rennyi is a very thin and fragile annual polypore that usually fruits alongside its asexual, yellowish to yellow-brown state with thick-walled chlamydospores (the margin of the illustrated fruitbody). The angular pores are white to cream, 3–4 per mm. Lacks cystidia. The hyaline spores measure 4–5(–5.5) × 2–2.5 µm. Forms a brown rot in decaying conifer wood, both in natural habitats and in buildings.

Oligoporus ptychogaster (page 893) also produces asexual spores, but from brown, cushion-shaped structures.

Widespread, rather rare in the nemoral zones, more common to the north; mainly in the autumn.

Anomoporia bombycina is a resupinate, soft-fleshed, white to pale violet annual polypore with 2–4 rounded to irregular pores per mm. The flesh is monomitic and the hyphae have clamps. The hyaline, amyloid, broadly ellipsoid spores measure 5–7 × 3–5 µm. Forms a brown rot in conifer wood.

Anomoporia kamtschatica ⋉ has irregular pores and small, almost spherical spores. *Anomoloma myceliosum* ▷▷ has white hyphal cords and produces a white rot.

Widespread in the hemiboreal–boreal zones, but virtually absent from the nemoral zones; mostly in the autumn.

Jan Vesterholt

The genus ***Trechispora*** includes cottony, monomitic, resupinate fungi that have either smooth, spiny or pore-shaped hymenophores (see also pages 997 & 1039). The spores are either smooth or ornamented with small spines, and the hyphae usually have ampullate swellings at the septa and crystals on the surfaces. They produce a white rot.

Trechispora mollusca is a ± cottony, white, resupinate annual polypore with a diffuse margin and 2–5, irregular pores per mm. The hyphae are up to 8 µm wide, clamped and swollen at the septa, and ± covered with small bipyramidal crystals. The hyaline spores are densely covered in short spines and measure 3.5–4 × 3–3.5 µm. Occurs on rotten wood, both deciduous and conifer.

Similar species have narrower hyphae in the hyphal cords and larger spores. *Trechispora hymenocystis* ▽ has rhomboid crystals and large vesicles in the hyphal cords. *Trechispora candidissima* ⋉ has rod-shaped crystals and no vesicles.

Widespread and fairly common; July–January.

Trechispora hymenocystis is a white, soft annual polypore with hyphal cords at the margin and 2–3 somewhat irregular pores per mm. The hyphae are swollen at the septa and there are large, balloon-shaped vesicles in the hyphal cords. The spores are rather densely covered in short spines and measure 4.5–5.5 × 3.5–4.5 µm. Occurs on rotten deciduous and conifer wood.

Trechispora mollusca △ and *T. candidissima* ⋉ have no vesicles in the hyphal cords.

Widespread and probably common, becoming scarcer towards the north; mostly August–November.

Anomoloma myceliosum is a thin annual polypore with a very fimbriate, white margin that has prominent hyphal cords. Has 1–4 rounded to irregular, pale cream pores per mm. The flesh is monomitic and there are no cystidia. The amyloid spores are almost spherical and measure 3–4 × 2.1–2.8(–3) µm. A white-rotter on conifer wood.

Anomoloma albolutescens ⋉ has yellow hyphal cords and yellow pores.

Widespread, but apparently very rare in the western lowlands and in the UK, with scattered occurrences in the hemiboreal zone and rare in the boreal zone; mainly August–November.

Jan Vesterholt

Sistotrema albopallescens is a soft, monomitic, white to yellowish annual polypore with 2–3 pores per mm. The basidia are clamped and have 6–8 sterigmata. The spores measure 3–4.5 × 2.5–3.5 µm (Q ≈ 1.1–1.2). Occurs on wood and litter. Produces a white rot.

Ceriporia reticulata ▷▷ has 4-spored basidia, allantoid spores and no clamps. *Porpomyces mucidus* ▷▷ has 4-spored basidia, clamps and smaller spores. Poroid species of *Trechispora* ◁ have echinulate spores. *Sistotrema dennisii* ⋉ is also poroid, but has allantoid spores. The poroid and spiny-toothed species of *Sistotrema* are complexes with numerous molecular species. The name *S. muscicola* ⋉ has been used for poroid species but belong to a group of spiny-toothed species. See more on *Sistotrema* page 996.

Widespread, but apparently rare, probably overlooked; autumn-winter.

Jan Vesterholt

Jan Vesterholt

Xylodon form resupinate fruitbodies with warty, spiny or pore-shaped hymenophores. Usually has constricted basidia. The hymenium has different types of cystidia and the hyphal system is monomitic or dimitic. All species produce a white rot. See also page 1044.

Xylodon subtropicus is a whitish to cream, resupinate, tough-fleshed annual polypore with 1–3 ± labyrinthine pores per mm. Lacks proper skeletal hyphae but the hyphal ends are covered in numerous crystals. Cystidia simple and capitate, heads up to 6 µm wide. Spores hyaline, smooth measure 4–5 × 3–4 µm. Occurs mostly on deciduous wood, particularly *Fagus* and *Quercus*; rarely on conifer wood.

Also known as *Hyphodontia radula* and *Schizopora radula*. *Xylodon paradoxus* (page 1044) is typically coarser and more dentate, has somewhat larger spores and is dimitic with proper skeletal hyphae. *Xylodon flaviporus* ▽ has more pores per mm and smaller spores.

Widespread and common; all year.

Jens H. Petersen/Thomas Læssøe

Xylodon flaviporus is a ± yellowish, resupinate, tough annual polypore with 3–5 irregular pores per mm. The cystidia are simple, capitate and often crystal-covered. Has skeletal-like hyphae and clamps. The smooth, hyaline spores measure 3.5–4.5(–5) × (2.5–)3–3.5 µm. Occurs on deciduous wood.

Also known as *Hyphodontia flavipora* and *Schizopora flavipora*. *Xylodon subtropicus* △ has coarser pores and slightly larger spores. Other similar species lack cystidia or skeletal-like hyphae.

A southern species that is almost exclusive to the nemoral zones, where it can be locally common; all year, but in best condition during the autumn.

Skeletocutis form resupinate to capped, mostly somewhat tough fruitbodies. They are dimitic with skeletal hyphae, and the generative hyphae usually have a crystal covering, mainly in the pores. The spores are very small, typically 0.5–1.5 µm wide, and allantoid. White-rotters. See also the capped species page 901.

Skeletocutis odora is a resupinate, tough, whitish to pale apricot-orange annual polypore with a somewhat fimbriate margin. It has (2–)3–5 somewhat irregular pores per mm. The flesh is dimitic and, when fresh, smells of garlic or stink bugs. Lacks cystidia, but has crystal-covered hyphal ends. The allantoid spores measure 4–5 × 1–1.5 µm. Forms a white rot in conifer wood, typically *Picea* in old forests with lots of dead wood.

Skeletocutis stellae ⋉ has narrower spores, smaller pores and smells less strongly; *S. papyracea* ⋉ does not have a disctinctive smell, has somewhat wider spores and has hyphae that partly dissolve in KOH.

Mainly an eastern (taiga) element in temperate Europe; all year.

Skeletocutis biguttulata is a white, resupinate annual polypore that is up to 2 mm thick and has 5–7 pores per mm. The allantoid, hyaline spores have two obvious drops and measure 4.5–6.2 × 1.2–1.6 µm. Occurs mostly on *Picea* wood, less common on other conifers. A white-rotter.

Skeletocutis subincarnata ⋉ has somewhat more irregular pores and lacks the prominent spore drops. *Sidera* ▷▷ has more allantoid spores.

Very rare in the nemoral zones, although possibly overlooked, but rather common farther north; all year.

Thomas Læssøe

Heinrich Holzer

See also **Antrodiella** on page 898.

Antrodiella parasitica is a resupinate, white annual polypore with 1–3 pores per mm and trimitic, tough flesh. Has lageniform to cylindrical cystidia. The hyaline spores measure 3–4 × 2–2.5 (–2.7) µm. Associated with fruitbodies of *Trichaptum* (page 914) and wood degraded by these polypores. A white-rotter.

The distinctive cystidia and the habitat make this species relatively easy to recognize, but see also *Skeletocutis carneogrisea* (page 901).

Very rare, but may be somewhat overlooked, increasing towards the north; perhaps all year.

Antrodiella niemelaei is a whitish, resupinate, tough annual polypore with 1–4, rather coarse, irregular pores per mm. The flesh is dimitic; has gloeocystidia. The hyaline, ellipsoid spores measure 2.8–4 × 1.8–2.2 µm. Obligately associated with wood decomposed by *Hymenochaetopsis tabacina* (page 973), typically in old *Salix* carrs. A white-rotter.

Frequently reported as *A. americana*. *Antrodiella faginea* ⋉ may also occur with *Hymenochaetopsis tabacina* but has 4–7 pores per mm. Other similar *Antrodiella* species lack gloeocystidia.

Rather widespread, rare, but also overlooked; mainly July–November.

Antrodiella citrinella is a ± lemon-yellow, resupinate to somewhat capped, tough annual polypore with 3–4(–5) pores per mm. Lacks cystidia. The hyaline spores measure 3–3.5 × 2–2.5 µm and have a large internal drop. Occurs on and around fruitbodies of *Fomitopsis pinicola* (page 867). A white-rotter.

Amyloporia xantha ◁◁ has allantoid spores and produces a brown rot.

Seems to be absent from the nemoral zones, but occasional in the hemiboreal–boreal zones; all year.

Cinereomyces lindbladii is a resupinate, ± pale grey, rarely ± cream annual polypore with a finely fimbriate margin. Has 3–4 pores per mm. The ± amyloid skeletal hyphae dissolves in KOH. Lacks cystidia but has clamps. The hyaline spores measure 5–7 × 1.5–2 µm. A white-rotter, mainly in conifer wood, rarely in deciduous wood.

Resupinate forms of *Bjerkandera adusta* (page 912) are darker, lack skeletal hyphae and have wider spores. Less grey forms recall many other resupinate polypores, but the KOH reaction is unique.

Widespread and rather common; all year, but mostly during late autumn and winter.

Sidera vulgaris is a cream, very thin annual polypore with 6–8 pores per mm. Has skeletal hyphae and small, pointed cystidia with crystals at the top or with loose, floating clumps of star-shaped crystals. The hyphae have clamps. The very allantoid spores measure 3–4 × 1–1.5 µm. Occurs on both conifer and deciduous wood. A white-rotter.

Sidera lenis ⋉ is very similar but has more crystal-bearing cystidia, somewhat larger spores and a more northerly distribution. Other species of polypore with crystal-covered cystidia have larger pores.

Widespread and rather common; all year.

Physisporinus, ***Rigidoporus*** and ***Emmia*** are polypore genera with rather wide spores and hyphae that lack clamps. Many species have small cystidia, sometimes capped with crystals. See also the perennial species of *Physisporinus* and *Rigidoporus* page 861.

Physisporinus vitreus is a waxy, ± greenish annual polypore. Occurs on vertical substrates and produces small, cap-like protuberances. Has 4–6 pores per mm. Stains somewhat brownish with age and after bruising. The flesh is monomitic and the hyphae lack clamps. The hyaline, smooth spores are almost spherical and measure 5–6 × 4–5 μm. Produces a white rot in wood of both deciduous and coniferous trees, mostly in damp habitats.

Physisporinus sanguinolentus ▽ is pure white when fresh and stains blood-red after bruising.

Widespread and common; mainly August–November.

Physisporinus sanguinolentus is an annual, pure white, resupinate polypore that stains blood-red and later brown after bruising. Has 3–4 pores per mm. The flesh is monomitic and lacks clamps. The spores measure 6–7 × 5–6 μm. Occurs on water-soaked wood, both deciduous and coniferous.

Apparently constitutes a species complex. *Physisporinus vitreus* △ may stain brownish but never blood-red when bruised; it is typically somewhat greenish and has smaller pores. Species of *Ceriporia* ▷▷ may also stain red, but they have oblong spores and lack cystidia.

Widespread and common; mainly August–November.

Rigidoporus obducens is a ± resupinate, pale cream, pale-fleshed, tough, annual to perennial polypore with 4–6 pores per mm. Has thick-walled, crystal-capped cystidia. The hyaline, ± thick-walled spores measure 3–4.5 × 2.5–3.5 µm. Occurs on trunks of various deciduous trees in forests and parks. A white-rotter.

Formerly treated as *Oxyporus obducens*. *Rigidiporus corticola* ▽ has gloeocystidia and larger spores. *Emmia latemarginata* ▽ also has larger spores but no gloeocystidia. Other similar polypores typically lack crystal-bearing cystidia.

Widespread, occasional; mainly late autumn–winter.

Rigidoporus corticola is a resupinate, mostly annual polypore with 2–4 pores per mm. Has crystal-capped cystidia and gloeocystidia. The spores are mainly seen floating in tetrads and measure (4.5–)5–6 × 3.5–4.5 µm. Occurs mostly on fallen *Populus* trunks, but also known from other deciduous hosts and from conifer wood.

Rigidoporus obducens △ has smaller pores and spores and no gloeocystidia. *Botryodontia millavensis* ⋉ has 1–2 pores per mm and occurs on conifers and *Philadelphus*.

Widespread and rather common in the hemiboreal–boreal zone but rare in the western lowlands; ± all year.

Emmia latemarginata is a resupinate, white to yellowish annual polypore with a somewhat fimbriate margin and 1–3 angular pores per mm. Has crystal-capped cystidia, but no gloeocystidia. The spores do not occur in tetrads and measure 5.5–7 × 4.5–6 µm. Occurs on various deciduous hosts. A white-rotter.

Was formerly treated in the genus *Oxyporus*. *Rigidoporus corticola* △ has gloeocystidia, the spores are mainly found in tetrads, and the fruitbodies are tougher.

A southern, rather rare species; all year.

Lucie Zíbarová

Jens H. Petersen/Tom Smidth

Species of ***Ceriporia*** form resupinate, ± waxy but fragile fruitbodies that mostly have or acquire pink colours. They are monomitic with wide, clampless hyphae and have smooth, ± allantoid spores. All are white-rotters. May be confused with other resupinate polypores and *Porotheleum fimbriatum* (page 1084), which is dimitic and has clamps.

The taxonomy of the genus is in a state of flux and species identification is difficult.

Jan Vesterholt

Ceriporia viridans is a rather soft, resupinate, mostly rather small annual polypore with pale cinnamon-brown colours, often with a greenish tinge. It typically slowly stains pinkish when bruised and with age. Has 3–5 pores per mm. The hyphae lack clamps. The cylindrical to allantoid spores measure 4–6 × 1.5–2 µm. Forms a white rot in various deciduous hosts, very rarely in conifers.

Ceriporia excelsa ⋉ has more expanded fruitbodies with somewhat coarser pores and rather shorter and wider spores that measure 3.3–4.3 × 2–2.5 µm. *Ceriporia aurantiocarnescens* ▽ has more vivid apricot colours and somewhat smaller spores.

Widespread and apparently rather common; mainly May–November.

Thomas Læssøe

Ceriporia aurantiocarnescens is a resupinate, distinctly apricot-orange annual polypore with 4–5 pores per mm. The somewhat bean-shaped spores measure 3.5–4(–4.3) × 1.8–2 µm. Forms a white rot in various deciduous hosts.

Ceriporia viridans △ is less vividly coloured and has somewhat larger spores. *Ceriporia excelsa* ⋉ should have somewhat wider spores and be more pink. A very difficult complex.

Rare or overlooked; mainly summer–autumn.

Ceriporia reticulata is a small, white annual polypore with 3–4 almost labyrinthine pores per mm. The hyaline, allantoid spores measure 7–9.5 × 2–3.5 µm. Occurs mostly on dead deciduous wood, but occasionally on old polypores or conifer wood.

The large, allantoid spores combined with a lack of clamps and the white colours make identification rather straightforward. Species of *Trechispora* ◁◁ are clamped, have spiny spores and hyphae with a swelling around the septa.

Widespread and common; mainly April–November.

Ceriporia purpurea is typically a rather small, resupinate, whitish to purple-red annual polypore with 3–4 pores per mm. Stains mostly slowly reddish when bruised. The allantoid spores measure 5–7 × 2–2.5 µm. Occurs on various deciduous hosts, including on attached dead branches (never on conifer wood).

Ceriporia bresadolae ⋉ is very similar but occurs on conifer wood. *Meruliopsis violacea* ▽ is also very similar but should have smaller spores. A difficult complex.

Widespread and occasional; mainly August–January.

Jens Maarbjerg/Jens H. Petersen

Meruliopsis violacea is a pale but strongly purple- to red-staining, white-margined annual polypore with 3–5 pores per mm and very large fruitbodies. The spores measure (3–)3.4–4.4 × (1.9–)2–2.4 µm. Occurs on both deciduous and coniferous wood.

Formerly also known as *Ceriporia tarda*. *Ceriporia purpurea* △ usually has smaller fruitbodies and larger spores. *Ceriporia excelsa* ⋉ is less strikingly coloured.

Mainly an eastern species, *e.g.* recorded in Finland and Estonia; summer–autumn.

Gloeoporus pannocinctus is a rather thick, waxy, greenish-tinged annual polypore with 6–8 very small pores per mm and a darker, somewhat gelatinous layer towards the substrate. The margin is pale and fimbriate. The flesh is monomitic but has thick-walled generative hyphae and obvious ± open clamps. The hyaline, allantoid spores measure 3.5–4.5 × 0.7–1 µm. Forms a white rot and very large fruitbodies on large, fallen trunks of deciduous trees, mostly *Fagus*, but seems to prefer trees that died standing rather than being blown down or felled.

Also known as *Ceriporiopsis pannocincta* and *Gelatoporia pannocincta*. A rather easily recognizable species, particularly due to the colour and the very narrow spores.

Widespread, and apparently becoming commoner in parts of temperate Europe; mostly July–November.

Niemelaea balaenae is a resupinate, cream, soft-fleshed annual polypore with 2–4 pores per mm. Has obvious, ± open clamps. The spores are ± thick-walled, have refractive contents, and measure 3.4–4.2(–4.8) × 2.5–3.2 (–3.8) µm; Qav. 1.3. Forms a white rot in deciduous wood, mostly *Salix*.

Niemelaea consobrina ⋉ is very closely related but has 1–2 pores per mm and a spore Qav. of 1.5. *Ceriporiopsis aneirina* ⋉ becomes browner and the spores are larger, up to 7 µm long. *Porpomyces mucidus* ▷▷ has hyphal cords and narrower spores (1.9–2.5 µm wide).

A rare and mainly northern species; all year.

Species of ***Ceriporiopsis*** are soft-fleshed, with delicate colours and obvious clamps. White rotters.

Ceriporiopsis gilvescens is an annual, ± reddish-tinged, resupinate polypore with 4–5 pores per mm. Has monomitic flesh and obvious clamps. The hyphae in the pores are covered in small crystals. The ± cylindrical spores measure 3.5–4.5(–5) × 1.5–2 µm. Forms large fruitbodies on large trunks of deciduous trees, mostly *Fagus*. A white-rotter.

Similar species of *Ceriporia* ◁◁ lacks clamps on the hyphae.

Mainly nemoral, rather rare; August–February.

Ceriporiopsis resinascens is a resupinate annual polypore that quickly stains brown and becomes waxy. Has 3–4 pores per mm. The flesh is monomitic and has obvious clamps. The hyaline spores measure 4–6(–8) × 2–2.6 µm. Forms a white rot in deciduous wood, mainly on attached branches in old *Salix* carrs.

Ceriporiopsis pseudogilvescens ⋉ has relatively wider spores, 3.5–6 × 2.5–3.1 µm. *Brunneoporus kuzyana* ⋉ occurs in the same habitat, but has larger spores.

Widespread and common in the northern parts, including southern Scandinavia, becoming scarcer farther south; all year.

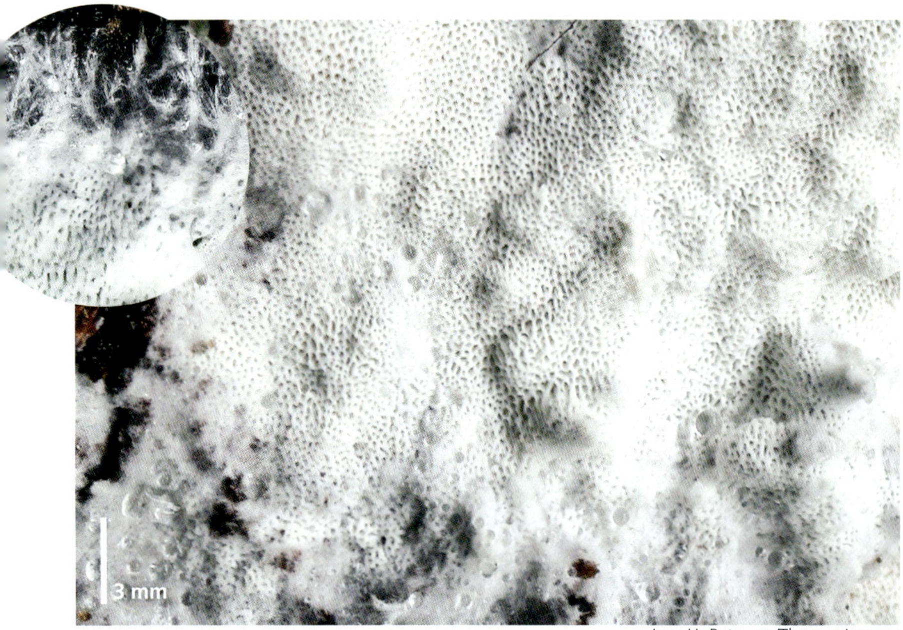

Jens H. Petersen/Thomas Læssøe

Porpomyces mucidus is a white to cream, easy to loosen, resupinate annual polypore with or without hyphal cords at the margin. Has 3–5 pores per mm. The hyphae have small, normal clamps and sometimes crystals. The hyaline spores measure (2.3–)2.5–3.5(–4) × 1.9–2.5 µm; Qav. 1.6. Forms a white rot in rather decomposed wood, both deciduous and coniferous.

Species of *Niemelaea* ⊲⊲ have large, open clamps and spores with refractive contents.

Widespread and rather common in the north, but becoming scarcer farther south; mainly August–October.

Junghuhnia nitida is a resupinate, buff to clay pink annual polypore that is easy to loosen from the substrate. The margin is pale and fimbriate. Has 5–7 angular to labyrinthine pores per mm. Dimitic, with obvious, long, thick-walled cystidia that are crystal-covered on the upper part. The hyaline spores measure 4–4.5 × 2–2.5 µm. Occurs on various deciduous hosts, mostly in forests on rich soils, rarely on conifer wood. A white-rotter.

Junghuhnia lacera ⋉ and *J. fimbriatella* ⋉ have more obvious hyphal cords and smaller spores: 3.8–4 × 2–2.2 µm and 2.3–3 × 1.9–2.2 µm respectively.

Widespread, common to occasional; mainly August–November.

Teppo Helo

Butyrea luteoalba is a resupinate, buff, rather firmly attached annual polypore with a pale margin. Has 4–6 small and angular pores per mm. Dimitic with skeletal hyphae and long, thick-walled cystidia with a crystal-covered upper part. The spores measure 4–5 × 1.5–2 µm. Occurs on conifer wood, mostly *Pinus*.

Junghuhnia collabens ⋉ also occurs on conifers but is typically darker. Other species of *Junghuhnia* △ may look very similar, but usually have larger pores and wider spores.

Mainly an eastern or montane species, absent from the western lowlands, including the UK; mainly autumn.

Inonotus obliquus mainly produces resupinate annual fruitbodies hidden below the bark on ± vertical surfaces of a standing host tree. Has 3–6 pores per mm. The fungus can be detected when the bark splits, and by the bright yellow spore-deposit. The sterile chaga-state on *Betula* (see lower images) is much more eye-catching than the fruitbodies. Has setae in the hymenium but not in the flesh. The spores measure 9–10 × 5.5–6.5 µm. Mostly on *Betula,* but also on *Fagus* and rarely on other deciduous trees. A white-rotter.

The similar *I. ulmicola* ⋉ has setae in the flesh and occurs on *Ulmus*.

Widespread and common, but rarely seen fruiting; all year (fertile in early autumn).

The strange black, almost carbonous cankers – the so-called chaga-state – contain very few asexual spores and the purpose of these prominent structures remains unknown. Chaga, crushed and used as a tea, has been used in traditional medicine in the northern part of Russia for hundreds of years.

Thomas Læssøe/Jens H. Petersen

Rosette-fungi and the like

This form group comprises species that, although not closely related, are collectively known as 'stipitate stereoids'. All the species in this group produce fan- to rosette-shaped or folded fruitbodies with a ± smooth hymenium on the under/outer side. The upper/inner side is mostly sterile. The fruitbodies are typically rather tough and range from very large in *Sparassis* to small in *Muscinupta*. Species of *Thelephora* form ectomycorrhiza, while the remaining species are presumed to be decomposers or moss parasites.

Other similar fungi:

– the clustered polypores may form large, rosette-shaped fruitbodies but they always have pores on the underside (page 846).

– capped stereoids are tough, mostly broadly attached to the substrate, and not stipitate (page 962).

– pleurotoids with reduced gills (*e.g. Gloiocephala* and *Arrhenia*) form soft-fleshed fruitbodies that usually have a veined hymenophore (page 60).

– cyphelloids do not normally form stems (page 1076).

– flattened, clavate species of *Ascomycota* are soft-fleshed with a hymenium on both sides and the spores are formed within asci (page 1350).

Further reading: 92, 142, 156, 163, 264.

Micro-drawings: spores first, followed by cystidia on the inside. cl. = clamps.

Approximate species number applies to temperate Europe.

Thelephora
9+ species
page 956

Podoscypha
1 species
page 955

photo Lukas Large

large; with numerous rosettes; with gloeocystidia and skeletal hyphae

very large; with numerous folds; without skeletal hyphae

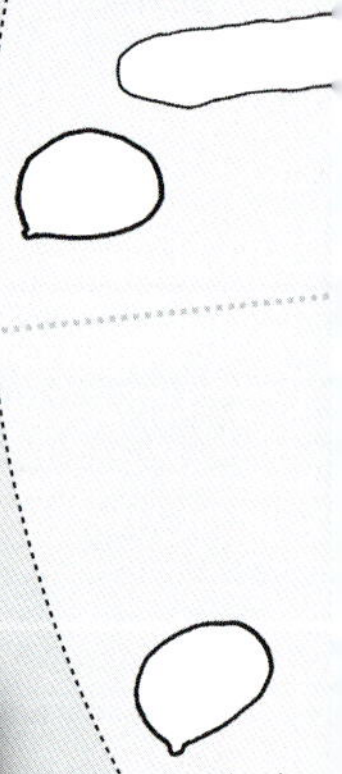

Sparassis
2 species
page 954

capped corticioid
page 962

stipitate stereoids
form fan- to rosette-shaped or folded fruitbodies with a hymenium on the ± smooth lower/outer side
medium-sized; with brown flesh, brown hyphae and brown, verrucose spores
+/- cl.
medium-sized to small; with white flesh and numerous, ± projecting cystidia
- cl.
Cotylidia
3 species
page 958
small; with white flesh, yellow colours and very small spores
- cl.
Stereopsis
1 species
page 959
medium-sized; with tough flesh and thick-walled cystidia
- cl.
I+
Stereum
(see page 962)
small and spatula-shaped; with projecting cystidia; on mosses
- cl.
Muscinupta
1 species
page 959
with skeletal hyphae
+cl.
other similar fungi
± tough
± fragile and soft-fleshed
clustered polypores
page 846
pleurotoids without gills
page 60
cyphelloids
page 1076
earthtongues and others
page 1350

Thomas Læssøe

Species of ***Sparassis*** have folded, large to very large annual fruitbodies with a ± one-sided hymenium. They are brown rotters.

•• ***Sparassis crispa*** has large to very large, highly branched and folded fruitbodies with ± curled, flat branches and a somewhat cauliflower-like appearance. It has a pleasant, aromatic smell. The hymenium is situated on the underside of the folds, and the flesh is monomitic with clamped hyphae. The spore-deposit is white. The hyaline, smooth, inamyloid spores measure (4.5–)5–7 × 3.5–5 µm. Forms a brown rot and fruits at the base of living or dead conifers, or on conifer stumps.

Sparassis brevipes ▽ has flatter, less curled, mainly ± zonate folds and typically decomposes *Quercus* wood. Microscopically the two species are almost identical. The cup-fungus *Peziza proteana* var. *sparassoides* (page 1314) may recall a small, young *S. crispa* but has a somewhat violet hymenium, occurs on sterilized soil (*e.g.* burnt ground), and forms spores in asci.

Widespread, occasional, almost absent from the boreal zone; July–January.

Jiří Polčák

•• ***Sparassis brevipes*** has large to very large fruitbodies with rather flat, not especially curled folds, which may be somewhat zonate. Microscopically mostly as *S. crispa*, but according to some sources it lacks clamps. A brown-rotter on various deciduous trees, especially *Quercus*, but has also been reported from *Abies*.

Sparassis laminosa and *S. nemecii* are somewhat more recent names for this species. *Podoscypha multizonata* ▷ also occurs on/with *Quercus* but is darker and has skeletal hyphae.

A southern species that reaches as far north as the Netherlands and north-east Germany, but recently also reported from Sweden, rare; autumn.

Podoscypha multizonata produces rosette-shaped, large to very large fruitbodies, up to 40 cm wide, with a rooting stem and white, tough flesh. The hymenium is smooth and situated on the underside of the folds, and the sterile upper side is ± smooth, zonate and pale brownish-red to darker brown. The flesh is dimitic with skeletal hyphae; it also has gloeocystidia in the cap cuticle and clamped hyphae. The inamyloid, smooth spores measure 4.8–6.2 × 3.8–4 µm. Occurs on roots of old *Quercus* and rarely *Fagus*, mostly in park-like habitats. Possibly parasitic but its precise biology is not known.

Cotylidia pannosa ▷▷ can look like a miniature version of *P. multizonata* but is not deeply rooting and has neither gloeocystidia nor clamps. *Sparassis brevipes* ◁ also occurs on *Quercus* but is paler and lacks skeletal hyphae.

Relatively widespread in the UK, but very rare in central and western continental Europe; summer–autumn.

Malcolm Storey

Thomas Læssøe

Thomas Læssøe

Thelephora is a genus with very variable fruitbodies from resupinate, shelf-, rosette- to bush-shaped or branched. All species have pale brown, angular-verrucose spores and brown, clamped hyphae. The genus is in need of a detailed review. All species are ectomycorrhizal. See also page 1152.

Thelephora terrestris is a brown, fan-shaped to imbricate *Thelephora* with a rugose-bumpy, felty upper side and, on specimens in active growth, a pale margin. The lower, fertile side is grey-brown with scattered warts. It may also occur in a completely resupinate form. The hyphae have clamps. The angular-verrucose, brown spores measure 8–12 × 6–9 µm. Occurs on ± acid, poor soils, including on soft, rotten wood; mostly with conifers.

Thelephora caryophyllea ▷ is typically smaller, almost funnel-shaped and has a smooth hymenium. *Thelephora atra* ▷ is darker and grows on clay-rich soils. Species of *Tomentella* and *Pseudotomentella* (pages 998 & 1035) are ± similar to the resupinate version of *T. terrestris*.

Widespread and common to very common; mainly August–November.

Thelephora atra is a dark, reflexed *Thelephora* with small, projecting caps that, at first, are yellowish-grey. The spores are ± spherical, measure 9–13 × 8–11 µm and have 1–2 µm long spines. Occurs on ± calcareous/ clay-rich soils, mostly in deciduous forests.

Darker grey than *T. terrestris* ◁ and has more rounded spores, but correct identification of species in this complex is very difficult. *Thelephora atrocitrina* ⋉ is possibly a closely related species that also occurs in the area.

Widespread, very rare, absent from the boreal zone; mainly June–September.

Thelephora caryophyllea is a rather small, dark, funnel-shaped *Thelephora* with a fringed-split margin and a grey-brown, ± smooth hymenium. The somewhat nodulose-angular spores measure 6.5–8.5 × 5–7 µm and have 1–1.5 µm long spines. Occurs on open, calcareous soils, mostly mycorrhizal with *Salix*.

Thelephora terrestris ◁ is more fan-shaped, more warty on the hymenium and has larger spores.

Widespread, occasional; July–November.

Jens Maarbjerg

Cotylidia form small, ± fan-shaped fruitbodies with long, projecting cystidia and smooth, inamyloid spores.

Cotylidia muscigena is a very small, fan-shaped to almost funnel-shaped *Cotylidia* with pale yellow-brown colours and highly projecting cystidia from all parts. Spores cylindrical and measure 6–8.5 × 2–2.5 µm. Occurs on mosses in open habitats, *e.g.* in dunes.

Cotylidia undulata ▽ is funnel-shaped and appears to sit on the soil (although still amongst mosses); it has smaller, ellipsoid spores, as does the fan-shaped *C. carpathica* ⋉.

Widespread, but rare or overlooked; mainly September–October.

Cotylidia undulata is a small, funnel-shaped *Cotylidia* with a fringed margin and pale yellow-brown colours. The hymenium is smooth. Has projecting, cylindrical cystidia. The ellipsoid spores measure 4–5 × 2–2.5 µm. Occurs on open soil amongst mosses, including those in the genus *Polytrichum*.

Cotylidia muscigena △ is smaller, more fan-shaped, sits directly on mosses and has cylindrical spores. *Cotylidia carpathica* ⋉ has ellipsoid spores but is fan-shaped. *Thelephora caryophyllea* ◁◁ has darker colours and ornamented, brown spores.

Widespread, rare or overlooked; mainly September–December.

Cotylidia pannosa is a rather small but fleshy, fan- to rosette-shaped *Cotylidia* with pinkish to pale brown colours. The hymenium is pale and smooth or wrinkled, while the upper side is radially felty. Has large cystidia in the hymenium. The spores measure 7–9 × 3.5–4 µm. Occurs on clay-rich soils, mostly in *Fagus* forests. The exact biology not understood.

Unlike the other species in the genus being more fleshy. *Podoscypha multizonata* ◁◁ is a larger, more zonate, tougher, rooting fungus with gloeocystidia.

Widespread in the nemoral zones, very rare but apparently increasing in some areas, absent from the boreal zone; September–January.

Stereopsis vitellina forms rather small, almost fan-shaped, yellow to yellowish-white, ± stemmed fruitbodies. The margin is mostly ragged and uneven. The sterile upper side is somewhat more fibrillose than the smooth hymenium. The flesh is monomitic and lacks cystidia, but the hyphae contain oil. The spores are inamyloid and measure 3–4 × 2–2.5 µm. Occurs on poor soils in conifer forests and plantations, usually in cavities, *e.g.* the entrance of fox earths, *etc*.

Unlikely to be confused with any other European basidiomycote, although there are many similar species in the tropics. It may recall *Spathularia flavida* (page 1358) but that species has a hymenium on both sides of the flattened part and forms spores in asci.

Nemoral to boreal, rare, most frequent in the boreal zone; August–December.

Thomas Læssøe

Muscinupta laevis forms very small, white, spatula-shaped fruitbodies. The hymenium is smooth and the sterile upper side is ± felty. The flesh is monomitic and lacks clamps. The cystidia are narrow and smooth, with a ± delimited head. The hyaline, smooth, inamyloid spores measure 4–5 × 2.5–3 µm. The fruitbody is firmly attached to mosses in the genus *Polytrichum* in damp forests.

Cyphellostereum pusiolum ⋉ is similar, but has no cystidia. Other similar moss-associated fungi, *e.g.* species of *Rimbachia* (page 1078), lack a stem and do not have such distinctive cystidia (see also amongst the cyphelloid fungi, page 1076). *Arrhenia spathulata* (page 64) is greyer, has different spores, lacks cystidia and is associated with mosses in the genus *Syntrichia*.

Widespread, but rather rare to rare; autumn.

Corticioids

This form group includes species with resupinate (flat) or reflexed (flat with a narrow protruding cap) fruitbodies and with a smooth, warty, spiny, toothed or veined hymenophore. Species with divided basidia or with spores that germinate with a replicate spore are shown on this wheel, but are treated with the resupinate jelly fungi (page 1162).

Some distinctive species can be told by macroscopical means. Colour, shape (mainly of the hymenophore) and ecology are the most important characters. The majority of the species, however, require microscopical study in order to reach an identification.

Phylogenetically the species are distributed among a series of orders that include fungi with completely different morphologies (boletes, agarics, polypores, *etc*.).

Most corticioids are decomposers, and the majority are white-rotters. Species in the *Boletales* (*e.g. Serpula* (page 984) and *Coniophora* (page 1034) produce a brown rot, and most members of the *Thelephorales*, *e.g. Tomentella* (page 998), are ectomycorrhizal. A handful of genera in other orders are also mycorrhizal.

The corticioid fungi is a large group with about 500 species in temperate Europe. Many rare or inconspicuous species are not included here, but representatives of most of the genera that occur in the area are included.

OTHER SIMILAR FUNGI:
– the rosette fungi have spathulate, cauliflower to rosette-shaped fruitibodies, with or without a ± well developed stem (page 952).
– thick-fleshed, resupinate hydnoids are treated on page 1054.
– resupinate polypores with labyrinthine pores have sterile pore margins (page 932).
– plant parasites on living stems and leaves, see *Exobasidiales* (page 1214) and *Taphrinales* (page 1622).

FURTHER READING: 36, 50, 67, 74, 75, 81, 99, 159, 163, 339.

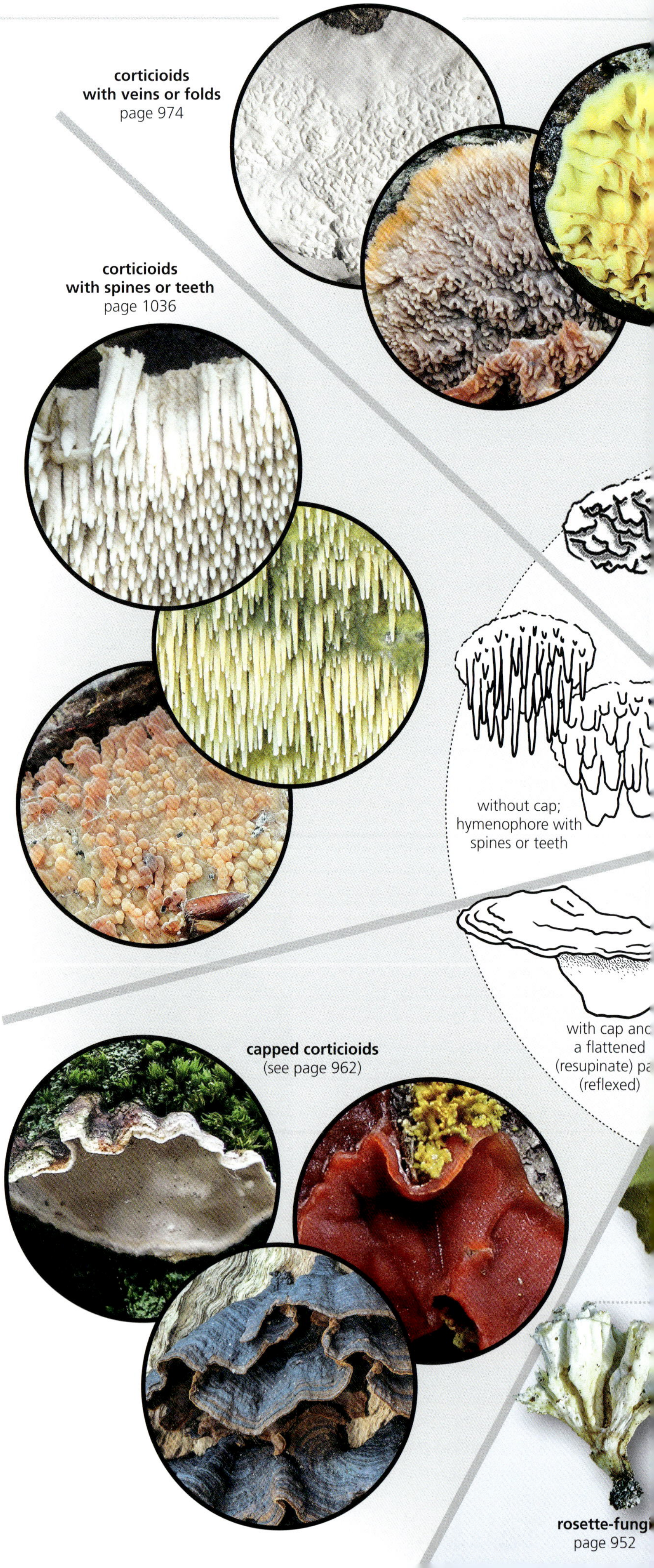

smooth, cottony or peeling, fragile corticioids
page 986

smooth, firm, whitish to greyish corticioids
page 1000

cottony or flaky, fragile

ithout cap; menophore folded or veined

without cap; hymenophore smooth or warty

gelatinous, waxy or leathery; whitish to greyish

corticioids
form ± flat fruitbodies that are found mainly on the underside or sides of dead wood or on herbaceous stems

smooth, firm, coloured corticioids
page 1012

gelatinous, waxy or leathery; coloured (not whitish to greyish)

other similar fungi

thick-fleshed hydnoids
page 1054

resupinate polypores

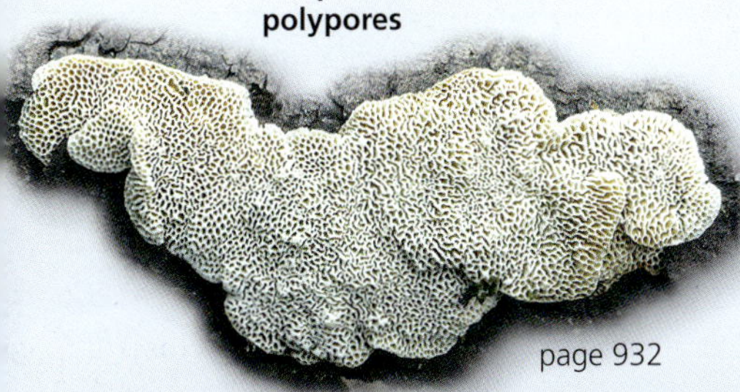

page 932

Exobasidiales
page 1214

Taphrinales
page 1622

Capped corticioids

This form group includes capped (reflexed/stereoid) corticioids with a smooth hymenophore. Capped genera with a wrinkled or echinulate hymenophore are shown in the blue part of the wheel and are treated under merulioid corticioids (page 974) and echinulate corticioids (page 1036).

Many species in this group have caps with colours, hairs and other useful characters. With experience quite a few can be recognized without the use of a microscope. All capped corticioids may start out without a cap, so always look for older fruitbodies. Phylogenetically the species are distributed in a number of groups that are not closely related, including orders dominated by corticioid, poroid or lamellate species.

Most capped corticioids are decomposers and form a white rot. *Serpula* (page 985), *Pseudomerulius* (page 981) and *Veluticeps* (page 967) are brown-rotters and species in the genus *Thelephora* (page 956) form ectomycorrhiza.

Other similar fungi:
– annual, capped polypores always have proper pores on the underside of the caps (check with a hand lens) (page 886).
– rosette-fungi either have a stem or narrow towards the base (page 952).

See also the overview and the references to further reading on page 960.

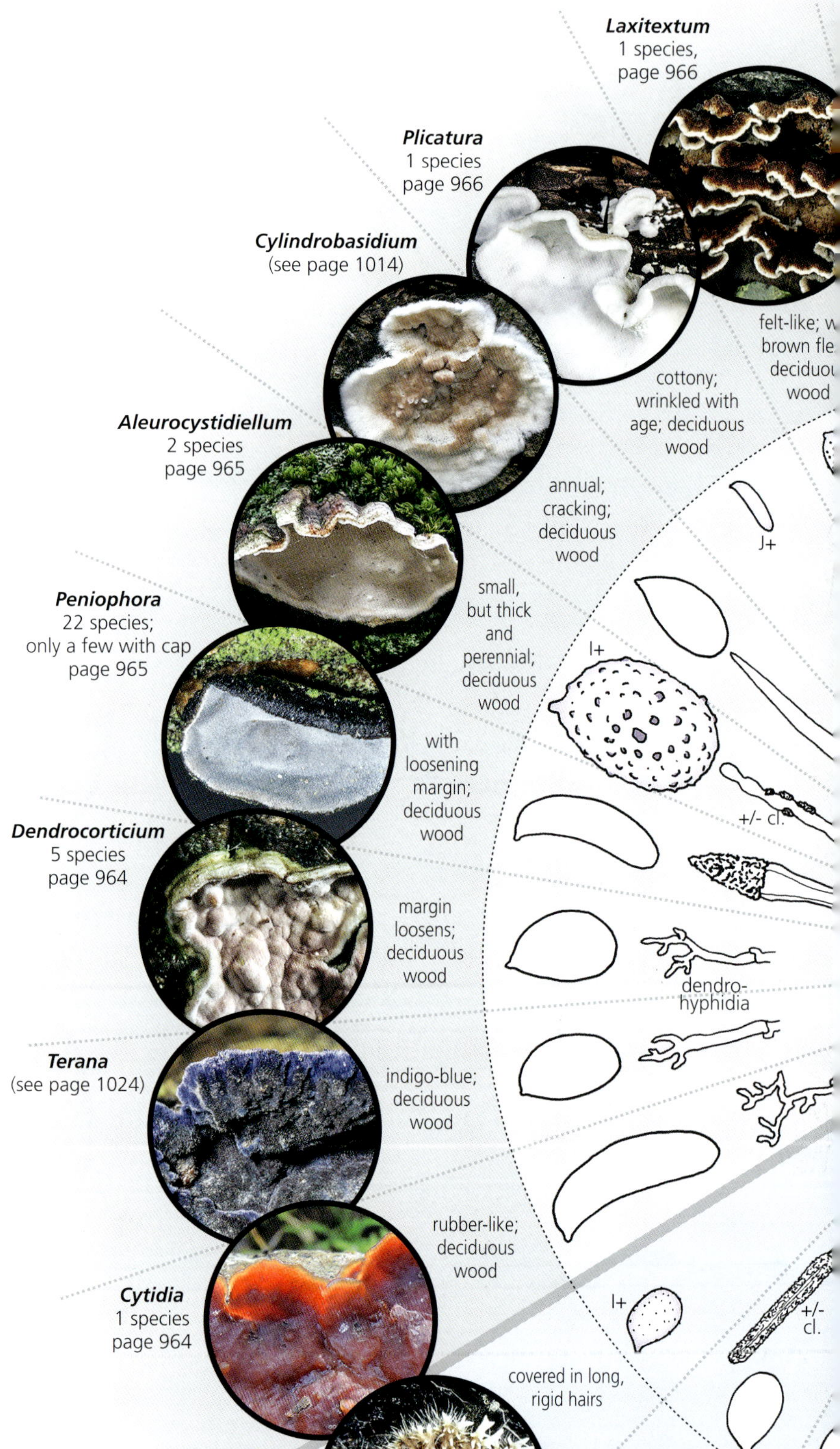

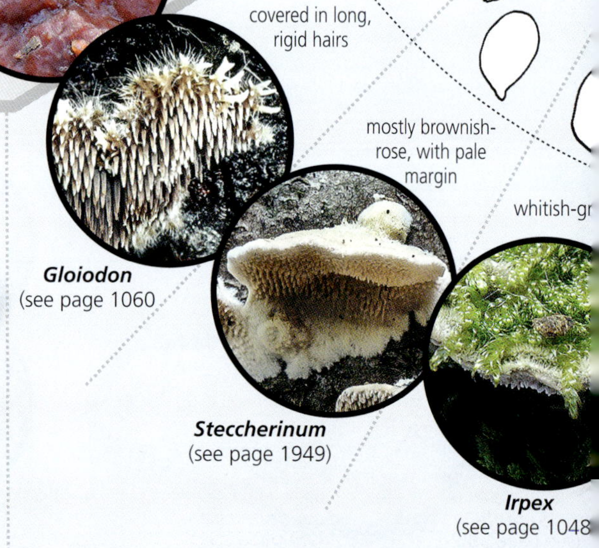

other similar fungi

annual, capped polypores
page 886

rosette-fungi
page 952

Micro-drawings: spores first, followed on the inside by cystidia. cl. = clamps.

Approximate species number applies to temperate Europe.

Jens H. Petersen/Thomas Læssøe

Cytidia salicina is a unique, thick, disk-shaped, dark scarlet to blood-red, gelatinous corticioid. When drying it becomes hard with a loosening margin (picture to the right) but revives after rewetting. Has numerous dendrohyphidia, but lacks cystidia; has clamps. The allantoid, smooth, inamyloid spores measure 12–18 × 4–5 µm. Occurs on *Salix* branches in wet scrub.

In consistency may recall *Peniophora quercina* ▷, but no species in that genus have scarlet to blood-red fruitbodies.

Widespread and rather common in the boreal zone, rather rare in the hemiboreal zone and nearly absent from the nemoral zones; all year.

Jens H. Petersen/

Dendrocorticium polygonioides is a rather thick, whitish, violet-tinged to pale brown corticioid with conspicuous cracks when dry and a pale, downy-hairy, loosening margin. Has coralloid dendrohyphidia and very long, narrow basidia. The hyphae have clamps and incrusted material. The egg-shaped spores measure 6–9 × 4–6 µm. Occurs on deciduous hosts, particularly *Salix*.

Corticium roseum (page 1023) is pinker and has larger spores; *C. boreoroseum* ⋉, on conifers, has hyphal cords. *Cylindrobasidium laeve* (page 1014) lacks dendrohyphidia.

Widespread in the hemiboreal–boreal zones of Scandinavia, absent from the nemoral zones; all year.

Peniophora quercina is a rather thick, tough, pale grey to dark greyish-violet *Peniophora* with loosening margins that reveal a black underside/cap surface. The hymenophore has an uneven, bumpy surface and is ± cracking. Has thick-walled, fusiform cystidia with crystals on the upper part. The allantoid, inamyloid spores measure 10–12 × 3–4 µm. Occurs on bark on attached, deciduous branches, mostly *Quercus*, but also on *Fagus* and other deciduous hosts.

Peniophora rufomarginata ⋉ also has a loosening margin but is more reddish, paler on the underside/cap, has smaller spores and occurs on *Tilia*. See more species of *Peniophora* page 1020.

Widespread and very common in the nemoral–hemiboreal zones; all year.

Aleurocystidiellum disciforme is a whitish to pale brown, disk-shaped, rather thick perennial corticioid with a ± obvious, narrow, protruding, ± pale brown cap. The flesh has many growth layers (inset image). Has constricted (moniliform) hyphidia. The amyloid, verrucose spores measure 15–18 × 10–14 µm. Occurs on bark of living or recently fallen older *Quercus* in damp forests, seemingly most often associated with *Quercus petraea*.

May recall a small, pale *Stereum rugosum* ▷▷ but that species stains red when bruised.

Widespread but rather rare in the hemiboreal zone, very rare in the nemoral zones; all year.

Jan Vesterholt/Jens H. Petersen

Plicatura nivea is a white to somewhat ochraceous, very soft, almost cottony, resupinate to capped corticioid with a smooth to somewhat veined, white to cream hymenophore. Lacks cystidia and clamps. The amyloid spores measure 4–5 × 0.5–1 µm. Almost exclusively on dead bark of *Alnus*, mostly *A. incana*.

Byssomerulius corium (page 977) is less cottony, has inamyloid spores and usually a more wrinkled hymenophore. See also *Plicatura crispa*, page 63.

Very common in the boreal zone, becoming scarcer farther south and absent from the nemoral zones; all year.

Laxitextum bicolor is a rather soft, felt-like, reflexed, dark red-brown to yellow-brown corticioid with a whitish to pale grey, smooth hymenophore. The flesh has a duplex structure with a thick, upper, dark brown felty zone and a compact, lower, paler zone (see inset image). Has long, ± cylindrical, oil-containing cystidia. The amyloid, verrucose spores measure 4–6 × 2.5–3.5 µm. Occurs on deciduous wood, mostly *Fagus* branches with soil contact, but also above ground on dead branches of e.g. *Salix*, and rarely on conifers.

There are a number of similar-looking fungi. Species of *Stereum* ▷▷ have a tougher flesh without a duplex structure and smooth spores. *Porostereum spadiceum* ▷ has larger, smooth spores and is more grey-brown. *Laurilia sulcata* ▷ has echinulate spores and produces perennial fruitbodies on *Picea*.

Widespread and rather rare to common, apparently rarer in Atlantic areas; mainly July–November.

The tiny ascomycote, *Capronia porothelia* (page 1612) is often seen as tiny black dots imbedded in the hymenium of *Laxitextum bicolor*.

Porostereum spadiceum is a reflexed corticioid with dull, grey-brown colours on both cap and hymenophore. The flesh is monomitic and 1–2 mm thick. Has brown, cylindrical cystidia with incrusted pigment. Spores smooth, inamyloid and measure 6–8 × 3–5 µm. Occurs on deciduous wood, particularly *Fagus*.

Species of *Stereum* ▷▷ have more vivid colours or redden when bruised. *Laxitextum bicolor* ◁ has a paler hymenophore and small, verrucose, amyloid spores.

Mainly a southern species that is rather frequent in the Netherlands, but very rare in the UK and in the Nordic countries; all year.

Thomas Læssøe

Laurilia sulcata is a leathery, perennial, reflexed corticioid up to 5 mm thick with a furrowed, zonate, brown upper side and a ± smooth hymenophore with a faint flesh-pink tinge. The cystidia are thick-walled, conical and have crystals. Has amyloid, echinulate, almost spherical spores measuring 5.5–6.5 × 5.5 µm. Occurs on fallen trunks of *Picea abies*.

Aleurocystidiellum subcruentatum ⋉ has much larger spores. *Laurilia* is closely related to the polypore *Bondarzewia* (page 851).

Mainly an eastern, boreal forest (taiga) species, absent from the nemoral–hemiboreal zones; all year.

Thomas Læssøe

Veluticeps abietina is a perennial, leathery, reflexed corticioid with 1–2 mm thick, zonate flesh, which is dimitic with brown skeletal hyphae. The cap is brown and hairy, and the hymenophore has a bluish-violet tinge and is minutely downy from projecting, very long, thick-walled cystidia; also with brown imbedded cystidia. Spores smooth, non-amyloid, and measure 9–13 × 4–5 µm. Forms a brown rot in conifer wood, mostly *Picea*.

Species of *Stereum* ▷▷ and a number of other stereoid fungi on conifer wood have amyloid spores and do not have brown skeletal hyphae.

Mainly a boreal species, also present in uplands in central Europe, but absent from the lowlands; all year.

Lucie Zíbarová

Lucie Zíbarová

Cystostereum murrayi is a large, reflexed, tough, perennial corticioid with a dark upper side and a pale brown, smooth to bumpy hymenophore that breaks into polygons when drying. Smells like coconut cake. Has thick-walled gloeocystidia. The spores are inamyloid, narrowly ellipsoid and measure 4.5–5.5 × 2–3 µm. Occurs on conifer wood; a white-rotter.

Two completely resupinate genera are similar: *Fibricium* ⋉ has no gloeocystidia and *Crustomyces* ⋉ has a warty hymenophore and dendrohyphidia.

Mainly a boreal species, but also in central European uplands, uncommon; all year.

Chondrostereum purpureum is a mainly annual, reflexed corticioid with a rather pale, hairy upper side and a ± purple, smooth underside. Has fusiform, projecting cystidia and distinctive inflated hyphal ends (vesicles) in the flesh. The almost cylindrical, inamyloid spores measure 6–8 × 2.5–3.5 µm. Grows on all kinds of deciduous wood and may occur as a parasite on *e.g. Prunus*; very rare on conifer wood. Mainly seen on the bark and cut ends of recently dead trunks.

Fresh specimens are hard to mistake, but old, worn specimens may look similar to *e.g.* species of *Stereum* ▷▷. *Veluticeps abietina* ◁◁ and species of *Amylostereum* ▷ may also have violet colours on the hymenophore but are found predominantly on conifer wood. *Laxitextum bicolor* ◁◁ normally has a paler under side that lacks violet, and small, amyloid spores.

Widespread and very common; all year.

Older fruitbodies of *Chondrostereum purpureum* are often attacked by the mould *Cladobotryum stereicola* ⋉.

Lucie Zíbarová

Amylostereum includes resupinate or reflexed, non-staining, leathery species with brown, thick-walled cystidia that have crystal caps and amyloid spores. They occur on conifers where they live in a symbiosis with wood wasps of the genera *Sirex* and *Urocreus*. The wasps store mycelium in special organs and transmit it to new trunks when egg-laying; the resultant larvae depend on the rot created by the fungus. See also the resupinate *Amylostereum laevigatum* page 1030.

Amylostereum areolatum is a rather large, perennial, reflexed corticioid with a dark, hairy upper side and a somewhat violet hymenophore. The flesh may have a dark line. Has conical, brown cystidia with incrusted pigment. The amyloid spores measure 6.5–7.5 × 2.5–3 µm. Forms a white rot in *Picea* and *Abies* wood. Considered harmful in forestry.

Amylostereum chailletii ▽ is more fragile (annual), less capped and lacks distinct cap felt and violet colours. *Chondrostereum purpureum* ◁ occurs on deciduous wood and has thin-walled, smooth cystidia.

Apparently mainly found in central Europe, occasional; all year.

Amylostereum chailletii is a reflexed, annual, rather dark brown, leathery, non-staining corticioid. The smooth to somewhat felty cap projects, at most, 1 cm out from the substrate and may be entirely absent. Has conical, brown cystidia with incrusted pigment. The amyloid spores measure 6–8 × 2.5–3 µm. Occurs on wood of *Picea* and *Abies*. Harmful in forestry.

Amylostereum areolatum △ is perennial and larger with a thicker felty layer on the cap. *Stereum sanguinolentum* ▷▷ is paler and stains red when bruised.

Widespread and common to occasional; all year.

Xylobolus frustulatus is a perennial, very tough, rather thick, whitish to pale clay-buff corticioid that recalls a jigsaw puzzle and, at most, has very narrow projecting caps. The flesh is monomitic with thick-walled hyphae and filled with echinulate-verrucose cystidia (acantohyphidia); lacks clamps. Spores amyloid and measure 4–5.5 × 3–4 µm. Occurs on *Quercus* wood in ancient woodlands and causes a characteristic, poroid rot (inset picture).

Stereum rugosum ▽ is also perennial, but reddens when bruised.

Widespread in the nemoral–hemiboreal zones, but mostly very rare and almost absent from the UK and parts of western Europe; all year.

Stereum form annual to perennial, very tough fruitbodies. Some species stain red or yellow when bruised. Has a monomitic hyphal system and numerous long gloeohyphae that break through the hymenium as cystidia, and smooth, amyloid spores. All are white-rottters on deciduous and conifer wood.

Stereum rugosum is a pale brown, leathery perennial corticioid that may form a narrow, dark brown cap but more often is completely resupinate. The flesh is zonate (inset picture) and very tough. The hymenophore stains when bruised. The amyloid spores measure 7–12 × 3–4.5 µm. Occurs on a number of deciduous hosts, particularly *Fagus*, *Corylus*, *Betula* and *Quercus*.

Stereum gausapatum ▷ also reddens, but is annual, somewhat more vividly coloured and mostly occurs on *Quercus*. *Stereum sanguinolentum* ▷ also reddens but grows on conifer wood.

Widespread and very common; all year.

Stereum rugosum is often attacked by jelly fungi in the genus *Phaeotremella* (page 1191).

Stereum sanguinolentum is an annual, rather thin-fleshed, somewhat dull-coloured, leathery corticioid that stains red when bruised. The amyloid spores measure 8–11 × 2.5–3.5 µm. Occurs on conifer wood, with *Picea*, *Pinus*, *Abies* and *Larix* as frequent hosts.

This is the only red-staining, leathery corticioid on conifer wood.

Widespread and very common; all year.

Stereum sanguinolentum is often attacked by jelly fungi, including *Phaeotremella foliacea* (page 1191) and *Naematelia encephala* (page 1192).

Stereum gausapatum is an annual, relatively thin, rather brown, white margined, resupinate to reflexed leathery corticioid that stains red when bruised. The spores are amyloid and measure 7–10 × 3.5–4.5 µm. Mostly associated with *Quercus*, mainly in open forests with many old trees.

Stereum rugosum ◁ is typically thicker (it is perennial) and more greyish. *Stereum sanguinolentum* △ occurs only on conifer wood.

Widespread in the nemoral–hemiboreal zones, common to occasional; all year.

Stereum hirsutum is a ± reflexed to completely resupinate, vivid apricot-orange, ± annual, leathery corticioid with a hairy, zonate cap. The hymenophore does not stain when bruised. The spores are amyloid and measure 6–7.5 × 2–3 µm. Occurs on rather fresh deciduous wood, frequently on piles of felled trunks, often *Quercus* and *Fagus*.

Stereum complicatum ⋉ is less than 0.6 mm thick and very pale; *S. subtomentosum* ▷▷ is larger, less resupinate, more zonate and less hairy; fresh specimens of that species stain yellow when bruised.

Widespread and very common; all year.

Jens Maarbjerg

Stereum subtomentosum is typically a capped, basally constricted, ± imbricate, vivid yellow-brown, only somewhat hairy, but very markedly zonate, annual–biennial leathery corticioid. When young and fresh it may stain yellow when bruised. Has narrow, pointed cystidia and large, cylindrical, thick-walled cystidia that are thinner-walled on the upper part. The spores are amyloid and measure 6–7 × 2–3 µm. Occurs on deciduous wood, typically larger trunks in damp forests; *Fagus* and *Alnus* are common hosts.

Stereum hirsutum ⊲⊲ has smaller, more strigose-hairy, non-staining fruitbodies. *Stereum insignitum* ▽ is even larger and almost funnel-shaped, and often more clay-buff.

Widespread and rather common, becoming scarcer towards the north; all year.

Stereum subtomentosum is quite often attacked by the violet flask fungus *Sphaerostilbella berkeleyana* ⋉ and rarely also by the discomycete *Hyphodiscus stereicola* (page 1390). The latter is found on the upper side, while the former mostly fruit on the hymenial side.

Jan Vesterholt

Stereum insignitum is a large, striking, fan- to almost funnel-shaped, very zonate, ± red-brown to clay-buff, almost smooth leathery corticioid that may stain yellow when bruised. Has cystidia with small, thick-walled protuberances (acanthohyphidia). The spores measure 4.5–7(–8) × 2–3(–3.5) µm. Occurs on trunks of deciduous trees.

Stereum subtomentosum △ has no acanthohyphidia.

A thermophilic species found only in the southern part of temperate Europe; all year.

Hymenochaete & ***Hydnoporia*** are characterized by brown setae that penetrate the hymenium (normally visible macroscopically with a good hand lens). There are also completely resupinate species (page 1032). *Inonotus* and the like (page 927) and *Phellinus* and the like (page 874) are close relatives.

Jens Maarbjerg/Jens H. Petersen

Hydnoporia tabacina is a thin and flexible, ± reflexed, zoned capped corticioid with projecting, brown setae on the hymenial side (hand lens required!). The setae are up to 100 µm long with small holes in the upper part. The smooth, inamyloid spores measure 4.5–7 × 1.5–3.5 µm. Occurs on decidious trees or shrubs, mostly *Salix* and *Corylus*.

Laxitextum bicolor ◁◁ is more felt-like, has no projecting setae and has small, amyloid spores. See also the resupinate species *Hydnoporia corrugata* (page 1033).

Widespread and common; all year. *Hydnoporia tabacina* is occasionally seen with the sepctacular parasite *Hypocreopsis lichenoides* (page 1526), and may also be attacked by *Trichoderma* species (page 1528).

The middle image shows a branch with a fruitbody seen from the underside.

Hymenochaete rubiginosa is a convex, imbricate, hard and stiff, very dark capped, perennial capped corticioid with a dark cinnamon hymenophore that has projecting brown setae (strong hand lens required!). The spores measure 4–6.5 × 2.5–3.5 µm. Occurs on old Quercus stumps, fallen branches *etc.*; also on *Castanea*.

Hymenochaete ulmicola ⋉ is a similar species with smaller fruitbodies that occurs on the bark of old living *Ulmus* trees. Completely resupinate species of *Hymenochaete* are treated on page 1032.

Widespread in the nemoral–hemiboreal zones, very common; all year.

Merulioid corticioids

This form group includes species with resupinate and reflexed fruitbodies that have a wrinkled or veined hymenophore. All such species were once included in the genus *Merulius*.

Some merulioid corticioids are very distinctive and can easily be recognized without the aid of a microscope, while others, *e.g.* species in *Leucogyrophana*, *Serpula* & *Pseudomerulius* are very difficult to separate without using micro-characters. The majority are white-rotters, but species related to the dry rot fungus *Serpula lacrymans* produce brown rots.

OTHER SIMILAR FUNGI:
- the wrinkled hymenophore grades into the spiny corticioids (page 1036).
- polypores with labyrinthine hymenophore (page 932) have sterile pore margins, whereas the species covered here have fertile margins.

See also the overview and the references to further reading on page 960.

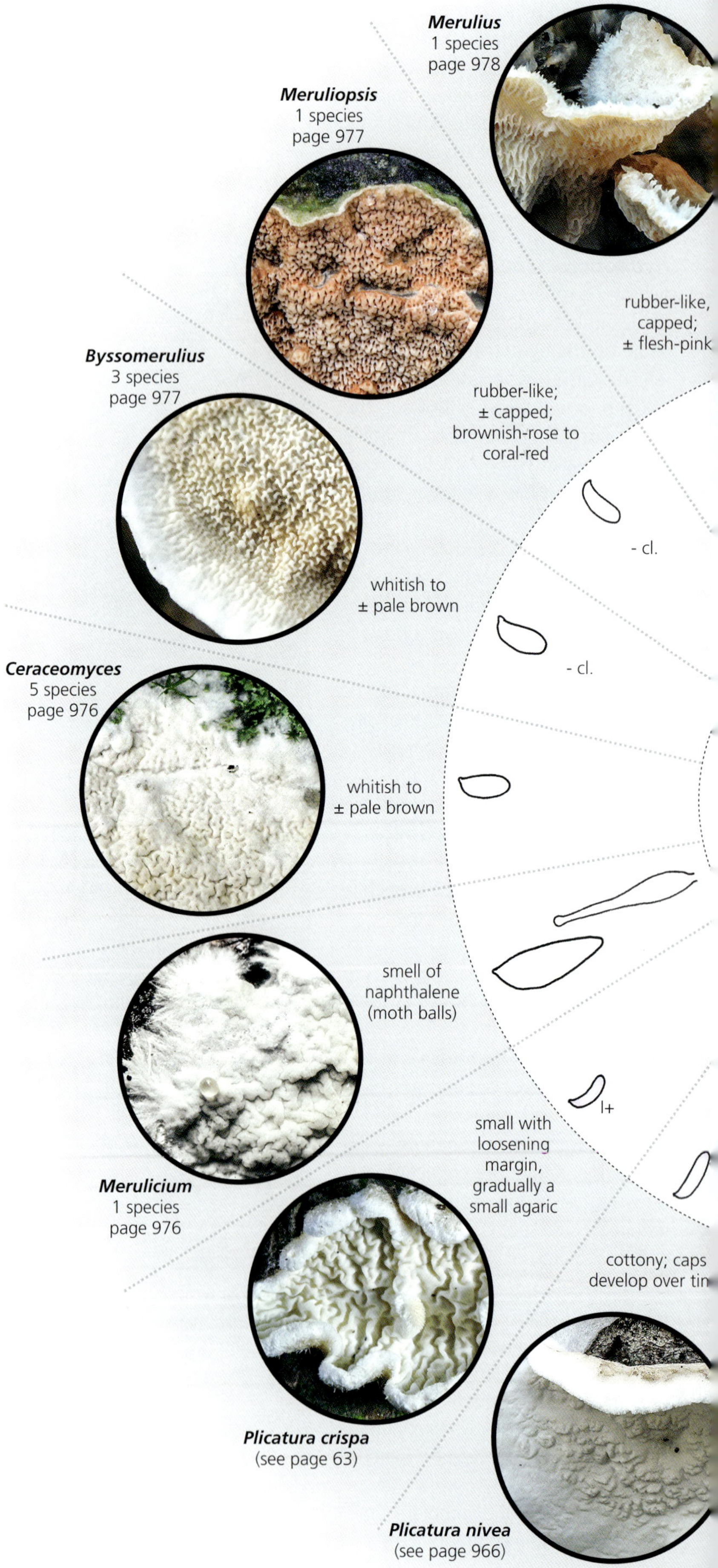

wrinkled hymenophore

hymenophore with radiating wrinkles

veined hymenophore

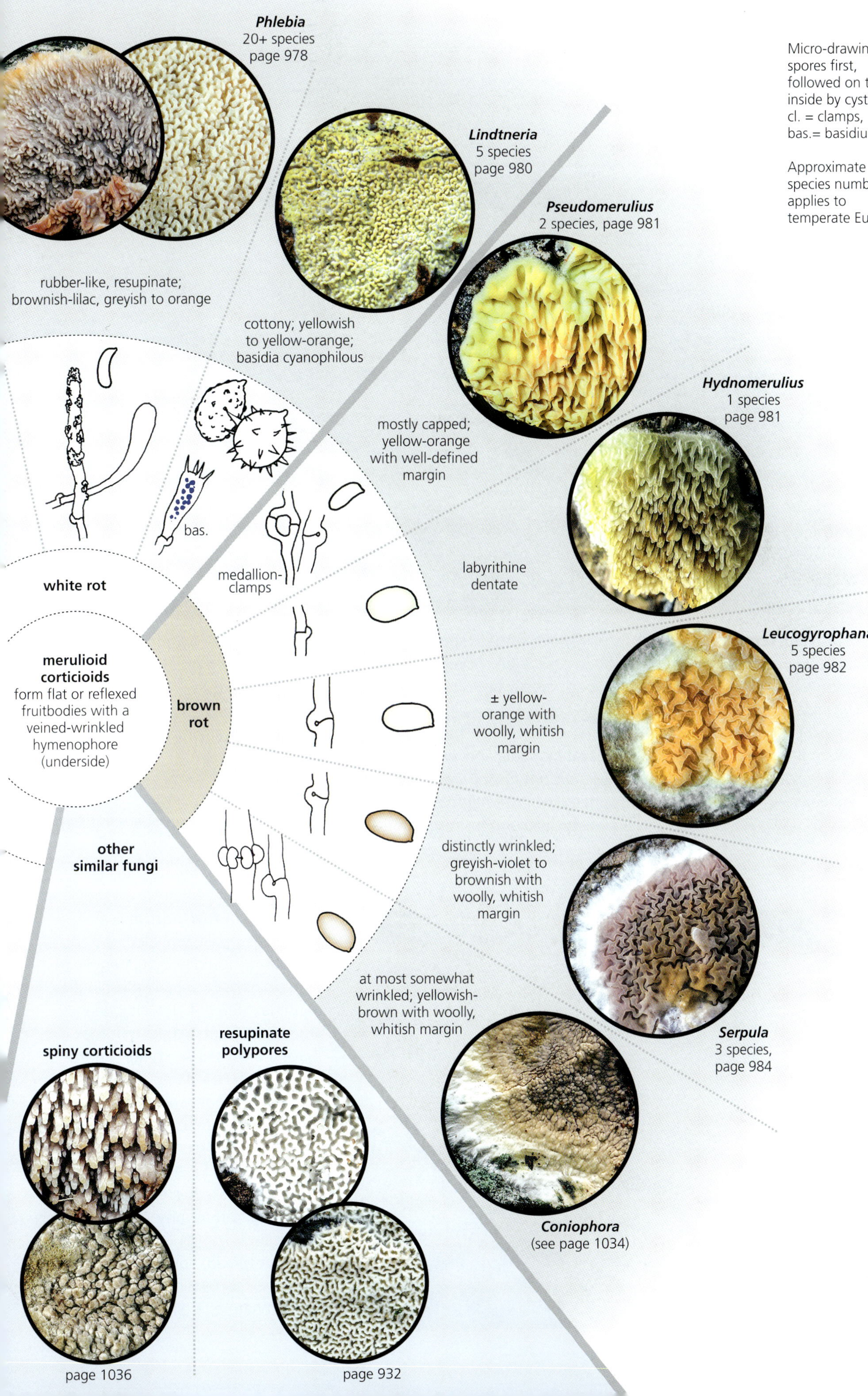

Micro-drawings: spores first, followed on the inside by cystidia. cl. = clamps, bas.= basidium.

Approximate species number applies to temperate Europe.

Thomas Læssøe

Ceraceomyces serpens is a resupinate, annual corticioid with a pale, somewhat fimbriate margin and a wrinkled-veined hymenophore in pale pastel colours ranging from cream through pale brown to greyish-violet. Lacks cystidia. The hyphae are thin-walled and have clamps and crystals. The inamyloid, smooth spores measure 4–5 × 2–2.5 µm. Occurs on dead wood, mostly deciduous; a white-rotter.

Phlebia rufa ▷▷ has large, clavate cystidia and allantoid spores. *Merulicium fusisporum* ▽ smells phenolic or like naphthalene, and has cystidia and fusiform spores. *Byssomerulius corium* ▷ is mainly reflexed with a projecting cap and has paler colours.

Widespread and common, becoming scarcer towards the north; most of the year.

Merulicium fusisporum is a resupinate, pale, wrinkled-veined annual corticioid with a phenolic to naphthalene-like smell. It is dimitic, with dextrinoid skeletal hyphae and lageniform cystidia that may have a brown encrustation on the upper part. Has clamps. The hyaline, fusiform, smooth spores measure 8–10 × 3.5–4 µm. Occurs deep in conifer brush wood on rich soil.

The dimitic flesh with dextrinoid skeletal hyphae makes the species very easy to recognize microscopically.

Widespread, apparently rare but probably overlooked due to its hidden nature; ± all year.

Byssomerulius corium is a typically reflexed annual corticioid with a white, downy-hairy cap and a whitish to pale brown, wrinkled hymenophore. The hyphae lack clamps. Lacks cystidia. The hyaline, smooth, inamyloid spores measure 5–7 × 2.5–3.5 µm. Occurs on branches from deciduous trees, rarely on conifer wood, often well above ground; a white-rotter.

Ceraceomyces serpens ◁ is not capped and has clamps. *Byssomerulius hirtellus* ⋉ lacks a cap and has cystidia and more vivid colours. *Merulius tremellosus* ▷▷ often has a more capped appearance, is more fleshy-gelatinous and pinker.

Widespread and very common, rare in the boreal zone; most of the year, although almost absent during the summer.

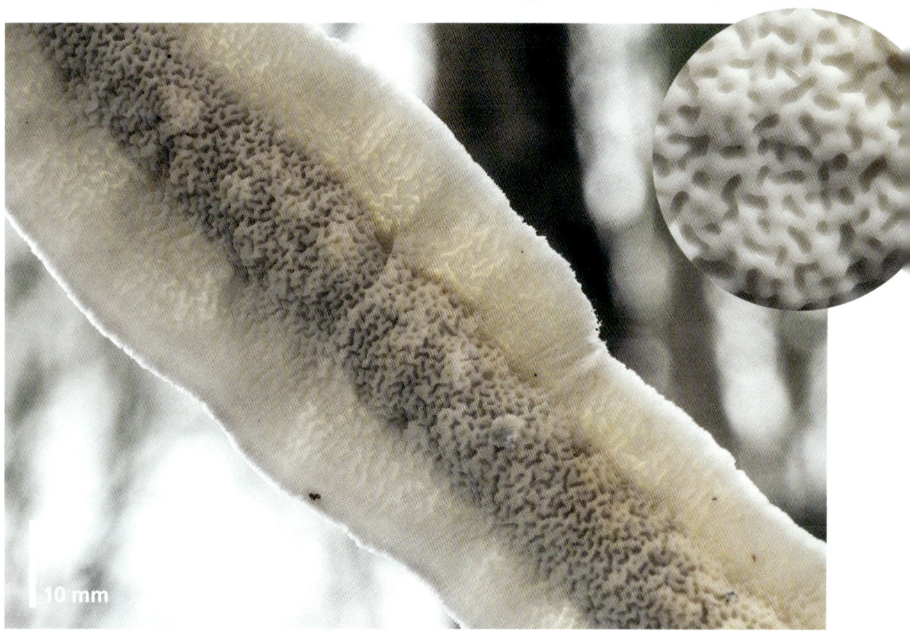

Meruliopsis taxicola is a reflexed to resupinate, ± rubber-like annual corticioid with a brownish-rose to coral-red, wrinkled-pored hymenophore, a white, downy margin and often a narrow, ± algae-covered cap. Lacks clamps and cystidia. Spores measure 4.5–6 × 1–1.5(–2) µm. Forms a white rot in *Pinus*, often above ground.

Gloeoporus dichrous (page 900) has clamps and occurs on deciduous wood. Species of *Ceriporia* (page 946) have ± well developed, typical pores. Species of *Leucogyrophana* ▷▷ are more fragile and are brown-rotters.

Widespread, rather common in the hemiboreal–boreal zones, rarer in Atlantic areas; all year.

Merulius and **Phlebia** are rather tough, somewhat gelatinous corticioids with a smooth, warty to wrinkled hymenophore. The wrinkles tend to be radiating. All species produce a white rot in deciduous or conifer wood. They are at their best during damp periods in late autumn and early winter.

Merulius tremellosus is a reflexed, rubber-like annual corticioid with a pale, felty-hairy upper side and a pinkish-buff to pinkish, radially wrinkled hymenophore. It may have a few cystidia-like projecting hyphae with crystals; has clamps. The allantoid spores measure 4–4.5 × 1–1.5 µm. Occurs on rotten wood of deciduous trees, often larger, decomposed trunks of *e.g. Fagus* and *Betula,* rarely on conifer wood.

Normally unmistakable. Other species with a similar hymenophore are not as thick and rubbery and have different spores.

Widespread and common; all year, except the summer months.

Phlebia rufa is a resupinate, annual corticioid with a ± clay-pink, wrinkled hymenophore. Has large, smooth, clavate cystidia. The spores are somewhat allantoid and measure 4.5–6.5 × 2–2.5 µm. Occurs on deciduous wood, with a preference for *Quercus*.

Ceraceomyces serpens ◁◁ is very similar but lacks the large clavate cystidia. *Phlebia radiata* ▷ has a more coarsely veined hymenophore with an evident radial structure, mostly orange colours and smaller spores.

Widespread and rather common in the nemoral–hemiboreal zones; all year, but mainly October-March.

Thomas Læssøe

Phlebia radiata is a resupinate, annual corticioid with a wrinkled-veined hymenophore and mostly rather vivid, orange to greyish-orange colours. It typically develops from circular fruitbodies with an evident radial structure (lower image). Later these fruitbodies merge to form larger, diffuse structures, which partly lose the radial pattern. Grows over mosses, bark fragments *etc.*, and may therefore appear ± capped. Has large, smooth clavate cystidia and clamps. The hyaline, smooth, inamyloid spores measure 4–5 × 1.8–2 µm. Occurs mainly on deciduous wood, *e.g. Fagus* and *Quercus*, but rarely also on conifer wood; a white-rotter.

Phlebia rufa ◁ does not have a distinct radial structure in the hymenophore, is usually less vividly coloured and has larger spores.

Widespread and very common in the nemoral–hemiboreal zones, becoming scarcer towards the north; most of the year, except during the summer months.

Jens Maarbjerg

Thomas Læssøe

Lindtneria species have cyanophilous spores and granules in the basidia. The fruitbodies have yellowish to orange-yellow colours. Occur on soil, litter, herbs or wood. Their biology is poorly understood.

Lindtneria trachyspora is a resupinate, vivid yellow-orange annual corticioid with a warty-veined-dentate to almost pored hymenophore. Has hyphal cords at the margin. Lacks true cystidia, but capitate hyphal ends occur. Has clamps. The basidia have cyanophilic internal drops. The ± spherical, highly ornamented spores stain dark blue in cotton blue and measure 7–9 µm. Occurs on calcareous soil amongst herbs in coastal scrub, open deciduous forests *etc.*, typically in sites with many other rare calciphilous species.

Other species of *Lindtneria* do not develop pores and have different spore shapes and ornaments or lack clamps.

Widespread in the nemoral–hemiboreal zones, very rare, but probably overlooked; July–November.

Thomas Læssøe

Lindtneria panphyliensis is a ± warty-veined-dentate, yellowish, annual corticioid with pale hyphal cords at the margin. Has clamps. The spores are fusiform with an irregular spiny ornament and measure (6–)7–8 (–10) × 4–6 µm. Occurs on soil and litter deep in herbaceous vegetation on open land or on barer forest floors.

Probably constitutes a species complex, since both morphology and ecology show much variation. *Lindtneria trachyspora* △ has ± spherical spores.

Widespread in the nemoral zones, rare, but apparently very much overlooked; July–December.

Lucie Zíbarová

Pseudomerulius aureus is a resupinate to reflexed, very vivid yellow-orange annual corticioid with a distinctly wrinkled hymenophore. The hyphae have clamps that are often large and open (medallion clamps). Lacks cystidia. The ± yellowish, ± thick-walled, ± dextrinoid, cyanophilous spores measure 4–4.5(–5) × 1.2–1.5 µm. Forms a brown rot in conifer wood.

Species of *Leucogyrophana* ▷▷ are less vividly coloured and have normal closed clamps. *Pseudomerulius montanus* ⋉ is less vividly coloured and has very small spores.

Widespread, occasional in the hemiboreal–boreal zones, rare in the nemoral zones; all year, except summer.

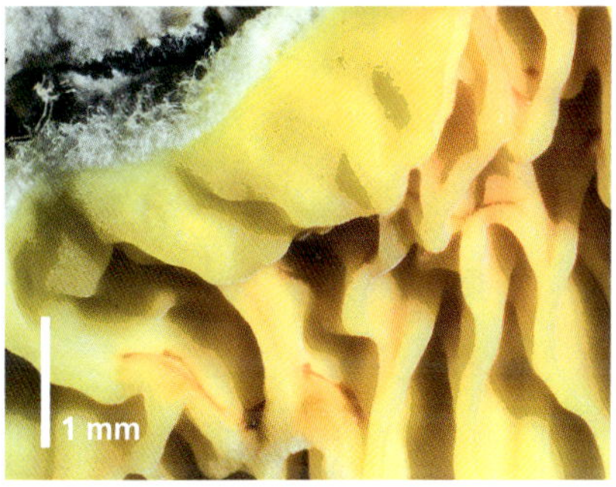

Hydnomerulius pinastri is a rather soft, toothed/wrinkled, pale yellow-brown, resupinate, brown-rotting, annual corticioid. Has thick-walled, somewhat dextrinoid, cyanophilous, yellowish spores measuring 5–6 × 3.5–4.5 µm. Known from rotten conifer wood, but also from *Juncus* in a cow-grazed meadow.

Sometimes accepted in the genus *Leucogyrophana* ▷▷ but the species in that genus are less dentate.

Widespread, but rare to very rare; mainly July–November.

Margaretha Liebmann

Species of ***Leucogyrophana*** generally have ± orange-yellow fruitbodies with a wrinkled hymenophore and whitish to yellowish spores less than 8 µm long. They are brown-rotters.

Leucogyrophana mollusca is a resupinate, annual corticioid with a highly wrinkled, ± brownish-yellow to brownish-orange hymenophore and a pale margin with thin threads. Has monomitic flesh, no cystidia and normal clamps. The spores are dextrinoid, smooth and ± thick-walled, and measure 6–7 × 4–4.5 µm. Forms a brown rot in decaying conifer wood, also occurs on construction timber inside buildings.

Other similar species of *Leuco-gyrophana* have smaller spores. *Serpula pulverulenta* ⋉ (also known as *L. pulverulenta*) has darker spores measuring 5.5–7 × 3.5–4.5 µm and is only found on timber in buildings within temperate Europe.

Widespread, occasional to rather rare, becoming scarcer towards the north; May–January.

Leucogyrophana sororia is a yellowish to apricot-orange, soft-fleshed annual corticioid with a wrinkled hymenophore that has a woolly, white margin. Lacks skeletal hyphae, and has thick-walled, yellowish, mostly non-dextrinoid spores that measure 4–4.5(–5) × 2.5–3 µm. Forms a brown rot in decaying conifer wood.

Leucogyrophana mollusca ⋉ has larger spores and *L. romellii* ⋉ has highly dextrinoid spores.

Widespread, rare to very rare; mainly July–November.

Leucogyrophana lichenicola is a vivid apricot-orange, annual corticioid that lacks skeletal hyphae. It may form sclerotia. The spores measure 4.5–5.5 × 3–3.5 µm. Occurs under and in tussocks of *Cladonia* subgenus *Cladina* (page 1486) and *Stereocaulon* ⋉ in dunes and heathlands.

This species is closely related to *L. romellii* ⋉, which has larger spores and occurs on woody substrates.

Very rare, but probably overlooked, within temperate Europe apparently only recorded from Denmark; August–October.

Thomas Læssøe

The genus ***Serpula*** forms resupinate, reflexed or capped fruitbodies with deeply wrinkled hymenophores. Young fruitbodies may be lilac-grey or pale yellow, but quickly become coloured orange-brown by the spores, which in species that occur outside buildings are 9–12 µm long. It is closely related to *Tapinella* (page 96). Brown-rotters.

Serpula himantioides is a ± olive-brown, highly wrinkled, resupinate annual corticioid up to 3 mm thick. Has skeletal hyphae about 2.5 µm wide. The rather thick-walled, pale brown, cyanophilous spores measure 9–12 × 5–7 µm. Occurs mainly on conifer wood with soil contact, but also on *e.g. Quercus* wood and sometimes found in houses. An aggresive brown-rotter.

Serpula lacrymans ▷ has thicker fruitbodies, is usually somewhat capped and has wider skeletal hyphae. Species of *Leucogyrophana* ◁◁ have smaller and paler spores and generally paler, more vividly coloured fruitbodies.

Widespread and rather common; mainly September–January.

Steen A. Elborne

Serpula lacrymans is a resupinate to distinctly capped annual corticioid with thick flesh and an olive-brown, highly wrinkled hymenophore. Has skeletal hyphae 2.5–6.5 µm wide. The smooth, brownish, cyanophilous spores measure 9–12 × 5–7 µm (see the spore-deposit below the fruitbody on the lower image). Occurs on trunks of conifers, but mostly in houses, where it is feared for its ability to penetrate brickwork and decay dry wood after having initiated growth on water-damaged wood. The degraded wood breaks into brown squares and completely lose the capacity to carry any weight.

Serpula himantioides ◁ has narrower skeletal hyphae and thinner, resupinate fruitbodies, and it is the dominating species out of doors. *Serpula pulverulenta* ⋉ has smaller spores, that measure 5.5–7 × 3.5–4.5 µm, and is only found in buildings.

Very rare in the wild, *e.g.* recorded in Denmark and the Czech Republic, but common in houses; mainly August–October (in the wild).

Steen A. Elborne

Fragile corticioids

This rather poorly defined form group includes resupinate corticioids where the fruitbody is relatively loosely attached to the substrate. They may be either cobweb-like, cotton-like or flaky and fragile. Species with thin, waxy or gelatinous fruitbodies are not included here, but amongst the firm, white or coloured corticioids with a ± smooth hymenophore.

Species in this group are best identified by using a compound microscope but macro-morphology may provide useful clues. Since all the species included have a loose hyphal structure it is easy to prepare flat preparations for use under the microscope – and as the hyphal structure can be seen clearly, they are much easier to work with than the waxy or gelatinous species.

Other similar fungi:
- whitish to pale grey corticioids with firm, tougher, waxy structures (page 1000).
- coloured corticioids with firm, tougher, waxy structures (page 1012).

See also the overview and the references to further reading on page 960.

Cobweb-like fruitbodies of *Botryobasidium* with 8-spored basidia in close up.

Loosely constructed fruitbodies of *Tomentella* with 4-spored basidia.

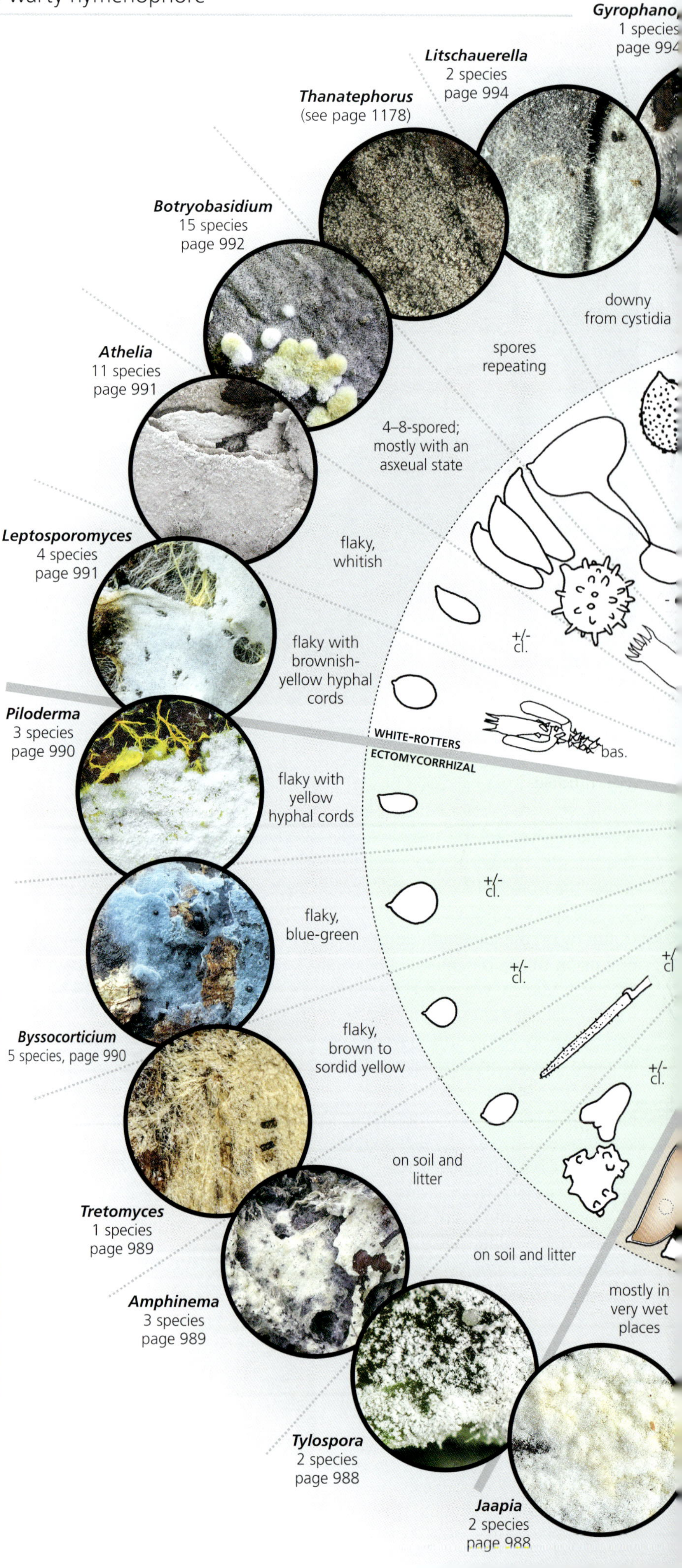

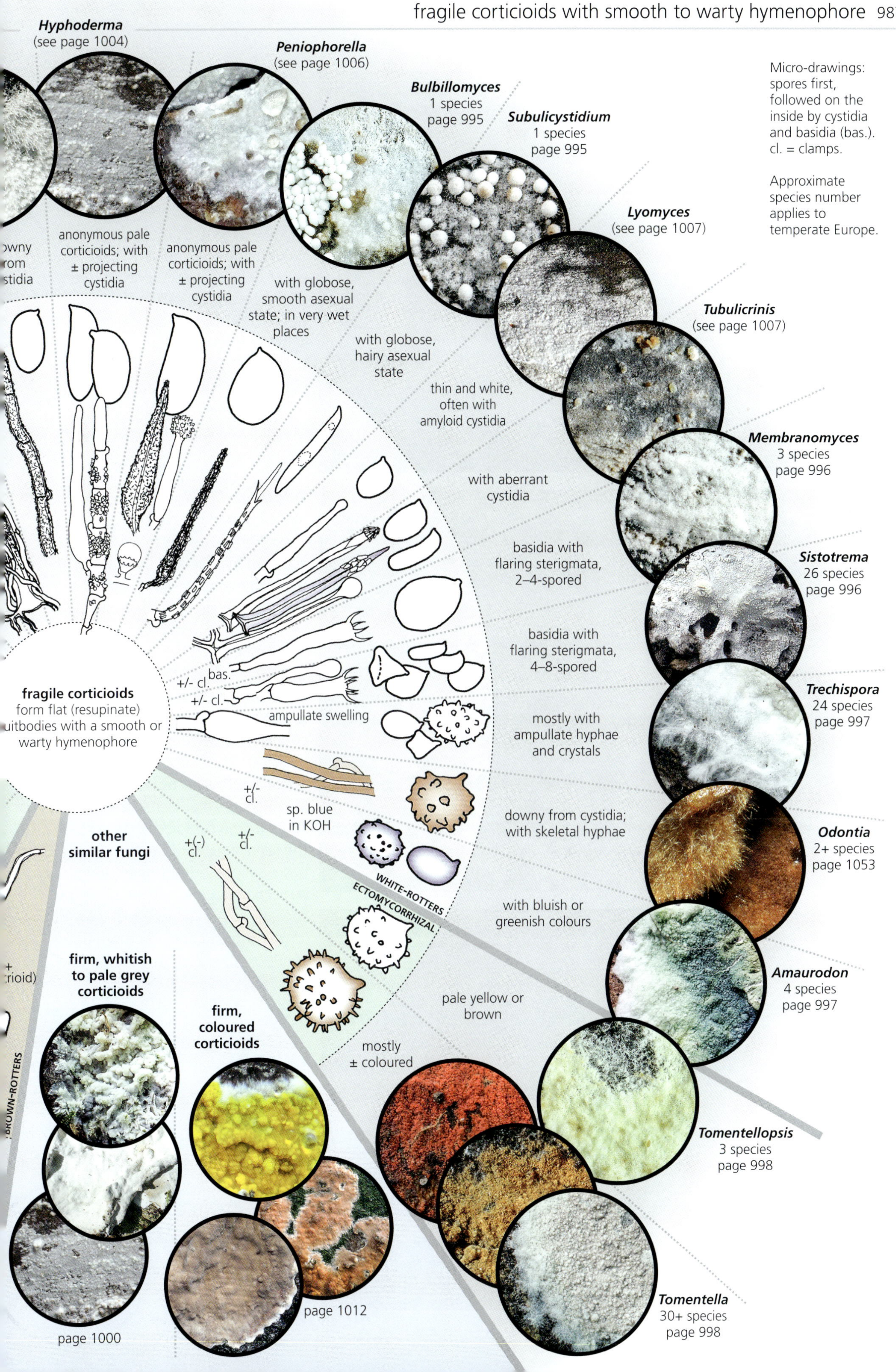
Hyphoderma
(see page 1004)
Peniophorella
(see page 1006)
Bulbillomyces
1 species
page 995
Subulicystidium
1 species
page 995
Micro-drawings: spores first, followed on the inside by cystidia and basidia (bas.). cl. = clamps.
Approximate species number applies to temperate Europe.
Lyomyces
(see page 1007)
Tubulicrinis
(see page 1007)
Membranomyces
3 species
page 996
Sistotrema
26 species
page 996
Trechispora
24 species
page 997
Odontia
2+ species
page 1053
Amaurodon
4 species
page 997
Tomentellopsis
3 species
page 998
Tomentella
30+ species
page 998
anonymous pale corticioids; with ± projecting cystidia
anonymous pale corticioids; with ± projecting cystidia
with globose, smooth asexual state; in very wet places
with globose, hairy asexual state
thin and white, often with amyloid cystidia
with aberrant cystidia
basidia with flaring sterigmata, 2–4-spored
basidia with flaring sterigmata, 4–8-spored
mostly with ampullate hyphae and crystals
downy from cystidia; with skeletal hyphae
with bluish or greenish colours
pale yellow or brown
mostly ± coloured
bas.
+/- cl.
+/- cl.
ampullate swelling
+/- cl.
sp. blue in KOH
+(-) cl.
+/- cl.
WHITE-ROTTERS
ECTOMYCORRHIZAL
fragile corticioids
form flat (resupinate) fruitbodies with a smooth or warty hymenophore
other similar fungi
firm, whitish to pale grey corticioids
firm, coloured corticioids
page 1000
page 1012
BROWN-ROTTERS

Jaapia ochroleuca is a ± yellowish, almost mould-like annual corticioid with long, almost cylindrical, smooth, thick-walled cystidia that have a thin-walled upper part. Has clamps. The dextrinoid, cyanophilous, thick-walled, fusiform spores measure 12–15(–18) × 4–6 µm. Occurs on wet, mostly conifer wood in marshes and other damp places; a brown-rotter.

Jaapia argillacea ⋉ differs by having only the central part of the spore plasma-filled and only this part being dextrinoid; the spores are even larger, 16–20 × 5–7 µm.

Widespread, probably rather overlooked; September–November.

Thomas Læssøe

Tylospora fibrillosa is a whitish, almost mould-like, resupinate annual corticioid with ± crystal-covered, clamped hyphae and no cystidia. The fruitbodies can be very large (lower image), *e.g.* along ditch banks. The very distinctive spores are triangular with small warts, and measure 6–7 × 5.5–6.5 µm. Occurs on humus and needle debris, mostly on ± acid, ± moist soil; ectomycorrhizal with conifers, particularly *Picea*.

Tylospora asterophora ⋉ forms similar fruitbodies but has three-lobed spores that lack warts. *Amphinema byssoides* ▷ occurs in the same environment, but usually has ± yellowish fruitbodies, long, projecting cystidia and small, simple spores.

Widespread and common, but overlooked; mainly August–January.

Amphinema byssoides is an ± isabelline, cottony annual corticioid that is minutely downy from very long, narrow, cylindrical, rough cystidia. The finely spiny hyphae have clamps. The inamyloid, smooth, almost cylindrical spores measure 4–5 × 2–3 µm. Occurs on litter and very rotten wood that is in contact with soil, mostly in conifer forests; it is an important ectomycorrhizal fungus in various types of woodland.

Species of *Tylospora* ◁ are typically more mould-like, have lobed spores and lack cystidia. *Botryobasidium isabellinum* ▷▷ has spherical, long-spined spores, as has the yellowish *Tomentellopsis echinospora* ▷▷. *Tretomyces* species ▽ have smaller, ± spherical spores.

Widespread and probably very common; mostly in the autumn.

Thomas Læssøe/Jens H. Petersen

Tretomyces microsporus is a sordid yellow, loosely built annual corticioid with orange-brown hyphal cords that have clamps at many septa. The spores are cyanophilous, somewhat thick-walled and measure 2–2.5 × 2–2.3 µm. Occurs in the litter or on very decomposed wood in *Pinus* forests and plantations on sandy soils; ectomycorrhizal (the inserted image shows ectomycorrhizal roots with visible external mycelium).

It macroscopically recalls *Amphinema byssoides* △. *Tretomyces lutescens* ⋉ has larger spores, measuring 3–3.5 × 2.7–2.9 µm, and hyphae that either have or lack clamps.

Very rarely recorded, but probably overlooked; autumn.

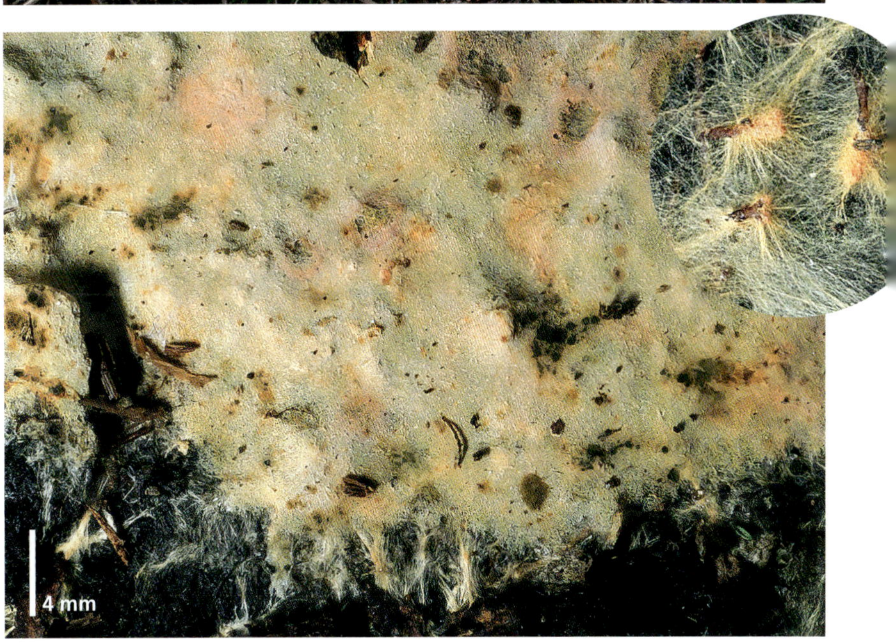

Byssocorticium is a genus of ± cobweb-like, brightly coloured corticioids with ± right-angled branchings on the hyphae and rather small, ± spherical spores. Ectomycorrhizal.

Byssocorticium atrovirens is a cobweb-like to cottony, greyish-blue annual corticioid with rather narrow, well-separated hyphae that have clamps at numerous septa. The thick-walled, almost spherical spores measure 3–4 µm in diameter and have a large internal oil drop. Occurs mostly on somewhat rich soil and usually in deciduous forests, *e.g.* with *Fagus*. The fruitbodies are typically formed in the leaf-litter.

Byssocorticium pulchrum ⋉ also has clamps but the spores are larger, 5.5–6 µm in diameter. There are also some further blue species that lack clamps. *Amaurodon mustialiensis* ▷▷ has larger, more ellipsoid spores and normally more greenish tinges; the spores are blue or stain blue in KOH.

Widespread, rather common to rare, becoming scarcer towards the north; mainly August–November.

Piloderma fallax has bright yellow hyphal cords and loosely attached, pale annual fruitbodies with crystal-bearing hyphae that lack clamps. Spores are hyaline, smooth, rather thick-walled, 3–3.5 × 2–3 µm. Occurs in the litter or in connection with very decomposed wood on sandy soils in coniferous forests, rarely in deciduous forests. Ectomycorrhizal.

Piloderma olivaceum ⋉ has darker fruitbodies. *Amphinema byssoides* ◁◁ does not have yellow hyphal cords. *Leptosporomyces fuscostratus* ▷ has more isabelline hyphal cords and is saprotrophic.

Widespread and common, but apparently rare in the nemoral lowlands; mostly August–November.

Leptosporomyces fuscostratus is a white, loosely attached annual corticioid with conspicuous yellow-brown hyphal cords and similarly yellow-brown hyphae. All septa have clamps. Lacks cystidia. The smooth, inamyloid spores measure 3.5–4 × 1.5–2(–2.5) µm.
Occurs on conifer wood and bark, particularly *Larix*.

Leptosporomyces montanus ⋉ has somewhat larger basidia and spores. Other similar species have ± hyaline hyphae.

Rather common and widespread, at least in part of the area (e.g. Denmark and Austria), but apparently rare in other parts; all year, but mainly August–September.

Thomas Læssøe/Jens H. Petersen

Athelia is a genus of thin, fragile, loosely attached resupinate corticioids. The hyphae typically branch at right angles and are often crystal covered. They are mostly white-rotters, more rarely parasites on lichens.

***Athelia epiphylla* s.l.** is a thin, chalk-white, rather loosely attached and built annual corticioid with rather wide, crystal-covered hyphae that have scattered clamps only in the lower part of the fruitbody. Lacks cystidia. The hyphae are 5–7 µm wide and the basidia 10–18 µm long. The narrowly ellipsoid, smooth, inamyloid spores measure 7–9 × 3–3.5 µm.
Occurs on litter and wood, mostly from deciduous trees, but also from ferns, *etc.*

There are a number of similar, closely related species.

Widespread and very common; mostly October–January.

Jens Maarbjerg

Thomas Læssøe

Thomas Læssøe

Athelia arachnoidea is a thin, mould-like, ring-forming annual corticioid with 2-spored basidia and needle-shaped crystals on the hyphae. Has clamps on the basal hyphae but not in the hymenium. The spores measure 8–10(–11) × 4.5–5.5 µm. Parasitizes *Lepraria* ⋉ and other lichens, but may also live as a decomposer on various substrates.

For a similar parasitic behaviour but on mosses, see *Chromocyphella muscicola* (page 1085) and species of *Trechispora* ▷▷. The 2-spored basidia and relatively large spores make this species quite easy to identify.

Widespread and rather common; mostly October–May.

Botryobasidium is a rather large genus of loosely constructed resupinate corticioids. The hyphae are typically branched at almost right angles. Many species have six or eight spores per basidium (see the image bottom left on page 986) and the spores and basidia usually stain in cotton blue (they are cyanophilous). All produce a white rot in wood.

Botryobasidium isabellinum is a somewhat mould-like, ± isabelline annual corticioid. The hyphae are very wide and lack clamps, and the basidia are 4-spored. The spores are hyaline, spherical, 7–10 µm in diameter, and have rather long spines. Occurs on litter and wood, in both deciduous and coniferous forests.

Also known as *Botryohypochnus isabellinus*. Other species of *Botryobasidium* ▷ typically have more than four sterigmata, different spore shapes and, at most, very indistinct warts on the spores. *Tomentellopsis echinospora* ▷▷ has smaller spores and shorter spines.

Widespread, occasional; mostly July–October.

Botryobasidium aureum is a somewhat mould-like, whitish annual corticioid that almost always occurs together with its cushion-shaped, cinnamon-buff, asexual state (dominating in the lower image). The asexual state often occurs by itself and produces large, yellow-brown, lemon-shaped conidia. The hyphae are wide with right-angled branchings and lack clamps; the basidia have six sterigmata. The spores are ± fusiform and measure 6–9 × 3–4 µm. Occurs on rotten wood of deciduous trees; very common on large, very decomposed trunks.

Botryobasidium capitatum ⋉ produces similar spores, but somewhat smaller conidia.

Widespread and common in the nemoral zones, becoming scarcer towards the north; mostly July–November.

Thomas Læssøe

Botryobasidium subcoronatum is a whitish to yellowish annual corticioid with smooth, wide hyphae that have sharp angles and clamps. Basidia mostly 6-spored, 20–25 µm long. The spores taper towards the top and measure 6–7.5 × 2.5–3.5 µm. Occurs mostly on ± brown-rotted conifer wood but is also found on deciduous wood.

Botryobasidium intertextum ⋉ lacks clamps at some septa and has spores that measure 7–10 × 1.5–2.5 µm.

Widespread and common; mainly March–November.

Gyrophanopsis polonensis is a chalk-white, fragile, somewhat cotton-like annual corticioid with highly projecting, cylindrical cystidia that have clamped septa and a grainy covering along their entire length. The ± thick-walled spores measure 6.5–9(–10) × 4–5 µm. Occurs on damp wood in marshes and bogs.

Also known as *Hyphoderma polonense*. *Hyphoderma setigerum* (page 1004) has the same type of cystidia but cylindrical spores that measure 7–11 × 3–4.5 µm and have an oil inclusion. *Subulicystidium longisporum* ▷ occurs in the same habitat but is thinner and has narrower cystidia with barbwire-like crystals. *Peniophorella pubera* (page 1006) is not as white and has more conical cystidia with a thick crystal cap.

Widespread and probably rather common, but easily overlooked in its wet habitat; April–November.

Litschauerella clematidis is a thin, whitish annual corticioid with highly projecting conical cystidia that are partly covered in a layer of thin hyphae. The ± spherical spores are finely rough in some mountants and measure 5–7.5 × 4.5–6.5 µm. Occurs on rather dry deciduous and conifer wood in scrub, plantations and forests.

The combination of spore shape and distinctive cystidia make this species easy to recognize.

Widespread in the nemoral–hemiboreal zones, but rare and possibly somewhat overlooked; all year.

Bulbillomyces farinosus is an annual corticioid with rather thin, ± smooth fruitbodies that, typically, are surrounded by an asexual 'bulbil-state' of small, ± spherical, smooth, air-filled, water-dispersed bodies ('bulbils'). The cystidia are covered in crystals and thick-walled, and with a hand lens may be seen to project from the hymenium. The smooth, thick-walled, inamyloid spores measure 6–10 × 4.5–6 µm. Occurs on wet deciduous wood, rarely conifer wood, even directly in water.

Subulicystidium longisporum ▽ has similar but somewhat smaller 'bulbils' that are covered in projecting hairs (cystidia) and may recall modern types of barbed wire.

Widespread and common in the nemoral zones, becoming scarcer towards the north; most of the year.

Subulicystidium longisporum is a white, thin and loosely built annual corticioid with projecting cystidia that are similar to modern barbed wire. The spores are cylindrical–fusiform and measure 10–15(–20) × 3.5–4.5 µm. Often seen together with a chalk-white, water-dispersed 'bulbil-state' that bears the same kind of projecting cystidia as in the fruitbodies. Occurs on damp wood, mostly deciduous, rarely on herbaceous litter in wet fields.

The 'bulbil-state' may recall that of *Bulbillomyces farinosus* △ but the projecting cystidia (hand lens required!) are diagnostic; *S. perlongisporum* ⋉ has even longer spores.

Widespread and common; most of the year, but rare in the spring.

Membranomyces delectabilis is a whitish to ± yellowish annual corticioid that lacks conspicuous hyphal cords. Cystidia and clamps are absent, but has large, oil-filled, 4-spored basidia with long, curved sterigmata. The ± thick-walled spores, 7–9 × 6–8 µm, contain refractive material. Occurs on wood and litter in carrs, park woodland and deciduous and coniferous forests.

Also known as *Clavulicium delectabile*. *Membranomyces spurius* ⋉ has hyphoid cystidia and is mostly 2-spored. *Clavulicium macounii* ⋉ has clamps, gloeocystidia, mostly 2-spored basidia and larger spores.

Widespread, but rare; mainly September–December.

Jens H. Petersen/Jan Vesterholt

Sistotrema species typically produce 6- or 8-spored basidia. The species form predominantly resupinate, loose-textured fruitbodies with a smooth, warty hymenophore (covered here), or a spiny hymenophore (page 1042) or a pore-shaped hymenophore (page 939). One species, *S. confluens* (page 1075), forms small, stemmed fruitbodies. Most species are white-rotters, but *S. confluens* is ectomycorrhizal.

Sistotrema brinkmannii is a rather thin, whitish annual corticioid with a grainy to finely warty surface that lacks hyphal cords. The hyphae have clamps and large crystals. The constricted basidia have six to eight sterigmata, and the curved, bean-shaped spores measure 4–5 × 2–2.5 µm. Occurs on rotten wood, litter, old polypores, *etc*.

There are a number of rather similar species of *Sistotrema* with similar basidia and spore size. However, the spore shape of *S. brinkmannii* is a particularly important identification character; macroscopically it also recalls various species of *Trechispora* ▷.

Widespread and very common; all year, less prominent during the summer.

Jan Vesterholt

Trechispora cohaerens is a white, loosely attached annual corticioid with conspicuous white hyphal cords. Has ampullate, swollen hyphae that have clamps, but no cystidia. The smooth spores are rather thick-walled, taper slightly towards the apiculus and measure 3.5–5 × 2.5–4 µm. Occurs on deciduous and conifer wood.

Trechispora confinis ⋉ looks similar but lacks hyphal cords and has smaller smooth spores that measure 3–3.5(–4) × 2–2.5 µm and are drop-shaped. *Trechispora silvae-ryae* ⋉ has less thick-walled but still smooth spores and dimitic flesh; *T. subsphaerospora* ⋉ has ± triangular spores. See also pages 938 & 1039.

Widespread in the nemoral–hemiboreal zones, rather common; all year, mainly August–December.

Amaurodon mustialaënsis is a blue to greenish, cottony annual corticioid with a paler, loosely built margin. The hyphae have clamps. The spore-deposit can be dark blue. The smooth, ellipsoid, thick-walled spores can be blue in water or stain blue in *e.g.* KOH; they measure 5–7 × 3.5–5 µm. Occurs on litter and wood that are in contact with calcareous soils, in both deciduous and coniferous forests.

Blue species of *Byssocorticium* ◁◁ have smaller or more spherical spores. See also page 1052 for the likewise bluish but more spiny *A. viridis*.

Widespread but very rare; mainly September–October.

Thomas Læssøe

Tomentellopsis echinospora is a buff to ± pale lemon-yellow to somewhat rose or greenish, cottony annual corticioid with a pale, loose margin. The hyphae lack clamps, and there are no cystidia. The spherical, almost hyaline spores measure 5–6 µm and have spines up to 1 µm long. Occurs on soil and litter, mostly in damp habitats in boggy forests and deciduous woodland.

Tomentellopsis pulchella ⋉ should have smaller spores and *T. submollis* ⋉ somewhat longer, less spherical spores. Other similar species may have clamps or more brown colours.

Widespread and probably rather common; July–November.

Tomentella is a genus of ectomycorrhizal, resupinate, mostly greyish, brownish or reddish corticioids that have distinctive ± spiny, generally somewhat angular, brown spores. Almost all species have clamped hyphae and most lack cystidia.

Tomentella is a large genus and species identification is fraught with difficulty.

Tomentella badia is a greyish to grey-brown, ± smooth annual corticioid that does not have a demarcated margin. The hyphae lack clamps, and there are no cystidia. The brown spores are an irregular, rounded shape, 8–11 µm in diameter, and have spines up to 2.5 µm in length. Ectomycorrhizal and occurs both in deciduous and coniferous forests.

Tomentella fuscocinerea ⋉ has more regularly shaped spores.

Widespread, but frequency not known; mainly July–November.

Tomentella atroarenicolor is a ± pinkish-buff, finely nodulose annual corticioid with ± yellow-brown, dimitic hyphal cords, hyphae with clamps and large, hyphae-like cystidia that measure 55–95 × 3–4 µm. The brown, irregularly shaped spores are 8–10 µm in diameter. Ectomycorrhizal in deciduous forests.

Other *Tomentella* species with cystidia have different, more distinct cystidial shapes.

Widespread, but frequency not known; mainly July–November.

Tomentella bryophila is a cinnamon-buff to brownish-orange, nodulose, peelable annual corticioid with a fringed margin. The hyphae have clamps and in the lower part of the fruitbody are brownish with somewhat incrusted pigment. The spores are evenly spherical, 7–11(–12) µm in diameter, and have long spines, up to 3 µm; the apiculus is amyloid. Ectomycorrhizal and found both with deciduous and coniferous trees.

More olive-coloured forms can perhaps be separated as *T. viridescens* ⋉; *T. ferruginella* ⋉ fits material with irregularly shaped spores.

Widespread and one of the most common species; June–November.

Tomentella lateritia is a dark coral-red to brownish-red, finely grainy annual corticioid. The lower hyphae are smooth and have clamps. The spores are almost triangular when seen in front view, measure 7–8.5 µm in diameter, and have spines up to 1 µm long. Ectomycorrhizal, typically with *Fagus* on calcareous soil, but probably also occurs with conifers.

Tomentella coerulea ⋉ is less red and has somewhat more irregularly shaped spores.

Widespread but rare in most areas, locally common in the nemoral zones; May–November.

Thomas Læssøe/Jens H. Petersen

Smooth, firm, ± whitish corticioids

This form group includes smooth, firm, relatively well-attached, resupinate corticioids with whitish or pale greyish hymenophores. The texture of the fruitbodies may be cheesy, waxy or leathery.

Most species in this group, also colloquially called 'white paint', can only be recognized with the aid of a microscope and suitable reagents.

Basidia attached to the side, rather than at the base, are termed pleural.

Other similar fungi:

– the distinction between this form group and the fragile, white corticioids is not clear-cut (page 986).

– dark grey and pinkish-buff species are treated amongst the attached, coloured corticioids (page 1012).

See also the overview and the references to further reading on page 960.

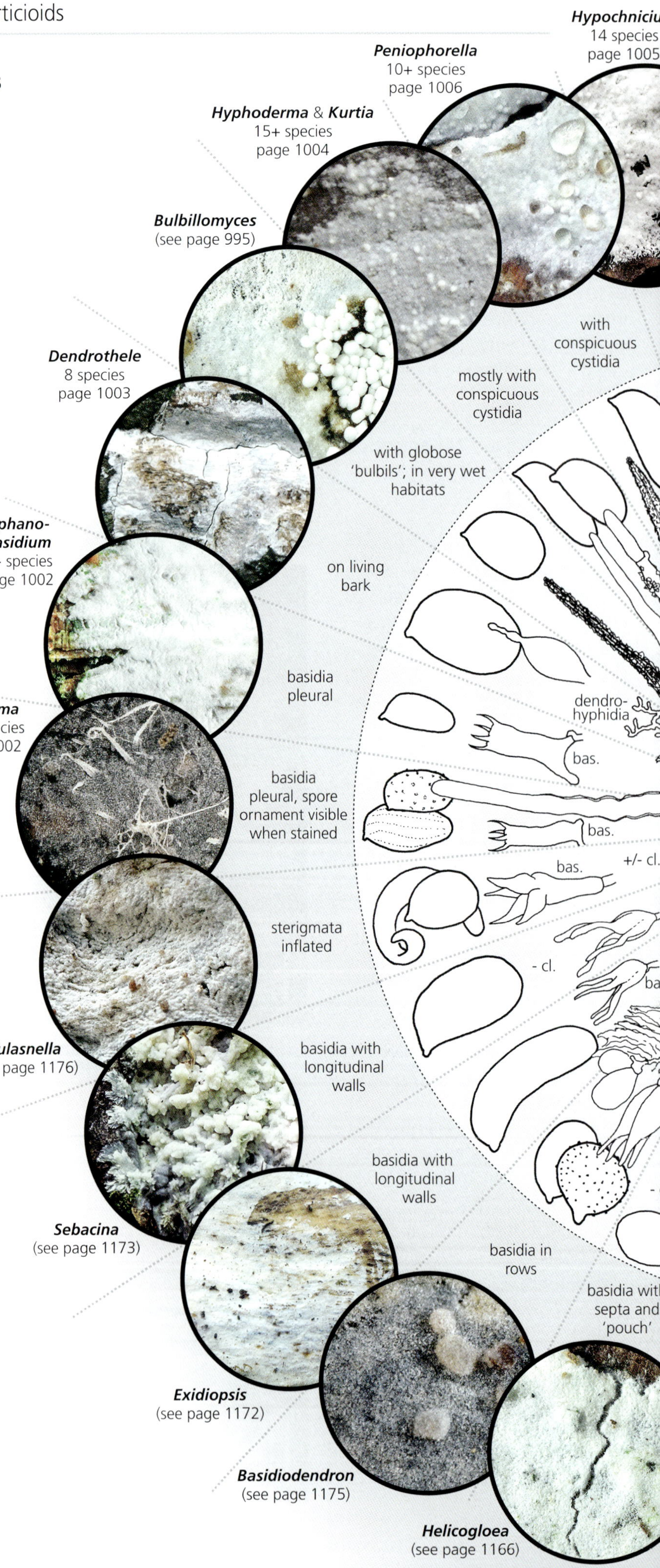

Micro-drawings: spores first, followed on the inside by cystidia.
cl. = clamps.
bas. = basidia

Approximate species number applies to temperate Europe.

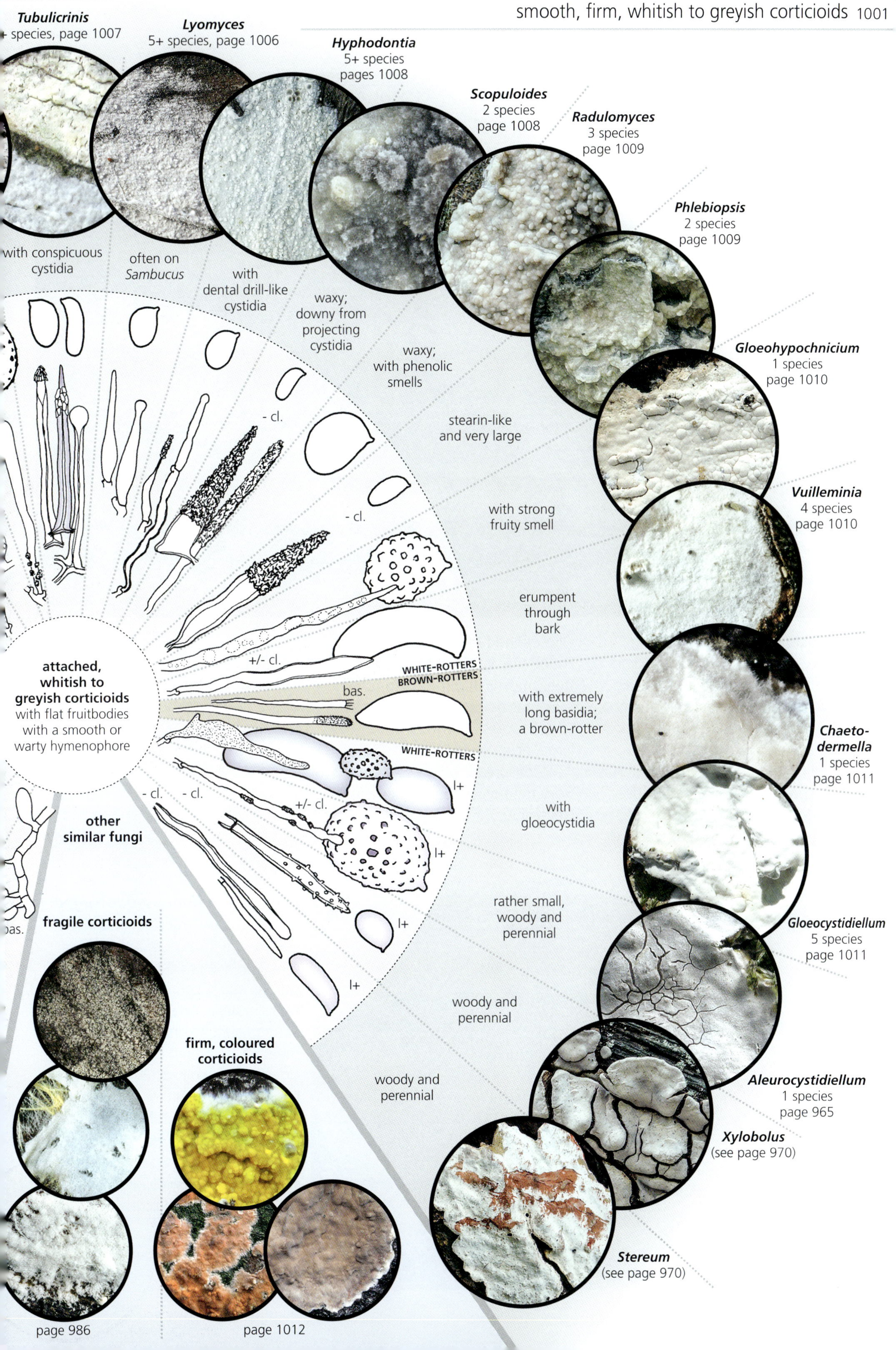
Tubulicrinis
species, page 1007
Lyomyces
5+ species, page 1006
Hyphodontia
5+ species
pages 1008
Scopuloides
2 species
page 1008
Radulomyces
3 species
page 1009
Phlebiopsis
2 species
page 1009
Gloeohypochnicium
1 species
page 1010
Vuilleminia
4 species
page 1010
Chaeto-
dermella
1 species
page 1011
Gloeocystidiellum
5 species
page 1011
Aleurocystidiellum
1 species
page 965
Xylobolus
(see page 970)
Stereum
(see page 970)
with conspicuous
cystidia
often on
Sambucus
with
dental drill-like
cystidia
waxy;
downy from
projecting
cystidia
waxy;
with phenolic
smells
stearin-like
and very large
with strong
fruity smell
erumpent
through
bark
with extremely
long basidia;
a brown-rotter
with
gloeocystidia
rather small,
woody and
perennial
woody and
perennial
woody and
perennial
- cl.
- cl.
+/- cl.
WHITE-ROTTERS
BROWN-ROTTERS
bas.
WHITE-ROTTERS
I+
- cl.
- cl.
+/- cl.
I+
I+
I+
attached,
whitish to
greyish corticioids
with flat fruitbodies
with a smooth or
warty hymenophore
other
similar fungi
bas.
fragile corticioids
firm, coloured
corticioids
page 986
page 1012

Xenasma pruinosum is a thin, chalky, pale grey annual corticioid with thick-walled cystidia that are widened towards the base and measure up to 130 × 10 µm, and thin-walled, short cystidia that have small protuberances at the tips. Has 4–6-spored, pleural basidia. The spores are finely verrucose in Melzer's reagent (smooth in KOH), narrowly ellipsoid and measure 6–7 × 3–4 µm. Found mainly on trunks of deciduous trees, mostly in old-growth forests.

Xenasma pulverulentum ⋉ has striate spores, and *X. rimicola* ⋉ has large, ellipsoid spores.

Widespread, but rare, ± absent from the boreal zone; mainly April–November.

Jens Maarbjerg

Aphanobasidium pseudotsugae is a thin, waxy, translucent annual corticioid with a ± smooth surface and a tar-like or phenolic smell. Has clamps and 4-spored, pleural basidia. Lacks cystidia. The ± fusiform to almond-shaped, inamyloid spores measure 5–8 × 3–4.5(–5) µm. Occurs on decorticated conifer wood in contact with soil.

Also known as *Phlebiella pseudotsugae*. The odour and the distinctive spore shape make the species rather easy to identify.

Widespread and probably very common; all year.

Dendrothele species are thin corticioids that grow exclusively on the bark of living trees. They tolerate desiccation and generate a new hymenium in damp conditions. They have rather large, smooth spores and dendrohyphidia.

Dendrothele acerina is a chalk-white, very thin annual corticioid with a rather sharp, delimited margin. When bruised it has a chlorine-like smell. Has very large, 4-spored basidia and basidia-like cystidia with a thin, undulating top and gnarled dendrohyphidia. The large spores are ellipsoid, inamyloid and thick-walled, and measure 10–12(–14) × 7–8(–9) µm; Q less than 2. Occurs mainly on bark of living *Acer campestre* and *A. platanoides*, very rarely on other hosts.

Dendrothele alliacea ⋉ has a larger spore Q (more than 2), and *D. wojewodae* ⋉ has a lower Q (less than 1.2). Other similar species lack cystidia and/or have 2-spored basidia or different spore shapes.

Widespread and common to occasional in the nemoral–hemiboreal zones; all year.

Dendrothele commixta is a very thin, chalk-white corticioid that looks like small, mouldy spots. Has 2-spored basidia, no cystidia and scarce dendrohyphidia. The spores are thick-walled, ellipsoidal/almond-shaped and measure 9–11 × 5–7 µm. Occurs on the bark of living old *Quercus* trees, mostly with other epiphytes, rarely on other hosts.

Rather easy to recognize due to the substrate, but should always be checked microscopically since it may resemble dead lichens.

Rather rare, but relatively widespread in the nemoral–hemiboreal zones, probably somewhat overlooked; all year.

Jens H. Petersen/Thomas Læssøe

Hyphoderma, ***Hypochnicium***, ***Kurtia*** and ***Peniophorella*** are rather anonymous genera of whitish, smooth to ± verrucose corticioids. They generally have rather large, smooth spores and basidia with a constriction in the middle. Many species have conspicuous cystidia, sometimes seen as a downy surface. All are white-rotters.

Hyphoderma setigerum is a resupinate, whitish to cream, smooth to indistinctly warty, downy-hairy annual corticioid with a diffuse margin. The down is formed of large, projecting, septate cystidia with crystals, each septum having a clamp. Spores ± cylindrical, inamyloid, smooth, have a large internal drop and measure 7–11 × 3–4.5 µm. Occurs mostly on deciduous wood; a white-rotter.

Gyrophanopsis polonensis (page 994) has narrower cystidia and smaller spores. *Candelabrochaete septocystidiatum* ▷▷ is typically somewhat more coloured, has narrower, allantoid spores and more densely septate cystidia that lack clamps.

Widespread, common; all year.

Jan Vesterholt

Kurtia argillacea is a soft, smooth to grainy, cream to greyish annual corticioid with a diffuse margin and a somewhat downy surface. It has long, tapering cystidia that measure up to 150(–200) × 20 µm and are distinctly swollen at the base, and usually have reddish sticky matter in the central part. The large, broadly allantoid, smooth, inamyloid spores measure 6–10 × 4–5 µm. Occurs on both conifer and deciduous wood; a white-rotter.

Several similar species are best told by different spore dimensions or slightly different cystidia.

Widespread and common; all year.

Hyphoderma roseocremeum is a resupinate annual corticioid with large fruitbodies, a diffuse margin and whitish to rose tinges. The surface is smooth to warty. Has clamps and long, cylindrical, smooth cystidia, up to 130 µm. The spores are cylindrical and oil-filled, and measure 8–12 × 3–4 µm. Occurs on deciduous wood, mostly *Fagus* and *Quercus*.

Hyphoderma medioburiense ⋉ has darker fruitbodies, larger spores and brownish secretions between the hyphae and on the cystidia. *Hyphoderma definitum* ⋉ has no rose tinges on the fruitbodies and shorter cystidia that measure 50–75 µm.

Widespread and rather common in the nemoral–hemiboreal zones, rare in the boreal zone; all year.

Hypochnicium geogenium is a somewhat waxy, white annual corticioid with a diffuse margin. Has clamps, large, cylindrical, smooth cystidia, and thick-walled, smooth, cyanophilous spores that measure 6–8(–9) × 4–4.5(–5) µm. Occurs predominantly on conifer wood, but also found on *e.g. Betula*.

Hypochnicium erikssonii ⋉ and *H. subrigescens* ⋉ have ± similar cystidia but spherical or almost spherical spores. The thick-walled spores separate *H. geogenium* from a number of species in other genera.

Widespread and rather common in the nemoral–hemiboreal zones, rare farther north; all year, mainly late in the season.

Jens Maarbjerg

Peniophorella pubera is a whitish to cream, minutely hairy annual corticioid with a diffuse margin. Has clamped hyphae and very large, fusiform, projecting cystidia that are completely covered by crystals. The smooth, inamyloid spores are somewhat curved and measure 7–10 × 3.5–5 µm. Occurs almost exclusively on deciduous wood, with *Fagus* being the dominant host in the nemoral zones.

Other similar species have different types of cystidia. The more warty-spiny *Scopuloides rimosa* ▷▷ has both fusiform and cylindrical cystidia and much smaller spores.

Widespread and common in the nemoral–hemiboreal zones; all year.

Jens Maarbjerg

Peniophorella praetermissa is a whitish, somewhat downy annual corticioid with a diffuse margin. Typically has three types of cystidia: a projecting type that is thickest above and below and often has crystals at the top; imbedded, pointed gloeocystidia; and short and goblet-shaped (so-called stephanocysts). The ± oil-filled spores measure 8–12 × 4–5 µm. Occurs on both deciduous and conifer wood.

A variable species that constitutes a species complex. *Peniophorella tsugae* ▽ and *P. pallida* ⋉ have lumps in the hymenium and no stephanocysts.

Widespread and common; all year.

Peniophorella tsugae is a whitish to greyish-cream annual corticioid with a diffuse margin; mostly with small, ± obvious brownish spots (hand lens required!). Has large, somewhat thick-walled, tapering cystidia and ± imbedded, shorter cystidia with swollen heads that have brown incrusted material. The ellipsoidal–tear-shaped spores measure 8–10 × 4–5 µm. Occurs mostly on conifer wood.

Peniophorella pallida ⋉ has narrower, different-shaped spores. *Peniophorella praetermissa* △ normally has goblet-shaped cystidia, so-called stephanocysts, and no brown lumps in the hymenium.

Rather southern and rare, but perhaps overlooked; mainly autumn.

Lyomyces sambuci is a ± chalk-white, almost smooth to irregular warty corticioid with large, diffuse fruitbodies; it is odourless. The hyphae have clamps. The cystidia are either slender and smooth with a small round head or shorter and tapering towards the tips. Has rather large scattered crystals in the flesh and the hymenium. The smooth, inamyloid, ± thick-walled spores measure 4.5–6(–7) × 3.5–4(–4.5) µm. Occurs mostly on wood of deciduous trees, primarily standing, decorticated *Sambucus* trunks, but also on dead stems of herbs in damp habitats; known from a wide range of hosts.

Probably forms a species complex. *Lyomyces incrustatus* ⋉ forms fruitbodies very low down in the litter in Salix carrs, *etc.* and has many more crystals, no tapering cystidia and somewhat larger spores. See also the more spiny *Lyomyces crustosus* (page 1045).

Widespread and very common, becoming scarcer in very northern regions; all year, but in best condition during the colder months.

The jelly fungus *Spiculogloea occulta* ⋉ has been found as a parasite in fruitbodies of *L. sambuci*.

Thomas Læssøe/Jens H. Petersen

Tubulicrinis subulatus is a rather thin, whitish to cream, minutely hairy annual corticioid. The 'hairs' consist of 'rooting', very thick-walled, ± amyloid cystidia that have a thin-walled, very pointed top and a cover of crystals just below. Has 4-spored, inamyloid basidia. The allantoid, inamyloid spores measure 6–8 × 1.5–2 µm. Occurs almost exclusively on conifer wood.

Tubulicrinis chaetophorus ⋉ has thicker-walled cystidia tops and wider spores. Other similar species with pointed cystidia have wider spores and narrower cystidia.

Widespread and probably rather common; all year.

Jan Vesterholt

Jens Maarbjerg

Hyphodontia alutaria is a smooth to warty, cream to pale cinnamon-buff annual corticioid with a floccose surface. Has numerous cystidia, some of which are shaped like dental drills and others having septa and swollen heads. The broadly ellipsoid spores measure 4.5–5 × 3–3.5 µm. Occurs mostly on conifer wood on the forest floor.

The combination of smooth to verrucose fruitbodies and the two types of cystidia makes the species easily recognizable microscopically. *Hyphodontia arguta* (page 1045) has distinctly spiny fruitbodies and somewhat wider spores.

Widespread and rather common; all year.

Scopuloides rimosa is a ± waxy, semi-translucent, warty-spiny annual corticioid with a diffuse margin. The dominating cystidium type is conical and covered in crystals, but there are usually some cylindrical, septate cystidia with crystals present. The inamyloid, smooth spores measure 3.5–4.5(–5) × 1.5–2(–2.5) µm. Occurs on damp deciduous wood, mostly on wet soils in *Alnus* carrs *etc*.

Also known as *S. hydnoidea*. *Peniophorella pubera* ◁◁ is similar, but has much larger spores and a different habit. The image shown includes a species of *Mollisia* (page 1436).

Widespread and probably common; most of the year, particularly in the summer.

Phlebiopsis gigantea is a large, diffuse, almost stearin-like annual corticioid with a rugose-warty surface. The hyphae lack clamps. Has large, projecting, thick-walled, fusiform cystidia, the upper half covered in coarse crystals. The smooth, inamyloid spores measure (5–)6.5–8 × 3–3.5 µm. Occurs on conifer wood, typically rather fresh stumps.

Phlebiopsis ravenelii ⋉ occurs on deciduous wood and has somewhat larger spores.

Widespread and common; mostly during winter.

Phlebiopsis gigantea is used as a biological control agent against harmful *Heterobasidion* species (page 856). Spores from *P. gigantea* are smeared on the cut ends of newly felled conifer stumps in order to reduce the risk of a major outbreak of rot caused by *Heterobasidion*.

Radulomyces confluens is a somewhat waxy, although quite soft, rather translucent, smooth to warty, cream to brownish-grey-tinged annual corticioid with a characteristic phenolic smell and a somewhat fringed margin. Has clamps and narrow hyphidia, but no true cystidia. The spores are smooth, ± thick-walled, broadly ellipsoid and measure 6–8.5 (–10) × 6–8.5 µm. Occurs on wood of deciduous trees, rarely conifers, at or just above soil level. Found on numerous hosts.

Radulomyces rickii ⋉ has almost spherical, very finely spiny spores; *R. molaris* (page 1051) has a similar smell, but a spiny hymenophore.

Widespread, very common; all year.

Thomas Læssøe

Gloeohypochnicium analogum is a rather thick, quite soft, smooth or rounded-warty, cream to cinnamon-buff annual corticioid with a penetrating, ± nauseous fruity smell that is almost impossible to remove from the fingers after having touched the fruitbody. Has clamps and ± cylindrical, large gloeocystidia. The broadly ellipsoid spores are verrucose, cyanophilous, inamyloid and measure 8–10(–12) × 6–8 µm. Occurs on the trunks of deciduous trees, mostly *Fagus*, usually in sites with long continuity and many fallen trunks.

The strong smell and the ornamented spores make this species easy to recognize.

Mainly confined to the nemoral zones with some records from the hemiboreal zone, rare to very rare; all year.

Vuilleminia cystidiata is a chalk-white, bark-splitting, almost dry annual corticioid with large, thick-walled cystidia that taper towards the tips. The allantoid spores measure 12–16 × 3.5–5 µm. Found mainly on attached *Crataegus* branches but also recorded on *e.g. Ilex*, *Cornus* and *Pyrus*.

The specific habitat and erumpent nature of the very pale fruitbodies should make it possible to identify this species in the field. Other species of *Vuilleminia* have more coloured fruitbodies (page 1025).

Widespread in the nemoral zones, less so in the hemiboreal zone, locally rather common to rare or absent; all year.

Chaetodermella luna is a thick, firm, ± cream annual corticioid with a somewhat fringed margin. The flesh is dominated by very long, up to 200 µm, cylindrical, thick-walled cystidia that have incrusted material. The basidia are also long and have clamps. The allantoid, inamyloid spores measure 10–15 × 4–5 µm. Produces a brown rot in *Pinus* wood and, farther south, in *Juniperus*.

Distinctive and unlikely to be misidentified, although where it occurs on *Juniperus* the darker *Amylostereum laevigatum* (page 1030) should be considered.

Widespread and common in the hemiboreal–boreal zones, but apparently absent o r nearly so from the nemoral zones; all year.

Gloeocystidiellum porosum is a cream to almost pinkish-buff, somewhat waxy, ± smooth annual corticioid with a diffuse margin. Has large, basally swollen gloeocystidia and clamped hyphae. The strongly amyloid, finely spiny, ellipsoid but flat-sided spores measure 5–6(–6.5) × 2.5–3(–3.5) µm. Produces a white rot in a range of deciduous hosts.

Gloeocystidiellum porosellum ⋉ lacks clamps; *G. clavuligerum* ⋉ has shorter and wider spores. *Gloiothele lactescens* (page 1029) is pale vinaceous-brown, exudes milk when fresh and tastes bitter.

Widespread and rather common; all year.

Smooth, firm, coloured corticioids

This form group includes resupinate corticioids with buff, brown, dark grey or vividly coloured, smooth to warty hymenophores. The texture of the fruitbodies varies from gelatinous, cheesy, waxy to leathery.

Some of the more brightly coloured species can be identified in the field, but the more anonymous require microscopic examination. It is difficult to make suitable preparations of the waxy species. The preparations are hard to make sufficiently thin and the tissues may be filled with lots of 'goo' that makes it difficult to make out the details.

OTHER SIMILAR FUNGI:

– species with completely white or pale grey fruitbodies (page 1000).

– species with fragile, cobweb-like or flaky fruitbodies (page 986).

– resupinate jelly fungi (with divided basidia) (page 1162).

See also the overview and the references to further reading on page 960.

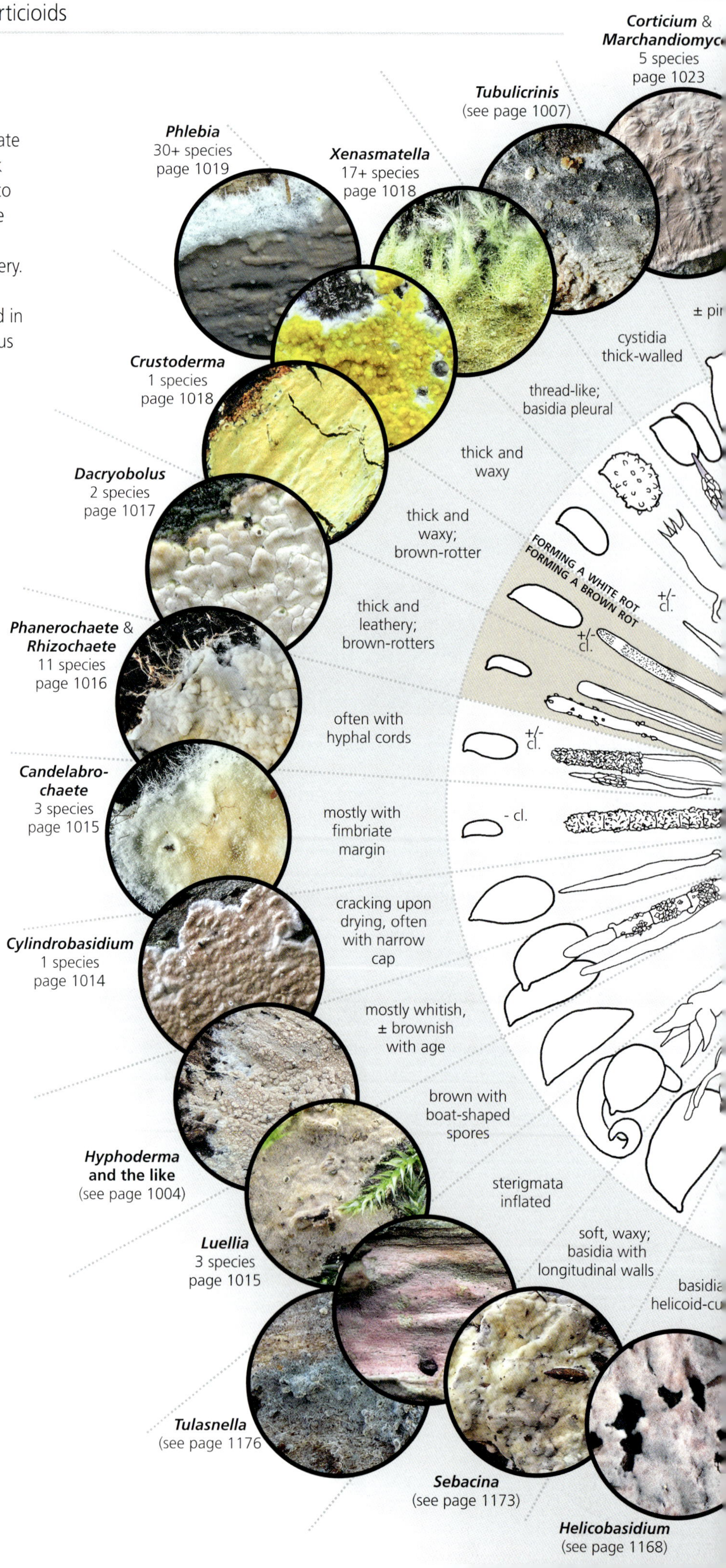

Micro-drawings: spores first, followed on the inside by cystidia.
cy. = cystidia
cl. = clamps.
bas. = basidia
I+ = amyloid

Approximate species number applies to temperate Europe.

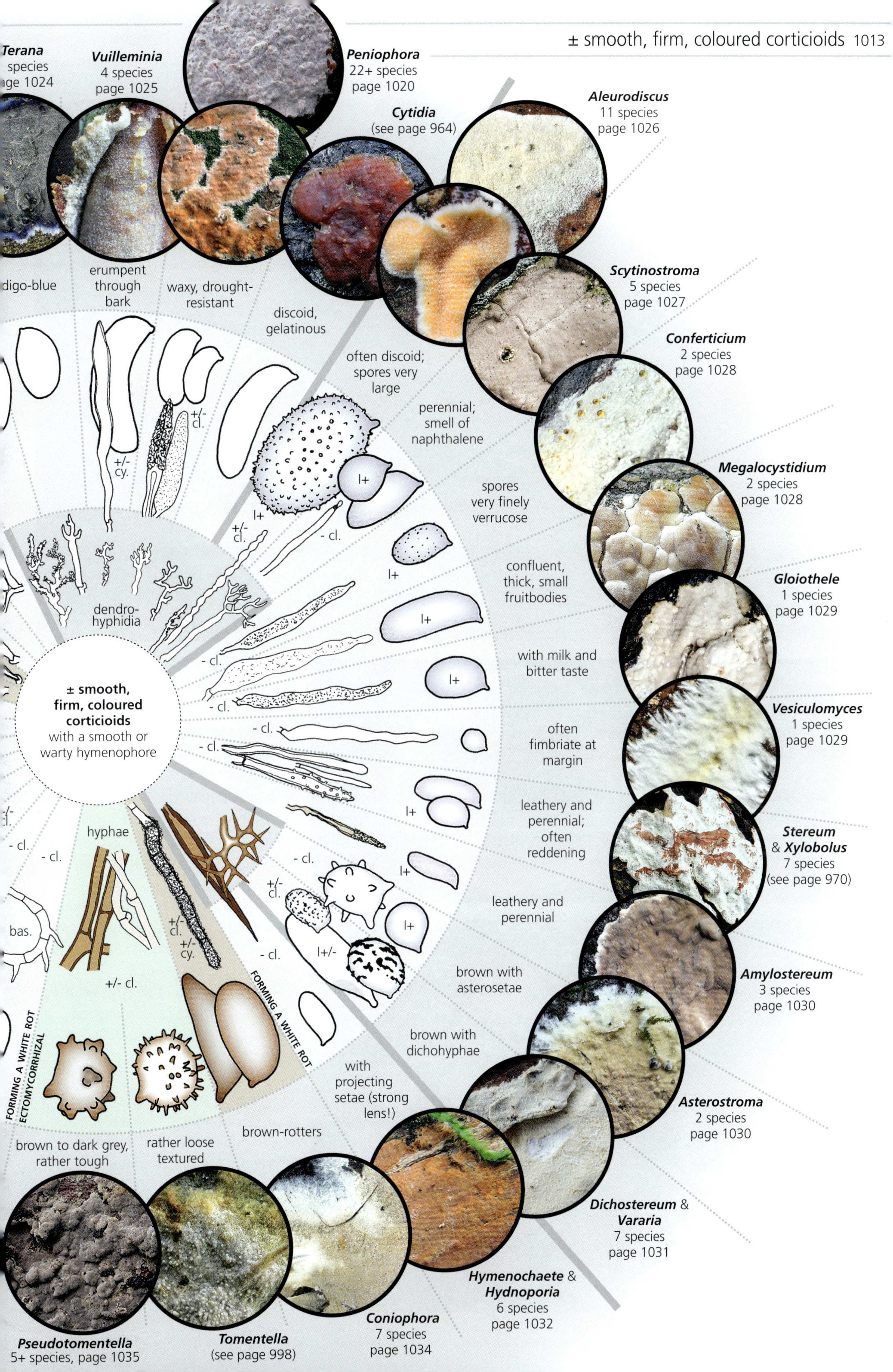

Terana
species
ge 1024
Vuilleminia
4 species
page 1025
Peniophora
22+ species
page 1020
Cytidia
(see page 964)
Aleurodiscus
11 species
page 1026
Scytinostroma
5 species
page 1027
Conferticium
2 species
page 1028
Megalocystidium
2 species
page 1028
Gloiothele
1 species
page 1029
Vesiculomyces
1 species
page 1029
Stereum
& Xylobolus
7 species
(see page 970)
Amylostereum
3 species
page 1030
Asterostroma
2 species
page 1030
Dichostereum & Vararia
7 species
page 1031
Hymenochaete & Hydnoporia
6 species
page 1032
Coniophora
7 species
page 1034
Tomentella
(see page 998)
Pseudotomentella
5+ species, page 1035
digo-blue
erumpent through bark
waxy, drought-resistant
discoid, gelatinous
often discoid; spores very large
perennial; smell of naphthalene
spores very finely verrucose
confluent, thick, small fruitbodies
with milk and bitter taste
often fimbriate at margin
leathery and perennial; often reddening
leathery and perennial
brown with asterosetae
brown with dichohyphae
with projecting setae (strong lens!)
brown-rotters
rather loose textured
brown to dark grey, rather tough
± smooth, firm, coloured corticioids
with a smooth or warty hymenophore
dendro-hyphidia
hyphae
bas.
FORMING A WHITE ROT
ECTOMYCORRHIZAL
FORMING A WHITE ROT
+/- cy.
+/- cl.
I+
- cl.
I+/-

Jens Maarbjerg/Jens H. Petersen

Cylindrobasidium evolvens is a ± pinkish-buff, honey-yellow or grey-olive annual corticioid that may develop small protruding caps with a pale upper side. Upon drying the fruitbody cracks into an irregular pattern and may loosen at the margin. Has clamps, long, pointed, imbedded cystidia and long, cylindrical basidia. The smooth, inamyloid spores are almost drop-shaped and measure 8–10 × 4–5 µm. Occurs mostly above ground level on attached deciduous branches and on piles of cut trunks, rarely on conifer wood.

Often recorded under the synonym *C. laeve*. May recall a thin-fleshed species of *Stereum* (page 970) but those often develop more pronounced caps and tougher fruitbodies. The ± drop-shaped spores of *C. evolvens* is a good distinguishing character.

Widespread and very common; all year, but in best condition in late autumn–winter.

Cylindrobasidium evolvens is sometimes attacked by the dark hyphomycete *Exosporiella fungorum* ⋉.

Luellia recondita is a pinkish-buff to pale grey-brown annual corticioid with a diffuse margin. It is often almost invisible due to its wood-like colours. Lacks cystidia, has clamps. Has brownish basal hyphae and clavate basidia. The hyaline, ± lemon-shaped spores measure (6–)7.5–9(–11) × 3.8–4(–4.5) µm. Occurs mostly on conifer wood, but also on fern litter, *etc.*

The name *Athelopsis recondita* is also used for this species. Other brown, smooth corticioids have different spore shapes. The similar *Luellia cystidiata* ⋉ has cystidia.

Widespread; occasional, but often overlooked; mainly August–March.

Candelabrochaete septocystidia is a ± buff to pinkish-buff annual corticioid with a thin, pale margin. It generally forms large fruitbodies. Has long, cylindrical, densely septate, crystal-covered cystidia. The septa lack clamps. The allantoid spores measure 4.5–6.5 × 1.5–2 µm. Produces a white rot on fallen trunks of deciduous trees, mostly in sites with large amounts of dead wood.

The very densely septate cystidia is a good diagnostic character, but the sharp-angled, thick-walled hyphae are also characteristic. *Candelabrochaete verruculosa* ⋉ has 4.2-6(-7) × 2-3(-4) µm large spores and repeating basidia.

Widespread in the nemoral and hemiboreal zones, but rare, although perhaps on the increase; probably most of the year.

Thomas Læssøe

Jens Maarbjerg

Phanerochaete & ***Rhizochaete*** are genera of ± smooth, resupinate corticioids with rather waxy fruitbodies that often have abundant hyphal cords at the margin. They tend to lack clamps (some may have clamps but only in the lower part of the fruitbodies). Most have prominent cystidia. All are white-rotters.

Phanerochaete velutina is a rather thick, large annual corticioid with eye-catching white hyphal cords and a downy hymenium from large, projecting, ± cylindrical, thick-walled cystidia, which are crystal-covered in the upper part. Most hyphae lack clamps but the basal hyphae may have several clamps at each septum. The almost cylindrical, smooth, inamyloid spores measure 5–7(–8) × 3–3.5(–4) µm. Occurs on deciduous branches in contact with soil; a white-rotter.

Rhizochaete radicata ▷ is darker and stains purple-red with KOH. Other smooth species of *Phanerochaete* have smaller, less projecting cystidia.

Widespread and common; all year.

Thomas Læssøe

Phanerochaete sanguinea is a cream to buff, ultimately reddish, annual corticioid that has hyphal cords at the margin and colours the substrate red. The hyphae are branched at right-angles and completely lack clamps. The cystidia are long, smooth, pointed and somewhat projecting. The hyaline spores measure 5.5–6(–7) × 2.5–3 µm. Occurs on both conifer and deciduous wood, where it forms a red-stained white rot.

Red-stained wood is a good identification character.

Widespread and rather common in the north to occasional towards the south; all year.

Rhizochaete radicata is a ± brownish-orange annual corticioid with prominent brownish-orange hyphal cords. The surface stains violet with a drop of KOH. Has thick-walled, pointed cystidia with crystals and no clamps. The hyaline, ellipsoid spores measure 4–5 × 2–3 µm. Occurs on deciduous and conifer wood in contact with soil.

Formerly treated as *Phanerochaete filamentosa*. The small spores combined with the KOH reaction make this species easy to recognize.

Widespread in the nemoral–hemboreal zones, occasional; mainly July–October.

Thomas Læssøe/Jens H. Petersen

Dacryobolus karstenii is a thick, almost leathery, cream to cinnamon-buff, ± perennial corticioid with a somewhat uneven but mostly quite smooth, although downy, surface that has a rather well-defined margin. The fruitbodies may become very large. Has a distinctly sour-aromatic odour, like that of *Fomitopsis pinicola* (page 867). Has clamps and long, smooth cystidia that originate in the flesh and are thick-walled at the base and thin-walled towards the tips. The hyaline, allantoid spores measure 4.5–6 × 1.2–1.5 µm. A brown-rotter in fallen trunks of conifers, mainly *Picea*.

Amylostereum chailletii (page 969) is more monochrome clay-buff and has pointed, crystal-covered cystidia tops. *Dacryobolus sudans* ⋉ has warts that exude drops at the top.

Widespread, occasional, apparently on the increase in parts of the nemoral zones; all year.

Thomas Læssøe/Jens H. Petersen

Jens H. Petersen/Thomas Læssøe

Xenasmatella vaga is a very thread-like, initially almost completely sulphur-yellow annual corticioid that, with age, develops a grey-olive, grainy hymenium. It stains reddish with KOH. Has clamps, but lacks cystidia. Has pleural, 4-spored basidia. The finely and evenly spiny, broadly ellipsoid spores measure (4.5–)5–5.5(–6) × 4–4.5 µm. Occurs typically on pieces of wood in the leaf-litter, in both deciduous and coniferous forests.

Also known as *Phlebiella vaga* and *P. sulphurea*. An easily recognizable species due to the vivid colour and the thread-like appearance of the fruitbodies.

Widespread and very common; mainly July–November.

Thomas Læssøe

Crustoderma dryinum is a firm, although thin-fleshed, pale yellow to pale orange-yellow or slightly greenish annual corticioid with very long, smooth, thick-walled cystidia. The hyphae have clamps. The ± fusiform, smooth, inamyloid spores measure 7–9 × 2.5–3.5 µm. Occurs on trunks of *Picea* and *Pinus*, but also on timber and burnt wood; a brown-rotter.

Species of *Coniophora* ▷▷ have a looser texture, different and brown spores, and usually acquire browner colours with age.

Mainly a hemiboreal–boreal species, occasional to rare; all year.

Jens H. Petersen/Thomas Læssøe

Phlebia is a large genus of waxy, mostly warty or wrinkled resupinate corticioids. In the microscope the hymenium is very dense, with narrow basidia and, usually, eye-catching cystidia. See also page 978.

Phlebia subochracea is a ± warty, ± yellow to orange-yellow annual corticioid with a ± fimbriate, pale margin. Has clamps and long, very pointed, smooth cystidia. The spores measure 5.5–6.5(–7) × 2.5–3 µm. It is adapted to habitats that are flooded during winter, and fruits when the wood is exposed during the summer; a white-rotter. It is often found together with *Mycoaciella bispora* (page 1051), *Mycoacia uda* (page 1050) and *Physisporinus sanguinolentus* (page 944), as well as various ascomycotes.

The habitat and very vivid colours make identification straightforward.

Widespread in the nemoral–hemiboreal zones and probably more common than the available records suggest; mostly July–October.

Phlebia lilascens is a very variable annual corticioid that changes colour from cream to greyish with tinges of red and lilac in relation to the water content of the waxy fruitbody. The surface is smooth to warty. Has clamps, but no cystidia. The narrowly ellipsoid, hyaline spores measure 4–4.5 × 2–2.5 µm. Occurs on fallen trunks and thick branches from a number of deciduous hosts.

Phlebia livida ⋉ has ± allantoid spores and may have upwards tapering cystidia.

Widespread, mainly in the nemoral–hemiboreal zones, occasional; most of the year.

Peniophora is a large genus of ± resupinate, smooth to verrucose corticioids with thick, waxy, drought-resistant, typically greyish, violet or reddish fruitbodies. Usually has conspicuous cystidia and sometimes also dendrohyphidia. Typically has allantoid, inamyloid spores. White rotters. See also page 965.

Peniophora limitata is a dark reddish-grey to ± brownish annual corticioid that typically has an almost black, well-delimited, non-loosening margin. It cracks upon drying. Has conical, crystal-covered, thick-walled cystidia. The hyaline, allantoid spores measure 8–12 × 3–3.5 µm. Occurs on bark, rarely wood, of *Fraxinus*, mostly on attached branches; rarely on other hosts *e.g. Syringa* and *Ligustrum*.

Peniophora quercina (page 965) and *P. rufomarginata* ⋉ have loosening margins.

Widespread and common in the nemoral–hemiboreal zones; all year.

Jan Vesterholt

Peniophora lycii is a ± pale greyish, cracking annual corticioid without a loosening margin. The surface is finely felty. The hymenium has branching dendrohyphidia and rounded, thick-walled cystidia. The spores are allantoid and measure 9–13 × 3.5–5 µm. Occurs on attached deciduous branches, mostly decorticated, in hedges and on older garden shrubs, *e.g. Syringa*, but also in forests.

The presence of dendrohyphidia and the distinctive cystidia are characteristic.

Very common and widespread in the nemoral zones, but nearly absent farther north; all year.

Often parasitized by *Tremella versicolor* (page 1193) and the pyrenomycete *Trichosphaerella decipiens* ⋉.

Peniophora pini is a thick, wrinkled-warty annual corticioid with reddish-grey to yellow-brown-tinged fruitbodies that have a thin, paler margin. Has hyphae that dissolve into a slime, and large, balloon-shaped gloeocystidia in the basal part of the fruitbody. It also has thick-walled, crystal-covered, conical cystidia. The allantoid spores measure 7–9 × 2.5–3 µm. Occurs on bark of attached *Pinus* branches, in forests and coastal plantations.

Peniophora pithya ⋉ has more coloured, permanent hyphae and usually occurs on *Picea*.

Widespread, occasional to rare; all year.

Peniophora polygonia is a rather thick, ± warty annual *Peniophora* with a greyish surface with orange flesh shining through; by scraping the surface the orange-red flesh becomes clearly visible. The fruitbodies are formed by confluent small, warty components. Lacks crystal-bearing cystidia but has large, clavate gloeocystidia and dendrohyphidia. The allantoid spores measure 9–12 × 2.5–4 µm. Occurs on bark of attached, corticated *Populus* branches, particularly *P. tremula*, rarely on *Salix*.

Peniophora rufa ⋉ also occurs on *Populus* but is redder and has crystal-covered cystidia.

Widespread and common in the nemoral–hemiboreal zones, becoming scarcer towards the north; all year.

Thomas Læssøe

Peniophora laeta is an apricot-orange to yellow-brown annual corticioid that develops under bark, and has long warts that enable it to break through. Has gloeocystidia and crystal-covered, conical cystidia. The allantoid spores measure 10–14 × 3.5–4.5 µm. Occurs on wood of attached *Carpinus* branches.

The erumpent nature and distinctive colour make this species easy to recognize. Other corticioids may subsequently form fruitbodies under the broken bark. Erumpent species of *Vuilleminia* ⊳⊳ have pale colours, no warts and very large spores.

Widespread within the range of the host (nemoral), locally common; all year.

Peniophora incarnata is a flesh-pink to peach-coloured, rather smooth annual corticioid with a somewhat paler, attached margin. Has large gloeocystidia, clamps and conical, crystal-bearing, thick-walled cystidia. The allantoid spores measure 8–12 × 3.5–5 µm. Occurs on wood and bark from a wide range of deciduous hosts, but mostly on attached dead branches; rare on conifer wood.

Peniophora laurentii ⋉ lacks clamps. *Peniophora erikssonii* ⋉ occurs on *Alnus* and has very large, ellipsoid spores. *Corticium roseum* ⊳ is more grey-rose and more felty on the surface; it lacks thick-walled, crystal-bearing cystidia.

Widespread and very common; all year.

Corticium and ***Marchandiomyces*** are genera of ± rose, resupinate corticioids with dendrohyphidia and very large basidia and spores. Most species lack cystidia. The fruitbodies are formed above ground, *e.g.* on attached, dead branches.

Corticium roseum is a grey-rose to brownish-rose annual corticioid with rather small, sharply delimited fruitbodies that have a somewhat felty margin and surface. Has clamps and numerous dendrohyphidia but lacks cystidia. The inamyloid, smooth, ellipsoid spores measure (18–)20–22 (–25) × 8–12(–14) µm. Occurs on bark of attached branches of *Salix*, rarely on *Populus*, mainly in old *Salix* carrs.

Corticium erikssonii ⋉, which occurs on *Alnus*, *Populus* (including *P. tremula)* and possibly *Ulmus*, has many small, grey-rose, ± loose, asexual 'bulbils' on the fruitbody. *Peniophora incarnata* ◁ is less felty and more peach-coloured; it also has crystal-bearing, thick-walled cystidia.

Widespread, common to occasional; all year.

Corticium roseum is sometimes parasitized internally by the jelly fungus *Celatogloea simplicibasidium* ⋉.

Marchandiomyces quercinus is a ± grey-rose annual corticioid with a somewhat wrinkled surface that forms small fruitbodies over old fruitbodies of *Colpoma quercina* (page 1480). Lacks cystidia but has very large, 2-spored basidia. The spores are very large and ellipsoid, measuring 20–30 × 12–15 µm.

Also known as *Laeticorticium quercinum*. *Peniophora incarnata* ◁ has cystidia with crystals, smaller, 4-spored basidia and allantoid spores.

Widespread, but easily overlooked, probably rather common, at least in parts of the nemoral–hemiboreal zones; February–June.

Terana caerulea is a spectacular resupinate to slightly reflexed, large, rather tough, bright blue, greyish-blue to almost blackish-blue annual corticioid with a pale to almost white, fimbriate margin. Lacks cystidia but has dendrohyphidia. The ellipsoid, smooth, inamyloid spores measure 7–9 × 4–6 µm. Occurs on deciduous wood, particularly *Fraxinus*, in damp forests with plenty of dead wood.

Also known as *Pulcherricium caeruleum* (or *coeruleum*). Almost impossible to mistake. Blue species of *Byssocorticium* (page 990) have very loosely constructed, cobweb-like fruitbodies and small, spherical spores. *Amaurodon* (page 997) has spores that stain blue in KOH but cotton-like textures. *Acanthophysellum lividocoeruleum* ⋉ is mostly found on conifer wood in the hemiboreal–boreal zones and has prominent 'hedgehog'-like cystidia.

Locally common in Atlantic areas of temperate Europe, including parts of the UK, the Netherlands and western Norway, but also present in central European uplands; all year.

Species of ***Vuilleminia*** form waxy fruitbodies below bark, which gradually dies and curls back. They have dendrohyphidia and very large basidia and spores. The fruitbodies tolerate desiccation. They are White-rotters. See also page 1010.

Vuilleminia comedens is an erumpent, pale, mostly ± lilac-tinged, waxy annual corticioid that tolerates desiccation. Lacks cystidia but has very large 4-spored basidia (inset image). The smooth, inamyloid, allantoid spores measure (15–)16.5–22.6(–24) × (4.2–)5.3–7.3(–8) µm; Qav. 3.5. Occurs on attached, deciduous branches, primarily on *Quercus*, but also on *Fagus*, *Alnus*, *etc.*

Vuilleminia alni was formerly regarded as a separate species, but growth experiments and DNA sequence data indicate that it is conspecific with *V. comedens*. *Vuilleminia megalospora* ⋉ has a Qav. of 2. *Vuilleminia coryli* ▽, which is found predominantly on *Corylus,* has long, ± cylindrical cystidia. *Vuilleminia cystidiata* (page 1010) occurs predominantly on *Crataegus* and is whitish with cystidia that taper towards the tip.

Widespread in the nemoral–hemiboreal zones and very common, becoming scarcer farther north; all year.

Vuilleminia coryli is a pale, pinkish-buff to greyish-lilac, waxy, erumpent annual corticioid with large, cylindrical cystidia that measure 90–120 × 6–9 µm, but which may be present in very limited numbers. The allantoid spores measure 17–23 × 5–6 µm. Occurs on the branches of a wide range of deciduous hosts but mainly *Corylus*, only very rarely on *Fagus* and *Quercus*.

Vuilleminia comedens △ is extremely similar and the presence of cystidia has to be checked carefully. *Vuilleminia cystidiata* (page 1010) is chalk-white and has acute cystidia.

Widespread and common in the nemoral–hemiboreal zones; all year.

Thomas Læssøe

Aleurodisus species generally form ± orange fruitbodies that have a ± grainy surface due to the very large, amyloid spores, which typically have spiny ornaments. They have large basidia with very flaring sterigmata, and often have dendrohyphidia. All are white-rotters. See also *Aleurocystidiellum disciforme* (page 965).

Aleurodiscus amorphus is a disk-shaped, ± confluent, peach- to apricot-orange annual corticioid with a pale, downy margin. Lacks clamps and cystidia, but has unbranched hyphidia with many constrictions. The spores are amyloid (inset image), spiny and very large, measuring 20–26 × 16–20 µm. Occurs almost exclusively on bark of *Abies*, mainly on attached, dead branches or newly cut branches.

Often parasitized by species of *Phaeotremella* (page 1192), which are seen as milky, gelatinous spots on the hymenium.

Widespread in the nemoral–hemiboreal zones, but rare in some parts, while quite common in others, e.g. in Danish plantations; mainly in the spring.

Aleurodiscus aurantius is a rather anonymous, buff-yellow to somewhat orange-tinged, ± felty annual corticioid. Lacks cystidia and clamps but has dendrohyphidia and 4-spored basidia. The spores are amyloid, spiny and large, measuring 16–22 × 12–15 µm. Occurs mainly on stems of *Rubus* and *Rosa*.

Easily identified due to the combination of amyloid, spiny spores and host species. *Aleurodiscus norvegicus* ⋉ has 2-spored basidia and is known from *Pinus* needles and stems of *Calluna*.

Found mainly in the nemoral zones, occasional to rare, but probably often overlooked; mainly May–December.

Aleurodiscus wakefieldiae is a very striking, rose-coloured annual corticioid with a finely felty, white, loosening margin and a somewhat uneven surface. Has clamps, warted hyphidia, gloeocystidia and 4-spored basidia. The spores are amyloid, spiny and very large, measuring 19.5–29 × 12.5–20 µm. Occurs on bark of deciduous trees, particularly *Quercus* and *Castanea*.

Pink species of *Peniophora* ◁◁ do not loosen from the substrate, have a different tinge and completely different micro-characters.

An Atlantic species, mainly recorded in the UK, where it can be locally common; all year.

Thomas Læssøe

Scytinostroma hemidichophyticum is a perennial, internally layered (inset image), widely expanded, pale buff corticioid with an incredibly strong smell of naphthaline (moth balls), which is often how it is detected. Has dendrohyphidia, skeletal hyphae and large basidia. The almost spherical, smooth, thin-walled, amyloid spores measure (4.5–)5–6.5(–7) × 4.5–6 µm. Occurs on trunks of deciduous trees and branches, mostly *Salix*, but known from a number of hosts; a white-rotter.

Scytinostroma portentosum, described from North America, is possibly the same species, in which case that name would take precedence. The very similar *Baltazaria galactina* ⋉ has smaller spores. Other strong-smelling corticioids are thinner and annual, lack internal layers and have different micro-characters.

Widespread, occasional, locally common; all year.

Thomas Læssøe/Jens H. Petersen

Jan Vesterholt

Conferticium ravum is a thick, effused, buff to cinnamon-buff, ± warty, sometimes internally zonate annual corticioid. Lacks clamps but has long gloeocystidia. The amyloid, verrucose spores measure 5.5–7.5 × 3.5–4.5 µm (the ornament is best seen on dead spores in Melzer's reagent). Occurs on trunks of *Populus tremula*; forms a white rot.

Conferticium ochraceum ⋉ is very similar, but may be recognized by the combination of the substrate (conifer wood) and its smooth spores. To the right in the image above a group of *Merismodes anomalus* fruitbodies (page 1090) can be seen.

Widespread, but apparently very rare, absent from most of the nemoral zones, with just a single record in Denmark; all year.

Megalocystidium leucoxanthum is a buff to cinnamon-buff, mostly rather thick annual corticioid, which is formed from confluent, small, wart-like fruitbodies. The surface may be pruinose, caused by the large spores (see inset image). Has clamps, very long gloeocystidia and large basidia. The ± allantoid, amyloid spores measure 12–20 × 4.5–7 µm. Occurs on deciduous wood, mainly *Salix* and *Populus*.

Megalocystidium luridum ⋉ has smaller, ellipsoid spores.

Widespread, occasional to rare; all year.

Gloiothele lactescens is an annual corticioid that produces large, smooth to rugose, pale pinkish-buff to pale vinaceous-brown, rather thick fruitbodies, which, when fresh, exude a milky liquid when cut (inset image). It has a bitter taste. Lacks clamps but has projecting hyphae (hyphidia) and large gloeocystidia that stain dark in sulfo-vanillin. The spores are smooth and somewhat amyloid, and measure 6–7 × 4–4.5 µm. Occurs on highly decomposed deciduous stumps and trunks, mostly in damp habitats with black, fertile soil.

Also known as *Gloeocystidiellum lactescens*. Species of *Megalocystidium* ◁ have larger spores and, at most, only taste very slightly bitter. They are not known to exude milky liquids.

Widespread and common in the nemoral zones, but rare in the hemiboreal zone; mainly summer–autumn.

Jens Maarbjerg

Vesiculomyces citrinus is a large, pale annual corticioid with a ± sulphur-yellow to straw-yellow tinge and a white, fimbriate margin. Lacks clamps, but has large, ± swollen gloeocystidia that taper towards the tip. The spores are almost spherical, smooth and amyloid, 4–5 µm in diameter. Occurs on fallen trunks of conifers.

Accepted in the genus *Gloiothele* by some mycologist. *Vesiculomcyces* has no sulfo-vanillin reaction in comparison to *Gloiothele* species, but the genera are phylogenetically closely related.

Widespread, common to rather rare; mostly during winter.

Jan Vesterholt/Jens H. Petersen

Amylostereum laevigatum is a rather thick, very firm, pale grey, well-defined, leathery, perennial corticioid that does not stain red when bruised. Has numerous brownish, very long cystidia in the flesh. The spores are cylindrical, smooth and amyloid, and measure 7–12 × 3–4 µm. Occurs on hard wood of *Juniperus*, *Taxus* and various conifers in the *Cupressaceae*.

Normally easy to recognize in the field, but young, thin fruitbodies require microscopic examination. Other species in the genus are ± pileate and usually occur on other conifers (page 969).

Widespread and rather commom, becoming scarcer in the north; all year.

Asterostroma laxum is a ± vinaceous-brown, rather loosely built annual corticioid with yellowish hyphal cords and a pale margin. The flesh hyphae branch repeatedly, but are not always regularly bifurcated, are thick-walled, brown and with pointed ends (so-called asterosetae, see inset image). Lacks clamps. The spores are ± spherical, smooth and amyloid, and measure 6–8 µm. Occurs mostly on conifer wood, usually in older plantations and forests.

Asterostroma medium ⋉ has verrucose spores, and various other ± similar corticioid species have clamps or more branching asterosetae. *Asterodon ferruginosus* (page 1052) has a spiny hymenophore with asterosetae.

Widespread, occasional to rare; mainly May–December.

Dichostereum boreale is a rather firm, warty, cream-coloured annual corticioid with a diffuse margin. It has cylindrical cystidia and numerous repeatedly bifurcating, thick-walled, pale brown, dextrinoid, somewhat pointed hyphae in the flesh (so-called dichohyphae). The spores are finely verrucose, ellipsoid and amyloid, and measure 4.5–6 × 3.5–4 µm. Occurs on conifer wood.

Dichostereum effuscatum ⋉ has almost spherical spores. Species of *Vararia* ▽ have smooth spores, and species of *Asterostroma* ◁ have much longer tips on the branching hyphae, which are less regularly bifurcated (asterosetae).

Boreal, occasional; all year.

Vararia ochroleuca is a cream to pale cinnamon-buff, rather loosely built annual corticioid with somewhat paler, ± distinct cords at the margin. The flesh has fine, repeatedly bifurcating, thick-walled, pale brown, dextrinoid, semi-acute hyphae (dichohyphae). The hyphae lack clamps; the gloeocystidia have an elongated neck and thick walls. The smooth, inamyloid spores measure 3–4 × 2.5–3 µm. Occurs on wood and litter from both conifers and deciduous trees; probably calciphilous.

Vararia gallica ⋉ has larger, fusiform, ± amyloid spores. *Vararia hauerslevii* ⋉ has larger, inamyloid spores; *V. investiens* ⋉ has clamps and *V. cremeoavellanea* ⋉ has smaller, ± ellipsoid, inamyloid spores.

Widespread, but very rarely recorded, possibly overlooked; mainly May–October.

Hymenochaete & **Hydnoporia** are recognized by their brown setae, which penetrate the smooth hymenium (visible with a good hand lens!). The spores are hyaline and smooth. All species are white-rotters. Capped species are covered on page 973.

Hymenochaete cinnamomea is a relatively thick, perennial, completely appressed corticioid with yellow-brown to cinnamon-brown colours. The flesh becomes layered and has smooth setae up to 180 µm long. The hyaline, smooth spores measure 5.5–6.5 × 1.5–2.5 µm. Forms a characteristic, pitted rot in wood of deciduous trees, mostly in older deciduous forests; rarely on conifer wood (*Juniperus*).

Hymenochaete jaapii ⋉ is rather similar but the flesh lacks layers and the setae are shorter, up to 65 µm long. *Hymenochaete fuliginosa* ▽ is darker, thinner and has shorter setae (40–100 × 6–8 µm); it grows on conifer wood, but also on *Quercus* and *Salix*. *Hydnoporia corrugata* ▷ is pale brown and cracks in a characteristic way upon drying.

Widespread, uncommon; all year.

Hymenochaete fuliginosa is a completely appressed, red-brown to almost purple-brown corticioid that usually lacks flesh and rarely cracks. Has short, projecting, smooth setae up to 100 µm long. The spores measure 4.5–6.5 × 1.8–3 µm. Occurs on the wood of conifers and *Salix*.

Hymenochaete subfuliginosa ⋉ occurs on *Quercus* and has smaller spores, 4.5–5.5 × 2.7–3.2 µm. *Hydnoporia corrugata* ▷ is paler, cracking and has setae with pits at the tip. *Hymenochaete carpatica* ⋉ occurs on the flaky bark of *Acer pseudoplatanus*.

Widespread and fairly common in the hemiboreal–boreal zones, but rare in the nemoral zones; all year.

Hymenochaete cruenta is a completely appressed, coral-red to carmine corticioid, which when fresh has a paler, yellowish margin. It has projecting, brown, smooth setae that measure 60–80 × 6–10 µm. The spores measure 6–8 × 2–2.5(–3) µm. Occurs on the bark of *Abies*.

A very easy species to recognize, as no other European *Hymenochaete* has such a vivid colour.

Found mainly in central Europe, where it is locally common, but extends north to the Netherlands and the UK, although extremely rare in both countries; all year.

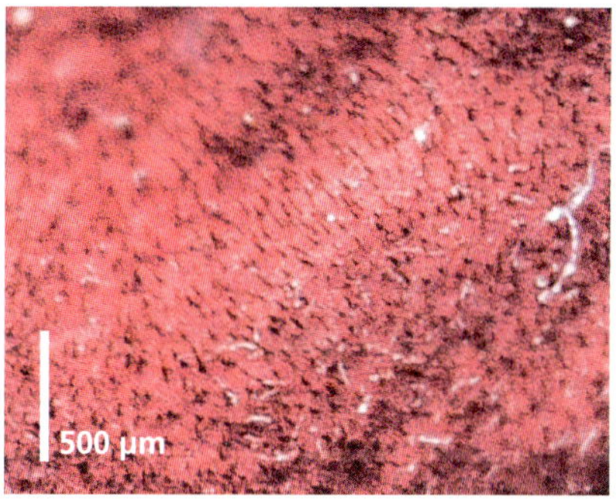

Hydnoporia corrugata is a completely resupinate, rather pale brown annual corticioid with short, projecting, brown setae that measure 40–100 × 6–14 µm and have a eroded tip. The spores measure 3.5–5 × 1–1.5 µm. Occurs in dry sites, on standing trunks of *e.g. Corylus*, *Rosa* and other ± thorny shrubs.

Other similar resupinate species are darker or have vivid colours and hardly crack. *Stereum rugosum* (page 970) reddens when bruised and does not have any brown setae.

Mainly nemoral and mainly Atlantic, locally common; all year.

May be parasitized by *Hypocreopsis rhododendri* (page 1527).

Thomas Læssøe

Coniophora is a genus of brown-spored, brown-rotting corticioids. It is related to the agaric genus *Tapinella* (page 96).

Coniophora puteana is a thick, ± rounded-warty, honey-yellow to olive-brown annual corticioid with a pale, somewhat fimbriate margin. Has hyphal cords. The surface is finely felty from projecting basidia (top inset image). Has several large clamps at each septum, but no cystidia. The brownish, smooth, rather thick-walled, only somewhat dextrinoid spores measure 10–15 × 7–9 µm. A brown-rotter on deciduous and conifer wood, and also occurs in houses where it may cause serious damage to timber.

Coniophora arida ▽ is thinner and has highly dextrinoid spores. *Coniophora olivacea* ▷ has large, cylindrical cystidia and *C. fusispora* ▷ has fusiform spores.

Widespread and common, becoming scarcer towards the north; mainly September–January.

Fruitbodies with large, yellow vesicles on the surface are parasitized by the fungus *Nodulisporium cecidiogenes* (lower inset image and page 1633).

Coniophora arida is a rather thin, pale-margined annual corticioid with grey-olive colours and hyphal cords. Lacks cystidia. The highly dextrinoid, yellowish spores measure 10–14 × 6–8 µm. Mainly a brown-rotter in conifer wood, mostly *Pinus*, in plantations and forests on sandy soils, rarely in deciduous wood.

Very similar to a thin version of *C. puteana* △, but should be recognizable by its paler, highly dextrinoid spores.

Widespread and fairly common; mostly September–January.

Coniophora olivacea is a grey-brown to olive-brown annual corticioid with pale hyphal cords and large, cylindrical, septate, ± crystal-covered cystidia (inset image). The ± yellow-brown, thick-walled, somewhat dextrinoid spores have a germ pore and measure 8–12 × 4–6 µm. A brown-rotter in both deciduous and conifer wood.

The cystidia are unique in the genus. Other corticioids with such cystidia have completely different spores.

Widespread, occasional; mainly October–January.

Coniophora fusispora is a cinnamon-buff to grey-olive annual corticioid that is dominated by hyphal cords. Lacks cystidia. The very dark, thick-walled spores are distinctly sinuous–fusiform and measure 14–20 × 5–8 µm. Occurs on timber, but also on herbs and tiny shrubs in open, sun-baked habitats, where the fruitbodies also spread out onto the soil.

The unique spores make this species easy to recognize.

Very rare with few scattered records in nature, but probably overlooked - also occurs indoors on construction timber; mainly July–October.

Thomas Læssøe

Pseudotomentella tristis s.l. is a brownish-black to almost blue-black, rather thick, almost fleshy annual corticioid, which may be loosened from the substrate. Lacks clamps. The hyphae and spores stain brown or ± bluish in KOH. The almost spherical to triangular, rather coarsely double-warted spores measure 6–9(–10) µm. Occurs mainly on litter and soil in fertile forests; ectomycorrhizal.

Probably constitutes a species complex. *Pseudotomentella atrofusca* ⋉ has smaller spores that measure 5–6(–6.5) µm.

Widespread and fairly common; mainly August–November.

Spiny corticioids

This form group includes resupinate and reflexed, thin-fleshed corticioids with a dentate to spiny hymenophore. A few species can be identified in the field, but in general the most important characters are the appearance of the spores and basidia, and in some cases the types of cystidia and hyphal structures, all of which require microscopy.

OTHER SIMILAR FUNGI:

– the hydnoids are more thick-fleshed and often have a cap and stem (page 1054).
– pendent clavarioids form 'clubs' directly on the substrate, rather than in connection with a subiculum or the flesh (page 1092).
– the merulioid corticioids have a ± veined hymenophore, which may partially split into rounded 'teeth' (page 974).
– the pores of annual, resupinate polypores may, with age, split into 'teeth' (page 932).

See also the overview and the references to further reading on page 960.

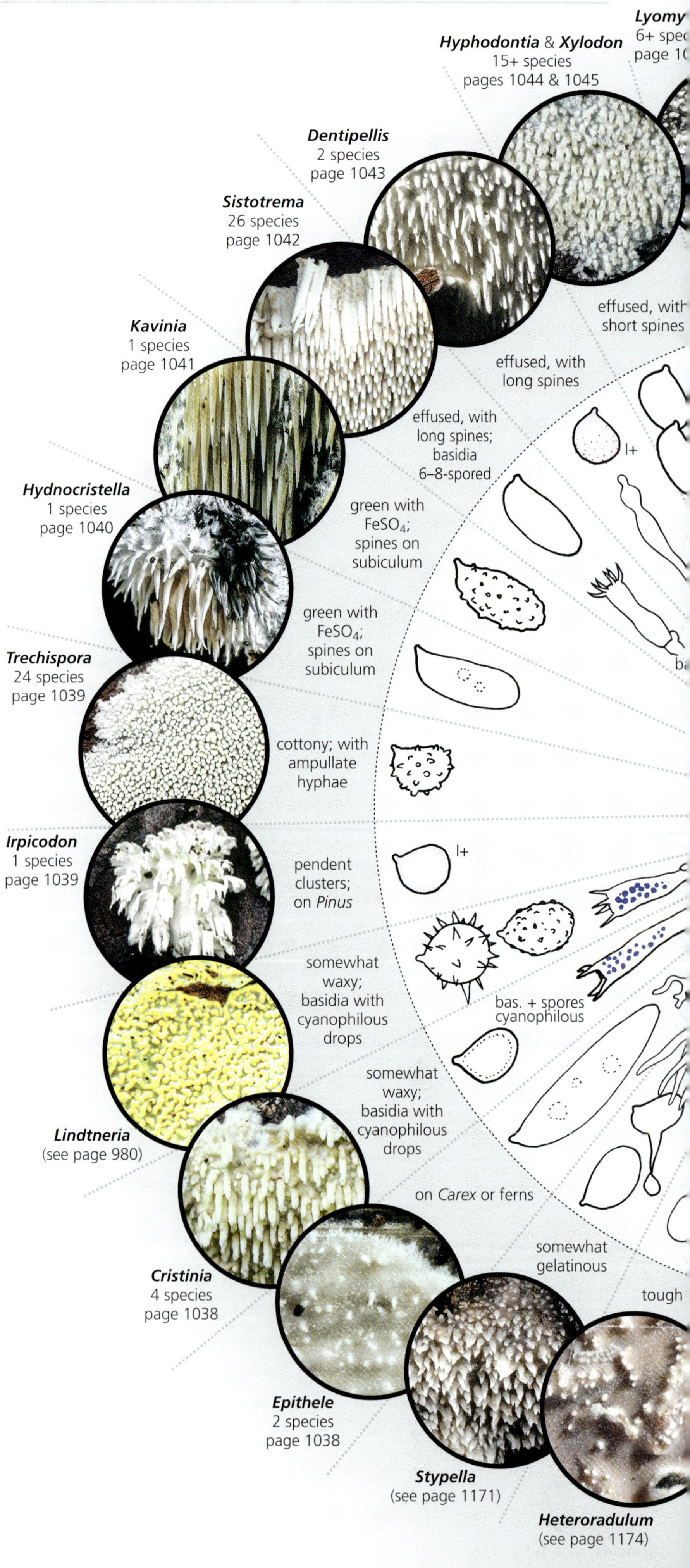

spines (round in section)

'teeth' (flattened)

Micro-drawings: spores first, followed on the inside by cystidia.
cl. = clamps.
bas. = basidia
I+ = amyloid
hy. = hyphae

Approximate species number applies to temperate Europe.

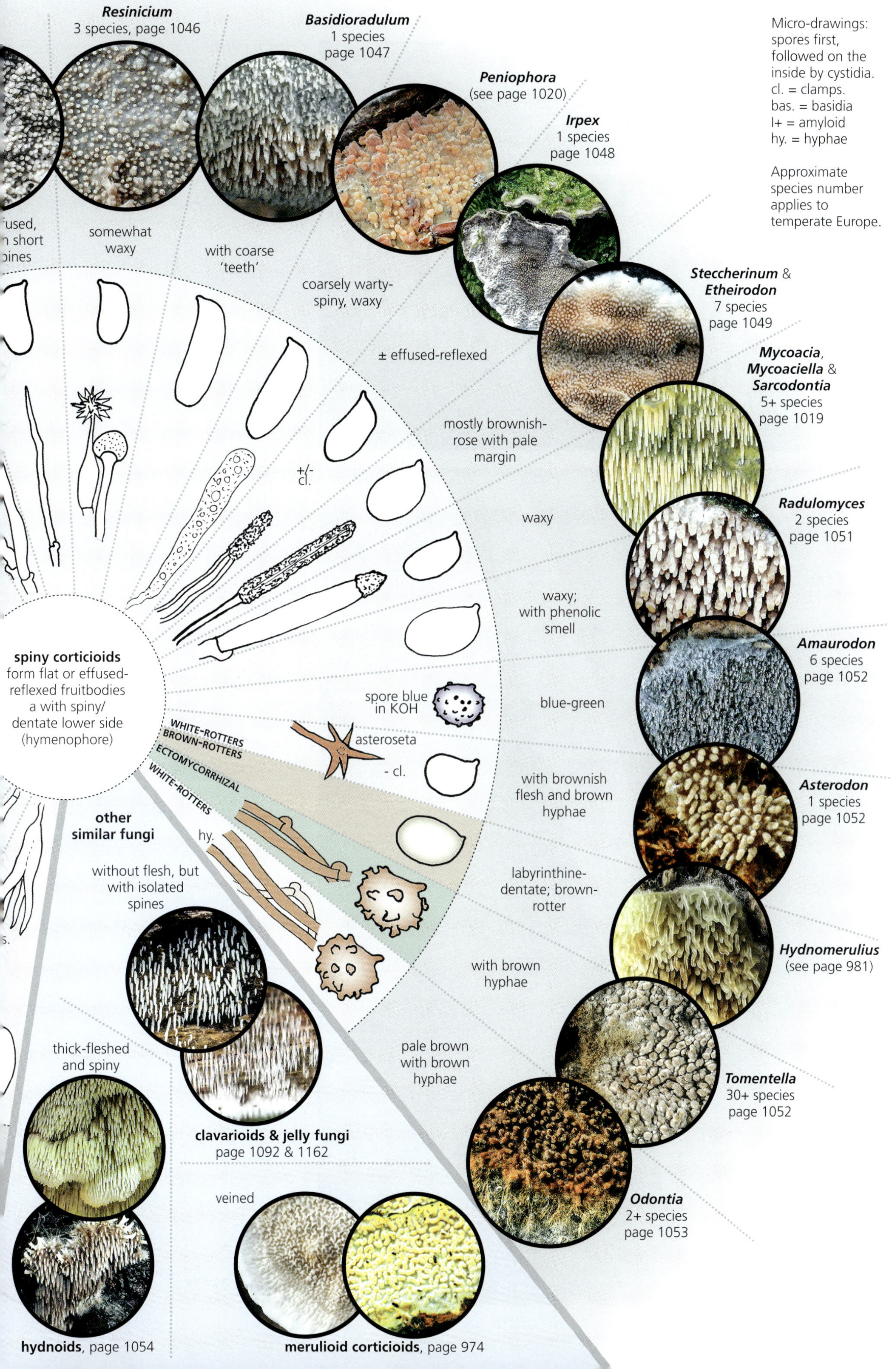

Cristinia eichleri is a yellowish, annual corticioid with ± obvious warts and up to 5 mm long, flattened spines; margin whitish and diffuse. It is monomitic and lacks cystidia. Has cyanophilous basidia and ± spherical, smooth, ± thick-walled spores, 5–7 µm in diameter. Occurs on soil, litter or wood in deciduous forests on ± calcareous soils.

Also known as *C. gallica*. *Cristinia helvetica* ⋉ is less dentate and has smaller spores, 3.5–4.5 × 3–3.5 µm. Species of *Lindtneria* (page 980) also have cyanophilous basidia but the spores are ornamented.

Widespread in the nemoral–hemiboreal zones, but rather rare; July–November.

Epithele typhae is a very characteristic, white to cream annual corticioid with rather scattered, short spines that consist of intermixed thin-walled hyphae. The hyphae have clamps. The very large, fusiform, thick-walled, inamyloid spores measure 16–30 × 5.5–8 µm. Occurs on various sedges and rushes, but predominantly at the base of living *Carex acutiformis*.

Distinctive and unlikely to be misidentified, although other white corticioids can also occur on marsh plants, *e.g. Lyomyces sambuci* (page 1007). *Epithele galzinii* ⋉ has smaller spores and occurs on ferns.

Probably common wherever large stands of *C. acutiformis* occur in the nemoral zone, with a few records from the hemiboreal zone; (May–) July–November.

Trechispora stevensonii is a spiny corticioid that also has scattered cotton-like tufts where asexual spores (conidia) are formed by fragmentation of the hyphae. It is monomitic with clamps and 4-spored basidia. Has flat crystals on the hyphae. The spores have small spines up to 0.5 µm high and measure 4–4.5 × 3–3.5 µm. Occurs on rotten wood of deciduous trees, mostly larger trunks, rarer on conifer wood.

The combination of obvious spines and a 'woolly' asexual state make this species relatively easy to recognize. Other conidia-forming species either have thick-walled conidia, smooth spores or skeletal hyphae. Amongst the spiny trechisporas that do not have an asexual state, *T. farinacea* ⋉ has strikingly short cells in the spines and *T. nivea* ⋉ has long cells. *Trechispora kavinioides* ⋉ has smooth spores. (See species with a smooth or pore-shaped hymenophore on pages 997 and 938.)

Widespread and rather common in the nemoral–hemiboreal zones, rare farther north; mainly June–November.

Irpicodon pendulus is a ± chalk-white, very delimited annual corticioid with long 'teeth' (± flat spines) and a tendency to form small caps. It has monomitic flesh and a few thick, refractive hyphae, but no cystidia. The somewhat allantoid, amyloid spores measure 4–5 × 2.5 µm. Occurs typically on dead branches high up in living *Pinus sylvestris*, where it produces a white rot.

Irpex oreophilus ⊳⊳ is rather similar but occurs on deciduous wood and the spores are inamyloid.

Widespread in the hemiboreal zone, but ± absent from the western lowlands; all year.

Johan Nitare

Hydnocristella himantia is a resupinate annual corticioid with almost white to brownish, very long spines that are situated on a subiculum or on a thin fleshy layer; it has a white, thread-like margin. Reacts green with $FeSO_4$ (lower image). The basal hyphae have cyanophilous warts. The spores are smooth and measure 8–12 × 3–5 µm. Occurs on wood and litter, mainly in old-growth woodlands on mineral-rich, mostly calcareous soils.

Kavinia alboviridis ▷ has verrucose spores and typically more olive-brown colours. *Dentipellis fragilis* ▷▷ has densely packed, white to cream spines, no $FeSO_4$ reaction and amyloid spores. Species of *Mucronella* (page 1096) have completely free, hanging spines and also lack a $FeSO_4$ reaction. Other similar species of corticioid form spines from a more well-defined flesh and the hymenium also covers the area between the spines.

Mainly widespread and fairly common in the hemiboreal zone, rare or absent from the nemoral and boreal zones; July–November.

Hydnocristella and **Kavinia** form fruitbodies with numerous spines suspended from a cobweb-like or more compact hyphal layer. They have ± ampullate, widened hyphae at some septa and stain green with $FeSO_4$. These genera are closely related to *e.g. Lentaria*, *Phaeoclavulina* and *Ramaria* (pages 1132, 1134, 1138). White-rotters.

Kavinia alboviridis is an annual corticioid with long, buff to pale olive-brown spines, a white subiculum and white threads at the margin. It stains green with $FeSO_4$ (lower image). All hyphae are smooth, but some irregular crystals may be present. The spores are somewhat tapering, have cyanophilous warts and measure 8–11 × 3–4.5 µm. Occurs on litter and wood, mostly from conifers, including *Juniperus*.

Hydnocristella himantia ◁ has smooth spores and spines that lack olive-brown shades. Other similar species of corticioid form spines from a more well-defined flesh, and the hymenium covers the flat parts of the fruitbody as well as the spines; they do not stain with $FeSO_4$.

Widespread but rather rare to very rare; mainly September–November.

Sistotrema raduloides is a white to cream annual corticioid with a diffuse margin and spines up to 4(–6) mm long. Has clamps and 6- to 8-spored basidia. Lacks cystidia. The smooth, narrowly ellipsoid to fusiform, inamyloid spores measure 7–9 × 2.5–3.5 µm. Occurs on very decomposed wood of deciduous trees, *e.g. Populus* or trunks of *Fagus*.

The combination of long spines and 6- to 8-spored basidia is unique. *Sistotrema muscicola* ⋉ has shorter spines and smaller spores. *Dentipellis fragilis* ▷ is very similar but has ± spherical, verrucose, amyloid spores and 4-spored basidia.

Widespread in the hemiboreal zone but uncommon, very rare in the nemoral zones; mainly August–October.

Dentipellis fragilis is a long-spined, white to cream annual corticioid that usually forms large fruitbodies with fimbriate margins. The spines are up to 15 mm long and may be somewhat flattened. Has gloeocystidia. The ± spherical, verrucose, amyloid spores measure 4–6 × 4–5 µm. Occcurs on large, fallen deciduous trunks, often *Fagus* or *Populus tremula*.

Dentipellis coniferarum ⋉ is a very similar species with smaller spores (3–3.5 × 2.5 µm). *Climacodon septentrionalis* (page 1056) has similar spines but forms numerous tiered caps. *Sarcodontia setosa* (page 1057) has yellow spines and a very strong smell. Young, shorter-spined specimens may recall *Sistotrema raduloides* ◁, but that species has inamyloid, smooth spores and 6- to 8-spored basidia.

Widespread, but rather rare to very rare in the nemoral–hemiboreal zones, and also very rare in the boreal zone; July–December.

Thomas Læssøe

Xylodon form resupinate fruitbodies with a warty, spiny or pore-shaped hymenophore. The basidia are usually constricted. The hymenium has different types of cystidia and the hyphal system is monomitic or dimitic. All species are white-rotters. See also page 940.

Xylodon paradoxus is a cream-coloured, dentate annual corticioid that usually has small, cap-like protuberances, and in all aspects is very variable. The hyphal ends have large, irregular crystals up to 3 µm wide, and are mostly capitate. The flesh is dimitic with a few skeletal hyphae. The smooth, inamyloid spores measure 5.5–6.5 × 3.5–4 µm. Occurs mainly on deciduous branches on the forest floor, less commonly on attached branches, mostly on *Fagus*; very rare on conifers.

Xylodon raduloides (page 940) has smaller spores and a hymenophore of labyrinthine pores. *Xylodon quercinus* ▽ has ± cylindrical spores and two types of cystidia. *Basidioradulum radula* ▷▷ has more robust 'teeth' and allantoid spores that measure 9–11 × 3–3.5 µm.

Widespread and common, becoming scarcer towards the north; all year, but in best condition during the autumn.

Xylodon quercinus is a coarse, irregularly dentate, cream-coloured annual corticioid with a diffuse margin and 'teeth' up to 3 mm long. Has both capitate and pointed cystidia. The spores are cylindrical to allantoid and measure 6–8 × 2.5–3 µm. Occurs on deciduous wood, often on newly fallen or attached *Quercus* branches.

Xylodon paradoxus △ lacks pointed cystidia and has a different spore shape. *Basidioradulum radula* ▷▷ has larger spores. *Radulomyces molaris* ▷▷ smells phenolic and has 6-8 µm wide spores.

Widespread and common in the nemoral and hemiboreal zones; all year, though in best condition in late autumn.

Xylodon asperus is an annual corticioid with rather widely spaced, short spines, 1–3 per mm, whitish to pale cinnamon-buff colours and a floccose surface. Microscopically, characterized by cylindrical cystidia that have a capitate, swollen top, and ellipsoid spores that measure 5–6 × 3.5–4.5 µm. Occurs on both deciduous and conifer wood.

There are many similar species of *Xylodon* that only can be separated with careful microscopy.

Widespread and common in the hemiboreal–boreal zones, less common farther south; all year.

Jan Vesterholt

Lyomyces crustosus is an initially white, later cream, scattered warty-spiny annual corticioid with a rather well-defined margin and spines up to 0.3 mm long. Has pointed cystidia in the hymenium and the flesh. The spores measure 5–6.5 × 2.5–3 µm. Occurs mostly on deciduous wood, and typically on dead, attached branches in damp woodlands, *e.g.* older *Salix* carrs.

Best told by the well-defined margin, short spines and presence of pointed cystidia. *Brevicellicium olivascens* ⋉ often occurs in the same habitat but lacks cystidia and has different spores.

Widespread and common; all year.

Thomas Kehlet/Jens H. Petersen

Hyphodontia arguta is a spiny, cream to pale, cinnamon-buff corticioid with a spiny hymenophore. Microscopically, characterized by a combination of numerous cystidia shaped like dental drills, and ellipsoid spores that measure 4.5–6 × 3.5–4 µm. Occurs on both deciduous and conifer wood.

Microscopically, the combination of spiny fruitbodies and distinctive cystidia shape make identification straightforward. Macroscopically, easily confused with *e.g.* species of *Sistotrema* ◁◁ and *Trechispora* ◁◁.

Widespread and rather common; all year, but in best condition during late autumn.

Resinicium bicolor is a ± waxy, finely, but obviously warty to short-spined annual corticioid with pale greyish to brownish colours; the margin is diffuse and the warts somewhat conical. Has clamps and small, pointed cystidia crowned with star-shaped crystals, and other cystidia with a large, swollen 'ball' on the capitate tip. The smooth, inamyloid, somewhat allantoid spores measure 5–8 × 2.5–3.5 µm. Occurs mostly on decorticated conifer wood on the forest floor; rarely on deciduous wood.

With experience identifiable in the field, and very easy microscopically due to the characteristic cystidia. *Resinicium furfuraceum* ⋉ can be smooth or warty but lacks 'star cystidia'; *R. pinicola* ⋉ has long spines and also lacks 'star cystidia' – both species occur on conifer wood.

Widespread and common to very common; all year.

Jens Maarbjerg

Basidioradulum radula is a whitish to dark cream, coarsely dentate annual corticioid with smooth 'teeth' (± flat spines) that are usually rounded at the tip. It generally has a smooth or fairly short-toothed margin, and the 'teeth' gradually become longer towards the centre. Has clamps and cylindrical, sinuous cystidia measuring 40–60 × 5–8 µm. The allantoid, smooth spores measure 8–10(–12) × 3–3.5 µm. Occurs typically on dead, attached branches of deciduous trees, particularly *Sorbus*, *Betula* and *Prunus cerasus*, rare on conifer wood.

Species of *Mycoacia* ▷▷ are often more yellow or brownish, while species of *Hyphodontia* ◁◁, *Xylodon* ◁◁, *etc.* usually have more floccose 'teeth' (hand lens required!).

Widespread and common; all year.

Irpex, ***Steccherinum*** and ***Etheirodon*** are three genera, which, together with the polypore genus *Junghuhnia* (page 950), are recognized by their thick-walled, crystal-covered cystidia. Many of the species have an apricot-orange hymenophore with a white, sterile margin. All are white-rotters.

Irpex lacteus is a effused-reflexed annual corticioid with a labyrinthine to dentate hymenophore and a hairy cap surface. The flesh is dimitic with skeletal hyphae, and the cystidia measure 80–150 × 4–9 µm, with the upper part covered by crystals. Lacks clamps. The smooth, inamyloid spores measure 5–6 × 2–3 µm. Occurs on deciduous trunks and fallen branches.

Steccherinum oreophilum ▽ has clamps and smaller fruitbodies. *Cerrena unicolor* (page 860) has clamps but lacks crystal-covered cystidia and forms more fleshy fruitbodies with labyrinthine pores.

Widespread in the hemiboreal zone, rare in the nemoral zones; all year.

Thomas Læssøe/Jens H. Petersen

Steccherinum oreophilum is a rather small, whitish, effused-reflexed annual corticioid with prominent 'teeth' and long, rigid hairs on the cap surface. The hyphae have clamps, and the cystidia are very long, pointed and crystal-covered. The spores measure 5–6.8 × 3–3.2 µm. Occurs on twigs of deciduous trees, e.g. *Betula*, *Populus* and *Ulmus*.

Irpex lacteus △ has larger fruitbodies and no clamps. *Irpicodon pendulus* ◁◁ occurs only on *Pinus* and has smaller, amyloid spores (4–5 × 2.5 µm).

Widespread, but apparently rare in most regions, possibly on the increase in the nemoral zones; all year but many spring records.

Thomas Læssøe/Jens H. Petersen

Steccherinum ochraceum is a resupinate to effused-reflexed annual corticioid with pale apricot orange colours and, in capped specimens, a felty, buff to cinnamon-buff cap surface. The margin is white and normally lacks prominent hyphal cords. The spines are up to 1 mm long. Has long, thick-walled cystidia that are crystal-covered in the upper part, and clamped hyphae. The smooth, hyaline, inamyloid spores measure 3–3.5(–4) × 2–2.5 µm. Occurs on deciduous wood, rarely on conifers.

Steccherinum bourdotii ▽ is more capped, has spines up to 2.5 mm long and spores that measure 4.5–5.2 × 3–4.5 µm. *Steccherinum robustius* ⋉ has spines up to 3 mm long and spores that measure 3.5–5 × 2.5–3.2 µm.

Widespread and common in the nemoral–hemiboreal zones, rare in the boreal zone; all year.

Steccherinum bourdotii is a distinctly capped annual corticioid with apricot orange, ± flat spines up to 2.5 mm long. Has clamps, and long cystidia with crystals. The spores measure 4.5–5.2 × 3–4.5 µm. Occurs on trunks of deciduous trees.

Capped forms of *S. ochraceum* △ have smaller caps, shorter spines and smaller spores. *Steccherinum robustius* ⋉ is ± resupinate and has even longer spines and somewhat smaller spores.

Widespread in the nemoral zones and increasing in frequency to the south, very rare in southern Scandinavia; all year.

Lucie Zíbarová

Etheirodon fimbriatum is a resupinate annual corticioid with very prominent hyphal cords; hymenium has ± wart-like spines and brownish-rose to reddish-grey colours. Has long, crystal-covered cystidia and clamped hyphae. The ellipsoid spores measure 3–3.5 (–4) × 2–2.5 µm. Occurs on deciduous wood with soil contact, mostly in damp woodlands, *e.g.* in stands of *Fraxinus*. Apparently also found on herbaceous stems in grasslands.

Similar species are white, cream to apricot-orange, with fewer or no hyphal cords.

Widespread and common, becoming scarcer towards the north; all year.

Mycoacia, ***Mycoaciella*** and ***Sarcodontia*** are characterized by their wax-like fruitbodies with cylindrical spines.

Sarcodontia uda is a sulphur-yellow to almost lemon-yellow annual corticioid with pointed spines up to 2 mm long that stains purple with KOH. Has fusiform cystidia, ± smooth hyphae in the tips of the spines and usually numerous, loose, rod-shaped crystals. The spores measure 5–5.5(–6) × 2–2.5 µm. Occurs on wood of deciduous trees, *e.g.* in temporarily dried out bogs and other damp places.

Mycoacia fuscoatra ▷ also stains purple with KOH, but has duller colours and highly encrusted hyphae in the tips of the spines. *Mycoacia aurea* ⋉ and *Mycoaciella bispora* ▷ do not stain purple with KOH; the latter has skeletal hyphae and longer spines. *Sarcodontia setosa* (page 1057) is larger, has spines up to 20 mm long, narrower spores and a strong sweetish-sour smell. *Phlebia subochracea* (page 1019) occurs in the same damp habitats, as does *Physisporinus sanguinolentus* (page 944, seen in the lower part of the middle image).

Widespread and common in the nemoral–hemiboreal zones, rare or absent from the boreal zone; mostly July–November.

Mycoacia fuscoatra is a cinnamon-buff to grey-brown, ± blackening annual corticioid with pointed spines that stains purple with KOH. It has pointed cystidia and numerous crystals on the hyphae in the tips of the spines. The spores measure 5–6 × 2–2.5 µm. Occurs on wood from various deciduous trees in damp woodlands.

The dull colour and pointed cystidia make this species fairly easy to recognize. *Sarcodontia uda* ◁ has brighter colours and ± smooth hyphae in the spines.

Widespread, ccasional; all year, mostly July–October.

Thomas Læssøe/Jens H. Petersen

Mycoaciella bispora is a straw-coloured to somewhat greyish, long-spined (3 mm or more) dimitic annual corticioid with skeletal hyphae that end in smooth, cystidia-like extensions. Basidia 2- to 4-spored. The spores measure 4–5 × 2.5–3 µm. Occurs on wood of deciduous trees in temporarily dried out bogs, *e.g.* in *Alnus* and *Salix* carrs.

Sarcodontia uda ◁ and *Mycoacia aurea* ⋉ have shorter spines and no skeletal hyphae. *Sarcodontia setosa* (page 1057) is larger, has spines up to 20 mm long and narrower spores.

Widespread in the nemoral–hemiboreal zones, but very rarely recorded; mostly July–November.

Thomas Læssøe

Radulomyces molaris is a dingy vinaceous-brown, coarsely flat-toothed annual corticioid with a ± phenolic smell. Lacks cystidia, but has clamps. The somewhat thick-walled, inamyloid spores measure 8–12 × 6–8 µm and have grainy contents. Occurs mainly on attached or recently fallen, mostly decorticated *Quercus* branches, rarely on other deciduous trees.

Basidioradulum radula ◁◁ has allantoid, narrower spores. *Radulomyces confluens* (page 1009) has a similar smell, but is smooth.

Widespread in the nemoral–hemiboreal zones, but more common in the hemiboreal zone; all year.

Asterodon ferruginosus is a sturdy, up to 5 mm thick, cinnamon-buff to orange-brown, warty-spiny annual corticioid with a whitish to orange-brown margin that has prominent hyphal cords. The dimitic flesh has brown asterosetae. The smooth, hyaline spores measure 5.5–7 × 4–5 µm. Occurs mostly on conifer wood, including in buildings, rarely on deciduous wood.

Odontia ferruginea ▷ occurs on deciduous wood, has double-warted spores and no asterosetae.

Widespread in the hemiboreal–boreal zones, occasionally found in buildings farther south; most of the year.

Thomas Læssøe

Amaurodon viridis is a grey-green to greyish-blue, spiny annual corticioid. Lacks cystidia, but has clamps. The spores are ± spherical, verrucose and almost hyaline, stain violet in KOH and measure 4–5.6 µm in diameter. Occurs on litter and wood in rich, warm scrub and deciduous forests.

The combination of spiny fruitbodies and verrucose, spherical spores are unique to this species. See also *A. mustialaënsis* (page 997).

Generally very local and rare, but common in e.g. coppiced *Corylus* stands in parts of western Norway; most of the year.

Tomentella italica is a whitish to cinnamon-buff spiny annual corticioid with up to 2 mm high ± spines/warts that are sometimes almost labyrinthine and confluent. Has monomitic hyphal cords. Lacks cystidia, but has clamps. The almost hyaline to pale brown, double-warted spores are somewhat irregular in outline, variable in shape and measure 7–9.5 × 6.5–7.2 µm. Occurs on rotten wood of deciduous trees.

Odontia ferruginea ▷ and *O. fibrosa* ▷ are more rust-brown. See smooth species of *Tomentella* on page 999.

Widespread in the nemoral–hemiboreal zones, but very rare; autumn.

Thomas Læssøe/Jens H. Petersen

Odontia comprises species that live as decomposers. Most have ± spiny, red-brown hymenophores.

Odontia ferruginea is a curry-yellow to orange-brown, long-spined or warty annual corticioid with a paler fimbriate margin. Has narrow hyphal cords. Lacks cystidia, but has clamps and skeletal hyphae. The brownish, double-warted spores measure 7–9(–10) µm in diameter and are somewhat irregular in shape. Occurs on rotten deciduous wood, on rich, mostly calcareous soils; saprotrophic.

Similar species in *Tomentella* form ectomycorrhiza; *T. italica* ◁ is much paler and has no skeletal hyphae. *Asterodon ferruginosus* ◁ has asterosetae, smooth spores and occurs on conifer wood.

Widespread, occasional to rare; May–November.

Odontia fibrosa is a red-brown annual corticioid with a ± warty-spiny, rarely smooth surface and prominent, dimitic, hyphal cords. The flesh has large, septate cystidia (see inset image), and, apart from a few hyphae on the subiculum, the hyphae lack clamps. The brown, spherical spores are double-warted and measure 6–8 µm in diameter. A decomposer, mostly in the litter layer in deciduous forests.

Microscopically, an easy species to recognize. *Odontia ferruginea* △ has no cystidia and the spines are more prominent.

Widespread, occasional; late summer–autumn.

Thomas Læssøe/Jens H. Petersen

Hydnoids

The hydnoids are defined here as either stemmed or thick-fleshed fungi with a spiny or dentate hymenophore. Fungi with thin, resupinate, spiny fruitbodies are treated under the form group corticioids (page 960).

The hydnoids constitute a form group of species that, for example, belong to the orders *Thelephorales*, *Cantharellales* and *Russulales*.

OTHER SIMILAR FUNGI:
– spiny species with flat (resupinate) fruitbodies and thin flesh are included under the corticioid fungi (page 1036).
– species with isolated spines that lack a common flesh are treated under the clavarioids (page 1092).

FURTHER READING: 50, 142, 163, 221, 229, 346.

Micro-drawings: spores and cystidia

Approximate species number applies to temperate Europe.

Auriscalpium
1 species
page 1060
tough;
on cones
I+
on
cones;
with
stem
Phellodon
7 species
page 1061
Hydnellum
25+ species
page 1064
tough to brittle;
with smell of curry
tough to brittle;
brown to coloured;
without smell
of curry
brittle; brownish
and mostly with
prominent scales;
without smell
of curry
Sarcodon
4+ species
page 1071
hydnoids
form hoof-
shaped, branched
or stemmed
fruitbodies with
a spiny underside
(hymenophore)
on soil;
with stem
brittle;
cap smooth and
pale or apricot-
orange
Hydnum
11 species
page 1072
other
similar fungi
small; with tooth-
like spines to
labyrinthine pores
small; with very
small, crowded
spines
Sistotrema
1 species with stem
page 1075
Mycorrhaphium
1 species
page 1075
thin-fleshed
isolated, hanging
spines
spiny corticioids
page 1036
clavarioids
page 1092

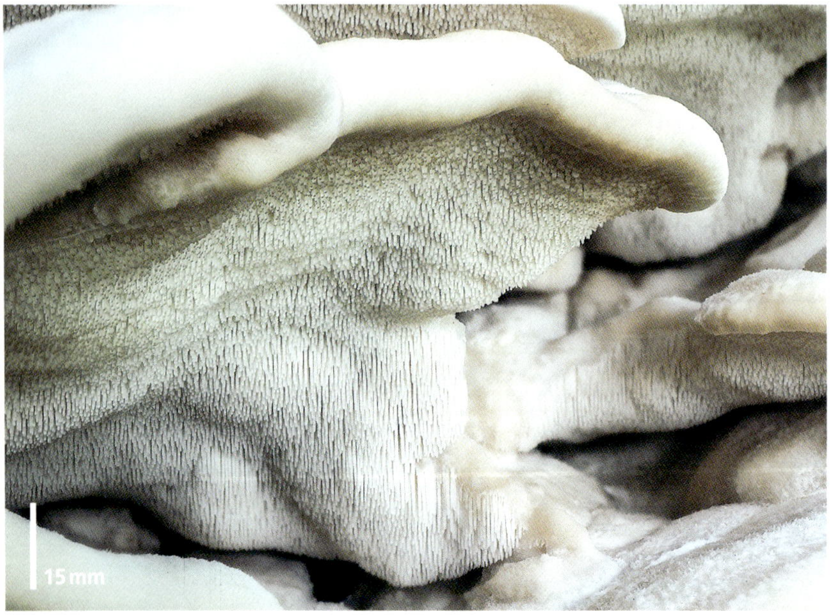

Climacodon septentrionalis is a huge, effused-reflexed, annual, yellowish-white, fleshy hydnoid that forms multiple rows of caps from a large, resupinate plate. It has broad, thick-walled cystidia. The ellipsoid to somewhat curved, smooth, inamyloid spores measure 3.5–6 × 1.5–2.5 µm. Occurs on living, sun-baked deciduous trees, to the west mostly on *Fagus*, but farther east preferring *Acer platanoides*; forms a white rot.

Distinctive and unlikely to be misidentified. *Spongipellis pachyodon* (page 906) is smaller, has coarser, flattened teeth and larger, almost spherical spores. *Sarcodontia setosa* ▷ is yellow and does not form multiple rows of caps. *Hericium cirrhatum* ▷▷ is smaller with irregular spines on the upper side of the caps; it has amyloid spores. *Climacodon septentrionalis* is related to *Sarcodontia setosa* ▷, *Phlebia* (pages 978 & 1019) and *Mycoacia* (page 1050).

A mainly eastern species, absent from the British Isles, the Netherlands *etc.*, with a few records in southern Scandinavia, but quite common in the Baltic Republics and also with scattered records in southern central Europe; mainly August–October.

Thomas Læssøe

Thomas Læssøe

Thomas Læssøe

Sarcodontia setosa is a large, pale yellow, resupinate to effused-reflexed, somewhat waxy hydnoid with up to 20 mm long, cylindrical spines and with up to 3 mm thick flesh. The smell is very strong, sweetish-sour, rather like butyric acid. Monomitic with clamps, but flesh also has very thick-walled, septate hyphae with ± pear-shaped tips (sclerocysts). The ± spherical spores are hyaline, smooth, inamyloid and measure 4–6 × 3.5–4.5 µm. Occurs on old *Malus* trees, rarely on other deciduous trees, forms a white rot.

Climacodon septentrionalis ◁ is even larger, whitish and has tiered caps and narrower spores. *Sarcodontia uda* (page 1051) have the same colour and texture, but are much thinner and have narrower spores. *Sarcodontia setosa* is related to *Climacodon septentrionalis* ◁, *Phlebia* (pages 978 & 1019) and *Mycoacia* (page 1050).

Widespread, but rare to very rare, absent from the boreal zone; mainly October–February, but may appear as early as July.

Hericium is a genus of large, ± white, wood-inhabiting, white-rotting hydnoids. The flesh has numerous gloeohyphae and the spores are hyaline, amyloid and usually verrucose.

●● ***Hericium erinaceus*** is a whitish to cream, sometimes pinkish-tinted, ± rounded *Hericium* with the lower side covered by 20–50 mm long cylindrical spines and with more irregular, short spines on the upper side. It has a mild taste and a somewhat fruity odour. The flesh is amyloid. Has gloeocystidia. The amyloid, finely ornamented, thick-walled spores measure 5–6.5 × 4–5.5 µm. Occurs on larger living or dead trunks, predominantly *Fagus,* at least in northern parts of the range, but also known from *e.g. Quercus*.

The three other *Hericium* species are typically more complex in structure, either being branched or having layered caps. *Hericium cirrhatum* ▽ has dextrinoid flesh.

Widespread, but very rare to rare in most areas, locally occasional; mostly August–November.

Holger Magdanz

●● ***Hericium cirrhatum*** is a large, whitish *Hericium* with numerous caps from a common base. The lower side has up to 15 mm long, downwards-pointing spines, while the upper side is covered by shorter, curved, irregular spines. The flesh is up to 5 mm thick, dextrinoid, and lacks a striking smell or taste. Spores hyaline, smooth, amyloid, measuring 3.5–4.5 × 2.5–3.5 µm. Occurs on various deciduous trees, often dead, predominantly *Fagus*.

Other species of *Hericium* △▷ are either finely branched or form spines from fewer large, rounded fruitbodies. *Climacodon septentrionalis* ◁◁ forms numerous tiered caps from a large, ± plate-like structure and is much larger with narrower spores.

Widespread, occasional to rather rare; June–December.

●● ***Hericium coralloides*** is a large, white to cream, very branched *Hericium* with each branch covered in pale, hanging spines up to 10 mm long. The flesh is amyloid and the minutely spiny, somewhat thick-walled spores measure 3.5–5 × 3–4 µm. Occurs on deciduous wood, particulary *Fagus* and *Betula*, both on standing trunks, which often have been killed by *Inonotus obliquus* (page 951), and on fallen trunks.

Hericium flagellum ▽ typically has somewhat longer spines and larger spores (5.6–6.3 × (4.4–)4.6–5.4 µm).

Widespread, occasional; June–February.

●● ***Hericium flagellum*** is a large, whitish to cream, heavily branched *Hericium* with up to 20 mm long, hanging spines. The flesh is amyloid, and the rather large, ornamented, amyloid spores measure 5.6–6.3 × (4.4–)4.6–5.4 µm. It apparently occurs exclusively on trunks of *Abies*.

Hericium coralloides △ has smaller spores, typically has shorter spines, and occurs on deciduous wood. The North American *H. abietis* ⋉ has been recorded once in England on timber that was presumed to have been imported.

Widespread in central Europe, reaching as far north as the Netherlands; late summer–autumn.

Jacob Heilmann-Clausen

Pekka Helo

Teppo Helo

Gloiodon strigosus is a stemless, but capped, rather small hydnoid with long, brownish spines and a brown, coarsely felty cap surface. It can rarely occur in resupinate forms. The flesh has a duplex structure and is dimitic with skeletal hyphae; it also has a few gloeohyphae that end in gloeocystidia. The spores are verrucose, hyaline, amyloid and measure 4.5–6 × 3.5–5 µm. Occurs typically on large, standing dead or fallen trunks of *Populus*, more rarely *Salix*, *Alnus* and other deciduous trees, mostly in old, damp, herb-rich forests.

Distinctive and unlikely to be misidentified, although the resupinate form may perhaps recall *Hydnocristella himantea* (page 1040). However, that species has completely different spores and stains green with $FeSO_4$. The resupinate *Mycoacia fuscoatra* (page 1051) has a waxy texture and different spores. The closely related *Auriscalpium vulgare* ▽ occurs on conifer cones.

Widespread in the hemiboreal–boreal zones, but absent from the nemoral zones and not recorded in the UK; most of the year.

Auriscalpium vulgare is a rather small, strigose-hairy hydnoid with a kidney-shaped cap that has crowded, rather long, pale brown to grey spines underneath. The stem is lateral, long and has brown hairs. Has gloeocystidia. The spiny, amyloid spores measure 4.5–5.5 × 3.5–4.5 µm. On ± buried *Pinus* cones.

Distinctive and unlikely to be misidentified; it is closely related to *Gloiodon strigosus* △.

Widespread and common; all year.

Stridvall/Jens H. Petersen

Phellodon (including *Bankera*) is a genus of hydnoids in the order *Thelephorales*. It includes rather brittle to tough species with pale, spiny spores. They have a curry-like smell. Species of *Sarcodon* ▷▷ and *Hydnellum* ▷▷ have coloured spores. All are ectomycorrhizal.

Phellodon fuligineoalbus is a brittle *Phellodon* that usually incorporates some of the litter layer in the cap surface. The cap is initially whitish, later more orange-brown to reddish-brown; it normally stays smooth during development. Has a curry-like smell. The hyaline, spiny-verrucose spores measure 4.5– 5.5 × 3–3.5 µm. Mycorrhizal with *Pinus* on sandy soils with a thin litter layer.

Phellodon violascens ▽ does not incorporate elements of the litter layer in the cap surface and has obvious violet-grey tinges; it is mycorrhizal with *Picea*.

Widespread, occasional to very rare, extinct in the Netherlands; August–November.

Phellodon violascens is a brittle finely-felty to somewhat scaly, whitish, later pale smoke-grey to violet-grey *Phellodon* that does not incorporate litter in the cap; the stem has similar colours and the spines are greyish. The flesh does not have zones and, in the stem, is reddish or somewhat purple-tinged. The odour is strong and curry-like. The hyaline spores are spiny and measure 4.5–5.5 × 4–4.5 µm. Mycorrhizal with *Picea*, mostly in ± calcareous sites.

Phellodon fuligineoalbus △ usually has litter incorporated in the cap surface and does not have violet-grey tinges; it is ± smooth and occurs with *Pinus*.

Widespread in the hemiboreal–boreal zones, occasional, but rare to absent from the nemoral zones; June–November.

Phellodon niger s.l. is a dark, mostly almost blue-black, zoned, tough, *Phellodon* with greyish-blue spines. The 1–2 cm thick stem has a dark, hard surface below the felt. The exposed flesh stains blue-green with KOH, and the flesh has a firm, dark centre (a duplex structure). The hyaline, spiny spores measure 3.6–4(–4.5) × 2.7–3.1(–3.5) µm; the spines are about 0.5 µm long. Ectomycorrhizal with deciduous trees, including *Fagus* and *Quercus*, and conifers, mostly in calcareous sites or on very poor sandy soils.

Constitutes a species complex with several, as yet, undescribed species that have different mycorrhizal partners. *Phellodon connatus* ▽ is paler, lacks thick felt on the surface and does not have a duplex structure.

Widespread, occasional to rather rare; July–November.

Phellodon connatus s.l. is a slender, tough, thin-fleshed *Phellodon* with a very zoned cap in shades of dark brown to purple-brown that has a white margin and pale brown-grey spines. The shiny, thin stem has no felty layer and the flesh has no hard centre and smells pleasantly spicy. Spores spiny, 3.5–4.5 × 3–4 µm; spines are up to 0.3 µm long. With conifers on poor sandy soils, rarely with deciduous trees.

Has also been called *P. melaleucus* and constitutes a species complex. *Phellodon tomentosus* ▷ has a thicker stem and paler colours. *Phellodon secretus* ⋉ is a small-spored species from northern, pristine *Picea* forests.

Widespread, occasional; July–November.

Phellodon tomentosus is a tough, rather pale, zoned *Phellodon* with brownish shades, and usually confluent fruitbodies that are almost smooth. The stem lacks a felty layer, except at the base and does not have a hard centre. The flesh does not stain, or only stains faintly, with KOH. The hyaline spores measure 3–4 × 2–3.5 µm and have spines 0.2 µm high. Ectomycorrhizal with conifers, mostly on sandy, poor soils.

Phellodon confluens ▽ is almost azonate, more felty on the surface and occurs with deciduous trees. Similar species of *Hydnellum* ▷▷ have brown spore-deposits.

Widespread, fairly common, occasional or very rare; July–November.

Phellodon confluens is a tough, brown, plain or somewhat zoned *Phellodon* with a very felty appearance and with protuberances mainly near the cap centre. The stem is mostly felty along its full length. The hyaline spores measure 3.5–4.5 × 3–4 µm and have spines up to 0.3 µm high. Mycorrhizal with deciduous trees on poor soils.

Phellodon tomentosus △ is more zoned, has smaller spores and is mycorrhizal with conifers.

Widespread in the nemoral–hemiboreal zones, occasional to rare; July–October.

Paul Hugill

Jens H. Petersen/Thomas Læssøe

Hydnellum is a hydnoid genus in the order *Thelephorales*. It includes rather brittle- to tough-fleshed species with brown, angular-verrucose spores. Species of *Sarcodon* ▷▷ are brittle and mostly have prominent scales. *Phellodon* ◁◁ species have pale spores. All are ectomycorrhizal.

Hydnellum species vary from convex when young to more depressed with age. The surface can be velvety-felty, more tufted-scaly or zoned to almost smooth. The flesh colour and taste are important characters.

***Hydnellum concrescens* s.l.** is a rather thin-fleshed, tough *Hydnellum* with ± confluent fruitbodies; the surface is initially somewhat tufted but quickly becomes smoother and radially fibrillose with ± obvious, concentric zones and brownish colours. The hyphae lack clamps. The spores are pale brown, have double-warts and measure 4.5–6 × 4–4.5 µm. Mycorrhizal with both deciduous trees and conifers mostly on sandy soils.

Constitutes a species complex with at least three macroscopically poorly defined and as yet unnamed species. *Hydnellum scrobiculatum* ⋉ does not have concentric zones on the cap and has larger spores. Other brown species of *Hydnellum* are mostly azonate.

Widespread, occasional to rare; July–November.

Hydnellum ferrugineum is a rather large and thick-fleshed, tough, azonate or radially fibrillose *Hydnellum*, which, when young, may secrete orange-red drops. It has a farinaceous smell and taste and brown flesh that becomes dark and then olive-brown with KOH (test a thin section). The hyphae lack clamps. The spores measure 5.5–6.5 × 3.5–4.5 µm. Mycorrhizal with *Pinus* and *Picea*.

Hydnellum peckii ▷ has a very bitter taste. *Hydnellum gracilipes* ⋉ and *H. cumulatum* ⋉ are far more thin-fleshed.

Widespread, occasional in the north but has declined in many nemoral areas; June–January.

Hydnellum spongiosipes is an azonate, persistently felty, brownish, rather thick-fleshed, tough *Hydnellum* that typically arises from a thick, brown, felty stem base. Usually has confluent caps. Flesh rather dark, and has a mealy smell and taste. Hyphae lack clamps. The brownish, double-warted spores measure 5.5–7 × 4.5–5.5 µm. Mycorrhizal with *Fagus* and *Quercus*.

Hydnellum ferrugineum ◁ occurs with conifers; *H. concrescens* ◁ is more thin-fleshed and concentrically zoned; and *H. compactum* ⋉ has paler, thicker, bitter-tasting flesh.

Widespread in the nemoral zones, but rare; mainly August–October.

Morten Christensen

Hydnellum mirabile is an azonate, rather thick, tough, hairy, at first curry-yellow, later sordid brown *Hydnellum* with yellowish to yellow-brown spines. The flesh is a rather pale yellow-brown and the taste somewhat bitter; it becomes very sponge-like in wet weather. The hyphae lack clamps. The verrucose, brown spores measure 5.5–6 × 4.5 µm. In conifer forests; mycorrhizal with *Pinus* and *Picea*, perhaps ± calciphilous.

Hydnellum geogenium ▷▷ is a more vivid yellow and thin-fleshed species.

Mainly hemiboreal, rare; late summer–autumn.

On IUCN's international red list.

Maroš Peiger

Hydnellum peckii is a thick-fleshed, tough, azonate or radially fibrillose *Hydnellum* that is at first very pale, but later red-brown; it exudes orange-red drops when young. The flesh is pale and very bitter-tasting. The hyphae have clamps. The coarsely verrucose spores measure 5–5.5 × 3.5–4 µm. Mycorrhizal with conifers, often *Pinus*, on sandy soils.

Hydnellum ferrugineum ◁ has a mild taste, is somewhat darker in the flesh and lacks clamps. *Hydnellum mirabile* △ and *H. compactum* ⋉, with deciduous trees, also taste bitter.

Widespread and rather frequent in the hemiboreal–boreal zones, rare or very rare in the nemoral zones; July–October.

Jan Vesterholt/Thomas Læssøe

Hydnellum geogenium is a confluent, slender, tough *Hydnellum* with lemon-yellow to curry-yellow colours, particulary on the felty stem. The spines are long and lemon-yellow to yellowish-grey. The hyphae have clamps. The brown, verrucose spores measure 4.5–5 × 3–3.5 µm. Often situated deep in the litter layer and is probably mycorrhizal with *Picea*. A good indicator of species-rich fungal communities, with many red listed species.

Distinctive and unlikely to be misidentified. *Xanthoporus syringae* (page 828) and *Gyrodon lividus* (page 795) may have similar colours, but have pores.

Widespread in the hemiboreal–boreal zones and occasional to rare, absent from the nemoral zones and northern parts of the boreal zone; late summer–autumn.

Hydnellum aurantiacum is a sturdy, tough *Hydnellum* with zoned, orange flesh in the thick stem and no blue colours. Has a mild taste. Lacks clamps. The brownish, ± double-warted spores measure 5.5–6.5 × 4.5–5 µm; the ornament is up to 1.5 µm high. Mycorrhizal with conifers on poor sandy or calcareous soils.

Hydnellum auratile ▷ is mycorrhizal with deciduous trees and conifers and has smaller fruitbodies, vivid orange flesh in the cap and a more orange cap surface.

Widespread and occasional in the hemiboreal–boreal zones, rare or absent from the nemoral zones; June–November.

Hydnellum auratile is a vivid brownish-orange to orange *Hydnellum* with orange, tough flesh that does not usually have darker zones. The hyphae lack clamps. The spores measure 5–6 × 3.5–4.5 µm and have coarse warts. Mycorrhizal with *Fagus* on calcareous soils, but also with conifers and other deciduous trees in calcareous sites.

Hydnellum aurantiacum ◁ is mycorrhizal with conifers and has paler, mostly dark-zoned flesh in the cap, which is orange-brown.

Widespread but local to rare in the hemiboreal–boreal zones, very rare to absent from the nemoral zones; mainly August–October.

Tapio Kekki/Thomas Kehlet

Hydnellum caeruleum is a sturdy, fleshy, tough, somewhat farinaceous-smelling *Hydnellum* with bluish and orange colours in the zoned flesh and on the cap. Has clamps. The spores measure 5.5–6 × 3.5–4.5 µm. Normally mycorrhizal with *Picea* and *Pinus*, but apparently also with *Abies* and *Fagus* (although such populations may prove to be different species).

Hydnellum suaveolens ▷ does not have orange in the blue flesh, while *H. aurantiacum* ◁ and *H. auratile* △ completely lack blue colours.

Widespread and occasional in the hemiboreal–boreal zones, very rare or absent from the nemoral zones – locally extinct; June–November.

Hydnellum martioflavum is a relatively slender, rather brittle, dark grey-brown, minutely downy *Hydnellum* with orange felt on the stem. Fresh specimens have a whitish cap margin. It has a farinaceous smell. The hyphae lack clamps. The brownish, verrucose spores measure 5–6.5 × 3.5–4.5 µm. Ectomycorrhizal with *Picea* in damp woodlands.

Distinctive and unlikely to be misidentified due to the orange felt on the stem.

Widespread but rather rare in the hemiboreal and southern boreal zones; late summer–autumn.

Jarkko Korhonen

Hydnellum suaveolens is a small, somewhat tough *Hydnellum* that has only bluish colours, including in the zoned flesh. The cap often incorporates elements of the litter layer. It has a strong odour of coumarin, aniseed or lavender. Hyphae with clamps. The ± double-warted spores measure 4–5 × 3–3.5 µm. Ectomycorrhizal with *Picea* in herb-rich older stands.

Hydnellum caeruleum ◁◁ is almost odourless or smells farinaceous and has orange flesh in the stem.

Widespread and occasional in the hemiboreal–boreal zones, virtually absent from the nemoral zones; late summer–autumn.

Hartmut Schubert

Hydnellum joeides is a rather brittle, pale, cream to cinnamon-buff, smooth *Hydnellum* with striking, lilac to greyish-blue flesh. The dark reddish-grey spines are deeply decurrent, even to the stem base. The brownish, ± double-warted spores measure 5.4–5.8 × 3.6–4.2 µm. Occurs in warm, rather rich deciduous forests, where it is ectomycorrhizal with *Quercus* and *Fagus*.

Other similar species of *Hydnellum*, e.g. *H. fuligineo-violaceus* ▷, occur in coniferous forests.

Widespread in the nemoral zones, but rare to very rare; late summer–autumn.

Hydnellum fuligineoviolaceum is a rather brittle, pale, very dark, felty to smooth *Hydnellum* with greyish-blue flesh. The taste is very bitter. The spores measure 5.4–6.5 × 4–5 µm. Ectomycorrhizal with *Pinus* on calcareous soils.

Hydnellum roseoviolaceum ⋉ also occurs with *Pinus* but only on poor soils, and is somewhat paler with a mild rather than bitter taste. *Hydnellum joeides* ◁ occurs with deciduous trees and has very deeply decurrent spines. *Hydnellum glaucopus* ▷▷ occurs with *Picea* and has paler flesh.

A rare hemiboreal species; late summer–autumn.

Jean-Luc Fasciotto

Hydnellum scabrosum is a rather brittle, pale, grey-brown to red-brown, rather finely concentrically scaly *Hydnellum* with blue-green colours in stem base. Very young fruitbodies are brownish-rose. With a farinaceous smell and a farinaceous but soon bitter taste. Without clamps. Spores brownish, verrucose, 6–7.5 × 3.5–5 µm. Mycorrhizal with *Pinus* on sandy soils.

A similar species with *Quercus* also passes as *H. scabrosum*; *H. glaucopus* ▷, with *Picea*, smells farinaceous with notes of liquorice or orange; *H. fennicum* ⋉, with conifers, smells of bitter almonds and *H. lepidum* ⋉, with deciduous trees, is farinaceous; *S. imbricatus* ▷▷ and *S. squamosus* ▷▷ both have clamps.

Widespread and occasional in the hemiboreal–boreal zones; autumn.

Leif Stridvall/Jens H. Petersen

Michael Krikorev

Hydnellum versipelle is a rather brittle, pale, felty to smooth, pinkish-buff to cinnamon-buff, non-staining *Hydnellum*. The odour is sweetish, recalling liquorice or vanilla; the taste is mild to bitter. The hyphae have clamps. The brownish, verrucose spores measure 4.5–5.5 × 3.5–4.5 µm. Ectomycorrhizal with *Picea* in forests on rather rich soils.

There are several similar species of *Hydnellum* with smooth to minutely scaly caps: *H. glaucopus* ▽ smells sweet but typically acquires ± violet spots and the fruitbody darkens when bruised; *H. fennicum* ⋉ smells of bitter almonds; *Sarcodon leucopus* ⋉ has a bitter taste and larger spores; *H. lundellii* ⋉ smells unpleasantly farinaceous and lacks clamps.

Widespread in the hemiboreal and southern boreal zones, but uncommon; late summer–autumn.

Michael Krikorev

Hydnellum glaucopus is rather brittle and vinaceous-brown or darker brown, but paler between the cracks. It is finely-felty to smooth and typically has violet-tinged spots. The stem may be bluish-grey at the base. It stains brown when bruised. Smell faintly farinaceous with notes of liquorice or orange; taste ± bitter. Sporer brownish, ± double-warted, 5–6 × 4–4.5 µm. Mycorrhizal with *Picea*.

May be confused with *H. versipellle* ⋉ but that species has paler colours. *Hydnellum scabrosum* ⊲⊲ always has a greenish-blue base, is more scaly and occurs with *Pinus* while *H. fennicum* ⋉ has a smell of bitter almonds. The farinaceous *H. lepidum* ⋉ is a deciduous woodland species.

Widespread in the hemiboreal–boreal zones, but virtually absent from the nemoral zones; late summer–autumn.

Sarcodon includes rather brittle-fleshed, mostly scaly species with brown, angular-verrucose spores. Species of *Hydnellum* ⪡ mostly have a smoother cap or tougher flesh and *Phellodon* ⪡ have pale spores. All are ectomycorrhizal.

Sarcodon imbricatus is a large *Sarcodon* with coarse, red-brown to purple-brown scales, usually with pale flesh visible between the scales. It tends to be ± funnel-shaped with coarse upright scales at the centre. The spines are greyish and decurrent. The flesh does not stain blue, even in the stem base. The taste is mild to bitter and the smell not distinct. The hyphae have clamps. The brownish, verrucose spores measure 7–8.5 × 5–5.5 µm. Ectomycorrhizal with *Picea* and possibly *Abies*, often in old damp woodlands and plantations on rather poor soils.

Sarcodon squamosus ▽ occurs with *Pinus* (and possibly *Abies*) and is darker, has a more inrolled cap margin and is less depressed; *Hydnellum scabrosum* ⪡ has a blue stem base and no clamps.

Widespread and common in the hemiboreal–boreal zones, but occasional to rare in the nemoral zones; July–November.

Sarcodon squamosus is a somewhat smaller and darker version of *S. imbricatus* △ with a prolonged inrolled cap margin and either no, or only a slightly, depressed cap centre. The scales are not as coarse. The hyphae have clamps and the spores measure 7–8.5 × 5–5.5 µm. Ectomycorrhizal with *Pinus* and possibly also with *Abies*, typically on poor, sandy soils.

The species on this page are used for dyeing wool and produce different colours. The rather similar *S. scabrosus* ⪡ lacks clamps.

Widespread and rather common in the hemiboreal–boreal zones, but rare to absent from the nemoral zones; mainly September–January.

Jens H. Petersen/Leif Stridvall

Hydnum is a genus in the *Cantharel-lales*. The species are recognized by their fragile fruitbodies with white, buff to apricot-orange colours, large basidia with rather flaring sterigmata, and smooth, hyaline to cream, spherical to ellipsoid spores. All are ectomycorrhizal. Molecular studies indicate that at least 11 species occur in temperate Europe.

•• ***Hydnum ellipsosporum*** is a relatively vivid, typically yellow-brown, apricot-orange to orange *Hydnum* with a tendency to produce flattened, slightly decurrent spines. The spores are ellipsoid, have a Q value of 1.55–1.75, and measure (8.5–)9.5–11 (–11.5) × (5–)5.5–7 µm. Occurs on rich soils in deciduous forests, often where *Craterellus cornucopioides* (page 47) might also be found.

Similar species of *Hydnum* have spines that are more spherical in section. *Hydnum ovoideisporum* ⋉ has, in addtion, a spore Q of 1.0–1.38, while *H. rufescens s.l.* ▽ and *H. umbilicatum* ss. auct. ▷ have spherical spores. *Hydnum vesterholtii* ▷ is paler with a spore Q of 1.27–1.3.

Mainly nemoral, occasional; July–January.

•• ***Hydnum rufescens* s.l.** is a ± apricot-orange, slender, non-umbilicate *Hydnum* with ± decurrent, spherical spines (in section). The ± spherical spores measure 7–8.5(–9) × 6–7.2(–7.5), av. 8–8.2 × 6.4–6.8, with a Qav of 1.19–1.28. Occurs on rich soils in deciduous forests or near deciduous trees in coniferous forests.

The name has been used collectively for all apricot-orange *Hydnum* species. *Hydnum rufescens* differs from some similarly coloured species by its almost spherical spores. *Hydnum umbilicatum ss. auct.* ▷ tends to be more gracile, usually has a central navel.

Widespread in the nemoral–hemiboreal zones, probably rather common; July–January.

•• ***Hydnum umbilicatum* ss. auct.** is a small, orange-yellow to ± apricot-orange, deeply umbilicate, ± zoned *Hydnum* with ± spherical spines (in section). Spores almost spherical, 7.8–9.2 × 7.4–8.8 µm, av 8.6 × 8 µm. Mostly in rather calcareous or clay-rich deciduous forests, mainly with *Fagus*.

Hydnum umbilicatum was described from New York and specimens from Europe may prove to be a different species. Other small apricot-orange hydnums are not deeply umbilicate. The paler *H. vesterholtii* ▽ is also small and umbilicate but has ellipsoid spores.

Widespread in the nemoral–hemiboreal zones, occasional; August–January.

•• ***Hydnum vesterholtii*** is a rather small, very pale, somewhat zoned and sometimes umbilicate *Hydnum* with spherical, mostly ± decurrent spines. The spores are broadly ellipsoid with a Q value of 1.27–1.30 and measure (7–)8–9(–9.5) × 6–7.5(–8) µm. On rich soils with *Fagus*, *Carpinus* and *Abies*.

Hydnum albidum ▷▷ is purer white, has smaller spores and is highly calciphilous. Other species form either larger or more apricot-coloured fruitbodies. *Mycorrhaphium pusillum* ▷▷ is even more fragile, the spines are more crowded and it has smaller spores.

Widespread, at least in the nemoral zones; autumn.

Jacob Heilmann-Clausen

•• ***Hydnum magnorufescens*** is a rather large *Hydnum* with somewhat subdued buff-yellow to pale apricot-orange colours and spherical spines. The surface discolours dark apricot-orange when touched or bruised. It has a ± pleasant, fruity smell. The spores measure (6.8–)7–8.5(–9) × (6.5–)6.8–8(–8.5) µm; Q value (1.0–) 1.04–1.14(–1.2); Qav. 1.06. Occurs on ± calcareous soils with deciduous trees, mostly *Quercus* and *Fagus*.

Very similar to *H. repandum* ▷▷, but is typically somewhat more reddish-tinged and calciphilous.

Apparently widespread in the nemoral zones; August–November.

Jacob Heilmann-Clausen

•• ***Hydnum repandum*** is a large, fleshy, mostly very pale, cream, buff or buff-yellow *Hydnum* with pale, ± decurrent, in section spherical spines. It can occur in completely white forms. The almost spherical spores measure (6.8–)7.5–8.5(–9.2) × (5.5–)6–7.8(–8) µm; Q 1.16–1.21. Occurs both with conifers and deciduous trees, typically on rather poor soils. Mycorrhizal with *e.g. Picea* and *Fagus*.

Hydnum magnorufescens ◁◁ is also large and fleshy but is more apricot-orange in colour, stains strongly when bruised and is more calciphilous. *Hydnum vesterholtii* ◁◁ has pale colours but is slender and has a zoned surface, and sometimes an umbilicate cap centre. The ellipsoid spores measure (7–)8–9(–9.5) × 6–7.5(–8) µm; Q 1.27–1.30.

Widespread, common to rather common or occasional; July–January.

Jan Vesterholt/Jens H. Petersen

•• ***Hydnum albidum*** is a rather small to medium-sized *Hydnum* with an almost chalk-white cap and cream, decurrent spines. The flesh stains somewhat, and the smell is pleasant. The spores are small, 4.5–5.5 × 3.5–4.5 µm. Occurs on warm, calcareous soils in older deciduous forests; mycorrhizal with *Fagus*.

White forms of other species of *Hydnum* may cause confusion but they have larger spores. *Hydnum albidum* was described from North America and it is unclear whether specimens from temperate Europe are the same species.

Mainly nemoral, with extensions into the hemiboreal zone, rare to very rare and absent in many areas; July–October.

Mycorrhaphium pusillum is a 1–2 cm wide, rather tough hydnoid with ± layered, whitish to pale pinkish-buff caps on thin stems. The lower side of the cap has small, very crowded, white spines, 7–8 per mm. Lacks clamps and cystidia. The inamyloid, smooth spores measure 3–3.5 × 1.9–2.2 µm and are formed on 1–2-spored basidia. Inhabits soils in boreal *Betula* forests.

Apparently related to decomposing polypores. *Sistotrema confluens* ▽ has spines that are less crowded and less regular. Similar species of *Hydnum* ◁ are larger, have larger and more distant spines and larger spores.

Recorded in Finland, but mostly east of temperate Europe; autumn.

Kaisa Tervonen

Kaisa Tervonen

Sistotrema confluens is a fleshy *Sistotrema* with cream to pinkish-buff, usually pale-margined, ± zoned caps, central to excentric stems and a decurrent, pale, spiny-dentate to almost pore-shaped hymenophore. The odour is unpleasant, spicy. The basidia are urn-shaped with 6–8 smooth, hyaline spores that measure 4.5–6 × 2–2.5 µm. Occurs on calcareous soils, both in coniferous and deciduous forests; it is considered to be ectomycorrhizal.

Species of *Hydnum* ◁ are more fleshy and have more pleasant smells; they also have larger and wider spores. *Mycorrhaphium pusillum* △ has spines that are much more crowded and regular, and 2-spored basidia. All other species of *Sistotrema* (pages 939, 996) are resupinate and are considered to be decomposers, and *S. confluens* should probably be placed in a separate genus.

Widespread, occasional, seems to have declined in parts of the nemoral zones; July–January.

Jan Vesterholt/Jens H. Petersen

Cyphelloids

This form group includes basidiomycotes with small cup-shaped, bell-shaped or tubular fruitbodies on various substrates. The largest fruitbodies are up to 10 mm wide (*Schizophyllum amplum*), while the smallest are only around 0.2 mm wide (*Pseudolasiobolus minutissimus*). The fruitbodies are generally positioned with the hymenium pointing downwards, allowing the spores to fall freely when released.

Most cyphelloids are reduced agarics, *i.e.* members of *Agaricales* that have become gill-less during the process of evolution. Some polypores, *e.g.* species of *Ceriporia* (page 946), may initially recall cyphelloids but, with age, the cups join to form 'normal' polypore fruitbodies.The numerous separated tubes in *Fistulina hepatica* (page 845) and the unusual 'gills' of *Schizophyllum commune* (page 62) can be interpreted as many cyphelloid fruitbodies joined on a common fleshy meta-structure. These species are also nested within the *Agaricales*.

OTHER SIMILAR FUNGI:

– many species of *e.g. Hyaloscyphaceae* may look similar, but the apothecia can point in all directions and they tend to look neater and less irregularly hairy. They form spores in asci (page 1372).

– some rosette-fungi recall large cyphelloids, but cyphelloids are never long-stemmed (page 952).

– pleurotoids are typically larger, with more well-developed gills (page 60).

FURTHER READING: 5, 6, 40, 156, 186.

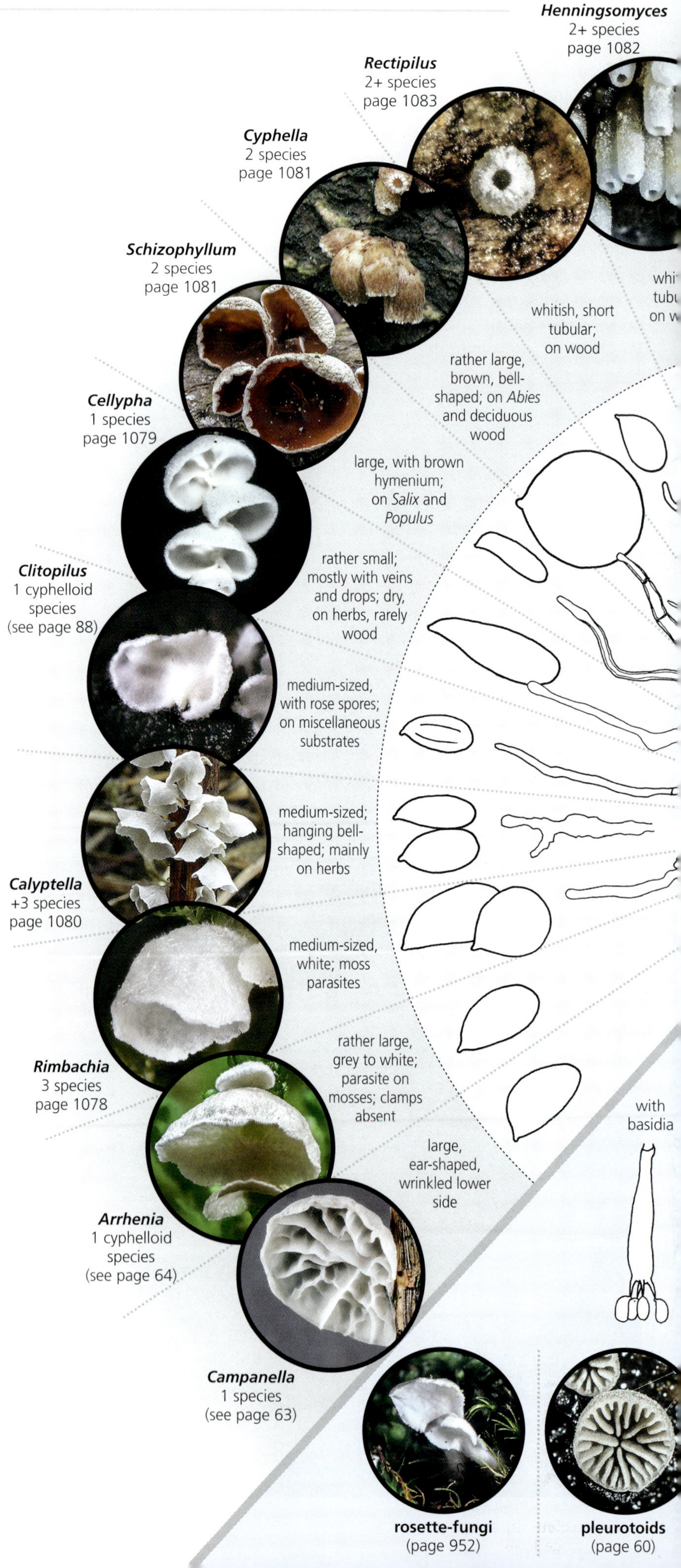

Micro-drawings: spores first, then surface hairs.

Approximate species number applies to temperate Europe.

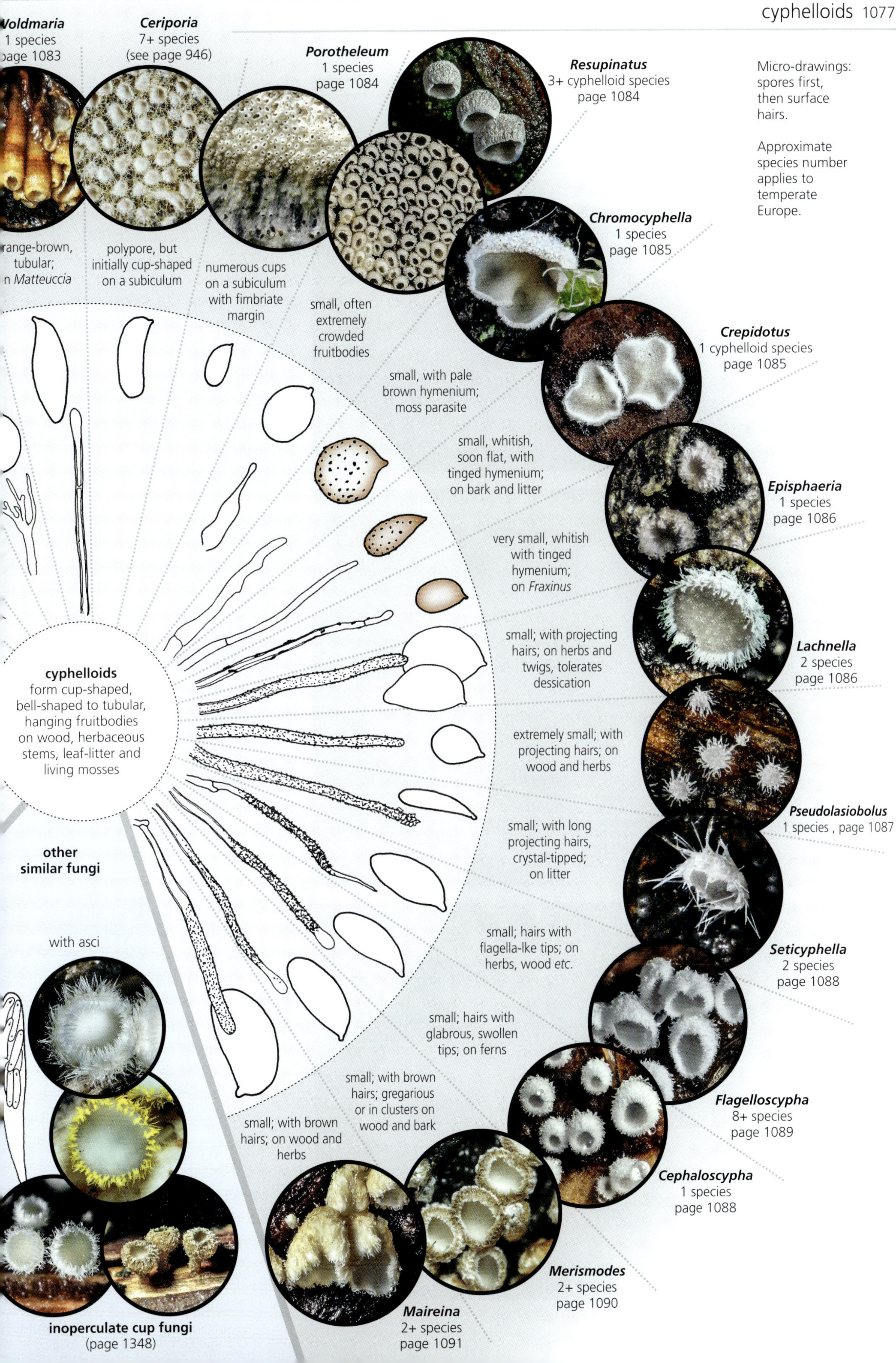

Rimbachia is a genus of medium-sized cyphelloids that parasitize mosses. Have clamps, hyphae-like hairs and hyaline, smooth, spherical to drop-shaped spores.

Rimbachia arachnoidea is a small, white cyphelloid with a smooth hymenophore. The upper side is finely-felty and attached on the back to mosses. The spores are almost spherical and measure 4–5(–6.5) × 4–5(–5.5) µm. Parasitizes acrocarpous mosses, *e.g.* species of *Mnium*, mostly in damp places in *Alnus* carrs *etc.* Usually site-faithful so may be found in the same spot year after year.

Rimbachia neckerae ▽ has more drop-shaped to ellipsoid spores and occurs on pleurocarpous mosses. *Rimbachia bryophila* ▷ has obvious rudimentary gills and spores that are longer than 5.5 µm. *Arrhenia retiruga* (page 65) is tinged grey, more open at maturity and lacks clamps.

Widespread in the nemoral–hemiboreal zones, occasional, but certainly overlooked; August–March.

Rimbachia neckerae is a small, white, finely-felty cyphelloid with a smooth hymenophore. The spores are drop-shaped to ellipsoid and measure 7.5–11 × (3.5–)5–7 µm. It parasitizes pleurocarpous mosses (such as *Ctenidium*, *Hylocomium* and *Ptilium*) in damp sites, *e.g.* in woodland springs.

Rimbachia arachnoidea △ also has a smooth hymenophore but occurs on acrocarpous mosses and has almost spherical spores.

Widespread in the nemoral–hemiboreal zones, but very rare or perhaps somewhat overlooked; mainly August–October.

Rimbachia bryophila is a cyphelloid that has obvious rudimentary gills. It is chalk-white and finely-felty on the upper side. The hyphae have clamps. The spores measure 6–7.5 × 5–6 µm. Occurs on various mosses, *e.g.* *Mnium* and *Rhynchostegium*, in *Alnus* carrs, *etc*.

The two other *Rimbachia* species ◁ have smooth hymenia, and *Arrhenia retiruga* (page 65) is more greyish and lacks clamps. *Clitopilus hobsonii* (page 88) has a reddish tinge on the gills and the spores have longitudinal ridges.

Widespread, but rare or somewhat overlooked; August–October.

Cellypha goldbachii is a rather small, white, mostly drop-covered, hairy, gregarious cyphelloid with a smooth to somewhat wrinkled hymenophore. The hairs are capitate. The spores are large, hyaline, ± fusiform, and measure 10–13.5 × 3.5–4 µm. Occurs mainly on grasses, sedges and rushes, mostly in dry places, *e.g.* along coasts, but may also occur on various lignicolous plants.

Species of *Calyptella* ▷▷ never have a tendency to be wrinkled, are mostly smooth and have smaller spores and hyphae-like hairs that lack a head; they mostly inhabit larger dicot herbs. Species of *Flagelloscypha* ▷▷ and *Lachnella* ▷▷ are mostly smaller with more evident marginal hairs that lack capitate tips. *Cephaloscypha mairei* ▷▷ has capitate hair tips but forms smaller fruitbodies on ferns. Other similar cyphelloids have brown, ± ornamented spores.

Widespread, mainly in the nemoral zones, rarely recorded but probably overlooked; May–November.

Thomas Læssøe/Jens H. Petersen

Calyptella is a genus of medium-sized cyphelloids with whitish to yellow colours. The upper side is smooth to minutely downy from hyphae-like, branching hairs. The spores are hyaline, smooth and ellipsoid to ± drop-shaped. Mostly on larger, dicot herbs, rarely on bark.

Calyptella capula is an almost chalk-white to somewhat translucent, bell-shaped, short-stemmed *Calyptella*. The margin is mostly somewhat floccose. The rudimentary stem is minutely downy. Lacks cystidia; has clamps. The hyaline, smooth spores measure 7–12 × 3.5–5 µm. Occurs on various standing year-old herbaceous stems, e.g. *Urtica* and *Epilobium*.

It is not easy to separate species of *Calyptella*. *Calyptella campanula* ▷ is often treated as a synonym of *C. capula*. *Calyptella gibbosa* ▽ is usually more greyish and smaller-spored, but intermediate specimens may make identification difficult. *Calyptella cernua* ⋉, on *Sambucus* bark, generally has very small and narrow cystidia between the basidia. Species of *Rimbachia* ◁◁ and *Arrhenia* (page 64) are stemless and occur on living mosses.

Widespread and rather common, at least in the nemoral zones, becoming scarcer towards the north; mainly May–November.

Calyptella gibbosa is a ± bell-shaped, pale grey, finely-felty, short-stemmed *Calyptella*. When dry it has a ± black stem. Lacks cystidia but the outer side has branching, hyphae-like hairs; has clamps. The spores measure 5–7 × 2.5–4.5 µm. Occurs on stems of various herbs, including ferns.

Calyptella capula △ should, in principle, have larger spores, up to 12 µm long, and possibly a somewhat different cap surface structure. *Arrhenia retiruga* (page 65) does not have a stem and lacks clamps.

There are scattered records, but the true frequency and distribution is not fully understood; mainly June–November.

Calyptella campanula is a ± bell-shaped, almost smooth, vivid yellow *Calyptella* with a short to very short, somewhat downy stem. The smooth, inamyloid spores measure 7–12 × 3.5–5 µm. Occurs typically on coarse, year-old herbaceous stems.

Calyptella capula ◁ is white, but possibly conspecific. *Calyptella cernua* ⋉ should have indistinct cystidia and should apparently be confined to *Sambucus* bark; it can apparently be either white or yellow.

Widespread and fairly common, at least in the nemoral zones; mainly June–October.

Schizophyllum amplum is a large, hanging, cup-shaped, tough, gelatinous cyphelloid with a white, felty surface and a somewhat veined, brownish hymenophore. It is tolerant of desiccation. The hyaline, smooth spores measure 7.5–10 × 2.5–3 µm. It is mainly associated with *Populus*, less so *Salix*, and is usually positioned high up on rather thin branches.

Cyphella digitalis ▽ occurs on *Abies* and has large, spherical spores. See also *Schizophyllum*, page 62.

Widespread in the nemoral zones, and at least locally common; all year.

Jacob Heilmann-Clausen

Cyphella digitalis is a rather large, brownish, bell- to thimble-shaped cyphelloid that usually occurs in clusters. The opening is pale and fringed with fine hairs. The hairs are capitate. The very large, almost spherical spores measure 16–19 µm. Occurs on bark of dead trunks and fallen branches of *Abies*, and also on attached, dead branches.

Schizophyllum amplum △ does not have a hairy opening and has completely different spores; it occurs on *Populus* and *Salix*. The rarely recorded '*Cyphella' ferruginea* ⋉ has large, spherical spores, 9.5–14 µm, and is only known from deciduous wood.

Widespread in central Europe; mostly during winter.

Zygmunt Augustowski/Morten Christensen

Henningsomyces is a genus of small, tubular, whitish cyphelloids with rather branched, hyphae-like hairs on the outside. Mainly in crowded groups and always on strongly decayed wood.

Henningsomyces candidus is a ± tubular, chalk-white, somewhat downy, white-spored cyphelloid. Lacks cystidia, and the outside has branching, thin-walled hairs. The spores are inamyloid, hyaline, smooth and measure 4.5–6 × 4–5 µm. Occurs both on large trunks of deciduous trees and on more decomposed, coarse, woody litter, both in deciduous forests and with conifers.

Henningsomyces puber ▽ is a nearly identical species with a somewhat more downy outer side that mostly occurs on rotten bark in damp woodlands, *e.g.* on *Betula*. Young, less tubular fruitbodies of *H. candidus* may recall species of *Rectipilus* ▷, but those mainly have unbranched hairs on the outside.

Widespread, rather common to rather rare; almost all year, but mainly August–October.

Henningsomyces puber is a whitish, crowded, ± tubular cyphelloid with a downy outer side. The spores are inamyloid, hyaline, smooth and measure 4–6.5 × 4–5.5 µm. Occurs mainly on coarse, rotten bark, *e.g.* from *Betula*, in damp woodlands.

Henningsomyces candidus △ is less downy, but the distinction between these two species is unclear and needs to be examined using modern tecniques.

Widespread, occasional, but rarely recorded; mainly August–November.

Rectipilus afibulatus is a very small, white, at first cup-shaped, but later somewhat more tubular cyphelloid with ± thick-walled, helical hairs that have rounded tips and no crystal cover. Lacks cystidia; hyphae lack clamps. The spores are drop-shaped, inamyloid, smooth, and measure 5–6.5(–7) × 3.2–4.2(–4.9) µm. Occurs on rotten bark of *Alnus* and *Salix*.

Rectipilus fasciculatus ⋉ has clamps. *Henningsomyces candidus* ◁ and *H. puber* ◁ have longer fruitbodies, shorter, ± branching hairs, clamps and different spores.

Only recorded a few times in Denmark, Sweden and the UK; May–September.

Woldmaria filicina is a rather large, ± tubular, crowded, yellow-brown cyphelloid with a paler opening and an appressed, felty-hairy outer side. The hairs are smooth and thick-walled with a thinner-walled tip. The spores are hyaline, smooth, fusiform and measure 12–13.5(–15) × 4–4.5 µm. Occurs only on dead frond bases of living specimens of the fern *Matteuccia struthiopteris*.

Also known under the name *Woldmaria crocea*.

Probably common and found wherever the host is found, but apparently increasing towards the north; all year, but in best condition during the autumn.

Porotheleum fimbriatum is intermediate between a resupinate polypore and a cyphelloid fungus. It is initially white, later cream, with prominent hyphal cords at the margin. Hyphal system is dimitic with skeletal hyphae and clamps. Cystidia absent. Spores hyaline, smooth, inamyloid, 5–6.5 × 3–3.5 µm. Occurs on larger deciduous branches and trunks, perhaps mainly on calcareous soils.

Also known as *Stromatoscypha fimbriata*. *Resupinatus poriaeformis* ▽ has globose spores and no hyphal cords. Species of *Ceriporia* (page 946) may also look similar, but lack clamps.

Widespread, but local and rare; mainly June–December.

Resupinatus can form small, greyish cyphelloid fruitbodies with spherical spores and simple hairs on the outside. See also species with gills page 70.

Resupinatus poriaeformis is an almost polypore-like *Resupinatus* with very crowded small, greyish cups, that are ± obviously connected by a subiculum. The almost spherical, smooth, inamyloid spores measure (4.5–)5–6(–8) × 4.5–5.5(–6) µm. Occurs on wood from various deciduous hosts.

Porothelium fimbriatum △ has hyphal cords, ellipsoid spores and skeletal hyphae.

Widespread, occasional; all year, but mainly late autumn.

Resupinatus griseopallidus is a small, cup-shaped, greyish, finely-felty, tough and gelatinous, almost solitary *Resupinatus* with a smooth hymenophore. The spherical, smooth spores measure 5–5.5 × 4.5–5 µm. Occurs mostly on twigs of *Juniperus* and *Syringa*.

There are no other cup-shaped, solitary, grey and gelatinous cyphelloids with spherical spores.

Very rarely recorded, but may be overlooked, mainly nemoral; all year.

Chromocyphella muscicola is a medium-sized, brown-spored cyphelloid with a white, downy outer side and a mostly smooth, brownish hymenophore. The spores are somewhat angular, have low warts and measure 8–10 × 7.5–8.5 µm with Q 1.1–1.4. Parasitic on mosses in damp fens, or on mosses on the bark of deciduous trees in damp forests where it may cause dead rings in the moss.

Crepidotus pallidus ▽ has almost flat fruitbodies, a somewhat higher spore Q and occurs on rotten litter and bark, not on mosses. *Episphaeria fraxinicola* ▷▷ has smooth, brown spores and occurs on bark. Similar discomycetes (page 1374) typically have longer hairs and point upwards.

Widespread in the nemoral zones, occasional to common; August–February.

Crepidotus pallidus is a small, very flat, brown-spored cyphelloid with somewhat irregular hairs at the margin and a pale brown hymenium. The hairs are smooth. The spores are finely echinulate and measure 6.5–9 × 3.5–5.5 µm; Q 1.5–2.2. Occurs on damp litter, including rotten leaves, bark and wood, including at the base of *Sarothamnus*.

Formerly accepted in the genus *Pellidiscus*. *Episphaeria fraxinicola* ▷▷ is smaller, more deeply cup-shaped, more hairy, occurs mainly on *Fraxinus* bark and has smooth spores. *Chromocyphella muscicola* △ occurs on mosses and is more cup-shaped. Hairy species of discomycetes (page 1374) are typically not as flat, have straighter hairs and form spores in asci.

Probably rather common and widespread, at least in the nemoral zones, but easily overlooked; all year, but apparently mainly in the autumn.

Episphaeria fraxinicola is a very small cyphelloid with dense, somewhat curly hairs, a white outer side and a pale brown hymenium. The hairs are thin-walled and covered with crystalline material. The brown spores are smooth and measure 6.5–8.5 × 4.5–6 µm. Occurs on bark, mostly twigs of *Fraxinus* – possibly on pyrenomycetous fungi in the bark.

Crepidotus pallidus ◁◁ is flatter, less hairy and has finely ornamented spores. *Chromocyphella muscicola* ◁◁ is bigger, has warty spores and occurs on mosses.

Probably rather common and widespread, but rarely recorded; all year.

Lachnella is a genus of desiccation-tolerant cyphelloids with straight, crystal-covered hairs that become thicker-walled towards the base. The hyaline, inamyloid spores are smooth. In dry weather the fruitbodies close and form small, hairy balls.

Lachnella alboviolascens is a rather large, drought-resistant cyphelloid that may be almost white to pale grey but only open during wet periods. Lacks cystidia. The ± cylindrical hairs are thick-walled, but thinner-walled towards the tip, and have a roughened surface. The spores are broadly almond-shaped, measuring 10–15 × 7.5–12 µm. Occurs on bark of deciduous trees, vines and large, coarse herbs.

Lachnella villosa ▷ has spores that are more lemon-shaped and somewhat narrower, and lance-shaped cystidia. *Lachnella alboviolascens* may resemble a species of *Lachnum* (page 1399) or *Trichopeziza* (page 1392) but these species form spores in asci.

Widespread and probably very common, but apparently ± absent from the boreal zone; all year.

Lachnella villosa is a small, drought-resistant, chalk-white, very hairy cyphelloid that only opens in wet weather; it has lance-shaped cystidia. The hairs are thicker-walled towards the base, roughened and ± cylindrical. The lemon-shaped spores measure 9–14 × 7–9 µm. Occurs on a wide range of coarse herbs, including *Urtica*, more rarely on the bark of woody plants.

Lachnella alboviolascens ◁ may also be whitish but has wider, more almond-shaped spores and lacks cystidia. Discomycetes (*e.g.* species of *Lachnum* (page 1399) or *Trichopeziza* (page 1392)) are very similar but form their spores in asci.

Probably common and widespread, but easily overlooked; all year.

Pseudolasiobolus minutissimus is a very tiny cyphelloid that recalls a very small, long-haired, somewhat untidy white discomycete (page 1374). The hyphae lack clamps. The hairs are somewhat pointed and covered with amorphous crystals. The smooth, ± drop-shaped, inamyloid spores measure 6–8 × 5–6 µm. Occurs on wood and stems of herbs in rather damp to very damp habitats.

Similar cyphelloids typically have either distinctly pointed or capitate hair tips and clamps. *Rectipilus afibulatus* ◁◁ also lacks clamps but is somewhat larger and has almost helical hairs.

Very rare or more likely very much overlooked, apparently mainly nemoral; known June–September.

The cyphelloid genus ***Seticyphella*** is characterized by its tiny fruitbodies that have a somewhat untidy appearance due to very long hairs that point in different directions. The hairs are covered by crystals, which are especially large at the tips.

Seticyphella tenuispora is a rather small, white, untidy-hairy cyphelloid with simple hairs that lack a flagellum and capitate tip. The spores measure (7–) 8.5–12 × 1.5–2.5 µm; Q value up to 4.6. On damp leaf-litter, *e.g. Salix*.

Seticyphella niveola ▽ has a spore Q value lower than 3.5.

Very rare or overlooked; April–October (perhaps all year).

Seticyphella niveola is a rather small, very long-haired, at the base constricted, chalk-white, white-spored cyphelloid with a simple hair tip. The spores measure 7.5–10 × 2–3.5 µm; Q 2.8–3.7. Occurs on leaf-litter in damp or moist habitats.

Seticyphella tenuispora △ has a spore Q of up to 4.5. Species of *Flagelloscypha* ▷ and *Cephaloscypha* ▽ either have flagellate or capitate hair tips. *Nochascypha filicina* ⋉ has a lower spore Q.

Very rare, but probably also very much overlooked; at least June–September.

Cephaloscypha mairei is a rather small, gregarious, rather long-haired, white cyphelloid that has capitate, smooth hair tips. The hyaline, smooth, inamyloid spores measure 8.5–11 × 3–4 µm. Occurs exclusively on fern litter (*Dryopteris*, *Athyrium*, *etc.*) in damp habitats.

The substrate and the capitate hair tips make identification easy. Macroscopically this species may be confused with other cyphelloids in the genus *Flagelloscypha* ▷.

Widespread, but rarely recorded, probably somewhat overlooked and seems to be more common towards the north; mainly September–October.

Flagelloscypha is a very difficult genus of small to very small, hairy, white cyphelloids. The hairs are typically crystal-covered, but naked at the base, and have a naked, flagella-like tip. The genus is poorly understood and a safe identification is often impossible. Some species have thin-walled hair bases and some have 2-spored basidia.

Flagelloscypha donkii is a small, hairy, white, cup-shaped, ± stemless, ± gregarious cyphelloid. The basidia are 4-spored. The hairs are thin-walled at the base and covered with pointed crystals, but have a naked flagella at the tip. The hyphae have clamps. The spores are narrowly drop-shaped, measuring 6–8 × 2.5–3.5 µm; Q 2–2.3. Occurs on leaves, dicot herbs and ferns.

Safe determination requires reference to specialist literature and considerable patience through the microscope.

Probably widespread, but rarely recorded; mainly autumn.

Flagelloscypha minutissima is a very small, very hairy, white, cup-shaped cyphelloid that usually produces small clusters of fruitbodies. The hair tips have a long, naked flagella, the middle of the hair is thick-walled and covered with pointed crystals, and the base is thin-walled. The basidia are 2-spored. The narrowly lemon-shaped spores measure 7–10.5 × 4–5.5 µm. Occurs on leaf- and twig-litter, mostly in damp habitats.

Various similar species of *Flagelloscypha* can only be ruled out by careful microscopy; most, however, are 4-spored.

Widespread and probably common; mainly June–November.

Jens Maarbjerg

Merismodes is a genus of small, brown, mostly gregarious or clustered, desiccation-resistant, white-spored cyphelloids with crystal-covered hairs that have naked, somewhat swollen tips. The species occur on bark and wood and may be situated high up on attached branches of deciduous trees. Some authorities prefer to use *Cyphellopsis* rather than *Merismodes*.

Merismodes anomalus is a highly gregarious, small, brown, at first narrowly urn-shaped but later more expanded cyphelloid with a felty-hairy outer side and long, brown, rough hairs, ending in a smooth, clavate tip. The smooth, hyaline spores are ellipsoid and measure 8–11 × 3.5–5 µm; Q 2–2.5. Occurs either on the bark of deciduous trees high up in the vegetation, *e.g.* on *Salix* bark, or on fallen trunks and branches of deciduous trees.

Merismodes fasciculata ▽ has allantoid spores that measure 7.5–11 × 2–3 µm; Q 3.5–4. A stemmed form of *M. anomalus* is also known as *M. stipitatus* ⋉.

Widespread, common and probably very common; all year.

Merismodes fasciculata is a rather small cyphelloid, which is erumpent in small, circular groups. The hairs are roughened but have smooth, clavate tips. The allantoid spores measure 7.5–11 × 2–3 µm; Q 3.5–4. Occurs on bark of living deciduous trees and scrub, mostly on *Salix* branches.

Also known under the name *M. confusa*. *Merismodes anomalus* △ normally occurs in diffuse, larger groups and has spores with a much lower Q value.

Widespread and probably very common; all year.

Species in the cyphelloid genus ***Maireina*** form small, brown or yellowish, hanging fruitbodies with completely crystal-covered hairs.

Maireina filipendula is a rather small, yellow-brown, silky-hairy cyphelloid with a pale, smooth hymenophore. The hairs are thick-walled, somewhat wider at the tip, and granulate. The hyphae have clamps, and there are no cystidia in the hymenium. The drop-shaped spores measure 5–5.9 × 2.5–3.1 µm. Occurs on dead, scale-like, basal leaves of living *Filipendula ulmaria*.

Maireina monacha ▽ has much larger spores, 10.5–15(–17) × 5.5–8.5(–10) µm. *Maireina callostoma* ⋉ has spores measuring 3–4.5(–5.5) × 2–3 µm and occurs on ferns and mosses, and *M. maxima* ⋉ has spores measuring 5–8 × 3–4.5 (–5) µm and occurs on wood. Species of *Merismodes* ◁ have hairs with naked, somewhat swollen heads. *Woldmaria filicina* ◁◁ is a ± yellow-brown, tubular cyphelloid that occurs at the base of the fern *Matteuccia struthiopteris*.

Currently only known from two sites in temperate Europe, one in Denmark and one in central Sweden; September–October.

Maireina monacha is a ± non-gregarious, brown cyphelloid with a whitish to pale grey, smooth hymenophore. The hairs are brown, roughened, and lack swollen tips. The large, smooth, hyaline spores measure 7.5–12(–16) × 5–7(–8) µm. Occurs on herbaceous stems and bark of deciduous trees, typically on dead attached branches.

Species of *Merismodes* ◁ are gregarious and have naked hair tips. *Maireina filipendula* △ has smaller spores and occurs on *Filipendula*. *Maireina monacha* strongly resembles species in the discomycete genus *Brunnipila* (page 1380).

Widespread, occasional; all year.

Clavarioids

Clavarioids includes *Basidiomycota* with normal, one-celled basidia and thread-like, awl-shaped, clavate or branched fruitbodies. Most clavate fruitbodies are negatively geotropic (they grow/point upwards), but some species with very small fruitbodies are ageotropic (the orientation is random) and species in the genus *Mucronella* ▷▷ are positively geotropic (they point downwards).

The clavate fruitbody is one of the simplest, most basic types of fungal fruitbody, and clavarioids have arisen numerous times through the evolution of the fungal kingdom. Clavate fruitbodies occur in *Ascomycota* (*e.g.* earthtongues), within the rust fungi, the jelly fungi and in numerous places amongst the remaining *Basidiomycota* with one-celled basidia, *e.g.* amongst the *Agaricales*.

The group of clavarioids with one-celled basidia includes about 20 genera, with around 200 species in temperate Europe.

Many small clavarioids live as decomposers of plant material, *e.g.* on stems of herbs or leaves. Some species of *Typhula* ▷▷ may parasitize plants. Amongst the larger clavarioids the genera *Ramaria* ▷▷, *Clavariadelphus* ▷▷ and *Thelephora* ▷▷ are ecto-mycorrhizals.

Like the hygrocyboid fungi (page 142), some or all species of *Clavaria*, *Clavulinopsis* and *Ramariopsis* ▷▷ may live in a not yet understood symbiosis with herbs (page 18). These species are mostly very demanding with regard to habitat, and, among other things, are sensitive to fertilizers, preferring habitats with a long continuity. They often share habitat with a number of rare hygrocyboids, entolomatoids, earthtongues – *e.g.* in unimproved grasslands or in damp forests on mull soils – and they are particularly good indicator species of especially biodiverse fungus habitats.

Continued on next page spread . . .

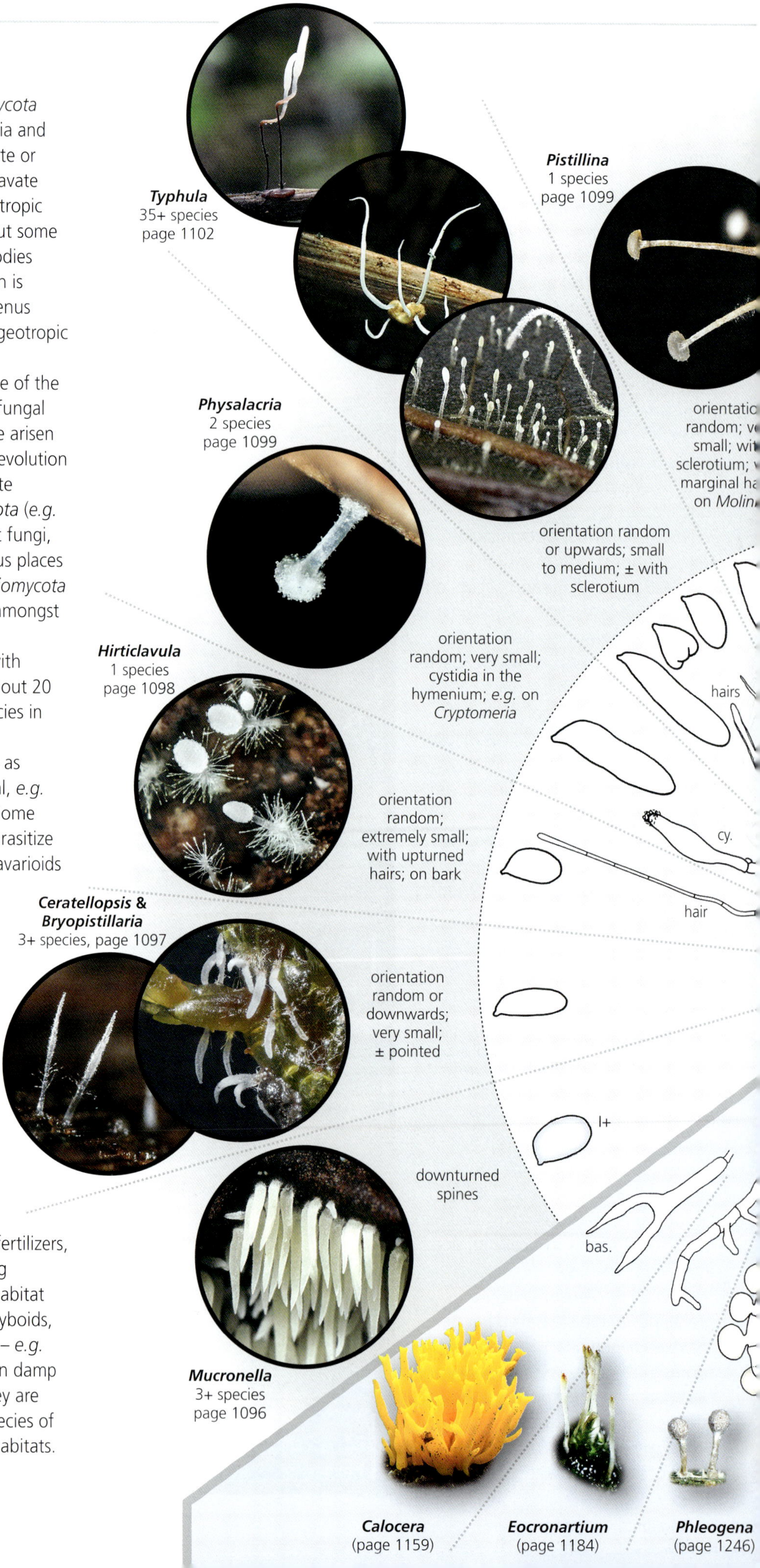

Micro-drawings: spores first, then basidia (bas.) and cystidia (cy.).

Approximate species number applies to temperate Europe.

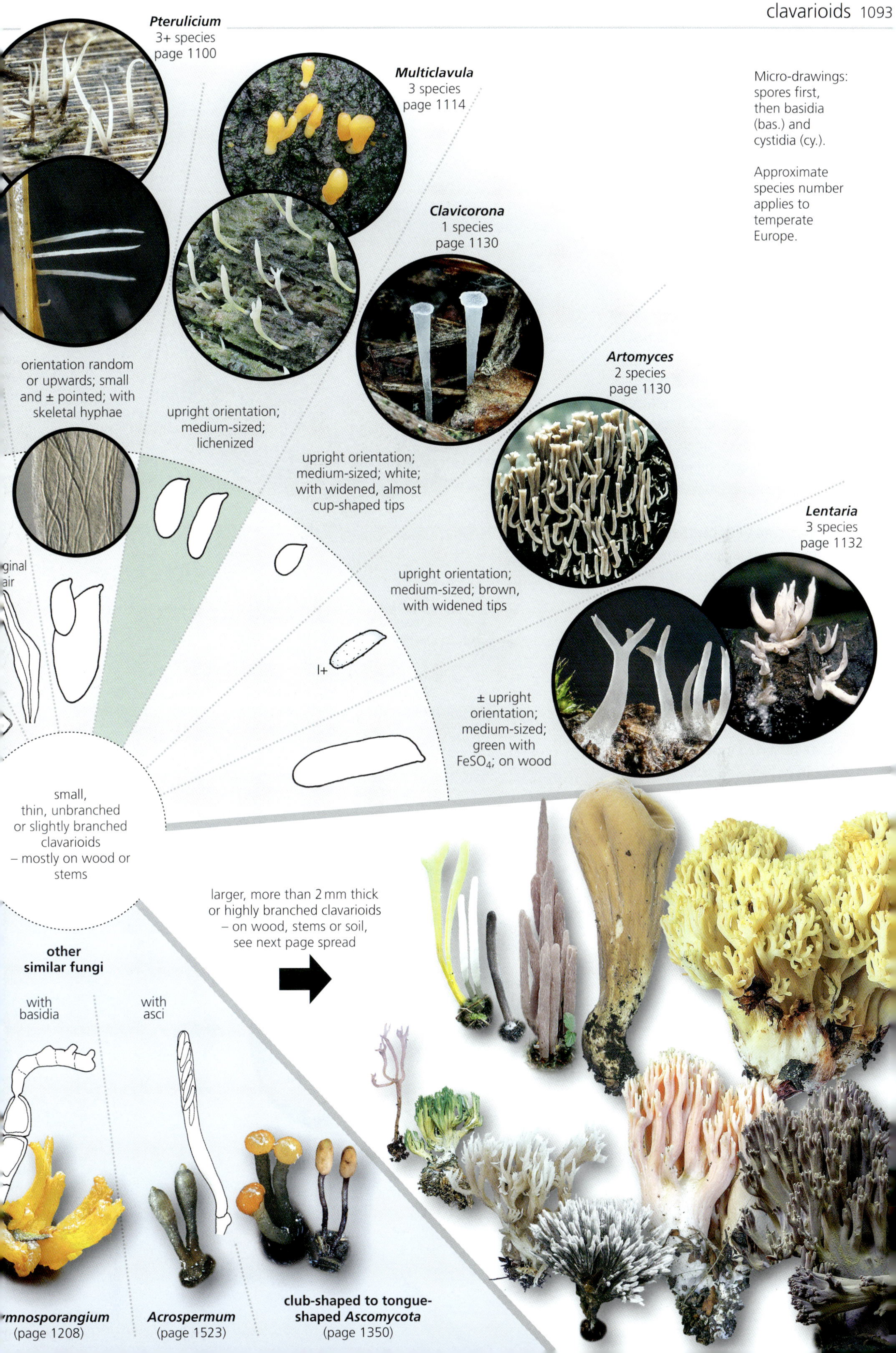

Clavarioids

continued from previous page spread . . .

OTHER SIMILAR FUNGI:
– species of *Calocera* look similar but have rubbery textures and tuning fork-shaped basidia (page 1159).
– greyish-white species of clavarioids with septate basidia are treated with the jelly fungi (page 1164).
– orange-yellow, gelatinous clavarioids with teliospores that germinate with septate basidia belong to the rust fungi (page 1208).
– *Ascomycota* with club-shaped apothecia are included with the earth tongues and others (see page 1350).
– *Acrospermum* is a small, dark, clavate *Ascomycota* on herbaceous stems with internal asci (see page 1523).
– *Ascomycota* with perithecia immersed in a clavate stroma are included with the pyrenomycetous fungi (see page 1510).
– for small, club-shaped fruitbodies with internal spore production and passive, dusty spore dispersal, see *Phloeogena*, asexual fungi and myce-tozoa (pages 1246, 1626 and 1646).

FURTHER READING: 31, 32, 38, 44, 63, 92, 100, 142, 143, 230, 232, 243, 244.

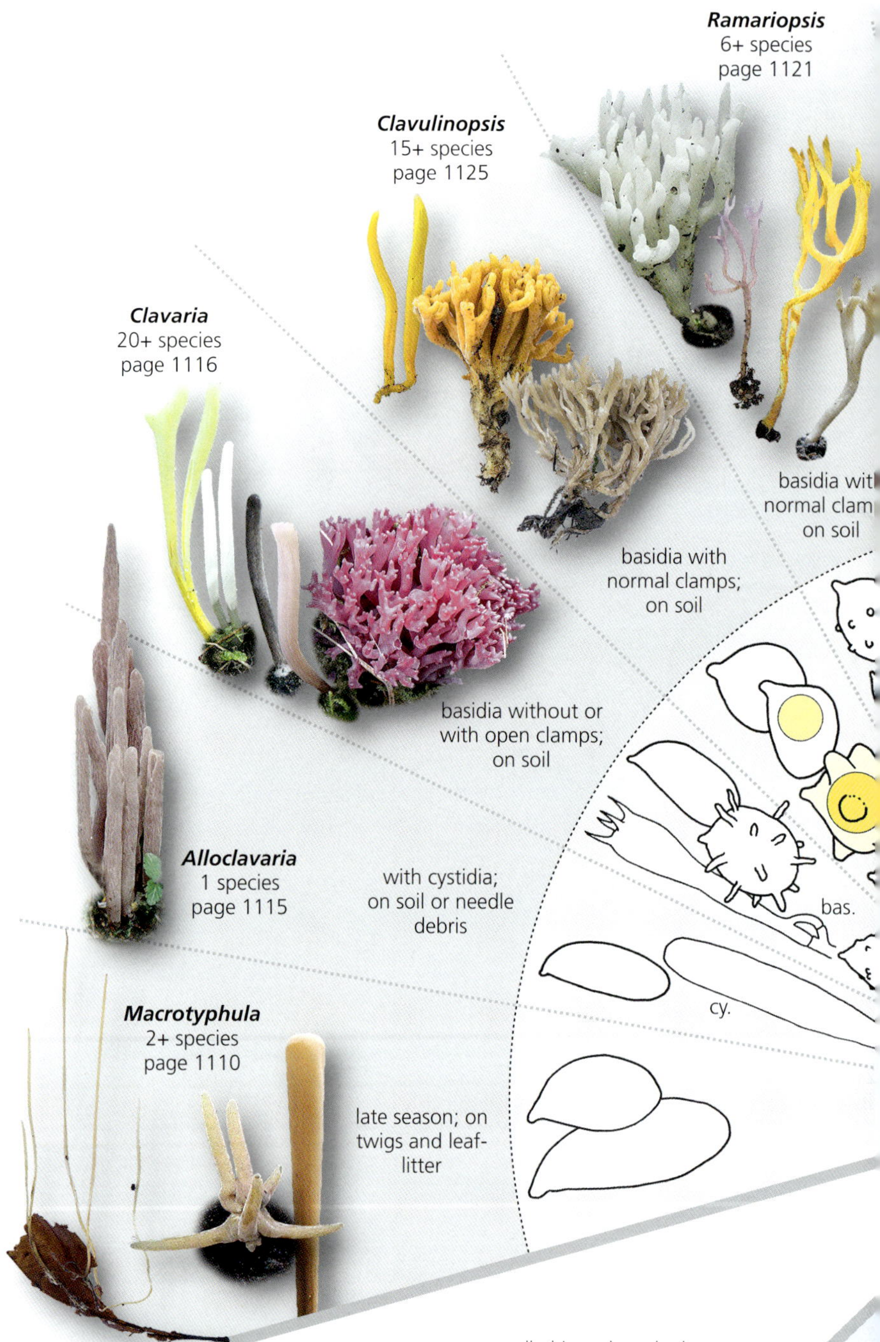

Micro-drawings: spores first, then basidia (bas.) and cystidia (cy.).

Approximate species number applies to temperate Europe.

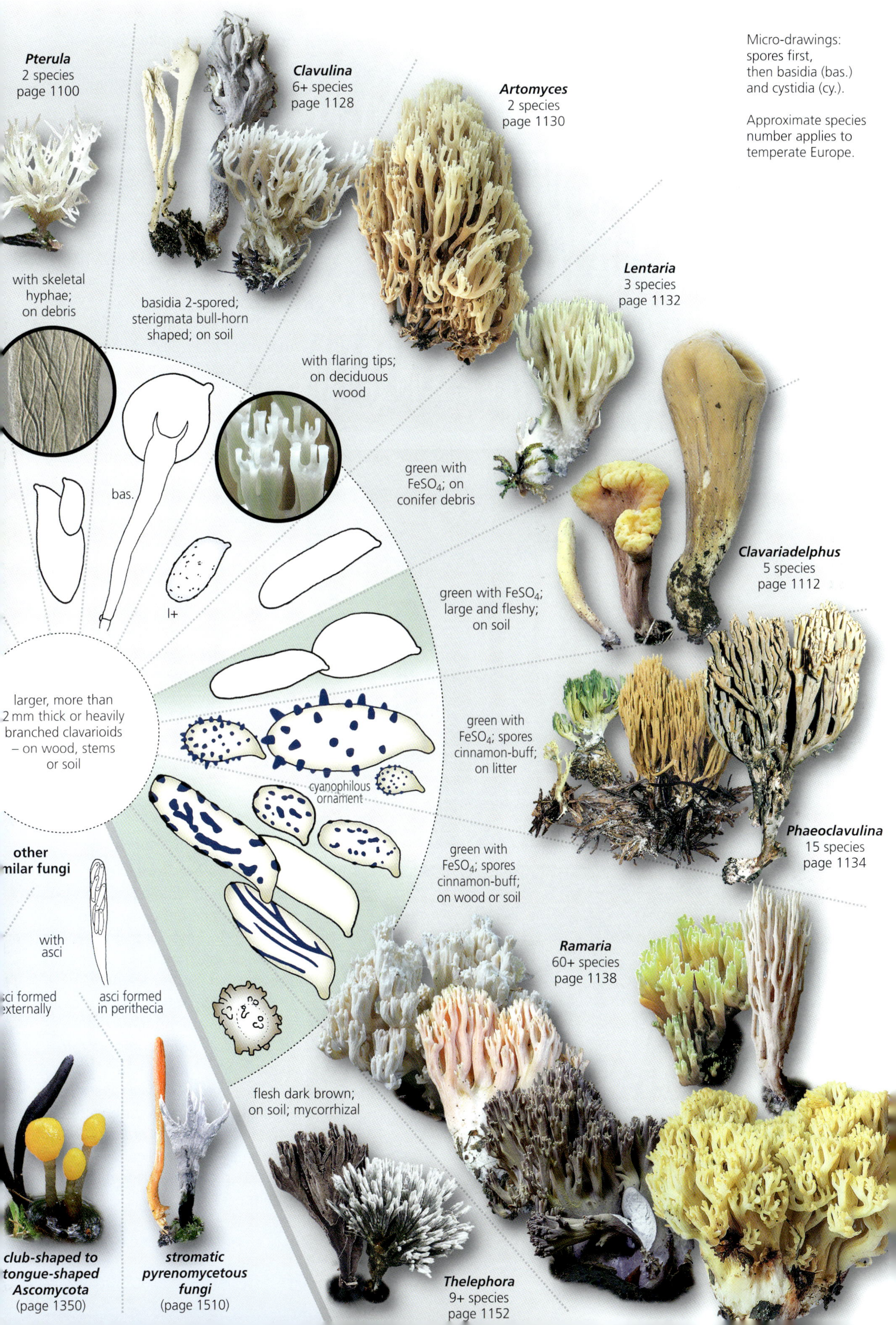

Species of ***Mucronella*** have small, downwards-turned, clavate fruitbodies (termed spines below) that grow in troops but do not have a basal, connecting flesh. The spores are smooth and ± amyloid. All occur on dead wood.

Hydnocristella and *Kavinia* (pages 1040–41) have browner spines that stain green with $FeSO_4$. *Sistotrema raduloides* (page 1042) and *Dentipellis fragilis* (page 1043) produce larger fruitbodies with more crowded spines on a thin subiculum or a thin flesh, and the latter has amyloid, ornamented spores. *Dentipratulum bialoviesense* ⋉ is similar, but has gloeocystidia.

Mucronella calva is a white to sordid yellow-brown clavarioid with hanging spines up to 1.5 mm long that appears directly from the substrate. The hyphae are thin-walled and have clamps. The amyloid, smooth spores measure 4–6 × 2.5–3.5 µm. Occurs on various substrates, but mostly rotten wood.

Mucronella flava ▷ is, even when young, distinctly yellow, and apparently has more amyloid spores (fruitbodies also stain dark blue in Melzer's, which does not seem to be the case with the white species of *Mucronella*).

Occasional, but easily overlooked; all year, but mostly mid-autumn.

Mucronella bresadolae is a clavarioid with hanging, white to yellowish, tapering, ± clustered spines up to 3 mm long; it sometimes has a ± visible thread-like subiculum. The amyloid spores measure 6–8.5 × 5.5–6 µm. Occurs on brown-rotted conifer wood or on old fruitbodies of *Osmoporus odoratus* (page 868).

Other species of *Mucronella* have smaller spores and occur in different habitats. Species of *Ceratellopsis* ▷ have smaller fruitbodies.

Widespread; mainly in the autumn.

Mucronella flava is a vivid yellow clavarioid with hanging, often clustered spines up to 2 mm long. Spores strongly amyloid and measure about 5 × 3 µm. On very rotten conifer wood, mostly in sandy plantations.

Mucronella calva ◁ is initially white and later becomes sordid yellow-brown. *Mucronella bresadolae* ◁ has longer, clustered spines, larger spores and is also white at first; it mainly occurs in connection with the brown-rotting polypore *Osmoporus odoratus* (see page 868).

Widespread, occasional to rare, and mainly reported from the nemoral zones; all year, but mostly late autumn.

Ceratellopsis aculeata is a very small, awl-shaped, monomitic, white clavarioid with fruitbodies with a ± random orientation. Lacks cystidia; hyphae have clamps. The smooth, ± amyloid spores measure 4–6 × 2–3 µm. Occurs mainly, or perhaps exclusively, on rotten parts of *Cladium* at lakesides and in marshes.

Ceratellopsis acuminata ⋉ has skeletal hyphae but similar spores. The also similar *Pterulicium gracilis* ▷▷ is dimitic with skeletal hyphae and has larger spores.

Probably widespread within the host range, but only a few records are known; mainly September–October.

Bryopistillaria sagittiformis is a very small, white clavarioid with a ± tapering, fertile part and a short, ± smooth stem. The tip can be pointed and sterile. Lacks cystidia; the hyphae have clamps and bipyramidal crystals. The amyloid spores measure (5.5–)7–8.5 × 3.5–4.5 µm. Occurs in damp habitats on rotten culms, mosses, bark *etc.*; its nutritional mode is not understood.

Mucronella calva ◁ has somewhat larger fruitbodies, typically grows on rotten wood and has more down-turned spines.

Widespread, but mainly reported from the nemoral zones; autumn.

Hirticlavula elegans is an extremely small, white clavarioid with a hairy stem; the hairs point upwards at a 45° angle and have ± swollen tips. The fertile part is ± egg-shaped and the fruitbodies point in all directions. The hyphae lack clamps and the hymenium does not have cystidia. The smooth, inamyloid spores measure 4.5–6.5 × 3–4.2 µm. Occurs on bark of deciduous trees, *e.g.* coarse, detached *Quercus* bark in damp forests.

The upwards-pointing hairs with swollen tips are unique to this species, and it is unlikely to be misidentified. However, it is easily overlooked as the fertile part is less than 0.26 × 0.2 mm.

Distribution poorly known, but reported from Denmark, Norway and the UK; May–November.

Physalacria cryptomeriae is a very small, chalk-white clavarioid with a stem and a globose, ± hollow head. The head is downy from projecting basidia and ± thin-walled cystidia. The spores measure 11–19 × 3–6 µm. Occurs mostly on *Cryptomeria* and usually on hanging dead branches, rarely on twigs of *Juniperus* and other members of the *Cupressaceae*

Physalacria stilboidea ⋉ occurs on leaves of *Hedera* and *Griselinia* in warmer parts of the UK; it has thick-walled, fusiform cystidia. *Physalacria* is a large genus in the tropics.

Widespread and apparently common where the host occurs; all year.

Pistillina brunneola is a very small clavarioid with nail-shaped fruitbodies that have a random orientation. The widened tip has a sterile lower side and projecting, cystidia-like hairs at the margin. The ± hairy stem becomes browner towards the base and arises from a small, elongated, dark, immersed sclerotium. The hyphae have clamps. The inamyloid, drop-shaped spores measure 8.5–12 × 3.5–5.2 µm. Occurs on decaying leaves of *Molinia* deep in dense stands of this grass in heathland habitats.

Also known as *Typhula brunneola*. Distinctive and unlikely to be misidentified, but hard to find. *Typhula berthieri* ▷▷ also occurs on *Molinia*, but has a more typical *Typhula* fruitbody with a somewhat hairy stem and a cylindrical fertile part on top.

Likely to be common, but is rarely reported, and mainly from Denmark; September–November.

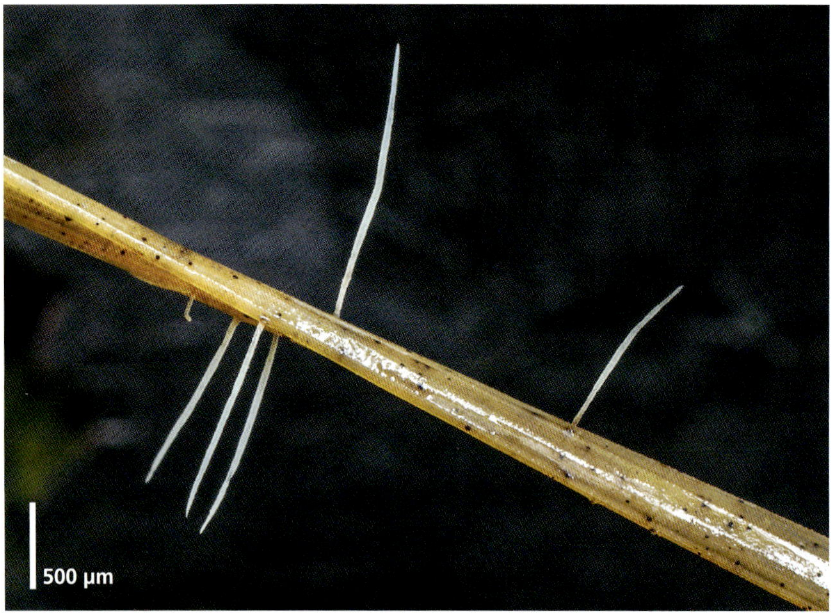

Pterula and ***Pterulicium*** are clavarioids with dimitic flesh that has skeletal hyphae which are very obvious microscopically. Species with small fruitbodies (*Peterulicium*) have no orientation while the larger *Pterula multifida* ▷▷ has upright branches. Relatively few species in temperate Europe, but many in the tropics.

Pterulicium gracilis is a small, white clavarioid with an almost thread-like appearance. The fruitbody ends in a pointed tip and the short stem may be brownish towards the base. The flesh is dimitic with skeletal hyphae. The basidia are 2-spored, and tapering cystidia may be present in the hymenium. The inamyloid, smooth spores measure 9–16 × 4–7 µm. Occurs on various herbaceous stems, both from dicots and monocots, and on other plant litter.

Pterula caricis-pendulae ▽ is characterized by its host, a more brown coloured stem with a sclerotium-like (jigsaw puzzle-like) surface and even longer spores that measure 12–17.5 × 4.8–6 µm. *Ceratellopsis acuminata* ⋉ has smaller spores, 4–6 × 2–3 µm.

Widespread and probably common, but infrequently reported; mainly May–October.

Ibai Olariaga Ibarguren

Pterula caricis-pendulae is a very small, white, acute clavarioid with a purple-brown-tinged stem that has a sclerotium-like surface. The flesh is dimitic with skeletal hyphae. Basidia 2-spored. The very large, inamyloid spores measure 12–17.5 × 4.8–6 µm. Occurs on rotten leaves of *Carex pendula*.

Pterulicium gracilis △ does not have a sclerotium-like stem surface and probably does not occur on *C. pendula*.

Poorly known but apparently widespread in the nemoral zones; spring–autumn.

Pterula multifida is a sordid whitish, highly branched and mainly trooping or fairy ring-forming clavarioid. The odour is unpleasantly phenolic. The flesh is dimitic with skeletal hyphae. Scattered, variably shaped cystidia may occur in the hymenium. The smooth to slightly ornamented, inamyloid spores measure 4.8–6.5 × 2.5–3.5 µm. Occurs on thick litter, mostly from conifers, at least in northern parts of temperate Europe, and typically on rich soils.

Pterula subulata ⋉ is more robust and darker, and acquires black branch tips with age; it has slightly wider spores. Pale, branched species of *Ramariopsis* ▷▷ have neutral smells, no skeletal hyphae, and the spores tend to be distinctly ornamented; they normally occur in other habitats. The asexual state of *Pseudonectria tilachlidii* (page 1631) is similar but produces asexual spores and occurs on rotten fungal fruitbodies.

Widespread, occasional to rare, apparently very rare in the boreal zone; mainly August–December.

Typhula is a large genus of small clavarioids that predominantly form fruitbodies from dead plant litter. Some species have random orientation, while others have upwards-pointing clubs. The hyphal structure is monomitic; the spores are amyloid or inamyloid, smooth and hyaline. The stem flesh is often dextrinoid. Most species have sclerotia that lie dormant through the winter. The surface of the sclerotia is formed of jigsaw-like plates. Some species are harmful to turf grasses or to other crops.

Typhula gyrans is a rather small, white *Typhula* that usually has a yellow-brown, immersed sclerotium. The stem has a covering of minute, acute hairs. The hyphae usually lack clamps. The sclerotium has a finely dotted surface when seen in the microscope. The spores are amyloid and measure 6–9 × 2.6–3.7 µm. Occurs on decaying leaves from deciduous trees, *e.g.* Alnus, in damp woodlands.

Probably forms a species complex of closely related taxa. Most other species of *Typhula* form larger fruitbodies. Other similar species of *Typhula* lack the dots on the sclerotial plates.

Widespread and probably very common; mainly September–November.

Typhula uncialis is a rather thick and almost cartilaginous *Typhula* that does not have a sclerotium. The flesh is ± gelatinous and the hyphae have clamps. It has clavate hairs on the stem. The inamyloid spores measure 5–9.6 × 2.5–4.5 µm. Occurs on coarse stems of herbs, *e.g.* from *Apiaceae* or *Epilobium*.

The rounded, tough clubs and lack of sclerotia make this species easy to identify. *Typhula quisquillaris* ▽ is very similar but is only found on *Pteridium*.

Widespread and common in the hemiboreal–boreal zones, but very rare in the nemoral zones; mainly August–October.

Typhula quisquiliaris is a white, rather tough *Typhula* with an evident, minutely hairy stem and with one club from each dark brown, immersed sclerotium. The amyloid spores measure 8.5–13 × 3.3–5 µm. Occurs on dead parts of *Pteridium*.

The combination of distinctive substrate and numerous, rather tough clubs make this species easy to recognize, even in the field.

Widespread and probably rather common; mainly September–October.

Typhula subhyalina is a small, nail-shaped, whitish- to pinkish-tinged *Typhula* that does not have a sclerotium. The well-defined stem has hairs 30–80 µm long. Lacks cystidia in the hymenium. The inamyloid spores measure 9–12 × 3.7–4.2 µm. Occurs mostly on rotten parts of *Typha*, but also on other monocot marsh plants.

Pistillina brunneola ◁◁ is smaller, has a sclerotium and occurs exclusively on *Molinia*. *Typhula capitata* ⋉ has a rounded head and typically more pinkish-tinged fruitbodies.

Rarely reported, but probably widespread, at least in the nemoral zones; May–October.

Typhula culmigena is a very small, white, clavate, short-stemmed *Typhula* that does not have a sclerotium. The inamyloid, almost heart-shaped spores are unique; they measure 5–6.5 × 4–5.7 µm. Occurs on dead parts of grasses, sedges (*Carex*) and *Typha*.

The lack of sclerotia and the distinctive spores should make identification straightforward.

Rarely reported but very easy to overlook and probably more common than records would indicate; mainly May–August.

Typhula berthieri is a small, whitish *Typhula* that arises from a sclerotium and has concentric rings on the sclerotial plates. The amyloid, ellipsoid to somewhat allantoid spores measure 8–12 × 3.5–4.5 µm. Occurs on litter of *Molinia*.

The concentric rings on the sclerotia plates is the best character for separating this species from other white typhulas on grasses. *Typhula culmigena* △ has heart-shaped spores. *Pistillina brunneola* ◁◁ is even smaller, with a well-defined, nail-shaped head.

Rarely reported but probably far more common than the records would indicate; mainly September–November.

Typhula sclerotioides is a slender, long, undulating, white to almost translucent *Typhula* that arises from somewhat oblong, almost black sclerotia that are gelatinous in the inner parts. The inamyloid spores measure 7.5–10 × 3.2–4 µm. Occurs on dicot herbs that tend to be decayed; the fungus then produces fruitbodies from the sclerotia.

Other similar species of *Typhula* have amyloid spores or different sclerotia.

As with many other typhulas, rarely reported, but probably rather common and widespread; late summer–autumn.

Typhula spathulata is a white to cream, clavate *Typhula* that usually forms several clubs from each of its dark brown, immersed sclerotia. The stems are finely hairy. The hyphae either have or lack clamps. The amyloid, ellipsoid to somewhat allantoid spores measure 5.6–10.5 × 3.2–5 µm. Forms sclerotia below the bark on various deciduous trees, but particulary *Fraxinus*.

Typhula sphaeroidea ⋉, apparently restricted to *Rubus*, has almost spherical, fertile heads and somewhat larger spores.

Widespread and probably rather common; August–November.

Typhula crassipes is a thin, white, worm-like, pointed *Typhula* that has several clubs arising from each large, somewhat irregularly shaped, brownish-yellow sclerotium. Typically has a rudimentary, ± smooth stem. Lacks clamps. The ellipsoid to drop-shaped spores measure 6.5–10 × 3.6–5.5 µm, and are either inamyloid, or at most only slightly, amyloid. Occurs on decaying dicot herbs.

The combination of large, yellowish sclerotia, each of which has several worm-like clubs, make this species easy to recognize.

Widespread, but apparently rather rare; June–October.

Jens H. Petersen/Thomas Læssøe

Typhula micans is a club-shaped to somewhat spatula-shaped, pink, rather small *Typhula* with a stem that lacks proper hairs, and which does not have sclerotia. The amyloid spores measure 9–12 × 4–6.5 µm. Occurs on coarse stems of herbs, *e.g.* umbels (*Apiaceae*).

Other pink *Typhula* species either have sclerotia or differently formed fruitbodies.

Widespread with scattered records, but probably not especially rare; mainly May–September.

Ibai Olariaga Ibarguren

Typhula incarnata is an elegant, slender, long-stemmed, pink *Typhula* that arises from pale brown sclerotia. The spores measure 8–10.5 × 3.7–4.8 µm and are either inamyloid or only slightly amyloid. Parasitic on grasses, where it causes symptoms known as grey snow mould; it can cause serious damage when it attacks sports fields, golf courses, *etc*.

Other pink *Typhula* species lack sclerotia, are smaller or are a different shape.

Widespread, but fruitbodies are rarely recorded, probably overlooked; late in the year.

Typhula lutescens is a slender, long-stemmed *Typhula*; the cylindrical fertile part typically has obvious yellow tinges. The clubs arise from yellow-brown sclerotia. The sclerotium plates either lack or have only indistinct dots. The inamyloid, somewhat sway-backed spores measure 7.7–10.9 × 2.6–3.6 µm. Occurs on leaf-litter, *e.g.* from species of *Populus* and *Fraxinus*.

Typhula ilicis ⋉, known from *Ilex* leaves, has very small, yellow fruitbodies; *T. chamaemori* ⋉ described from *Rubus chamaemorus* may be a distinct species.

Widespread, but rarely recorded; mainly September–October.

Typhula caricina is a white, slender, rather tall *Typhula* with a distinctly hairy, very long stem and an immersed, orange-brown, oblong sclerotium. The sclerotial plates lack ornament. The spores are almost cylindrical, amyloid and measure 10.5–13(–15) × 3.4–4.5 µm. Occurs on rotten parts of *Carex* in dense *Carex* stands in marshes, *etc*.

Similar species of *Typhula* have relatively wider spores and occur on other hosts.

Widespread, but rarely recorded, although most likely rather common; mainly October–November.

Typhula athyrii is a slender, white *Typhula* with brown sclerotia from which a single club emerges. The amyloid spores measure 8–14 × 3.4–4.4 µm; Q 2.4–2.8. Occurs on decaying ferns; known from *Athyrium* and *Dryopteris*.

Typhula olivascens ▽ has a higher spore Q and somewhat olive-tinged fruitbodies. It is known from *Athyrium* and *Matteuccia*. *Typhula quisquiliaris* ◁◁ is a more clavate, tougher species that only occurs on *Pteridium*.

Widespread, but rarely recorded; mainly September–October.

Typhula olivascens is a very slender, long, slightly olive-tinged *Typhula* that has dark brown sclerotia from which a single club emerges. The hyphae have clamps. The inamyloid spores measure 11.5–16.5 × 3.5–6 µm; Q 2.7–3.4. Occurs on *Athyrium* and *Matteuccia*.

This species complex requires more detailed study. *Typhula athyrii* △ does not have olive tinges and has a lower spore Q value.

Widespread, but the few records are mainly from the hemiboreal–boreal zones; late summer–autumn.

Typhula erythropus is a rather large, long-stemmed *Typhula* with a white, cylindrical fertile part and a long, red-brown and minutely hairy stem (strong hand lens required!) that arises from a somewhat elongated, red-brown, erumpent sclerotium. The amyloid spores measure 5–7.5 × 2.3–3.4 µm. Occurs on leaves from deciduous trees, including from petioles, *e.g.* from *Fraxinus*.

Typhula neglecta ▽ is almost identical but has somewhat smaller fruitbodies that have a completely smooth stem. *Typhula buxi* ⋉ occurs on *Buxus* and has spherical sclerotia.

Widespread and common, becoming scarcer towards the north; mainly September–November.

Typhula neglecta is a rather slender *Typhula* with a white, cylindrical fertile part and a smooth, shiny, red-brown stem. The inamyloid or slightly amyloid, ± sinuous spores measure 9–13.5 × 2.5–4 µm. Occurs on leaf-litter from deciduous trees.

Typhula erythropus △ is typically larger with a minutely hairy stem (strong hand lens required!).

Widespread but rare, although probably overlooked; mainly September–November.

Species of ***Macrotyphula*** form long, brownish clubs with or without sclerotia. All are decomposers of wood and litter.

Macrotyphula phacorrhiza is a very long, pale brown clavarioid with a single fruitbody that develops from a lens-shaped, pale brown sclerotium. The inamyloid spores measure 10–16 × 4.5–5.5 µm. Occurs on debris from deciduous trees, mostly in damp woodlands, *e.g.* in stands of *Fraxinus*, and similar places.

Macrotyphula juncea ▽ is very similar, but is attached directly to the leaf-litter by its hairy stem bases (it lacks sclerotia).

Widespread and probably common; mainly September–November.

The sclerotia of *Typhula phacorrhiza* can be parasitized by the inoperculate discomycete *Episclerotium sclerotipus* (page 1362).

Macrotyphula juncea is a very thin, long, pale brown clavarioid with a pale-haired, sterile base that binds the leaf-litter together (inset image). The ± almond-shaped, inamyloid spores measure 6–12 × 3.5–6 µm; Q 1.5–2.1. Occurs on thick, damp, leaf-litter in *Fraxinus* stands, *etc*.

Macrotyphula phacorrhiza △ is very similar, but the fruitbody emerges from a pale brown, lens-shaped sclerotium. *Macrotyphula tremula* ⋉, which occurs on species of the fern *Athyrium*, is apparently rather similar but has spores measuring 11–13 × 3.4–4.5 µm with a Q of 2.9–3.2.

Widespread and common; September–December, mainly late in the season.

Macrotyphula fistulosa is a medium-sized to large, pale brown, hollow, slender clavarioid that is wider towards the tip. The hymenium lacks cystidia and the ellipsoid, inamyloid spores measure 12.5–17.5 × 6–7.5 µm. Occurs on ± buried sticks from deciduous trees in woodlands.

The variety *M. fistulosa* var. *contorta* (bottom image) is a short, twisted, smaller version of the main variety that is found on dead, attached branches, mostly *Betula*, *Corylus*, *Alnus* and *Salix*. It is often considered a species in its own right based om the broader spores 14–18.5 × 6.5–9.5 µm and its presumed different host range. *Macrotyphula juncea* ◁ is smaller (at most 1 mm thick) and occurs on leaf-litter. The ectomycorrhizal *Clavariadelphus pistillaris* ▷▷ is much larger and occurs on soil.

Widespread and rather common; mainly October–January.

Jan Vesterholt

Jens Maarbjerg

Jan Vesterholt

Clavariadelphus is a genus of large clavarioids with large, smooth, white to somewhat yellowish, inamyloid spores. Flesh stains green with $FeSO_4$. All are probably ectomycorrhizal.

Clavariadelphus ligula is a rather slender to fairly broad, clavate, clay-coloured to yellow-brown, up to 10 cm tall clavarioid, which may have indistinct greenish tinges. The stem base typically has hyphal cords. Tastes ± bitter. Hyphae have bipyramidal crystals. The ± sinuous spores measure 9–14 × 3–4.5 µm. Occurs on soil and litter in conifer forests, probably mycorrhizal.

Clavariadelphus sachalinensis ⋉ has larger spores, 17–22 × 4–6 µm. The flesh of *C. helveticus* ▽ stains clay-pink when cut.

Widespread and fairly common in the hemiboreal–boreal zones, very rare in the nemoral zones; late summer–autumn.

Clavariadelphus helveticus is a rather compact *Clavariadelphus* with pointed, initially whitish fruitbodies that later become pinkish-buff. The surface and flesh stain clay-pink when bruised. The hyphae have needle-shaped crystals on the surfaces. The spores measure 9–12.5 × (5–)5.5–6.5(–7) µm. Ectomycorrhizal with *Fagus* and *Abies*.

Clavariadelphus ligula △ is more gracile, has larger spores, differently shaped crystals and occurs in conifer forests. Some mycologists appear to synonymize the two names under *C. ligula*. *Clavariadelphus pistillaris* ▷ is pale yellow when young and also has differently shaped crystals on the hyphae.

Poorly known in temperate Europe, but apparently fairly widespread in southern Sweden – and, as the name implies, it was described from Switzerland; autumn.

Clavariadelphus truncatus is a very robust, flat-topped, yellow-brown clavarioid that stains vivid red with KOH. Young specimans are ± orange yellow. The hyphae have needle-like crystals on the surfaces. The spores measure (8.8–)9.5–12.5(–13.5) × (4.8–)5.5–7 µm. Ectomycorrhizal with conifers.

In theory, only confusable with deformed specimens of *C. pistillaris* ▽ but the fruitbodies of that species are only pale yellow when young, it stains orange with KOH, the hyphae lack needle-shaped crystals, and it occurs with deciduous trees.

Widespread and fairly common in the boreal zone, less so in the hemiboreal zone, and absent from the nemoral zones; late summer–autumn.

Clavariadelphus pistillaris is a very large, pinkish-buff to vinaceous-brown clavarioid with a rounded top; initially, the colour is pale yellow. The surface becomes darker by bruising and stains orange with KOH. The flesh stains clay-pink when bruised and tastes bitter. The hyphae are finely spiny. The spores measure (8.5–)9.5–13(–16.5) × (4.5–)5.5–7.2 (–8) µm. Ectomycorrhizal with deciduous trees on rich, mostly calcareous soils, in the nemoral zones mainly with *Fagus*, farther south mostly with *Quercus,* and in the north presumably with *Populus*; often fruits from mole or vole burrows.

Clavariadelphus helveticus ◁ has more slender fruitbodies, is whitish when young and has needle-shaped crystals on the hyphae.

Widespread, rather common to quite rare, becoming scarcer towards the north; August–November.

Multiclavula is a genus of small, lichenized clavarioids. The fruitbodies arise from a grainy, green thallus and can be found either on poor soils or on wood.

Multiclavula vernalis is a rather small, orange-yellow, lichenized clavarioid. The short stem is whitish. The basidia are 4-spored. The hyaline, smooth, inamyloid spores measure 8–12 × 2.5–3.5 µm. Found in troops during spring on damp, peaty soil, *e.g.* with *Drosera*, on a mat of green algae with a network of fungal hyphae (a thallus).

May be confused with species of the non-lichenized genera *Clavulinopsis* ▷▷ and *Clavaria* ▷, but these occur mainly from summer to autumn. *Mitrula paludosa* (page 1361) has asci and usually a more obvious stem, and occurs in very wet habitats in spring and early summer.

Widespread, but rather rare to very rare; April–May.

Multiclavula corynoides is a cream or slightly pinkish lichenized clavarioid that, with age, typically acquires a couple of branches. Basidia 4- to 6-spored. The spores measure 5.6–8.2 × 2.2–3.5 µm. Arises from a grainy, green thallus on peaty soil.

Other whitish, small clavarioids are not associated with a grainy, green thallus. *Multiclavula mucida* ▷ occurs on trunks of deciduous trees.

Mainly a boreal species, with scattered occurrences in the hemiboreal zone; summer–autumn.

Multiclavula mucida is a simple, upwards-tapering, whitish to pinkish-buff, rather small, ± tough-fleshed, lichenized clavarioid that may eventually develop a couple of branches at the tip. Arises from a fine, grainy, green thallus. Basidia 4- to 6-spored. The spores measure 4.5–7.7 × 1.8–3.2 µm. Occurs on damp trunks of deciduous trees, mostly *Populus tremula*, in areas with high precipitation.

Distinctive and unlikely to be misidentified. The similar-looking *M. corynoides* ◁ occurs on peaty soil.

Widespread, but uncommon and local, very rare in the nemoral zones; summer–autumn.

Alloclavaria purpurea is a large, violet to greyish-violet, white-spored clavarioid with slender, clavate, pointed, ± hollow fruitbodies that usually occur in large troops, and within these troops, mainly in small clusters. The hymenium bear large, up to 130 × 10 µm, cylindrical, thin-walled cystidia; the hyphae lack clamps. The spores measure 7–12 × (3–)4–5 µm. Usually occurs in rather damp, ± calcareous, conifer forests, but may also occur in more open areas in the mountains; perhaps mycorrhizal.

The smaller *Clavaria fumosa* ▷▷ is found mainly in unfertilized grasslands, does not have cystidia and is more greyish.

Widespread in the hemiboreal–boreal zones, occasional; late summer–autumn.

Ania Kubiak

Clavaria species form rather small and fragile fruitbodies with inflated hyphae that lack clamps. *Clavaria fragilis* ◁, *C. rosea* ▽, *C. amoenoides* ▷▷, *C. fumosa* ▷▷ and *C. zollingeri* ▷▷ are completely clampless, but other species have open clamps at the basidia bases. All are probably biotrophic with plants in the same way as waxcaps, but more research is needed. *Clavaria* is polyphyletic and should be split into smaller genera.

Clavaria fragilis is a rather large, white, very fragile, mostly clustered and trooping *Clavaria*. Basidia lack clamps at the base. Spores broadly ellipsoid to ovoid, smooth, measure 4–6.5 × 2.8–4.5(–5) µm. Occurs on light, ± calcareous or clay-rich soils in scrub and unfertilized grasslands.

Clavaria falcata s.l. ▷ is typically somewhat shorter, has a clamp at the base of the basidia and larger spores.

Widespread, occasional; June–December.

Clavaria rosea is an unbranched, fragile, mostly strikingly pink clavarioid. The clampless hyphae are somewhat inflated, and the basidia are also unclamped. The ellipsoid spores measure 5–8 × 2.5–3.5 µm. Occurs in calcareous scrub together with other clavarioids, or in ± calcareous, unfertilized grasslands.

Clavaria incarnata ▷ has open clamps at the base of the basidia and tends to be less vividly coloured.

Widespread, but rare to very rare; June–November.

Johan Nitare

Clavaria falcata s.l. is a white to greyish-white, thin to rather thick, rather fragile clavarioid with a ± well-delimited stem. The basidia have clamps that are open and usually appear as a downturned curved peg. The spores measure (6.5–)7.5–10.5 (–11) × (4–)4.5–7.5(–8.5) µm, and a percentage may have ± cylindrical protuberances, giving a star-shaped appearance. Occurs on clay-rich soils in grasslands and scrub, along watercourses inside forests, *etc*. It is not especially sensitive to chemical fertilizers, and is one of the first species to appear once applications have stopped.

This species belongs to a poorly understood, not yet fully revised complex that has a large variation in spore shape and in the general appearance of the fruitbodies. *Clavaria tenuipes* and *C. daulnoyae* are included here in *C. falcata s.l.* *Clavaria crosslandii* ⋉ has smaller spores and *C. fragilis* ◁ lacks clamps at the base of the basidia.

Widespread and common; mainly July–December, but also known from spring.

Thomas Læssøe

Clavaria incarnata is an unbranched, flesh-pink to pink, fragile clavarioid with a ± delimited, smooth, shiny stem. The basidia have open clamps at the base. The spores measure 6.6–7.1 × 4.1–4.9 µm. Occurs in scrub and ± calcareous, unfertilized grasslands, usually together with other clavarioids, entolomatoids, *Geoglossum*, *etc*.

Clavaria roseoviolacea ⋉, known from Slovakia, has larger spores measuring 6–10.5 × 6–9 µm. *Clavaria rosea* ◁ does not have clamps, even at the base of the basidia.

Widespread, but rather rare or overlooked; July–November.

Thomas Læssøe

Clavaria flavipes is a rather slender, yellowish clavarioid with a strongly coloured, shiny stem. The hyphae lack clamps. Basidia have a large, open clamp. The almost spherical, always smooth spores measure 6–8 × 5.5–7.5 µm; Qav. below 1.1. Occurs on ± calcareous soil in scrub and unfertilized grasslands.

Clavaria flavostellifera ▽ is similar, but has narrower spores (Q above 2). *Clavaria argillacea* ▷ occurs on peaty soils with *Empetrum*. Yellow clavarioids in the genus *Clavulinopsis* ▷▷ typically have less translucent stems and have clamps on the hyphae.

Widespread, but rare; mainly September–November.

Thomas Læssøe

Clavaria flavostellifera is a rather slender, yellow clavarioid with a short, smooth, shiny, yellow stem. Hyphae lack clamps; basidia have a large, open clamp. The smooth spores are mixed with a few thick-walled, ± spiny spores that have refractive contents. The normal, smooth spores measure 5–7.5(–8) × 4.5–6(–6.5) µm; Qav. 1.20–1.56. Occurs on ± calcareous soils in scrub and unfertilized grasslands.

Clavaria flavipes △ always has smooth spores with a lower Q value. *Clavaria amoenoides* ▽ lacks clamps, tapers towards tip and is pale yellow.

Widespread, very rare or somewhat overlooked; mainly September–October.

Johan Nitare

Clavaria amoenoides is a pale yellow, usually clustered, ± upwards-tapering, worm-like, mostly twisted clavarioid with a short, pale stem. The hyphae and basidia lack clamps. The smooth spores measure 6–8 × 3.5–4.5 µm. Occurs in unfertilized grasslands of high conservation value, typically together with a number of other red-listed grassland fungi.

Clavaria flavipes △ and *C. flavistellifera* △ have more rounded, less twisted clubs, open clamps at the base of the basidia, and are mostly more vividly coloured.

Widespread, occasional to very rare; July–October.

Clavaria argillacea is a pale yellow, mostly somewhat compressed, slender, with age ± wrinkled clavarioid with a smooth, clearly delimited, more vivid yellow stem. The hyphae lack clamps; basidia have large, open clamps. The spores measure 9–13 × 4.5–5.5(–6) µm; Qav. 2–2.4. Occurs either singly or in small clusters in relatively dry heathlands, where it forms biotrophic relationship with *Empetrum*.

Clavaria sphagnicola ▽ is very similar but occurs in *Sphagnum* mires; the spores are shorter with a Qav. of 1.5–1.6. Other yellow clavarioids belonging to the genus *Clavaria* ◁ do not occur in acid heathland habitats with *Empetrum*. *Clavulinopsis luteoalba* ▷▷ may occur in nutrient-poor habitats but is more orange and has clamped hyphae.

Widespread and rather common; mainly September–December.

Clavaria sphagnicola is a pale yellow clavarioid with a vivid yellow, smooth stem. The hyphae lack clamps; basidia have a large, open clamp. The spores measure (7.5–)8.5–10.5(–11.5) × (5–)5.5–6.5(–7) µm; Qav. 1.5–1.6. Occurs in *Sphagnum* in mires, but probably associated with species in the *Ericaceae* that grow in the same environment.

Clavaria argillacea △ is very similar but has a higher spore Q and occurs on dry soils with *Empetrum*; it may also have a stronger tendency to fruit in clusters.

Rarely reported, but possibly present in many *Sphagnum* bogs and mires; mainly September–November.

Thomas Læssøe

Thomas Læssøe

Clavaria greletii is a rather slender, pale lead-grey to soot-coloured *Clavaria* with a rounded top and somewhat darker, shiny stem. Hyphae lack clamps; basidia have an open clamp. Spores almost spherical to ellipsoid, smooth, (6.4–)7–9(–9.5) × (5.6–)6.1–7.8(–8.7) µm; Q 1.13–1.20. In calcareous, unfertilized grasslands and on small, dry mounds in wet fens.

Other dark clavarioids do not have clamps at the base of the basidia, and some have ornamented spores or other spore shapes. Species of *Geoglossum* (page 1350) form spores in asci.

Widespread in the nemoral–hemiboreal zones, but very rare; mainly September–October.

Clavaria asperulispora is a rather slender, soot-coloured *Clavaria*. The hyphae and basidia lack clamps. The thick-walled, verrucose spores measure (3.7–)4.2–5.3(–5.5) × (3.4–) 3.7–4.9(–5.1) µm; Q 1.04–1.16. Occurs in unfertilized grasslands of high conservation value, including dune slacks with *Salix*.

Clavaria atrofusca ⋉ has a different spore shape and a higher Q of 1.64–1.81. Other dark clavarioids have smooth spores.

Very rare, found mainly in the nemoral zones; mostly September–October.

Thomas Læssøe

Clavaria fumosa is a large, clustered, fragile, pale reddish-grey to somewhat flesh-pink clavarioid. It completely lacks clamps and cystidia. The smooth, drop- to almond-shaped spores measure (6–)6.1–7(–7.4) × (3.2–)3.4–3.9(–4.1) µm. Occurs in unfertilized grasslands of high conservation value, more rarely in damp woodlands.

Alloclavaria purpurea ◁◁ has cystidia in the hymenium and more obvious purple colours. *Clavulinopsis umbrinella* ▷▷ is more branched and has clamped hyphae.

Widespread, occasional to rather rare; July–November.

Clavaria zollingeri ss. auct. is a highly branched, pale-spored, strikingly purple-violet clavarioid with rather rounded branch tips. It completely lacks clamps and cystidia. The almost spherical, smooth, hyaline spores measure 5–6.5(–7.2) × 3.8–5.2 µm. Occurs in unfertilized grasslands with a high fungal diversity, particularly of waxcaps.

There are also collections (often in northern parts of the area) with a higher spore Q that probably represent a separate species. *Ramariopsis pulchella* ▽ is a more slender fungus with small, spiny spores.

Widespread, occasional in the hemiboreal–boreal zones, but rare to very rare in the nemoral zones; July–November.

Species of ***Ramariopsis*** are normally ± branched and occur in grasslands and on rich soil in forests. All have clamped hyphae and normal clamps at the base of the basidia. The spores are ± ornamented with warts or spines that are so small that a high quality oil immersion lens is required to see them. All are presumed to form a biotrophic symbiosis with plants.

Ramariopsis pulchella is a slender, very striking, ± lavender, rather sparsely branched clavarioid with pointed branch tips. The hyphae have clamps and the surface has tetragonal crystals. The spores have very small, scattered spines and measure (2.8–)3–4(–4.1) × (2.2–)2.3–3(–3.1) µm. Occurs in scrub and unfertilized, typically calcareous, grasslands, usually deep down in the turf.

Very pale specimens may be confused with other gracile *Ramariopsis* ▷▷ species. *Clavaria zollingeri* △ has more robust, fleshy fruitbodies and larger, smooth spores.

Scattered records, but probably rather overlooked; June–October.

Thomas Læssøe/Jens H. Petersen

Thomas Læssøe

Ramariopsis robusta is, compared to other *Ramariopsis*, a large and robust, white to cream species with numerous branches from a rather thick stem. It has clamps, but lacks conspicuous crystals on the hyphae. The drop-shaped to almost spherical spores measure (3.5–)4–5.2 × 3.1–4.3 µm; Qav. 1.26, and have small, well-defined spines up to 0.7(–1.5) µm long. Occurs in old, unfertilized grasslands, typically associated with a number of other clavarioids and a range of waxcaps.

Ramariopsis robusta has only recently been separated from *R. kunzei* ▽, which generally forms smaller, more elegant fruitbodies and has more spherical spores.

Widespread, occasional, but the precise distribution is unclear due to confusion with *R. kunzei*; mainly July–October.

Ramariopsis kunzei is a gracile, whitish clavarioid with a rather thin, felty stem. It has clamps, but lacks conspicuous crystals on the hyphae. The almost spherical spores measure 3.5–5.5 × 3–5; Qav. 1.15, and have small, well-defined spines up to 0.7 µm long. Occurs on soil in scrub and in damp forests, probably also in old, unfertilized grasslands.

Ramariopsis robusta △ can be separated by its larger, more robust fruitbodies and somewhat more oblong spores (Qav. 1.26); it is a typical grassland fungus. The 2-spored *R. bispora* ⨯ has discoid crystals. *Ramariopsis tenuicula* ▷ has even more gracile fruitbodies that become more coloured with age, and has smaller spores with indistinct spines.

Widespread, occasional, but the distribution is unclear due to confusion with *R. robusta*; mainly July–October.

***Ramariopsis arbuscula* nom. prov.** is an apparently undescribed *Ramariopsis* that is characterized by its very small, densely branched, chalk-white fruitbodies that typically occur in troops rather deep in the vegetation. The spores are almost spherical and very finely spiny, (2–)2.2–2.8 (–3.1) × 1.7–2.5 µm. On calcareous soil in scrub and in herbaceous vegetation.

Ramariopsis minutula ⋉, described from France, *R. kunzei* ◁ and *R. robusta* ◁ have larger spores. Other *Ramariopsis* species have different colours or branching patterns.

Known from a handful of southern localities in Norway and Denmark; August–November.

Ramariopsis tenuicula is a rather slender and sparsely branched, white to grey-brown-tinged clavarioid. The hyphae have clamps and usually small, discoid crystals. The finely spiny spores measure 3–4.5(–5) × 3–3.5(–3.8) µm; Qav. 1.3. Occurs in scrub, mull-rich deciduous forests, deep in herb cover or in the turf of unfertilized grasslands.

Resembles *R. kunzei* ◁ but more slender and with duller colours. Other similar species of *Ramariopsis* have almost smooth spores.

Widespread and probably not rare, but easily confused with other species; mainly July–November.

Thomas Læssøe

Ramariopsis luteoochracea is a slender, ivory to ± buff clavarioid with one or two branchings. The hyphae have clamps and bipyramidal crystals. The smooth or nearly smooth spores measure 3.3–5 × 2.3–3 µm; Qav. 1.5. Occurs deep in herbaceous vegetation on mull-rich soils, both on open land and in scrub and forests.

Ramariopsis tenuicula △ and *R. kunzei* ◁ have more spherical spores (Qav. 1.26–1.3) with more obvious spines.

Probably widespread, but rarely reported, possibly somewhat overlooked; August–October.

Ramariopsis crocea is a relatively open-branched, strikingly yellow to orange-yellow clavarioid with pointed branch tips. The taste and smell are insignificant. The small, finely spiny spores measure (3–)3.5–4.5 × 2.8–3.5(–4) µm. Occurs in scrub, abandoned fields, rich deciduous forests and unfertilized grasslands.

Clavulinopsis corniuculata ▷ is less vividly coloured, tastes farinaceous and has larger, smooth spores.

Widespread, occasional to rare; mainly July–November.

Ramariopsis citrina is a very slender, finely branched, pale lemon-yellow clavarioid with pointed branch tips. The finely spiny spores measure 3.5–4.5 × 2.8–3.3 µm. Occurs on soil and herbaceous litter, *e.g.* from *Geranium*, on ± calcareous soils, in ± dry, open places, *e.g.* exposed coastal scrub or drier mounds in marshes, generally deep down in the turf.

Resembles *Ramariopsis crocea* △ but more slender and paler; it has been suggested that the two are conspecific.

Very rare, but probably also somewhat overlooked; mainly August–October.

Species of ***Clavulinopsis*** are unbranched to branched clavarioids, mostly with yellow to orange colours. Both hyphae and basidia have clamps and the spores are generally smooth, spherical, ellipsoid or star-shaped. Species belonging to the genus *Clavaria* may look very similar but normally have more shiny, translucent stems and completely lack clamps on the hyphae; they may, however, have open clamps at the base of the basidia.

Clavulinopsis corniculata is an orange-yellow to apricot-orange, branched, farinaceous-tasting clavarioid with pointed to rounded tips. The hyphae have clamps. The smooth, hyaline spores measure 5–6.5(–7.5) × 4.5–6.5 (–7) μm. Occurs in unfertilized grasslands, but also in scrub and on damp, black soils in woodlands; it is not a good indicator of high conservation value grasslands.

Ramariopsis crocea ◁ is typically more vividly coloured and gracile and does not taste farinaceous; it has smaller, finely ornamented spores. Species of *Phaeoclavulina* ▷▷ have coloured spore-deposits and ornamented spores. *Calocera viscosa* (page 1159) has very rubbery fruitbodies, completely different spores, tuning fork-like basidia and occurs on conifer wood.

Widespread and rather common; mainly August–January.

Clavulinopsis umbrinella is a pale, yellowish grey, vinaceous-brown to reddish-grey, rather branched, white-spored clavarioid with rounded tips. The hyphae have clamps. The smooth spores measure 5–7 × 4.5–6.5 μm. Occurs in unfertilized grasslands of high conservation value and always with other rare grassland fungi.

Ramariopsis kunzei ◁◁ is paler and has ornamented spores, and, like other similar species of *Ramariopsis*, has smaller spores.

Widespread, occasional to rare, becoming scarcer towards the north; mainly August–November.

Clavulinopsis helvola is an unbranched, yellow clavarioid with ± rounded tips and often, ± pronounced longitudinal furrows. Clamps are present. The spores are essentially spherical but have numerous large protuberances, giving an almost star-shaped appearance; they have a large, internal, golden drop and measure 5–6.5(–7.5) × 4.8–5.5(–6.5) µm. Occurs in unfertilized grasslands, more rarely in scrub on mineral-rich soils or in damp woodlands on black soil.

Other yellow, unbranched clavarioids are best told by their spore size and shape; they lack the prominent protuberances.

Widespread and common – one of the most common clavarioids in temperate Europe; mainly August–December.

Bo Levesen

Clavulinopsis fusiformis is a vivid yellow, highly clustered clavarioid with somewhat flattened, inflated, fusiform fruitbodies and a bitter taste. It has clamps. The smooth, almost spherical spores have a prominent apiculus and measure 5.5–7.5 × 4.5–6(–7) µm. Occurs in unfertilized grasslands.

The spores have to be checked carefully to rule out other yellow clavarioids that can occasionally occur in clusters: *C. laeticolor* ▷ has almost triangular spores with a distinct apiculus; *C. helvola* △ has spores with long protuberances; *C. luteoalba* ▷ and species of *Clavaria* have a small apiculus on the spores.

Widespread and rather common in western, Atlantic parts of temperate Europe, but rare elsewhere; mainly August–October.

Clavulinopsis luteoalba is an orange-yellow clavarioid that grows singly or in small clusters. It often has flattened tips and may branch once near the tip. It can either have or lack an evident farinaceous or mouldy smell and taste. The hyphae have clamps. The smooth, almond-shaped spores have a small apiculus and measure 5.5–7(–8.5) × 3–4.5 µm. Occurs in unfertilized, mossy grasslands or on mineral-rich soils in scrub and deciduous woodlands; usually on rather poor soils, *e.g.* with *Hieracium pilosella*.

Other yellow *Clavulinopsis* species are typically less orange and have less flattened tips, but *C. luteoalba* is best told by its spore shape and small apiculus. Yellow species in the genus *Clavaria* ◁◁ typically have a bright yellow, delimited stem and lack clamps on the hyphae, but may have an open clamp at the base of the basidia.

Widespread and rather common; mainly August–November.

Clavulinopsis laeticolor is a yellow, unbranched clavarioid that usually occurs singly or in loose clusters. Odourless but may have a bitter taste. Has clamps. The ± ellipsoid to almost triangular spores have a prominent, somewhat lateral apiculus and measure 5–7.5(–8.5) × (3.2–)4–6.5 µm. Occurs in unfertilized grasslands or on mineral-rich soils in scrub and deciduous forests.

Clavulinopsis helvola ◁ has spores with protuberances; *C. luteoalba* △ has ± ellipsoid spores with a small apiculus; and *C. fusiformis* ◁ has spherical spores.

Widespread and rather common; mainly July–November.

Jens Maarbjerg

Jens Maarbjerg

Clavulina is a genus of medium-sized, ± branched clavarioids that have basidia with two, well-spaced, horn-like sterigmata and large, almost spherical spores. Molecular studies have established that a further range of hard-to-identify species also occurs. All are ectomycorrhizal.

***Clavulina coralloides* s.l.** is a somewhat untidy, multi-branched, white, grey to vinaceous-brown clavarioid with pointed, ± serrated tips. The fruitbodies sometimes have more rounded tips and may become darker. Has clamps and 2-spored basidia. The hyaline, spores are smooth and measure (6.8–)7.5–9.5 (–10.5) × (5.5–)6–8(–8.8) µm; Q 1–1.35. Mycorrhizal with a number of hosts, especially on mineral-rich soils in forests, *Salix* scrub, *etc.*

This species is often treated under the name *C. cristata*. *Clavulina cinerea* ▷ is darker grey and does not have serrated apices. *Clavulina incarnata* ⋉ has a pinkish tinge and ± cylindrical cystidia in the hymenium. *Clavulina rugosa* ▷ is whitish, either not or only slightly branched, and has rounded tips. Species of *Phaeoclavulina* ▷▷ and *Thelephora* ▷▷ have coloured spore-deposits. *Tremellodendropsis tuberosa* (page 1197) has tougher, more rubbery fruitbodies and the basidia split at the top.

Widespread and very common; June–January.

Fruitbodies of *C. coralloides* with dark grey stems have typically been parasitized by the pyrenomycetous fungus *Helminthosphaeria clavariorum* (see lower image and page 1596). The dark surface normally holds the dark, septate conidia of the parasite, and the blackish, hairy perithecia are only evident occasionally.

Clavulina cinerea is a rather large, multi-branched clavarioid with grey colours and mostly ± rounded, branch tips; it branches from the base. The hyaline, smooth spores measure 7.1–10.6 (–11.6) × 5.6–8.4 µm; Q 1.2–1.4. Mycorrhizal with deciduous trees and conifers, often on rich soils.

Clavulina reae ⋉ is similar but has a more well-defined stem. *Clavulina amethystina* ⋉ is more violet-grey and the spores are more oblong; Q 1.5–1.6. Greyish forms of *C. coralloides* ◁ have ± serrated apices.

Widespread, rather common; June–December.

Clavulina rugosa is a rather slender, simple or slightly branched, distinctly longitudinally grooved, fairly large clavarioid with an evident stem and ± chalk-white colours that may darken slightly with age. It has 2-spored basidia and almost spherical, hyaline spores that measure (8–)9–11.5(–13) × (6.5–)7.5–10 µm; Qav. 1.14–1.30. Ectomycorrhizal, mostly with conifers on clay-rich or mull soils, rarer with deciduous trees.

Very branched forms can be told from the similarly pale-coloured *C. coralloides* ◁ by the distinct stem, more rounded tips and larger spores.

Widespread and common; mainly June–January.

Clavicorona taxophila is a small, whitish, ± trumpet-shaped clavarioid. The lower, sterile part is somewhat more translucent than the fertile part and finely floccose. The widened top is indistinctly dentate at the margin. Lacks cystidia. The basidia are 4-spored, and the hyphae have clamps. The hyaline, finely ornamented, inamyloid, broadly ellipsoid spores measure 3–4 × 2.5–3 µm. Occurs on soil and litter, usually in *Salix repens* scrub in dune slacks, rarely in *Juniperus* scrub, on *Taxus* litter or under exotic conifers.

Highly distinctive and unlikely to be misidentified.

Widespread, but rarely reported and easily overlooked, at least locally quite common; mainly September–November.

Artomyces is recognized by the widened branch tips that typically have the appearance of a serrated crown. Spores amyloid and ± finely, ornamented (× 1,000 lens required!). Occurs on wood.

Artomyces cristatus is a clay-buff to vinaceous-brown, unbranched or slightly branched, medium-sized clavarioid. The crown-like tips are somewhat paler. The amyloid, ± smooth or very finely ornamented spores measure 5–6.5 × 2–2.5 µm. Occurs on old, rotten trunks of conifers in coniferous forests.

Artomyces pyxidatus ▷ is larger, more branched, paler and occurs on deciduous wood.

Found in the hemiboreal–boreal zones, rare; late summer–autumn.

Artomyces pyxidatus is a very distinctive, rather large, multi-branched, pale-spored clavarioid with branches that end in crown-like tips. The branching pattern is candelabrum-like with several more branches from each branch point. The hyaline spores are amyloid, finely ornamented and measure 4–5 × 2–3 µm. Occurs on dead trunks of deciduous trees, mostly on *Populus tremula* and other *Populus* species.

Artomyces cristatus ◁ is smaller, darker, less branched and occurs on conifer wood. *Ramaria stricta* ▷▷, which, like *A. pyxidatus*, occurs on deciduous wood, may have a similar stature but does not display crown-like tips and the spore-deposit is buff. *Lentaria subcaulescens* ▷▷ is also similar and occurs on deciduous wood, but lacks the crown-like tips, is smaller and has larger, smooth, inamyloid spores.

Widespread and rather common in the hemiboreal and boreal zones, rare in the nemoral zones but is apparently becoming more common; mainly June–October.

Jens H. Petersen/David Boertmann

Lentaria is a genus of branched clavarioids with smooth, hyaline, inamyloid or amyloid spores and a green flesh reaction with $FeSO_4$. It belongs in the *Gomphales* together with *e.g. Ramaria* ▷▷ and *Phaeoclavulina* ▷▷. Molecular studies have demonstrated that species of *Lentaria* are mixed with species of *Clavariadelphus* ▷▷, *Hydnocristella* and *Kavinia* (pages 1040–41) and that the genus should probably be split into several further genera.

Lentaria byssiseda is an untidy-looking clavarioid fungus. The colours are a diluted cream, greenish to flesh pink. The surface colours green with $FeSO_4$. The large, somewhat sinuous, amyloid, smooth spores measure 12–24 × 3.5–5 µm. Occurs in calcareous habitats on twigs or other litter under *Juniperus* or various other conifers; less common on deciduous wood.

Lentaria dendroidea ▽ occurs on litter and wood but is more similar to a *Phaeoclavulina* ▷▷ and has smaller spores. Species of *Clavulina* ◁◁ occur on soil and have ± spherical, inamyloid spores.

Mainly a northern, hemiboreal–boreal species where it is occasional, very rare in the nemoral zones and apparently absent from the UK; mainly August–November.

Lentaria dendroidea is medium-sized, branched to highly branched clavarioid with a thick, white, felty stem and pale yellow branches. The inamyloid, smooth spores measure 7–11 × 3–4 µm. Occurs on debris in ± calcareous woodland, both with deciduous trees and conifers, often connected to decomposed wood.

May recall species of *Ramaria* ▷▷ and *Phaeoclavulina* ▷▷, but these have coloured, ± ornamented spores (but see the smooth-spored *R. schildii* ▷▷). Other species of *Lentaria* are far more gracile and extremely pale.

Only found in hemiboreal–boreal Scandinavia, rare; late summer–autumn.

Lentaria subcaulescens is a pale, whitish to somewhat blue-tinged, almost translucent, loosely branched clavarioid that may stain yellowish with age. The hyphae have clamps. The ± amyloid, smooth, hyaline spores measure 5–7.5 × 3–4 µm. Occurs on large rotten trunks of both deciduous and coniferous trees, and on old decaying sawdust, *etc*.

There is currently no consensus on how many species should be recognized in this group of clavarioids; a broad approach has been taken here. Other names that are in use for similar material are: *L. albovinacea*, *L. delicata* and *L. epichnoa*. *Lentaria afflata* ⋉ is also similar but lacks clamps. White species of clavarioids in the genus *Ramariopsis* ◁◁ also look similar to *L. subcaulescens* but have finely ornamented, inamyloid spores and normally occur on soil or fine plant litter.

Widespread, rare; mainly October–November.

Jacob Heilmann-Clausen

Species in the genus ***Phaeoclavulina*** are all decomposers of deciduous or needle debris that form ± medium-sized fruitbodies and have curry-yellow, brownish to greenish colours. The drop-shaped spores are spiny, often rather small and cinnamon-buff. The hyphal cords have various types of surface crystals. Fruitbody surfaces stain green with $FeSO_4$. The genus was formerly included in *Ramaria* ▷▷, which currently mostly includes lignicolous and mycorrhizal, often more fleshy species.

Phaeoclavulina abietina is a medium-sized, decomposing, multi-branched clavarioid with yellowish-brown and greenish colours. The hyphal cords are pale and have star-shaped crystals on the surfaces. The drop-shaped, finely spiny spores measure 6.5–9.5 × 3.5–4.5 µm; Q 1.9–2.4. Occurs on needle debris from *Picea* and *Abies*, mostly on rich soils.

The rarer *Ramaria apiculata* ▷▷ has large, ellipsoid, verrucose spores that measure 7–12 × 4–5.5 µm; it occurs on conifer wood. In southern parts of Europe a few very rare species with larger fruitbodies and spores are found, *e.g. R. ochraceovirens* ⋉ and *R. ochrochlora* ⋉.

Widespread and rather common; July–January.

Phaeoclavulina corrugata is a medium-sized, decomposing, multi-branched clavarioid with undulating, pale pinkish-buff branches and a peppery taste. The hyphal cords have star-shaped crystals on the surfaces. The drop-shaped, spiny spores measure 4–6.5 × 2.5–3.5 µm; Q 1.5–1.9. Occurs on needle debris of *Picea*, *Juniperus* and *Abies*.

Also known as *Ramaria myceliosa* – a more recent name. *Phaeoclavulina flaccida* ▷ is similar but does not taste peppery and has somewhat thicker branches and larger spores.

Rather rare; mainly August–November.

Phaeoclavulina flaccida is a medium-sized, decomposing, multi-branched, ± cinnamon-buff clavarioid. The hyphal cords have hyphae covered in star-shaped crystals. The finely spiny spores measure 6.5–9.5 × 3–5 µm; Q 1.7–2.2. The most common habitat is probably needle beds under *Picea*, *Abies* and *Juniperus*, but it also occurs on debris from deciduous trees (lower image) and in coastal grasslands (middle image); more than one species could be involved.

The smaller *Phaeoclavulina roellinii* ⋉ seems to be very closely related or perhaps identical; it is associated with dry, open, ± calcareous habitats and should have flatter branch tips, a dark KOH reaction on the surface and somewhat more vivid colours. *Phaeoclavulina flaccida* can easily be confused with *P. eumorpha* ▷▷ but that species tends to produce coarser fruitbodies with somewhat larger spores; it also has irregular rather than star-shaped crystals on the hyphal cord hyphae. *Phaeoclavulina corrugata* ◁ has a peppery taste, and both *P. corrugata* and *P. decurrens* ▷▷ have smaller spores.

Widespread and fairly common; mainly July–November.

Phaeoclavulina decurrens is a medium-sized to rather large, decomposing, multi-branched clavarioid with rather vivid, curry-yellow to yellow-brown colours. The hyphal cords have star-shaped crystals on the surface. The drop-shaped, finely spiny spores measure 4.5–7.2 × 2.7–4 µm; Q 1.6–2.0. On debris from deciduous trees, mostly on calcareous soils.

Phaeoclavulina eumorpha ▽ may be of a similar size but has larger spores, lacks star-shaped crystals and occurs on needle debris. *Ramaria stricta* ▷▷ has verrucose spores and occurs on wood.

Widespread, occasional, more or less absent from the boreal zone; mainly August–November.

Phaeoclavulina eumorpha is a medium-sized, decomposing, multi-branched clavarioid with ± yellowish-brown colours. It has irregular crystals on the surface of the hyphal cord hyphae. The finely spiny spores measure 7–12 × 3–5 µm; Q 1.8–2.3. Occurs mainly in fairy rings on needle beds in plantations and forests with *Picea* or *Abies*, or with *Juniperus*.

Mostly confused with other *Phaeoclavulina* species, but those that are similar have star-shaped crystals on the hyphal cords; *P. flaccida* ◁◁ is also somewhat more gracile and has smaller spores. *Phaeoclavulina decurrens* △ is typically larger, a more vivid colour, has star-shaped crystals, smaller spores and occurs in deciduous forests. Specimens of *P. eumorpha* can rarely have olive-green colours, and then may be confused with *P. abietina* ◁◁, but that species also has star-shaped crystals on the hyphal cords.

Widespread and common; July–January.

Phaeoclavulina macrospora is a medium-sized to rather large, decomposing, multi-branched clavarioid with at first buff-yellow to curry-yellow, but later more grey-brown colours. All parts colour dark brown to black when bruised and the fruitbodies dry black. The flesh tastes bitter and has 2-spored basidia. The ± ellipsoid spores measure 12–20 × 5–8 µm and have spines up to 2 µm long. Occurs in calcareous, thermophilic habitats with a high diversity of rare herbs and fungi, usually with *Juniperus*, *Pinus* or *Corylus* in the vicinity.

Also known as *Ramaria broomei* and *R. nigrescens*. The combination of large, spiny spores and 2-spored basidia is unique to this species. Macroscopically, it may resemble *Ramaria testaceoflava* ▷▷ but that species stains more greyish-violet when bruised, the spores are ellipsoid, verrucose and smaller, and it is ectomycorrhizal with conifers.

Known from Norway (*e.g.* coastal areas in the south), Sweden (*e.g.* Öland's alvar), the UK and Germany, absent from the boreal zone; late summer–autumn.

Thomas Læssøe

Thomas Læssøe

Species of ***Ramaria*** have cinnamon-buff, ± cyanophilic, ellipsoid to almost sinuous spores that mostly have a warty or ± striate ornament – only a few species have completely smooth spores. The spore ornament is also cyanophilic and can only be studied in detail after staining with cotton blue. Many species produce large, fleshy fruitbodies and all stain green on the hymenium with $FeSO_4$. The fruitbodies vary tremendously with age: initially, they are mostly pale, then become darker and more coloured on the middle branch parts but remain pale at the tips; finally, the middle branch parts become brownish from the cinnamon-buff spores that accumulate particularly in the branch forks. The genus currently includes both decomposers and ectomycorrhizals and is phylogenetically polyphyletic, being mixed with species of *Gomphus* (page 54), *Kavinia/Hydnocristella* (page 1040–41) and similar genera. Undoubtedly, *Ramaria* will have to be split into further genera (see also note under *R. botrytis* ▷▷).

Ramaria suecica is a medium-sized, multi-branched clavarioid with a mild taste and pale pinkish-buff to pale pinkish colours. The hyphae have clamps. The verrucose spores measure 7–10.5 × 3–5 µm; Q 2.1–2.5. Occurs on needle debris, mostly from *Juniperus*, possibly also on *Salix* debris, often in coastal scrub.

Ramaria gracilis ▷▷ has somewhat finer branches, smells of aniseed and has skeletal hyphae in the branch flesh. *Ramaria stricta* ▷ has skeletal hyphae in the hyphal cords and occurs on wood. *Ramaria suecica* is only distantly related to the other species of *Ramaria*.

Fairly common in the hemiboreal–boreal zones but rare to very rare in the nemoral zones; mainly August–November.

Ramaria stricta is a medium-sized to rather large, multi-branched clavarioid that, when young, is pale yellow, but which quickly becomes more pinkish-buff to brownish-rose with pale yellow tips. Flesh and branches stain red-brown when bruised. Smells spicy with faint aniseed components. Occurs both in rather gracile, open branched forms and in very compact forms (f. *compacta*, see lower image). Hyphae have clamps and hyphal cords have skeletal hyphae. The verrucose spores measure 7.5–10.5 × 4–5 µm; Q 1.9–2.2. Occurs on deciduous wood, including wood chips and sawdust.

Wood-inhabiting species of *Ramaria* can be difficult to identify. *Ramaria concolor* ⋉ lacks yellow colours and occurs on conifer wood. A similar, but undescribed, species occurs on *Ammophila* in dunes. *Ramaria apiculata* ▷▷ has green colours and lacks skeletal hyphae. A fourth species, *R. rubella* ▷▷, with reddish-brown colours, has very short spores.

Widespread and common in the nemoral–hemiboreal zones, but very rare in the boreal zone; mainly August–November.

Josef Christan

Ramaria rubella is a rather large, clay-pink, multi-branched clavarioid. The hyphae have clamps, and skeletal hyphae are present in the basal hyphal cords. The rather short, verrucose spores measure 6.5–9 × 4–5.5 µm; Q 1.3–1.6. Occurs on well-decayed conifer wood in damp woodlands.

The rare *R. concolor* ⋉ also occurs on conifer wood but has longer spores; Q 1.9–2.2. *Ramaria stricta* ◁◁ also has longer spores, the branch tips are yellowish and it occurs on deciduous wood.

A rare, hemiboreal–boreal species that is absent from the nemoral zones; late summer–autumn.

Thomas Læssøe

Ramaria apiculata is a medium-sized, multi-branched clavarioid with greenish colours in the branch forks and on the branch tips. The hyphae have clamps; lacks skeletal hyphae. The verrucose spores measure 7–12 × 4–5.5 µm; Q 1.8–2.2. Occurs on conifer wood, perhaps with a preference for calcareous soils, mainly in old, species-rich conifer forests

This is the only lignicolous *Ramaria* that completely lacks skeletal hyphae. *Ramaria tsugina* ⋉ is very similar, but has hyphal cords with skeletal hyphae. The more gracile *Phaeoclavulina abietina* ◁◁ occurs on needle beds and has small, spiny, drop-shaped spores.

Widespread, occasional in the hemiboreal–boreal zones, very rare in the nemoral zones; September–October.

Ramaria gracilis is a medium-sized, delicately multi-branched clavarioid with pale pinkish-buff colours and a strong smell of aniseed. The flesh has skeletal hyphae and the hyphae have clamps. The verrucose spores measure 5–7.5 × 3–4.5 μm; Q 1.5–1.9. Occurs mostly on needle beds under *Picea*, mainly on rich, often calcareous soils.

The combination of a distinct aniseed smell and skeletal hyphae in the flesh is unique within this genus. *Ramaria suecica* ◁◁ may look similar but usually has somewhat thicker branches, lacks an aniseed-like smell and does not have skeletal hyphae in the flesh. *Ramaria stricta* ◁◁ smells at most only very faintly aniseed-like and occurs on wood in deciduous forests; only the hyphal cords have skeletal hyphae. Species of *Pterula* ◁◁ also have skeletal hyphae but have smooth, white spores and different smells.

Widespread, occasional to rather rare in the boreal zone becoming scarcer towards the north; mainly August–November.

Jan Vesterholt

The following species of *Ramaria*, unlike the preceding species, are all supposedly ectomycorrhizal, and all are typically more fleshy.

Ramaria testaceoflava is a medium-sized clavarioid with at first pale yellow (below), later cinnamon-brown to orange-brown branches (above). The flesh stains greyish-violet where bruised and tastes bitter. The hyphae have clamps and are ± covered with rod-shaped crystals. The spores are verrucose, have spherical warts and measure 8.5–15 × 4–6 µm; Q 2.0–2.6. Mycorrhizal with *Picea* on richer soils in old-growth forests.

The greyish-violet reaction in the flesh and the rod-shaped crystals are unique characters. *Phaeoclavulina macrospora* ◁◁ stains blacker when bruised, has very large, spiny spores and occurs mainly in more open habitats, sometimes with *Juniperus*.

Widespread in the hemiboreal–boreal zones; autumn.

Ramaria fumigata is a medium-sized, grey-brown clavarioid with violet colours on the stem, branches and tips. The hyphae have clamps. The verrucose spores measure 8–13 × 4–6 µm; Q 1.9–2.3. Mycorrhizal with *Fagus*, mostly in places with other rare mycorrhizal fungi.

Ramaria fumigata is one of few species of fungi in temperate Europe to have violet colours. *Ramaria fennica* ⋉ is identical but occurs in old-growth conifer forests on calcareous soils. *Ramaria fumigata* is almost the same colour as the closely related *Gomphus clavatus* (page 54). Species of *Clavulina* ◁◁ may have greyish to violet-grey colours but all have 2-spored basidia with large, ± spherical, smooth and hyaline spores.

Widespread in the nemoral zones, but rare; August–October.

Ramaria karstenii is a medium-sized, mycorrhizal clavarioid with cinnamon-brown to grey-brown colours. The hyphae lack clamps. The spores measure 7.5–11.5 × 4–6.5 µm (Q 1.6–2.0) and have small, sinuous warts. In conifer forests (probably mostly *Picea*) on calcareous soil.

The closely related *R. spinulosa* ⋉ occurs in deciduous woodlands. Similar wood-inhabiting species of *Ramaria* have clamps and usually also have skeletal hyphae in the hyphal cords. Species of *Phaeoclavulina* ◁◁ have clamps and spiny, drop-shaped spores.

Mainly in the hemiboreal–boreal zones, rather rare; autumn.

Katriina Bendiksen

Ramaria sanguinea is a fleshy, pale yellow, densely branched, fairly large clavarioid with a rather narrow and compressed branching pattern and wine-red surface spots. The hyphae lack clamps. The spores have spherical to oblong warts, mostly in rows, and measure 7.5–12 × 4–5.5 µm; Q 1.9–2.4. Mycorrhizal with *Fagus* on mineral-rich soils, more rarely associated with *Picea*.

Ramaria eosanguinea ⋉ should have fewer spots and the hyphae have clamps. *Ramaria krieglsteineri* ▷ has slightly warmer colours, lacks red spots and has very long spores. *Ramaria brunneicontusa* ▷ has somewhat longer spores and the hyphae have clamps.

Widespread, but mostly in the nemoral–hemiboreal zones, rather rare to very rare; mainly August–October.

Ramaria brunneicontusa sensu auct. is a fairly large, fleshy, multi-branched clavarioid with pale yellow branches that lack red spots. The hyphae have clamps. The spores measure 9.5–16 × 4–6 µm; Q 2.2–2.8. It is mycorrhizal with *Fagus* on rich soils in thermophilic habitats.

Ramaria sanguinea ◁ has rather similar colours, but has wine-red spots at the base, smaller spores and lacks clamps. *Ramaria krieglsteineri* ▽ has even narrower spores; Q 3.1–3.5. *Ramaria flavobrunnescens* ⋉ has smaller fruitbodies, smaller spores (7–10 × 3–5 µm; Q 2.0–2.4) and the hyphae have clamps. There are several similar species in temperate Europe, *e.g. R. flava* ⋉, where the warts of the spores are only visible after colouring but not in outline, and *R. flavoides* ⋉ and *R. pallido-saponaria* ⋉ with partly gelatinous flesh. Of these the former has clamps whereas the latter is without. Unfortunately there is confusion about the name *R. brunneicontusa*, as Scandinavian authors often apply the name to a golden-orange species growing with conifers.

Widespread but rare; mainly August-October.

Ramaria krieglsteineri is a fairly large, fleshy, multi-branched clavarioid with pale yellow branches. The hyphae lack clamps. The spores are very slender, measuring 9–18 × 3.7–5.5 µm; Q 2.5–3.2. Ectomycorrhizal with *Fagus* on rich, ± clayey soils.

The name *R. longispora* has been applied to this species. It may be confused with other pale yellow species of *Ramaria*, particulary *R. brunneicontusa* △, but that species has shorter, wider spores and the hyphae have clamps.

Widespread in the nemoral–hemiboreal zones, rare; mainly August–October.

Ole Martin

Johan Nitare

Ramaria flavescens is a large, fleshy, multi-branched clavarioid, which at first is orange-yellow, but later the branches become pale apricot-orange, while the stem top and branch tips remain a more orange-yellow colour; finally, the entire fruitbody becomes ± pinkish-buff. The major branches may become hollow with age. The hyphae have clamps. The spores are covered with distinct, mostly somewhat sinuous warts and measure 8–13.5 × 3.5–6 µm; Q 2.1–2.7. Occurs on rich, often clayey soils; ectomycorrhizal with both deciduous trees and conifers.

This species belongs to a difficult complex of ± orange species of *Ramaria* that used to be lumped with *Ramaria aurea* ⋉. More recently, some North American names have come to be used, particulary *R. largentii* ⋉. Molecular studies of this group have demonstrated that the suite of species in North America is mostly different from the suite of species in Europe.

Widespread in the nemoral–hemiboral zones, occasional to very rare; mainly August–October.

Ramaria boreimaxima is a very large, thick-stemmed, fleshy, multi-branched clavarioid that at first is orange-yellow. The branches quickly become apricot-orange, while the stem apex and branch tips remain a more orange-yellow colour. Finally, the entire fruitbody becomes duller and ± pinkish-buff. The hyphae have clamps. The spores have fine warts ± in rows, and measure 10–12 × 4–5 µm; Qav. 2.4. Occurs in dry habitats; mycorrhizal with *Pinus*.

This species belongs to a difficult complex of ± orange species of *Ramaria*. The most common species is probably *R. flavescens* ◁, which has paler colours, a thinner stem and spores with coarser, non-aligned warts; it tends to occur in damper habitats. There are further similar species.

A hemiboreal–boreal species, found mainly in Norway, Sweden and Finland; August– September.

Johan Nitare

† ***Ramaria formosa*** is a rather large, multi-branched clavarioid that branches from low down. The branches are flesh-pink and the tips pale yellow. The hyphae have clamps. The spores have coarse, sinuous warts and measure 8.5–15 × 4–6.5 µm; Q 1.8–2.4. Mycorrhizal with *Fagus* on rich soils, mostly in places with many species of *Russula*.

The almost stemless fruitbodies and diluted colours are good field characters. *Ramaria fagetorum* ▷ has an obvious stem, pronounced rose branches with pale yellow tips, smaller spores and lacks clamps. Other species in the group include *R. subbotrytis* ▷, *R. ignicolor* ▷ and *R. neoformosa* ⋉, all of which lack clamps.

Widespread in the nemoral–hemiboreal zones, rare to very rare; mainly August–October.

Ramaria schildii is a fairly large, multi-branched clavarioid with pale, flesh-pink branches that may acquire violet spots; the tips are pale yellow. The hyphae have clamps. The spores are long and smooth, measuring 10–15.5 × 3.5–5 µm; Q 2.8–3.3. Mycorrhizal with conifers.

The smooth spores are unique within this genus, but are also found in the ramarioid *Lentaria dendroidea* ◁◁, which has smaller fruitbodies, whitish spores and is a decomposer of needles.

Mainly in the hemiboreal–boreal zones, rare; late summer–autumn.

Ramaria fagetorum is a fairly large, fleshy, multi-branched clavarioid with rose branches and pale yellow tips. It has a well-delimited stem. The hyphae lack clamps. The spores have warts in rows and measure 6.5–11.5 × 3–5 µm; Q 2–2.6. Mycorrhizal with *Fagus* on mor-like, mineral-rich soils, mostly in sites with many species of *Russula*.

May be confused with *R. formosa* ◁ but that species is a paler flesh-pink and branches from the base; it also has somewhat larger spores and the hyphae have clamps.

Mainly nemoral, rather rare; July–October.

Ramaria ignicolor is a medium-sized, multi-branched clavarioid with rose, mostly hollow branches and yellow tips. The hyphae lack clamps. The small, broad spores have small, spherical or sinuous warts and measure 6–10 × 4–6.5 µm; Q 1.3–1.6. Mycorrhizal with *Picea* and *Pinus*.

The similar *R. fagetorum* △ has extremely large branches and longer, narrower spores measuring Q 2–2.6.

Mainly hemiboreal–boreal, rare; late summer–autumn.

Josef Christan

Ramaria subbotrytis is a medium-sized, fleshy, multi-branched clavarioid with bright rose branches and branch tips. The hyphae lack clamps. The spores are coarsely verrucose and measure 7.5–12 × 3–4.5 µm; Q 2.2–2.8. Ectomycorrhizal with deciduous trees.

This is one of very few species of *Ramaria* in temperate Europe to have both rose branches and rose tips. The name is based on North American material, and it is likely that material from Europe may prove to be a separate species.

Widespread in the nemoral–hemiboreal zones, rare; late summer–autumn.

Rudolf Markones

† ***Ramaria pallida*** is a large, fleshy, multi-branched clavarioid with whitish, cream to pale pinkish-buff colours, and sometimes a slightly lilac tinge on the branch tips. With age, the whole fruitbody is coloured cinnamon-buff by the spores. It may show small, greyish-violet spots, and always has a ± spicy, not especially pleasant smell of coffee grains, used cloth or similar. The hyphae lack clamps, and the spores are rather short, with coarse, sinuous warts and measure 8–13.5 × 4.5–6.5 µm; Q 1.7–2.2. Predominantly ecto-mycorrhizal with *Fagus* on ± clay-rich soils, mostly in sites with many other ectomycorrhizal species, *e.g.* other *Ramaria* species, species of *Cortinarius* subgenus *Phlegmacium* (page 698) and *Russula* species (page 372).

Ramaria botrytis ▷ is rather similar but has a ± sweetish smell, flesh-pink to rose tips, clamped hyphae and long, striate spores.

Widespread in the nemoral–hemiboreal zones, occasional; mainly June–October.

•• ***Ramaria botrytis*** is a rather large to large, fleshy, multi-branched clavarioid with pale branches and flesh-pink to rose branch tips. It has a mild taste and a ± sweetish smell. The hyphae have clamps. The spores have a spiralling, striate ornament and are long, measuring 11–18 × 4–6 µm; Q 2.4–3.1. Mycorrhizal, mostly with *Fagus* on mor-like, mineral-rich soils with *e.g.* other *Ramaria* species, species of *Cortinarius* subgenus *Phlegmacium* (page 698) and *Russula* species (page 372), but also in dry forests with *Pinus*.

May superficially recall *R. pallida* ◁ but that species does not have rose branch tips, lacks clamps and the spores are short and verrucose. *Ramaria botrytis* has at least two very similar 'sister species' in temperate Europe: *R. rubripermanens* ⋉ and *R. rubrievanescens* ⋉. The former has spores that measure 9–13 × 3.5–5 µm; Qav. 2.6, the latter has spores measuring 9–15.5 × 4–7 µm; Q 2.1–2.6. As these spores are only marginally smaller and wider than in *R. botrytis*, safe identification without molecular data is almost impossible.

The species with rose tips and spiralling stripes on the spores form a separate phylogenetic unit, and since *R. botrytis* is the type species of *Ramaria*, it and its sister species may end up as the only true species of *Ramaria*, the others having to be reassigned to new genera.

Widespread in the nemoral–hemiboreal zones, occasional; mainly July–October.

Species of ***Thelephora*** can be resupinate, effused-reflexed or fan-shaped (see page 956) or ± branched-clavarioid. All have brown hyphae and brown, angular-verrucose spores. In comparison to *Phaeoclavulina* ◁◁ and *Ramaria* ◁◁, clavarioid *Thelephora* typically have ± flat branches and often hymenium on one side only; they form ectomycorrhiza.

Thelephora anthocephala is a medium-sized, somewhat branched, ± upright, odourless clavarioid with flat branches. The flesh may become greenish in KOH. The brown, angular-verrucose spores measure (7–)8–10(–11) × (5–)6–8.5 µm and have spines (0.5–)1–1.5(–2) µm long. Occurs on rich soils, mostly in deciduous forests.

Thelephora penicillata ▽ is almost appressed to the substrate and typically very pale. *Thelephora palmata* ▷ is taller and more *Ramaria*-like; it smells of rotten cabbage.

Widespread in the nemoral–hemiboreal zones, occasional; mainly July–October.

Thelephora penicillata is a very split and very flattened *Thelephora* that typically has pale branch tips which gradually become darker towards the centre of the fruitbody. The flesh stains green in KOH. The brown, angular-verrucose spores measure 7–10 × 5–7(–8) µm and have spines 1–1.5 µm long. Occurs on ± rich soils in deciduous forests and parks.

Thelephora anthocephala △ may be found in the same sites but is more upright. *Thelephora terrestris* (page 956) occurs in poorer, often sandy habitats, and is less split.

Mainly nemoral–hemiboreal, occasional; July–November.

Thelephora palmata is a medium-sized, branched clavarioid with flattened, dark brown branches and a strong smell of rotten cabbage. The spores are brown, angular-verrucose and measure 8–12 × 7–9 µm. Mycorrhizal with conifers on mull or clay soils.

Recalls a brown *Phaeoclavulina* or *Ramaria* ◁◁ but the branches of those species are more spherical in section and they never smell of rotten cabbage. Furthermore, species of *Phaeoclavulina* have spiny or low verrucose, non-angular spores and inflated, hyaline hyphae. *Thelephora palmata* is separated from other branched *Thelephora* species ◁ by its quite long stem, upright branches and smell.

Widespread and rather common; mainly August–November.

Dacrymycetales

The *Dacrymycetales* is a natural group characterized by tuning fork-shaped basidia, mostly septate spores that bud off microconidia, and by the predominantly orange-yellow, gelatinous to rubbery fruitbodies. Mature spores are best studied from deposits on slides. Some species form simple asexual spores (arthrospores) from structures that are similar to the fruitbodies, but softer.

The individual genera are poorly delimited in relation to each other, and the largest genus, *Dacrymyces,* is not monophyletic.

All species are wood-decayers and form a brown rot. Some species, *e.g. Dacrymyces stillatus,* can cause serious damage to timber, including wooden window frames.

FURTHER READING: 50, 142, 163, 265.

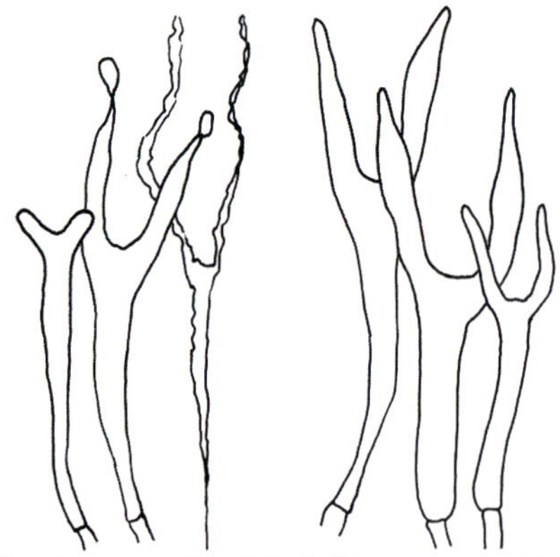

Development of basidia (left); types of basidia (right).

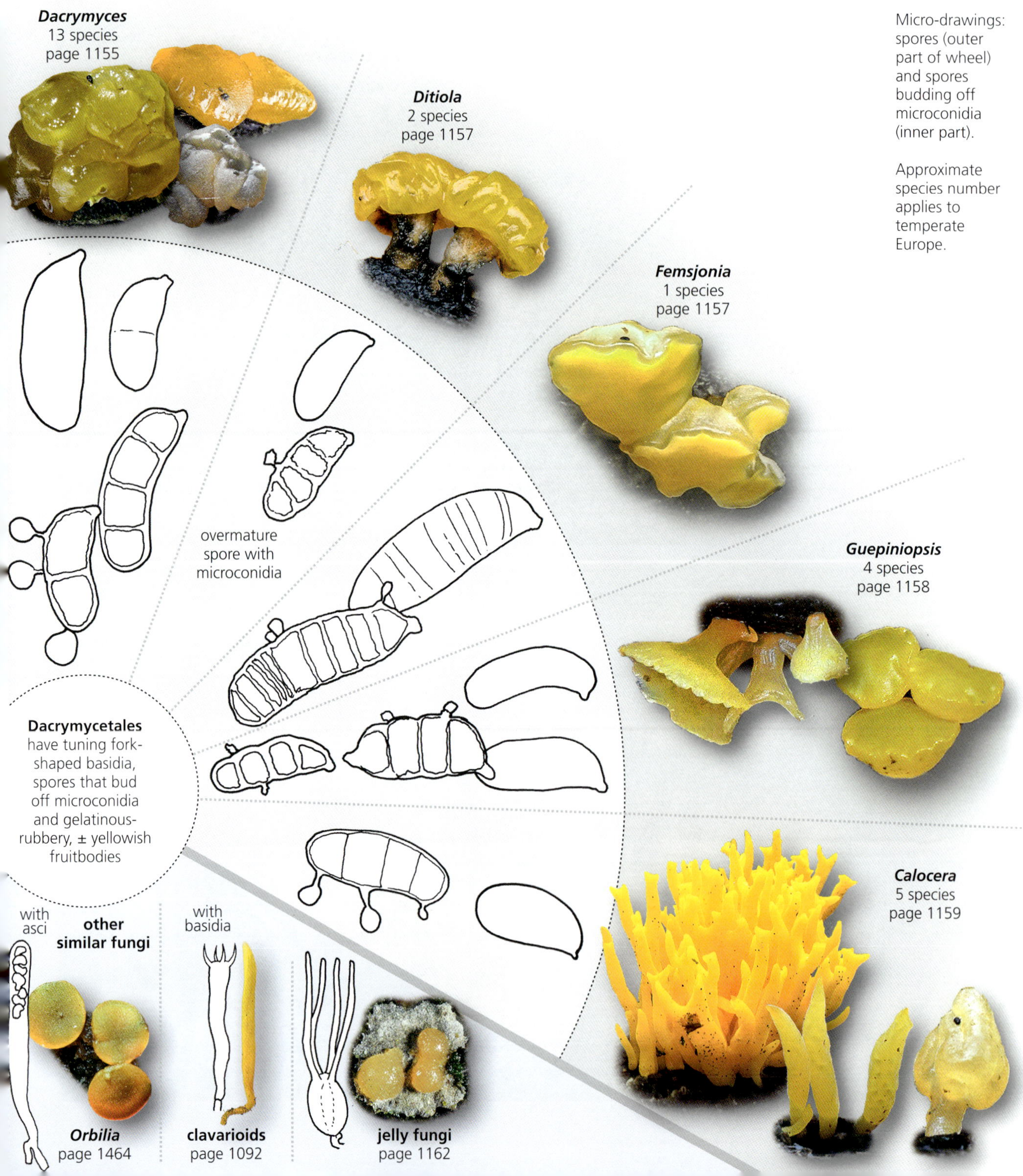

Micro-drawings: spores (outer part of wheel) and spores budding off microconidia (inner part).

Approximate species number applies to temperate Europe.

Dacrymyces form small, cushion-shaped to turbinate fruitbodies that may aggregate in resupinate structures. The spores have up to nine septa. All are wood-degraders and brown-rotters.

Dacrymyces stillatus often occurs in both a sexual and an asexual state on the substrate. The asexual state is reddish-orange and soft, and can easily be squashed (lower image), while the sexual state is adnexed, translucently yellow and firmer, more rubbery-gelatinous. The hyphae lack clamps. The basidia are tuning fork-shaped. The thick-walled spores are flattened on one side, at maturity have three thick septa and measure 12–14 × 3.5–4 µm; they bud off small, globose conidia from each cell. Occurs on moist, rotten wood of coniferous and deciduous trees.

May be difficult to separate from *D. lacrymalis* ▽ but that species lacks the prominent asexual state and has somewhat thinner-walled spores.

Widespread and very common; all year.

Dacrymyces lacrymalis is bright yellow, translucent, wrinkled, narrowly attached, medium-sized, *Dacrymyces* that does not have a proper stem. The spores have three somewhat thickened septa and measure 12–14 × 3.5–4 µm. Occurs mostly on decomposed wood from deciduous trees, more rarely on conifer wood.

The similar *Dacrymcyes stillatus* △ is usually accompanied by a reddish asexual state.

Widespread and common; all year.

Thomas Læssøe

Dacrymyces chrysospermus is a golden-orange, rather large *Dacrymyces* that is typically ± turbinate. The hyphae lack clamps. The basidia are tuning fork-shaped and do not have a basal swelling. The 7–9-septate spores measure 17–23 × 4–8 µm. Occurs on wood and bark of conifers, mainly *Pinus*.

Guepiniopsis estonica ▷▷ has wider spores (8–12 µm), and more swollen basidia. *Dacrymyces chrysospermus* is sometimes interpreted as a species with clamps but here interpreted as without.

Widespread, uncommon; all year.

Thomas Læssøe

Dacrymyces enatus is a *Dacrymyces* with a ± resupinate, uneven surface and dirty, pale brown colours. The hyphae have clamps. The up to 3-septate spores measure 9–14 × 3–4.5 µm. Occurs on coarse, fallen wood, *e.g.* trunks of *Quercus*, but also known from conifers.

Other resupinate species of *Dacrymyces* have more vivid, yellowish colours. Corticioids do not have tuning fork-shaped basidia and are normally less gelatinous.

Mainly found in the nemoral zones, rare; probably all year.

Dacrymyces macnabbii is a mostly pale, whitish to yellowish, ± turbinate *Dacrymyces* with somewhat confluent fruitbodies. It has open clamps and numerous branched hyphae (hyphidia) amongst the basidia. The mainly one-celled spores measure 10.5–13 × 3.5–4 µm. Occurs on conifer wood.

Dacrymyces punctiformis ⋉ has normal closed clamps and fewer hyphidia in the hymenium.

Widespread in the nemoral–hemiboreal zones, rare or overlooked; probably all year.

Lucie Zibarova

Femsjonia peziziformis is a large, striking, golden, somewhat stemmed and almost discoid dacrymycetalean fungus with a ± chalk-white outer side. It has tuning fork-shaped basidia and clamps. The 1–15-septate spores measure 25–32 × 8–10 µm. A brown-rotter on wood and bark, mainly *Quercus*.

The combination of fruitbody shape and substrate should make identification straightforward. It is also known under the name *Ditiola peziziformis*. *Guepiniopsis estonica* ▷▷ also has spores with many septa but occurs on conifer wood.

Widespread, locally rather common, but very rare in most regions; mainly August–December.

Ditiola radicata has a distinct stem, a rooting base and very tough flesh. The fruitbody has a convex, yellow, fertile head and a minutely hairy lower side and stem. It has a distinctive chewing gum-like smell. Lacks clamps. The 1–3-septate spores measure 10–12 × 3–4 µm. A brown-rotter on conifer wood, mostly on sun-exposed *Pinus* in sandy plantations.

Dacryonema rufum ⋉ has a reddish stem, smaller, finely granulate fruitbodies and spores with one septum. It may occur as sterile stems.

Widespread, occasional; all year.

Jacob Heilmann-Clausen

Guepiniopsis includes mostly rather stipitate to conical dacrymycetalean fungi.

Guepiniopsis buccina is a short-stemmed, rooting, almost discoid fungus with a wrinkled outer side that has thick-walled, clavate cells, a crenulate margin and golden colours. The hyphae lack clamps; the basidia are shaped like slender tuning forks. Spores 11–14.5(–16) × 4–6 µm, with three thick septa at maturity. On deciduous hosts.

Other *Guepiniopsis* species are found on conifers.

Mainly in southern and western parts of temperate Europe, rare; probably all year.

Seppo Kytöharju

Guepiniopsis alpina is an at first almost clavate, later turbinate, tough, yellow to orange-yellow dacrymycetalean fungus with a hairy outer side. The hyphae have clamps and clavate, thick-walled hairs. The thick-walled, 1–3-septate spores measure 15–17.5(–20) × 5–6 µm. The microconidia are ± cylindrical, 4.5 × 1.5 µm. Occurs on *Pinus* wood.

Also known as *Heterotextus alpinus* ⋉. Similar species of *Dacrymycetales* have wider spores, typically with more septa.

Widespread in the hemiboreal–boreal zones but almost absent from the nemoral zones; mainly in spring.

Guepiniopsis estonica is a cushion- to disc-shaped, golden dacrymycetalean fungus with rather wide, tuning fork-shaped basidia, the hyphae lack clamps. The ellipsoid, thin-walled, up to 9(–10)-septate spores measure 18–24(–26.5) × 8–12 µm. Occurs on conifer wood.

Also known as as *Dacrymyces estonicus*. *Dacrymyces chrysospermus* ◁◁ has narrower spores (4–8 µm). Other similar species of *Dacrymyces* have clamps.

Widespread, rather rare; probably most of the year.

Calocera form clavate or branching, mostly orange-yellow very rubbery fruitbodies. The spores are, at maturity, up to 4-celled and produce microconidia. To ensure that mature spores are studied it is advisable to prepare spore deposits, *e.g.* on a microscope slide.

Thomas Læssøe

Calocera cornea is a slender, yellow, rarely pale yellow, normally unbranched, ± acute *Calocera* that usually has a rather indistinctly delimited, short stem. The basidia are tuning fork-shaped. Lacks clamps. The mature spores develop only one septum (best seen from a spore-deposit) and measure 7.5–9.5 × 3–3.5 µm. Found mainly on deciduous wood, with *Fagus* the dominant host in the nemoral zones.

Calocera glossoides ▷▷ is typically more wrinkled, less slender and develops three septa on the mature spores. *Calocera furcata* ▽ also develops three septa on mature spores but occurs on conifer, rather than deciduous wood.

Widespread and very common; mainly June–March.

Calocera furcata is a rather small, slightly branched, yellow *Calocera* that usually has several ± flattened main branches from the base. The branch tips are ± pointed. The spores are 3-septate (best seen from a spore-deposit) and measure 10–14 × 3–4.5 µm. Occurs on conifer wood, mainly *Pinus*.

Branched forms of *C. cornea* △, which can occasionally occur on conifer wood, are best recognized by the mature spores having only one septum.

Widespread, occasional to common; mainly August–January.

Thomas Læssøe

Calocera glossoides is a rather small, unbranched, ± tapering, typically wrinkled-furrowed, yellow to pale yellow *Calocera* with a well-delimited, short, sometimes paler stem. The allantoid spores have, at maturity, three septa (best seen from a spore-deposit) and measure 10–14 × 3–4.5 µm. Occurs on deciduous wood, mostly on decorticated *Quercus* branches, mainly in sites with a long continuity.

Calocera cornea ◁◁ has fruitbodies with stems that are less distinctly delimited, is usually smoother, and the mature spores only develop one septum.

Widespread in the nemoral–hemiboreal zones, but rare to very rare, except in the UK, where it is apparently more common; all year, but most records from October.

Thomas Læssøe

Calocera pallidospathulata is a ± unbranched, pale yellow *Calocera* with very irregular, compressed and rounded-clavate fruitbodies. The stem is translucent and almost white. The up to 3-septate spores measure 10–15 × 3.5–5 µm. Mainly on conifer wood.

Calocera cornea ◁◁ is yellower and more slender, and the spores are 1-septate at maturity. *Calocera furcata* ◁◁ is typically branched with pointed branch tips and more colourful.

Common in the UK, where probably introduced from North America, but likely to be spreading to other parts of temperate Europe, *e.g.* the Netherlands; all year.

Calocera viscosa is a rubbery, very tough and pliant, large, rooting, orange-yellow to reddish-yellow, two- to four-times-branched *Calocera* that usually has bifurcated tips to the branches. The hyphae lack clamps and the basidia are tuning fork-shaped. The somewhat curved spores either have no septa, or only one, and measure 8.5–10.5 × 3–4.5 µm. Occurs on conifer wood that is usually buried.

The very rubbery texture makes this species easy to identify. Other species of *Calocera* have much smaller fruitbodies with either fewer or no branches. Yellow clavarioids in the genera *Clavulinopsis*, *Ramariopsis* and *Clavaria* (page 1094) have fruitbodies that are much more fragile.

Widespread and common; June–January.

Jelly fungi

This form group includes species that mostly have septate, long, sterigmate basidia and spores that can usually produce a copy of themselves (repeating to form secondary spores). Many species are gelatinous and have basidia imbedded in a protective gel. The sterigmata find their way from within the fruitbody to the surface, where the spores are formed.

The fruitbodies are highly variable in shape and colour, from flat (resupinate), verrucose or spiny, to large, folded, brain-like, clavate, branched or cap-like. One group of jelly fungi does not even produce obvious fruitbodies, but occurs either as parasites inside other fungal fruitbodies or forms almost invisible films on, for example, wood.

As defined above (aberrant basidia and repeating spores), the jelly fungi is a form group that also includes fruitbodies that are not jelly-like, and phylogenetically they usually belong to the subphylum *Pucciniomycotina* and the orders *Auriculariales*, *Cantharellales*, *Sebacinales* and *Tremellales*, all of which are within the subphylum *Agaricomycotina* (page 13).

Jelly fungi are decomposers or parasites on plants or other fungi. A group of species are part of the so-called 'orchid mycorrhiza', where orchids parasitize the fungi involved.

Jelly fungi includes more than 50 genera, with more than 200 species in temperate Europe. In this section, 30 genera are covered by the two wheels, including the most showy species. The majority of the genera not treated in this publication form resupinate, mostly invisible fruitbodies

continued on next page spread . . .

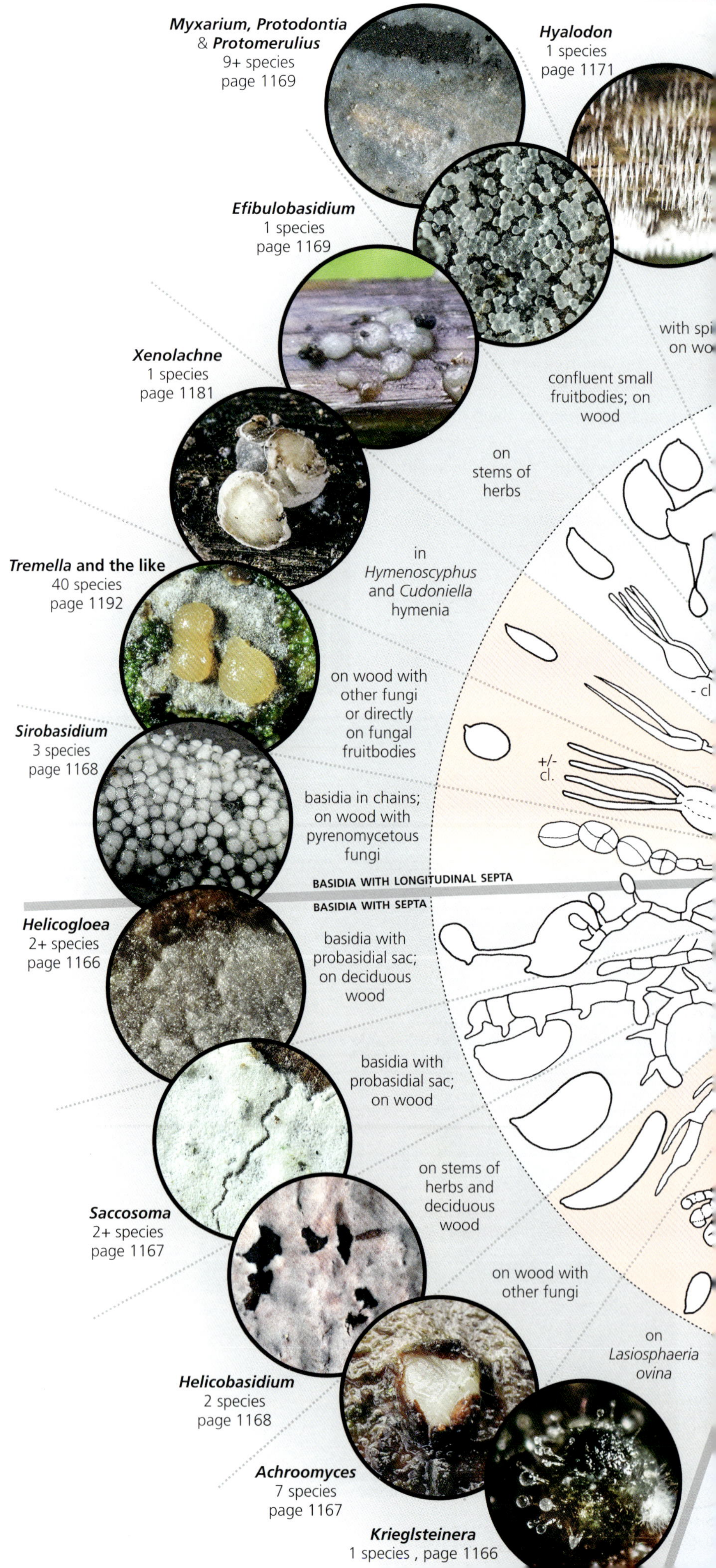

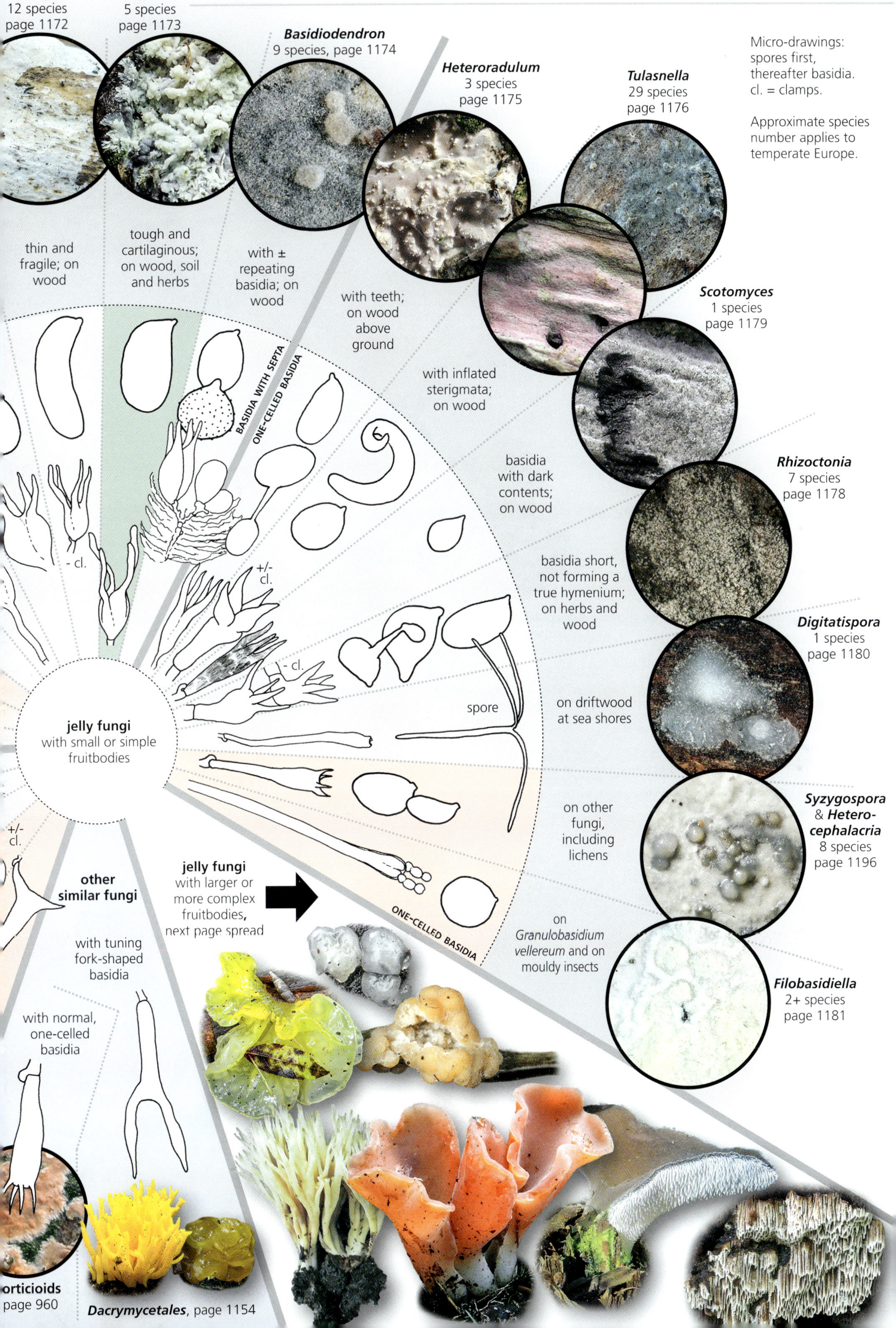
Micro-drawings: spores first, thereafter basidia. cl. = clamps.
Approximate species number applies to temperate Europe.
Exidiopsis 12 species page 1172
Sebacina 5 species page 1173
Basidiodendron 9 species, page 1174
Heteroradulum 3 species page 1175
Tulasnella 29 species page 1176
Scotomyces 1 species page 1179
Rhizoctonia 7 species page 1178
Digitatispora 1 species page 1180
Syzygospora & *Hetero-cephalacria* 8 species page 1196
Filobasidiella 2+ species page 1181
thin and fragile; on wood
tough and cartilaginous; on wood, soil and herbs
with ± repeating basidia; on wood
with teeth; on wood above ground
with inflated sterigmata; on wood
basidia with dark contents; on wood
basidia short, not forming a true hymenium; on herbs and wood
on driftwood at sea shores
on other fungi, including lichens
on *Granulobasidium vellereum* and on mouldy insects
BASIDIA WITH SEPTA
ONE-CELLED BASIDIA
ONE-CELLED BASIDIA
- cl.
+/- cl.
- cl.
spore
+/- cl.
jelly fungi with small or simple fruitbodies
other similar fungi
jelly fungi with larger or more complex fruitbodies, next page spread
with tuning fork-shaped basidia
with normal, one-celled basidia
orticioids page 960
Dacrymycetales, page 1154

Jelly fungi

continued from previous page spread . . .

or are parasites within other fungi or plants, *e.g.* the genera *Bourdotigloea*, *Endoperplexa*, *Hauerslevia*, *Herpobasidium*, *Heteroacanthella*, *Insolibasidium*, *Kriegeria*, *Naohidea*, *Occultifur*, *Oliveonia*, *Phragmoxenidium*, *Pseudostypella*, *Renatobasidium*, *Serendipita*, *Spiculogloea* and *Trimorphomyces*.

Other similar fungi:
– species of *Dacrymycetales* are also gelatinous but are mostly yellow and have characteristic, tuning fork-shaped basidia. The spores are mostly septate and do not repeat. However, conidia are often produced directly from the spores (page 1154).
– resupinate jelly fungi can be very difficult to tell apart from corticioid fungi; those, however, have one-celled basidia and spores that do not repeat (page 960).
– *Aporpium* may be confused with resupinate polypores (page 932).
– *Pseudohydnum* may be confused with hydnoids, but hydnoids are not gelatinous (page 1054).
– *Tremellodendropsis* and *Eocronartium* may be confused with clavarioids (page 1092).
– some *Ascomycota*, *e.g. Ascotremella*, *Ombrophila* and *Bulgaria*, superficially recall jelly fungi, but they form spores in asci (page 1364).

Further reading: 50, 70, 100, 142, 163, 176, 189, 190, 192, 222, 268, 269, 270, 271, 306, 307.

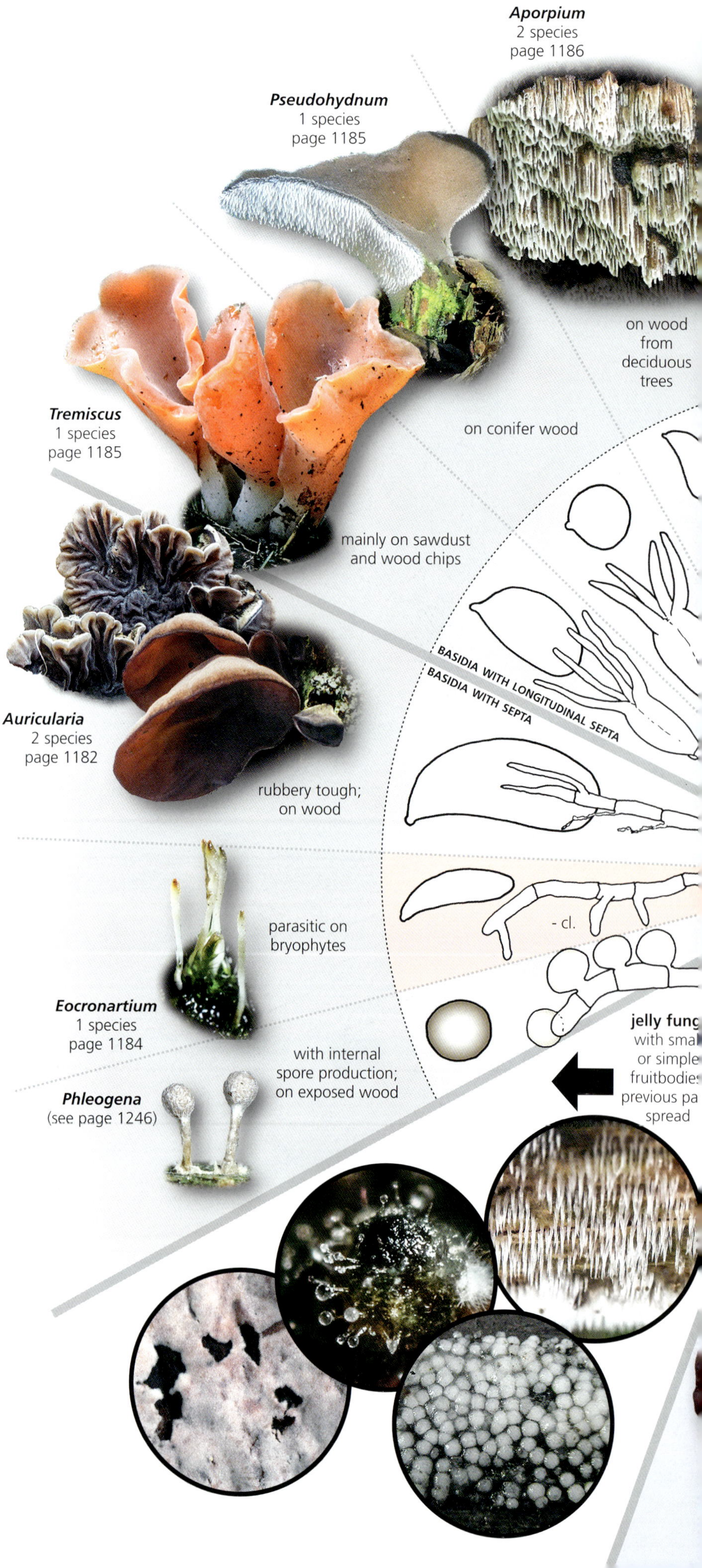

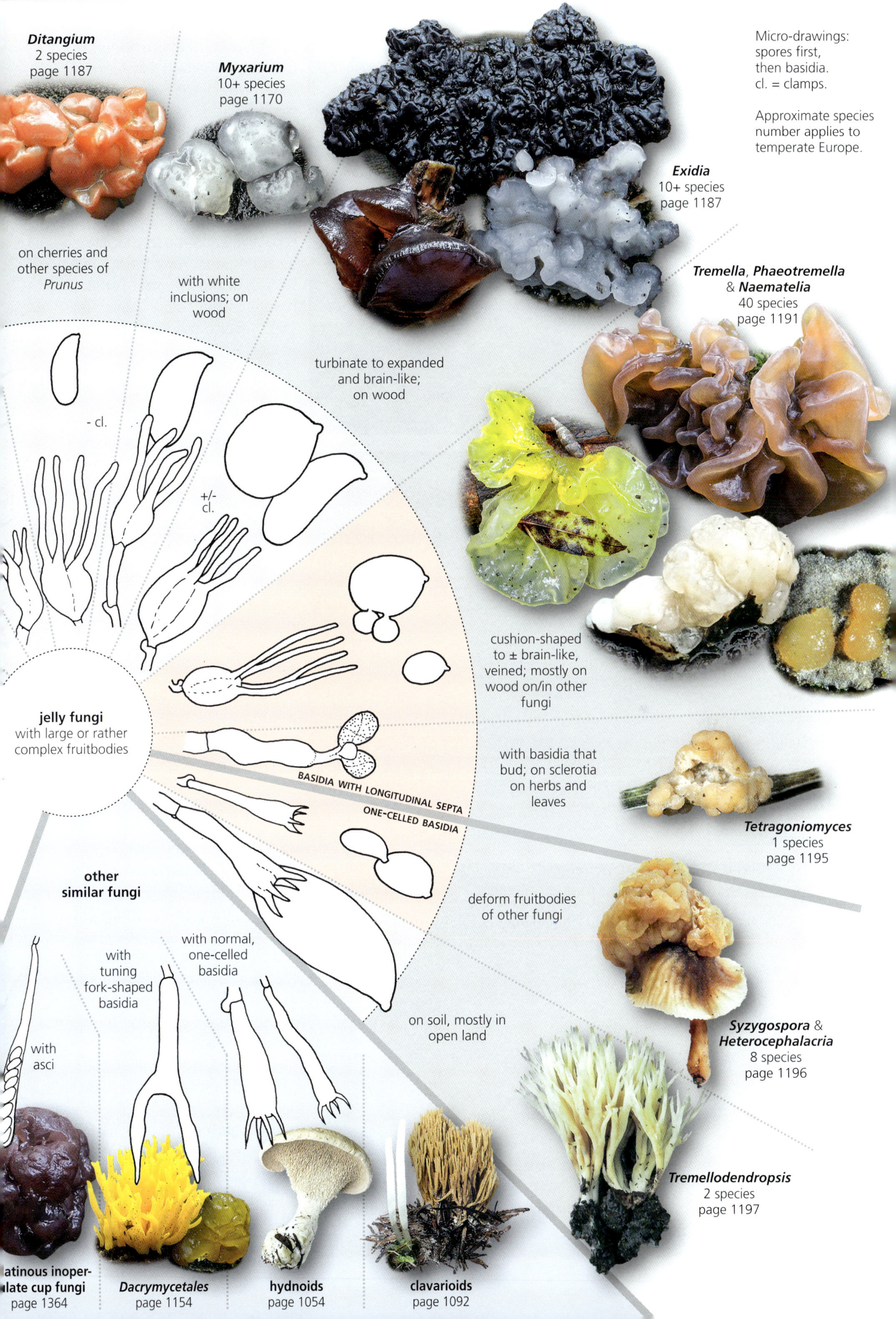
Micro-drawings:
spores first,
then basidia.
cl. = clamps.
Approximate species number applies to temperate Europe.
Ditangium
2 species
page 1187
on cherries and other species of Prunus
Myxarium
10+ species
page 1170
with white inclusions; on wood
turbinate to expanded and brain-like; on wood
Exidia
10+ species
page 1187
Tremella, Phaeotremella & Naematelia
40 species
page 1191
cushion-shaped to ± brain-like, veined; mostly on wood on/in other fungi
- cl.
+/- cl.
jelly fungi
with large or rather complex fruitbodies
BASIDIA WITH LONGITUDINAL SEPTA
ONE-CELLED BASIDIA
with basidia that bud; on sclerotia on herbs and leaves
Tetragoniomyces
1 species
page 1195
deform fruitbodies of other fungi
Syzygospora & Heterocephalacria
8 species
page 1196
on soil, mostly in open land
Tremellodendropsis
2 species
page 1197
other similar fungi
with asci
with tuning fork-shaped basidia
with normal, one-celled basidia
atinous inoper-
late cup fungi
page 1364
Dacrymycetales
page 1154
hydnoids
page 1054
clavarioids
page 1092

Krieglsteinera lasiosphaeriae is a highly aberrant jelly fungus that forms clusters of transversely septate basidia on the tips of thick-walled, colourless, translucent hyphae that have a clamp at a median septum. The budding spores measure 4.5–8 × 2–3.5 µm and end up in a slimy head. Spore dispersal is passive and probably facilitated by arthropods. Parasitic on the hairy surface of the pyrenomycetous fungus *Lasiosphaeria ovina* (page 1592), which occurs on deciduous wood.

A distinctive, although cryptic fungus that should cause no identification problems. The hairs of this parasite are projecting and shiny in comparison to the non-shiny, felty, white hairs of the host.

Widespread and probably rather common, but easily overlooked; April–November, perhaps most frequent in early summer.

***Helicogloea lagerheimii* s.l.** is a film-like, gelatinous fungus with large, allantoid, transversely septate basidia that have a basal, sac-like cell (a probasidium). The hyphae lack clamps. The almond-shaped, ellipsoid to loaf-shaped, one-celled, repeating spores measure 10–15 × 5–8 µm. Occurs on rotten deciduous wood.

Described from Ecuador, and belongs to a species complex that is poorly understood, with species such as *Bourdotigloea vestita* ⋉ having more swollen hyphae. *Saccosoma farinaceum* ▷ forms thicker, cottony fruitbodies; species of *Achroomyces* ▷ do not have a sac-like probasidium.

Widespread and probably rather common; all year.

Achroomyces disciformis is a firm, when dry cartilaginous jelly fungus with ± discoid, erumpent fruitbodies. Although the fruitbodies are almost invisible in dry weather, they can be made visible by placing infected branches of *Tilia* in a bucket of water. The basidia are 2–4-celled and have transverse septa. The ± allantoid, hyaline, one-celled, repeating spores measure 27–36 × 6–7.5 µm. Occurs on thin, rather fresh, sometimes still attached, dead *Tilia* branches, perhaps associated with pyrenomycetous fungi.

A distinctive species that is unlikely to be misidentified. *Myxarium nucleatum* ▷▷ and *M. hyalinum* ⋉ are common on the same substrate but are not erumpent; they have longitudinally divided basidia.

Widespread in the nemoral–hemiboreal zones, but often overlooked; all year.

Saccosoma farinaceum is a very cottony white jelly fungus that could be mistaken for sterile mycelium, but consists of numerous sinuous, transversely divided, 4-celled basidia, which have a huge sac-like appendage at the base (a probasidium) and hyphae with clamps. The repeating spores measure 11.5–20.5 × 8–14.5 (–15.8) µm. Mostly on conifer wood.

Helicogloea lagerheimii ◁ and species of *Bourdotigloea* ⋉ mainly produce thinner, gelatinous fruitbodies; species of *Bourdotigloea* also often have cystidia.

Widespread, occasional but easily overlooked; probably all year.

Thomas Læssøe

Thomas Læssøe

Helicobasidium purpureum is a strange, tough and thread-like, purple jelly fungus that usually occurs as sterile, felty mycelium. Only occasionally can the helical, transversely septate, 4-celled basidia be found within the felt. The hyphae are rather thick-walled, up to 8 µm wide, and lack clamps. The one-celled, repeating spores measure 8–14 × 4–8 µm. Found mostly on herbaceous stems, including living ones, but may also occur on the underside of rotten wood, mainly on rich soils. Also known as a parasite of rust fungi (*Tuberculina* ⋉), and may form an asexual state with globose conidia.

Helicobasidium longisporum ⋉ has spores that measure 11–32 × 3.5–7 µm. Sterile material cannot be identified with certainty.

Widespread, but rarely reported; most of the year.

Margaretha Liebmann

Sirobasidium albidum is a whitish jelly fungus with confluent, small, milky-white fruitbodies and 4-celled, longitudinally divided, ovoid to spherical basidia, which are formed in chains of up to 8. The upper basidia are unclamped, easily loosens, and act as progagules. Sexual spores are rarely seen. Occurs on stromata of pyrenomycetous fungi of the genus *Eutypa* (page 1569), *e.g.* on *Acer*.

Other species of *Sirobasidium* either have differently shaped basidia, shorter chains of basidia, or red-brown colours.

Very rare or overlooked, only recorded in the nemoral zones; probably mainly in late winter and spring.

Efibulobasidium albescens forms small, translucent cushions that consist of clampless hyphae, 2–4-spored longitudinally divided basidia and ± branched hyphidia. The repeating spores are ellipsoid to curved drop-shaped and measure 15–21 (–25) × 5.5–7.5 µm. Occurs on the stems of herbs.

Easily identified microscopically, but macroscopically a very anonymous fungus. The asexual *Chaetospermum chaetosporum* (inset image) occurs on the same substrates and is possibly conspecific – the asexual name has priority over the sexual name.

Widespread in the nemoral–hemiboreal zones; mainly August–October.

Myxarium*, *Protodontia *and* ***Protomerulius*** are genera of small, ± translucent jelly fungi that usually form aggregations of grain-like, cushion-shaped to spiny fruitbodies that often coalesce into larger, ± resupinate structures. They have longitudinally divided basidia with a short to long stem, hyphidia and repeating spores. Cystidia may be present.

Myxarium crystallinum produces crowded, shiny grain-like fruitbodies that only coalesce slightly, and which are virtually invisible in dry conditions. Lacks cystidia, but has hyphidia and basidia that are divided once longitudinally to become 2-spored. The repeating, almost spherical to broadly ellipsoid spores measure 5–9.5 × 4.5–6.5 µm. Occurs on rotten wood, usually deciduous.

Ofella glaira ⋉ and other species of *Myxarium* have 4-spored basidia.

Widespread, mainly nemoral, occasional; mainly May–November.

Thomas Læssøe/Jens H. Petersen

Protomerulius dubius is a resupinate, ± translucent jelly fungus with a minutely hairy surface caused by the numerous, projecting, thick-walled, up to 300 µm long, cylindrical cystidia. Internally, the fruitbodies have prominent star-shaped crystals and longitudinally divided, 2- or 4-spored basidia. The very variable, repeating spores measure (3.5–)4–8.5(–9.5) × (2.5–)3–4.5 µm. Found mostly in old growth forests on rotten wood from deciduous trees.

The internal crystals and distinctive cystidia make this species easy to identify microscopically.

Widespread, occasional; probably most of the year.

Myxarium podlachicum quickly coalesces to form resupinate, ± translucent, gelatinous fruitbodies with ± obvious, whitish internal grains (inset image). It has pointed, scattered, up to 60 µm long cystidia in the hymenium and 4-spored basidia. The repeating spores measure 6–8.5 × 3–4 µm. Occurs on rotten deciduous wood.

Similar species of *Myxarium* lack pointed cystidia, but *Protomerulius dubius* △, for example, has cylindrical cystidia. *Myxarium nucleatum* ▽ and *M. hyalinum* ⋉ have white inclusions but the fruitbodies are thicker and the spores larger.

Widespread and common, at least in the nemoral zones; all year.

Myxarium nucleatum is a rather large cushion-shaped to somewhat confluent, almost translucent to milky jelly fungus with internal, white crystalline grains that are very visible in dried material. Has clamps and hyphidia. The longitudinally divided basidia have a clamped septum above a stem hypha. Spores ± allantoid, repeating, measure 10–15 × 3.5–5 µm. Occurs on fallen and attached dead branches of deciduous trees.

Myxarium hyalinum ⋉ is pale brownish. *Exidia thuretiana* ▷▷ and whitish species of *Tremella* ▷▷ do not have white inclusions and usually have different spore shapes.

Widespread and very common; at its best during mild winter periods.

Protodontia subgelatinosa is a jelly fungus with short spines that are held together by a thin, coherent layer that also is covered with hymenium. Hyphidia are present, but no cystidia. The repeating spores measure 4–7.5 × 3.5–5.5 µm. Usually occurs on deciduous wood, mainly in forests with a long continuity and plenty of dead wood.

Myxarium legonii ⋉ has narrower spores, 4–6.5 × 2.5–3 µm. *Hyalodon piceicola* ▽ has unconnected spines up to 3 mm long and lacks hyphidia.

Widespread, occasional; mainly August–November.

Hyalodon piceicola is a spiny jelly fungus with up to 3 mm long, hanging, pointed, occasionally slightly branching, white to yellowish spines that are crowded but do not have a connecting tissue. When dried-out, the spines become firm and horn-like in texture. Lacks cystidia and hyphidia. The basidia are divided longitudinally and 4-spored. The repeating spores measure 4–6 × 3.5–4.5 µm. Occurs on rotten conifer wood.

Protodontia fascicularis ⋉ should have spines in clusters and somewhat larger spores. Other species of *Protodontia* △ usually have basally connected, shorter spines and branching hyphidia. *Kavinia* (page 1040), *Mucronella* (page 1096) and *Dentipellis* (page 1042) have similar fruitbodies but one-celled basidia and ± amyloid, non-repeating spores.

Found mainly in the boreal zone, where occasional; mostly in autumn.

Thomas Læssøe

Exidiopsis includes thin, mostly greasy-gelatinous, resupinate jelly fungi with hyphidia, longitudinally divided basidia and repeating spores.

Exidiopsis effusa is a thin, whitish, pale lilac or somewhat greyish, greasy-gelatinous jelly fungus which is so fragile that even a fingerprint will expose the substrate (upper image). It has longitudinally divided, 4-celled basidia and clamps, but lacks cystidia. Spores ± allantoid, repeating, measure 10.5–17 × 3.5–5 µm. Found mostly on decorticated, fallen, deciduous branches. During periods of relatively high humidity and temperatures around zero, it may form hair-like ice crystals, a so-called goblin's beard.

Exidiopsis calcea ▷ is similar microscopically but has dry, chalk-white fruitbodies and is found mostly on attached branches. *Exidiopsis grisea* ⋉ forms darker fruitbodies on conifer wood.

Widespread and common; mainly during mild winter periods.

Jan Vesterholt

Exidiopsis calcea is a ± chalk-white, almost dry jelly fungus. The hyphae have clamps. Lacks cystidia. The basidia are divided longitudinally and are 4-spored. The ± allantoid, repeating spores measure 12.5–16.5 × 5–6.5 µm. Occurs on both deciduous and conifer wood, mostly on attached, dead branches.

Probably constitutes a species complex. *Exidiopsis peritricha* ⋉ is more brownish and is cracking upon drying. *Exidiopsis effusa* ◁ has a more gelatinous flesh. The somewhat similar species of *Vuilleminia* (page 1025) typically develop under loosening bark and have one-celled, very large basidia with short, curved sterigmata.

Widespread and rather common; all year.

Thomas Læssøe

Sebacina incrustans is a very firm, tough-cartilaginous and thick-fleshed jelly fungus that produces fruitbodies from soil level upwards onto whatever will support it, including living herbs. The hyphae lack clamps; the basidia are longitudinally divided and 4-celled. Lacks skeletal hyphae and resting spores in the flesh. The somewhat bean-shaped, repeating spores measure 10–12.5 × 6.5–7 µm. Occurs on soil, litter, *etc.* in deciduous and coniferous forests; forms ectomycorrhiza.

There are many poorly known and mostly undescribed species in this genus, and it is difficult to identify any species of *Sebacina* with certainty. However, *S. dimitica* ⋉ has skeletal hyphae in the flesh, and *S. helvelloides* ⋉ and *S. epigaea* ⋉ have star-shaped resting spores (chlamydospores) in the flesh.

Widespread and rather common; mainly August–October.

Basidiodendron is a genus of jelly fungi with longitudinally divided basidia, and where new basidia are often formed on top of the old ones, creating stacks of collapsed basidia with the active ones on top. Many species have cystidia and ± globose, sometimes ornamented spores. All are decomposers of wood.

Basidiodendron trachysporum **s.l.** is a thin, somewhat chalky, greyish corticioid jelly fungus with stacks of collapsed basidia. The hyphae have clamps. Has gloeocystidia that measure 40–50 × 4–8 µm. The spherical, finely echinulate spores measure 6.5–9 µm. Occurs on rotten conifer and deciduous wood.

Macroscopically recalls a species of *Botryobasidium* (page 992). The closely related *Basidiodendron eyrei* ▽ and *B. diminutum* ⋉ both have smooth, ± spherical spores.

Widespread and fairly common; most of the year, although mainly September–November.

Jens H. Petersen/ Thomas Læssøe

Basidiodendron eyrei is a rather thin, pale brownish, almost smooth corticioid jelly fungus. It has gloeocystidia, and the basidia are in dense stacks. The smooth spores are almost spherical but tend to to be wider than long, measuring 4.3–5 × 4.6–5 µm. Occurs on rotten wood, usually deciduous.

Basidiodendron diminutum ⋉ has similar smooth spores but they are longer than wide, measuring 5–6 × 3.5–5 µm. *Basidiodendron trachysporum* s.l. △ has finely echinulate spores, and *B. radians* ⋉ and *B. rimosum* ⋉ have larger spores.

Widespread, probably rather common, but infrequently recorded; most of the year.

Heteroradulum deglubens is a resupinate, firm-fleshed jelly fungus with a well-defined, ± free margin and a ± spiny-warty hymenophore. The colour ranges from pale with grey-rose tinges to almost orange-brown. The hyphae are thick-walled and have indistinct clamps. The basidia are 4-spored, only partly divided longitudinally from above, and have thick, finger-like sterigmata. The repeating spores measure 15–20 × 5.5–8.5 µm. Occurs on deciduous wood, often on attached *Fraxinus* branches high up in the canopy.

Also known as *Eichleriella deglubens*. The very similar *H. kmetii* ⋉ becomes capped with age, and can also be told by the presence of brownish skeletal hyphae. *Heteroradulum deglubens* also recalls various corticioids, but all have non-divided basidia and non-repeating spores.

Widespread in the nemoral–hemiboreal zones, occasional; all year.

Thomas Læssøe

Jens Maarbjerg

Tulasnella is a genus of jelly fungi characterized by swollen, almost balloon-shaped sterigmata and one-celled basidia. The spores repeat, and occur in a very wide range of shapes. Many species produce either no fruitbody or a very cryptic fruitbody, often within other fungal fruitbodies. Various asexual states are known and a number of species are involved in so-called 'orchid mycorrhiza'.

Tulasnella violea is a magenta, thin, smooth, corticioid jelly fungus that lacks cystidia. The rather wide hyphae (2–5 µm) lack clamps, and the almost spherical, repeating spores measure 5.5–8(–10) × 5–6.5(–8.5) µm. Occurs on decomposed wood of deciduous trees, rarely conifer wood.

Other clampless *Tulasnella* species usually have inconspicuous or completely invisible fruitbodies.

Tulasnella subglobispora ⋉ has clamps and greyish fruitbodies.

Widespread and rather common; most of the year.

Tulasnella eichleriana is a very thin, greyish corticioid jelly fungus with clampless hyphae and small, almost spherical, repeating spores that measure 3–4.5(–6) × 2.5–3(–4) µm. Occurs on rotten wood and litter in deciduous and coniferous forests.

Tulasnella cystidiophora ▷ also lacks clamps and has the same spore shape, but has gloeocystidia. *Tulasnella violea* △ and the whitish *T. albida* ⋉ have larger spores.

Widespread and probably common; all year.

Tulasnella cystidiophora is a thin, somewhat mould-like or corticioid, pale grey jelly fungus with clampless hyphae and gloeocystidia up to 100 µm long. The spherical spores measure 4–6(–7) µm. Occurs on decomposed wood.

Similar species of *Tulasnella* with spherical spores either have clamps (*e.g. T. hyalina* ⋉) or no cystidia (*T. eichleriana* ◁ and others).

Mainly nemoral, infrequently recorded but perhaps overlooked; mainly September–March.

Tulasnella tomaculum is a very inconspicuous, greyish jelly fungus that lacks clamps. The ± allantoid, repeating spores measure 4–7 × 2.5–3 µm. Occurs on rotting wood in deciduous forests.

Tulasnella danica ⋉ has spores that are more distinctly allantoid and larger, measuring 10–15(–17.5) × 3–4 µm. *Tulasnella permacra* ⋉, *T. allantospora* ▽ and *T. curvispora* ⋉ have clamps. A species of either *Orbilia* or *Hyalorbilia* (pages 1464–1467) is also present in the image.

Widespread in the nemoral–hemiboreal zones, probably rather common; all year.

Tulasnella allantospora is a very thin, greyish-white jelly fungus. Microscopically characterized by somewhat narrow, clamped hyphae and rather small, allantoid spores that measure 6.5–9(–10) × 2.5–3 µm. Occurs on rotten wood, predominantly deciduous.

Tulasnella curvispora ⋉ has longer spores measuring 10–14(–18) × 2.5–4 µm. Other similar species of *Tulasnella* either have different spores or lack clamps.

Widespread, but infrequently recorded; probably all year.

Thomas Læssøe

Jens H. Petersen/Thomas Læssøe

Rhizoctonia is a genus of loosely constructed, ± corticioid jelly fungi with one-celled basidia that have two to eight, large, conical sterigmata. The spores repeat and several layers of basidia may develop. Decomposers, or parasites on wood, herbs and litter. Also involved in so-called 'orchid mycorrhiza'. Many of the species were formerly accepted in the genera *Thanatephorus* and *Ceratobasidium*.

Rhizoctonia solani is a thin, pale grey, mould-like jelly fungus. Hyphae lack clamps and up to 12 µm wide; basidia one-celled, barrel-shaped with long, projecting sterigmata. The repeating, ellipsoid spores measure 8–12 × 5–6 µm. It occurs mostly in an asexual state ('*Moniliopsis solani*'). Parasitic on dicot herbs, *e.g.* potato.

Also known as *Thanatephorus cucumeris*. There are several similar species with other spore shapes, *e.g. R. fusispora* ▽, or that have larger spores, *e.g. R. amygdalispora*.

Widespread, but rarely reported, probably common; mainly in summer.

Rhizoctonia fusispora is a pale, mould-like jelly fungus with one-celled basidia. The repeating, fusiform to lemon-shaped spores measure 8.5–15(–18) × 5–10 µm. Occurs on soil, litter, rotten wood, *etc.*, in deciduous and coniferous forests.

Rhizoctonia solani △ has ellipsoid spores, but otherwise is very similar. Various other *Rhizoctonia* species also differ by their spore shape.

Widespread and probably rather common; April–December.

Rhizoctonia ramicola is a thin, pale, mould-like jelly fungus with barrel-shaped, 4-spored basidia that have very long sterigmata. The hyphae are rather wide and lack clamps. The repeating, somewhat pointed spores measure (6–)7–9(–11.5) × (3–)4–5 (–6.5) µm. Occurs on soil and litter in deciduous and coniferous forests.

Also known as *Ceratobasidium cornigerum*. *Rhizoctonia stridii* ⋉ has spores that measure 6–7.5 × 3–3.5 µm, and *R. anceps* ⋉ has ± ellipsoid spores measuring (7–)9–12.5 × 5–6 µm.

Widespread and presumably common; most of the year.

Jan Vesterholt

Scotomyces subviolaceus is a rather dark, ± greyish-violet, resupinate jelly fungus with a greyish surface. The basidia are one-celled, have long, sinuous sterigmata and a large clamp at the base. With age, the wide hyphae and basidia become filled with a violet-grey pigment. The repeating, one-celled, smooth spores measure 6–8 × 4–5.5 µm. Mostly on ± decomposed conifer wood.

A distinctive species that is unlikely to be misidentified.

Widespread but rarely reported; probably all year.

Thomas Læssøe

Jørgen Koch

Teppo Rämä

Teppo Rämä

Digitatispora marina is a resupinate to cushion-shaped, translucent, whitish jelly fungus with very unusual four-armed spores that are loosened from the basidia by water movement. The spores are formed on one-celled basidia that lack sterigmata. The fungus fruits on marine driftwood in the upper tidal zone.

A distinctive but rather obscure fungus that should be easy to identify microscopically. Only a few other basidiomycotes are adapted to life in salty water. One example is *Nia vibrissa* (page 1234), but that species produces internal spores in small, hairy balls. Ascomycotes are, however, well represented in the marine environment. Both freshwater and marine fungi often produce spores with long appendages, which enable the spores to float and then settle better on their appropiate substrates.

Distribution and frequency not well understood, but probably widespread; likely all year.

Xenolachne longicornis is a jelly fungus that does not form proper fruitbodies, but instead parasitizes the fruitbodies of inoperculate discomycetes, *e.g. Cudoniella* and *Hymenoscyphus* (pages 1418 & 1421). The 2-spored basidia have a longitudinal septum, long sterigmata and clamped hyphae that also produce haustoria. The repeating spores measure 8.5–12.5(–13.5) × 3–3.5(–4) µm.

The combination of habitat and distinctive basidia is unique.

Widespread, but rarely reported; mainly April–October.

Thomas Læssøe

Filobasidiella lutea forms a pale yellowish, felty-hairy covering on the white to pink corticioid *Granulobasidium vellereum* ⋉. The basidia are very long and narrow, (200–)240–290(–310) × 12–13 µm, and have a clamp at the base; they are also somewhat widened at the tip, where there are short, longitudinal, incomplete septa (basidia can be seen in profile on top of the host hymenium in the lower image). Spores are sessile, spherical and measure (5.5–)6–6.5(–7) µm – developed in four chains.

A very distinctive fungus.

Widespread in the nemoral zones, but rarely recorded; mainly September–March.

Jens Maarbjerg

Auricularia form extremely cartilaginous, effused-reflexed to ear-shaped fruitbodies on wood. The basidia have transverse septa. The cartilaginous texture makes it almost impossible to see the basidia in their entirety with a microscope.

• ***Auricularia auricula-judae*** is a cartilaginous, ± ear-shaped, brown, rarely completely white jelly fungus with a minutely downy upper side and a smooth to veined, somewhat paler lower (hymenial) side. It dries to a horn-like texture but completely regains its shape when re-wetted. Basidia are large with three transverse septa. The one-celled spores measure 13–15 × 5–6 µm. Occurs on deciduous wood, both fallen, decorticated trunks and on bark of standing trees, particularly *Sambucus nigra*.

A very distinctive species that is unlikely to be misidentified. Very young fruitbodies may recall species of *Peziza* (page 1313) but those have very fragile flesh and the hymenium points upwards.

Widespread and common in the nemoral zones; all year.

Thomas Læssøe

Auricularia mesenterica is an effused-reflexed, rather thick-fleshed, rubbery to cartilaginous jelly fungus with a very hairy-felty, rather pale, ± zoned upper side and a greyish, veined lower (hymenial) side. The one-celled, hyaline spores measure 15–18 × 5–6 µm. Occurs on fallen trunks of deciduous trees, mostly *Ulmus*.

The upper side may recall a species of *Stereum* (page 970) or *Trametes* (page 918), but the combination of texture and a veined hymenophore make it unmistakable.

Widespread in the nemoral–hemiboreal zone, but probably declining due to lack of suitable *Ulmus* trunks; all year.

Thomas Læssøe

Eocronartium muscicola is a small, slender, tough, white, tapering clavarioid jelly fungus, which turns yellowish when drying out. The hyphae lack clamps and the basidia are 4-spored with three transverse septa. The ± banana-shaped, repeating spores measure 22–25 × 5–6.5 µm. It is a parasite on bryophytes in forests and marshes, and is known from at least 21 moss hosts, including *Rhytidiadelphus triquetrus*.

All other small, white clavarioids (page 1116) are more fragile, have one-celled basidia and are not parasitic on living bryophytes. *Eocronartium muscicola* is related to rust fungi (page 1198), despite their lack of proper fruitbodies.

Widespread, occasional in northern parts of temperate Europe, but rare in the nemoral zones; July–October.

Tremiscus helvelloides is a ± funnel-shaped, pink to orange, rather firm and thick-fleshed jelly fungus with a pale, short stem and an initially smooth, but later wrinkled hymenophore. The basidia are longitudinally divided and 2–4-spored. The ± ellipsoid, repeating spores measure 15–17.5 × 11–12.5 µm. Occurs on conifer wood, mostly sawdust, wood chips or ± buried sticks in damp, typically calcareous sites, *e.g.* road sides in forests. It is probably native to the Alpine region or boreal areas, but has found its way to the nemoral zones, maybe via imported timber.

A very distinctive species that is unlikely to be misidentified. The generic name *Guepinia* is also in use, but this is an older name for a genus of plants in the *Brassicaceae*.

Widespread, rather rare, but apparently spreading; mainly August–November.

Jørgen Christiansen/Jens H. Petersen

Pseudohydnum gelatinosum is a very distinctive, fleshy, highly gelatinous jelly fungus with spatula- or tongue-shaped, short-stemmed fruitbodies that have a minutely hairy, grey-brown upper side and a lower hymenophoral side with pale, crowded spines. The basidia are longitudinally divided and 2–4-spored. The repeating spores measure 12.5–16 × 8.5–12.5 µm. Typically occurs on decomposed conifer stumps.

A very distinctive fungus that is unlikely to be misidentified. All capped hydnoids (page 1054) are non-gelatinous.

Widespread and common; July–March.

Aporpium is a genus of grey-brown, polypore-like jelly fungi with longitudinally divided basidia.

Aporpium macroporum is a polypore-like jelly fungus with (1–)2–3(–4) pores per mm. The flesh is dimitic and the spores measure 5–7 × 2.7–3.7 µm; Qav. 1.9. Occurs on deciduous wood, mostly *Populus tremula*.

Aporpium canescens ▽ is very similar but has smaller pores and narrower spores (Qav. 2.3). *Aporpium macroporum* may be confused with various resupinate polypores, *e.g. Niemelaea* and *Ceriporiopsis* (page 949), but these have one-celled basidia.

Has a very eastern distribution within temperate Europe, rather rare; mainly autumn.

Thomas Læssøe

Aporpium canescens is a poroid jelly fungus with rather fine pores, 4–6 per mm, and a pale, somewhat woolly margin. The tube layer discolours brownish when bruised. It has a dimitic hyphal system, and the flesh has clamps. The spores measure 5.1–6.9 × 2.2–2.9 µm; Qav. 2.3. Occurs on wood of deciduous trees.

The very similar *A. macroporum* △ has wider spores (Qav. 1.9) and fewer ((1–)2–3(–4)) pores per mm. *Aporpium caryae* ⋉ is a very similar North American species, the name for which was formerly used erroneously for European material.

Widespread, but rarely recorded; mainly autumn.

Ditangium cerasi is a ± turbinate to somewhat confluent, pale orange to orange-rose jelly fungus that occurs together with its urn-shaped asexual state (inset image). The hyphae lack clamps and have longitudinally divided basidia. The cylindrical to almost allantoid, repeating spores measure 8.5–11 × 2.5–3.5 µm. Conidia measure 6–7.5 × 1–2 µm. Occurs on bark of deciduous trees, possibly exclusively on *Prunus* species.

Ditangium incarnatum ⋉, on *Salicaceae*, has smaller fruitbodies and shorter spores. Other somewhat similar species of *Exidia* ▽ and *Dacrymyces s.l.* (page 1154) do not have a prominent asexual state and usually have clamped hyphae.

Widespread in the taiga towards the north and the east of temperate Europe, but absent from the nemoral zones; most of the year.

Karel Tejkal/Sava Krstic

Exidia form turbinate or cushion-shaped, ± brain-like folded fruitbodies with longitudinally divided basidia. All are wood decayers. Species of *Phaeotremella* ▷▷ and *Tremella* ▷▷ are similar but are typically more folded and are parasites. *Ascotremella faginea* (page 1366) has asci.

Exidia thuretiana is an often ± disc-shaped, ± veined jelly fungus with blue-grey to yellow-grey tinges and a thin margin. The ± allantoid, repeating spores measure 15–20 × 4.5–6.5 µm. Occurs on deciduous wood, often newly fallen *Fagus* branches, but also on other deciduous trees.

The less firm, smaller and more translucent *Myxarium nucleatum* ◁◁ and the brownish *M. hyalinum* ⋉ have white, firm, crystalline bodies inside the fruitbodies.

Common and widespread, becoming scarcer towards the north; mostly in winter.

Thomas Læssøe

Exidia nigricans is a ± brain-like, typically confluent, rather thick jelly fungus with small warts on the hymenium. The colour is translucently grey-brown to almost tar-black. The basidia are divided longitudinally. The allantoid, repeating spores measure 11–15 × 3.5–4.5 µm. Occurs mostly on wood and bark from a number of deciduous trees, rarely on conifer wood, typically on rather fresh wood, both on standing trees and on dead attached branches and fallen wood.

Also known as *E. plana* and, confusingly, as *E. glandulosa* ▷. However, *E. glandulosa* is more turbinate, has a distinctly, felty upper side and very obvious warts on the hymenium; it mainly occurs on *Quercus*. *Exidia pithya* ▽ is thinner and only occurs on *Picea* and other conifers. The form *E. nigricans* f. *lobata* ⋉ has larger spores that measure 20–22 × 4–5 µm.

Very common and widespread; all year but in best condition in late autumn and during mild winter periods.

Exidia pithya is a thin, confluent, irregularly rugose, tar-black jelly fungus that in dry weather dries to an almost paper-thin covering. The allantoid, repeating spores measure 11–14 × 4.5–5 µm. Occurs mostly on fallen *Picea* trunks, but also on other conifers.

Exidia nigricans △ is very similar but thicker and more veined; it occurs predominantly on deciduous wood.

Widespread and very common; all year, but at its best in late autumn and during mild winter periods.

Exidia glandulosa is a very dark, thick- and tough-fleshed, ± turbinate jelly fungus with a hairy, dark brown to brownish-black upper side and a more grey-black, finely warty lower (hymenial) side. The basidia are 4-spored and divided longitudinally. The allantoid, repeating spores measure 12.5–16 × 3.5–5.5 µm. Grows in crowded groups on bark of attached or newly fallen *Quercus* branches, occasionally on other hosts, *e.g. Corylus*.

Also known as *E. truncata*. *Exidia nigricans* ◁ does not have a hairy upper side and has smaller warts on the hymenium; it also forms more confluent fruitbodies. See also the ascomycote *Bulgaria inquinans* (page 1371).

Common and widespread; all year but at its best in the cooler months.

Exidia candida is a rather large, brain-like, veined, ± marginally fimbriate, rather pale yellow-brown to red-brown jelly fungus. The ± allantoid, repeating spores measure 10–14 × 3–4.5 µm. Occurs mostly on recently dead *Betula* trunks in damp forests.

Formerly known as *E. cartilaginea*. *Exidia thuretiana* ◁◁ is less coloured and typically smaller and more gelatinous. *Exidia repanda* ▷▷ also occurs on *Betula* but is more cushion-shaped, a more pure brown and has smaller fruitbodies with a smooth margin.

Widespread and fairly common in the hemiboreal–boreal zones, rare in the nemoral zones; probably all year.

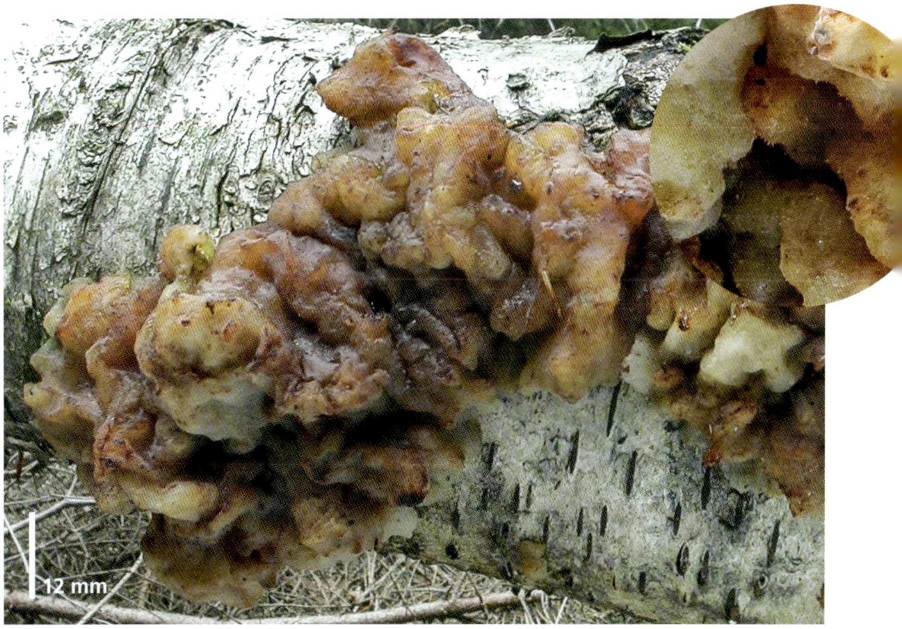

Thomas Læssøe

Exidia repanda is a ± cushion-shaped to confluent, red-brown jelly fungus with a rounded, smooth margin. The ± allantoid, repeating spores measure 12–14 × 2.5–3.5 µm. Occurs almost exclusively on recently dead *Betula* trunks and branches.

Exidia candida ⊲⊲ also occurs on *Betula* but has paler fruitbodies with a fimbriate margin. *Exidia saccharina* ▽ occurs on conifer wood; *E. recisa* ▽ is more turbinate and occurs on attached *Salix* branches.

Widespread and rather common in the hemiboreal–boreal zones, but very rare in the nemoral zones; probably all year, but at its best during mild winter periods.

Exidia recisa is a rather small, ± turbinate, mostly hanging, brown, finely rugose jelly fungus. The allantoid, repeating spores measure 11–15 × 2.5–3.5 µm. Occurs on attached dead *Salix* branches, mostly in wet marshes, rarely on other deciduous hosts.

Exidia brunneola ⋉ occurs on *Populus* and has larger spores that measure 16–24 × 3–5 µm.

Very common and widespread; all year, but rarely evident June–August.

Thomas Læssøe

Exidia saccharina is a rather compact, somewhat brain-like, reddish-brown jelly fungus. The ± allantoid, repeating spores measure 11–13.5 × 3.5–4 µm. Mostly associated with *Pinus* bark, often in coastal plantations in the nemoral zones, rarely on other conifer hosts.

Phaeotremella foliacea ▷ is more lobed and occurs in association with *Stereum sanguinolentum* (page 971). *Exidia repanda* △ looks similar, but grows almost exclusively on *Betula*.

Widespread across temperate Europe but especially common in the hemiboreal–boreal zones; all year.

Naematelia, ***Phaeotremella*** and ***Tremella*** are highly gelatinous jelly fungi with longitudinally divided basidia. All are fungal parasites and obtain their nutrients via haustoria that penetrate the hyphae of their hosts.

Phaeotremella foliacea is a large, lobed, pale brown to dark brown, highly gelatinous jelly fungus. The repeating spores measure 5.3–9.1 × 4.7–8.5 µm. It parasitizes *Stereum sanguinolentum* (page 971) which always occurs on conifer wood, but the host is often not visible.

Two similar species of *Phaeotremella* are found on *Stereum rugosum* (page 970), which occurs on deciduous wood: the rather small *P. fimbriata* ⋉, which is initially dark but quickly blackens; and the larger and paler *P. frondosa* ▽. *Exidia saccharina* ◁ also occurs on conifers but is more brain-like and has larger spores. *Ascotremella faginea* (page 1366) is purple-brown, very brain-like and has asci in the hymenium.

Widespread and rather common; all year, but rarely evident May–August.

Phaeotremella frondosa is a large, lobed, orange-brown jelly fungus. The repeating, smooth, one-celled spores measure 6.4–10.2 × 5.1–8.7 µm. It parasitizes species of *Stereum* (page 970) on deciduous wood.

Phaeotremella fimbriata ⋉ occurs on the same hosts but is smaller and darker. *Phaeotremella foliacea* △ parasitizes *Stereum sanguinolentum* (page 971) on conifer wood. *Exidia candida* ◁◁ and *E. repanda* ◁ occur on deciduous wood, are more cushion-shaped to brain-like and have larger spores.

Widespread and rather common; all year, but rarely evident May–August.

Phaeotremella simplex is a small, rather colourless, sticky, cushion-shaped jelly fungus. The hyphae lack clamps and the basidia are 2-spored. It has spherical, repeating spores that measure 6.5–8 µm, but the fungus is mostly asexual. It parasitizes *Aleurodiscus amorphus* (page 1026), and fruits on top of the host's hymenium.

Phaeotremella mycetophiloides ⋉ parasitizes the same host but has clamps and is 4-spored.

Widespread in the nemoral zones, occasional to rare; mostly found in late winter–early spring.

Naematelia encephala is a very firm, cartilaginous, brain-like, whitish to pinkish jelly fungus that has an inner white core of host tissue. The repeating spores measure (6–)7.4–11(–12) × (5–)6.2–9.9(–11.6) µm. It parasitizes *Stereum sanguinolentum* (page 971) which always occurs on conifer wood.

A distinctive species that is unlikely to be misidentified, but see also the smaller and more translucent *Myxarium nucleatum* ⊲⊲ which has white, firm crystalline bodies inside the fruitbodies.

Widespread and common; all year, but rarely evident May–July.

Naematelia aurantia is a medium-sized, yellow to yellow-orange, veined and lobed jelly fungus. The repeating spores measure 5.5–9 × 4.5–7 µm. It parasitizes species of *Stereum*, mostly *S. hirsutum* (page 971) which occurs on deciduous wood.

Tremella mesenterica ▷ has larger spores and basidia, and parasitizes species of *Peniophora* (page 1020).

Occurs in southern and western parts of temperate Europe, has not been recorded in Scandinavia; all year.

Tremella mesenterica is a fairly large, bright yellow to orange-yellow, veined and lobed jelly fungus that dries to a more orange colour. Occurs both in an asexual state and a sexual state, or as a combination. The basidia are divided longitudinally. The repeating spores measure (8–)9.8–14.9(–17.4) × (7.4–)7.9–13.5(–15.6) µm. It parasitizes various species of *Peniophora* (page 1020), but is not directly associated with the fruitbodies of the host.

Naematelia aurantia ◁ is very similar but parasitizes species of *Stereum* (page 970) and has much smaller spores (5.5–9 × 4.5–7 µm).

Widespread and very common: all year, but rarely evident May–July.

Tremella versicolor is a very small, disc- to cushion-shaped, ± amber-coloured jelly fungus. It occurs mostly in an asexual state, but occasionally produces longitudinally divided basidia. It fruits directly on the surface of *Peniophora* fruitbodies, e.g. *P. lycii* (page 1020).

May superficially recall a species of the ascomycote genus *Orbilia* (page 1464), but *T. versicolor* is more cushion-shaped and uneven in size, and less translucent.

Widespread and at least common in the nemoral zones, but easily overlooked; all year.

Tremella globispora is a small, almost globose jelly fungus that at first is almost colourless, but later becomes somewhat darker, and which dries to almost black. The longitudinally divided basidia have a basal stem. The repeating spores measure (5.4–)5.9–9.6(–10.2) × (5.4–)5.9–10.2(–12) µm. It parasitizes species of *Diaporthe* (page 1575) and is typically found on thin, attached branches of *Quercus*.

Tremella indecorata ⋉ has somewhat larger fruitbodies (3–10 mm as opposed to 1–3 mm in *T. globispora*) and basidia that usually lack a stem.

Widespread in the nemoral–hemiboreal zones, probably common; mostly found during mild winter periods.

Tremella exigua is a rather small, very dark, greenish, olive-grey to black, cushion-shaped to confluent jelly fungus. The basidia are clavate, and predominantly have oblique rather than longitudinal septa. The repeating spores measure 6.5–11 × 5.8–8 µm. It probably parasitizes species of *Diaporthe* (page 1575) and/or bitunicate pyrenomycteous fungi, *e.g.* on *Berberis* and *Sarothamnus*.

Distinctive and unlikely to be misidentified, provided it is checked microscopically.

Widespread in the nemoral–hemiboreal zones, rarely recorded but easily overlooked; probably all year.

Tetragoniomyces uliginosus is a very distinctive and unusual jelly fungus. The fruitbodies are ± buff and brain-like and somewhat hollow. The verrucose basidia measure 10–12(–15) × 7–10 µm, are thick-walled and divided into four cells; they do not produce spores but act instead as propagules and resting structures. It fruits on grasses and leaf-litter in damp places, but is probably a parasite on sclerotia produced by other fungi growing in the same areas.

The unique basidia make this species easy to recognize microscopically.

Widespread, at least in the nemoral zones, and probably rather common; mainly June–December.

Thomas Læssøe

Jens Maarbjerg

Syzygospora tumefaciens forms irregular, brain-like galls on the cap of the host. The hyphae have clamps. The spores measure 6–8(–9.5) × 3–4.5 µm and bud off yeast-like cells. It parasitizes species in the *Gymnopus dryophilus* (page 290) complex.

Syzygospora effibulata ⋉ also forms galls on *Gymnopus* or *Rhodocollybia*, but it lacks clamps.

Widespread in the nemoral zones, rather rarely reported; autumn.

Syzygospora pallida forms small, milky to translucent protuberances on the surface of the host hymenium. It has so-called zygoconidia that are formed in connection with clamp formation. The basidia are one-celled and 4–6-spored. The spores measure 8.5–9 × 5.5–6.5 µm and may produce conidia. A parasite, probably exclusively on *Phanerochaete sordida* (see page 1016).

The host and general appearance are good characters, but it should always be checked microscopically.

Distribution and frequency not well understood but, like the host, probably widespread and common; mainly May–October.

Heterocephalacria bachmannii forms small, pale to dark brown, cushions at the branch tips of the host; they are gelatinous in wet weather but hard in dry weather. The basidia are mostly 4-spored. The spores measure 7–9 × 4.5–6 µm. It parasitizes species of *Cladonia* (page 1486).

Also known as *Syzygospora bachmanii*. Could possibly be confused with the fruitbodies of the host, but these contain asci.

Widespread and probably common; all year.

Tremellodendropsis form branched, tough, *Clavulina*-like fruitbodies with basidia that have longitudinal septa only at the top. Occur on soil, but the biology is not fully understood.

Tremellodendropsis tuberosa is a rather tough, branched clavarioid jelly fungus. The branches are ± flattened, pointed and somewhat fimbriate at the tips. The basidia have incomplete, longitudinal septa at the top. The hyphal system is dimitic. The large, almond-shaped to fusiform, smooth, hyaline spores measure 12–20 × 5–9 µm. Occurs on soil, mostly in calcareous sites, in scrub, old, unfertilized grasslands, *etc.*

Tremellodendropsis helvetica ▽ is smaller, paler and has smaller spores. Branched species of *Thelephora* (page 1152) have brown, ornamented spores. Similar clavarioids, *e.g. Clavulinopsis umbrinella* (page 1125), are more fragile, have more rounded branch tips and much smaller spores. Species of *Clavulina* (page 1128) have ± spherical spores, while species of *Lentaria* (page 1132) occur on wood. Furthermore, all these genera have one-celled basidia that have no hint of septation.

Widespread in the nemoral–hemiboreal zones, rather rare to rare; mainly August–January.

Tremellodendropsis helvetica is a slender, pale clavarioid jelly fungus. The small spores measure (6–)8–11 (–13) × 3.5–4.8(–5) µm. Occurs on soil in calcareous, thermophilic scrub and deciduous forests.

Tremellodendropsis tuberosa △ is darker, more robust and has markedly larger spores. Clavarioids in the genus *Ramariopsis* (page 1121) have even smaller spores that are finely ornamented.

Widespread in the nemoral zones but rarely recorded; late summer–autumn.

Thomas Læssøe

Thomas Læssøe

Johan Nitare

Rusts and smuts

This group includes a number of biotrophic plant parasites that do not form proper fruitbodies; all belong to the *Basidiomycota*. There are about 30 genera of rusts and 20 of smuts in temperate Europe; the aim here is only to provide an overview of this group.

Rusts have a particularly complicated life-cycle that may involve up to five types of spore, and in many cases a change (shift) in host. The spore types can be summarized as follows. **Type 0** (spermogonia with spermatia): a basidiospore (Type IV) germinates on a suitable host and an internal mycelium develops from which tiny, flask-shaped spermogonia (n) are formed in spots on the host's surface. The spermatia may then transfer nuclei to suitable compatible hyphae on other spermogonia. **Type I** (aecidia with aecidiospores): the dikaryotic hyphae that result from fertilization with spermatia may then produce ± discoid aecidia in which dikaryotic (n+n) aecidiospores are formed. **Type II** (uredia with uredospores): in host-shifting species the aecidiospores infect a second host and a new dikaryotic mycelium develops from which uredia and uredospores are formed (n+n). **Type III** (telia with teliospores): a Type II mycelium finally forms telia with thick-walled teliospores (n+n) that typically overwinter. The following spring the teliospores germinate with 4-celled basidia after fusion of the two nuclei and a subsequent meiosis. **Type IV**: basidiospores (n) are then produced and can reinfect the first host.

Smuts and *Microbotryales* have much simpler life-cycles, infecting just one host and finally producing spores externally or internally. They can, for example, transform the inner parts of a flower to spores or may break out through stems, roots, seeds or leaves.

Both rusts and smuts are serious pathogens on crop plants.

Other similar fungi:
– numerous fungi and fungoid organisms live as parasites on plants, some of which may form structures that resemble those from rusts and smuts, see images in the blue section.

Micro-drawings: teliospores (III) and basidia (bas.).

0: spermogonia
I: aecidia
II: uredia
III: telia

Approximate species number applies to temperate Europe.

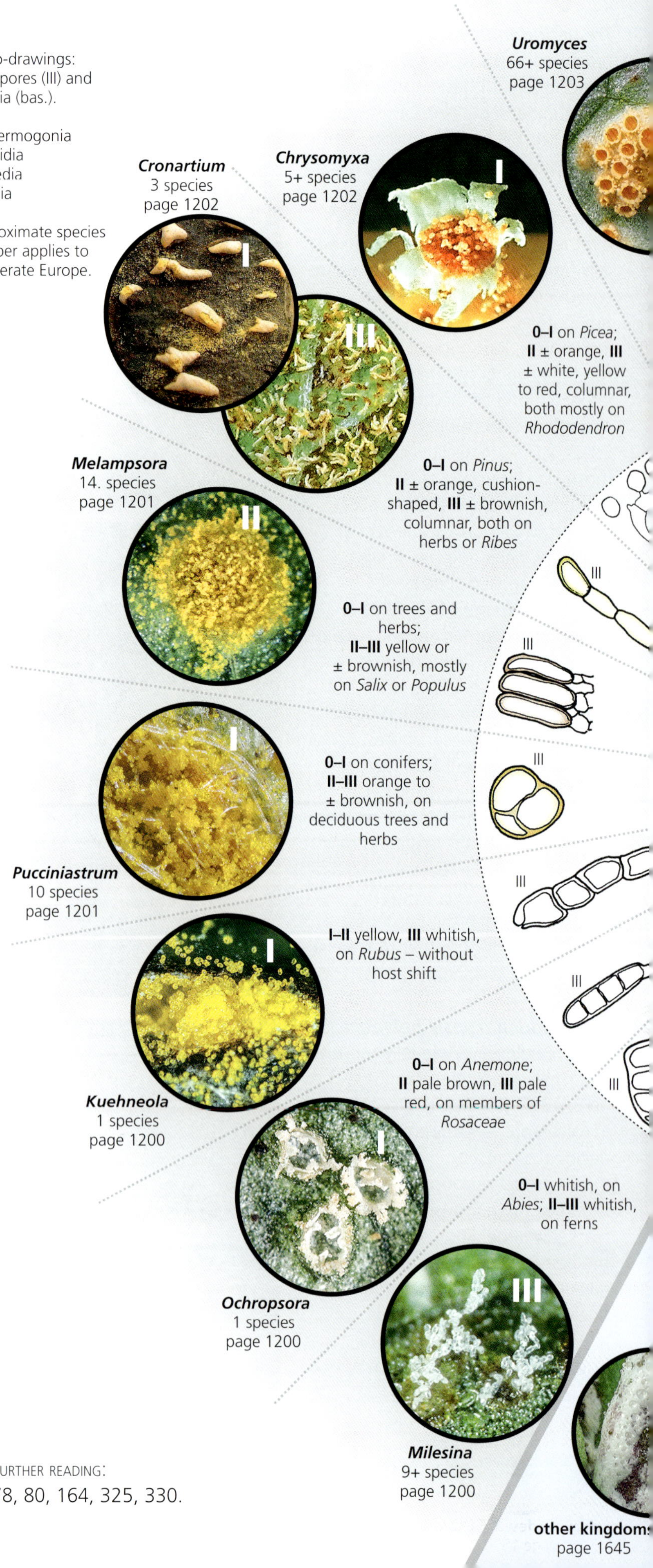

Further reading:
78, 80, 164, 325, 330.

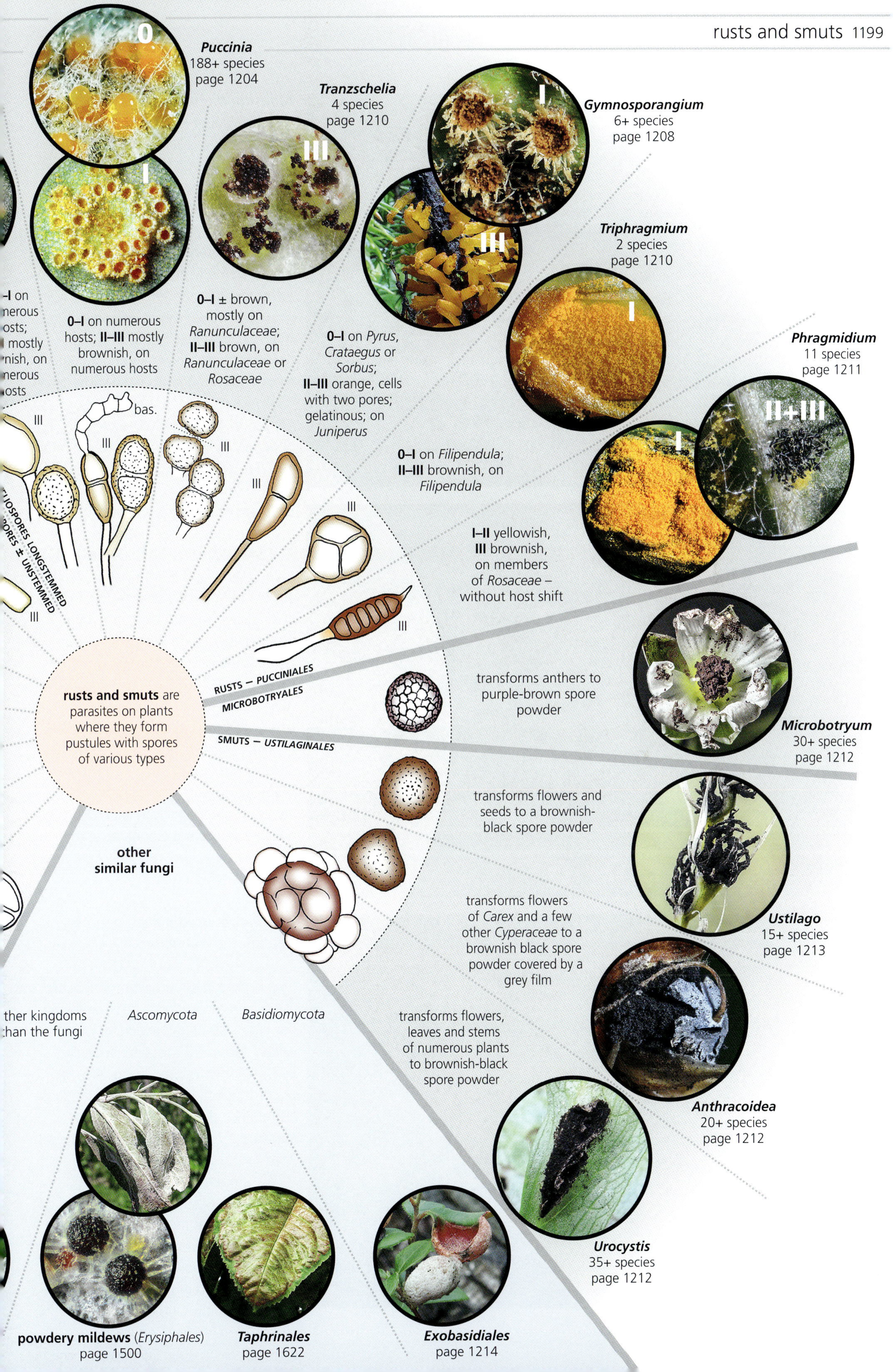

Puccinia
188+ species
page 1204
Tranzschelia
4 species
page 1210
Gymnosporangium
6+ species
page 1208
Triphragmium
2 species
page 1210
Phragmidium
11 species
page 1211
-I on
nerous
osts;
mostly
nish, on
nerous
osts
0–I on numerous hosts; II–III mostly brownish, on numerous hosts
0–I ± brown, mostly on Ranunculaceae; II–III brown, on Ranunculaceae or Rosaceae
0–I on Pyrus, Crataegus or Sorbus; II–III orange, cells with two pores; gelatinous; on Juniperus
0–I on Filipendula; II–III brownish, on Filipendula
I–II yellowish, III brownish, on members of Rosaceae – without host shift
bas.
TIOSPORES LONGSTEMMED
ORES ± UNSTEMMED
rusts and smuts are parasites on plants where they form pustules with spores of various types
RUSTS – PUCCINIALES
MICROBOTRYALES
SMUTS – USTILAGINALES
other similar fungi
transforms anthers to purple-brown spore powder
Microbotryum
30+ species
page 1212
transforms flowers and seeds to a brownish-black spore powder
Ustilago
15+ species
page 1213
transforms flowers of Carex and a few other Cyperaceae to a brownish black spore powder covered by a grey film
Anthracoidea
20+ species
page 1212
transforms flowers, leaves and stems of numerous plants to brownish-black spore powder
Urocystis
35+ species
page 1212
ther kingdoms
than the fungi
Ascomycota
Basidiomycota
powdery mildews (Erysiphales)
page 1500
Taphrinales
page 1622
Exobasidiales
page 1214

Milesina dieteliana is a white-spored rust with spore types 0/I on needles of *Abies* and types II/III on species of *Polypodium*. The type I state is seen as very small pale yellow blisters along the midrib of the needle. Spore type II is seen as small white chains of stipitate, echinulate uredospores measuring 23–48 × 16–26 µm. Type III spores are formed within the fern cells.

There are species of *Milesina* that occur on *Blechnum*, *Dryopteris*, *Asplenium*, *Athyrium* and *Polystichum*, all of which appear to have spore types 0/I on *Abies*.

Distribution poorly known, but generally considered rare in temperate Europe; most of the year.

Ochropsora ariae is a host-shifting rust with spore types 0/I on *Anemone* and II/III on *Sorbus,* and possibly also on *Malus*, *Pyrus* and *Prunus padus*. Type I aecidia are situated on pale, elongated host leaves, and are very pale, small and crowded, with fimbriate margins; type I spores are pale brown and measure 19–28 × 16–22 µm. Teliospores (type III) form below the epidermis and finally become cylindrical and 4-celled.

The only rust with spore types II/III on *Sorbus*, but many have 0/I. *Tranzschelia pruni-spinosae* ⋉, on *Anemone,* has aecidia more than 1 mm wide and the host leaves are not elongated.

Widespread and probably common; mainly March–August.

Kuehneola uredinis is a non-host-shifting rust on *Rubus*, *e.g. R. fruticosus s.l.* and *R. idaeus*. The type I spore state, which probably acts as an overwintering state, is similar to the type II state and produces orange-yellow spores. Multicelled (typically 5–7), pale teliospores (spore type III) are produced in very limited quantities and do not overwinter. Type I spores measure 19–23 × 18–20 µm, and type II 21–27 × 16–19 µm.

A number of *Phragmidium* ▷▷ species also occur on *Rubus*, but have large quantities of violet-black teliospores; the uredospores are stemmed and have obvious pores.

Widespread in the nemoral zones, occasional; July–October.

Pucciniastrum always has spermogonia (spore type 0) and aecidia (type I) on conifers. Teliospores (type III) are ± brown, stemless and multicellular.

Pucciniastrum areolatum forms spore types 0/1 on *Picea abies* cones, where it develops crowded aecidia with pale spores that measure 20–30 × 16–22 µm; it has a strong, sweetish odour. Spore types II and III are found on *Prunus*; type II measure 15–21 × 11–15 µm.

Habitat and appearance make the 0/1 spore-state unmistakable.

Widespread and common in the hemiboreal–boreal zones, less so in the nemoral zones; all year, but mainly March–June.

Pucciniastrum agrimoniae is only known from spore types II/III on *Agrimonia*. Type II spores are bright orange and measure 18–21 × 14 µm. Type III are brown, form between type II, have oblique septa, are typically 4-celled and measure about 30 × 21–30 µm.

Other *Pucciniastrum* species have spore types 0/I on conifers, and *P. agrimoniae* may also have such a state on *Abies*; *P. epilobii* ⋉ causes damage on cultivated *Abies*. There are no other rusts on *Agrimonia*.

Widespread, mainly in the nemoral zones, common in western parts of temperate Europe; mainly August–September.

Melampsora populnea forms spore types 0/I on *Larix* and *Pinus* needles, but also on *Corydalis*, *Chelidonium majus* and *Mercurialis*. Types II and III occur on leaves of *Populus*; type II spores measure 15–25 × 11–18 µm.

The taxonomy of this species is very complex and it may be split into *Melampsora laricis* ⋉ on *Larix*, *M. magnusiana* ⋉ on *Corydalis*, *M. pinitorqua* ⋉ on *Pinus*, and *M. pulcherrima* ⋉ and *M. rostrupii* ⋉ on *Mercurialis*. There are no other rusts on *Corydalis* and *Mercurialis*, but rusts with larger spores (type II) are found on *Populus* (although only *Melampsora populnea* occurs on *P. tremula*).

Widespread and common; most of the year.

Niels V. Mogensen/Julia Kruse

Cronartium ribicola is a host-shifting rust with spore types 0/I on 5-needled *Pinus* and types II/III on various species of *Ribes*, including *R. nigrum* and *R. uva-ursi*. Overwinters in thickened *Pinus* branches. Type I spores measure 22–29 × 18–20 µm and type II 21–25 × 13–18 µm. Type III spores form multicelled orange columns, germinate quickly with basidia.

This is the only rust on 5-needled *Pinus*, but a number of other rusts occur on 2-needled species of *Pinus*. *Puccinia ribis* × also occurs on *Ribes* but produces brown bladders, and there are two rust species with 0/I spore-states.

Widespread in the nemoral–hemiboreal zones; all year.

This species occurs in two forms, one is native to the Alps, the other invaded Europe from Asia during the 19th century. The latter form is especially aggressive to *Pinus strobus*, a species native to North America. It was introduced there in 1910 via *Pinus strobus* seedlings from France, and has subsequently caused huge problems in American pine forests.

Chrysomyxa ledi is a host-shifting rust. It parasitizes *Rhododendron* species, e.g. *R. tomentosum* (syn. *Ledum palustre*), where it forms spore types II and III. Spore types 0/I are found on needles of *Picea*, where aecidia are formed that produce a white, floccose peridium and orange spores. Type II spores measure 22–27 × 16–22 µm; type III form multicelled chains, are hyaline, and measure 20–30 × 10–14 µm.

Puccinia rhododendri × also occurs on *Rhododendron* and has coloured teliospores; the situation on *Picea* is more complex.

Eastern hemiboreal–boreal with extensions to eastern parts of the nemoral zone; spring–autumn.

Uromyces is recognized by having one-celled teliospores with a long stem.

Uromyces limonii is a non-host-shifting rust that forms spore types 0–III on *Limonium vulgare*. Spore type I measures 21–32 × 18–26 µm and type II 22–32 × 20–28 µm; type III is dark, one-celled and measures 24–50 × 14–25 µm.

There are no other rusts on *Limonium*.

Widespread in the nemoral zones, probably present in most stands of *Limonium*; summer–autumn.

Thomas Læssøe

Uromyces ficariae is a non-host-shifting rust that forms spore types II–III on *Ficaria verna/Ranunculus ficaria*, with (II-)III in brown, well-defined, somewhat oblong pustules. Type III spores are one-celled, smooth-walled and measure 22–38 × 18–26 µm.

Uromyces dactylidis s.l. ▽ also occurs on *Ficaria verna/Ranunculus ficaria* but forms spore types 0/I.

Widespread, mainly in the nemoral zones, common; spring.

Thomas Læssøe

Uromyces dactylidis is a host-shifting rust with spore types 0/I on *Ficaria/Ranunculus* and II/III on grasses. The type I spores are pale yellow, 17–25 µm in diameter and finely echinulate; type II measure (14–)20–32 × (14–)18–25 µm; and type III are one-celled, shiny brown and measure 17–30 × 14–24 µm.

Uromyces ficariae △ occurs on *Ficaria verna* but only has type II–III spores. Unfortunately, on other species of *Ranunculus* it is not possible to separate *Puccinia magnusiana* ⋉ from the type I state of *U. dactylidis*.

Widespread and very common; spring–autumn.

Puccinia is recognized by the long-stemmed teliospores with a predominantly 2-celled spore body. It is by far the largest genus amongst the rusts.

Puccinia pulverulenta is a non-host-shifting rust that forms spore types 0–III on *Epilobium*. The aecidia are bright yellow, disc-shaped and have distinctly lobed margins. Type II spores measure 20–23 × 16–19 µm. Type III pustules are up to 1 mm in diameter and have blackish-brown, (1–)2(–3)-celled spores that measure 22–30 × 15–18 µm.

Puccinia epilobii ⋉ only forms type III spores on *Epilobium* and the pustules are less than 0.25 mm in diameter.

Widespread, mainly nemoral, common; spring–summer.

Puccinia sessilis is a host-shifting rust with spore types 0/I on monocot hosts, e.g. *Allium*, *Arum*, *Maianthemum*, *Convallaria, Polygonatum*, *Paris*, *Dactylorhiza*. Spore types II/III form on *Phalaris*, type II having scattered pores that measure 20–28 × 18–24 µm.

Other rusts on *Phalaris* include: *Puccinia graminis* (see *P. striiformis* ▷▷); *P. striiformis* ▷▷, which has type II/III states in rows, and *P. coronata* ⋉, which has type III spores that have protuberances. Two species of *Melampsora* ◁◁ can occur on *Dactylorhiza,* but have a type I spore-state and lack a peridium. *Puccinia hordei* ⋉, which is almost identical to *P. sessilis,* and a *Melampsora* ◁◁ species that lacks a peridium, can occur on *Allium*.

Widespread, common; spring–autumn.

Puccinia urticata is a host-shifting rust that forms conspicuous swellings (galls). Spore types 0/I form on the stems and leaves of *Urtica*, and spore types II/III on *Carex*. Type I spores measure 16–27 × 12–20(–24) µm; type III spores are 2-celled and measure 35–66 × 14–24 µm.

Puccinia iridis ⋉ also occurs on *Urtica* and has type I spores, but does not form galls on the host. A number of other species of rust can be found on *Carex*, with *P. caricina* ⋉ being very similar to *P. urticata*.

Widespread, rather common, becoming scarcer towards the north; spring–autumn.

Puccinia punctiformis is a non-host-shifting rust that occurs on *Cirsium arvense*. In the 0 spore-state it has a nauseous-sweetish smell. Spore type I is similar to type II and measure 21-28 µm in diam. Type III spores are 2-celled, brown, finely rough and measure 26–42 × 17–25 µm. It is systemic within the host, and even present in the root system. Other rusts on *Cirsium* lack the characteristic smell and often some of the spore states. *Puccinia cnici* ⋉ is found mostly on *Cirsium vulgare*, and *P. calcitrapae* ⋉ is more common on *Arctium* and *Carduus*.

Widespread and very common, becoming scarcer towards the north; spring–autumn.

It has been suggested that *Puccinia punctiformis* could be used as a bio-control agent against *Cirsium arvense*, which can be invasive and is often regarded as an agricultural weed. Although various experiments have been carried out worldwide, none have proved to be effective. New experiments in Canada are designed to show whether sufficient systemic infections can be obtained by adding teliospores to leaf rosettes in the autumn.

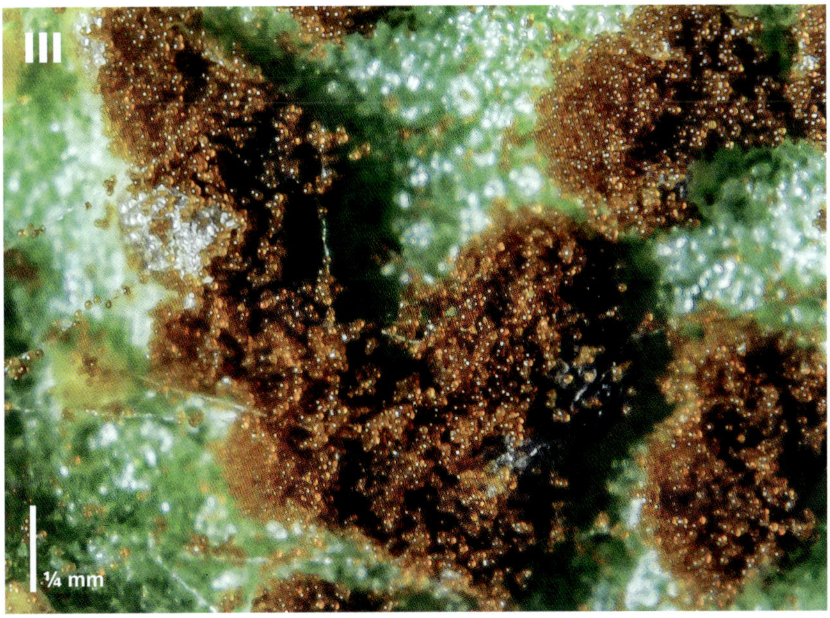

Wikipedia

Puccinia striiformis was, until recently, considered to be a non-host-shifting rust, found only on *Berberis*. However, experiments have shown that it may in fact be a host-shifting species, with some strains of the fungus forming spore types 0/I on species of *Berberis* and spore types II/III on a range of grasses, including cultivated species. The uredia (spore type II) are situated in characteristic stripes on leaves and stems and type II spores measure 15–30 × 12–24 µm. The mainly 2-celled teliospores (spore type III) measure 30–75 × 16–24 µm. particularly in cereal crops.

Puccinia graminis ⋉ can be identifed by the aequatorial pores on the uredospores; *P. striiformis* and *P. recondita* ⋉ have scattered pores; the latter also produces its uredia in scattered pustules rather than in long yellow stripes.

Widespread and a serious economic problem in cereal farming; summer–autumn.

The very similar *Puccinia graminis* has been known as a very serious pathogen on wheat, rye *etc.* since the middle ages. In the early 1800s the Danish schoolmaster N.P. Schøler came up with the idea that there was a connection between the very extensive use of *Berberis* as a hedging plant and outbreaks of rust on rye. In 1865, the German mycologist Heinrich Anton de Bary finally proved that it was indeed *Puccinia graminis* that was infecting the two different plant genera. In 1903 a law was passed in Denmark that outlawed *Berberis*, and required that it should be exterminated. This law was abandoned in 1992 due to the relative rarity of *Berberis* and the rust resistance of modern varieties of grain having reduced the problem. The law may, however, now have to be revoked.

Puccinia violae is a non-host-shifting rust that is only found on species of *Viola* and has spore types 0–III. The aecidia have prominent white margins and orange spores that can be found on all parts of an infected plant, 20–28 × 17–23 µm. The type III spores are 2-celled, brown, finely ornamented and measure 20–40 × 15–23 µm.

Puccinia fergussonii ⋉ occurs on *Viola* but has only type III spores, which are smooth.

Widespread and common; spring–autumn.

Puccinia phragmitis is a host-shifting rust with spore types 0/I on leaves (not stems) of *Polygonaceae*, e.g. *Rumex*, *Fallopia* and *Persicaria*; spore types II/III are found on *Phragmites*. The aecidia (spore type I) are pale, disc-shaped and typically in circular groups on somewhat thickened, reddish spots. The telia (spore type III) are black and form rows on leaf sheaths and leaves; the type III spores are 2-celled, have a long stem and measure 45–65 × 16–25 µm.

Puccinia magnusiana ⋉ also occurs on *Phragmites* and has uredospores with scattered pores; *Puccinia graminis* ◁ can occur on *Phragmites* but has thinner spore walls (1–1.5 µm compared with 3–4 µm in *Puccinia phragmitis*). *Uromyces polygoni-avicularis* ⋉, which occurs on *Polygonaceae*, has smaller, more irregular spores. *Uromyces acetosae* ⋉ has aecidia on stems and on both sides of a leaf.

Widespread and very common; most of the year.

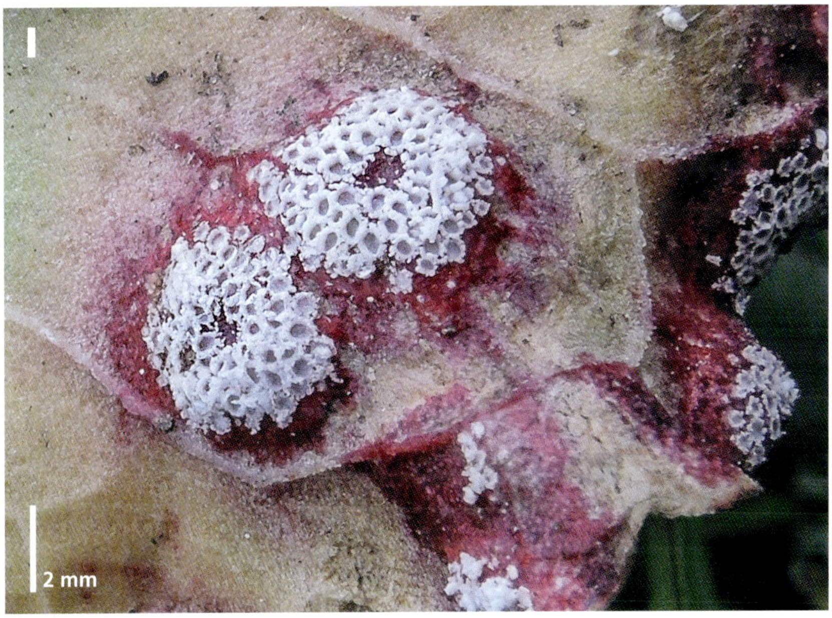

Ole Martin/Jens H. Petersen

Species of ***Gymnosporangium*** typically occur as systemic infections on species of *Juniperus* where they form ± gelatinous, prominent, ± orange telia and often galled host tissues.

Gymnosporangium clavariiforme is a host-shifting rust that causes thickenings on trunks of *Juniperus communis*. It occurs as a systemic infection, from which it forms large, ± cylindrical, yellow-orange telia (spore type III) in spring. Type II absent. Types 0/I mostly on leaves and fruits of *Crataegus*, more rarely *Aronia*, *Cotoneaster*, *Cydonia*, *Pyrus* and *Sorbus*; at first horn-shaped and aggregated, later split and very floccose. Type I spores 22–30 × 18–26 µm; type III 2-celled, 40–120 × 12–22 µm, form a gelatinous matrix.

Gymnosporangium cornutum ▷ also occurs on *Juniperus communis* but has more *Tremella*-like telia and causes either no or only moderate deformations on young branches; *G. gaeumanni* ⋉ occurs on the needles. Other species of *Gymnosporangium* occur on different species of *Juniperus*.

Widespread and very common in natural stands of *Juniperus communis*; mainly visible March–August.

Gymnosporangium sabinae is a host-shifting rust with spore types 0/I on *Pyrus* and type III on deformed branches of various *Juniperus* species, but not on *J. communis*. The telia are large, flattened, conical and quickly become sordid yellow-brown; type III spores measure 38–48 × 20–30 µm and have two pores in each cell. The 0/I spore-states are formed in connection with orange to bright red shiny spots on *Pyrus* leaves; spore type I is found in crowded groups, mostly on the underside of the leaves, and, with age, acquires a grid-like extension of the aecidia; the type I spores measure 25–32 × 22–29 µm.

Gymnosporangium tremelloides ⋉ forms flat to cushion-shaped telia on deformed branches and trunks of various *Juniperus species,* including *J. communis*. *Gymnosporangium clavariiforme* ◁ forms clavate, long telia on deformed branches and trunks of *J. communis*; *G. cornutum* ▽ produces telia on young, green shoots causing either no or only moderate deformations.

Widespread in the nemoral–hemiboreal zones, has increased in recent times; spring–autumn.

Gymnosporangium cornutum is a host-shifting rust with spore types 0/I on species of *Sorbus*, where first punctate spermogonia and later horn-like aecidia are formed; type I spores measure 20–29 × 18–25 µm. Not known in a type II-state. Spore type III occurs on young, green, non-galled branches of *Juniperus*, where gelatinous, brownish-orange telia are formed; the spores are stemmed, 2-celled, measure 32–52 × 18–28 µm and have one or two pores.

Other species of *Gymnosporangium* form telia on the more woody parts of *Juniperus*.

Widespread and common; mainly July–September (spore types 0/I).

Tranzschelia anemones is a non-host-shifting rust that occurs on *Anemone* and *Thalictrum* where it forms dark brown to black type 0 and type III spore-states. Teliospores (spore type III) are mostly 2–3-celled (but quickly split to become to one-celled), measure 26–55 × 15–30 µm and have a very thick, brown, warty wall.

Puccinia singularis ⋉ and *P. pulsatillae* ⋉ occur on *Anemone* but have non-splitting, 2-celled teliospores.

Widespread and fairly common, becoming scarcer towards the north; on *Anemone* mainly April–May, on *Thalictrum* mainly July September.

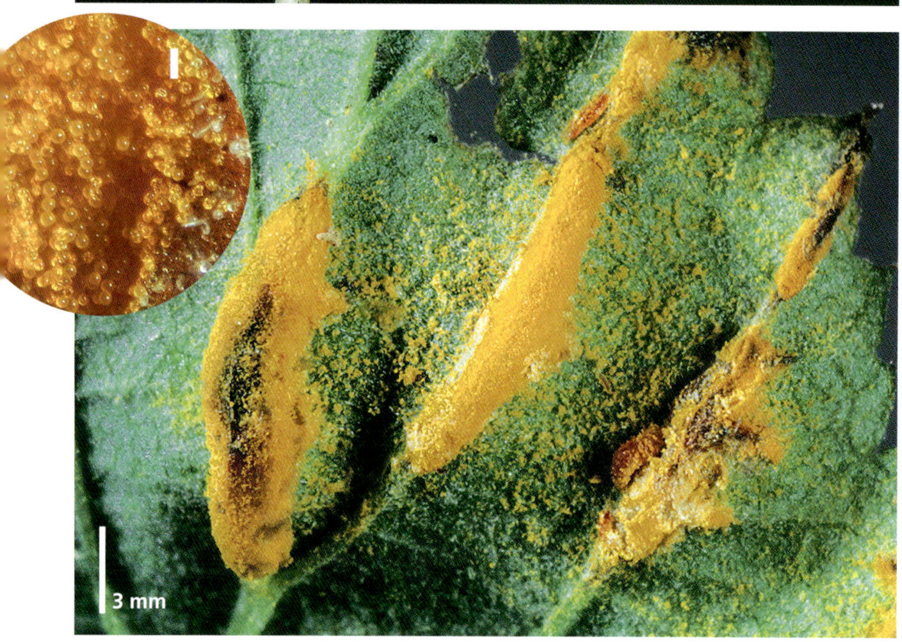

Triphragmium ulmariae is a non-host-shifting rust that occurs on *Filipendula ulmaria* with spore types 0–III. The type I state is similar to the type II uredo state and produces very conspicuous orange-yellow spore masses on leaves and stems. Type I spores measure 25–28 × 18–21 µm and type II measure 20–30 × 18–25 µm. Type III spores are 3-celled and 35–49 µm in diameter.

Triphragmium filipendulae ⋉ occurs on *Filipendula vulgaris*. No other rusts occur on either of these two hosts.

Widespread and common; summer.

Phragmidium is a non-host-shifting rust genus that occurs on *Rosaceae*. It has multicelled, dark teliospores with a basal stem.

Phragmidium mucronatum is a non-host-shifting rust with 0–III spore types occurring on *Rosa* (both on wild species and cultivated forms). The teliospores are dark, verrucose, long-stemmed, 6–9-celled and measure 67–110 × 32–36 µm.

Phragmidium fusiforme ⋉ has 8–14-celled teliospores. *Phragmidium rosa-pimpinellifoliae* ⋉ has a conical, pointed tip to the teliospores. *Phragmidium mucronatum* and *P. tuberculatum* ⋉ have rounded teliospore tops with sharply delimited tips; *P. tuberculatum* also has verrucose aecidiospores (they are very finely echinulate in *P. mucronatum*).

Widespread and common, becoming scarcer towards the north; summer–autumn.

Thomas Læssøe

Phragmidium bulbosum is a non-host-shifting rust that occurs on leaves and shoots of *Rubus*, including *R. caesius*, *R. fruticosus* and *R. corylifolius*. The teliospores are black, coarsely verrucose, 4–7-celled and do not have constrictions; they measure 60–110 × 30–36 µm and have a stem up to 160 µm long and up to 20 µm wide at the base.

Phragmidium violaceum ⋉ has 3–5-celled teliospores. *Phragmidium rubi-idaei* ⋉ occurs only on *R. idaeus*, and *P. acuminatum* ⋉ only on *R. saxatilis*.

Widespread and common in the nemoral–hemiboreal zones; summer–autumn.

Microbotryum stellariae is a smut-like parasitic fungus that attacks and transforms anthers of *Stellaria* to a violet-black spore mass. The spores are spherical with a net-like ornament and measure approx. 8 µm in diameter.

The same symptoms can be seen on other *Caryophyllaceae*, such as *Silene*, and are caused by the closely related *M. violacea* ⋉. Similarly, *M. scabiosae* ⋉ causes symptoms on *Scabiosa.*

Widespread and common; spring–early summer.

Jens H. Petersen/Thomas Læssøe

Anthracoidea arenariae is a smut that attacks *Carex* fruits and seeds. The spore mass is covered at first by a greyish film, which later disintegrates to reveal a dark brown mass of smut spores. Spores spherical, finely roughened, with walls of uneven thickness, and measure 13–23 × 11–19 µm. Occurs in fruits of *Carex arenaria*, also known from *e.g. C. leporina*.

No other fungus transforms the fruits of this host to a black, initially covered, spore mass. A range of similar species of *Anthracoidea* occur on other species of *Carex*.

Widespread, probably common within the range of the host; summer–autumn.

Urocystis eranthidis is a smut that causes swellings on leaves and stems of *Eranthis*, which later crack to reveal a large mass of black smut spores. The spores are multicelled; there is a dark cell at the centre surrounded by pale cells 20–40 µm in diameter.

There are no similar fungi on *Eranthis*.

Widespread and common, at least locally; spring.

Ustilago is a large genus of smuts that transform flowers to dark spore powder. Many are serious pathogens on crop plants.

Ustilago tritici is a smut that attacks *Triticum* and transforms the seeds to an exposed olive-brown spore mass (naked smut). The finely echinulate spores are ± spherical, yellow-brown to olive-brown, have one paler side and measure 5–9 µm in diameter.

Can cause very serious crop losses, and is also known as *U. nuda* f. sp. *tritici*. *Tilletia caries* ⋉ is a more covered smut, where the spore mass is partly hidden by the host's awns, lemmas, *etc*. The spores have a reticulate pattern and are 12–17 µm in diameter.

There are similar smut species on *Hordeum* and *Avena*.

Widespread, occasional; becoming scarcer towards the north.

Ustilago maydis is a smut that causes swellings of the seeds and cobs in *Zea mays*; the infected parts become lead-grey and may reach 20 cm in size. The thin covering of the host then splits to disclose a powdery mass of dark smut spores; stems and leaves may also be infected. The ± spherical, echinulate spores measure 7–13 × 7–11 µm.

There are no other similar fungi on *Zea mays*.

Widespread and probably rather common, *e.g.* in unsprayed stands planted as game cover; autumn.

The smut-infected cobs are considered a delicacy in Mexico (huitlacoche) and can even be found in a tinned version.

Exobasidiales

The order *Exobasidiales* has just one genus of biotrophic parasites in temperate Europe. It partly deforms shoots and leaves of the host, which is also stained red, and forms a whitish covering of hymenia that have long, one-celled basidia on some surfaces. Some species form infections only in a few leaves, while others produce systemic infections. The species in temperate Europe occur on shrubs and dwarf shrubs of the *Ericaceae*.

Farther south the related parasite *Laurobasidium lauri* ⋉ produces horn-like protuberances on *Laurus* trunks.

OTHER SIMILAR FUNGI:
- the ascomycote *Taphrina* occurs predominantly on trees and may form witch's brooms (page 1622).
- the powdery mildews (*Erysiphales*) are more mealy and, with age, form small, spherical fruitbodies with asci (page 1500).
- some organisms that resemble fungi, e.g. *Albugo*, have neither basidia nor asci (page 1645).

FURTHER READING: 80, 100, 164, 209.

Taphrina and others
page 1622

powdery mildews (*Erysiphales*)
page 1500

other kingdoms
page 1645

other similar fungi

Exobasidiales
is parasitic on species of *Ericaceae* that become deformed (galled) and red-stained

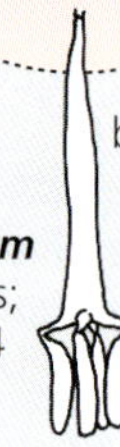

basidium with spores

Exobasidium
25+ species;
page 1214

Exobasidium oxycocci occurs within *Vaccinium oxycoccos* and *V. microcarpum*. The annual shoot of the host stretches and becomes pale, and finally is completely covered by erumpent basidia. The curved, 1–3-septate spores measure 12–15 × 3–3.5 µm.

Exobasidium rostrupii ⋉ occurs on the same host but only causes symptoms on the leaves, which acquire bright red spots on the upper side and have spore production underneath.

Widespread and probably rather common; summer–autumn.

Exobasidium vaccinii occurs on leaves of *Vaccinium vitis-idaea*. The leaves discolour red and become thick, and spores are formed on basidia on the mealy white lower sides. May also attack flower buds and young shoots, which become misshapen. The curved, 1–3(–7)-septate spores measure 11–19 × 2–4 µm; germinates with microconidia.

Exobasidum splendidum ⋉ and *E. juelianum* ▽ occur on the same host but form larger, systemic infections that include both leaves and stems.

Widespread and very common where the host occurs; summer–autumn.

Thomas Læssøe

Exobasidium juelianum is systemic and discolours the shoots of *Vaccinium vitis-idaea* bright red; it also causes dwarf growth. The spore-producing pale tissue is seen mostly under the leaves. The spores are either one-celled, or have one septum, and measure 9–14 × 2–4 µm.

Exobasidium splendidum ⋉, on the same host, is an even redder, more northern, 2-spored species with curved, 0–1-septate spores that measure 20–27 × 6–11.5 µm; *E. vaccinii* △ causes localized infections of single leaves, but not of whole shoots.

Widespread, common in the hemiboreal–boreal zones, rarer farther south; summer–autumn.

Exobasidium uvae-ursi infects plants of *Arctostaphylos uva-ursi*. The parasitized plants typically produce dense, annual shoots with numerous dark red leaves and stems. The hymenium is formed on the underside of the leaves. The curved, 1–3-septate spores measure 15–22 × 5–6 µm; germinates with microconidia.

Exobasidium sydowianum ⋉ occurs on the same host but the infection is limited to just tiny areas on the leaves.

Widespread in the hemiboreal–boreal zones, occasional, rare in the nemoral zones; summer–autumn.

Julia Kruse

Bird's nest fungi

Bird's nest fungi belong to the *Basidiomycota*, and produce basidia and spores within special egg-like structures called peridioles. In *Sphaerobolus* the single peridiole is shot actively away from the fruitbody. In other genera the small peridioles are held within cup-shaped structures, and are 'splashed out' by heavy raindrops. Traditionally, bird's nest fungi are included in the gasteroid form group.

All species in the group are decomposers and occur typically on large herbaceous stems, wood or dung.

OTHER SIMILAR FUNGI:
– some operculate discomycetes may superficially resemble bird's nest fungi, but they do not have peridioles and the spores are borne in asci in a hymenium inside the cup (page 1278).
– *Stictis* may superficially recall *Sphaerobolus* but lacks peridioles and forms the spores in asci in a gelatinous hymenium (page 1472).

FURTHER READING: 50, 92, 156, 241.

Micro-drawings: spores.

Approximate species number applies to temperate Europe.

Sphaerobolus stellatus **s.l.** is a unique, approximately 2 mm wide gasteroid fungus that opens in a star-like fashion, after which a ± orange-brown ball (a peridiole) is shot several metres away. During the opening an orange colour is seen (upper image), and after shooting a pale projecting 'balloon' is seen (middle image). Spores and basidia are held within the peridiole. The hyaline, thick-walled spores measure 6–10 × 5.5–6.5 µm. Occurs on litter, very decomposed wood, and old dung, mainly in open habitats.

In Europe there are at least two species. *Sphaerobolus stellatus* has a massive peridiole and a gelatinous outer wall whereas *S. iowensis* ⋉ has a chambered peridiole and no gel. Only one other European genus, *Pilobolus* (page 1642), has the ability to propel a spore ball several metres, but the mechanism used by the two genera is quite different. In *Sphaerobolus* the propelling force is created by tensions between six different wall layers. At maturity, one of the layers under the peridiole swells, and the resulting tension results in an inversion of the upper layer and the peridiole is shot away. *Pilobolus* propel the spore ball by creating an internal pressure within the head that eventually 'explodes' and delivers the necessary force.

Widespread and common; mainly June–December.

Jan Vesterholt

Jan Vesterholt

Thomas Læssøe/Jens H. Petersen

Mycocalia denudata is a rather small (1–2 mm wide), well-hidden fungus with an irregular, white fruitbody that has a thin skin (peridium) which holds the rather numerous, yellowish to pale red-brown peridioles. The peridium is very ephemeral. The hyaline spores measure 5.5–8.5 × 4.5–7 µm. On damp, rotten straw, often in stands of *e.g. Juncus effusus*, but also on old manure.

Mycocalia duriaeana ⋉ also has red-brown peridioles but occurs in drier habitats. *Nidularia deformis* ▽ has larger (5–15 mm wide), fruitbodies, darker peridioles and occurs on wood. *Mycocalia minutissima* ⋉ and *M. sphagneti* ⋉ have fruitbodies that have just one, very dark peridiole.

Widespread, rather rare, but probably greatly overlooked; July–October.

Thomas Læssøe

Nidularia deformis is an irregularly splitting, 5–15 mm wide, cushion-shaped, initially white, but later pale yellow-brown to brown gasteroid fungus with many red-brown, lens-shaped peridioles. The hyaline spores measure 5–9.5 × 4–6 µm. Occurs on conifer and deciduous wood, including wood chips, mostly in sandy plantations or in swampy forests. Also occurs in sawmills and similar places.

Mycocalia denudata △ is much smaller, has yellowish peridioles and occurs on damp herbs. *Mycocalia duriaeana* ⋉ is also smaller but has similar, dark-coloured peridioles; it occurs in dry habitats.

Widespread, occasional; mainly August–December.

Crucibulum crucibuliforme is a barrel-shaped bird's nest fungus with a pale, downy outer side, a yellow-brown, almost smooth lid and white, lens-shaped peridioles (seen in longitudinal section of the fruitbodies in bottom image), that are attached to a small 'string' (funiculus). The hyaline spores measure 6–10 × 3.5–5 µm. Occurs on wood chips and other kinds of wood, mostly in rather exposed sites, along roads, *etc*.

Species of *Cyathus* ▷▷ have taller, more flaring fruitbodies, and either very dark peridioles or a striate-furrowed inner side.

Widespread and common; mainly July–January.

Cyathus form cup-shaped fruitbodies that typically are at least twice as high as wide, and contain ± grey peridioles that are attached to a small 'string' (funiculus). At maturity, heavy rain splashes the peridioles several metres away.

Cyathus striatus is a brownish, velvety-hairy bird's nest fungus with a pale lid and a furrowed-striate, grey and shiny inner side, and rather pale, greyish, lens-shaped peridioles that are attached by an elastic 'string' which can stretch many centimetres. The spores measure 12–21 × 7–12 µm. Typically occurs in troops on wood and wood chips that are in contact with the soil, often in rather damp sites and mainly in deciduous forests.

Within temperate Europe, an unmistakable species.

Widespread, common, becoming scarcer towards the north; mainly July–January.

Cyathus stercoreus is a bird's nest fungus with a hairy lid, a smooth inner side and very dark peridioles. The outer surface of the fruitbody is covered by long, rigid hairs. The spores measure 23–33 × 17–30 µm. Occurs on old dung – both from larger animals and from rabbits and hares – in open, dry habitats.

Cyathus olla ▽ is associated with coarse herbs, the hairs are less prominent, and it has paler peridioles and much smaller spores.

Mainly recorded in the nemoral zones, rare; can probably be found almost all year, but possibly mainly during the autumn.

Flemming Rune/Christian Glabasnia (Wikipedia)

Cyathus olla is a distinctly bell-shaped bird's nest fungus with a skin-like lid, a smooth inner side and mouse-grey, lens-shaped peridioles. The outer surface is felty, but not covered by long, rigid hairs. The spores measure 9–12(–14) × 6.5–8.5 µm. Typically occurs amongst coarse herbs in open areas or stubble fields.

Cyathus stercoreus △ can be recognized by its characteristic substrate, and by its very dark peridioles, paler covering of rigid hairs and much larger spores. *Crucibulum crucibuliforme* ◁◁ is yellower and has barrel-shaped fruitbodies with paler peridioles.

Widespread, occasional but probably very much overlooked due to its seldom-visited habitat, becoming scarcer towards the north; mainly June–March.

Stinkhorns

The stinkhorns and their allies (*Phallales*) is a monophyletic group of fungi that form spores internally in a bulb-shaped 'witch's egg'. At maturity, the egg opens to reveal a slimy, dark spore mass. At the same time the fruitbodies emit a nauseating smell of carrion or faeces that attracts flies and other insects, which then disperse the spores. The spore mass also contains sugars as a reward for the insects. All species live as decomposers and are traditionally included in the gasteroid form group.

The insect-aided dispersal limits the potential distance that spores can travel compared to normal wind-disseminated spores. A consequence, it has been possible to follow the relatively slow spread of some of the introduced species in this order, *e.g. Clathrus archeri*.

Other similar fungi:
– some basidiomycote truffles form structures similar to witch's eggs, but these stay underground (page 1256).

Further reading: 50, 92, 156, 241.

Species of ***Mutinus*** are small stinkhorns with a not very clearly delimited fertile area on top of the stem. Has red and orange colours.

Mutinus caninus is a very slender stinkhorn with orange colours under an olive-black spore mass, as well as part way down the stem. The white witch's egg is elongated and has orange colours when sectioned. It is not as foul-smelling as *Phallus impudicus* ▷▷. The smooth, dark spores measure 4.5–6.5 × 1.8–3 µm. Occurs on wood chips, on and around old stumps, and in similar places, in parks, deciduous forests and with conifers.

Mutinus ravenelii ▽ has red colours under the spore mass and normally also down the stem; it mainly occurs in gardens. *Mutinus elegans* ⋉ is very pointed, has vivid orange colours on the stem and the spore mass is even less well delimited.

Widespread and common; mainly June–January.

Mutinus ravenelii is a slender stinkhorn that tapers towards the top, does not have a well-delimited 'head', and has red colours under the olive-black spore mass; it also usually has red tinges down the stem. The witch's egg is slender and white, and has red colours when sectioned. Not as foul-smelling as *Phallus impudicus* ▷▷. The dark, sooty spores measure 5–7 × 1.8–2.5 µm. Sometimes occurs in large troops, typically on compost and disturbed soil in gardens and parks.

Mutinus caninus △ is ± orange under the spore mass.

Rare and meteoric in its occurrence – probably introduced from North America; June–October.

Jacques Landry

Species of ***Phallus*** are rather large stinkhorns that have a spore mass on a bell-shaped or cap-like 'head'. In temperate Europe all have whitish stems and white or purplish skin (peridium) on the witch's egg.

Phallus impudicus is a very characteristic phalloid fungus with a completely white witch's egg that has a thick gelatinous zone just under the skin (peridium). The olive spore mass is located on a well-delimited cap-like 'head'. A short net is hidden underneath the cap, and may occasionally hang free; such veiled forms are called *P. impudicus* var. *pseudoduplicatus* ▷, or, in error, *P. duplicatus*. The witch's egg is connected to a system of very thick, tough hyphal cords that smell of radish. The foul-smelling spore mass covers a honeycomb-like structure that is revealed once flies and other insects have dispersed the spores (inset image). The smooth, dark spores measure 3.2–6 × 2–3.2 µm. Occurs on and around rotten stumps, in deciduous forests and with conifers, rarely in more open habitats, including dunes.

Phallus hadriani ▷ has a peridium (and a witch's egg) that quickly stains grey-rose to purple when exposed or bruised.

Widespread and very common; June–January.

Phallus impudicus* var. *pseudo-duplicatus is a variety that has a rudimentary veil hanging from below the fertile 'head'.

It has also been described from North America and called *P. impudicus* var. *togatus*, but this is a synonym of *Dictyophora duplicata*. Other mistaken names include *P. indusiatus* and *Dictyophora indusiata*.

Variety *pseudoduplicatus* is found in the same places as the normal variety, *P. impudicus* var. *impudicus*, but is much rarer.

Phallus hadriani is a large stinkhorn with a white stem and a well-delimited 'head' that has a rather wide top. It is mainly characterized by its grey-rose to purple witch's egg. The 'head', with its foul-smelling, olive-brown spore mass, has a coarse, honeycomb-like structure underneath. The smooth, dark spores measure 4.5–6.5 × 1.8–3 µm. It produces thick hyphal cords in stands of *Ammophila* in white dunes, but very rarely may occur in more ruderal habitats.

Phallus impudicus ◁ can also occur in dunes, probably on buried wood, but has a permanently white peridium and witch's egg.

Widespread along exposed, sandy coasts, rare inland, rather common in the nemoral zones but absent from the hemiboreal–boreal zones; mainly June–October.

Jens H. Petersen/Jan Vesterholt

Species of ***Clathrus*** are stemless, ± red stinkhorns with free branches or net-like structures. All have a dark brown to black, extremely foul-smelling spore mass.

Clathrus archeri is a highly distinctive, red-armed stinkhorn that breaks out of an ovoid, almost white witch's egg. The very foul-smelling, dark spore mass is situated on top of the red, backwards-curving arms. The dark, smooth spores measure 4.5–8.5 × 2.5–3.5 µm. Occurs amongst litter, on wood chips and compost in deciduous forests, parks, gardens and grasslands.

Unmistakable when mature, but the witch's egg may recall that of *C. ruber* ▽ and *Aseroë rubra* ▷.

Widespread and locally common in the nemoral zones; mainly August–November.

Jan Vesterholt

Clathrus ruber is a bright red, latticed stinkhorn that arises from a greyish-white, spherical witch's egg. The slimy, very foul-smelling spore mass is located on the inside of the lattice. The spores measure 4–6 × 1.5–3.2 µm. Occurs on soil and wood chips in warm sites.

The witch's egg may recall that of *Clathrus archeri* △, but when fully developed the fungus is unmistakable. *Ileodictyon cibarium* ⋉ is also latticed, but is completely white.

Rather widespread in the nemoral zones, *e.g.* in the UK and the Netherlands and farther south, but absent from most of Scandinavia; all year, but less evident in cold winters.

Aseroë rubra is a spectacular stinkhorn with a pinkish-tinged witch's egg, from which a pink, cylindrical stem emerges crowned with bright red, tapering, projecting 'tentacles'. The extremely foul-smelling spore mass is situated on the inner part of the tentacles. The smooth, dark spores measure 4.5–7 × 1.7–2.5 µm. Occurs on soil amongst litter in mixed forests on slightly acidic soils.

A unique fungus that is very unlikely to be confused with anything else.

Native to Australia. In Europe apparently only known from two sites near London in England. It was probably introduced via footwear or via imported wood or living plants; summer–autumn.

Thomas Læssøe

Lysurus cruciatus is a rather small stinkhorn with a pale witch's egg, from which a white stem emerges. The stem is crowned with five to eight converging, short, ± yellowish arms with a dark spore mass on the inside. The smell is unpleasant. The smooth, dark spores measure 4–6 × 1.5–2.5 µm. Occurs on disturbed, nutrient-rich soils.

A distinctive species that, within temperate Europe, is unlikely to be misidentified.

Probably introduced to Europe. There are scattered records from the nemoral zones, never recorded from the hemiboreal–boreal zones; mainly summer–autumn.

José Ignacio Fernández Simón

Puffballs and similar gasteroid fungi

This form group includes basidiomycotes where the spores are produced inside the fruitbody and form a powder at maturity. The species may either be ball-shaped, with or without a stem, or have a star-like appearance when mature. The latter may open and close depending on moisture – they are hygroscopic. The surface 'skin' is termed a peridium. Some species have a sterile elongation of the stem within the spore-containing ball – a columella. Others may have a short, wide stem, which can be massive or have spongy flesh.

The spores are hydrophobic and dispersed by the splash from raindrops. Almost all species have dark, thick-walled spores – probably an adaptation to prolonged exposure to sun and wind. The spores may be mixed with elastic, branched, thick-walled hyphae – a capillitium. A spore mass with all elements it is termed a gleba.

Traditionally, all the species included here were placed in the artificial class *Gasteromycetes*, along with the bird's nest fungi, stinkhorns and basidiomycote truffles. Phylogenetically the majority belong within the agarics (*Agaricales*). *Battarraea*, *Montagnea* and *Chlorophyllum agaricoides* are intermediate types that in some respects look similar to closely related agarics, but the gills are transformed to a dusty spore mass at maturity. Such forms are also called secotioids after the genus *Secotium*, where they were formerly placed.

OTHER SIMILAR FUNGI:
– truffles have tuber-like fruitbodies that develop below ground (page 1256).
– cleistothecial fungi are asco-mycotes with tiny, 0.2–3 mm wide, closed fruitbodies (page 1500).

FURTHER READING: 92, 134, 135, 136, 137, 138, 142, 156, 162, 241, 320, 365.

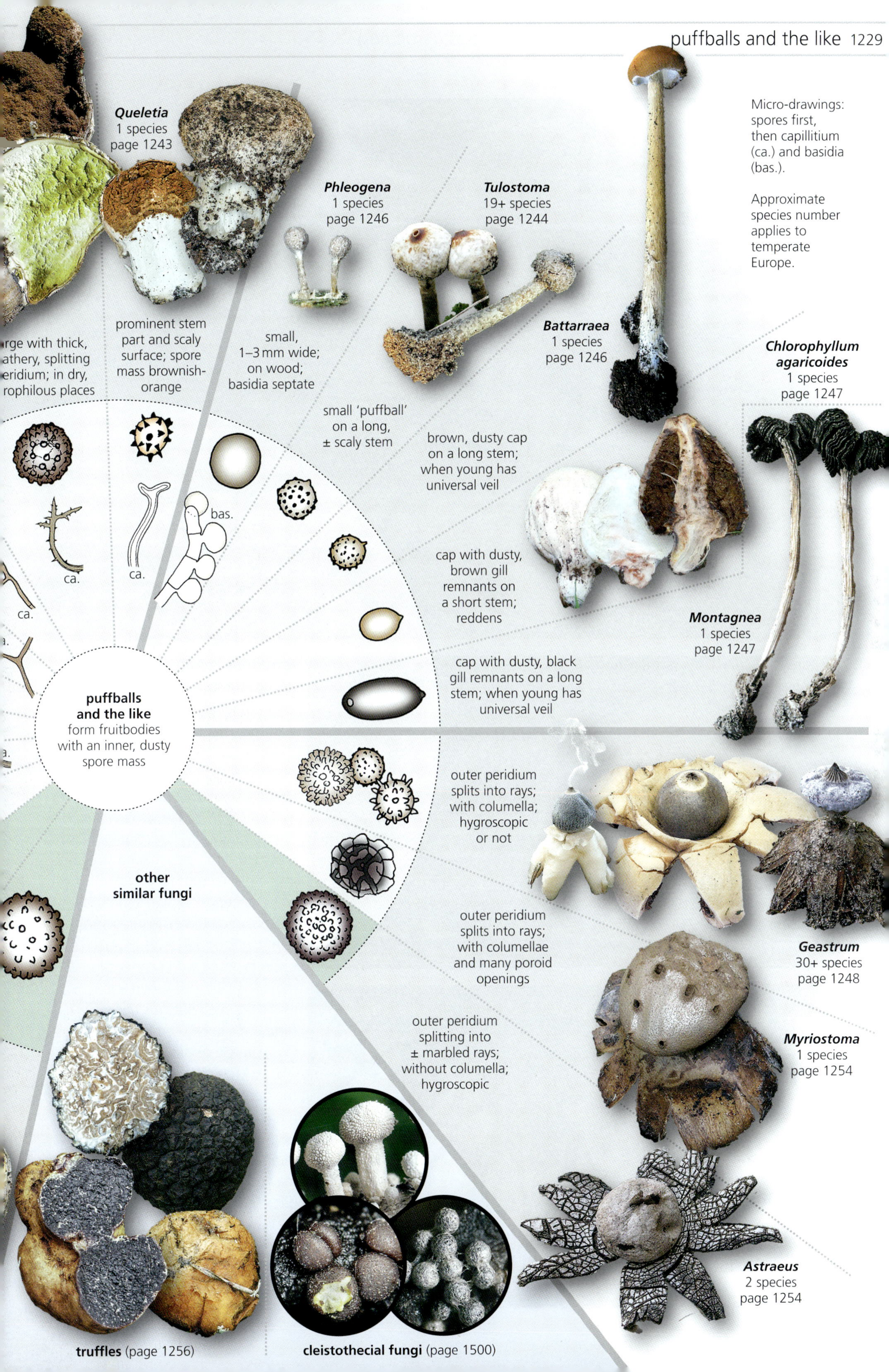
Micro-drawings: spores first, then capillitium (ca.) and basidia (bas.).
Approximate species number applies to temperate Europe.
Queletia
1 species
page 1243
Phleogena
1 species
page 1246
Tulostoma
19+ species
page 1244
Battarraea
1 species
page 1246
Chlorophyllum agaricoides
1 species
page 1247
Montagnea
1 species
page 1247
Geastrum
30+ species
page 1248
Myriostoma
1 species
page 1254
Astraeus
2 species
page 1254
rge with thick, athery, splitting eridium; in dry, rophilous places
prominent stem part and scaly surface; spore mass brownish-orange
small, 1–3 mm wide; on wood; basidia septate
small 'puffball' on a long, ± scaly stem
brown, dusty cap on a long stem; when young has universal veil
cap with dusty, brown gill remnants on a short stem; reddens
cap with dusty, black gill remnants on a long stem; when young has universal veil
outer peridium splits into rays; with columella; hygroscopic or not
outer peridium splits into rays; with columellae and many poroid openings
outer peridium splitting into ± marbled rays; without columella; hygroscopic
ca.
ca.
ca.
bas.
puffballs and the like
form fruitbodies with an inner, dusty spore mass
other similar fungi
truffles (page 1256)
cleistothecial fungi (page 1500)

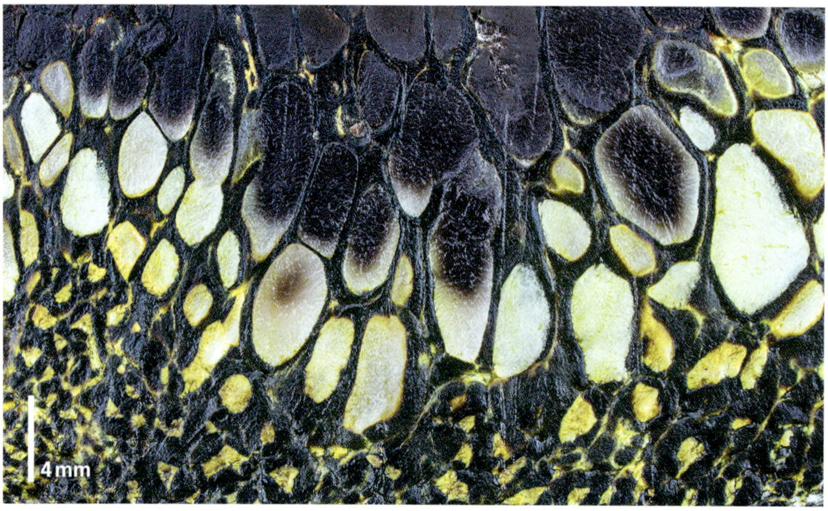

Pisolithus capsulifer is a very peculiar fungus. Initially it has large underground fruitbodies that eventually emerge from the substrate to form a large stem with a broad head; this then finally disintegrates to reveal a brown spore powder. Young fruitbodies in section (lower images) have abundant yellow pigment and crowded, firm, pea-shaped structures with internal spore production. The spiny, spherical, brown spores measure 7–8 µm in diameter. Ectomycorrhizal with *Pinus* on very nutrient-poor sand or on exposed soil in connection with mine spoil (lignite, kaolin, *etc.*).

The name *Pisolithus tinctorius* has frequently been used for this species. The name *P. arrhizus* has also been in use. Molecular studies show that the genus includes at least 14 species, many of which are southern and occur with *Eucalyptus*. However, in temperate Europe at least three species may occur with conifers (and possibly deciduous trees).

Widespread but rare in all zones; mainly June–February.

Scleroderma is a genus of ectomycorrhizal, puffball-like fungi (earthballs) with a rather thick peridium and a blackish, marbled spore mass that has a ± metallic smell. The spores are spherical, dark and ornamented. Some species have clamped hyphae. All may cause mild poisonings.

† ***Scleroderma verrucosum*** is a rather large, irregular, minutely dark-scaled earthball with a distinct root-like base and a thin, reddening peridium. The hyphae do not have clamps. The rather small, spiny, spherical spores measure 8–10 (–11) µm (without the ornament). Ectomycorrhizal with *Quercus* and other deciduous trees.

Scleroderma areolatum ▽ is very similar but has smaller fruitbodies, more regular scales and a less pronounced root-like base. The spores are larger, typically more than 11 µm in diameter. *Scleroderma septentrionale* ▷▷ grows in sandy habitats. Species of *Lycoperdon* ▷▷ do not redden in the flesh.

Common and widespread, except in the boreal zone; July–November.

Thomas Læssøe

† ***Scleroderma areolatum*** is a small earthball with a distinct, but rather short, root-like base, a thin, reddening peridium and very obvious, small, regular, dark, adpressed scales on a pale yellow-brown background. The hyphae lack clamps. The spherical spores have rather short spines and measure 10–13 µm (without the ornament). Ectomycorrhizal with *Quercus*, rarely other deciduous trees, mainly on clay-rich soils.

Scleroderma verrucosum △ is bigger with a larger root-like base and less regular peridial scales. The spores are less than 11 µm in diameter. See also *S. cepa* ▷▷.

Common and widespread in the nemoral–hemiboreal zones; July–December.

Thomas Læssøe

† ***Scleroderma citrinum*** is a sturdy, thick-skinned, coarsely scaled, mainly ± yellowish earthball that almost, or completely, lacks a stem-like base. The hyphae have clamps. The spores are spherical, have an incomplete, raised reticulum on the surface and measure 10–12 µm (without the ornament). Ectomycorrhizal with deciduous trees, mainly *Fagus*, *Quercus* and *Betula*.

Scleroderma bovista ▽ normally has an evident, but short, stipe-like base, is not as yellow and has less obvious scales; the spores have a complete reticulum. The two species also have different habitat preferences; *S. citrinum* prefers acidic soils, while *S. bovista* prefers disturbed, rich soils.

Widespread and very common, but rather rare in the boreal zone; mainly July–November, but partially rotten fruitbodies can be found year-round.

Thomas Læssøe

† ***Scleroderma bovista*** is a large earthball with a rather thin peridium, a ± scaly surface and a small, stem-like base. Rarely, it occurs totally underground. The hyphae have clamps. Spores have a complete reticulum and measure 9–13 µm (without ornament). Ectomycorrhizal on rich soils, mainly with *Quercus*.

Scleroderma citrinum △ has an incomplete reticulate spore ornament. *Scleroderma cepa* ▷ is browner and has long spines on the spores. *Scleroderma septentrionale* ▷ is very similar, but normally has a very well-defined stem-like base; it occurs in sand.

Rather common in the nemoral–hemiboreal zones; June–January.

Jan Vesterholt

† ***Scleroderma cepa*** is a potato-like earthball that lacks a stem but has a thick bundle of hyphal cords at the base; the pale yellow-brown to reddish-brown surface is somewhat cracking. The peridium lacks clamps and, when in a fresh condition, is thicker than 1 mm. The spores have curved spines about 2 µm long and 10–12 µm in diameter (without the ornament). Ectomycorrhizal in warm sites, mainly with *Quercus*, *e.g.* in parks and gardens.

Scleroderma bovista ◁ has a reticulate spore ornament and the hyphae have clamps. *Scleroderma areolatum* ◁◁ and *S. verrucosum* ◁◁ have more distinct, dark, peridial scales and a prominent root-like base; the spores have shorter spines (less than 1 µm).

Widespread in the nemoral–hemiboreal zones, occasional to rare; July–November.

† ***Scleroderma septentrionale*** is a large, 'sandy soil' earthball with a very pronounced root-like base. It can be almost completely buried. The hyphae have clamps. The spores have a complete reticulum and measure 9–15 µm (without the ornament). Ectomycorrhizal with *Pinus*, and perhaps *Salix*, in dunes and other sandy habitats.

Scleroderma bovista ◁ has a less evident root-like base, somewhat smaller spores and paler scales. *Scleroderma verrucosum* ◁◁ and *S. areolatum* ◁◁ lack clamps, have spiny spores and are more scaly.

Widespread, but rarely recorded; mainly August–October.

Thomas Læssøe

Jørgen Koch

Nia vibrissa is a small, pale brown, hairy, gasteroid fungus. The ± ellipsoid spores have a long appendage at one end, a sterigmal remnant in the other end, and four, long appendages emerging from the centre. The spores measure 7–16 × 4.5–8 µm. Occurs on marine driftwood.

There are no similar species. Only a very few basidiomycotes, *e.g. Digitatispora marina* (page 1180), form fruitbodies on marine driftwood.

Distribution and frequency not well known; probably all year.

Thomas Læssøe

Disciseda form small, puffball-like fruitbodies that, at maturity, invert to reveal an opening that originally was positioned at the attachment point.

Disciseda bovista produces rather large, somewhat compressed fruitbodies. The spores are relatively large, measuring 5–7 µm in diameter, and strongly ornamented with spines. Occurs on ± exposed soil in steppe-like habitats.

The spores of *D. candida* ▽ measure 4–5 µm and the ornament is finer; *D. hyalothrix* ⋉ has larger spores with a distinct ornament of flattened spines.

Widespread in the nemoral zones except in UK, rare; August–October.

Thomas Læssøe

Disciseda candida produces relatively small, greyish, somewhat compressed fruitbodies with a rather large opening. The small, finely spiny spores measure 4–5 µm in diameter. Occurs in warm, sun-exposed grasslands, mainly on somewhat disturbed soil, *e.g.* in parking areas.

Disciseda bovista △ has larger, more distinctly ornamented spores (5–7 µm in diameter), and *D. hyalothrix* ⋉ has even larger spores with an ornament of flattened spines.

Widespread in the nemoral zones but absent from the UK; mainly August–October.

Bovista form small, puffball-like fruitbodies that open via a pore at the top. The outer peridium is normally shed at maturity. The capillitium is mainly branched, flexible and has thin, pointed ends (a so-called *Bovista*-type capillitium).

Thomas Læssøe

Bovista plumbea is a medium-sized puffball that does not have sterile tissue (a subgleba) at the base. The fruitbody is at first chalk-white (upper image), but the outer peridium is soon lost, and the inner, lead-grey peridium becomes visible (lower image). The fruitbody opens via a somewhat irregular, floccose pore. The capillitium is branched and elastic. The almost spherical to ellipsoid, very finely ornamented spores measure 4–6 × 4–5.5 µm and have a 8–15 µm long, ± straight sterigmal remnant. Occurs in dry grassland, mainly in coastal areas; tolerant of fairly high levels of fertilizer application.

Bovista nigrescens ▽ and *B. graveolens* ⋉ have larger, at maturity browner, fruitbodies. They also have more spherical spores, and may have a recurved sterigmal remnant.

Widespread and common; mainly June–October, but may be found year-round.

Jens Maarbjerg/Thomas Læssøe

Bovista nigrescens is a large, at maturity smooth and ± dark brown puffball that has a somewhat irregular opening. The warty spores measure 4–6 µm and have a straight to slightly curved sterigmal remnant that is up to 10 µm long. Typically found in fertilized grassland, often together with species of *Agaricus*.

Bovista graveolens ⋉ looks very similar but has a long, sharply bent sterigmal remnant on the spores. *Bovista plumbea* △ has a grey surface at maturity, almost smooth, more ellipsoid spores and a ± straight sterigmal remnant.

Widespread and common; mainly June–October, but may found year-round.

Anne Storgaard

Bovista aestivalis is a ± pear-shaped puffball with obvious and extensive yellow-brown sterile tissue at the base (a subgleba), and tiny convergent spines or granules on the peridium. The base of the fruitbody is often tinged orange. Spores spherical, lack sterigmal remnants, finely ornamented and measure 3.5–4.5 µm. Occurs in dry grassland or in warm sites in open forests.

May recall *Lycoperdon lividum* ▷▷, but that species has a spongy, ± stem-like subgleba and is never orange at the base. The similar *B. paludosa* ▽ occurs in rich fens.

Widespread, occasional; mainly September–November.

Thomas Læssøe

Bovista pusilla is a small, non-rooting puffball that lacks sterile tissue (a subgleba) at the base. When mature, dark brown with a faint violet tinge and a simple, ± floccose opening. Spores 4.5–5.5 µm in diameter with a 3–9 µm long sterigmal remnant. Occurs on dry, sandy, calcareous soils.

Bovista tomentosa ⋉ is partly immersed in the substrate and is somewhat larger; the spores are not completely spherical and have a longer sterigmal remnant; *B. furfuracea* ⋉ has a fragile capillitium and no sterigmal remnant; and the alpine *B. limosa* ⋉ has a projecting opening.

Widespread, rather rare to fairly common; mainly autumn.

Jean-Marc Moingeon

Bovista paludosa is a medium-sized, grainy to smooth, eventually grey-brown puffball with compact, grey-brown sterile tissue at the base (a subgleba). The fruitbody has a simple opening. The ± spherical, very finely ornamented spores measure 4–5 µm and have a sterigmal remnant up to 15 µm long. Occurs in rich fens, *e.g.* with *Schoenus*.

The combination of habitat and microscopical characters make this species easy to recognize.

Widespread, but rather rare, *e.g.* absent in the Netherlands and Denmark; late summer–autumn.

The puffball genera ***Lycoperdon*** & ***Apioperdon*** form small to medium-sized fruitbodies that open via a pore at the top or through decay of the upper part of the peridium. Typically have spongy (chambered) tissue in the stem (a subgleba), which is ± solid in *Apioperdon*. Capillitium of branched, but not particularly pointed hyphae.

Apioperdon pyriforme is a ± long-stemmed, pale grey-brown, brown or purple-brown puffball that quickly looses its rather short peridial spines (the lower ones are more permanent than those at the top). It has pale, ± solid stem flesh (a subgleba) and distinct, white hyphal cords. The almost smooth spores measure 3.5–4.5 µm and have no sterigmal remnants. Occurs on wood, sometimes buried, mainly from deciduous trees and in many different environments.

Similar species of *Lycoperdon* have chambered stem flesh (a subgleba). *Lycoperdon perlatum* ▽ may occur on wood but has conical spines that leave a fine, reticulate pattern on the surface after detachment. *Lycoperdon dermoxanthum* ⋉ has a ± solid subgleba but the shape of the fruitbody is different, and, in comparison to *A. pyriforme*, the capillitium has many more pores.

Widespread and very common; August–November, but can be found all year in an old, weathered condition.

Lycoperdon perlatum is a rather large, ± pear-shaped, pale puffball with a small, round opening. The conical spines leave a distinct reticulate pattern after detachment. The scars are surrounded by persistent grains (inset image). Spores 3–4 µm; without sterigmal remnants. On soil and litter, rarely on wood, in both open and forested habitats.

Lycoperdon nigrescens ▷▷ has convergent, longer and darker spines. *Lycoperdon excipuliformis* ▷▷ is similar when young, but the entire upper peridium disintegrates at maturity.

Widespread and very common; mainly August–November.

Lycoperdon nigrescens is a dark grey- to chestnut-brown, ± pear-shaped puffball with groups of rather long (up to 3 mm), convergent spines and a small, round opening. The detached spines leave a fine reticulate pattern. The spores measure 3.5–4.5 µm and lack sterigmal remnants. Occurs on ± acidic soils and peat in open and wooded habitats.

Lycoperodn perlatum ◁◁ is paler and has one central conical spine or several compacted spines in each spine-group. *Lycoperdon echinatum* ▽ has spines up to 5 mm long and is also reticulate at maturity.

Widespread and common; mainly August–November.

Thomas Læssøe/Jens H. Petersen

Lycoperdon echinatum is a grey-brown, medium-sized to rather large puffball with convergent spines up to 5 mm long, which, when detached, leave a very distinct, dark, coarse reticulum. The spores measure 4.5–6 µm. Occurs on ± calcareous clay soils in deciduous forests.

Lyoperdon nigrescens △ has shorter, darker spines and is a smaller puffball from habitats on poor soils. *Lycoperdon caudatum* ⋉ also has long spines but is paler with a very indistinct reticulum.

Widespread in the nemoral–hemiboreal zones, occasional, very rare in the boreal zone; mainly August–November.

Lycoperdon umbrinum is a ± orange-brown, medium-sized, densely spiny puffball with convergent spines that do not leave a pattern when detached; the mature gleba is olive-brown. Capillitium has scattered pores. Spores finely ornamented, 4–4.5(–5) µm; in preparations not mixed with sterigmal remnants. Occurs on soil, typically with conifers.

Lyoperdon nigrescens △ has a reticulate surface. *Lycoperdon molle* ▷ has a coarser spore ornament, loose sterigmal remnants, and occurs on richer soils.

Widespread, occasional to rather common; mainly August–November.

Lycoperdon molle is a relatively large, rather pale, short-stemmed puffball with small grains and small, fragile, ± convergent spines. The capillitium is elastic with only a few pores. The spores are rather coarsely verrucose and measure 4–4.5–(5) µm – in preparations mixed with free sterigmal remnants. Mainly in ± open forests on rich soils.

Lycoperdon decipiens ⋉ and *L. atropurpureum* ⋉ have an even coarser spore ornament. The former has numerous pores in the capillitium and the latter has a very dark capillitium.

Widespread, occasional; mainly August–October.

Enrique Rubio Domínguez

Lycoperdon lividum is a pear-shaped, rather small puffball with distinctly chambered, brownish sterile tissue at the base (a subgleba). Peridium has distinct, but small, conical warts or small spines; later almost smooth. The capillitium has numerous large and spherical pores. Spores almost smooth, 4–4.5 µm, lack loose sterigmal remnants in preparations. Occurs on sandy, unfertilized soils in open habitats.

Most similar species have more distinct spines. *Lycoperdon dermoxanthum* ⋉ has a very pale subgleba.

Widespread, common, often coastal; September–December.

Lycoperdon mammiforme is a very distinctive, rather large puffball with loose flakes and rose powder on young fruitbodies. The spores measure 4–5 µm and are mixed with sterigmal remnants in preparations. Occurs on mineral-rich soils in open deciduous forests, mainly places with numerous other rare species, *e.g.* phlegmacia (*Cortinarius* subgenus *Phlegmacium* (page 698)).

Lycoperdon molle △ does not have loose flakes or pink colours.

Nemoral–hemiboreal, rare; mainly August–October.

Jens H. Petersen/Thomas Læssøe

Lycoperdon pratense is a rather small, flat-topped puffball with a skin-like membrane separating the basal sterile tissue (the subgleba) from the gleba (see arrow on inset image). The surface is mealy-spiny and, at maturity, has a large, irregular opening. Lacks a true capillitium. The spores measure 3–4 µm. Occurs in fertilized grasslands, including lawns; forms fairy rings.

Often reported as *Vascellum pratense*. The larger *L. utriforme* ▽ has a true capillitium with comma-shaped pores. *Lycoperdon lividum* ◁◁ has a true capillitium and lacks a separating membrane.

Widespread, common; mainly July–November.

Lycoperdon utriforme is a very large puffball with a wide stem that becomes scaly with age. At maturity the upper peridium disintegrates to produce a bowl-like fruitbody. The capillitium has comma-shaped pores. The almost spherical to ellipsoid, smooth spores measure 4–5 µm. Occurs in sandy grasslands.

Formerly placed in the genera *Calvatia* and *Handkea*. *Mycenastrum corium* ▷▷ is more 'thick-skinned' and lacks the distinct sterile base; it has a thorny capillitium. *Lycoperdon pratense* △ is much smaller.

Widespread, rather common; mainly June–December, but can be found in weathered condition all year.

Thomas Læssøe

Lycoperdon excipuliforme is generally a tall, thick-stemmed puffball with a disintegrating peridium. The surface on young specimens is finely spiny. The capillitium has comma-shaped pores. The spores are spiny and measure 4–4.5 µm; in preparations they are mixed with loose sterigmal remnants. Occurs on ± acidic soil in forests and plantations, rarely in more open habitats.

Short-stemmed forms may recall *Lycoperdon molle* ◁◁ but that species has a capillitium with spherical pores. *Lycoperdon utriforme* ◁ is wider, more scaly and has smooth spores.

Widespread, common; mainly August–November, but can be found in weathered condition all year.

Thomas Læssøe

Lone Melchior Hvidegaard

Calvatia fragilis is a large puffball with a whitish, smooth to floccose outer peridium. The gleba is, at maturity, purple-brown or dark violet. The capillitium is fragile and has numerous pores and septa. The spherical, coarsely verrucose spores measure 4.5–6 µm in diameter. Occurs in dry, warm habitats, in open forests, *etc*.

Also known as *C. cyathiformis*. The similar *C. candida* ⋉ has an olive-brown to grey-brown gleba.

Found in southern parts of temperate Europe, not recorded in the Nordic countries, rare; mainly August–November.

Enrique Rubio Domínguez

• ***Langermannia gigantea*** is a very large to gigantic puffball with a smooth surface; it lacks or almost lacks sterile basal tissue (a subgleba). The outer, skin-like peridium disintegrates at maturity and leaves a rather long-lasting, floccose, brown ball filled with spores. The capillitium is fragile and has many spherical pores. The almost spherical, finely roughened spores measure 3.5–5.5 µm. Occurs on rich soil on lawns, on cultivated fields, in hedgerows, *etc*. It is often found amongst nettles (*Urtica*) and may form large fairy rings, as in the upper image.

Mycenastrum corium ▷ has a very thick peridium that often splits into thick rays; it also has distinctive coarsely spiny capillitial elements.

Widespread and fairly common, becoming rarer towards the north; mainly June–November, but may be found in weathered condition all year.

Mycenastrum corium is a large, very thick-skinned puffball that lacks sterile basal tissue (a subgleba) but has a thick hyphal cord at the base. It may open almost star-like and disintegrates very slowly. It has a coarsely spiny capillitium. The large, reticulate spores measure 8–11 µm in diameter. Occurs in dry, sun-exposed, nutrient-rich habitats, often calcareous, *e.g.* on cow- or horse-grazed fields.

Langermannia gigantea is not nearly as thick-skinned and, with age, completely loses the peridium; the capillitium lacks spines.

Nemoral, with only a few records in the hemiboreal zone, rare to very rare and absent from the British Isles; mainly August–October, but present in weathered condition all year.

Johan Nitare

Queletia mirabilis is a ± clustered, erumpent, rather large puffball-like fungus with a ± flattened top that has a collar-like base. The gleba is orange-brown, and is exposed by an irregular splitting of the peridium. Stem massive, white. Capillitium thick-walled, ± unbranched and fragile. Basidia absent at maturity. Spores brown, coarsely spiny and measure 6–8.5 × 5.8–7.5 µm. Occurs on very decomposed wood and soil.

A highly distinctive fungus. *Dictyocephalos attenuatus* ⋉, known from Spain, has a brown stem, persistent basidia and a different spore ornament.

Nemoral, very rare; summer–autumn.

Lucie Zibarova

Tulostoma is a genus of long-stemmed puffballs with a distinct opening at the top. Mainly appear late in the year, and typically associated with mosses.

Tulostoma brumale is a rather pale, relatively slender *Tulostoma* with a brown zone around the projecting opening. The stem has relatively pale, brown scales. The thick-walled capillitium has hourglass-shaped swellings around the septa, and a fine grainy surface. The rough spores are almost spherical and measure 4–6 µm in diameter. Occurs on sandy, ± calcareous soils, typically in dunes, but also in gravel pits, *etc.* – always associated with mosses.

Tulostoma kotlabae ▷ is paler and lacks a brown zone around the opening. The capillitium is obliquely widened at the septa and lacks incrusted grainy material. *Tulostoma niveum* ⋉ is very pale and small, and the spore ball is somewhat elongated (not globose). *Tulostoma fimbriatum* ▽ and *T. winterhoffii* ⋉ have non-projecting openings, darker stems and lack swellings around the septa.

Widespread in the nemoral–hemiboreal zones; all year, but rarely evident during May–August.

Jens H. Petersen/Thomas Læssøe

Tulostoma fimbriatum is a sturdy, thick-stemmed *Tulostoma* with a fimbriate, non-projecting opening. The capillitium has no swellings in connection with the septa. The spores measure 4–6 µm in diameter. Occurs on sandy, ± calcareous, mainly somewhat disturbed soils, in dunes, grasslands, gravel pits, *etc.*

Tulostoma winterhoffii ⋉ is very similar but has spores of uneven size and swellings around the septa.

Almost exclusively nemoral, occasional, rarer towards the west; most of the year, but more common from late autumn to spring.

Tulostoma kotlabae is a rather slender, pale *Tulostoma* with a projecting opening that lacks a brown zone. The stem is pale brown and finely scaly. The capillitium is smooth and usually obliquely widened at the septa. The spores measure 4–5.5 µm in diameter. Occurs with mosses on sandy, calcareous soils in dunes or in similar sandy habitats.

Tulostoma brumale ◁ has a brown zone around the opening and symmetrical swellings around the septa. *Tulostoma niveum* ⋉ has an elongated spore ball.

Mainly nemoral, very rare to rare; mostly during winter.

Thomas Læssøe

Tulostoma melanocyclum is a dark-stemmed *Tulostoma* with a projecting, smooth opening surrounded by a brown zone. The capillitium has simple, non-swollen septa. The spores measure 4–6 µm in diameter, av. 5.2–5.4 µm. Occurs with mosses on ± calcareous sand in dunes, *etc.*

Tulostoma brumale ◁ has a paler stem and hourglass-shaped swellings around the septa. *Tulostoma calcareum* ⋉ is more robust, more orange-brown on the stem and has slightly smaller spores, av. 4.7–5 µm. There are other similar species farther south.

Nemoral–southern hemiboreal, very rare; mainly during winter.

Rudolf Markones

Tulostoma pulchellum is a distinctive *Tulostoma* with a membranous outer peridium and a white inner spore ball. The well-defined, somewhat raised opening is fimbriate. The capillitium lacks swellings. The spores are coarsely verrucose to almost striate and measure 4–6 µm in diameter. Occurs on ± disturbed, sandy, calcareous soils.

Tulostoma fimbriatum ◁ has a fimbriate but less obvious opening, and no distinct, semi-persistent outer peridium.

Only known from southern parts of temperate Europe, rare; mainly during winter.

Hartmut Schubert

Thomas Læssøe/Jens H. Petersen

Phleogena faginea is a small clavate, slime mould-like, dry and tough basidiomycote with a smell of lovage. It has passively dispersed, dusty spores formed on basidia that have transverse septa but no sterigmata. The brown, smooth, thick-walled spores measure 8–10 µm in diameter. Occurs mainly on bark and wood of deciduous trees, especially on recently dead, standing, dry trunks of *Fagus*.

Superficially may recall various slime moulds (page 1646) but they are not as tough, are smaller and typically have ornamented spores.

Widespread, but mainly nemoral, occasional; most of the year.

Jan Vesterholt

Battarrea phalloides is a very tall, reduced, so-called secotioid agaric with a convex head that is initially covered by remnants of a universal veil, with another part of the veil forming a volva at the stem base. The fruitbody originates from a white witch's egg. The capillitium has a spiral-shaped ornament. The spore mass is yellow-brown. The ± spherical, verrucose spores measure 4–6.5 µm in diameter. Occurs in sandy, warm habitats in scrub or open forests.

A highly distinctive species that cannot be mistaken for any other gasteroid fungi in temperate Europe.

Nemoral, *e.g.* known from England, Germany, Poland and farther south, rare; summer–autumn.

Jan Vesterholt

Chlorophyllum agaricoides is a reduced, so-called secotioid agaric that is closely related to agaricoid members of the genus *Chlorophyllum* (see page 341). The fruitbody has either a short or almost absent stem and a ± pear-shaped, white cap. The flesh reddens. The spore mass is olive-brown and becomes dusty at maturity. Lacks a distinct smell or taste. The spores are smooth, thick-walled, have a germ pore and measure 7–10 × 6–8 µm. Occurs on soil in dry, steppe-like grassland.

A distinctive fungus that is unlikely to be misidentified.

Known from Öland (Sweden) and farther southwards and eastwards, very rare; summer–autumn.

Kees Bas

Montagnea arenaria is a gasteroid (secotioid) fungus that has a close relationship with *Coprinus* (see page 522). The fruitbodies are initially a closed ball immersed in sand. After expansion the long stem has a buried volva at the base. The gills curl up at maturity. The spores are dark brown, thick-walled, very variable in size, 7–16 × 5–10 µm, and have a distinct germ pore; they are not actively shed. Occurs in sandy, warm habitats.

A distinct species but possibly the same as *M. radiosa*, in which case that is the valid name. *Montagnea* is so closely related to *Coprinus* that the two genera may have to be lumped.

A southern species within temperate Europe with no records from the British Isles, the Netherlands or the Nordic countries, rare; summer–autumn.

José Ignacio Fernández Simón/Jan Vesterholt

Geastrum (earthstars) initially produce onion-like structures, which later crack open from the top. The inner spore ball is then lifted up by the recurving rays. The opening is a pore, and it can be smooth, fimbriate or furrowed; it may be located within a well-delimited circular zone, although such a zone can be absent. The spores are spherical, warty-spiny and brown. Decomposers on ± calcareous soils or on decaying compost, *etc*.

Geastrum michelianum is a very large, fleshy earthstar with an outer peridium that typically splits into two parts, with the inner layer forming a collar (see arrow) at the base of the spore ball. It has a fairly faint delimited zone around the fimbriate opening. The unopened fruitbody resembles a brown, scaly onion. Spores 4.5–5.5 µm. On ± disturbed rich soil, compost, *etc.*, in woodlands, parks and gardens.

Formerly known under the name *G. triplex* but that is a tropical species. *Geastrum melanocephalum* ▷▷ is closely related but easily recognized by its naked spore mass; *G. saccatum* ⋉ and *G. lageniforme* ⋉ lack the inner collar, are more felty on the outside, smaller and paler, and typically have narrower, more tapered rays.

Widespread in the nemoral–hemiboreal zones, where rather common to rare, rare in the boreal zone; mainly September–November.

Geastrum fimbriatum is a medium-sized, rather pale earthstar with a sessile, smooth spore ball; the fimbriate opening is not within a well-delimited zone. The surface of the fruitbody incorporates soil particles, *etc*. The finely spiny spores measure 3–4 µm in diameter. Occurs on mull soils in deciduous woodland and in conifer plantations.

Geastrum lageniform ⋉ and *G. saccatum* ⋉ are similar but have a well-delimited zone surrounding the opening. *Geastrum rufescens* ▷ is reddish on the outer peridium, with a short-stemmed spore ball when dry.

Widespread, occasional, becoming scarcer towards the north; mainly August–November.

Geastrum rufescens is a rather compact, fleshy earthstar, which, in fresh condition, is reddish on the inner sides of the rays. The fimbriate opening is not within a delimited zone. When dry a short stem holds the inner spore ball. The spores are relatively large, 4–5 µm in diameter. Occurs on ± calcareous, mainly sand-mixed mull soils in forests and scrub.

Geastrum fimbriatum ◁ lacks the reddish colour, has spores that are only 3–4 µm in diameter and fruitbodies that are normally smaller and less fleshy.

Predominantly nemoral–hemiboreal, occasional; mainly August–November.

Thomas Læssøe/Ole Martin

Geastrum coronatum is a large, 7–12-rayed, rather dark, ± non-hygroscopic earthstar that, when dry, has a short-stemmed spore ball. It can either lack or have an indistinct delimited zone around the fimbriate opening. The spore ball is finely grainy to completely smooth. Spores rather coarsely verrucose, 4.5–5.5 µm in diameter. Occurs on mull soils in gardens, scrub and forests.

Geastrum pseudolimbatum ▷▷ is separated by its well-delimited opening zone and hygroscopic peridium rays.

Nemoral–hemiboreal, rather rare; mainly August–November (old fruitbodies seen all year).

Jens Maarbjerg

Geastrum minimum s.l. is a very small, 6–9-rayed, non-hygroscopic earthstar with a well-delimited zone around the fimbriate, somewhat projecting opening. The spore ball has a short stem when dry and a mealy surface. The spores measure 4.5–5.5 µm in diameter. Occurs on sandy, ± calcareous soils in dry habitats.

The name *Geastrum minimum* encompasses at least two ± identical species, *G. granulosum* and *G. marginatum*, but is no longer applied in a strict sense. *Geastrum schmidelii* ▷▷ is similar and occurs in similar habitats, but it has a distinctly furrowed-plicate opening.

Nemoral–hemiboreal, occasional; mainly in the autumn.

Thomas Læssøe

Thomas Læssøe

Geastrum pseudolimbatum is a medium-sized, somewhat hygroscopic earthstar (the rays curl somewhat over the spore ball in dry weather and expand again in wet conditions). It has a projecting, fimbriate opening within a rather well-delimited zone. The outer surface of the rays incorporates soil particles. When dry, the spore ball is short-stemmed and the surface somewhat rough. The spores measure 4.5–5.5 µm in diameter. Occurs typically in open, sandy, steppe-like vegetation.

Geastrum coronatum ◁◁ has a smooth or finely grainy spore ball, a ± non-delimited opening and rays that are not as pointed and, at most, only very slightly hygroscopic. *Geastrum corallinum* ▽ is hygroscopic but the spore ball is sessile.

Nemoral–hemiboreal, very rare; mainly in the autumn.

Flemming Rune

Geastrum corollinum is a small, hygroscopic earthstar (the rays curl over the spore ball in dry weather and expand again in wet conditions). It darkens with age and becomes very tough. The zone surrounding the fimbriate opening is fairly well-delimited and the spore ball is sessile. The spores measure 4–4.5 µm in diameter. Occurs in dry, warm habitats, in open scrub, *etc.*

Geastrum floriforme ⋉ is also hygroscopic but does not have a well-delimited opening zone, is typically even smaller, and the spores are somewhat larger (up to 5.5 µm) and more distinctly ornamented.

Nemoral–hemiboreal, rare; mainly September–November.

Thomas Læssøe

Geastrum fornicatum is a very large earthstar, that 'stands tall' with, typically, four steep rays from an immersed bowl-like base. The spore ball has a chimney-like, fimbriate opening that is not within a well-delimited zone. The spore ball is short-stemmed in dry conditions and has a collar-like base. The spores measure 3.5–4.5 µm in diameter. På tør, ofte kalkholdig bund, sjældent indvendig i hule stubbe og stammer.

A highly distinctive species that is unlikely to be misidentified. *Geastrum quadrifidum* ▽ similarly has a basal 'bowl' but is much smaller. *Geastrum britannicum* ⋉ is also smaller and has a dark zone surrounding the opening.

Widespread in the nemoral zones, very rare; mainly August–November.

Geastrum quadrifidum is a small, rather upright earthstar typically with four rays and a membranous, basal 'bowl' in litter. Has a distinct delimited zone around the projecting, fimbriate opening and a finely pruinose surface on the spore ball. Young fruitbodies are very pale. Spores 4–5 µm in diameter. Occurs on needle debris in calcareous forests, rarely in deciduous forests.

Geastrum minimum s.l. ⊲⊲ has no basal 'bowl' and normally more than four rays. *Geastrum britannicum* ⋉ has more rays and a very dark zone around the opening.

Mainly nemoral–hemiboreal with scattered occurrences in the boreal zone, occasional to rare; mainly August–November.

Thomas Læssøe

Jens H. Petersen/Jens Maarbjerg

Geastrum pectinatum is a medium-sized, 6–10-rayed, non-hygroscopic earthstar with a pointed, projecting, furrowed opening that does not have a delimited surrounding zone. The somewhat mealy spore ball is furrowed at the base, lacks a collar (lower inset image), and has a distinct stem when dry or when old and weathered. A collar-like structure may be present at the base of the stem. The strongly ornamented spores measure 3.5–4.5 µm in diameter. Occurs mainly on needle debris in conifer forests and in *Juniperus* scrub, rarely in deciduous woodlands; also on *Formica* anthills.

Geastrum striatum ▽ is very similar but has an evident collar at the base of the spore ball, which also has a more obvious whitish covering. Other similar species of *Geastrum* have either a sandpaper-like surface to the spore ball or a distinct delimited opening zone.

Widespread, fairly common to occasional; mainly August–November.

Jens Maarbjerg

Geastrum striatum is a rather slender, non-hygroscopic earthstar with a short-stemmed, whitish, mealy spore ball that has a collar at the base. The opening is furrowed-plicate, long and pointed. The stem of the spore ball is not visible in young, fresh specimens. The spores measure 4–4.5 µm in diameter. Occurs on rich, often calcareous soils, mainly on needle debris in parks and gardens, but also amongst litter in hedgerows.

Geastrum pectinatum △ is superficially very similar but the spore ball is darker and lacks a downturned collar.

Nemoral–hemiboreal, fairly common to rare; mainly in the autumn.

Geastrum pseudostriatum is a rather small earthstar with a rough spore ball surface, a long, tapering, furrowed opening, a many-rayed outer peridium and no basal 'bowl'. The spores measure 4.5–5.5 µm (av. 4.8). Occurs mainly in dry, calcareous, steppe-like habitats.

Geastrum berkeleyi ⋉ is a larger woodland species; *G. campestre* ⋉ is very similar but is somewhat hygroscopic and has larger spores (av. 5.3 µm).

Mainly nemoral, and found mostly in south-eastern parts of temperate Europe, rare to very rare; usually September–November.

Morten Christensen

Geastrum elegans is a rather small and pale earthstar with a furrowed opening that does not have a delimited surrounding zone. The spore ball is sessile. Occurs in sandy ± calcareous habitats.

May recall *G. fimbriatum* ◁◁ in colour and size but that species does not have a distinctly furrowed opening and occurs in woods rather than open country. *Geastrum schmidelii* ▽ occurs in similar habitats but has a delimited opening zone and, when dry, a short-stemmed spore ball.

Nemoral–hemiboreal, most frequent in coastal dunes, rather rare; mainly autumn.

John Breugelmans

Geastrum schmidelii is a small, 6–9-rayed, non-hygroscopic earthstar with a furrowed-striate, pointed opening within a well-delimited zone. The almost smooth, brownish to brown-grey spore ball has a short stem when dry. The spores measure 4.5–5.5 µm. Occurs in sandy, mainly calcareous, grasslands, typically just behind white dunes; rare away from the coast.

Geastrum pseudostriatum △ has a roughened spore ball surface; *G. elegans* △ has a sessile spore ball and *G. minimum s.l.* ◁◁ has a fimbriate opening.

Nemoral–hemiboreal, occasional; mainly October–January.

Thomas Kehlet/Ole Martin

Geastrum melanocephalum is a distinctive, very large earthstar that resembles an unpeeled, brown onion. It has thick flesh and an exposed spore mass that does not have a peridium. The spores measure 3.5–4.5 µm in diameter. Occurs in warm habitats, typically on rich mull soils, in scrub, open forests, parks, *etc*.

The naked spore mass makes this species unmistakable and formerly led to placement in its own genus, *Trigaster*. *Geastrum michelianum* ◁◁ is very similar but the spore mass is covered by a peridium.

Nemoral–hemiboreal, rare; mainly in the autumn.

Jens H. Petersen /Thomas Læssøe

Myriostoma coliforme has a somewhat shiny spore ball with many round, simple openings. The outer peridium is many-rayed, and the spore ball is attached by many short stems. The spiny-reticulate spores measure 3.5–4 µm in diameter. Occurs in scrub in very warm habitats on rich, ± calcareous soils.

Distinctive and unlikely to be misidentified.

Mainly nemoral, very rare, but with an apparently healthy population in dunes in the Netherlands; mainly in the autumn, but may be present in weathered condition all year.

Nikolay Palazov

Thomas Læssøe

Astraeus hygrometricus s.l. is a *Geastrum* 'lookalike'. The speckled rays are hygroscopic (curl over the spore ball in dry weather). The spore ball is finely felty with a simple opening. The verrucose, brown spores measure 7–10 µm in diameter. Ectomycorrhizal with *e.g.* evergreen *Quercus* in warm habitats, but also with conifers.

This genus is not well studied within temperate Europe. Another similar species, *A. pteridis*, has been recorded in Norway and Denmark and it is possible that further records from the area belong to that species rather than with *A. hygrometricus*. *Astraeus* is a member of *Boletales* and fairly closely related to *Scleroderma* ◁◁. Young, unopened fruitbodies of several *Astraeus* species are eaten in South-east Asia.

Nemoral and mainly found in southern parts of temperate Europe; all year, but in best condition in the autumn.

Jan Vesterholt

Truffles

Truffles constitute a form group where the tuber-like fruitbodies are formed ± underground (hypogaeus) and spore dispersal is passive. This strategy has arisen multiple times during the evolution of the fungal kingdom, both amongst the *Ascomycota*, *Basidiomycota* and *Zygomycota* and within the *Glomales* (A-mycorrhiza-forming fungi), where truffle-like asexual structures can be found. The fungal wheels shown here have the basidiomycote and other non-ascomycote truffles on the first page spread and the ascomycotes (the 'true truffles') on the second.

Almost all truffles are ectomycorrhizal. They are dispersed by animals, and various mammals are attracted by the pungent odours. Boar, deer, rodents, *etc.* dig the fruitbodies out of the soil and eat them. Some rodents hoard truffles. Spores from some of the species tolerate passage through the gut. The odours vary from species to species, and some may resemble pheromones. It is probably compounds of the latter nature that make true truffles an exclusive ingredient in fine cuisine.

The spores are formed internally and the mechanism for active release has been lost. Within the basidiomycote truffles this means that the sterigmata on the basidia and the apiculus on the spores are absent or have changed, and that the spores are typically more symmetrical than in basidiomycotes with active dispersal. Within the ascomycotes the cylindrical ascus has typically evolved into a ± balloon-like shape without special structures, but in some species the cylindrical shape has been maintained. The ascospores tend to be extremely large (20–50 µm).

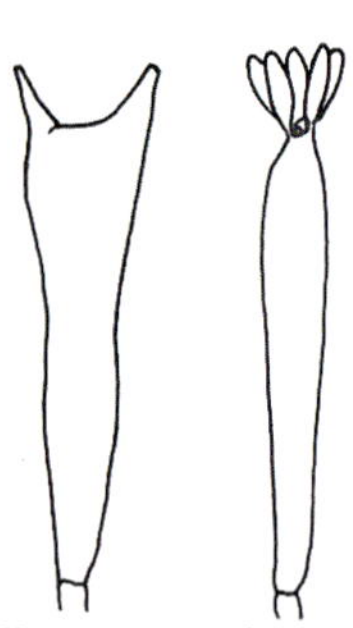

False truffle basidia: left, a 2-spored basidium from *Hymenogaster*; right, a 6-spored basidium without sterigmata from *Hysterangium*.

Micro-drawings: spores.

Approximate species number applies to temperate Europe.

How to find truffles (truffling)

Some truffles may form fruitbodies at soil level with the top clearly visible; this is often the case with, for example, species of *Rhizopogon* and *Choiromyces*. However, as the majority form completely immersed fruitbodies in litter or rotten wood, or in soil down to 10–15 cm, sometimes deeper, they have to be dug out (*e.g.* with a small hand rake). Successful truffling requires a certain 'feel' for the ecology of the species. One simple prerequisite is the presence of suitable mycorrhizal partners, mainly trees and shrubs, *e.g. Corylus*, *Tilia*, *Fagus*, *Quercus* or ectomycorrhizal conifers. Relatively light, mull soils can often be rewarding to rake through, mainly in places with a warm microclimate. Rake the surface and keep an eye out for any tuberous object (alas, many will be fungal primordia, pebbles, old nuts, *etc.*). Some truffles form potato-sized fruitbodies, but many are small, down to the size of a pea. If you think you have found a truffle, cut it through with a sharp blade and check the internal structure with a hand lens; if veined or spongy you are in luck.

A shortcut can be taken by looking out for animal scrapes, *e.g.* from deer or squirrels – or, even better, train a dog!

Other similar fungi:
- witch's eggs may look similar but do not grow underground (page 1222).
- puffballs and the like typically become dusty at maturity and are normally not immersed in soil (page 1228).
- cleistothecial fungi produce tiny fruitbodies with tiny asci (page 1500).

Further reading: 7, 98, 163, 197, 238, 239, 240, 303, 317, 318.

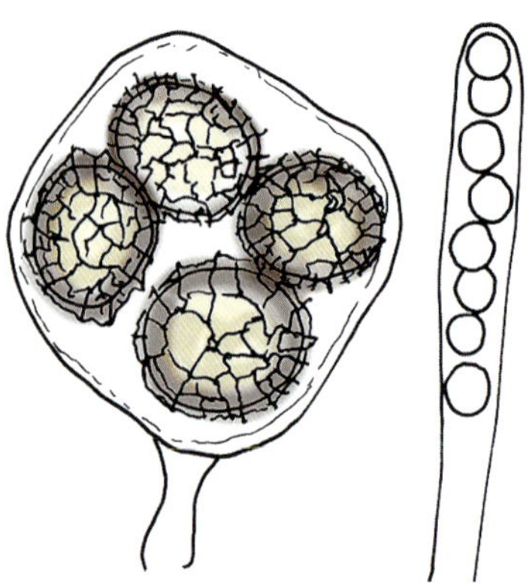

Types of truffle asci: left, a balloon-shaped type from *Tuber aestivum*; right, a cylindrical ascus from *Hydnocystis*.

Micro-drawings: spores.
Approximate species number applies to temperate Europe.
Hydnotrya
5 species
page 1270
Pachyphlodes
5+ species
page 1271
Hydnobolites
1 species
page 1271
Choiromyces
1 species
page 1272
Tuber
14 species
page 1273
photo: Flemming Rune
Elaphomyces
17+ species
page 1276
ascomycote truffles are typically folded/veined internally, except the more uniform *Elaphomyces* that eventually become powdery
other ruffles
typically
ber-like,
m-like or
ny inside
other similar fungi
with asci; fruitbodies tiny
cleistothecial fungi
page 1500

Johan Nitare

Geastrum flexuosum is a very deviating earthstar that stays underground. Fruitbody has a thick, pale peridium and numerous white hyphal cords; the inside has a columella from which capillitial threads radiate out. The spore mass (gleba) is black and powdery at maturity. Mature fruitbodies crack open in an irregular fashion. The verrucose, dark brown, spherical spores measure 4–5 µm in diameter. Occurs in black mull in warm, deciduous forests.

This species is closely related to *G. corollinum* (page 1250) but was formerly treated as *Radiigera atrogleba*. The real *R. atrogleba* ⋉ should now be referred to as *Schenella simplex* and is a more southern species that does not occur in temperate Europe.

Only known from one Swedish site in the hemiboreal zone; all year, but in best condition in the autumn.

Lucie Zíbarová

Phallogaster saccatus is a pale, basidiomycote truffle with a distinct stem and a gelatinous, veined, olive-green gleba. The stem is short and thick, and has thick, white hyphal cords from the base. The spores are smooth and measure 4–5 × 1.5–2 µm. Occurs in damp coniferous and mixed forests, in soil and rotten wood. Ectomycorrhizal.

Rather closely related to the stinkhorns in the *Phallales* (page 1222), but belongs with *Hysterangium* ▷ in the neighbouring order *Hysterangiales*; *Hysterangium* differs, amongst other things, by not having a stem.

Only known from southern and south-eastern parts of temperate Europe, very rare; spring–autumn.

Thomas Læssøe

Hysterangium calcareum is a ± irregularly shaped basidiomycote truffle surrounded by an abundant, white, ± stringy mycelium. Gleba ± olive-green and rubbery firm. The white, loosening peridium is up to 0.6 mm thick. The ± hyaline, smooth spores measure 11–18 × 4–5 µm. In calcareous soils in warm deciduous forests. Ectomycorrhizal.

Related to *Phallogaster* ◁ and belongs to the *Hysterangiales*, a sister order to the *Phallales*. The very abundant surface mycelium and narrow spores are characteristic of this species.

Mainly nemoral, very rare; summer–autumn.

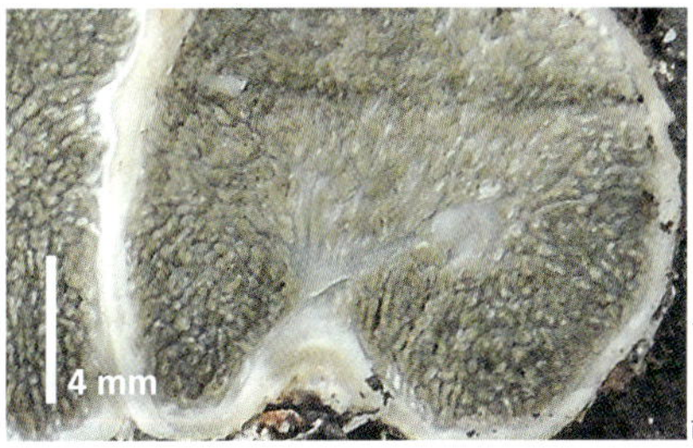

Thomas Læssøe

Gautieria morchelliformis is a highly veined and folded, yellow-brown basidiomycote truffle that lacks an outer peridium. It has a ± well-defined white stem and a columella. Smells rather foul. The spores are ornamented, ellipsoid, truncate at the end and measure 15–28 × 8–12 µm. Occurs in rich, ± calcareous soil, typically in deciduous forests. Ectomycorrhizal.

Related to *Ramaria* in the *Gomphales* (page 1138). *Gautieria otthii* ⋉ is similar but has smaller spores. There are also other closely related species of *Gautieria*.

Widespread in the nemoral zones with extensions to the hemiboreal zone, rare; summer–autumn.

Thomas Læssøe

Rhizopogon is a truffle genus with rubber-foam-like gleba and rather small, dark, smooth spores. It is closely related to the bolete genus *Suillus* and is ectomycorrhizal with conifers.

Rhizopogon obtextus is a buff to yellow-brown basidiomycote truffle with darker hyphal cords on the outside. The gleba is at first firm and finely chambered, later soft, almost liquid and dark olive-green. The smell is at first fruity, later ± rotten and foul. The surface has no reaction to KOH or $FeSO_4$. Basidia clavate to cylindrical. The spores are truncate at one end and measure (4.8–)5–8.5(–11) × 2–4 µm. Occurs mainly in sandy soils; mycorrhizal with *Pinus*.

Also commonly known as *R. luteolus*. *Rhizopogon verii* ⋉ is very similar and is also widespread in temperate Europe; it has thick-walled, lageniform basidia. *Rhizopogon occidentalis* ⋉ has spores that are not truncate, and initially has a paler peridium that stains red when bruised and with KOH. *Rhizopgon roseolus* ▽ lacks hyphal cords on the peridium and also stains red.

Widespread and common; June–November.

Thomas Læssøe

Rhizopogon roseolus is a rather small, pale yellowish *Rhizopogon* that does not have conspicuous hyphal cords on the surface. It reddens when bruised and becomes very foul-smelling with age. It has a simple, thin peridium with parallel hyphae. The spores measure (5–)6–8(–10) × (2–)2.5–4(–5) µm and are only very slightly truncate. Occurs on ± calcareous soils with *Pinus*.

Rhizopogon rubescens ⋉ is sometimes recognized as a separate, more intensely coloured species; *R. abietis* ⋉ has a loose, intricate peridium and truncate spores.

Widespread; occasional; June–October.

Sclerogaster hysterangioides is a firm, white basidiomycote truffle with a white, stringy, covering. Gleba grey-green, small-chambered, later yellowish; interior has a whitish columella. The verrucose, yellowish, spherical spores measure 4–5 µm. It is not clear whether the species is ectomycorrhizal or not; in mixed forest with ± calcareous soils.

Closely related to *Geastrum* (page 1248). The more yellowish *S. compactus* ⋉ has slightly larger spores and less mycelial covering. Species of *Hysterangium* ◁◁ have ± fusiform spores.

Mainly nemoral, either very rare or overlooked; summer–autumn.

Thomas Læssøe

Octaviania asterosperma is a medium-sized basidiomycote truffle, which at first is completely white and silk-like, but quickly turns bluish to black. The brownish gleba is chambered and white-veined. Smell is sweetish to curry-like. Spores spherical, yellow-brown, with triangular spines, 13–16.5 µm (including the spines). Occurs in ± calcareous soils in deciduous forests. Ectomycorrhizal.

Related to *Leccinum* (see p. 762). *Octaviania vacekii* ⋉ has less prominent spines, but still a very high ornament. Other *Octaviania* species may have striate spore ornaments.

Widespread in nemoral–hemiboreal zones, occasional; June–October.

Chamonixia caespitosa is a white, quickly blue-staining, mostly rather superficially positioned basidiomycote truffle with at first a whitish, later darker, chambered gleba. The ribbed, somewhat lemon-shaped, red-brown spores measure 18–20 × 12–15 µm. Mycorrhizal with *Picea*, mostly in calcareous sites with a high groundwater table. Often found on or inside rotten wood; ectomycorrhizal.

Related to the bolete genus *Leccinum* (see p. 762). Distinctive and unlikely to be misidentified.

Mainly hemiboreal–boreal, rare; summer–autumn.

Stanislav Glejdura

Jens H. Petersen/Johan Nitare

Melanogaster form ± foul-smelling basidiomycote truffles with a black, slimy gleba. Related to *Paxillus* (page 757). Spores smooth, brown and mostly ± fusiform. Ectomycorrhizal.

Melanogaster ambiguus is a grey-brownish basidiomycote truffle with a black/white-marbled, soon slimy gleba. Smell is unpleasant. Spores constricted one end, ± lemon-shaped, 14–20 × 8–10.5(–12) µm. In rich soil in deciduous forests, gardens, *etc*.

Other species of *Melanogaster* have differently shaped, smaller spores.

Widespread in the nemoral–hemiboreal zones, rather common; April–October.

Thomas Læssøe/Jacob Heilmann-Clausen

Melanogaster broomeanus is a large, firm basidiomycote truffle with yellow-brown colours and a dark, pale-veined, soon slimy gleba. Smells sweetish-nauseous. The smooth, tick-shaped spores measure 6–8 × 3–4 µm; Qav. 1.8–2.1. Ectomycorrhizal with various deciduous trees on clay or mull soils, often in rather disturbed places, along forest roads, *etc*.

Melanogaster variegatus ▽ has ellipsoid spores. The remaining species of *Melanogaster* have longer, usually differently shaped, spores.

Widespread in the nemoral–hemiboreal zones, rather common to occasional; June–November.

Jan Vesterholt

Melanogaster variegatus is a ± orange-brown basidiomycote truffle with a strong, fruity smell. The spores are ellipsoid and measure 6–9 × 3–6 µm; Qav. 1.3–1.7. Occurs in mull, mostly in disturbed sites with various deciduous trees.

The similarly small-spored *M. broomeanus* △ has tick-shaped spores with a higher Q value (1.8–2.1). *Melanogaster ambiguus* △ has large, lemon-shaped spores.

Mainly nemoral, somewhat rare to rather common in southern parts of temperate Europe; spring–summer.

Hymenogaster is a large and difficult genus of basidiomycote truffles with ± fusiform, mainly ornamented, brown spores. *Hymenogaster* is related to *Naucoria* (page 628). Ectomycorrhizal.

Hymenogaster niveus is a rather large, completely white *Hymenogaster* with a chambered, pale grey-brown gleba. The yellowish, somewhat lemon- or flounder-shaped spores typically measure 13–18 × 8–12 µm and have a wrinkled-warty, loosening outer wall and a very short apiculus. Occurs on rich soil with various deciduous trees, often in gardens and parks.

Descomyces albus ⋉ is somewhat similar but has a more complex peridium, the spores are more verrucose, and it occurs with *Eucalyptus*.

Widespread, but exact distribution and frequency not well understood, probably rather common; mainly August–December.

Thomas Læssøe

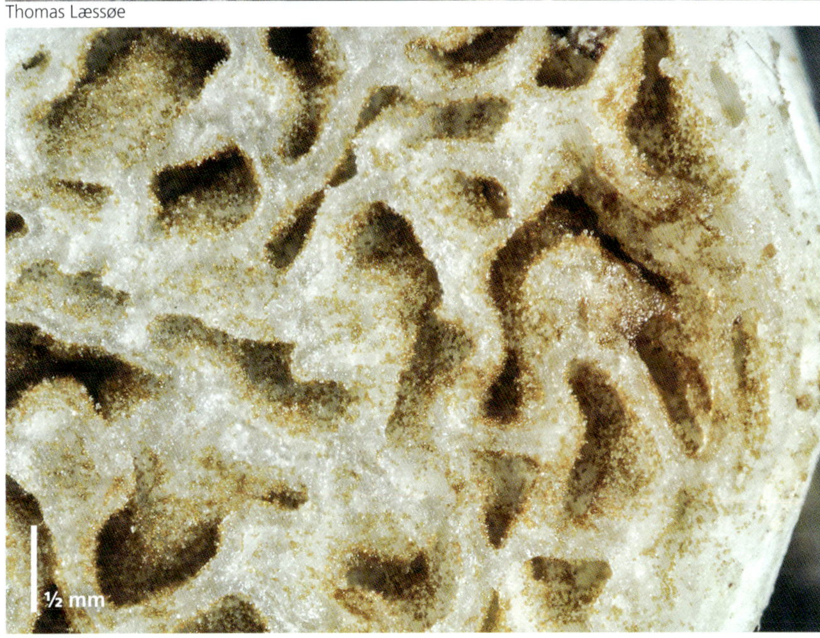

Hymenogaster luteus is a fairly distinctive *Hymenogaster* with a yellow gleba and smooth, fusiform to lemon-shaped spores that lack a loosening outer wall. Occurs on rich, mull soils; associated with various deciduous trees.

Hymenogaster bulliardii ⋉ has a somewhat darker gleba and darker spores that are more distinctly lemon-shaped.

Mainly nemoral, rare or perhaps overlooked; summer–autumn.

Thomas Læssøe

Stephanospora caroticolor is a vivid orange-yellow, medium-sized basidiomycote truffle with a chambered, non-slimy gleba. The cyanophilous, long-spined, ellipsoid spores have a peculiar, truncate end and measure 10–16 × 7–12 µm. Occurs on rich, ± calcareous soils in thermophilic woodlands; ectomycorrhizal with deciduous trees.

A very distinctive species that is unlikely to be misidentified. It is related to the corticioid genus *Lindtneria* (page 980) within the *Agaricales*.

Widespread in the southern parts of the nemoral zones, rare; summer–autumn.

Russula candida is a medium sized, whitish basidiomycote truffle with a cream, chambered, non-slimy gleba and a ± evident pale 'inner stem' (columella). The spherical, spiny, amyloid spores measure 8.5–11(–12.5) × 7–9(–11) µm. They have a truncate apiculus. In rich soils with deciduous trees, in parks and open deciduous forests; ectomycorrhizal.

Also known as *Macowanites candidus*, *Sclerogaster candidus* and *Elasmomyces krjukowensis*. *Russula candidissima* ⋉ is a very similar species with interconnected warts on the spores. *Lactarius stephensii* ⋉ is a similar *Lactarius*-truffle with white milk and a buffish peridium, which reach Öland in Sweden in the north but mainly occurs further south.

Widespread in the nemoral-hemiboreal zones; rare to very rare; July–October.

Endogone is an ectomycorrhizal member of the phylum *Zygomycota*.

Thomas Læssøe

Endogone flammicorona is a white, firm zygomycote truffle with a yellowish, juicy to milky gleba. The spores are surrounded by a strange projecting, veined structure (hence the name '*flammicorona*') and measure 80–140 µm in diameter. Rather superficial in the soil, in both deciduous and coniferous forests.

Endogone lactiflua ▽ has a less strongly veined outer structure on the spores, but a denser, more finely spun covering.

Widespread but rarely recorded; summer–autumn.

Thomas Læssøe

Endogone lactiflua is a chalk-white, firm zygomycote truffle with a milky, yellow-brown gleba. The large spores have a dense, stringy covering and measure 70–200 µm in diameter. Occurs rather superficially in both deciduous and coniferous forests.

Endogone flammicorona △ has a more projecting, veined spore surface structure, but otherwise is very similar macroscopically.

Widespread, probably rather common; summer–autumn.

Glomus macrocarpum recalls a truffle but is asexual and belongs in *Glomeromycota*/*Glomales* (see pages 12–13). The truffle-like structure contains numerous very large, spherical, long-stemmed, thick-walled, asexual resting spores that measure 85–230 µm in diameter. Superficially in turf, deep in grass tussocks and in other herb-rich places. A-mycorrhizal.

Glomus microcarpum ⋉ has somewhat smaller spores that measure 30–50 µm in diameter. *Glomus fuegianum* ⋉ has spores that are grouped in clusters and measure 30–80 µm in diameter.

Widespread, probably common; mainly July–November.

Johan Nitare

Geopora is an ascomycote genus with both truffles and normal cup fungi (page 1304). Related to *Humaria* (page 1308). Ectomycorrhizal.

Geopora cooperi is a rather large truffle with a veined, pale brown interior and a ± brown-felty surface. The asci are cylindrical and have a lid (operculum). Spores smooth, broadly ellipsoid, 18–22 × 12–14 µm. Rather superficially on ± calcareous soil in coniferous and deciduous forests.

Recalls species of *Hydnotrya* ▷▷ but those have ornamented spores.

Widespread, rare or perhaps overlooked; summer–autumn.

Jan Vesterholt

Geopora clausa is a ± clay-buff, rather small, hollow to somewhat veined truffle with a minutely hairy-felty to minutely scaly surface. It has a pleasant smell. The asci are cylindrical but do not have a functional lid. The smooth, ellipsoid spores measure 19–25 × 13–18 µm. Occurs both in sandy *Pinus* forests and in more scrub-like habitats on sandy soil.

Geopora cooperi △ is more veined, more hairy and has asci with a distinct lid. Species of *Genea* ▽ have ornamented spores.

Occurs in southern parts of temperate Europe, rather rare; summer–autumn.

Caroline Hobart/Thomas Læssøe

Genea includes partly underground, ± folded ascomycote truffles with ± cylindrical asci. Ectomycorrhizal.

Genea hispidula is a small, ± orange-brown, typically almost superficial truffle that is covered in long, rigid hairs, has a distinct opening on the top and a non-folded interior. The asci are cylindrical. The ellipsoid spores are covered in coarse, ± rounded warts and measure 29–45 × 20–33 µm. Occurs in deciduous forests and scrub on ± calcareous soils.

Other species of *Genea* are either not as hairy or are strongly folded.

Widespread, at least in the nemoral zones, rather common; mainly September–October.

Genea verrucosa is a dark grey-brown, ± folded truffle with a warty surface. May be enveloped in a white mycelium formed by a parasitic basidiomycote. The asci are up to 220 µm long, ± cylindrical and contain a row of eight large spores that measure 23–29(–33) × 21–24(–27) µm and have an up to 3 µm high ornament of irregular, ± truncate warts. Occurs in ± calcareous soil in thermophilic deciduous forests and scrub.

The more folded *G. fragrans* ⋉ and *G. sphaerica* ⋉ have longer asci (more than 260 µm) and, typically, longer spores.

Widespread in nemoral–hemiboreal zones, rather rare; mainly July–October.

Thomas Læssøe

Hydnocystis bombycina is a small, pale, compact, internally white/pale brown, veined truffle. The surface is minutely hairy. The asci are cylindrical, and the spherical, smooth spores measure 17–23 µm in diameter. Occurs in deciduous and coniferous forests on mull soils; the ectomycorrhizal status is currently in doubt.

Closely related to *Geopyxis* (page 1311). Similar truffles with spherical spores have obvious spore ornaments.

Widespread in the nemoral–hemiboreal zones, but rare; summer–autumn.

Johan Nitare

Paurocotylis pila is a very unusual smooth, ± coral-red truffle. The interior is coarsely folded. The asci are ± cylindrical and typically 4–6-spored. The spores are spherical, smooth and measure 17–20 µm in diameter. Superficially on disturbed soil in gardens and parks. Possibly a decomposer rather than mycorrhizal.

Related to *Geopyxis* (page 1311). A highly distinctive species that is unlikely to be misidentified.

Introduced from New Zealand to the UK (presumably with soil) where it is now widespread towards the north – so far not recorded elsewhere in Europe; most of the year.

Clive Shirley/Jerry Cooper

Thomas Læssøe

Hydnotrya is a genus of ascomycote truffles with ± folded, but non-solid fruitbodies and with very large, pale brown, ornamented spores. *Hydnotrya* is related to *Gyromitra* (page 1285). Ectomycorrhizal.

Hydnotrya tulasnei is a rather large, ± smooth, buff, greyish-brown to dark red-brown truffle; the inside is strongly folded but retains some cavities. The asci are more or less irregular and have scattered to somewhat more organized rows of spores. The spores are hyaline to brown, spherical, and have a coarse, irregular warty ornament up to 7 µm high; they measure 22–32 µm in diameter (without the ornament). Occurs mostly in deciduous forests, but also with conifers – mostly in sandy, ± calcareous soils.

Hydnotrya bailii ⋉ apparently has smaller fruitbodies, cylindrical asci with spores in one row, and is perhaps strictly associated with *Picea*. Other species of *Hydnotrya* are ± hollow and the spores have other ornaments or are a different size.

Widespread and fairly common; mainly July–October.

Hydnotrya michaelis is a pale brown, somewhat irregularly shaped truffle with a pale and hollow interior. The spores are ellipsoid, have a non-striate, thick ornament of interconnected rounded warts and measure 26–34 × 18–23(–25) µm (without the ornament). Fruits at the soil surface in association with conifers on ± sandy soils.

The other hollow species of *Hydnotrya*, *H. confusa* ⋉ and *H. cubispora* ⋉, have striate spore ornaments and the latter has almost box-shaped spores.

Widespread, but rather rare; mainly April–September.

Balsamia polysperma is a rather small, brownish-orange, warty, firm-fleshed truffle with a solid, pale, veined interior. The asci are balloon-shaped and have clustered spores that are ellipsoid, smooth, hyaline and measure 21–26(–31) × (12–)13–15 (–18) µm. Fruits underground, and mainly mycorrhizal with deciduous trees in ± calcareous soils.

Also known as *B. platyspora*. It is related to *Helvella* (page 1288). *Balsamia vulgaris* ⋉ is very similar but has longer spores.

Widespread in the nemoral–hemiboreal zones, rather rare to rare; summer–autumn.

Enrique Rubio Domínguez

Pachyphlodes nemoralis is a medium-sized, brown, flat-warty ascomycote truffle with a ± yellowish 'navel'. The gleba is ± yellowish to greenish and veined/folded. The spores are spherical, have a twisted, low, confluent ornament and measure 16.8–19 µm. On rich soils in deciduous forests – usually fruiting rather deep in the soil.

Pachyphlodes is related to e.g. *Peziza* (page 1313) and *Plicariella* (page 1322). *Pachyphlodes citrina* ⋉ has a more discrete spore ornament; *P. melanoxantha* ⋉ has coarser warts on the peridium.

Widespread in the nemoral–hemiboreal zones, probably uncommon; summer–autumn.

Thomas Læssøe/Jens H. Petersen

Hydnobolites cerebriformis is a small, pale, whitish to yellow-brown, ± irregular, brain-like truffle. The asci are broadly ellipsoid, mostly 8-spored, inamyloid and disorganized. Lacks paraphyses. The spores are spherical, have a coarse, reticulate ornament and measure 18–25 µm (without the ornament). In deciduous forests and damp scrub, in mull soils or inside very decomposed wood with soil contact.

Hydnobolites is phylogenetically related to *Peziza* (page 1313). Species of *Tuber* ▷▷ are mostly more regularly rounded and have fewer spores in the asci.

Widespread, but rare, perhaps overlooked; summer–autumn.

Thomas Læssøe

Ebbe Wedel Eriksen

Choiromyces venosus is a large to very large, somewhat irregularly shaped truffle with almost potato-like colours and a pale marbled/veined gleba. The odour is very heavy and somewhat nauseous. The spores are ± spherical, spiny, and measure 16–22 µm. Mycorrhizal with deciduous trees, *e.g. Fagus* and *Quercus*, but also known from conifer forests; on clay-rich, ± calcareous soils.

Choiromyces is closely related to *Tuber* ▷. *Tuber magnatum* ⋉ is similar, but has a more agreeable smell and the spores have a coarse reticulum. Species of *Scleroderma* (page 1231) have metallic smells and quickly turn almost black and pulverulent internally. *Choiromyces* has been claimed to be mildly poisonous but more recent data indicate that this is not the case.

Widespread in the nemoral–hemiboreal zones; rare but locally occasional; summer–autumn.

Thomas Læssøe

Tuber is a large genus of ectomycorrhizal ascomycote truffles that have fruitbodies with a ± solid, distinctly veined gleba and strong smells at maturity. The spiny or reticulate spores are large and pale brown. The famous edible truffles, *T. magnatum* ⋉ and *T. melanosporum*, are southern species.

• ***Tuber aestivum*** is a completely black, rather large, coarsely warty truffle with a greyish, marbled/veined gleba. It has a strong truffle aroma. The spores have a reticulate ornament and measure 24–40 × 21–29 µm. Thermophilous and often found in sun-baked sites with various deciduous trees in clay-rich, mostly calcareous soil.

Tuber mesentericum ⋉ has a basal 'navel' and at maturity a phenolic smell. *Tuber macrosporum* ▷▷ has larger spores (45–75 × 30–50 µm) and smaller warts on the peridium. *Tuber brumale* ⋉ has a spiny ornament.

Widespread in the nemoral zones north to Gotland in Sweden, occasional to rare; June–December.

Christian Lange

• ***Tuber macrosporum*** is a red-brown, warty/scaly truffle with a white/orange-brown-marbled gleba. The asci are balloon-shaped and 1–3(–4)-spored. The ellipsoid spores have a fine reticulum with 7–9 cells across the spore, and measure 45–75 × 30–50 µm. A thermophilic species from rich, calcareous soils, in connection with *Quercus*, *Tilia*, *etc*.

The combination of fine red-brown peridial scales and very large spores make the species easy to recognize.

Widespread in the nemoral zones, reaching as far north as the southern parts of Denmark but mainly more southern, rare; August–October.

Enrique Rubio Domínguez

Tuber maculatum is a buff-yellow to pinkish-buff, rather small to medium-sized truffle with a pale, marbled/veined gleba. The peridium consists of an outer, waxy, compact layer of angular, brownish cells and an inner more hyphal layer. The spores have a reticulate ornament with 4–8 meshes lengthwise, and measure 25–47(–55) × 21–40 µm (without the ornament). In mull or clay soils, mostly in deciduous woodland, but also with conifers.

Tuber puberulum ▷ and others may have ± distinct hairs on the peridium and no wax-like outer layer.

Widespread in the nemoral–hemiboreal zones, occasional; July–November.

Tobias Bøllingtoft

***Tuber rufum* s.l.** is a rather small to small, cinnamon-brown to pale orange-brown, ± smooth truffle with a cinnamon-brown and white marbled gleba. The spores are spiny and measure 25–32(–40) × 18–25 (–29) µm (without the ornament). Mycorrhizal with deciduous trees in mull, often ± calcareous soils – in woodlands, parks and gardens.

Most similar *Tuber* species have a reticulate spore ornament. *Tuber rufum* is both morphologically and genetically rather variable and is here treated in a rather broad sense.

Widespread in the nemoral–hemiboreal zones, occasional; June–November.

Enrique Rubio Domínguez

Tuber rapaeodorum is a ± pale orange-brown, rather small, smooth truffle with white veins in the dark brown gleba. The outer peridium is a dense layer of 6–30 µm wide cells (pseudoparenchymatic), the inner layer is thick and consists of intricate of 4–5 µm wide hyphae. Peridial hairs are rare or lacking; asci typically 3–4 spored. The ellipsoid spores measure 37–60 × 28–40 µm; Q 1.3–1.5. Occurs in a rather wide range of habitats, mostly with deciduous trees.

Tuber maculatum ◁ has a layer of angular, thick-walled cells in the outer peridium. *Tuber puberulum* ▽ is more hairy.

Widespread, rare to occasional; summer–autumn.

Tuber puberulum is a small, very pale truffle with a pale brown, pale-veined gleba. The peridium is very thin and covered with tufts of pointed, rather long, slightly thick-walled hairs. Asci mostly 1–3-spored. The spores are almost spherical, Q 1.18–1.35, and measure 28–55 µm in diameter. Occurs in light mull soils, mainly in deciduous forests.

Tuber rapaeodorum △ and *T. borchii* ⋉ have a higher spore Q, a thicker peridium and fewer hairs.

Widespread, becoming scarcer towards the north, occasional; summer–autumn.

Tobias Bøllingtoft

Thomas Læssøe

Thomas Læssøe

Thomas Læssøe

Species of ***Elaphomyces*** have a thick, outer peridium and a dense, uniform spore mass. Belongs to the *Eurotiales* together with many cleistothecial fungi (page 1500). Ectomycorrhizal. Some species are parasitized by species of *Tolypocladium* (page 1518).

Elaphomyces muricatus is a rather large, cinnamon-buff to yellow-brown, coarsely warty *Elaphomyces* with a marbled (in section) peridium and a dark, pulverulent spore mass. The red-brown, spherical spores have an ornament of dense, short, curved spines and measure (16–)18–21(–23) µm. Usually occurs in rather rich soils, often calcareous, in both deciduous and coniferous forests, but also in poorer habitats.

Elaphomyces muricatus var. *reticulatus* ⋉ has larger spores, up to 28 µm. *Elaphomyces decipiens* ⋉ has radiating pale spots in the peridium; the peridium of *E. quercicola* ⋉ is covered in tapering, long spines and is enveloped in a hyphal layer containing soil and litter; the spores have more isolated spines. *Elaphomyces barrioi* ⋉ has uneven peridial spines, a darker peridium and more truncate spines on the spores; *E. violaceoniger* ⋉ has a very dark peridium and a very irregular spore ornament of confluent warts and spines.

Widespread, becoming scarcer towards the north, common; all year.

Elaphomyces granulatus is a rather large to very large, brown *Elaphomyces* with rather small peridial warts and a white to yellowish (in section) uniform peridium. The spores are red-brown, spherical, measure 20–29 µm, and have an ornament of thick, up to 2 µm high, ± convergent warts. Occurs in poor, usually sandy, soils, mostly in conifer forests and *Fagus* stands.

Elaphomyces asperulus ▷ has a bluish to wine-red-tinted peridium (as seen in section).

Widespread, becoming scarcer towards the north, rather common; all year.

Elaphomyces asperulus is a brownish, finely warty *Elaphomyces* with a bluish to wine-red (in section) uniform peridium. The spherical spores measure 22–30 µm and have short, rod-shaped warts that form an irregular, reticulate pattern. Occurs in both deciduous and coniferous forests, and in a range of soil types.

Elaphomyces hassiacus ⋉ has a dense, uniform spore ornament.

Widespread, becoming scarcer towards the north, rather common to occasional; all year.

Thomas Læssøe

Elaphomyces maculatus is a rather large, black and smooth *Elaphomyces* that typically has traces of a greenish mycelium on the surface. The spores, 28-42 µm, are brown, spherical, and have a very dense, slightly radiating and grouped ornament of tiny rods. Occurs in ± calcareous soils, *e.g.* with *Quercus* and *Corylus*.

Elaphomyces leveillei ⋉ has smaller spores, 18–25 µm, and a more fragmented spore ornament.

Widespread in the nemoral–hemiboreal zones, rare and not recorded in some countries; all year.

Thomas Læssøe

Elaphomyces cyanosporus is a dark, indistinctly warty *Elaphomyces*. The gleba is greyish-blue and has a strong smell, very like a truffle in the genus *Tuber*. The peridium is pale (in section) and fairly uniform. The spores are bluish, spherical, measuring 23–26 µm, and have a 1.5–2.5 µm high, reticulate ornament. Occurs with deciduous trees; thermophilous.

Elaphomyces persoonii ⋉ has larger fruitbodies (30–70 mm; 15–30 mm in *E. cyanosporus*), and darker spores with a more complex reticulate ornament.

Recorded in the UK and farther south, apparently very rare; all year.

Enrique Rubio Domínguez

Operculate cup fungi

The operculate cup fungi are characterized by a small lid (an operculum) at the top of the asci, and all operculate species are classified in the order *Pezizales*. The open fruitbodies (apothecia, see page 30) have the hymenium on the surface/inner side of the cap or cup. They are often rather large and may be ± cup-shaped, either have or lack a stem or, as in the morels, be more folded or pitted.

The hymenium is a mixture of asci and paraphyses. The spores are mostly large and may be ornamented with *e.g.* a raised reticulum or with warts or spines. The ends of the spores may have appendages.

Some species are ectomycorrhizal, while others are decomposers of dead organic material, *e.g.* dung, wood or debris. A third group is parasitic, mainly on bryophytes and hepatics.

The dung-inhabiting (coprophilous) species often produce very small fruitbodies compared to the ectomycorrhizal species. The smallest fruitbodies are found in the genera *Ascozonus*, *Coprotus* and *Thelebolus* and can be less than 0.5 mm in diameter. The species in these three genera also have aberrant asci, where the top cracks instead of opening with a lid, and are classified in an order of their own, the *Thelebolales*.

OTHER SIMILAR FUNGI:

– inoperculate cup fungi have asci with an internal plug-like structure or no structure at all. They are mostly smaller, typically 0.5–5 mm in diameter, and mainly occur on dead herbaceous stems, fallen leaves or dead wood. Club-shaped inoperculate cup fungi (page 1348) may resemble small morels (page 1280) and often occur on soil.

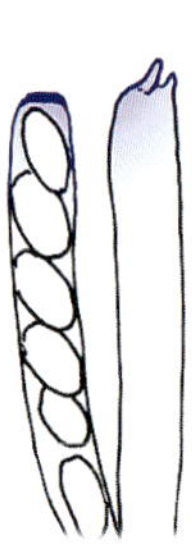

Amyloid asci with lid in a *Peziza* before and after spore discharge.

operculate cup fungi with yellow, orange, red to pinkish hymenium
page 1328
operculate cup fungi
have open fruit-bodies (apothecia) and asci with a lid (an operculum)
other similar fungi
asci without operculum, but often with an inner apical apparatus
inoperculate cup fungi
page 1348

Morels and the like

All species belong to the operculate cup fungi, *i.e. Ascomycota*, and have an exposed hymenium and asci that open with a lid (an operculum). All have inamyloid asci. They produce the largest fruitbodies (apothecia) within the ascomycotes – species of *Morchella* ▷ and *Gyromitra* ▷▷ can reach more than 20 cm in height. The hymenophore is often highly folded, resulting in the hymenium having a large surface area.

Morchella may form mycorrhiza-like associations with plants, *e.g.* with *Rosaceae*, *Ulmus* or herbs. *Gyromitra* is saprotrophic, while *Helvella* is ectomycorrhizal.

Other similar fungi:
– some species in the group 'earthtongues and the like' may look similar to tiny morels. These, however, have inoperculate asci (page 1350).

See also the main wheel, page 1278.

Further reading: 101, 180, 266, 301 328.

Morchella is a genus of large, operculate cup fungi with a pitted and hollow head on a hollow stem. The spores are large and smooth. The genus probably includes both mycorrhizals and saprotrophs.

●● ***Morchella esculenta*** is a large, at maturity ± buff *Morchella* with a hymenophore of large, rather regular honeycomb-like cells. The hollow top may be conical or almost globose, and the stem length and thickness is very variable. It has a pleasant, spicy odour. The smooth spores measure 18–22(–27) × 12.5–14 µm, and they have numerous external drops at the poles but none internally. Occurs on rich, ± calcareous soils – often with *Ulmus*, but also found with *e.g. Fraxinus* and *Populus*; mycorrhizal.

Morchella esculenta is part of difficult species complex. *Morchella vulgaris* ▽ is darker with more irregular hymenophoral cells. *Morchella americana* ⋉, found in central Europe, was probably recently introduced from North America. *Morchella steppicola* ⋉ occurs in the steppes of Eastern Europe.

Widespread, occasional; April–June.

●● ***Morchella vulgaris*** is a relatively large to large, rather dark *Morchella* with a hymenophore that has a rather irregular honeycomb-like pattern of cells with thick walls. The smooth spores measure 18–22(–27) × 12.5–14 µm and have numerous external drops at the poles, but none inside. Occurs on ± calcareous soil in deciduous forests, parks, gardens and dunes, may occur with a range of partner trees, *e.g. Crataegus*, *Malus*, *Fraxinus* and *Sorbus*; mycorrhizal.

Morchella americana ⋉, which was apparently introduced to Europe from North America, is very similar and probably requires sequencing in order to confirm identification. *Morchella esculenta* △ is typically paler and the cells in the hymenophore are more regular.

Widespread, occasional; mainly April–June.

Jean-Marc Moingeon

Jean-Marc Moingeon

Jean-Marc Moingeon

•• **The *Morchella conica/elata* complex** includes species with distinct longitudinal ridges on the hymenophore divided by horizontal ridges, and which have a ± conical shape to the fertile part. The stem is almost smooth and nearly white. The fruitbodies are hollow, like species in the *M. esculenta* group ◁◁. The hymenium colour ranges from pale grey through purple-grey to dark grey-brown and can vary depending on maturity.

Both macro- and microscopically the species in this complex are almost identical and identification without DNA data requires first-hand experience of many sequenced collections. In temperate Europe the following species are known to occur: *M. deliciosa* ⋉, *M. eohespera* ⋉, *M. eximiodes* ⋉, *M. exuberans* ⋉, *M. importuna* ▽ and *M. purpurascens* (top image was determined as *M. purpurascens*). Note that the two classic and very old names in the group, *M. conica* and *M. elata*, are missing from the list; the use of these names has still to be formally agreed. All species have smooth spores that measure 17–25 × 10–15 µm, and occur in many types of woodland, but mostly on calcareous soils; some produce mass fruitings after forest fires.

Widespread as a group, occasional; March–June.

•• ***Morchella importuna*** is a member of the conical, saprotrophic *Morchella* group △. The hymenium is ± grey-brown, has 12–20 vertical ridges and numerous horizontal ridges, creating a laddered appearance. The smooth spores lack drops and measure 18–24 × 10–13 µm. Occurs on recently laid wood- and bark-chip beds, often in towns and cities, but may also occur after fires in conifer forests; it is often characterized as a ruderal.

Possibly the same species as originally described as *M. elata* ⋉, but this has yet to be resolved.

Widespread in the nemoral–hemi-boreal zones, occasional; April–June.

•• ***Morchella semilibera*** is a small, thin-fleshed *Morchella* with a, ± conical, usually rather dark fertile part (the upper image depicts rather young pale apothecia), and a thin-fleshed, grainy, hollow stem. The margin around the lower part of the head is free of the stem. The asci are cylindrical and 8-spored. The ellipsoid spores have external oil drops at the poles and measure 22–26(–30) × (12–)15–16(–18) µm. Occurs on ± clay-rich, disturbed ground, mostly rather damp places, *e.g.* under the leaves of *Petasites*.

This is the only European species of *Morchella* with a free margin around the head. In *Verpa* ▷▷ the fertile part is even more free, only attached at the very top of the stem.

Nemoral–hemiboreal, occasional; mainly April–May.

Verpa is a small genus of morel-like species where the stem is attached to the cap as a clapper is to a bell. One species has a wrinkled hymenophore; the other is nearly smooth. Both have large, smooth spores and the asci are either 2- or 8-spored. Assumed to be decomposers but may possibly form a kind of mycorrhiza.

• ***Verpa bohemica*** is a yellow-brown, veined-wrinkled *Verpa* with an almost free fertile part. The asci are 2-spored. The spores are narrowly ellipsoid, somewhat asymmetrical and measure 55–87 × 17–20(–22) µm. Thermophilic, on sandy, ± calcareous mull soils, typically with *Populus*.

Also known as *Ptychoverpa bohemica*. *Morchella semilibera* ◁◁ has a free margin but the fertile part and the stem connect less than halfway towards the top and it has 8-spored asci.

Widespread, found mainly in continental areas and very rare in the lowlands of north-western Europe; mainly March–May.

• ***Verpa conica*** is an almost non-veined to somewhat wrinkled-veined, rather dark *Verpa* with a thimble-shaped fertile part that is narrowly attached to the stem. The asci are 8-spored. The spores have external oil drops at the poles and measure 21–25 × 11–15 µm. Occurs on rich soils in open deciduous forests and scrub or in more open habitats, *e.g.* in coastal grasslands.

Verpa bohemica △ is wrinkled like *Morchella semilibera* ◁◁, and has 2-spored asci. Species of *Helvella* ▷▷ are more saddle-shaped.

Widespread, occasional, becoming scarcer towards the north; mainly April–May.

Gyromitra form cup-shaped or folded, rather brain-like apothecia, and usually have a furrowed stem that lacks a complete cavity. Spores are smooth or ornamented and have large appendages at the poles. Probably predominantly decomposers. Many or all species contain the poisonous compound gyromitrin. See also the cup-shaped *Gyromita* on page 1292.

††**Gyromitra esculenta** is a brain-like, red-brown to purple-brown *Gyromitra* with a short, white, thick stem. The smell is spicy and pleasant. The ellipsoid-fusiform, smooth, hyaline spores have two small drops and measure 20–24 × 10–11.5 µm. Occurs on soil and wood chips close to conifers, typically in *Pinus* forests and plantations.

Gyromitra splendida ⋉ is a northern, longer-stemmed species with larger, more fusiform spores measuring 23–30 × 11.5–14 µm. *Gyromitra gigas* ▷▷ has larger, sometimes more yellow-brown fruitbodies and spores with rounded appendages at the poles.

Widespread, occasional; February–May.

††**Gyromitra gigas** is a very large and heavy, brain-like, dark cinnamon-buff, brownish-orange to purple-brown *Gyromitra*. The spores measure 26.5–30.5 × 11.5–15 µm and have a finely reticulate ornament and rounded appendages at the poles. Occurs on both rich and poorer soils, near decaying conifer wood.

Gyromitra esculenta ◁◁ does not have spore appendages and is typically smaller and darker when young. *Gyromitra fastigiata* ▽ has a more angular or jester's/fool's cap-shaped appearance, occurs with deciduous trees and has coarse spines at the spore poles.

Widespread, rather rare to very rare, becoming scarcer towards the north; February–July.

Thomas Læssøe

††**Gyromitra fastigiata** is a very large, veined-grooved, irregularly shaped, sometimes somewhat resembling a jester's or fool's cap, brownish-orange to orange-brown *Gyromitra* with a thick, pale stem. The spores have a distinct reticulate ornament, conspicuous spines at the poles and measure 26.5–36.5 × 12–14 µm. Occurs on calcareous soils in deciduous forests, *e.g.* around fallen trunks of *Fagus*.

Gyromitra gigas △ occurs with conifers and has a more rounded shape; the spores have rounded appendages at the ends.

Widespread in the nemoral zones with extensions into the hemiboreal zone but very rare, absent from most of the north-western lowlands, including the British Isles; May–June.

††***Gyromitra infula*** is a long-stemmed *Gyromitra* with a purple-brown, jester's/fool's cap-shaped fertile part. The smooth, ellipsoid spores measure 20–21 × 7.5–9 µm. Occurs on wood, wood chips and sawdust, mostly from conifers.

Gyromitra ambigua ⋉ has larger, more fusiform spores and a darker stem. *Gyromitra fastigiata* ◁ is larger, more irregularly veined, has larger spores and occurs in connection with deciduous wood; it is a spring rather than autumn species.

Widespread and common in the hemiboreal–boreal zones, rather rare in the nemoral zones; mainly September–November.

†?***Gyromitra korshinskii*** is a very distinctive *Gyromitra* with a very wrinkled, partly rose stem and a rounded-veined, grey-brown fertile part that has a whitish underside. The spherical, smooth spores measure 8–12 µm. It typically occurs on moss-covered trunks in coniferous forests and on large trunks of *Populus tremula*, but also known from garbage dumps.

Also known as *Pseudorhizina sphaerospora*. A very unusual species that is unlikely to be misidentified.

Found mainly in the boreal zone, very rare; summer.

Heinrich Holzer

Helvella is a genus of either cup-shaped or more folded cup fungi that mostly have an evident, and usually pitted-veined, stem. The spores are rather large, oblong and smooth or, rarely, finely verrucose. Most species are probably ectomycorrhizal. See also cup-shaped species of *Helvella* page 1298.

Helvella crispa is a large, pale *Helvella* with an irregular, ± saddle-shaped fertile part attached to a longitudinally grooved, ± cream stem. The lower side of the fertile part is minutely downy and pale pinkish-buff. The asci are 270–320 µm long. The ellipsoid, smooth spores measure 18–21 × 11–12.5 µm. Occurs on ± rich, mostly disturbed soil, but always close to ectomycorrhizal trees or shrubs.

Helvella leucophaea ⋉ is closely related but has a more golden-brown hymenium. *Helvella lactea* ⋉ is similarly pale, but is more saddle-shaped with a smooth lower side, and is shorter and more broad-stemmed; the asci are also shorter.

Widespread, common; August–December.

Helvella elastica is a tall, slender *Helvella* with a cylindrical, white stem and a greyish, saddle-shaped fertile part, the margin of which is inturned and ± appressed to the stem. The sterile lower side is minutely downy. The flesh in the fertile part is partly gelatinous. The spores measure 18–21 × 11–13.5 µm. Occurs on rich, mostly calcareous soils in coniferous and deciduous forests.

Helvella danica ⋉ is shorter-stemmed, almost bell-shaped, occurs in clusters, flesh not gelatinous and spores measure 19.5–22.4 × 12.2–13 µm; *H. carnosa* ⋉ is similar to *H. danica* ⋉, but is darker, more saddle-shaped and minutely downy. *Helvella levis* ⋉ has a freer, fertile margin and wider spores that measure 17–21 × 13.5–17 µm; *H. bicolor* ⋉ also has a freer, projecting margin and a darker hymenium (brownish-black when dry).

Widespread and rather common; July–December.

***Helvella lacunosa* s.l.** is a variable *Helvella*, both in size and colour, but is usually nearly black (rarely pure white). The irregular ± saddle-shaped fertile part is attached to the whitish to greyish, longitudinally furrowed stem in 3–4 places. In dry material, the dark hymenium contrasts with the paler stem and sterile part of the top. The smooth spores measure 15–19 × 9.5–12 µm. It is assumed to form mycorrhiza with a range of different deciduous and coniferous trees, mostly on rather disturbed, moist soils. May also form fruitbodies on rotten wood.

Helvella lacunosa belongs to a very difficult species complex. *Helvella pallescens* ⋉ is very pale and long-stemmed; *H. atra* ⋉ is small and dark-stemmed; while *H. nigra* ⋉ is a large and grey with veins on the hymenium. *Helvella pseudolacunosa* ⋉ is not considered to be European, but Danish material was sequenced and cited in the original account; it is, perhaps, a synonym of one of the species already mentioned.

Widespread and common (in a broad sense); May–December.

Helvella fallax is a grey-black to black, rather slender-stemmed and saddle-shaped *Helvella*. The stem and the sterile part of the top are distinctly downy and the stem is pale at the base. The smooth spores measure 17–19 × 10–11 µm. Occurs on mostly calcareous, damp soils in woodlands of various types.

Other similar species are not as dark and have a more pronounced contrast between the dark hymenium and pale stem. *Helvella pulla* ⋉ is somewhat paler.

Widespread, rather rare; June–November.

Jens Maarbjerg

Operculate cup fungi with dull colours

Apothecial ascomycotes with disc- to cup-shaped apothecia, with or without stems, and with whitish, greyish, brownish, violet, or black hymenia. The asci have a lid (an operculum) or, in *Thelebolus* and *Ascozonus,* a cracking top.

OTHER SIMILAR FUNGI:
– operculate cup fungi with yellow, orange, red to pink colours are treated in a separate wheel (page 1328).

See also the general wheel, page 1278.

FURTHER READING: 34, 49, 51, 52, 53, 73, 101, 108, 109, 118, 233, 301, 310, 329, 349, 358.

Micro-drawings: spores first, followed by paraphyses (p), asci (asci) and hairs (h).

Approximate species count applies to temperate Europe.

Gyromitra form cup-shaped or folded apothecia, mostly with a distinctly ridged, ± solid stem. Spores smooth or ornamented; sometimes with large appendages at the ends. Probably predominantly decomposers. Many species contain the poisonous compound gyromitrin. See also the brain-like *Gyromitra* on page 1285.

†? ***Gyromitra parma*** has ± cinnamon-buff to cinnamon-brown, discoid apothecia with a short, furrowed, pale stem. Spores broadly ellipsoid, measure 26–30 × 11.5–13.5 µm; have a reticulate ornament and finger-like, appendages at the ends, 2 µm long. On trunks of *Fagus* in calcareous habitats, mainly in very species-rich sites.

Also known as *Discina parma*. *Gyromitra leucoxantha* ⋉ occurs on soil and has larger spores with truncate appendages that measure 32–39.5 × 13–15 µm (without the appendages). The darker *G. accumbens* ⋉, also on soil, has even larger spores, 35–45 × 12–15 µm. *Gyromitra ancilis* ▽ has pointed appendages at the spore ends and is more red-brown; it occurs predominantly on conifer wood.

Scattered in the nemoral zones, very rare; April–June.

†? ***Gyromitra ancilis*** has discoid, at first somewhat cup-shaped, but later almost convex apothecia with a ± wrinkled, dark cinnamon-brown to red-brown surface and a paler, somewhat furrowed lower side. The reticulate spores have pointed appendages at each end and measure 26.5–30.5 × 11.5–14.5 µm (without the appendages). Mainly on conifer stumps, predominantly *Picea*, often, but not necessarily, in ± calcareous sites.

Also known as *G. perlata* and *Discina perlata*. *Gyromitra warnei* ⋉ has smaller spores. Other species of discoid *Gyromitra* occur on deciduous wood.

Widespread, occasional; March–May.

• ***Disciotis venosa*** is a large, yellow-brown to cinnamon-brown fungus with very fragile, but rather thick flesh, and a wrinkled-veined surface. It has a distinct nitrous/chlorine smell. The sterile lower side is ± sordid cream. Asci inamyloid. The smooth spores measure 22–25 × 12–14 µm and have external oil drops at the ends. Occurs on rich, damp soil in open forests, parks and gardens – often in stands of *Petasites*.

Discoid species of *Gyromitra* ◁ are tougher and do not smell nitrous.

Widespread in the nemoral–hemiboreal zones, occasional; February–June.

Thomas Læssøe

Rhizina undulata has large, purple-brown, irregularly shaped, rather tough apothecia that initially have a very pale margin. The underside is attached with thick root-like hyphal cords (see inset image) – the generic name is a reference to these cords. The spores are very finely ornamented, the ends having acute appendages that are thickened at the base, and measure 27–33 × 10–11.5 µm (without the appendages). Occurs mostly on burnt soil and wood in coniferous forests, but also on very poor, sandy soils; it is considered to be parasite on conifers.

Distinctive and unlikely to be misidentified.

Widespread, occasional; mainly June–January, but can be found all year.

Tobias Bøllingtoft

Jens H. Petersen/Thomas Læssøe

Sarcosphaera coronaria has large, thick-fleshed apothecia with a stellate opening but is initially globose and very pale lilac (lower image). When mature and open, the greyish-violet to purple hymenium is exposed. The asci are 8-spored, cylindrical and amyloid. The rough, ellipsoid spores measure 15–16.5 × 8–10 µm. Occurs in and on calcareous soils, both in deciduous woodland and in coniferous forests and plantations, presumed to be ectomycorrhizal.

Distinctive and unlikely to be misidentified. It is related to *Peziza* (page 1313).

Widespread in the nemoral–hemiboreal zones, and even found in the boreal zone, but rare and local; April–July.

Michal Mikšík

Urnula craterium has large, grey-brown, leathery, urn-shaped apothecia that have a serrated margin and a minutely scaly, paler outer surface. The smooth, ellipsoid spores do not have drops and measure 29–35 × 13–14 µm. Occurs on ± buried *Corylus* branches on light, rich, ± calcareous soils.

Urnula hiemalis ⋉ is more expanded, occurs on soil and has a darker purple-brown surface and somewhat smaller spores.

Widespread in the hemiboreal and southern boreal zones, including uplands in central Europe, but absent from the nemoral zones; spring.

Sarcosoma globosum is a large, brown, scaly, 'heavy', thick-fleshed and very gelatinous operculate cup fungus. It matures very slowly through the winter and spring, and during this process loses part of its internal liquid and becomes more wrinkled. The hymenium is dark grey-brown to olive-brown. The asci are 8-spored and inamyloid, and the narrow paraphyses are typically brown. The spores are smooth, ellipsoid, have small drops at the ends, and measure 20–30(–34) × 7–10(–13) µm. Occurs in damp *Picea* forests with a thick moss carpet, *e.g.* in older grazed forests, often along streams. Its precise ecology is not fully understood, but it is assumed to be a decomposer.

Very distinctive and unlikely to be misidentified.

Occurs from southern Norway and up through Scandinavia into parts of the Baltic republics and eastern Europe; January–May.

The species is included on the IUCN international red list.

Thomas Læssøe

Michal Mikšík

Michal Mikšík

Pseudoplectania form dark, finely-felty apothecia and spherical spores. Decomposers of wood, needle debris and *Sphagnum*.

Pseudoplectania nigrella has blackish, when immature shiny, cupulate apothecia with a felty surface that is covered in helical dark hairs. The spherical, smooth spores measure 9–11 µm and are positioned asymmetrically within a colourless slime coat. Occurs on needle debris and rotten wood of *Picea*.

Pseudoplectania episphagnum ▷ occurs in wet mires directly on *Sphagnum*; it has smaller fruitbodies. *Pseudoplectania melaena* ▽ has apothecia with a stem, somewhat larger spores, paraphyses that are twisted and downturned at the top, and occurs on wood of *Abies*. The recently described *P. lignicola* ⋉ has spores that lie centrally within a colourless slime coat; it is also more prone to growing on wood. The somewhat smaller, inoperculate *Piceomphale bulgarioides* (page 1413) occurs on *Picea* cones, but has asci with an amyloid apical plug and ellipsoid spores.

Widespread and rather common; October–June, maturing during the spring.

Jean-Marc Moingeon

Pseudoplectania melaena has rather large, black, relatively long-stemmed apothecia with a somewhat hairy outer side. The paraphyses are bishop's crook-shaped. The spherical, smooth spores measure 11.5–13 µm and may have a large, lateral slime coat. Occurs on wood of *Abies*.

Also called *P. vogesiaca*. *Pseudoplectania nigrella* △ has straight paraphyses, somewhat smaller spores and the apothecia are usually sessile. *Pseudoplectania lignicola* ⋉ has smaller spores that lie centrally within a slime coat.

Widespread in central and eastern Europe, uncommon; spring.

Pseudoplectania episphagnum has rather small, almost sessile to short-stemmed, black apothecia. The base has long, dark hairs. The spherical spores measure 11–12 µm. Occurs in wet mires, directly on *Sphagnum*.

Until recently, referred to in error as *P. sphagnophila*. *Pseudoplectania nigrella* ◁ does not occur on *Sphagnum* and has somewhat larger fruitbodies.

Widespread in the nemoral–hemiboreal zones, rare to very rare; mainly May–June.

Thomas Læssøe

Plectania melastoma has very distinctive, initially brownish-orange, cupulate, ± short-stemmed, tough, thick-fleshed apothecia with an almost black hymenium. The margin is furrowed-scaly, and the base emerges from a dark hyphal mat. The apothecia darken with age, and with KOH dissolve into a very distinct, purple pigment. The asci are inamyloid. The smooth, fusiform spores measure 21–25 × 9–12.5 µm and have an indistinct, colourless slime coat. Occurs on needle debris and other litter in ± calcareous forests and in more open heathlands with small woody perennials.

Species of *Urnula* ◁◁ are not as rust-coloured and have larger, ± ellipsoid spores.

Widespread, becoming scarcer towards the north, rather rare to very rare; mainly April–June.

Desmazierella acicola is a very distinctive, small, dark operculate cup fungus with flat apothecia and long, brownish-black setae that emerge from the outer surface, the margin and from the hymenium. The paraphyses are strongly branched towards the tips, brown and finely spiny; they are longer than the asci. The ellipsoid, smooth spores have a colourless slime coat and measure 16–19 × 8–11 µm. It also has a characteristic brown, hairy asexual state with conidiophores in a circular (verticillate) pattern. Occurs on needle debris of *Pinus*.

The poorly known *Picea*-associated *D. piceicola* ⋉ has much larger, ornamented, fusiform spores that measure 48–52 × 11–12 µm.

Widespread, common, but apothecia can be difficult to locate in some years; February–June (apothecia).

Helvella form cup-shaped or more folded apothecia, usually with a pitted-furrowed stem. Many species, perhaps all, are ectomycorrhizal. See also the folded species of *Helvella* (page 1288).

Helvella leucomelaena has grey-brown, short- and broad-stemmed, cup-shaped apothecia with rounded furrows at the base. The surface is minutely downy. The smooth, ellipsoid spores measure 20–24 × 11–13 µm. Ectomycorrhizal with *Pinus* on ± calcareous, mostly clay-rich soils.

Helvella confusa ⋉, with *Picea* and perhaps *Abies*, is typically smaller with a more obvious stem; *H. solitaria* ▷▷ is larger, has an evident stem and occurs mainly with deciduous trees.

Widespread, occasional; January–June.

Helvella acetabulum has medium-sized to large, ± clay-buff apothecia, which are somewhat paler when young. It has low, rounded, forking ribs running from the base onto the cupulate part. The smooth, ellipsoid spores measure 16–19 × 11–12 µm. Ectomycorrhizal, typically with deciduous trees on ± calcareous, mostly rather damp soils.

Helvella costifera ⋉ and *H. calycina* ▽ are typically more strongly veined on the surface, more greyish and occur from June onwards. There are several further similar species of *Helvella* that have a more northern/alpine distribution.

Widespread and locally rather common; March–June.

Thomas Læssøe

Helvella calycina has rather large, grey-brown apothecia with a somewhat paler, highly veined surface, on which, with age, numerous transverse veins develop. The smooth, ellipsoid spores measure 15–19 × 10–13 µm. Occurs mostly on calcareous soils in deciduous and coniferous forests; ectomycorrhizal.

Helvella costifera ⋉ has hardly any transverse veins and a greyer hymenium. *Helvella acetabulum* △ is browner and the surface is less markedly but more sharply veined; it also occurs earlier in the year.

Widespread, rather local and rare – much confused with *H. costifera*; July–November.

Helvella macropus has grey, pale-haired, very long-stemmed apothecia with a cupulate top and a greyish hymenium; the outer surface is downy-hairy. The stem is rather thin and does not have furrows. The finely verrucose, fusiform spores measure 20–26 × 9–12 µm. Occurs on rich soil or very rotten wood in damp deciduous forests; probably ectomycorrhizal.

Also known as *H. bulbosa*. *Helvella fibrosa* ⋉ has ellipsoid, smooth spores.

Widespread, common; mainly June–October.

Helvella solitaria has rather dark, rather large apothecia that have a lengthwise-furrowed stem and ribs that barely reach the cupulate to saddle-shaped upper part. The outer surface is minutely downy and the hymenium a dark grey-brown. The smooth, ellipsoid spores measure 17–19 × 11–13.5 µm. Occurs on rich soils in forests and parks; probably ectomycorrhizal.

Also known as *H. queletii*. *Helvella confusa* ⋉ is a smaller, *Picea*-associated species. Other species with a furrowed stem either have veins on the surface of the cup, are a different colour, or are more distinctly saddle-shaped.

Widespread, occasional to rare; May–October.

Helvella corium has rather small, almost black apothecia with an evident stem that can either be somewhat furrowed or lack furrows. The surface has paler tufts of hairs, particularly at the apothecial margin. The smooth, ellipsoid spores measure 18–22 × 11–13 µm. Occurs mostly on calcareous, damp soil; ectomycorrhizal with *Salix*.

In the nemoral zones it should be easy to recognize this species, but in more alpine-arctic areas it can occur together with several other very similar species of *Helvella*.

Widespread, occasional to rare; April–October.

Jan Vesterholt

Helvella terrestris has small, greyish, minutely hairy, cupulate to compressed apothecia. The hymenium is grey-brown, purple-black or black. The spores are very large, measuring 50–60 × 10–15 µm, fusiform and smooth; using an SEM a minute warty ornament can be seen. Occurs on sandy soils in riverbeds, *etc.*, often with the similarly operculate cup fungus *Trichophaea hybrida* ⋉.

Also known as *Pindara terrestris*. A very distinctive species that is unlikely to be misidentified.

Widespread but very rare and apparently absent from the nemoral zones; autumn.

Maroš Peiger

Wynnella silvicola has orange-brown, narrowly ear-shaped, tough-fleshed apothecia with a darker brown hymenium. The paraphyses are straight. The smooth spores have a large drop and measure 24–26 × 14–15 µm.

Has often been reported under the wrongly applied name *W. auricula* – a synonym of *Otidea leporina* ▷▷. Species of *Otidea* are softer-fleshed and have spores with two drops. Some mycologists treat this species as a *Helvella* △, where it then occupies an isolated position.

Widespread, but very rare, known from Sweden but mostly seen in the southern, upland parts of temperate Europe; autumn.

Jean-Marc Moingeon

Otidea form sideways-slit to cone-shaped, rather soft-fleshed fruitbodies. All European species have smooth spores. Ectomycorrhizal. See also the yellow to orange species of *Otidea* on page 1332.

Otidea leporina is a rather small *Otidea* with somewhat narrow, yellowish-brown to yellow-brown, sideways-slit apothecia. Paraphyses ± curved at the tips. The ellipsoid spores have two drops and measure (12–)12.5–14(–15) × 7–8.5 µm. Occurs on needle debris amongst moss in ± rich coniferous forests and plantations.

Otidea brunneoparva ⋉ has similar wide spores but is much darker and the paraphyses are hooked at the tips. A small group of similar species of *Otidea* have an obvious grainy covering on the outside, and shorter spores, see *e.g.* *O. tuomikoski* ▽.

Widespread and rather common; August–October (–November).

Jens H. Petersen/Jarkko Korhonen

Otidea tuomikoskii is a very slender, yellow-brown, sideways-slit *Otidea* with a grainy/warty covering on the outside; some warts are more than 85 µm high (inset image). When dry, the felt at the base is ± cinnamon-buff. The flesh (in section) stains ± yellow in KOH. Spores (9.5–)10–11(–12) × 5.5–6.5(–7) µm. Occurs on needle beds on ± calcareous soil in coniferous forests (mostly *Picea*) and plantations.

Otidea nannfeldtii ⋉ has somewhat lower warts (less than 85 µm high) and the flesh does not stain in KOH; the basal felt is paler. *Otidea papillata* ⋉ is wider.

Widespread, especially in northern parts, rare or overlooked; mainly September.

Otidea alutacea is a rather large, wide *Otidea* with yellowish-brown to grey-olive colours. It does not have conspicuous lumps on the basal mycelium. The spores have two drops and measure (13.5–) 14.5–16.5 (–17.5) × 6.5–7.5(–8) µm; Qav. range 2.1–2.2. Typically with deciduous trees, mostly on ± calcareous soils or sand.

A rather poorly understood species complex. *Otidea cochleata* ⋉ is sometimes recognized based on its larger spores; *O. platyspora* ⋉ has obviously larger spores and blackens somewhat when bruised; *O. bufonia* ▽ has dark lumps on the basal mycelium.

Widespread, occasional; July–November.

Otidea bufonia is a rather large, wide, sideways-slit, often two-coloured *Otidea* with an at first greyish, later brown outer side, and a grey-olive, later brown hymenium. It has striate incrustations on the hyphae. The fusiform spores have two drops and measure (12–)13–16.5(–18) × 6–7.5(–8) µm. Occurs on mulchy soils with deciduous trees and with *Abies*.

Otidea mirabilis ▽ is very similar but usually has distinct lilac tinges on the outside when young, and the hyphae lack a striate covering.

Widespread, occasional; mainly August–October.

Jens H. Petersen/Thomas Læssøe

Otidea mirabilis is a wide, sideways-slit, lilac/blue-tinged *Otidea*. The spores have two drops and measure (13–)13.5–16(–17) × 6–7 (–7.5) µm. Occurs on mulchy soil, mostly or exclusively with conifers.

Otidea bufonia △ lacks the lilac or blue shades on the outside and has conspicuous incrustations on the hyphae.

Mainly in eastern parts of the nemoral–hemiboreal zones, very rare or overlooked; August–October.

Jens Maarbjerg

Geopora form more or less immersed fruitbodies with brown, sinuous, ± branching hairs on the outside. The spores are smooth and have one large or two smaller drops. Ectomycorrhizal. See also the truffle-like species of *Geopora* page 1268.

Geopora arenicola is a rather small *Geopora* that usually remains hidden in the soil but, at maturity, has a rather wide, somewhat split opening. The outside is brown and loosely hairy. The ± uniguttulate, ellipsoid spores measure (18–)20–25.5(–26.5) × (11.5–)12–15 µm. Occurs in sandy and sand-mixed clay-rich soils with *Salix* and other deciduous trees.

Geopora foliacea ▷ is more superficial and has narrower spores with strikingly parallel sides. *Geopora sumneriana* ▽ fruits in the spring and has somewhat fusiform spores. *Geopora cervina* ▷ and *G. tenuis* ⋉ form smaller and rather superficial apothecia; the former has fusiform spores, while the latter has broadly ellipsoid spores.

Widespread and probably rather common; May–November.

Thomas Kehlet

Geopora sumneriana is a large *Geopora* that splits into a star-like pattern. The outside is brown and hairy; the hairs have septa and are brown and pliable. The hymenium is cream. The somewhat fusiform spores usually have two drops and measure 30–33(–37) × 14–15 (–15.5) µm. Occurs on soil in connection to *Cedrus* in parks, cemeteries, etc.

A distinctive species that by its ecology and fruiting period should be easy to identify. The similarly large-spored *G. nicaensis* ⋉ occurs in the autumn with other partners.

Widespread in the nemoral zones, mainly in southern parts, occasional to rare; February–May.

Geopora foliacea is a medium-sized, partly superficial, usually rather irregular and somewhat splitting *Geopora* with a brown-haired to scaly outer side and greyish to pinkish-buff hymenium. The ± cylindrical spores measure (23–)25–27(–28.5) × (11–) 12–14(–15) µm. Occurs with deciduous trees on ± clay-rich soils.

The cylindrical spore shape is unique within the genus. *Geopora arenicola* ◁ is more immersed in the substrate. *Geopora tenuis* ⋉ and *G. cervina* ▽ have smaller apothecia, and *G. sumneriana* ◁ occurs in the spring with *Cedrus*.

Nemoral–hemiboreal, rare; September–November.

Geopora cervina is a small, moderately immersed *Geopora* with a brown, ± hairy outer side. The hymenium is greyish. The ± fusiform, usually unilate spores measure 21–26.5 × 10.3–12.8 µm. Occurs in sandy-mulchy soils with deciduous trees.

Geopora tenuis ⋉ has broadly ellipsoid spores measuring 18.5–23 × 9.5–12 µm. *Geopora foliacea* △ has wider, ± cylindrical spores; the apothecia are also somewhat larger. Other similar, small, brown-haired operculate cup fungi have stiffer, setose hairs.

Widespread, occasional to rare or overlooked; July–October.

Tricharina gilva is a rather vividly coloured, pale brownish-orange, up to 10 mm wide, rather shallow operculate cup fungus with a scaly-hairy outer side; hairs are pale brown, thin-walled, and clustered. The smooth, ellipsoid spores measure 16–18 × 8.5–10 µm. On ± moss-covered burnt ground in deciduous and coniferous forests.

Similar species typically occur on unburnt soils, but the genus is poorly researched. *Tricharina cretea* ⋉ apparently has a pale grey hymenium and stiffer hairs, making the margin dentate; *T. ochroleuca* ⋉ and *T. praecox* ⋉ have deeper apothecia and less vivid yellow-brown colours.

Distribution poorly known; May–September.

Trichophaea has up to 800 µm long, pointed, brown, setose hairs.

Trichophaea woolhopeia is a small, gregarious, pale grey cup fungus with rather flat apothecia that have brown, clustered marginal, setose hairs which are pointed and up to 260 µm long. The broadly ellipsoid, smooth, uniguttulate spores measure 20–22.5 × 16–18 µm. Occurs on mulchy, mostly calcareous soils in deciduous and coniferous forests, probably ectomycorrhizal.

The smooth, broad spores are distinctive.

Widespread and probably rather common; July–November.

Trichophaea gregaria is a medium-sized, gregarious, greyish *Trichophaea* with rather long, dark brown, projecting hairs up to 780 µm long. The smooth to rough, fusiform spores measure 20–27 × 10–13 µm. Occurs on sandy soil and litter, but also on burnt ground. It is presumed to be ectomycorrhizal.

May constitute a species complex, and the name *Trichophaea hybrida* is also in common use. The combination of large, fusiform spores and long hairs make the complex easy to recognize.

Probably widespread and rather common; July–October.

Trichophaea hemisphaerioides is a gregarious, medium-sized *Trichophaea* with a pale greyish hymenium and rather short, up to 400 µm long, brown setae on the outside and at the margin. The finely ornamented spores have two drops and measure 13–16 × 7–8 µm. Occurs on burnt ground in deciduous and coniferous forests.

Trichophaea abundans ⋉ has the same ecology but smaller apothecia (up to 5 mm compared with up to 12 mm in *T. hemisphaerioides*) and shorter hairs. Species of *Geopora* ◁◁ have larger spores and less pointed, usually flexible hairs.

Widespread, occasional; March–October, potentially all year.

Trichophaeopsis bicuspis is a small cup fungus with rather few, long, setose hairs and a pale hymenium. The hairs are up to 1,000 µm long, and the hairs at the base of the apothecium are forked at the base, with the longest 'prong' pointing upwards and the shorter 'prong' downwards (inset image). The asci are 8-spored. The smooth, broadly ellipsoid spores measure 15–17 × 9–12 µm and have many small drops at the poles. Occurs on ± soil-mixed litter in damp deciduous woodlands, *e.g.* on fallen leaves and twigs of *Populus tremula*.

The similar *T. tetraspora* ⋉ occurs on cow dung and has 4-spored asci.

Distribution and frequency poorly known, but probably uncommon; August–September.

Sphaerosporella brunnea is a sessile, dark orange-brown operculate cup fungus with a somewhat fimbriate margin. The outside has flexible, up to 165 µm long brown hairs, but the hairs at the margin are shorter, paler and very pointed. The paraphyses are imbedded in a brown, sticky mass. The spherical spores have one large drop and several small drops and measure 14–16.5 µm. Occurs on mossy burnt ground, but supposedly ectomycorrhizal.

The combination of a brown hymenium, spherical spores and short, brown hairs is unique.

Widespread, occasional to rare; June–November.

Humaria form rather large, cupulate apothecia with 1–2 mm long, pointed, brown, setose hairs and finely verrucose spores that have two drops. Ectomycorrhizal.

Humaria hemisphaerica is a large, cup-shaped *Humaria* with a greyish hymenium and a densely brown-haired surface. The hairs are stiff, pointed and measure up to 1.5 mm at the margin of the apothecium. The ellipsoid spores have two drops, measure 20–27 × 10–13 µm, and have irregular warts up to 2 µm high. Occurs on soil or on very decomposed wood or litter; ectomycorrhizal.

The combination of large, ornamented spores and apothecia (up to 3 cm wide) make this species easy to recognize. *Humaria solsequia* ▽ is the same shape but distinctly yellow-brown and has smaller spores.

Widespread, common; May–November.

Mouldy, whitish apothecia of *H. hemisphaerica* are parasitized by *Hypomyces cervinigenus* ⋉ – which is mainly seen in the asexual state and is very common.

Miloš Villaris

Humaria solsequia is a medium-sized, cupulate operculate cup fungus with vivid, yellow-brown colours and stiff, brown surface hairs up to 1(–2) mm long. The tips of the paraphyses have orange contents. The spores are narrowly ellipsoid, smooth to very finely verrucose (at ×1,000 in cotton blue), have two large drops and measure 15–18 × (6.5–)8–9 µm. Occurs on soil or rotten wood, mostly in calcareous sites.

Also known as *H. aurantia*. Colour between the brown and orange cup fungi; may recall a large *Cheilymenia* (page 1344) but those species usually have multi-rooted apothecial hairs.

Southern and eastern within temperate Europe, rare; autumn.

Leucoscypha leucotricha has white, almost translucent, long-haired, deeply cupulate apothecia. The large spores have two drops, a ± distinct ornament of low warts, and measure 24–41.5 × 10–15.5 µm. Occurs in deciduous forests, deep in litter in contact with ectomycorrhizal roots, which it probably parasitizes.

There is some doubt as to how this species varies morphologically. Some mycologists recognize specimens with large spores as *L. erminea* ⋉.

Widespread, but becoming scarcer towards the north, probably rather common but easily overlooked in its hidden habitat; mainly May–October.

Arpinia fusispora has ± clustered, white-flocculose, buff apothecia with a narrowed, almost stem-like base. May discolour pinkish with age. Asci 8-spored, inamyloid. Paraphyses ± straight, somewhat widened above. The smooth spores have two drops, are somewhat fusiform, with a Q-value above 1.8, and measure 14–16 × 6–7 µm. Occurs in damp woodlands on ± calcareous soils.

Arpinia luteola ⋉ has shorter spores. Other species have more ellipsoid spores. All are rare. Very young *Otidea* ◁◁ apothecia may look similar but the paraphyses have ± bent tips.

Mainly a boreal species, generally very rare and local; summer–autumn.

Jan Vesterholt

Tarzetta cupularis is a short-stemmed, scaly-downy, pale cinnamon-buff operculate cup fungus with a dentate margin of adhering, thin-walled, brownish hairs. Paraphyses are unbranched. The ellipsoid smooth spores usually have two drops and measure 21–25 × 11–15 µm. Occurs on rich soils with various deciduous trees, *e.g.* with *Corylus* in gardens and scrub; ectomycorrhizal.

The species' delimitation is poorly resolved and the number of recognized species varies considerably. *Tarzetta catinus* ⋉ can be separated by being more stemmed, more fine-grained and more robust, but is included by some in *T. cupularis*. Amongst the sessile species, *T. spurcata* ⋉ has branching paraphyses, swollen at the tip, and *T. gaillardiana* ⋉ has very small, up to 5 mm wide apothecia. A version of *T. cupularis* with small spores, (19–)20–23 × (9.5–)10.5–12.5 (–13) µm, is known as *T. scotica* ⋉.

Widespread and common; April–December.

Sepultariella patavina is a rather small, somewhat immersed, sordid apricot cup fungus with a pale, finely serrated apothecial margin. The hyaline hairs are somewhat thick-walled and blunt-tipped. The ± fusiform spores have two drops and measure 24–30 × 12–13 µm. Occurs on ± calcareous, sandy soils. Possibly ectomycorrhizal.

Sepultariella semi-immersa ⋉ is very similar, but has smaller spores measuring (18.2–)21.9–23 (–26.7) × (9–)11.1–11.7(–13.4) µm.

Widespread in the nemoral–hemiboreal zones, rather rare, but probably overlooked; June–October.

Geopyxis carbonaria is a rather small, distinctly wine glass-shaped, yellow-brown operculate cup fungus with a pale fimbriate-serrated margin. The asci are inamyloid and 8-spored. The smooth, somewhat fusiform spores do not have drops and measure 12.5–16.5 × 7–10 µm. Occurs on rather recently burnt soil in coniferous forests; apparently biotrophic but the exact ecology is not fully understood.

Geopyxis delectans ⋉ is ± discoid and sessile. *Geopyxis alpina* ⋉ is also sessile but is cup-shaped and occurs on unburnt soil

Widespread, occasional; mainly April–October.

Pseudombrophila cervaria is a greyish-brown, almost turbinate operculate cup fungus with a narrow, dark, raised, ± serrated apothecial margin. It has simple, hyaline hairs and the asci are inamyloid. There is purple pigment in the flesh and between the paraphyses. The ellipsoid, smooth spores do not have drops and measure 14.4–18.7 × 7–9.3 µm; Q 1.8–2.3. Occurs on dung of various larger mammals.

Pseudombrophila theioleuca ⋉ lacks the purple pigment, is a paler yellow-brown and has shorter spores.

Distribution poorly known, uncommon; June–October.

Pseudombrophila ripensis is a large species of *Pseudombrophila* that develops from an immersed, dark sclerotium. The apothecia are at first spherical and somewhat grainy-scaly, but become deeply cupulate with a yellow-brown hymenium and a white, dentate margin. The spores do not have drops and measure 13.5–17.7 × 8.8–9.6 µm. Occurs on cow dung in dry grassland, more rarely on dung of sheep or moose (*Alces alces*).

The large sclerotia are unique in this genus and amongst dung-inhabiting operculate cup fungi.

Nemoral, very rare, but probably somewhat overlooked; mainly May.

Lajos Boros

Smardaea planchonis is a dark violet, rather small operculate cup fungus with a ± smooth margin. The asci are inamyloid and 8-spored. The spores are spherical, measure 11–12 µm, and, seen through a light microscope, mostly smooth. Occurs on soil, typically amongst litter from various conifers, mostly from *Cupressaceae*, *e.g.* *Cupressus* and *Juniperus*.

There are a number of other species of *Smardaea* that can only be identified with the aid of a microscope. *Smardaea amethystina* ⋉ has fusiform, coarsely verrucose spores, while *S. ovalispora* ⋉ has ellipsoid, smooth spores.

Nemoral, very rare; typically late in the autumn.

Nico Dam

Thomas Læssøe

Marcelleina persoonii is a small, violet, quickly flattened operculate cup fungus with a ± smooth margin; the violet pigment is situated within the paraphyses. The asci are inamyloid and 8-spored. The spherical spores have a ± complete reticulum and measure 10–12 µm. Occurs on ± calcareous, sometimes burnt soil – mostly in rather damp sites.

Marcelleina rickii ⋉ has low, short crests on the somewhat smaller spores. *Smardaea planchonis* △ has smooth spores, and the violet pigment is found all over. *Ionopezia gerardii* ▷ has fusiform spores.

Widespread but becomes scarcer towards the north, rather rare; July–October.

Peziza s.l. is a large and complicated group, mostly with rather large cup-shaped, rarely folded apothecia. The asci are amyloid above or along their full length. The spores are smooth, verrucose or striate, ellipsoid to fusiform. The genus *Peziza* has recently been split into many smaller genera, which include both decomposers and ectomycorrhizals.

Geoscypha tenacella is a medium-sized, pale violet to brownish-violet pezizoid. The asci have an amyloid zone around the operculum. The ellipsoid spores have two drops, are finely verrucose and measure 12–15 × 7–8 µm. Occurs amongst mosses on fairly recently burnt ground.

Also known as *Peziza subviolacea*. Other species on burnt ground include: *Daleomyces petersii* ⋉, which has spores that have two drops and oblong warts; "*Peziza*" *lobulata* ⋉, which has smooth spores that lack drops, and "*Peziza*" *moseri* ⋉, which likewise has smooth spores but with two drops at maturity. Further similarly coloured species occur on calcareous soils.

Widespread, probably rather common; spring–autumn.

Ionopezia gerardii is a small, soon flattened, violet pezizoid. It is easily recognized by the unusual, fusiform, longitudinally ribbed spores (× 1,000 magnification required), which measure 25–34 × 9–12 µm. Occurs on clay-rich, mostly calcareous soils in damp woodland; probably saprotrophic.

Other small, violet operculate cup fungi have differently shaped and smaller spores. *Ionopezia gerardii* is more related to species of *Marcelleina* ◁ than to other pezizoids.

Widespread, rather rare, but probably somewhat overlooked; July–October.

Phylloscypha boltonii is a rather large pezizoid with a sordid violet hymenium and a finely warty outer side. The asci are amyloid at the top. The finely verrucose, ellipsoid spores have two drops and measure 15–18 × 7.5–9 µm. Occurs on clay-rich, usually calcareous soils in forests.

Geoscypha tenacella ⊲⊲ occurs on burnt ground and has smaller spores. *Daleomyces exogelatinosus* ▽ has almost smooth spores. *Ionopezia gerardii* ⊲⊲ is small and disc-shaped with fusiform, finely longitudinally ribbed spores.

Mainly nemoral, rare; mostly September–October.

Daleomyces exogelatinosus is a rather small to medium-sized, ± violet, ± smooth pezizoid with a very thin, outer gelatinous layer. The asci are amyloid at the top. The very finely verrucose spores (almost smooth in a normal microscope) have two drops and measure 16.3–18.8 × 8.8–10 µm. Occurs on damp, almost bare, very calcareous soil.

Geoscypha tenacella ⊲⊲ occurs on burnt ground and lacks the gelatinous layer. The larger, more floccose-verrucose *Pachyella celtica* ⋉ has a more irregular and more obvious spore ornament.

Nemoral, apparently rare; September–November.

Thomas Læssøe

Daleomyces phillipsii is a large, pale violet, cauliflower-like pezizoid. The asci have an amyloid top. The spores have two drops, fine, irregular, oblong warts and measure 10–13 × 6–7 µm. Occurs on burnt or sterilized soil.

Species of *Sparassis* form rather similar but larger fruitbodier. They are basidiomycotes (page 954).

Nemoral, widespread but rare; mainly in the autumn and during mild winter periods.

Galactinia saniosa is a blackish-blue, rather small to medium-sized, soon flattened pezizoid that, when bruised, exudes a very striking, blue latex. The asci are mainly amyloid in the upper part. The spores have irregular, coarse warts and measure 15–17 × 7–9 µm. Occurs on damp, clay-rich soil in deciduous and mixed forests or on very decomposed wood; ectomycorrhizal.

No other operculate cup fungus with bluish latex has an almost black hymenium. *"Peziza" phlebosphora* ▽ has a ± olive-brown hymenium and pale blue latex.

Rather rare; July–October.

"_Peziza_" _phlebospora_ is a medium-sized brown to olive-brown pezizoid with a pale bluish latex (only seen in young, fresh specimens). The spores measure 17–19(–20) × 8–10 µm and have an ornament of oblong, sinuous warts. Occurs on burnt ground and sandy soils in forests and plantations, mostly where the mineral soil is exposed; probably ectomycorrhizal.

"Peziza badiofusca" ⋉ exudes latex but is smaller and the spores measure 13–15 × (8–)9–10 µm. *Galactinia saniosa* △ has darker latex and lacks olive colours. *Legaliana badia* ▷▷ and *L. limnaea* ▷▷ do not contain latex.

Nemoral–hemiboreal, rarely recorded; mainly August–October.

Paragalactinia michelii is a medium-sized pezizoid with a violet-tinted hymenium and a more yellow-brown outer side. The flesh stains yellow when cut. The asci stain diffusely blue in iodine reagents. The spores have two drops, irregular warts and measure 16–18 × 7.5–8.5 µm. On clay-rich soils in deciduous forests; ectomycorrhizal.

The combination of yellow stain and brownish-violet colours should make identification straightforward. Two other yellow-staining species, *Paragalactinia succosa* ▷▷ *and P. succosella* ▷▷, have yellowish or greyish hymenia and wider spores.

Nemoral–hemiboreal, occasional to rather common; July–October.

Paragalactinia succosa is a cinnamon-buff to olive-grey, medium-sized, distinctly cupulate pezizoid with a somewhat paler outer side. The flesh contains latex and stains bright yellow when cut. The amyloid asci become diffusely blue along their full length in iodine reagents. The ellipsoid spores have two drops, irregular, isolated warts and measure 19–22 × 10–12 µm. Occurs on mulchy, mostly ± calcareous soils in deciduous forests; ectomycorrhizal.

Paragalactinia succosella ▽ is smaller, greyer on the hymenium and more flattened at maturity; the spores are somewhat smaller. *Paragalactinia michelii* ⊲⊲ has violet shades on the hymenium and narrower spores.

Widespread and common in the nemoral–hemiboreal zones, becoming scarcer in the boreal zone; mainly July–October.

Paragalactinia succosella is a rather small, often soon flattened, grey-brown pezizoid with flesh that contains some yellow-staining latex. The asci have a diffuse blue colour in iodine reagents. The spores have one or two isolated drops, low, irregular warts and measure 16.5–19 × 8.5–10 µm. Occurs on clay-rich, mostly calcareous soils in forests; ectomycorrhizal.

Paragalactinia succosa △ is larger, more cinnamon-buff, more cupulate and has larger spores with two drops and a higher ornament. *Paragalactinia michelii* ⊲⊲ has violet shades on the hymenium and narrower spores.

Nemoral–hemiboreal, occasional; mainly July–October.

Legaliana limnaea is a medium-sized to large pezizoid with a ± olive-brown hymenium and a somewhat more red-brown, warty-grainy outer side. The asci have a strong blue colour in the upper part in iodine reagents. The ellipsoid spores have two drops, low, comma-shaped warts and measure 17–20 × 8–10 µm. It typically occurs on moist soil, mostly with *Salix* and *Alnus*. Ectomycorrhizal.

Legaliana badia ▽ is darker, occurs in drier habitats, and has a reticulum on the spores. "*Peziza*" *phlebospora* ⊲⊲ exudes a pale blue latex when bruised.

Probably rather common in most of temperate Europe; July–October.

Legaliana badia is a rather large, red-brown to purple-brown, somewhat clustered pezizoid that, when young, has a ± obvious olive tint on the hymenium. The flesh has no coloured latex. The asci become intensely blue in the upper part in iodine reagents. The spores have two drops, an incomplete reticulum and measure 17–19.5 × 7–9.5 µm. Occurs on acid, peaty soil, mostly in coniferous forests and plantations; ectomycorrhizal.

Legaliana limnaea △ grows more moist and have comma-shaped warts on the spores. "*Peziza*" *depressa* ⋉ have rounded warts or spines on the spores and occurs on richer soils. "*Peziza*" *phlebospora* ⊲⊲ has more or less the same olive shades but, young, fresh, specimens have a pale blue latex.

Widespread, common; June–December.

Thomas Læssøe

Elaiopezia polaripapulata is a greenish-yellow to olive-brown, thick-fleshed pezizoid with a darker, grainy outer side. The asci are amyloid and mainly stain only in the upper parts. The coarsely and irregularly warty spores have two drops, are ellipsoid with ± truncate ends, and measure 17.5–20 × 8.8–10 µm. Occurs on deciduous wood or piles of wood chips in deciduous forests; a decomposer.

Other similar species, *e.g.* *Phaeopezia apiculata* ▽, have distinct spines or coarse warts at the spore ends.

Nemoral, very rare; mainly June–September.

Ján Komár

Phaeopezia apiculata is a rather small, ± olive pezizoid with a grainy outer side. The densely verrucose spores measure 18–24 × 8–10 µm and have conical, acute appendages at both ends – hence the name *apiculata*. Occurs on decomposed wood in damp woodlands; a decomposer.

Other similar species do not have the characteristic appendages on the spores or prefer different habitats. *Elaiopezia obtusapiculata* ⋉ has more rounded spore appendages, and *E. polaripapulata* △ has lower, confluent warts at the spore ends.

Nemoral–hemiboreal, very rare; mainly June–October.

Karen Hansen

"Peziza" subcitrina is a rather small, soon flattened, rather thick-fleshed, pale yellow pezizoid. The asci are amyloid and stain mainly around the operculum. The smooth spores do not have drops, have a lateral, colourless slime coat and measure 17.5–20 × 8.8–10 µm. Occurs on rich soils, including compost, in damp forests, gardens *etc.*; a decomposer.

"Peziza" buxea is either an older synonym or a similar, somewhat darker species.

Mainly nemoral, rarely reported; mostly September–October.

Peziza ammophila is a cupulate to almost tulip-shaped, red-brown pezizoid that usually splits into rays at the apothecial margin, and has a very fragile stem that roots deeply into sand. The asci are amyloid and stain a deep blue around the operculum. The spores are smooth, lack drops and measure 16–17 × 9–10 µm. Occurs deep in sand in white dunes, typically in sparse *Ammophila* vegetation; a decomposer.

Peziza ammophila f. *macrospora* ⋉ has larger spores that measure (17–)18–20(–21) × (9–)10–13 (–13.5) µm. Species of *Geopora* ◁◁ are distinctly brown-haired on the outer side.

Widespread in the nemoral–hemiboreal zones, occasional; July–November.

Peziza arvernensis is a very large, yellow-brown to cinnamon-brown pezizoid with a paler, floccose outer side; it can either have or lack a stem. The finely but distinctly verrucose spores (warts up to 0.5 µm high) lack drops and measure 16–17.5 × (7–) 9–9.5 µm. Occurs on soil mixed with sawdust, soiled wood chips, *etc.*, not on large trunks; a decomposer.

This species belongs to a difficult complex. *Peziza varia* ▷▷ has a finer spore ornament (× 1,000 required!). *Peziza vesiculosa* ▷▷ has completely smooth spores and occurs on nutrient-rich substrates.

Widespread in the nemoral–hemiboreal zones, occasional; April–October.

Karen Hansen

Peziza varia is a medium-sized to very large, at first pale yellow, later honey-yellow to clay-buff, usually somewhat stemmed pezizoid. The margin is typically finely serrated and the outside ± grainy. The amyloid asci stain an intense blue around the operculum. Spores lack drops and have a very fine ornament that can only be observed at high magnification (× 1,000 required); they measure 14.5–17.5 × 8–10.5 µm. Occurs on large, very decomposed trunks of deciduous trees, on thick piles of wood chips, in damp cellars (lower image) and even on dung. In the top image it is associated with various slime moulds (see page 1646). The middle image depicts rather young and somewhat drought-damaged apothecia.

This species belongs to a difficult complex, and numerous names have been or are in use, *e.g. P. cerea*. *Peziza arvernensis* ◁◁ has a spore ornament that is visible at × 400 magnification.

Widespread, becoming scarcer towards the north, common; mainly May–November in the wild.

Peziza vesiculosa is a clay-buff, often ± clustered, thick and fragile, deeply cupulate pezizoid with a somewhat paler, grainy-warty outer side. The amyloid asci stain an intense blue around the operculum. The completely smooth spores do not have drops and measure 20.5–24 × 11–12.5 µm. Occurs on various forms of compost, middens, fertilized stubble fields, old hay stacks, *etc.*, but not on wood; a decomposer.

Most other brown, large *Peziza* species have ornamented spores or a completely different ecology. *Peziza varia* ◁ soon becomes less cupulate and has smaller spores. *Peziza fimeti* ▽ is much smaller, less cupulate and has smaller, but similarly smooth spores.

Widespread, rather common, becoming scarcer in the boreal zone; mainly May–November.

Peziza fimeti is a rather small, ± clay-buff pezizoid with a serrated-grainy margin. The paraphyses usually have somewhat swollen and bent tops. Asci amyloid, staining around the operculum. The smooth to nearly smooth spores measure 18–22 × 9.5–12.5 µm. Occurs both on rich soil and directly on dung, *e.g.* from cows; a decomposer.

Peziza alcis ⋉ occurs on moose (*Alces alces*) dung, and some undescribed species occur *e.g.* on beaver (*Castor*) dung or *Phragmites* culms; all have smaller spores. *Peziza vesiculosa* △ is a large, fleshy, cupulate species.

Widespread, rather common; mainly May–November.

Thomas Læssøe

Thomas Læssøe

Plicariella flavovirens is a cupulate, almost black operculate cup fungus with a green tint on the hymenium and conical warts on the outside. The amyloid asci stain mostly in the upper part. The spherical spores have an ornament of thin, truncate, regular warts and measure 14–15.5 µm. Occurs on calcareous, mineral soils in deciduous forests. Presumed to be ectomycorrhizal.

Plicariella scabrosa ⋉ has a coarser spore ornament. Species of *Plicaria* ▽ have similarly ornamented, spherical spores but are thinner-fleshed, less verrucose and occur on burnt ground.

Nemoral, very rare; July–October.

Plicaria trachycarpa is a very dark, almost black, ± flattened operculate cup fungus with a downy margin. The spherical, hyaline spores measure 10–11 µm and have 1 µm high, rounded warts. Occurs on burnt ground in forests.

Plicaria carbonaria ⋉ has a more irregular ornament up to 2 µm high. *Plicaria acanthodictya* ⋉ has a reticulum on the spores and *P. endocarpoides* ⋉ has smooth spores.

Widespread, apparently rare; all year, except during periods with frost.

Adelphella babingtonii is a rather small, disc-shaped operculate cup fungus with a ± buffish, somewhat convex hymenium. Flesh is gelatinous. Asci amyloid; best seen in living material. The ellipsoid, smooth spores measure 17–23 × (9–)10.6–13.7 (–16) µm and have two large drops and numerous small ones. Occurs on water-soaked wood in streams, at lake margins, in wooden water barrels, *etc.*

Formerly named *Pachyella babingtonii. Pachyella punctispora* ⋉ has finely echinulate spores. *Psilopezia nummularia* ⋉ has larger spores and inamyloid asci. *Thecotheus rivicola* ▷▷ has spore appendages.

Widespread, rather common; mainly May–October.

Ruben Walleyn

The genera **Pachyella** and **Adelphella** have somewhat thick-fleshed fruitbodies with an outer gelatinous layer, and generally occur in very wet habitats. *Adelphella* ◁ has somewhat amyloid asci and *Pachyella* strongly amyloid asci. The very similar genus *Psilopezia* ⋉ has inamyloid asci; there are further differences in the flesh anatomy.

Pachyella peltata is a large, wrinkled-veined, brown operculate cup fungus with a paler outer side. The yellow-staining flesh is up to 3 mm thick and has an outer gelatinous layer with radiating thin hyphae. Asci amyloid. The smooth spores usually have two drops and measure 27–30 × 13–15 µm. Occurs on damp wood in wooded fens, *etc.*

The smaller *Adelphella babingtonii* ◁ is disc-shaped and has smaller spores. Brown species of *Peziza* ◁◁ lack the outer gelatinous layer. *Pachyella clypeata* ⋉ is a very similar species from North America, and this name has been used in error in Europe.

Central Europe, rare; summer–autumn.

Pachyella violaceonigra is a rather large operculate cup fungus with dark violet colours. The outside has a gelatinous layer with thin, radiating hyphae. Asci amyloid. The ellipsoid spores have two drops and fine warts; they measure 19–27 × 11–13 µm. Occurs on damp wood in *Alnus* carrs, *etc.*

Violet species of *Peziza* ◁◁ do not have an outer gelatinous layer with radial hyphae. *Pachyella peltata* △ is brown and has smooth spores.

Nemoral, rare; summer–autumn.

Enrique Rubio Domínguez

Species of ***Ascobolus*** form small fruitbodies with somewhat amyloid, projecting asci and dark spores that usually have a furrowed or cracking wall. There are numerous species on dung and other organic material, including burnt soil.

Ascobolus immersus is a very small, yellowish, somewhat immersed, almost turbinate *Ascobolus* that does not have an evident margin. It has 1–10(–40) amyloid, very projecting, 8-spored asci. The very large, somewhat longitudinally furrowed, violet spores have a thick slime coat and measure 54–67 × 32–38 µm. The spores are ejected as one cohesive projectile. Occurs on dung from herbivores, particularly sheep.

Other species of *Ascobolus* typically have more asci per fruitbody and smaller, often differently ornamented spores.

Widespread, common; most of the year.

Ascobolus sacchariferus is a whitish, ± barrel-shaped *Ascobolus* with well-defined apothecia. The purple, longitudinally striate spores measure 16.5–20.5 × 8.5–10.5 µm. Occurs on deer dung.

Ascobolus albidus ⋉ has larger spores, measuring 25.5–29.5 × 12–13.5 µm, and occurs on dung from various herbivores. *Ascobolus carletonii* ⋉ has smaller spores, measuring 13–17 × 7.5–8.5 µm. *Ascobolus degluptus* ⋉ has larger spores that measure 27.5–35 × 16–18 µm. *Ascobolus brassicae* ⋉ has spherical spores. There are many more similar species of *Ascobolus*.

Distribution and frequency are not well understood, but probably not rare; most of the year.

Ascobolus furfuraceus is a rather large, yellow-brown *Ascobolus* with a well-defined, ± serrated-scaly margin. The violet spores have relatively parallel, ± anastomosing stripes and measure 22.5–25.5 × 11.5–13 µm. Occurs on dung from various herbivores.

Ascobolus crenulatus ⋉ is yellow-green-tinted and has smaller spores, measuring 12.5–15 × 7–8 µm.

Widespread and common; most of the year.

Ascobolus carbonarius is a large, almost black *Ascobolus* with a serrated margin. The ellipsoid, ± purple, irregularly ornamented spores measure 20.5–25.5 × 11.5–14 µm. Occurs on burnt ground with scattered mosses.

Ascobolus geophilus ⋉ may also occur on burnt ground but has a finer spore ornament.

Widespread, occasional to rather rare; most of the year.

Saccobolus beckii is a small *Saccobolus* with very small, almost hyaline, discoid apothecia. The spores are violet, measure 17–23 × 8.5–10 µm and have a 1–3 µm high coarsely verrucose ornament; they are 'glued' together in a ball that measures 42–60 × 18–24 µm and has sticky appendages at the ends.

Saccobolus verrucisporus ⋉ has smaller spores, measuring 14–16 × 8–9 µm and similar spore balls; *S. obscurus* ⋉ has a finer, lower spore ornament (up to 0.5 µm); *S. depauperatus* ⋉ has almost smooth spores. There are many other similar species.

Distribution and frequency are not well understood; most of the year.

Thecotheus has small fruitbodies with amyloid, projecting, operculate asci and hyaline spores. Decomposers of dung and plant litter.

Thecotheus pelletieri is a very small, turbinate to barrel-shaped, almost colourless operculate cup fungus. Asci are amyloid, projecting and 32-spored. The smooth, hyaline spores have a slime coat, are ellipsoid to fusiform, and measure 32–40 × 15–22 µm. Occurs on dung from herbivores.

Other similar species have 4–8-spored asci.

Widespread, probably not rare; all year.

Thecotheus rivicola is a greyish, rather small operculate cup fungus with a ± evident margin and a flattened to convex hymenium. Asci amyloid, projecting. The paraphyses are slender and branched. The spores are smooth, ellipsoid to fusiform, measure 16.4–19 × 8.3–9.8 µm, and have rounded appendages at both ends and a loosening slime coat. Occurs on water-soaked wood at lake margins, *etc*.

Adelphella babingtonii ◁◁ is browner and lacks spore appendages. Other species of *Thecotheus* occur on dung and have smaller fruitbodies.

Nemoral, apparently very rare; July–October.

Ascozonus woolhopensis is a very small, translucent-white, turbinate cup fungus with a somewhat hairy outer side. The small, hyaline hairs have a few septa. The inamyloid asci have a ring-shaped thickening and open by a slit down into the thickening. It has 64 fusiform spores that measure 11–15 × 3–5 µm. Occurs on dung both from carnivores and herbivores.

Some mycologists include specimens with a variable number of spores per ascus in a broader concept of the species. *Ascozonus cunicularis* ⋉ has non-septate hairs; *A. monascus* ⋉ has only one ascus per fruitbody.

Widespread, probably common; in the cooler months of the year.

Thelebolus has inamyloid asci that crack in the top (no operculum), and which may contain a very large number of spores. The fruitbodies are tiny and occur on dung.

Thelebolus stercoreus is a very small (max. 0.3 mm wide), smooth, yellow-brown ascomycote with a varying number of asci in each apothecium, from just one (lower image) to 50. Similarly, each ascus has a variable number of spores, from 32 up to several thousand, and opens by a slit or crack. The spores are hyaline, smooth and variable (up to 10 × 5 µm). Occurs on dung from birds and various other carnivores and herbivores.

Thelebolus microsporus ▽ is up to 1 mm wide and has asci that are 8-spored. Species of *Trichobolus* ⋉ are similar but have very long, projecting hairs. Species of *Ascozonus* ◁ are completely white and downy.

Distribution and frequency are poorly known; mainly in the cooler months of the year.

Thelebolus microsporus is a small (up to 1 mm wide), yellow-brown, smooth ascomycote with up to 100 asci per apothecium. The asci are inamyloid, 8-spored and open by a slit. The paraphyses are swollen above and have yellow contents. The smooth, hyaline spores measure 7–10 × 3.5–4.5 µm. Occurs on dung, including from birds.

Thelebolus stercoreus △ is smaller and has from 32 to several thousand spores per ascus.

Widespread and probably common; mostly in the cooler months of the year.

Yellow, orange, red to pink operculate cup fungi

Fungi with cup- to goblet-shaped or flattened apothecia, yellow, orange, red or pinkish hymenium, and asci with a lid (an operculum) or, in the genus *Coprotus*, with a slit-like opening mechanism. All genera, apart from *Iodophanus* and *Peziza*, have inamyloid asci.

OTHER SIMILAR FUNGI:
– operculate cup fungi with whitish, greyish, brownish, violet or black hymenia are found in the wheel on page 1290.

See also the general wheel, page 1278.

FURTHER READING: 33, 34, 39, 49, 73, 101, 119, 198, 199, 200, 233, 292, 359, 360, 361, 363, 364.

Micro-drawings: spores first, followed by paraphyses (p), asci and hairs (h).

Approximate species number applies to temperate Europe.

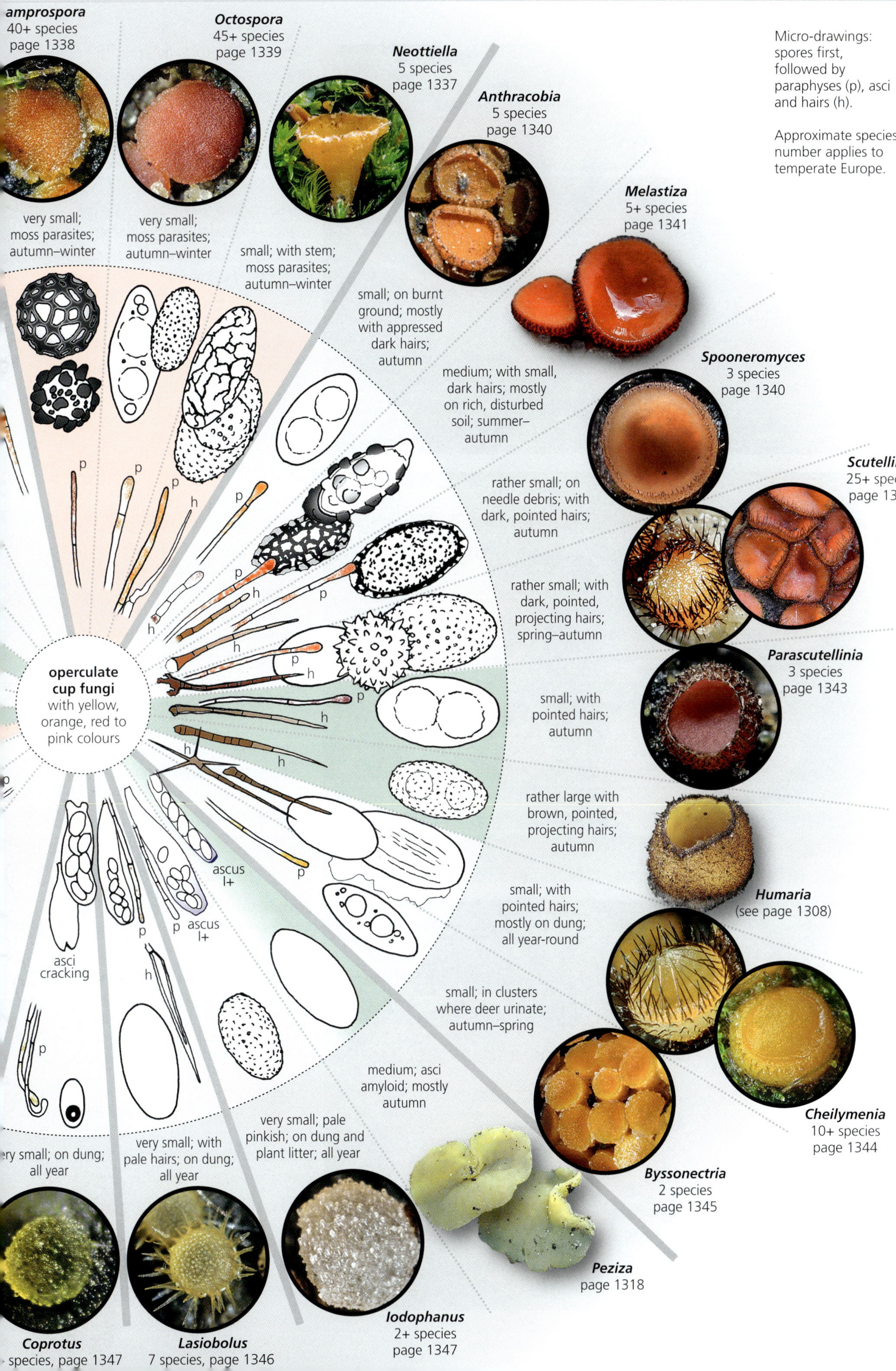

Sarcoscypha form large, bright scarlet or rarely orange fruitbodies with large, smooth, almost cylindrical spores in 8-spored, inamyloid asci. All fruit in winter–spring and are decomposers on wood.

Sarcoscypha austriaca is a large, cupulate, scarlet, tough-fleshed *Sarcoscypha* with a somewhat paler, dense felt on the outside; can also rarely occur in an orange form. The outside hairs are corkscrew-curled (microscope required). The living spores have many internal small drops, are concave at the ends and measure (22–)26–40(–50) × (11.5–)12–15 (–16) µm. Occurs on wood of deciduous trees (*e.g. Salix*), mostly on wood buried in ± moist soil in damp woodlands and scrub.

Sarcoscypha coccinea ▽ has ± straight hairs on the outer side. The living spores also have multiple drops but are rounded at the ends (may be slightly concave in dead spores); it prefers drier, highly calcareous habitats and the main host is *Fagus*. *Sarcoscypha jurana* ⋉ also has straight hairs but the spores have two drops and truncate or somewhat concave ends; it has a preference for *Tilia*.

Widespread in the nemoral–hemiboreal zones, occasional; (November–)January–April(–May).

Morten Christensen

Thomas Læssøe

Sarcoscypha coccinea is a scarlet, ± stemmed, tough *Sarcoscypha* with a paler, somewhat hairy outer side. The hairs are ± straight. The spores have many small drops, are typically rounded at the ends (may be slightly concave in dead spores) and measure (24.5–) 27–43(–51) × (10–)11–14 (–14.5) µm. Occurs on dry, calcareous soil, mostly on *Fagus* wood, but also on other hard woods.

Sarcoscypha austriaca △ has corkscrew-like hairs and *S. jurana* ⋉ has spores with two drops. Both have truncate or somewhat concave spore ends.

Widespread, nemoral, uncommon to locally common; (September–) February–April.

Microstoma protractum is a spectacular, long-stemmed, almost tulip-shaped, scarlet, operculate cup fungus with a dentate, somewhat hairy apothecial margin. Occurs mainly in clusters, with the upper parts joined to a common, deeply rooting stem. The spores are thick-walled and measure 40–45 × 15–20 µm. Occurs on ± buried sticks on rich soils in temporarily flooded riverbeds, *Alnus* carrs, *etc*.

A highly distinctive species that is easy to identify.

Found mainly in continental parts of Europe, absent from oceanic parts of the nemoral zones; spring.

Jiří Polčák

Pithya form rather small, ± orange apothecia and the asci have spherical spores. Decomposers on conifers.

Pithya cupressina is a small, orange, ± sessile, operculate cup fungus with a smooth margin and a ± downy, paler outer side. The spherical, double-walled, smooth spores measure 10–11 µm. On attached, dead twigs of mostly introduced *Juniperus* species in gardens, but also on *J. communis* and other members of the *Cupressaceae*.

Pithya vulgaris ▽ has larger apothecia, larger spores and occurs mainly on *Abies*.

Mainly nemoral–hemiboreal, occasional; mostly October–May.

Pithya vulgaris is rather small, yellow-orange, disc-shaped, tough-fleshed operculate cup fungus with a smooth margin. The outside is paler and minutely hairy at the base. The asci are inamyloid and 8-spored. The paraphyses have yellow contents. The spherical, smooth, double-walled spores measure (11–)13–14(–15) µm; outer wall dissolves in KOH. Occurs mostly on recently dead *Abies* branches (sometimes also attached), rarely on other conifers.

Pithya cupressina △ has smaller apothecia, smaller spores and occurs mainly on *Juniperus*.

Mostly nemoral, common to rare; December–May.

Jaroslav Malý

Caloscypha fulgens is a deeply cupulate cup fungus with an apricot-orange hymenium and a more yellowish, scurfy outer side that quickly discolours turquoise or greyish-blue. The spherical spores are smooth and measure 6–7 µm. It also occurs in an asexual state (known as *Geniculodendron pyriforme*). Occurs on ± calcareous soils, typically in coniferous forests, *e.g.* with *Abies*, but also in deciduous forests. It can also parasitize conifer seeds.

Distinctive and unlikely to be misidentified.

Widespread but very sporadic, mainly nemoral–hemiboreal to southern boreal; February–May.

Otidea form sideways-slit to cone-shaped, soft-fleshed apothecia. Asci inamyloid and 8-spored. All European species have smooth spores and typically narrow paraphyses with downturned tips. Ectomycorrhizal. See also the browner species of *Otidea* page 1302.

Otidea onotica is a large, tall, side-slit, ± orange-yellow to apricot-orange *Otidea* that, when young, may have pink tinges on the hymenium (lower image), sometimes occurring as spots. The outer side is somewhat mealy-floccose. The mycelial felt at the base stains yellow with KOH. The paraphyses bend downwards at the tips. The spores have two drops and measure (11–)12–13.5(–14) × (5.5–)6–7 µm. Occurs on rich soils in both deciduous and coniferous forests; ectomycorrhizal.

Other species of *Otidea* are browner or bright yellow and lack pink spots; they are typically smaller.

Widespread, rather common; August–November.

Otidea concinna is a vivid yellow, side-slit *Otidea* with a paler hymenium. The outside usually completely lacks veins or ribs. The spores have two drops and measure (10–)10.5–12 × (5–)5.5–6.5 µm with a Q-value at 1.8–2. Occurs on mulchy, mostly calcareous soils in deciduous and coniferous forests; ectomycorrhizal.

Can easily be confused with *O. minor* ⋉ but that species has somewhat narrower spores measuring (10–)10.5–12.5(–13) × (4.5–)5–5.5(–6.2) with a Q-value at 2.1–2.3; it also usually has veins/ribs on the outer side of some of the apothecia within a fruiting mycelium, and the outer side is not as intensely coloured. *Otidea phlebophora* ▽ has a distinctly veined and ribbed outer side.

Nemoral–hemiboreal, rather rare; June–October.

Otidea phlebophora is a vivid yellow, usually non-slit *Otidea* with a distinctly veined and ribbed outer side. The spores have two drops and measure (9.5–)10–11.5(–12) × (4.5–)5–5.5(–6.2) µm, Q 2–2.1. Occurs on ± calcareous soils in deciduous and coniferous forests; ectomycorrhizal.

Otidea concinna △ has ± the same colours but is side-slit and lacks the prominent ribs on the outer side. *Otidea minor* ⋉ is usually less veined/ribbed and is somewhat paler and side-slit.

Nemoral–hemiboreal, rare; July–November.

Thomas Læssøe

Thomas Læssøe

Sowerbyella has yellow to orange apothecia with a root-like 'stem'.

Sowerbyella radiculata is a ± greenish-yellow, somewhat wrinkled *Sowerbyella* with a downy, whitish outer side and a root-like extension. Spores coarsely ornamented with warts and wrinkles that almost form a reticulum, measure (12–)14.5–16(–18) × (6.5–)7.5–9 μm. Occurs on needle debris or litter from deciduous trees on ± calcareous soils.

Sowerbyella imperialis ▽ has more yellow to orange yellow apothecia and isolated small warts on the spores.

Widespread, uncommon to rare; July–November.

Sowerbyella imperialis is a large, yellow to orange-yellow *Sowerbyella* with long root-like extensions. The outer side and hymenium are almost the same colour. The spores have an ornament of mostly isolated, rather small warts and measure 12.5–14 × 6–7.5 μm. Occurs on litter in calcareous coniferous forests (*e.g. Abies*), possibly also in deciduous forests.

Sowerbyella polaripustulata ⋉ occurs with deciduous trees and has a more echinulate spore ornament; other similar species have different spore dimensions and ornaments.

Nemoral–hemiboreal, very rare; August–November.

Alain Brissard

Sowerbyella rhenana is a spectacular, usually clustered, apricot-orange *Sowerbyella* with a whitish, short and thick, sometimes branched stem and a somewhat felty-hairy or flocculose outer side. The spores have a reticulate ornament and measure 18–22(–25) × 9–11(–12) μm. Occurs on calcareous soils in deciduous and coniferous forests.

Other species of *Sowerbyella* △ are a brighter yellow and typically have larger apothecia. Species of *Aleuria* ▷ are almost sessile.

Nemoral–hemiboreal, rare; autumn.

Aleuria is a genus of yellowish to orange operculate cup fungi that are mostly sessile. The spores have reticulate or verrucose ornaments. Species of *Sowerbyella* ◁ are similar but have stems or root-like extensions. Species of *Melastiza* ▷▷ have a more hairy outer side, especially at the margin, and the hairs tend to be darker.

Aleuria aurantia is typically a very large, sessile, orange, operculate cup fungus with tiny, white floccules on the outer side. At maturity the apothecia become very expanded and somewhat irregular in shape. It typically occurs in crowded groups. The asci are inamyloid and 8-spored. The ellipsoid spores have a uniform, rather coarse reticulum and measure 13–15 × 7.5–9 µm (without the ornament). Occurs on disturbed, gravelly, often clay-rich soils along roads in parks and forests.

Typical specimens are very distinctive, but small specimens may be confused with other *Aleuria* species, species of *Melastiza* ▷▷ or *Sowerbyella* ◁.

Widespread, common; July–January.

Aleuria luteonitens is a rather small *Aleuria* with bright yellow colours. The outer side has inconspicuous, several-celled, thin-walled, rather short hairs. The spores have a reticulate ornament of meshes 0.5–2.3 µm long, which extends into points at the ends of the spores; they measure 14–16 × 6–8.5 µm (without the ornament). Occurs on clay-rich soils.

Aleuria bicucullata ⋉ has a coarser spore ornament with larger meshes, 2.8–6.3 µm, and more prominent hairs on the outer side. *Sowerbyella reguisii* ⋉ is a very similar species or perhaps just a synonym.

Widespread but rarely reported; autumn.

Enrique Rubio Domínguez

Pulvinula convexella has rather small, sessile, flattened apothecia and orange to red colours. The margin is smooth and sharply delimited. The paraphyses are branched and curve downwards at the tips. Asci inamyloid and 8-spored. The spherical, smooth spores have many drops and measure 16–18.5 µm. Occurs on clay-rich, damp mineral soils.

Pulvinula constellatio ⋉ is smaller and pinker and has spores with one large and numerous small drops; *P. miltina* ⋉ has smaller spores; and *P. carbonaria* ⋉ occurs on burnt ground. *Pulvinula alba* ⋉ is small and whitish.

Widespread, occasional; June–November.

Miladina lecithina is a vivid orange-yellow, disc-shaped, rather small operculate cup fungus with a smooth outer side. The asci are inamyloid and 8-spored. The ellipsoid, finely verrucose spores have many drops and measure 20–22.5 × 12–15 µm. Occurs on wood in unpolluted streams in deciduous forests.

The only other bright yellow discoid ascomycote on wood in streams is the inoperculate *Vibrissea flavovirens* (page 1455), which is smaller, pale yellow and has thread-like spores.

Widespread, mainly nemoral, but uncommon; May–November.

Pyronema omphalodes is a confluent, small, smooth, operculate cup fungus with rose to flesh-pink apothecia that are formed on a ± evident white subiculum. The smooth spores do not have drops and measure 14–16.5 × 8–9.5 µm. Occurs on newly burnt ground or on other kinds of sterilized soil.

Pyronema domesticum ⋉ is very similar but has fine, pointed hairs on the apothecia and larger spores measuring 16–20 × 10–12 µm.

Widespread, occasional; most of the year.

Ramsbottomia asperior is a rather small, apricot-orange, almost smooth operculate cup fungus with an indistinct margin. The asci are inamyloid and 8-spored. The almost spherical spores have spines 1–2 µm long and measure 16.5–17 × 13–15 µm. Occurs on damp mineral soils in forests and scrub.

Ramsbottomia crechqueraultii ⋉ and *R. macracantha* ⋉ have completely spherical spores. The latter has very long (3–6 µm) spines on the spores.

Widespread, rare or overlooked; July–October.

Neottiella rutilans is a stemmed, wine glass-shaped, apricot-orange, small to medium-sized operculate cup fungus with a white, fringed-serrated apothecial margin and a somewhat felty outer side. The stem is ± immersed in the host. The asci are inamyloid and 8-spored. The paraphyses have orange-coloured drops that stain green in iodine reagents. The spores have an incomplete reticulum and measure 22–27 × 13–15 µm. Occurs on *Polytrichum* mosses on poor, sandy soils, *e.g.* in dunes and heathlands.

The rather similar *Neottiella hetieri* ⋉ occurs on mosses on burnt ground and has smooth spores. *Neottiella vivida* ⋉ has a warty spore ornament and the northern *N. aphanodictya* ⋉ has a very fine, reticulate spore ornament (× 1,000 magnification required) and smaller spores. Species in the moss parasite genera *Lamprospora* ▷▷ and *Octospora* ▷▷ are sessile and do not have pale marginal hairs.

Widespread and rather common; (August–)October–December.

Lamprospora & ***Octospora*** are two genera of moss (or rarely liverwort) parasites that produce orange to reddish, mostly very small apothecia. *Lamprospora* has ± spherical spores and *Octospora* has ± oblong spores; both have inamyloid asci. These genera include numerous hardly discernible species that, mostly, are strictly associated to certain host genera.

Lamprospora retispora is an apricot-orange *Lamprospora* with a more yellowish, dentate margin. The almost spherical spores have a 0.5 µm high reticulum with 8–12 visible meshes and measure (16–)18–20(–22) × (13–)15–16 (–17) µm. Occurs on mosses in the genus *Syntrichia, e.g. S. ruralis*.

A number of other *Lamprospora* occur on other mosses and have somewhat different spore sizes and ornaments. *Lamprospora dictydiola* ⋉ occurs on *Tortula muralis* and has smaller spores that measure 14–16 µm. *Lamprospora tortulae-ruralis* ⋉, also on *Syntrichia ruralis*, has a more irregular, incomplete reticulum.

Nemoral, rare or probably overlooked; mostly in the cooler months of the year.

Lamprospora rugensis is a small, dark apricot-orange *Lamprospora* with a somewhat raised, warty apothecial margin. The spherical spores measure (15–)16–18(–20) µm and have an up to 1 µm high reticulum with 5–9 visible meshes. Occurs on species of moss in the genus *Bryum*.

A range of similar species occurs on other host mosses.

Widespread, but rarely reported; mainly June–October.

Octospora humosa is a large, vivid orange to reddish *Octospora* with a distinct, somewhat serrated margin. The smooth spores have one or two drops, are almost cylindrical and measure 19–23 × 11–13 µm. Occurs on sandy soils in association with mosses in the genus *Polytrichum*.

Octospora leucoloma ⋉ occurs on *Bryum argenteum* and has fusiform, smooth spores. *Octospora roxheimii* ⋉ occurs on *Funaria hygrometrica* on burnt ground.

Widespread, occasional; all year, but mainly recorded in the cooler months.

Thomas Læssøe

Octospora rubens is a small, reddish *Octospora* with a well-defined margin. The ellipsoid, smooth spores measure 15–18 × 10–12 µm. Occurs on the moss *Ceratodon purpureus*.

Octospora rustica ⋉ occurs on the same host but has smaller spores that measure 13–16 × 9–11 µm. *Octospora melina* ⋉ may also occur on the same host but has verrucose spores that measure 16–18.5 × 10.5–13 µm.

Widespread, rare, or probably overlooked; mainly in the cool months of the year.

Octospora ithacaensis is a small, yellowish *Octospora* that can either have or lack pale hairs at the base. The asci are 8-spored. The ellipsoid-fusiform, finely verrucose spores have two drops and measure 19–22 × 10–11 µm. Occurs on liverworts in the genus *Marchantia*.

Sometimes placed in the genus *Neottiella* ◁◁. This species is unlikely to be confused with any other fungus that occurs on *Marchantia*.

Nemoral, very rare or overlooked; mainly August–October.

Anthracobia melaloma is a sordid yellow-brown *Anthracobia* with pale brown, adpressed hairs on the outside, especially at the margin. The ellipsoid, smooth spores have two drops and measure 14–22 × 7–10 µm. Occurs on rather recently burnt ground.

Very similar to the other species in this genus but, according to some mycologists, is less vividly coloured compared to *A. macrocystis* ⋉ and paler, more yellow-brown than *A. maurilabra* ⋉ and the very dark *A. subatra* ⋉. The latter two species have hairs with many septa, while other species have hairs that either lack septa, or have, at most, two septa. *Anthracobia rehmii* ⋉ has large, fusiform spores that measure (19–)25–29(–34.5) × (6.5–)9–11 (–12) µm.

Widespread, occasional to rare; most of the year.

Jens H. Petersen/Thomas Læssøe

Spooneromyces laeticolor is an apricot-orange, cupulate operculate cup fungus with scattered, brownish, projecting hairs. The hairs are pointed, thick-walled, non-rooting and many-celled. The ellipsoid spores measure 16–23.5 × 8–10.5 µm and have a rather fine ornament of ± confluent warts and fine spines at the poles. Occurs on needle debris.

Spooneromyces helveticus ⋉ has coarsely reticulate spores, and *S. velenovskyi* ⋉ (syn. *S. microsporus*) has much smaller spores, 14–16 × 5.5–7 µm, with a finer ornament. Species of *Scutellinia* ▷▷ have ± rooting hairs.

Mainly nemoral–hemiboreal, rare or perhaps somewhat overlooked; August–November.

Melastiza produce orange to orange-red apothecia with adpressed, pale or dark hairs on the outer side. The asci are inamyloid and 8-spored. The spores have a reticulate or warty ornament that is higher at the poles. All species are decomposers.

Aleuria ◁◁ lacks brown hairs. *Anthracobia* ◁ has adpressed, pale brown hairs, but smooth spores. *Spooneromyces* ◁, *Scutellinia* ▷▷ and *Humaria* (page 1308) have pointed, projecting, brown hairs. *Cheilymenia* ▷▷ is more yellow.

Melastiza cornubiensis is an orange-red, soon flattened *Melastiza* with short, adpressed, dark hairs on the margin and outer side. The ellipsoid spores measure 16–17 × 7–10 µm and have both a reticulate, rather fine ornament and drawn out, spiny spore ends. Occurs on clay-rich, gravelly, disturbed mineral soils, both in forests and in more open habitats.

Also known as *M. chateri*. *Melastiza contorta* ▽ is more cupulate and pale-haired. *Melastiza flavorubens* ⋉ does not have a distinct reticulum on the spores, and *M. carbonicola* ⋉ has a coarser reticulum and larger protuberances at the poles; it grows on burnt ground, but also on bare, calcareous soils.

Widespread, rather common; (May–)September–January.

Melastiza contorta is a deeply cupulate, strikingly apricot-orange *Melastiza* with small, whitish to pale brown hairs on the outer side, mainly at the margin. The spores have coarse warts up to 7 µm high, and measure 23–26.5 × 12–15 µm. Occurs on soil in deciduous and coniferous forests.

Aleuria aurantia ◁◁ may be floccose, but has hardly any hairs on the outer side, and has smaller spores. Other species of *Melastiza* have adpressed, darker hairs. *Humaria solisequia* (page 1308) has projecting hairs up to 2 mm long at the margin and finely verrucose spores.

Widespread, rather rare; autumn.

Scutellinia is a large, difficult genus of ± orange-red operculate cup fungi with pointed, brown, rooting hairs that usually have several root branches. Asci inamyloid. Spores spherical to oblong, usually ornamented. Decomposers, mostly on rotting wood or exposed mineral soils.

Species of *Cheilymenia* ▷▷ may also have rooting hairs but are typically smaller, have brighter orange to yellow colours and mostly occur on dung or rich soils.

Scutellinia scutellata is an orange-red to scarlet *Scutellinia* with marginal hairs that are dark brown and very long, up to 1.6(–3) mm. The hairs have several roots. The ellipsoid spores measure 17.5–23 × 10.5–13 µm and have rather small warts that partly coalesce and are up to 0.9 µm high. Mainly on water-soaked, rotten wood in boggy forests, but may also fruit directly on damp soil and litter.

This species and *S. crinita* ⋉ form a difficult complex. There are several further, very similar species, e.g. *S. subhirtella* ⋉ with somewhat shorter hairs and a more regular spore ornament, and *S. hirta* ⋉, on large trunks of deciduous trees, which has shorter, thicker hairs and spores measuring 22–25.5 × 10–13.5 µm.

Widespread, common; mainly May–November.

Jens H. Petersen/Thomas Læssøe

Scutellinia trechispora is a scarlet *Scutellinia* with brown hairs of different lengths on the margin and outer side. The longest hairs are up to 2 mm long and have a split, root-like base. The spherical spores have truncate warts up to 4 µm high and measure 15–17.5 µm. Occurs on ± calcareous, moist soils in forests.

Amongst the round-spored species, *S. paludicola* ⋉ has larger spores (more than 25 µm) and shorter hairs. *Scutellinia minor* ⋉, *S. hyperborea* ⋉ and *S. barlae* ⋉ have a lower ornament (less than 2 µm high), and *S. minor* ⋉ does not have completely spherical spores.

Widespread, occacional to rather rare; mainly June–October.

Scutellinia superba is a rather small, ± apricot-orange *Scutellinia* with short, brown hairs (up to 360 µm long). The broadly ellipsoid spores have ± confluent warts up to 0.5 µm high and measure 19–24 × 11–16 µm; the outer wall loosens when the spores are heated in cotton blue. Occurs on open, sandy, moist soils.

The similar *S. mirabilis* ⋉ has narrower spores (a higher spore Q) and *S. torrentis* ⋉ has shorter spores.

Widespread, rare or overlooked; May–December.

Scutellinia setosa is a very small, mostly pale apricot-orange *Scutellinia* with very long, marginal hairs in relation to the size of the apothecia. The hairs have several roots. The ± smooth spores are multiguttulate and measure 19–22 × 11–13 µm. Occurs on very rotten wood of deciduous trees, mostly large trunks.

The very small fruitbodies (max. 2 mm) and the ± smooth spores are the best separating characters.

Mainly nemoral, rare; June–February.

Parascutellinia carneosanguinea has rather small, cupulate, dark grey-rose apothecia with an outer covering of appressed, brown, non-rooting, multicellular hairs. The finely verrucose spores have two drops, measure 22–29 × 12–15 µm. Occurs on sandy, damp soils and litter in ± calcareous, temporarily inundated habitats; ectomycorrhizal.

Parascutellinia arctespora is perhaps an older, valid name for this species. *Parascutellinia fuckelii* ⋉ can apparently be recognized by an apricot-orange hymenium and completely smooth spores; it occurs on wood.

Widespread, rare or overlooked; July–October.

Cheilymenia produce yellow to reddish-orange, rather small apothecia that usually have projecting, pale brown, ± rooting hairs. A few species are smooth. The spores are ellipsoid and smooth, verrucose or finely striate, and may have a loosening outer wall. Occurs mostly on rich soils or dung.

Scutellinia ◁◁ is typically more orange-red and often occurs on wood or on mineral soils. Species of *Spooneromyces* ◁◁ do not have rooting hairs and have spores with an ornament of sinuous warts or a reticulum that is attenuated at the poles.

Cheilymenia vitellina is a large, yellow *Cheilymenia* with brown-tinged, scattered marginal hairs up to 0.7 mm long. The ellipsoid spores may have a loosening outer wall and measure 14–17 × 8–9 µm; with high magnification (× 1,000) a fine striate ornament can be seen. Occurs on mulchy soil, mostly in *Urtica* stands along forest roads.

Also known as *C. oligotricha*. *Cheilymenia theleboloides* ⋉ occurs on plant litter and has shorter, indistinct hairs up to 0.25 mm long. *Cheilymenia crucipila* ⋉ has an orange hymenium.

Widespread, occasional; mainly June–November.

Cheilymenia fimicola is a rather small, long-haired, typically apricot-orange *Cheilymenia* with rooting hairs. The ellipsoid, smooth spores have a loosening outer wall and measure 18–22.5 × 10–12 µm. Occurs on herbivore dung, typically cowpats.

Cheilymenia raripila ⋉ has larger spores, measuring 23–26.5 × 13–16.5 µm. *Cheilymenia coprinaria* ⋉ has narrower spores, 8–9.5 µm wide. *Cheilymenia stercorea* ▷ has 'star-shaped' hairs on the outside and *C. pulcherrima* ⋉ has non-rooting hairs.

Widespread, rather common; most of the year.

Cheilymenia stercorea is a long-haired, ± apricot-orange *Cheilymenia* with hairs that have short, acute, 'star-shaped' branches at the base (see inset image). The smooth spores do not have drops and measure 16.5–20 × 8–10 µm. Occurs on herbivore dung.

Cheilymenia crucipila ⋉ has similar hairs but occurs on soil and has verrucose spores. Other species of *Cheilymenia* that occur on dung have unbranched hairs.

Widespread, fairly common to occasional; most of the year.

Cheilymenia granulata is an almost coral-red to reddish-orange, mostly smooth *Cheilymenia* with a finely serrated apothecial margin and fine grains on the outside. The spores measure 14.5–17 × 7–9 µm and have very indistinct longitudinal striae (× 1,000 magnification required). Occurs on herbivore dung, particularly cowpats.

Due to the lack of setose hairs this species was formerly placed in a separate genus, *Coprobia*.

Widespread, very common; most of the year.

Byssonectria terrestris has clustered, orange-yellow, smooth apothecia, typically on a subiculum. Margin is dentate. Paraphyses bent at the tips. Spores fusiform, smooth, with two drops, ± thickened at the poles, 18.4–25.6 × 8–9.6 µm. On needle debris, *etc.*, where deer have urinated.

The very similar *B. fusispora* ⋉ is perhaps less associated with deer and apparently has an indistinct subiculum, straight paraphyses and larger spores, 24–29 × 7–11 µm. *Pseudombrophila guldeniae* ⋉, which has brown apothecia, may fruit in between *Byssonectria* apothecia.

Widespread, occasional; October–May.

Lasiobolus has small fruitbodies with one-celled, ± hyaline hairs and smooth, hyaline spores. Occurs on dung.

Lasiobolus ciliatus is a yellowish- to reddish-tinged *Lasiobolus* with projecting, setose hairs up to 500 µm long. Asci cylindrical; spores usually in one row. The spores measure (19.4–) 20.4–25.2(–25.6) × 10–12.1(–12.5) µm, and have a slime coat but no drops. Occurs mostly on cow and horse dung.

Also known as *L. papillatus*. *Lasiobolus cuniculi* ⋉ has more clavate asci; spores are in two rows. *Lasiobolus intermedius* ▽ has smaller spores.

Widespread, common; most of the year.

Lasiobolus ruber is a typically orange-red to orange-yellow *Lasiobolus* that has hairs up to 950 µm long. The asci are cylindrical. The ± fusiform spores measure 23.7–28.3(–31) × (10–)10.5–12 µm. Occurs on *e.g.* deer and cow dung.

Best told by the very long hairs and very long, fusiform spores. *Lasiobolus macrotrichus* ⋉ has more ellipsoid, shorter spores, and *L. diversisporus* ⋉ has shorter hairs and spores that are yellow at maturity.

Widespread, but rarely recorded; most of the year.

Lasiobolus intermedius is a relatively large, up to 0.7 mm wide, orange to orange-yellow *Lasiobolus* with setose hairs up to 630 µm long. The asci are cylindrical and the spores are in one row. The spores are ellipsoid and measure (14.1–)14.7–16.2(–16.8) × 8.9–9.9 µm. Occurs on dung from a range of herbivores.

Lasiobolus microsporus ⋉ has even smaller spores.

Widespread, rare or overlooked; most of the year.

Iodophanus carneus is a small operculate cup fungus with cushion-shaped, rose- to orange-tinged apothecia that have somewhat projecting asci at maturity (see inset picture). The asci are diffusely amyloid, and the tips of the paraphyses have numerous small drops. The ellipsoid to somewhat fusiform spores have a fine, warty to wrinkled outer wall and measure 15–20 × 7.5–10.5 µm. Occurs mostly on herbivore dung, but also on nutrient-rich soils, soiled textiles, *etc*.

Iodophanus hyperboreus ⋉ apparently has less fusiform spores with somewhat larger warts. *Iodophanus testaceus* ⋉ is apparently more orange and has wider spores. There is considerable disagreement amongst mycologists as to how species in *Iodophanus* should be treated, and a dedicated molecular study is required.

Widespread, probably common; most of the year.

Coprotus granuliformis is a rather small, whitish to yellowish *Coprotus* that lacks both hairs and elongated cells at the apothecial margin. The 8-spored, cracking/slitting asci are broadly clavate, 40–50 × 15–25 µm, and the paraphyses are widened to 5–8 µm at the tips. The hyaline, smooth, ellipsoid spores do not have drops and measure 11–12.5 × 6–7.5 µm. After some time in a preparation, the spores develop a large, refractive air bubble. Occurs on herbivore dung.

Coprotus breviascus ⋉ has narrower paraphyses; *C. aurora* ⋉ has almost cylindrical asci and curved paraphyses; and *C. ochraceus* ⋉ has larger spores.

Widespread and probably common; most of the year.

Inoperculate cup fungi

The inoperculate cup fungi is a form group of ascomycotes with open fruitbodies (apothecia) where spore release is mostly controlled by an inner, cork-like structure at the ascus tip. The name refers to the lack of an operculum (a lid on the top).

The apothecia are mostly small (less than 2 mm wide) and cushion- or disc-shaped to cupulate, and numerous species occur on wood or herbaceous stems. A few genera produce larger or ± club-shaped apothecia. Species in the *Sclerotiniaceae* produce apothecia from overwintered sclerotia; many of these are plant parasites.

The hymenium is in most cases made of asci and paraphyses. The spores are typically hyaline, small and smooth.

This form group is very large with nearly 300 genera and perhaps around 1,000 species in temperate Europe. The generic delimitation will undoubtedly undergo many changes in the coming years, and many more species will be described. Here we have included 139 genera with 256 species.

OTHER SIMILAR FUNGI:

– the operculate cup fungi have asci that open with a lid (an operculum). The fruitbodies are mostly larger and many species occur on soil or wood. A group of species on dung have very small apothecia (page 1278).

– the apothecial lichens may have fruitbodies that can look very similar to species in this form goup, but most have obvious thalli with algae or cyanobacteria (page 1484).

– cyphelloids are small basidiomycotes with mainly downturned, cupulate or ear-shaped fruitbodies (page 1076).

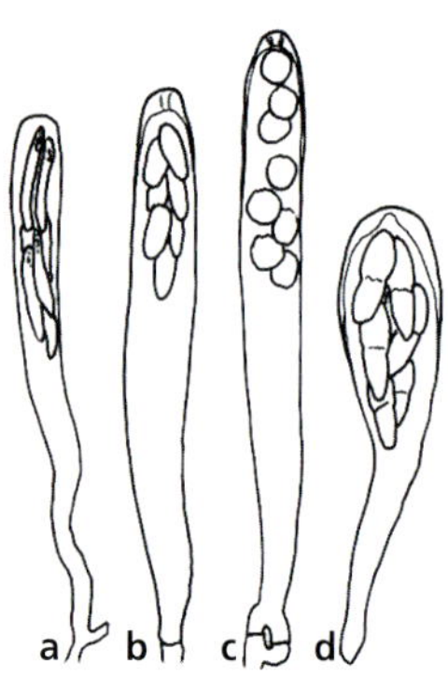

asci with a: simple top; b: cylindrical apical apparatus and simple base; c: cylindrical apical apparatus and crozier at the base; d: thick-walled top.

low, smooth to felty inoperculate cup fungi
(low = less than 2 mm above the substrate, page 1426)
erumpent inoperculate cup fungi
(opening by a lid, with teeth or slit-like, page 1468)
smooth, cupulate, ostly long-stemmed; ot or hardly gelatinous
smooth, discoid, sessile to short-stemmed, gelatinous or not
slit-like or with lid or teeth
inoperculate cup fungi
(open fruitbodies and asci without lid)
other similar fungi
with asci
with basidia
without asci or basidia
operculate cup fungi
(pages 1278)
apothecial lichens
(page 1484)
cyphelloids
(page 1076)
asexual fungi
(page 1626)

Earthtongues and the like

This form group includes the species of inoperculate cup fungi that have clavate, tongue-shaped, spatulate or morel-like fruitbodies.

Almost all species in the group have asci with an internal apparatus near the tip that is blown out and inverted during spore ejection. This apparatus often stains blue in iodine. The genus *Neolecta* (page 1363) deviates by having asci that split open (no apparatus) and by the lack of paraphyses among the asci.

Many species are decomposers but earthtongues, for example, possibly form biotrophic associations with plants in the same way as the hygrocyboid fungi (*Hygrocybe* and others) are thought to do (page 18).

OTHER SIMILAR FUNGI:
– clavarioids may have very similar fruitbodies but are basidiomycotes, *i.e.* the spores form on basidia (page 1092).
– jelly fungi and *Calocera* may be clavate in shape but are tougher and have septate or tuning fork-shaped basidia (page 1154 & 1162).
– stromatic pyrenomycetous fungi such as, for example, *Cordyceps*, *Trichoderma* and *Xylaria*, have the flesh filled with perithecia (see page 30) and the openings can be seen as small dots on the surface (page 1510).
– long-stemmed inoperculate cup fungi have a rather well-defined disc with hymenium (page 1404).

See also the general wheel, page 1348.

FURTHER READING: 14, 21, 49, 101, 166, 167, 168, 277.

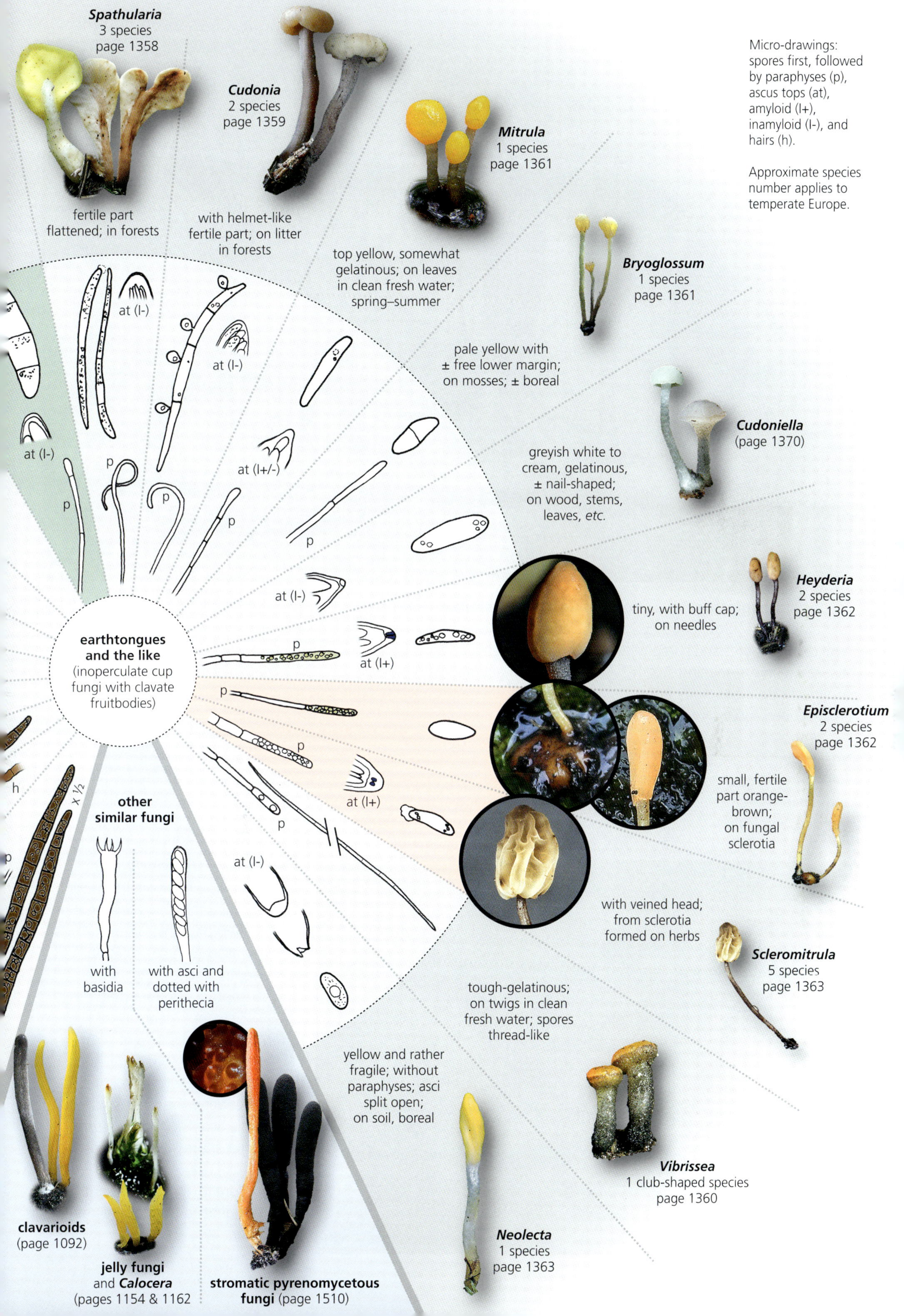

Micro-drawings: spores first, followed by paraphyses (p), ascus tops (at), amyloid (I+), inamyloid (I-), and hairs (h).
Approximate species number applies to temperate Europe.
earthtongues and the like (inoperculate cup fungi with clavate fruitbodies)
Spathularia 3 species page 1358
fertile part flattened; in forests
Cudonia 2 species page 1359
with helmet-like fertile part; on litter in forests
Mitrula 1 species page 1361
top yellow, somewhat gelatinous; on leaves in clean fresh water; spring–summer
Bryoglossum 1 species page 1361
pale yellow with ± free lower margin; on mosses; ± boreal
Cudoniella (page 1370)
greyish white to cream, gelatinous, ± nail-shaped; on wood, stems, leaves, etc.
Heyderia 2 species page 1362
tiny, with buff cap; on needles
Episclerotium 2 species page 1362
small, fertile part orange-brown; on fungal sclerotia
with veined head; from sclerotia formed on herbs
Scleromitrula 5 species page 1363
tough-gelatinous; on twigs in clean fresh water; spores thread-like
Vibrissea 1 club-shaped species page 1360
yellow and rather fragile; without paraphyses; asci split open; on soil, boreal
Neolecta 1 species page 1363
at (I-)
at (I-)
at (I-)
at (I+/-)
at (I-)
at (I+)
at (I+)
at (I-)
p
p
p
p
p
p
p
p
p
p
h
x ½
other similar fungi
with basidia
with asci and dotted with perithecia
clavarioids (page 1092)
jelly fungi and Calocera (pages 1154 & 1162)
stromatic pyrenomycetous fungi (page 1510)

Species of ***Geoglossum*** have smooth to scaly stems, ± brown, mostly multicellular spores and paraphyses that may curve at the tips. Asci I+. Apothecia are often white from an infection caused by *Hypomyces papulosporae* ⋉. A number of smaller genera have been split from *Geoglossum* based on *e.g.* spore and paraphysis characters, e.g. *Hemileucoglossum* ▷.

Geoglossum fallax is a tall and slender earthtongue, typically with a contrast between an almost black hymenium and a more brownish, distinctly scaly stem. Paraphysis tips are pear-shaped and imbedded in brown matter. Released spores are one-celled and smoky-grey, and overmature spores are 7–12-septate and brown; released spores measure (66–)71–100 (–105) × 5–7 µm; av. length 86 µm. Occurs mostly in unfertilized grasslands and scrub together with hygrocyboids (p. 142), *etc.*

Apothecia of *G. elongatum* ⋉ have the same appearance but the spores are shorter (av. length 61 µm based on released spores); *G. starbaeckii* ▽ has brown-walled paraphyses.

Widespread and rather common; August–December.

Geoglossum starbaeckii is a slender, dark, ± brown earthtongue with a scaly stem. The somewhat downcurved paraphyses are brown-walled with constrictions at the septa. The ± brown spores are 7–9-septate and measure 65–85 × 5–6 µm. Occurs in unfertilized grasslands and scrub.

The spores are somewhat shorter than *G. fallax* △, which has brown matter between the more swollen paraphysis tips; *G. vleugelianum* ⋉ has shorter spores with up to 7 septa.

Widespread, rather rare, but more frequent to the north; June–November.

Geoglossum cookeanum is a sturdy, ± black earthtongue that typically has a flattened fertile part and a somewhat felty stem surface without contrasting scales. The paraphyses are brown with many short, broad, constricted cells in the upper part. The spores soon become brown within the asci, are uniformly 7-septate, and measure (50–)56–80(–90) × 5–7 µm. In unfertilized grasslands and scrub, mainly on sandy, calcareous soils.

Geoglossum simile ⋉ prefers boggy habitats and has less swollen, only partly constricted cells in the paraphyses.

Widespread and locally rather common; August–December.

Thomas Læssøe

Geoglossum umbratile is a slender, dark, ± smooth-stemmed earthtongue. The stem may be browner than the hymenium. Asci 8-spored. The brownish paraphyses are slender, with downturned tips. The spores soon become brown, are 7-septate and measure (51–)56–80(–85) × 4.5–7 µm. Occurs in unfertilized grasslands, including in older, unfertilized lawns, and in scrub. On image with *Clavulinopsis corniculata* (page 1125).

Geoglossum cookeanum △ and *G. simile* ⋉ have broader spores and more swollen paraphysis cells; *G. montanum* ⋉ has 4–6-spored asci.

Widespread, fairly common; June–December.

Thomas Læssøe

Hemileucoglossum littorale is a rather short, mostly somewhat twisted-flattened, black earthtongue. The stem has ± setose, dark hairs with a rounded tip. The spores are ± hyaline, 0–3(–10)-septate and measure (36–)50–62(–88) × (4–)5–6.5(–8) µm. Fruits with *Littorella* in dried-out oligotrophic lakes. Sometimes occurs in association with *Hygrocybe phaeococcinea* (page155) and with other earthtongues.

The closely related *H. pusillum* ⋉ occurs on slopes with seepages of calcareous water. At least one species of *Geoglossum* ◁ has been found with *Littorella*.

Nemoral, very rare or overlooked; August–October.

Thomas Læssøe

Trichoglossum hirsutum is a rather variable black earthtongue with a stem covered by long, setose hairs (inset image) and also with projecting setae from the hymenium. The fertile, upper part is mostly capitate. Setae very long, black and pointed. Asci I+, 8-spored. The spores soon become brown (within the asci), are 15-septate and measure (86–)96–140(–165) × 6–7 µm. Occurs in unfertilized grasslands, marshes, scrub and forests, usually together with various clavarioids (pages 1116–1127) and entolomatoids (page 448).

Earthtongues belonging to the genus *Trichoglossum* are best told by their stiff-haired stems and the dark setae in the hymenium. *Trichoglossum variabile* ⋉ has 9–14 septate spores; *T. tetrasporum* ⋉ has 4-spored asci; and *T. walteri* ⋉ has 7-septate spores. *Leucoglossum leucosporum* ⋉ is very similar to a *Trichoglossum*, but has hyaline spores.

Widespread, fairly common; August–January.

Apothecia of *Trichoglossum* and other earthtongues are often white from an infection caused by *Hypomyces papulosporae* ⋉.

Glutinoglossum glutinosum is a dark brown earthtongue with a very slimy, smooth stem. The paraphyses are thin with abruptly swollen, pear-shaped tips. Spores are pale brown, 1–3-septate and measure (46–)56–75(–85) × 4–5 µm. Occurs in unfertilized grasslands, scrub, *etc*.

Also known as *Geoglossum glutinosum*. The slimy stem is characteristic of this genus. Although *Geoglossum difforme* ⋉ also has a slimy stem, the spores are 15-septate. The newly described but widespread *Glutinoglossum heptaseptatum* ⋉ has 7-septate spores; *G. triseptatum* ⋉ has 3-septate but slightly longer spores.

Widespread, fairly common; July–December.

Sabuloglossum arenarium is a tar-like, irregularly shaped, dry and ± smooth earthtongue. The spores are hyaline, one-celled, multiguttulate and measure (21–)26–35(–45) × 4–6 µm. Occurs on sand in grey dunes, together with *Empetrum* and usually *Clavaria argillacea* (page 1119).

Also known as *Corynetes arenarium* and *Geoglossum arenarium*.The combination of habitat and the hyaline, one-celled, rather short spores make this species easy to identify. *Hemileucoglossum littorale* ◁◁ occurs with *Littorella* and has longer, multicellular spores.

Widespread, rather common to rare, mostly coastal; August–December.

Jan Vesterholt

Sarcoleotia globosa is a small earthtongue with a brown, floccose-scaly stem and a convex, brownish-black head- to disc-shaped fertile part with a rounded-inturned margin. Asci I+, 8-spored. Paraphyses brown and curved at the tips. Spores ± fusiform, hyaline, 0–1-septate, 21–33 × 2.5–4.5 µm. Occurs on mossy, sandy soils, *e.g.* in heathlands.

Other similar earthtongues are either completely black or more tongue-shaped. *Thuemenidium atropurpureum* ▽ is brown but has multicelled spores.

Widespread, very rare in the nemoral zones, more frequent in the north; (May–) August–November.

Thuemenidium atropurpureum is a dark purple-brown to almost black, rather large and twisted fungus with a minutely scaly stem. Asci I+, 8-spored. Paraphyses imbedded in brown matter and swollen at the tips. Spores hyaline, have many drops, are 0–6(–9)-septate and measure 20–35 (–39) × 5–6 µm. Occurs in unfertilized, mostly calcareous grasslands.

Geoglossum fallax ◁◁ and *G. elongatum* ⋉ have much longer spores. The poorly known *Maasoglossum aseptatum* ⋉ has a smooth, black stem and non-septate spores.

Widespread, rather rare, becoming scarcer towards the north; August–November.

Knud Knudsen

Microglossum is a genus of clavate, mostly brightly coloured, dry inoperculate cup fungi with one-celled, hyaline spores and straight paraphyses. The spores have a flat and a curved side. Occurs in unfertilized, often calcareous grasslands and on rich soils in scrub and forests.

Microglossum griseoviride is a grey-green to dark yellow-green *Microglossum* with a distinctly scaly stem. Asci I+, 8-spored. The spores are hyaline, one-celled, guttulate, have a ± flat side and a convex side, and measure (14–)16–20(–23) × 4–5 µm. Occurs on relatively well-drained forest soils – typically in *Fagus* or *Quercus* forests.

The very similar *M. viride* ▽ is more yellowish on the fertile part, and typically occurs with liverworts on peaty soils in damp or wet habitats; it has somewhat longer and wider spores. *Microglossum tenebrosum* ▷ and other species of *Microglossum* (except *M. viride* ▽) have smooth stems; *M. tenebrosum* also has shorter spores.

Widespread, at least in the nemoral zones, occasional; August–October.

Knud Knudsen

Microglossum viride is a green to yellow-green *Microglossum* with a distinctly scaly stem. The hyaline spores have drops, a ± flat side, and measure 18.5–20.5 × 5–5.5 µm. Occurs on moist, peaty soil, mostly together with large, non-leafy liverworts.

Microglossum griseoviride △ is a more greyish-green and has narrower spores; it grows in drier, forested habitats. *Microglossum tenebrosum* ▷ has a smooth stem and shorter spores.

Widespread, occasional, probably more common in northern parts than in the nemoral zones; mainly August–October.

Microglossum tenebrosum is a smooth-stemmed, ± twisted and acute *Microglossum* with rather dark turquoise to violet-brown hymenial colours and a rather short, ± turquoise stem. Asci I+, 8-spored. Spores measure (10–)11.8–15.4(–20) × (3–)3.7–4.5 (–5.5) µm, Qav. 3.4; have a few large drops. Occurs on calcareous soils in ± open habitats.

Belongs to a difficult complex of ± olive to turquoise, smooth-stemmed species. The spore Q values are: *M. nudipes* ⋉ 3.9, *M. clavatum* ⋉ 3.8, *M. pratense* ⋉ 3.7, *M. truncatum* ⋉ 3.8 and *M. parvisporum* ⋉ 3.05.

Distribution poorly known, mainly nemoral, rare; autumn.

Thomas Læssøe

***Microglossum olivaceum* s.l.** is a ± olive to brownish-olive, smooth-stemmed *Microglossum*. Asci I+, 8-spored. Spores measure (11–)12.5–18(–21.5) × 3.8–4.7 × 4.6–5.3 µm. Occurs mostly in unfertilized, mainly calcareous, grasslands.

Recent DNA sequence analyses has caused an explosive increase in the number of *Microglossum* species recognized. Currently, it is hard to delimit *M. olivaceum* precisely, although it is the oldest available name for a *Microglossum* with a smooth stem.

Widespread, rare; late summer–autumn.

***Microglossum fuscorubens* s. auct.** is a ± coral-red to flesh-pink *Microglossum* with a rather short, usually slightly paler stem that may have an olive tinge at the base. Asci I+, 8-spored, with croziers. Spores have 4–6 large drops and measure (13.7–)15.7–19.5(–22.3) × (3.3–)4–4.8 µm. Occurs on calcareous soils in grasslands, scrub and forests.

See note under *M. olivaceum* △. The French mycologist, Boudier, originally described a duller-coloured fungus under this name, but it is currently used for this brighter-coloured species.

Widespread, at least in the nemoral zones, rare; mainly September–November.

Spathularia form spatula-like fruitbodies with narrow, almost thread-like spores and similar twisted paraphyses. Occurs mostly on needle debris in forests and plantations.

Spathularia flavida is an inoperculate cup fungus with a flattened, spatula-shaped fertile part that has straw-yellow to yellow colours and a paler stem. The fertile part becomes wrinkled with age, while the stem remains ± smooth. Asci I-, 8-spored; spores in a parallel bundle. The paraphyses are strongly curved-sinuous and narrow. The spores are hyaline, tapering and almost thread-like, and measure 35–62 × 2–2.5(–3.5) μm. Overmature spores become septate and produce egg-shaped conidia from the cells. Occurs on needle debris, typically from *Larix*. Sometimes seen with *Trichoderma leucopus* (page 1528) – a possible parasite on both *Spathularia* and *Cudonia* ▷.

Spathularia rufa ▽ is red-brown-tinged and lacks any yellow; *S. nigripes* ⋉ has a grey- to violet-tinged stem and also lacks yellow colours.

Widespread, occasional but rare in parts of the north-western lowlands of temperate Europe; June–November.

Spathularia rufa is a rather pale, reddish-cinnamon to buff *Spathularia* with a ± transversely wrinkled hymenium; the stem is somewhat darker. The spores are thread-like and measure 30–50 × 1.5–2 μm. Occurs on litter amongst mosses in damp *Picea* forests and upland *Betula* forests.

Spathularia flavida △ is microscopically almost identical, but has distinctive yellow colours. *Spathularia nigripes* ⋉ has a grey to violet-tinged stem.

Mainly boreal, occasional, absent from the nemoral zones; summer–autumn.

Cudonia form rather small, firm, stipitate fruitbodies with a convex, strongly inrolled, pale fertile part. Spores narrow, multicelled, and at maturity germinate with conidia. Mostly on needle debris, sometimes in fairy rings.

Cudonia confusa is a rather monochrome, cinnamon-brown to grey-brownish *Cudonia*, although the stem may be somewhat darker than the hymenium. Asci I-, 8-spored. Paraphyses thread-like, curved at the tips. The spores are hyaline, multicelled, taper at one end and measure 30–50 × 2–2.5 µm; they may produce conidia directly from the cells. Grow mostly in fairy rings on needle debris, but also known to occur in deciduous forests. Sometimes in the company of *Trichoderma leucopus* (page 1528), which may be a parasite on *Cudonia* and *Spathularia* ◁.

Cudonia circinans ▽ is very similar but has a more evident contrast between a paler hymenium and a darker stem.

Widespread and rather common in the hemiboreal–boreal zones but rare in the nemoral zones and absent from some lowland countries; August–October(–December).

Thomas Læssøe

Jens Maarbjerg

Cudonia circinans is a contrasty *Cudonia* with a cream to pinkish-buff fertile part and a darker, red-brown or greyish stem. Asci I- and 8-spored. The multicelled spores are attenuated towards one end and measure 30–50 × 1.5–2.5 µm; may produce conidia. Mostly in fairy rings on needle debris, rarely on soil and litter in deciduous forests. Sometimes in the company of *Trichoderma leucopus* (page 1528), a possible parasite on *Cudonia* and *Spathularia* ◁.

Cudonia confusa △ is recognized by the less contrasty apothecia.

Widespread and rather common in the hemiboreal–boreal zones but rare in the nemoral zones; July–October.

Leotia lubrica is a long-stemmed inoperculate cup fungus with a rubber-like texture. The fertile head is ± olive with a downcurved/ inrolled, free margin. The stem is yellowish and finely grainy. It may stain green when dry, and in wet weather may be covered with a slime layer (inset image). Asci I-, 8-spored. Paraphyses branched, with somewhat widened tips. The spores are fusiform, have many drops, are 5–7-septate and measure 18–27 × 5–6 µm. Occurs on soil in damp forests and forested bogs; forming arbutoid mycorrhiza.

Several European species of *Leotia* apparently hide under the name *L. lubrica*; amongst other things, one has a yellow stem when dry and one a green stem. It is somewhat unclear what hides under the name *L. atrovirens* ⋉ (described from France), but the name is currently used for a distinct North American species.

Widespread and common; mainly July–November.

Specimens with somewhat sticky, dark green heads (image to the right – from Bhutan) are probably parasitized by *Hypomyces leotiicola* ⋉, which rarely (or perhaps never) has perithecia in Europe.

Vibrissea truncorum is a long- and ± dark-stemmed *Vibrissea* with a lens-shaped, yellow to orange-yellow fertile head. Asci I+, 8-spored. Spores are thread-like, 16-celled and measure 200–250 × 1–1.5 µm; they have a loosening slime coat. On dead wood in running, clean fresh water.

A distinctive species. Other species of *Vibrissea* (page 1455) are sessile or very short-stemmed and less orange on the hymenium. *Mitrula paludosa* ▷ has a differently shaped fertile head, a paler stem and completely different spores.

Mainly hemiboreal–boreal, occasional to rather common; mostly spring–summer.

Mitrula paludosa is a very distinctive, long- and white-stemmed, ± water-soaked/gelatinous inoperculate cup fungus with a ± striking, orange-yellow, capitate fertile part that has an adpressed, free margin. Old, late-season specimens may acquire more dingy colours (lower image). The stem base may be swollen into a cotton-like ball. Asci I+, 8-spored. Paraphyses straight and not widened at the tips, strongly branching. The spores are hyaline, somewhat clavate, one-celled and measure 10–19 × 2.5–3 µm. Occurs in clean, often slow-moving fresh water, amongst rotten leaves, *Alnus* catkins and mosses.

A distinctive species that is unlikely to cause identification problems. *Bryoglossum gracile* ▽ is a more northern, less water-dependent species associated with mosses in fens; *B. rehmii* ⋉ is even more terrestrial and occurs on mosses in conifer forests, mainly in the boreal zone.

Widespread, occasional, locally common; February–July(–August).

Bryoglossum gracile is a long-stemmed ascomycote with a thin, pale yellow, slightly scaly stem and an egg-shaped, pale yellow to buff, fertile head with a free margin. Asci I+, 8-spored. Spores are hyaline, 1–2-celled, 10.6–13 × 2.4–3.5 µm, and may be finely punctate. On mosses in fens, including *Paludella* fens.

Bryoglossum rehmii ⋉ is also rather northern but occurs on mosses in conifer forests and is duller-coloured. *Mitrula paludosa* △ occurs in very wet habitats, has a whiter stem when young, and the top has a less free margin.

Mainly boreal, occasional; summer–early autumn.

Heyderia cucullata is a small, pale brown inoperculate cup fungus with a long, downy stem and a helmet-like, pale pinkish-buff fertile part with a free margin. The asci are 8-spored and have a crozier at the base. The spores are hyaline, one-celled, pointed, banana-shaped and measure 11.5–16(–18.5) × 2.3–3(–3.5) µm. Occurs on decaying needles of *Picea*, or rarely *Abies* or *Pinus*.

Heyderia pusilla ⋉ is somewhat smaller, does not have croziers and occurs on needles of *Pinus*.

Rarely found, but probably rather common; mainly late autumn.

Thomas Læssøe

Episclerotium sclerotipus is a pale yellow-stemmed inoperculate cup fungus with a longitudinally furrowed, elongated, helmet-like, pale orange-yellow fertile part. Asci 8-spored, have croziers, I+. The hyaline, fusiform spores have a loosening slime coat and measure 5.5–9 × 1.5–2.5 µm. Occurs on the large, brown sclerotia from *Typhula phacorrhiza* (page 1110) in forested bogs, *etc.*

The somewhat smaller *E. sclerotiorum* ⋉ occurs on sclerotia from *Sclerotinia trifoliorum* (see *Sclerotinia* page 1416).

Nemoral, very rare, but probably also highly overlooked; mainly October.

Scleromitrula calthicola is a small, long- and brown-stemmed inoperculate cup fungus that arises from a brown sclerotium and has a *Morchella*-like, pinkish-buff fertile part. Asci I+, 8-spored and have croziers. The spores are hyaline, one-celled and measure 6.5–10 × 1.5–2.5 µm. Occurs on sclerotia in damp litter amongst tall herbs; known from *Caltha palustris* and *Iris pseudacorus*. It apparently also occurs on *Filipendula ulmaria*.

Scleromitrula spiraeicola ⋉ has smaller spores, measuring 5–7 × 1.5–2.7 µm, but is otherwise very similar and occurs in stands of *Filipendula*.

Nemoral, very rare or perhaps overlooked; June–July.

Neolecta vitellina is a white-stemmed, ± clavate ascomycote with a yellow fertile part. The hymenium is unique, consisting exclusively of asci that lack croziers and no paraphyses; the ascus tips are thickened and I-. Spores hyaline, one-celled and measure 6–8.5 × (3–)3.5–4.5 µm. Occurs on soil around living *Picea abies* in boreal coniferous forests; apparently parasitic.

Spathularia flavida ⊲⊲ has a flattened fertile part and the hymenium has paraphyses. Species of *Calocera* (page 1159) occur on wood and are rubbery. Clavarioids (pages 1126–1127) are fragile and have basidia. *Hypocrea leucopus* (page 1528) forms a stroma with small ostiolar dots from immersed perithecia.

Mainly boreal, rather rare; late summer–autumn.

Thomas Læssøe

Large, gelatinous inoperculate cup fungi

A form group with often quite large, discoid, turbinate or brain-like, highly gelatinous apothecia. The spore release is controlled by an internal, cork-like structure in the ascus top. The species in this group are predominantly wood-decayers.

OTHER SIMILAR FUNGI:
– grades into the group of small, gelatinous, inoperculate cup fungi. It can be difficult to determine whether or not a tiny cup fungus has gelatinous matter in the tissues, and it is important to check the wheels for both the short- and long-stemmed inoperculate cup fungi (pages 1404 & 1426).
– gelatinous, tongue-shaped species are included on page 1350.
– gelatinous, operculate cup fungi are included on page 1290.
– jelly fungi have basidia, see page 1162. However, they can superficially resemble species of *Ascotremella* and *Ombrophila* and microscopy may be required.

See also the general wheel, page 1348.

FURTHER READING: 21, 49.

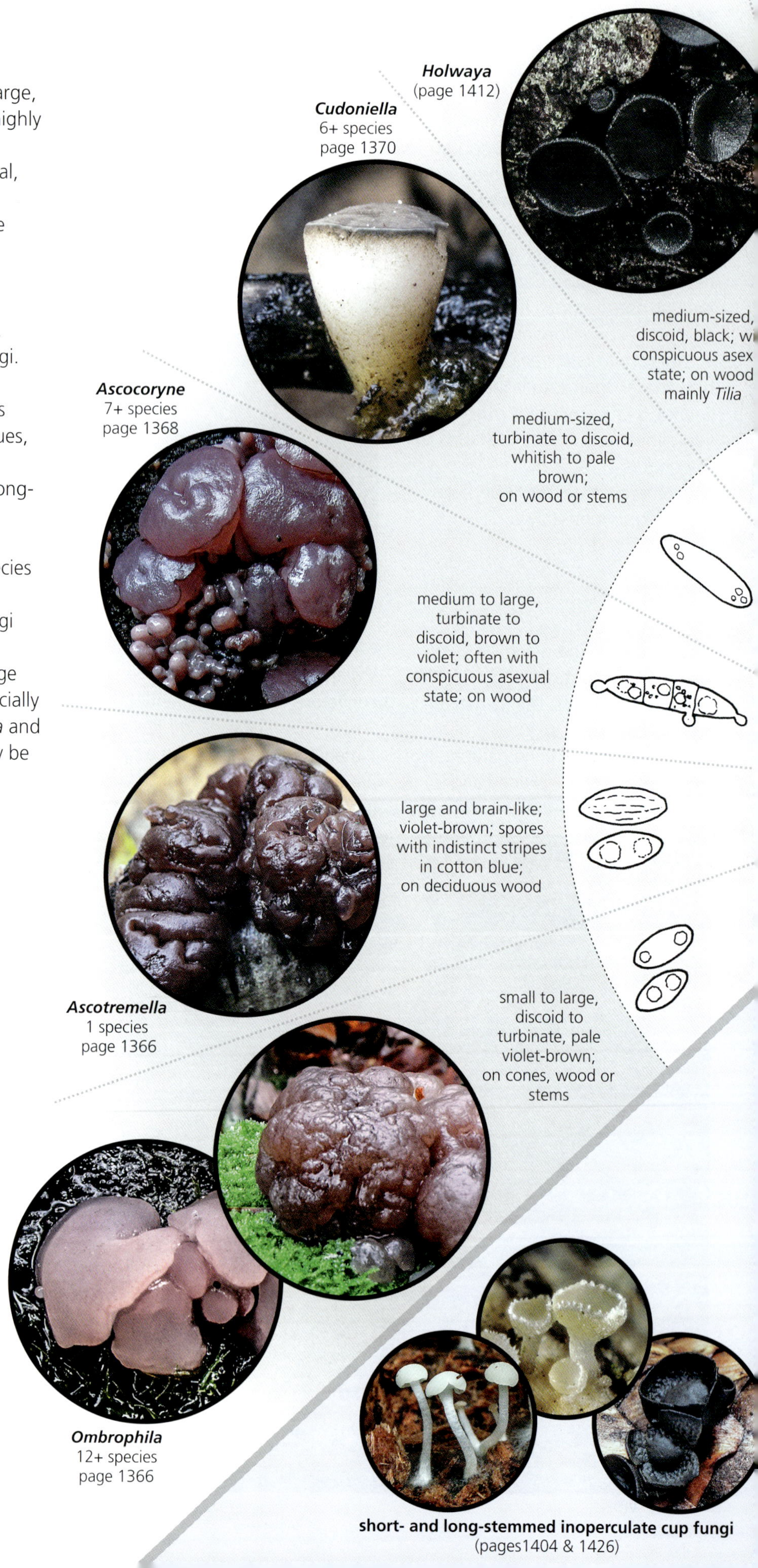

Micro-drawings: spores outermost, followed by paraphyses (p).

Approximate species count applies to temperate Europe.

Ionomidotis
(page 1407)
medium to large, greenish-yellow to black; stains dark in KOH (ionomidotic reaction); on deciduous wood
Bulgariella
1 species
page 1370
medium-sized, black; only slightly gelatinous; on deciduous wood
Bulgaria
1 species
page 1371
large, purple to black; highly gelatinous (like jellied candy); on deciduous wood
Vibrissea
7 species
(page 1455)
small and only moderately gelatinous; with thread-like spores; on deciduous wood in fresh water
Orbilia
(page 1464)
small and at most moderately gelatinous; with small spores and swollen paraphyses; on wood or stems
p
large, gelatinous inoperculate cup fungi
with brain-shaped, turbinate to discoid apothecia
other similar fungi
with asci
with basidia
asci inoperculate
asci operculate
earthtongues and the like
(page 1350)
operculate cup fungi
(page 1290)
jelly fungi
(page 1162)

Ascotremella faginea is a rather firm, although highly gelatinous, brain-like, ± wine-red, inoperculate cup fungus. Asci I+, cylindrical, 8-spored. Spores hyaline, very delicately longitudinally striate (at × 1,000), ± broadly fusiform, (7.7–) 8.1–9.5(–9.8) × 3.8–4.1 µm. May occur in a similar but asexual form. Occurs mainly on corticated branches from deciduous trees.

Ombrophila pura ▽ is paler, more turbinate and has smooth spores. '*Neobulgaria*' *pura* var. *foliacea* ▽ is often confused with *A. faginea*.

Mainly nemoral–hemiboreal, occasional; (June–)September–December(–March).

Ombrophila forms ± gelatinous, pale lilac to pale grey apothecia on wood, cones, stems or other dead organic material. Spores bi- or multi-guttulate, rather small, ellipsoid to slightly curved and smooth.

Ombrophila pura is a large, turbinate, later convex, flesh-pink to brownish-vinaceous, highly gelatinous, often clustered *Ombrophila*. The asci are I+ and 8-spored. The bi-guttulate, ellipsoid, smooth spores measure 7.5–9 × 3.5–4.5 µm. Occurs mainly on recently fallen *Fagus* trunks and thicker branches, rarely on other hosts.

Although until recently treated as belonging in *Neobulgaria*, molecular studies have confirmed a placement in *Ombrophila*. '*Neobulgaria*' *premnophila* ⋉ (described from Norway) should apparently produce smaller, more yellowish apothecia and occurs on *Picea*. The variety '*Neobulgaria*' *pura* var. *foliacea* (lower image) is brain-like and can easily be mistaken for *Ascotremella faginea* △ but that species normally has darker apothecia and the spores have faint striations.

Mainly nemoral with extensions into the hemiboreal–boreal zones, common to rare; (July–) September–February.

Thomas Læssøe

Ombrophila violacea is a rather small, up to 7 mm wide, convex, ± vinaceous, sessile to short-stemmed *Ombrophila* with a very gelatinous inner flesh. The asci are I+. The smooth, bi-guttulate spores measure 6.5–9 × 3.5–4.5 µm. Occurs in very damp habitats, on rotten cones, wood, *etc*.

There is a group of similar, poorly defined *Ombrophila* species, many of which tend to be less gelatinous. '*Cudoniella' rubicunda* ⋉ occurs on *Pinus* cones and has spores that measure 7–8(–9) × 2,6–3 µm.

Widespread but the exact distribution and frequency is poorly known; mainly August–October.

Ombrophila pileata is a very pale, very long-stemmed, rather small *Ombrophila*. The asci are I+ and 8-spored. The large, fusiform but flat-sided or curved, multi-guttulate spores measure (14.5–)18–27–33) × 3.3–4.2 µm. Occurs on decaying *Carex* leaves in boggy habitats.

Various similar *Ombrophila* species have shorter spores. *Ombrophila lacustris* ⋉ has strongly curved spores and also occurs on *Carex*. *Cudoniella clavus* ▷▷ has smaller spores measuring 10.5–15 × 3.5–5 µm and occurs on rotten, wet wood.

Widespread, rare or overlooked; mainly August–September.

Ombrophila rivulorum is an almost sessile, up to 8 mm wide, very pale, partly gelatinous *Ombrophila*. Asci I+. It has yellow crystals in the flesh and two types of spores: very oil-filled measuring 15–18.5 × 4.2–4.8 µm; and less oil-filled, 12–16 × 3.5–4.2 µm. Occurs on roots and leaves of *Salix* and *Populus* in damp habitats.

Cudoniella clavus ▷▷ has only one type of spore measuring 10.5–15 × 3.5–5 µm and occurs on wet deciduous wood.

A number of similar *Ombrophila* species are mainly more gelatinous and occur in different habitats.

Widespread, but rarely recorded; mainly July–September.

Ascocoryne form firm, somewhat gelatinous, mostly wine-red to purple apothecia that are often accompanied by a club- or spatula-like asexual state. The spores are ellipsoid, fusiform to banana-shaped, and often become septate when overmature; they germinate with small conidia. Occur on wood or rooting into *Sphagnum*.

Ascocoryne sarcoides is a gregarious to clustered, lilac to wine-red, ± stemmed *Ascocoryne* that is usually accompanied by a spatula-shaped asexual state. The paraphyses have gradually widening tips. Asci are I+. Spores ellipsoid, have two large and many small drops and measure 9–21(–26) × 3–5.5 µm; when overmature, 2–4-celled and ellipsoid to ovoid conidia 2–3 × 1–2 µm are produced in chains. Occurs mainly on only slightly decayed deciduous wood, rarely on conifer wood.

Ascocoryne inflata nom. prov. ▷ has a more brownish-lilac tinge, the spores are more pointed and the paraphyses have globose tips; *A. cylichnium* ▽ has larger spores filled with small drops and produces single conidia on the spores. *Ascocoryne albida* ▷ is often accompanied by a pestle-shaped, often rather dark asexual state.

Widespread, common; August–February.

Ascocoryne cylichnium is a rather large, solitary or only slightly clustered *Ascocoryne* that is never found accompanied by an asexual state. Paraphyses gradually widen at the tips. Asci I+. Spores have many small but no large drops, measure 18–33 × 4–5.6 µm; when overmature, 3–6-celled with single conidia produced directly from the spore wall. Occurs mainly on rather strongly decayed deciduous wood, rare on conifer wood.

Ascoryne sarcoides △ mainly occurs with its spatula-shaped asexual state and the spores have large as well as small drops.

Widespread, common; August–March.

Ascocoryne albida is a greyish-vinaceous to grey-purple *Ascocoryne* that typically occurs with its pestle- or matchstick-shaped asexual state or often only in this state (lower image). The spores have rounded ends and measure 9–19 × 3.5–4.8 µm. The asexual state has onion-shaped conidiophores that may resemble basidia. Occurs mainly on rotten deciduous wood, rarely on conifer wood or on cones.

Also known as *A. solitaria*. The asexual state of *A. sarcoides* ◁ is paler and ± flattened at the tips.

Probably widespread and rather common, but often overlooked and rarely reported; mainly October–December.

Benny T. Olsen

Ascocoryne inflata nom. prov. is a wine-red to brownish-purple *Ascocoryne* with ± discoid apothecia. The paraphyses have a globose head. Asci I+. The rather pointed, banana-shaped spores have large drops and measure 14–29 × 4.4–6 µm; when over mature, 2–3-celled and conidia are produced in chains. Occurs on deciduous wood.

Yet to be formally named. *Ascocoryne cylichnium* ◁ has gradually widening tips to the paraphyses and small drops in the spores.

Widespread, at least in the nemoral zones, occasional; mainly October–November.

Jean-Marc Moingeon

Ascocoryne turficola is a deeply rooting, pale olive-brown to wine-red *Ascocoryne*, typically with the stem and hymenium having contrasting colours. Paraphyses cylindrical and have many drops. Asci I+. The fusiform spores have many drops and measure (12–)15–20(–21) × (4–)–7 µm; they may have a coating of gel and may produce single conidia. Occurs in raised bogs in *Sphagnum* but probably in connection with *Cyperaceae*.

Also known as *Sarcoleotia turficola*. A very distinctive species.

Widespread, very rare in the lowlands of north-western Europe, increasing but not common towards the north; mainly August–November.

Cudoniella clavus is a ± distinctly stemmed, pale, non-reddening *Cudoniella* with a convex hymenium. Stem downy and often dark at the base. Asci faintly I+. Paraphyses have many drops. Spores 1–2-celled and asymmetrical, measuring 10.5–15 × 3.5–5 µm. Occurs mainly on deciduous wood and litter in wet habitats or directly in clean, ± flowing fresh water.

The somewhat reddening *C. tenuispora* ⨯ is more turbinate or barrel-shaped, up to 25 mm wide and usually has a concave hymenium. *Cudoniella acicularis* (page 1418) occurs on old *Quercus* stumps and is not gelatinous.

Widespread, rather common to occasional; mainly April–July.

Thomas Læssøe

Bulgariella pulla is a ± clustered, rather dark brownish-grey to slightly olive-tinged inoperculate cup fungus with a somewhat raised, convex hymenium. Paraphyses thread-like with dark, grainy contents. Asci I-. The dark brown, one-celled spores have many drops and measure 10–15(–17) × 7.5–9 µm. Occurs on deciduous wood, mainly *Betula*, but also *Fagus*.

Bulgaria inquinans ▷ is a larger, warty, much more gelatinous fungus. Species of *Ionomidotis* (page 1407) are less gelatinous and have smaller spores.

Widespread, but found mainly in the hemiboreal–boreal zones and very rare in the lowlands of north-western Europe; late summer–autumn.

Bulgaria inquinans is a large, often clustered, thick-fleshed, purple-brown to black inoperculate cup fungus with at first cupulate, later convex apothecia, and a shiny hymenium that produces a black stain from deposited spores. The outer side is scurfy-squamulose, and the gelatinous fleshy is marbled (middle image). The asci are I+ and 8-spored, with 4 dark spores at the top and 4 colourless spores below. The dark spores are blackish-brown, have a germ slit and measure 10–16 × 6–7.5 µm. Occurs on recently fallen, corticated *Quercus* and *Fagus* trunks, and is probably an endophyte in the living trees.

A highly distinctive species. The basidiomycote *Exidia glandulosa* (page 1189) has 4-celled basidia, a softer texture, a velvety outer side and a warty hymenium; it mainly occurs on thinner branches.

Widespread, but mainly nemoral–hemiboreal, common; almost all year, but in best condition during the autumn.

Downy to hairy inoperculate cup fungi

This form group of small inoperculate cup fungi has the outer side, and especially the margin, covered by projecting hairs. They are mainly decomposers and occur predominantly on herbaceous stems, fallen leaves, cones and wood.

The group contains many genera, and the appearance of the hairs (the shape, size, colour, crystal-covering, guttulation, *etc.*) is a very important distinguishing character. Some of the more distinctive species can be recognized in the field, but identification mostly requires the use of a microscope.

The majority of the species are classified within the *Hyaloscyphaceae*, and many have long paraphyses with lanceolate or tapering tips that project above the asci (see figure below).

Other similar fungi:

– cyphelloid fungi (page 1076) have small, hanging fruitbodies. They are *basidiomycotes* and can be recognized by their basidia and spore morphology – the spores having a small appendage (apiculus) where they were attached to the sterigmata. They are often not as elegant as the hairy inoperculate cup fungi and the hairs are more disorderly.

– some other inoperculate cup fungi have hairs but they are short and adpressed, making the apothecia look felty rather than hairy (a hand lens or dissecting microscope is required) – see the general wheel to the inoperculate cup fungi, page 1348.

Further reading: 21, 22, 23, 49, 78, 101, 115, 116, 160, 253, 319.

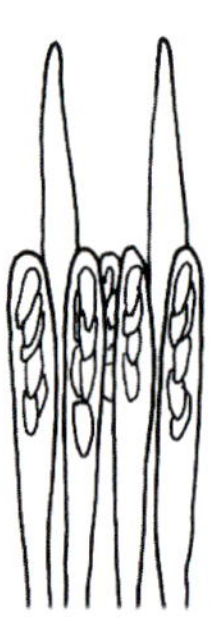

A typical hymenium in the genus *Lachnum* with asci and lanceolate, projecting paraphyses.

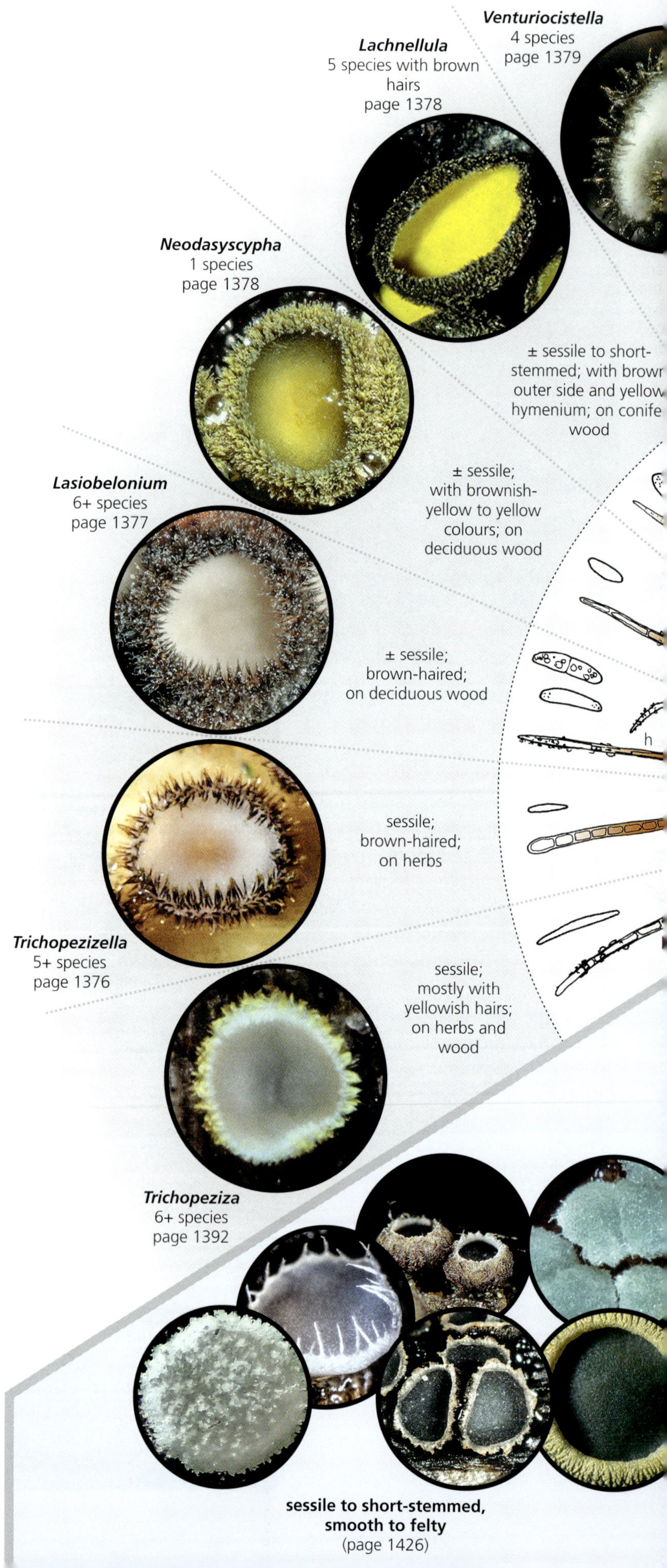

Micro-drawings: spores first, followed by hairs (h).
Approximate species number applies to temperate Europe.
± sessile; ark-haired th two hair types; on ood, leaves and herbs
Brunnipila
11 species , page 1380
sessile to short-stemmed; pale brown; on herbs and leaves
Cejpia
(page 1435)
sessile with downy grey hymenium and dark basal hairs; on *Poaceae*, and *Cyperaceae*.
Dennisiodiscus
(page 1451)
sessile with greenish-grey hymenium and dark orange hairs; on *Poaceae* and *Carex*
Perrotia
2 species
page 1382
± sessile; with orange-yellow to orange hairs; on deciduous wood or *Poaceae*
Albotricha
6 species
page 1382
± sessile; with white or coral hairs; on herbs or deciduous wood
± sessile on subiculum; mostly with yellowish to orange hairs; mostly on wood
Arachnopeziza
10 species
page 1383
h
± hairy inoperculate cup fungi – species with dark hairs
other similar fungi
with asci
with basidia
species with pale hairs,
see next page spread
high, sessile to long-stemmed, smooth to felty
(page 1404)
cyphelloids
(page 1076)

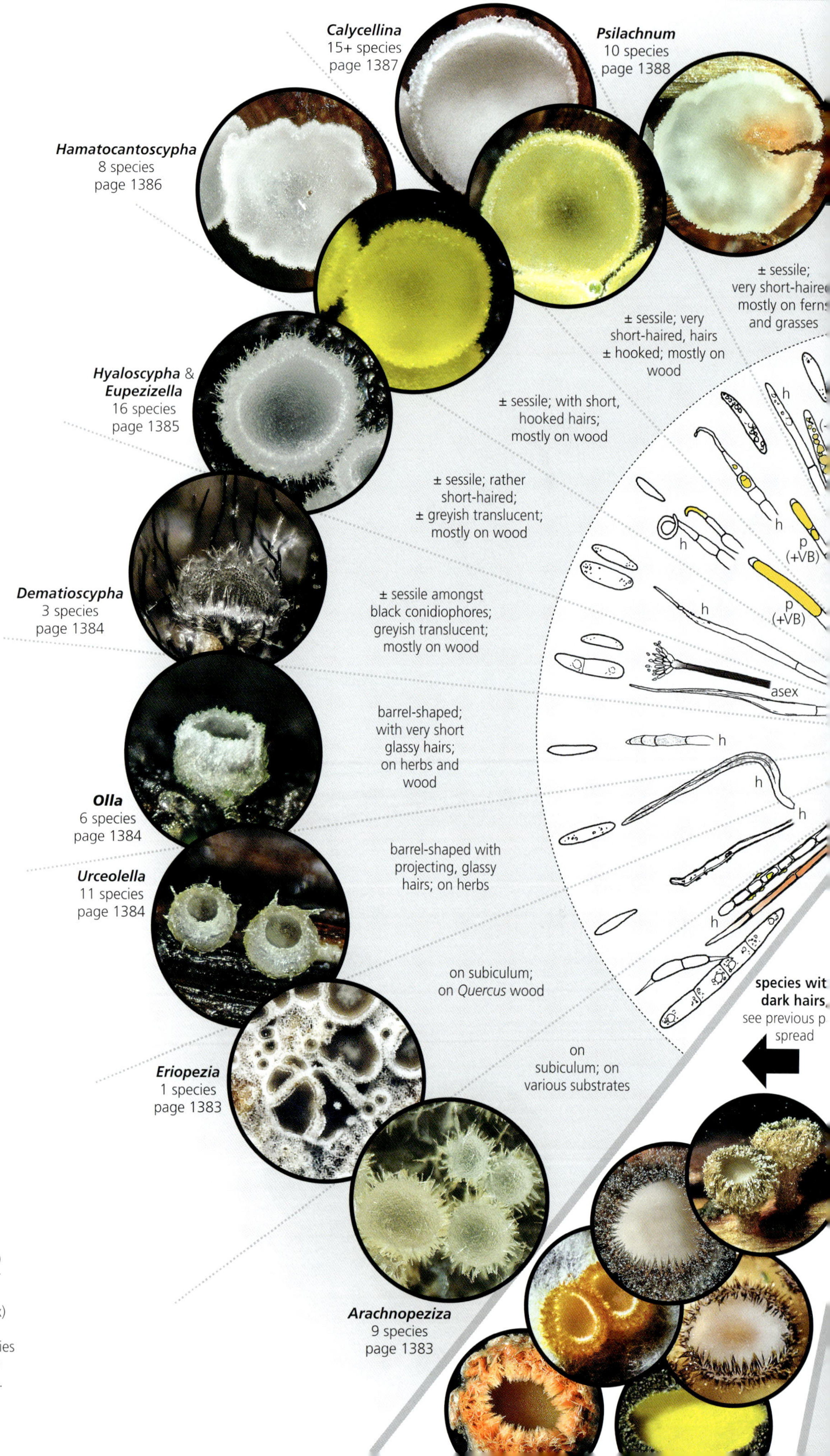

Micro-drawings: spores first, followed by hairs (h), paraphyses (p) refractive vacuolar bodies (VB) and asexual state (asex)

Approximate species number applies to temperate Europe.

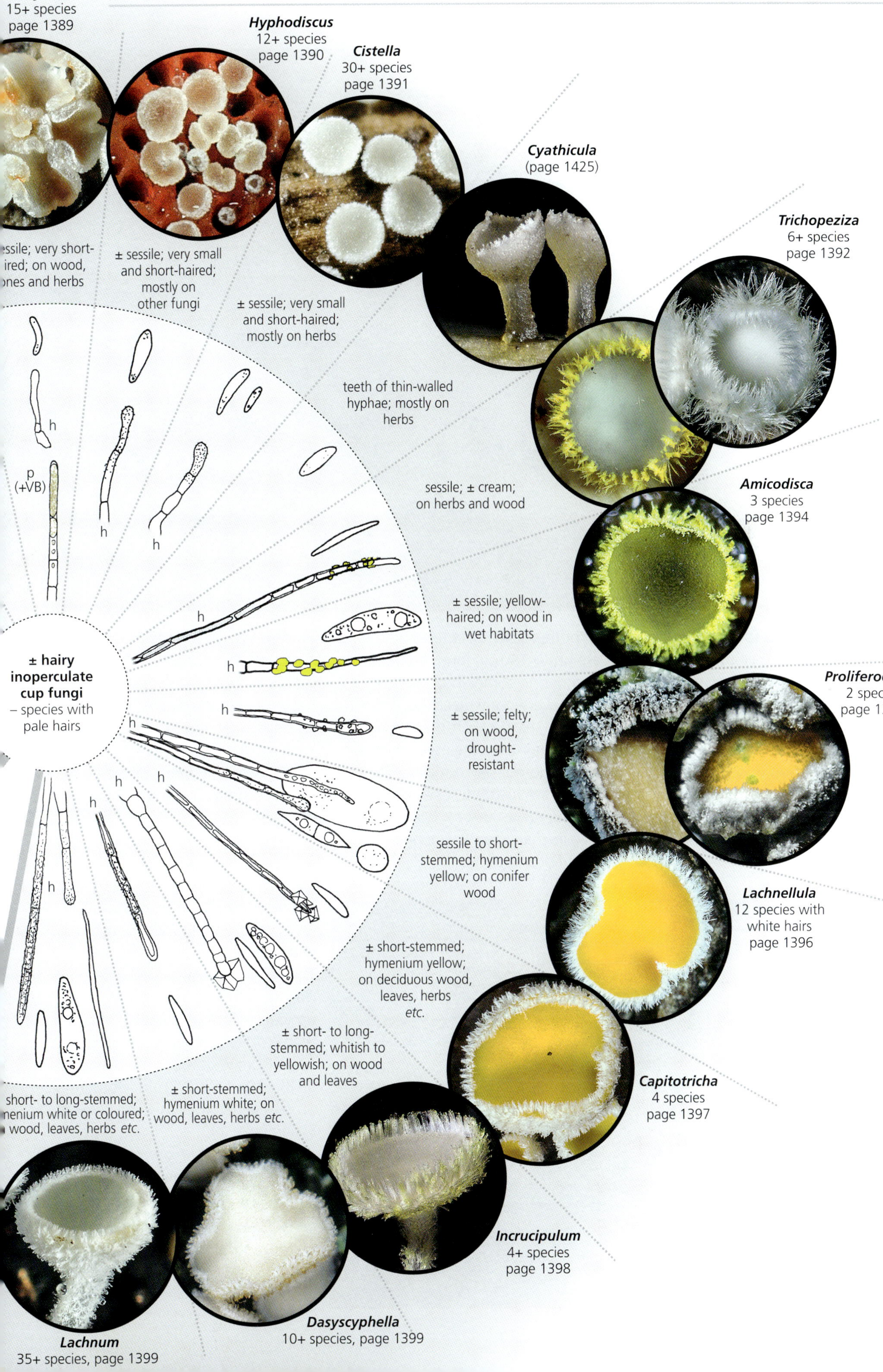
± hairy inoperculate cup fungi – species with pale hairs
Calycina
15+ species
page 1389
…ssile; very short-…ired; on wood, …nes and herbs
Hyphodiscus
12+ species
page 1390
± sessile; very small and short-haired; mostly on other fungi
Cistella
30+ species
page 1391
± sessile; very small and short-haired; mostly on herbs
Cyathicula
(page 1425)
teeth of thin-walled hyphae; mostly on herbs
Trichopeziza
6+ species
page 1392
sessile; ± cream; on herbs and wood
Amicodisca
3 species
page 1394
± sessile; yellow-haired; on wood in wet habitats
Proliferodiscus
2 species
page 1395
± sessile; felty; on wood, drought-resistant
Lachnellula
12 species with white hairs
page 1396
sessile to short-stemmed; hymenium yellow; on conifer wood
Capitotricha
4 species
page 1397
± short-stemmed; hymenium yellow; on deciduous wood, leaves, herbs etc.
Incrucipulum
4+ species
page 1398
± short- to long-stemmed; whitish to yellowish; on wood and leaves
Dasyscyphella
10+ species, page 1399
± short-stemmed; hymenium white; on wood, leaves, herbs etc.
Lachnum
35+ species, page 1399
short- to long-stemmed; …menium white or coloured; … wood, leaves, herbs etc.
p
(+VB)
h

Trichopezizella form rather small apothecia with pale hymenia and straight, brown, smooth hairs that may be so scattered that the surface in between can be seen. Occur on stems of herbs.

Trichopezizella nidulus is an almost sessile, brown, stiff-haired *Trichopezizella* with a greyish hymenium. The thick-walled, smooth, rather distant, rooting hairs are up to 150 µm long, multicelled, rounded to somewhat widened and have slightly paler tips. The paraphyses are fusiform and wider and longer than the asci, which are I+ and do not have croziers. The spores are fusiform, one-celled and measure (7.2–)8–11 × 1.5–1.9 µm. Occurs mainly on *Polygonatum multiflorum* stems.

Also known as *Lasiobelonium nidulum. Trichopezizella rubroguttata* ▽ occurs on dicot stems and has more crowded hairs, somewhat longer spores and the asci have croziers. *Trichopezizella relicina* ⋉ has longer hairs and larger spores.

Widespread, common; March–October.

Trichopezizella rubroguttata is a sessile, brownish *Trichopezizella* with yellowish drops on the hairs. The hairs are crowded, multicelled, clavate, ± brown, thick-walled and up to 150 µm long. Asci I+, 8-spored, with croziers. Paraphyses lanceolate, up to 6 µm wide. Spores one-celled, 8–13 × 1.5–1.8 (–2.2) µm. Occurs on coarse herbaceous stems, *e.g.* of *Epilobium*.

Trichopezizella nidulus △ is smaller and mainly occurs on *Polygonatum*.

At least nemoral but probably more widespread, rarely reported, but probably overlooked; mainly May–June.

Species of ***Lasiobelonium*** form rather large apothecia with pale hymenia, and have long, often sinuous or curly, ± brown hairs that usually have a fine crystalline covering. Occur on deciduous wood or stems of *Epilobium hirsutum*.

Lasiobelonium variegatum is a brownish, sessile, pale brown-haired *Lasiobelonium*. The hairs are straight to curled, thick-walled, multicelled, brownish with hyaline tips, and up to 170 µm long close to the apothecial margin and longer lower down. They have a pale crystalline covering, mostly on the upper parts. Paraphyses narrowly lanceolate, up to 4 µm wide. Asci I+, 8-spored, 57–70 × 5–6 µm, with croziers. At maturity the spores are 2-celled and measure 8–14(–17) × 2–3 µm. Occurs on deciduous wood, perhaps mostly *Populus tremula* and other species of *Populus*.

Lasiobelonium lonicerae ⋉ has somewhat shorter hairs and shorter spores. *Lasiobelonium barbatum* ▽ has lanceolate paraphyses but they are wider.

Seemingly mainly nemoral, occasional; mainly March–June.

Lasiobelonium barbatum is a sessile, grey-brown *Lasiobelonium* with a brown, felty covering and a pale hymenium. Hairs 200–270 × 4–5 µm, thick-walled, brown and have incrusted material; end cells clavate and colourless. Asci I+ (red), 50–80 × 4–7.5 µm, with croziers. Paraphyses lanceolate, 2.5–5 µm wide, project 15–20 µm. Spores narrowly ellipsoid, 2-celled, 9–12 × 2–3 µm. On deciduous wood, mostly on attached twigs of *Lonicera*.

Also known as *Trichopezizella barbata*. *Lasiobelonium belanense* ⋉ has larger spores and a brown subiculum. Other similar inoperculate cup fungi have cylindrical to narrowly lanceolate paraphyses.

Widespread in the nemoral–hemiboreal zones, occasional; mainly May–July.

Neodasyscypha cerina is a leathery, ± curry-yellow, rather large, short-stemmed inoperculate cup fungus with a densely hairy outer side; the innermost hairs are greyish and the outermost more yellow-brown. It can dry out and revive. The flesh is ± gelatinous. A purple pigment dissolves in KOH. Hairs up to 150 × 4 µm, basally thick-walled and have a grainy covering. Paraphyses cylindrical to narrowly lanceolate and project 5–10 µm above the asci. Asci I+, 8-spored. Spores very small, ellipsoid to fusiform, one-celled and measure 4–5.5 × 2–2.5 µm. Occurs on ± decorticated deciduous wood, *e.g.* attached *Quercus* and *Fagus* branches but also on cut surfaces of stumps.

The combination of dry habitat, colours, small spores and purple pigment make this species very distinctive.

Widespread in the nemoral–hemiboreal zones, occasional; September–April, mainly in cooler periods.

Peter Schirmer

Lachnellula flavovirens is a striking contrasty, rather large *Lachnellula* with a bright yellow hymenium and a densely hairy, dark grey-brown outer side. Hairs are somewhat pointed, up to 300 µm long and densely covered in coarse grains. Asci I-, 8-spored. Spores one-celled, hyaline, 8–12 × 3–4 µm. On conifer twigs.

Lachnellula pini ⋉ occurs on cankers on *Pinus* and has larger spores. *Lachnellula fuscosanguinea* ⋉ is a decomposer on *Pinus* and has larger spores. Various other *Lachnellula* species have more pointed hairs. See also the white-haired species on page 1396.

Mainly in southern parts of temperate Europe; most of the year.

Venturiocistella form very small apothecia with some short, rounded hairs and some long, dark, pointed hairs; both may have roughened walls. Occurs on bark, leaves of deciduous trees or grasses.

Venturiocistella pini is a very small inoperculate cup fungus with dark, projecting hairs and a pale hymenium. It has two types of hairs: one measures 50–100 × 3.5–5.5 µm, is pointed, thick-walled, blackish-brown and has a rough base; the other measures 20–50 × 4–5.5 µm, is cylindrical, thin-walled, pale brown and has a rough upper part. Asci I+, brown but blue with addition of KOH, 8-spored. Spores fusiform, one-celled and measure 10–14 × 1.5–2.5 µm. Occurs amongst algae and lichens on bark of attached *Pinus* branches. Shown on the lower image with *Peniophora pini* (page 1021).

A highly distinctive species.

Rarely reported, but perhaps widespread and common; apparently most of the year.

Jan Vesterholt

Venturiocistella diversipila is a very small inoperculate cup fungus with a pale hymenium that has a few projecting, seta-like hairs and some shorter, pale hairs (inset image). The seta-like hairs are thick-walled, blackish-brown, 75–195 × 4.5–7 µm and basally rough. Asci I-, 8-spored. Spores fusiform and measure 6–8.5 × 1.5–2 µm. Occurs on leaves of *Betula* and *Quercus* in the litter layer (see also the image of *Calycellina lachnobrachya* page 1387).

Venturiocistella venturioides ⋉ occurs on leaves of *Vaccinium uliginosum* and *V. heterotricha* ⋉ occurs on *Deschampsia cespitosa*.

Distribution and frequency poorly known; August–January.

Species of ***Brunnipila*** form rather small, short-stemmed apothecia with pale hymenia and yellowish-brown hairs. The hairs are mostly rough-walled and typically have large crystals at the tips. Occur on wood, leaves or herbaceous stems.

Brunnipila clandestina is a distinctly stemmed, clay-buff, ± adpressed felty-hairy *Brunnipila*. The hairs are brown, cylindrical, 80–140 × 3–5 µm, ± thick-walled, rough-walled, multicelled and have crystal-covered, somewhat widened tips. Asci I+, 8-spored, without croziers. The paraphyses are lanceolate, up to 5 µm wide and project up to 30 µm above the asci. The one-celled spores measure 6–9 (–11) × 1.5–1.8 µm. Occurs on year-old, dead *Rubus idaeus* canes.

Brunnipila calyculiformis ▽ is very similar but occurs on wood of deciduous trees. *Brunnipila cannabina* ⋉ occurs on coarse herbaceous stems and the asci have croziers.

Widespread, rather common; mainly March–July.

Brunnipila calyculiformis is a short-stemmed, brown-haired *Brunnipila*. The hairs measure 150–250 × 4–5 µm, are multicelled, rough-walled and have crystals at the tips. Asci I+. The paraphyses are lanceolate, 4–5 µm wide and project up to 25 µm above the asci. The narrowly ellipsoid, one-celled spores measure 8–12 × 2 µm. Occurs on deciduous wood, *e.g. Corylus*.

The somewhat shorter-spored *B. clandestina* △ is very similar but only occurs on *Rubus idaeus*.

Known from few records, but probably overlooked; May.

Brunnipila fuscescens is a short-stemmed, clay-buff *Brunnipila*. The brown, cylindrical, thick-walled, multicelled hairs measure 60–100 × 3–7 µm and usually have a crystal-covered pointed tip. Asci I+, 40–45 × 3–4 µm, with croziers. Paraphyses lanceolate, 3–5 µm wide and project 10–20 µm above the asci. The fusiform spores measure 6–9 × 1.5–2 µm. Occurs on *Fagus* leaves and cupules in the litter layer.

Brunnipila brunneola ⋉ grows on *Quercus* leaves and acorns and has somewhat larger spores. A number of other inoperculate cup fungi also occur on *Fagus* leaves, *e.g. Rutstroemia sydowiana* (page 1420), *Hymenoscyphus epiphyllus* (page 1460), *Coccomyces tumidus* (page 1471), *'Lachnum' patulum* ⋉, *Incrucipilum ciliare* (page 1398) and *Scutoscypha fagi* ⋉.

Widespread in the nemoral zones, very common; March–September.

Brunnipila palearum is a rather long-stemmed, brownish *Brunnipila*. The brown, cylindrical, rather thick- and rough-walled, multicelled hairs measure 50–70 × 3–5 µm, and either have or lack crystals on the tips. Asci I+, 45–50 × 4 µm, without croziers. Paraphyses lanceolate, 4–5 µm wide, project up to 15 µm. Spores fusiform and measure 7–14 × 1.5–2 µm. Occurs on rather thin grass culms.

There are several similar species of *Brunnipila* on other substrates. *Albotricha albotestacea* ▷▷ is short-stemmed and redder.

Widespread, at least in the nemoral–hemiboreal zones, common; mainly May–July.

Albotricha albotestacea is a long-haired *Albotricha* with vinaceous-brown hairs and a whitish hymenium. The tapering, multicelled, rather thick-walled, smooth hairs measure 170–230 × 4 μm. Asci I+, 8-spored, with croziers. Paraphyses lanceolate, project 20–30 μm above the asci. Spores fusiform, one-celled and measure 7–15 × 1.5–2 μm. On dead grass culms, *e.g. Phragmites*, *Phalaris* and *Calamagrostis*.

Brunnipila palearum ◁◁ occurs on similar substrates but is longer-stemmed and clay-buff.

Widespread in the nemoral–hemiboreal zones and probably rather common; May–October.

Perrotia flammea is a sessile, orange-red, drought-resistant species with a pale yellow-brown hymenium. Hairs are thick-walled, multicelled, cylindrical, measure 150–200 × 4–5 μm, and have small, superficial pigment lumps. Asci I-. Paraphyses cylindrical, non-projecting. Spores curved, cylindrical, 1–2-celled and measure 10–14 × 2.5–3 μm. Occurs on exposed deciduous wood.

Arachnopeziza aurelia ▽ is a similar colour but has completely different spores and a white subiculum.

Mainly hemiboreal–boreal, occasional; most of the year.

Arachnopeziza aurelia is a striking orange *Arachnopeziza* with either a rather limited, or a more pronounced, whitish to yellow subiculum. When fully expanded, the hymenium is pale orange, up to 5 mm wide and has ± agglutinated, pointed, yellow-orange hairs at the margin. Asci I+. Spores 11–20 × 2.5–4 μm, 2–4-celled with pointed appendages at the ends. Occurs on litter, including woody debris, particularly from *Quercus*.

Perrotia flammea △ has similar colours but no subiculum and completely different spores. *Arachnopeziza trabinelloides* ⋉ has 2-celled spores and hardly any subiculum.

Nemoral–hemiboreal, occasional; from early spring to summer.

Arachnopeziza form hairy apothecia on a subiculum.

Arachnopeziza aurata is a straw-yellow, rarely white *Arachnopeziza* on a pale subiculum and with ± cylindrical hairs and paraphyses. Asci I+, 8-spored, without croziers. Spores up to 8-celled and measure 50–75 × 1.5–3 µm. Occurs on rotten wood of deciduous trees, e.g. *Quercus* and *Fagus*.

Other pale species of *Arachnopeziza* and *A. aurelia* ◁ have smaller spores; *A. eriobasis* ⋉ recalls *Crepidotus pallidus* (page 1085) and has small, 2-celled spores.

Nemoral–hemiboreal, rather common; most of the year.

Arachnopeziza obtusipila is a small, pale *Arachnopeziza* with yellowish hairs and yellowish subiculum. The hairs are long, multicelled, blunt-tipped and have yellow pigment lumps. Spores are 4-celled and measure 26–38 × 3.4–4 µm. Occurs on deciduous wood and on conifer wood and cones.

Arachnopeziza aurata △ has longer spores; *Resinoscypha variepilosa* ⋉ has much smaller spores that measure 4–7 × 1.8–2 µm; and *A. delicatula* ⋉ has swollen hair tips and very small apothecia.

Nemoral, rare or overlooked; apparently most of the year.

Eriopezia caesia has a grey-brown to somewhat olive hymenium with white, ± felty marginal hairs and a whitish subiculum. Hairs cylindrical to irregular or can even have branches. Asci I+, 8-spored. Paraphyses cylindrical. The one-celled, fusiform to clavate spores measure 4–8 × 1.5–2 µm. Occurs on decorticated, split or cut *Quercus* wood, rarely on other types of wood.

The colour recalls species of *Mollisia* (page 1436), but these have spherical flesh cells, different hairs and typically a browner subiculum.

Widespread in nemoral–hemiboreal zones, rather common; all year.

Urceolella crispula is a very small, whitish, barrel-shaped inoperculate cup fungus. The hairs project at a tangent, measure 60–135 × 3.5–6 µm and are pale, very thick-walled, one-celled, I-, and have a narrow central cavity. The hair walls dissolve in 5% KOH (not in 2%). Asci I+, 8-spored, 20–43 × 3–5.5 µm. Spores narrowly ellipsoid to cylindrical and measure (5–)6–10(–11) × 1.5–2 µm. Occurs on dead herbaceous stems, *e.g. Apiaceae* and *Fallopia*.

Urceolella aasii ⋉, known from *Urtica* stems, is more yellow- to orange-tinged.

Widespread, but rarely recorded; mainly March–June.

Olla millepunctata is a ± barrel-shaped, very small, whitish to pale grey *Olla* with a white margin. Hairs are one-celled, 15–60 µm long, cylindrical, glass-like, dextrinoid, solid in the upper part and do not dissolve in 5% KOH. Asci I+, 8-spored, 20–45 × 3.5–6 µm. Spores ellipsoid, one-celled and measure 4–7.5 × 1.5–2 µm. Occurs on coarse herbs and *Rubus idaeus* canes.

Also known as *Hyalopeziza millepunctata*. *Olla scrupulosa* ⋉ occurs on deciduous wood; *Unguicularia aspera* ⋉ occurs on *Osmunda*; and other similar species occur on *Poaceae* and *Cyperaceae*.

Mostly nemoral, very common; mainly May–June.

Dematioscypha dematiicola is a small, greyish species with pale, partly solid, 110 × 3–5 µm, tapering hairs that do not have external lumps of pigment. Asci I-, 40–54 × 4–7.5 µm, 8-spored. Paraphyses cylindrical. Spores 7–9.5 × 2.5 µm, narrowly ellipsoid, 1–2-celled. Occurs on deciduous wood surrounded by what is assumed to be its asexual state, a dark hyphomycete, '*Haplographium catenulatum*', with hyaline conidia.

Other species either have larger spores or thin-walled hairs with external pigment lumps; *D. richonis* var. *olivacea* ⋉ is greenish.

Nemoral–hemiboreal, probably common; most of the year.

Hyaloscypha has thin, pointed, mainly one–few-celled hairs. A difficult genus.

Hyaloscypha fuckelii* var. *fuckelii is a ± sessile, thin-haired, white *Hyaloscypha*. The hairs are pointed and measure 30–60(–90) × 2–6 µm. Asci I+, 8-spored, 25–44 × 4.6–6.5 µm, with croziers. Paraphyses cylindrical. Spores ± fusiform, 6–10 × 1.9–2.3 µm. Occurs on deciduous wood, rarely on conifer wood, mosses or grass culms.

Hyaloscypha fuckelii var. *alniseda* ⋉ has asci that lack croziers, and *H. leucoconica* ▽ has thicker and longer hairs.

Mainly nemoral, probably common; most of the year.

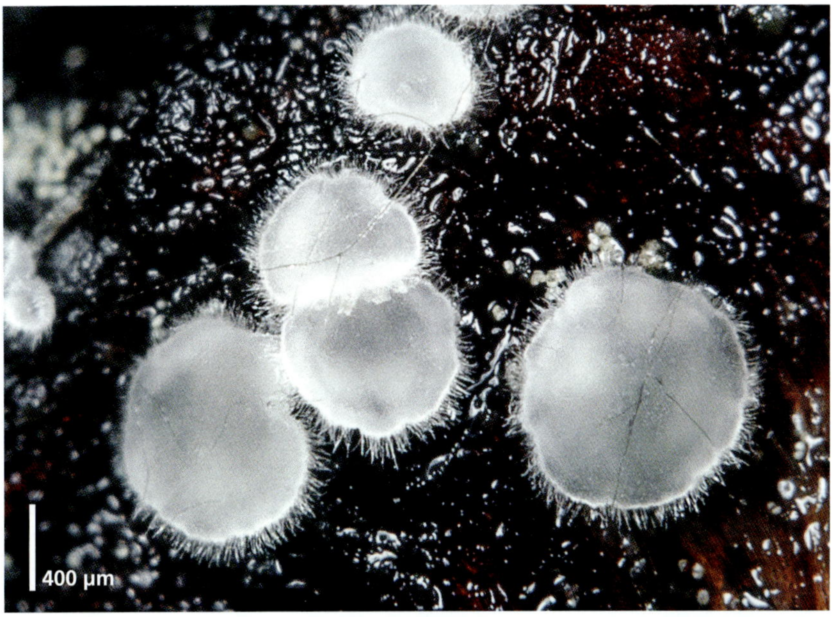

Hyaloscypha leuconica* var. *leuconica is a sessile, white to yellowish, long-haired *Hyaloscypha*. The tapering hairs measure 40–250 × 3–9 µm and are ± thick-walled. Asci I+, 22–50 × 4.5–6.8 µm, without croziers. Paraphyses cylindrical. The ± cylindrical spores measure 5.5–10 × 1.5–3 µm. On deciduous wood, rarely on conifer wood, *Calluna*, cones, catkins or cupules.

Hyaloscypha leuconica var. *bulbopilosa* ⋉ has asci with croziers, but is otherwise as var. *leuconica*; *H. fuckelii* △ has shorter, thinner-walled hairs.

Nemoral–hemiboreal, probably rather common; mostly in spring.

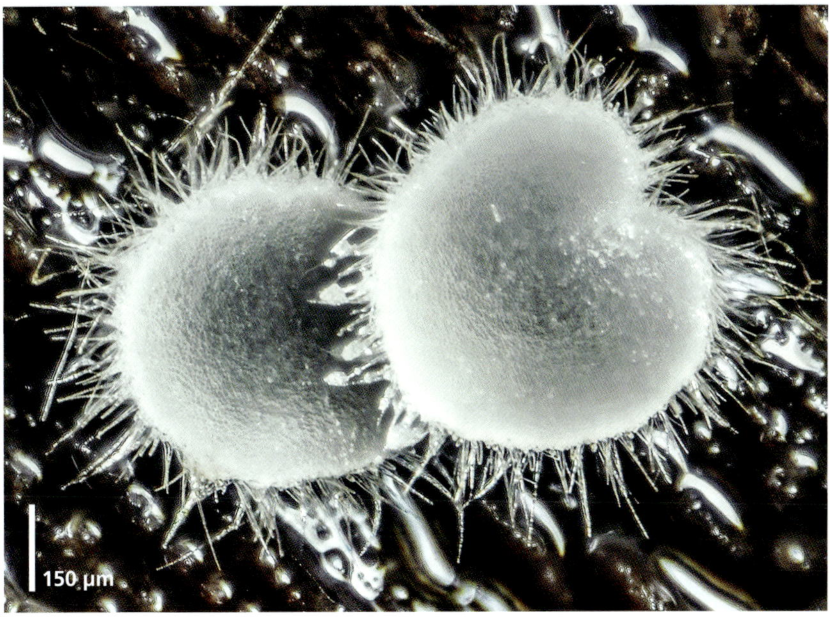

Eupezizella aureliella is a ± sessile *Hyaloscypha* with yellow-brown pigment lumps on the hairs (a strong lens required or check in water preparations). The hairs are narrow above, thin-walled and measure 25–60 × 2.5–5 µm. Asci I+, 35–54 × 5–8 µm, with croziers. Paraphyses cylindrical. The narrowly ellipsoid spores measure 5.5–10 × 2–3 µm. Almost exclusively on conifer wood.

Eupezizella britannica ⋉ has somewhat larger spores and *E. roseoguttata* ⋉ has rose pigment lumps on the hairs.

Widespread, common; all year.

The genera ***Hamatocanthoscypha*** and ***Calycellina*** form small, very short-stemmed apothecia; the paraphyses usually contain yellow, refractive vacuolar bodies (VBs). The hairs are short (strong lens required!). In *Hamatocanthoscypha* the top hair cell is ± pointed and hooked. In *Calycellina* the hairs are ± tapering, flagella-like and sinuous.

Hamatocanthoscypha uncipila is a short-stemmed, ± bright yellow, very short-haired *Hamatocanthoscypha*. The non-projecting hairs are cylindrical to attenuated, measure 10–35 × 2–3.8 µm and have yellow contents and hooked tips. Asci I+, 8-spored, 34–59 × 5.3–7.3 µm, with croziers. The paraphyses are cylindrical, 1–2.6 µm wide and have yellow contents. The fusiform to clavate spores measure 5–10.3 × 2–3 µm. Occurs mostly on conifer wood and cone scales.

Hamatocanthoscypha straminella ⋉ (perhaps the same as *H. sulphureocitrina*) has asci that are I- and occurs on deciduous wood.

Mainly nemoral, occasional; autumn.

Hamatocanthoscypha laricionis* var. *laricionis is a small, white, downy *Hamatocanthoscypha*. Hairs are hooked, cylindrical to slightly tapering, thin-walled and measure 15–28 × 2–4 µm. Asci I+, 8-spored, 22–30 × 4–5 µm, with croziers. Paraphyses cylindrical. Spores are clavate and measure 4.8–7 × 1–2 µm. Mostly on cones, rarely on bark, needles, *Lycopodiaceae* or leaves from deciduous trees.

Hamatocanthoscypha l. var. *minutissimus* ⋉ has shorter asci that do not have croziers. Similar species occur on *Polytrichum* and ferns.

Widespread and probably rather common; most of the year.

Calycellina ulmariae is a sessile to very short-stemmed buff to yellow *Calycellina*. The hairs are pointed, flagella-like, sinuous, smooth, thin-walled and measure 10–44 × 2–4 µm. The asci are I+, 4-spored and measure 29–39 × 4.5–6.5 µm. Paraphyses cylindrical. The fusiform, one-celled spores measure 12–13 × 2 µm. Occurs on *Filipendula ulmaria* stems from the previous year.

Often treated as a species of *Phialina*. There are several similar species of *Calycellina* on other substrates.

Nemoral–hemiboreal, probably common; mainly May–September.

Calycellina lachnobrachya is a very small, nearly sessile, rather vivid yellow *Calycellina* with fairly 'untidy' hairs. The hairs are pointed, flagella-like, sinuous, ± thin-walled and measure 18–64 × 2.5–4 µm. The asci are I+, predominantly 4-spored and measure 28–52 × 5–8 µm. Paraphyses cylindrical. Spores fusiform, 1–2-celled and measure 11–19 × 2–3.5 µm. Occurs on leaves of deciduous trees. Here seen together with *Venturiocistella diversipila* (page 1379).

Calycellina foliicola ⋉ and other species of *Calycellina* usually have 8-spored asci.

Widespread, but rarely recorded; July–October.

Calycellina chlorinella is a ± sessile, translucent greenish-yellow, very minutely downy *Calycellina*. The hairs are somewhat paraphysis-like and intergrade with the hymenial elements at the margin. It has spherical cells in the flesh. Asci mostly I+, 8-spored, 40–50 × 3.5–4 µm, with croziers. Paraphyses ± cylindrical, 1–2 µm wide. Spores ± allantoid and measure 5–7(–8) × 0.7–1 µm. Occurs on dark-stained herbaceous stems, e.g. *Urtica* and *Fallopia*.

The faint green colour and the round-celled, *Mollisia*-like flesh make this species fairly easy to identify.

Nemoral, probably common; mainly August–October.

Psilachnum has, like the species on preceding page spread, refractive vacuolar bodies (VBs) in the paraphyses. The apothecia are small, usually yellowish, and have short, not especially differentiated or tapering, sometimes bent hairs. Occur on ferns, *Equisetum*, *Cyperaceae*, *Poaceae*, other herbs or wood.

Psilachnum chrysostigmum is a ± sessile, white *Psilachnum* that stains yellow or reddish when bruised. The short, smooth, one-celled hairs measure 30–50 × 2.5–3 µm. Asci I+, 8-spored, 31–37 × 4–5 µm, with croziers. Paraphyses not obviously projecting, 2 µm wide, cylindrical to somewhat fusiform, guttulate. Spores fusiform to clavate, 6–7 × 1.5–2 µm, with small drops at the poles. Occurs on dead petioles of various ferns, including *Dryopteris* and *Athyrium*.

Psilachnum versicolor ⋉, on *Asplenium scolopendrium*, has spores with two drops; *P. pteridigena* ⋉, on *Pteridium*, has spores that measure 8–12 × 2.5–3 µm; and *P. inquilinum* ⋉, on *Equisetum*, is yellowish.

Nemoral–hemiboreal, very common; March–November.

Psilachnum rubrotinctum is a ± sessile, greyish-white to yellowish *Psilachnum* that immediately stains orange-red in ammonia vapours (inset image). Hairs gradually taper to 60 × 5 µm, 4–5-celled, smooth, thin-walled. Asci I+, 4-spored. Paraphyses tapering, 1.5–3 µm wide. The fusiform spores measure 6–7 × 1.5–2 µm. Described from stems of *Filipendula ulmaria*.

Psilachnum micaceum ⋉ has 8-spored asci and hairs that ± collect together to form teeth; known from *Filipendula* and *Cirsium*.

Nemoral, rarely reported; most of the year.

Calycina form sessile or short-stemmed apothecia with smooth or extremely short-haired margins (strong lens required!), and the paraphyses have vacuolar bodies. Occur on wood, cupules, cones or stems of herbs. See also the completely smooth species of *Calycina* on page 1458.

Calycina conorum is a ± sessile, usually clustered, whitish to cream *Calycina* that stains orange to reddish after bruising and on drying. The outside is minutely downy from blunt-tipped hairs. The hyphae in the flesh have ± glassy, thickened walls. Asci I+ reddish in Lugol, 8-spored, 45–60 × 4–6 µm, without croziers. Paraphyses cylindrical, 2–3 µm wide and do not project. The spores are curved, one-celled and measure 6–10 × 1.5–2.5 µm. Occurs on *Pinus* cones.

Calycina venceslai ⋉ occurs on conifer wood and has ellipsoid spores measuring 6–7.5 × 2.5–3.5 µm.

Widespread and fairly common, at least in the nemoral zones; mainly February–June.

Calycina vulgaris is a rather short-stemmed, whitish to cream or reddish, short-haired *Calycina*; the hairs have rounded tips. Asci I-, 8-spored, 35–45 × 4–5 µm, with croziers. Paraphyses ± cylindrical, with refractive contents in upper parts. The allantoid, one-celled spores measure 5–9 × 1.5–2 µm. Erumpent through holes in bark of deciduous trees, mostly on attached *Salix* branches in willow carrs.

Various similar species of *Calycina* have larger and typically differently shaped spores.

Nemoral–hemiboreal, probably rather common; mostly during winter.

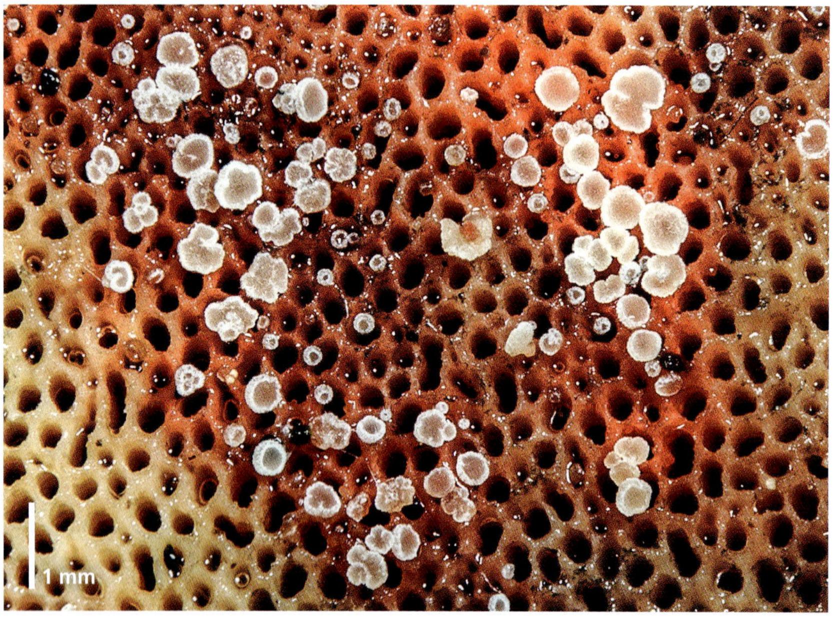

The genera **Hyphodiscus** and **Cistella** form small, finely downy apothecia with short, blunt-tipped, mostly somewhat clavate, finely verrucose hairs. Occur on wood, stems, leaves, needles or fungal fruitbodies.

Hyphodiscus hymeniophilus is a yellowish-white to greyish-white, small *Hyphodiscus* with a mostly somewhat undulating apothecial margin. The clavate, roughened hairs measure 15–40 × 3–5 µm. Asci I+ (reaction somewhat variable), 8-spored, 30–40 × 5 µm, with croziers. Paraphyses do not obviously project above the asci. Spores fusiform, hyaline, one-celled and measure 4–6 × 1.5–2 µm. Occurs mostly on old polypores or corticioids, which are stained in a pinkish or reddish colour – mainly on *Antrodia serialis* (page 896), rarely on similar stained, fungus-infected wood.

Various similar species of *Hyphodiscus* are not associated with pink- or red-stained substrates. Red-stained polypores are not necessarily infected with *H. hymeniophilus*.

Widespread, occasional, but overlooked; most of the year.

Hyphodiscus stereicola is a blue-green-tinged, clustered *Hyphodiscus*. The hairs are grey-green, ± straight, somewhat tapering, thin-walled, 2–4-celled, measure 31–62 × 2–2.8 µm and have loose lumps on the surface. Asci I-, 8-spored, with croziers. The spores measure 7.6–11.5 × 2.6–4 µm. Occurs on the upper, hairy side of *Stereum subtomentosum* (page 972).

The distinctive substrate and unusual colour make this species easy to identify. Other greenish *Hyphodiscus* species, *e.g. H. viridipilosus* ⋉, on wood, have smaller spores measuring 3–4 × 1.4–1.9 µm.

Within temperate Europe only known from Estonia; most of the year.

Cistella acuum is a short-stemmed, small, white *Cistella*. The hairs are thin-walled, clavate, 2–5-celled, roughened towards the upper part and measure up to 44 × 6.5 µm. Asci I+, 8-spored, 25–30 × 4–5 µm, with croziers. The paraphyses are cylindrical, 1.5–2 µm wide and project somewhat above the asci. The ± fusiform, one-celled spores measure 4–5.5(–6.8) × 1.5–2 µm and have small drops at the poles. Occurs on needles, rarely twigs and cones, of various conifers, including *Pinus*, *Picea* and *Abies*, but also on *Cupressaceae*.

Cistella conorum ⋉ is apparently a very similar, but completely sessile species on *Picea* cones. *Cistella* is a large and difficult genus with many species that occur on *e.g.* monocot and dicot herbs, ferns, leaves from deciduous trees and wood.

Widespread and very common; most of the year.

Cistella grevillei is a sessile, pale buff to pale flesh-pink, rather large *Cistella*. The hairs are whitish to pale yellowish, several-celled, ± thin-walled, measure 30–75 × 2.5–4.5 µm and are clavate with roughened tips. Asci I+, 8-spored, 43–56 × 4.5–7 µm. Paraphyses narrowly lanceolate, 1.8–3.8 µm wide and project 4–12.5 µm above the asci. Spores cylindrical to fusiform, 8.3–10.1 × 1.8–2.3 µm. Occurs on coarse herbaceous stems.

Cistella hungarica ⋉ has somewhat shorter spores and asci.

Widespread and very common; most of the year, but mainly March–May.

Trichopeziza form rather large, felty-hairy apothecia. The hairs are usually coloured yellowish from external pigment deposits. Mostly on coarse herbaceous stems, rarely on wood.

Trichopeziza leucophaea is a sessile *Trichopeziza* with a whitish, greyish to slightly rose hymenium. The marginal hairs are vivid yellow, the hairs farther down ± brown (inset image). They are ± cylindrical to gradually attenuated, measure 140–225 × 3–4.5 µm, have small pigment lumps and rather thick-walls, and are multicelled; the contents do not usually change colour in KOH. Asci I+, 8-spored, 55–80 × 4–6 µm. Paraphyses lanceolate, 4–5 µm wide and project 15–25 µm. Spores curved-cylindrical, one-celled and measure 11–18 × 1.5–2 µm. On *Apiaceae*, *Urtica* and other coarse herbaceous stems.

Other yellow-haired species in this genus on coarse herbaceous stems, including *T. violascens* ⋉, have hair contents that stain violet in KOH. *Trichopeziza sulphurea* ▽ occurs on similar substrates but has larger spores measuring 25–35 × 2–3 µm. *Trichopeziza discolor* ⋉ also occurs on herbs but has even longer spores measuring 40–65 × 1.5–2.5 µm. *Trichopeziza subsulphurea* ▷ occurs on wood and has shorter spores.

Widespread and common; (March–) April–June(–October).

Trichopeziza sulphurea is sessile with a pale hymenium and bright yellow hairs that are cylindrical to attenuated, multicelled, rather thick-walled and measure 150–225 × 3–4.5 µm. Asci I+, 8-spored, 80–100 × 6–8 µm. Paraphyses lanceolate, 4–5 µm wide and project 5–20 µm above the asci. The curved-cylindrical, 1–4-celled spores measure 25–35 × 2–3 µm. Occurs mostly on *Urtica* stems, rarely on other coarse herbaceous stems.

Trichopeziza leucophaea △ has shorter, one-celled spores (see above for other similar species).

Widespread but mainly nemoral–hemiboreal, common; most of the year, records predominantly in the autumn.

Trichopeziza subsulphurea is a sessile *Trichopeziza* with a whitish to pale grey hymenium. The upper and marginal hairs are bright yellow, those lower down are ± brown. The hairs are thick-walled, multicelled, have yellow lumps on the surface and measure 100–150 × 3–4 µm. Asci I+ (somewhat), with croziers. Paraphyses cylindrical, do not project above the asci. Spores one-celled and measure 6.5–11.5 × 2–3 µm. On deciduous wood, *e.g.* decorticated *Ulmus*, stump surfaces, *etc.*

Other similar species typically occur on coarse herbaceous stems and have larger spores.

Widespread but rarely recorded; mainly October–March.

Trichopeziza mollissima is a very pale cream, sessile *Trichopeziza* with long, white hairs. The hairs are ± thick-walled, tapering, blunt-tipped and measure 150–225 × 3–4.5 µm. Asci I+, 8-spored, 50–70 × 3.5–5 µm. Paraphyses lanceolate, 3.5–7 µm wide and project 15–25 µm above the asci. The fusiform, one-celled spores measure 8–13 × 1.5–2 µm. Occurs on coarse herbaceous stems, *e.g. Epilobium*, *Urtica* and *Apiaceae*.

The name *T. mollissima* has also been in use for species with yellower hairs, treated here under the name *T. leucophaea* ◁. *Trichopeziza discolor* ⋉, mostly on stems of *Urtica*, is not as pale, usually has yellow lumps on the hairs and has much larger spores measuring 40–65 × 1.5–2.5 µm.

Widespread and common; mainly April–August.

Amicodisca is a small genus of relatively small inoperculate cup fungi with rather short, yellow-green, tapering hairs. All occur on deciduous wood, mostly in wet habitats.

Amicodisca virella is a rather small, up to 0.6 mm wide, mostly sessile *Amicodisca*. The projecting, sulphur- to greenish-yellow hairs are crowded and fused, which creates a furrowed margin. The hymenium is greyish. The hairs are thin-walled, somewhat pointed, distantly septate and have external yellow pigment lumps. The paraphyses are cylindrical and guttulate. Asci up to 75 µm long. Spores one-celled, ± ellipsoid-fusiform and measure 17–25 × 4–6 µm (from dead material) and 18–28 × 5–7.2 µm (from living spores). Occurs in clustered groups on very damp or wet deciduous wood, typically in winter-flooded sites.

Amicodisca svrcekii ▽ has smaller spores (av. 7.9 × 2.3 µm – measured from dead material), and *A. groenlandica* ⋉ has longer asci, up to 116 µm.

Widespread but rarely recorded; summer–autumn.

Amicodisca svrcekii has sulphur-yellow to greenish-yellow hairs in tufts/stripes on the outer side. The hymenium is greyish. The hairs are thin-walled, somewhat pointed and have external yellow pigment lumps. The one-celled, hyaline spores measure 7.2–8.6 × 2–2.8 µm (from dead material) and 11.5–16 × 3–3.5 µm (from living spores). Occurs on damp or wet deciduous wood.

The small spores distinguish *A. svrcekii* from the two other species in this genus.

Widespread but rarely recorded; summer–autumn.

Proliferodiscus form drought-resistant, densely hairy apothecia, and have ± yellowish hymenia, cylindrical, granulate hairs and violet reactions with KOH.

Proliferodiscus pulveraceus is an up to 1.5 mm wide, drought-resistant, whitish to pale grey, almost woolly *Proliferodiscus* with a yellow hymenium. New apothecia are quite often formed on top of older ones; they only open in very damp weather. The hairs are cylindrical, somewhat thick-walled, distantly septate and finely granulate. Paraphyses thread-like. Asci I+, 8-spored. The flesh stains violet in KOH. The one-celled, hyaline spores measure 5.6–11 × 1.8–2.7 µm (from living spores). The asexual state forms greyish pycnidia, which also stain violet in KOH. Occurs on rather hard, typically decorticated, mostly sun-exposed wood from deciduous trees.

Species of *Lachnellula* ▷▷ have other spore shapes, and species of *Lachnum* ▷▷ typically have lanceolate paraphyses and are less drought-tolerant. Species of *Trichopeziza* ◁◁ and *Albotricha* ◁◁ do not have yellow hymenia.

Nemoral–hemiboreal, occasional; all year.

Proliferodiscus tricolor is an up to 2.5 mm wide, ± greyish-blue *Proliferodiscus* with densely agglutinated hairs and a pinkish-buff to somewhat rose hymenium. The hairs are granulate, cylindrical and rather densely septate. Asci I+. The pigment on and in the hairs and in the flesh becomes ± blue in KOH. Spores 1–2-celled and measure 9–14 × 2.8–3.7 µm. On bark of living *Quercus*, usually more than 1 metre up, often with lichens, liverworts, *etc.*

The habitat and colours are unique, although some species of *Trichopezizella* ◁◁ and the arctic/alpine *Solenopezia leucostoma* ⋉ may also have bluish hairs.

Nemoral–hemiboreal; rare; all year.

Lachnellula form rather large apothecia with yellow to orange hymenia and mostly white, rarely dark (page 1378), dense hairs. The hairs are granulate but, in contrast to the rather similar genus *Capitotricha* ▷, lack large crystals on the hair tips. Mostly decomposers but also parasites on conifers. *Lachnellula willkommii* ⋉ causes *Larix* canker.

Lachnellula calyciformis is a short-stemmed, 1–4 mm wide, yellow to orange, rarely cream-yellow *Lachnellula* with white marginal and outer side hairs. The hairs are cylindrical-sinuous, 3–4 µm wide, hyaline, multicelled, drop-filled, somewhat thick-walled, up to 180 µm long and have blunt tips. Asci I-, 8-spored, 45–63 × 4–5.5 µm, without croziers. Paraphyses are cylindrical, 1.5 µm wide and projecting. Spores are one-celled, ellipsoid, have two drops and measure 5–7 × 2.5–3 µm. Mostly on twigs and branches of *Abies*, but also on *Pinus* and *Picea*.

Lachnellula resinaria var. *resinaria* ▷ has smaller spores measuring 2.5–4.5 × 1.7–2.5 µm; *L. subtilissima* ⋉ has somewhat longer, aguttulate, fusiform spores measuring 6–9 × 2–2.5 µm. Many similar species of *Lachnellula* have larger spores.

Widespread, common; most of the year, except during summer.

Lachnellula suecica is a short-stemmed, orange *Lachnellula* with white marginal and outer side hairs. The hairs are granulate, cylindrical and 2.5–4 µm wide. Asci I+ (red), rarely I-, 50–80 × 5–8.5 µm, with croziers. Paraphyses cylindrical, 2–2.5 µm wide, projecting. The spores are spherical to almost spherical or egg-shaped and measure 4–6.5 µm in diameter. Occurs on bark of *Larix*, *Picea*, *Pinus* and *Juniperus*.

Lachnella resinaria var. *calycina* ⋉ also has almost spherical spores but they measure 2.3–3.5 × 2–2.8 µm.

Mainly hemiboreal–boreal, occasional to common; most of the year.

Lachnellula resinaria* var. *resinaria is a short-stemmed, pale yellow *Lachnellula* with white marginal hairs. The hairs are cylindrical, rough and measure 70–125 × 2–3 µm. Asci I-, 8-spored, 38–45 × 3–4 µm. Paraphyses cylindrical, 1.5–2 µm wide, projecting. The broadly ellipsoid spores measure 2.5–4.5 × 1.7–2.5 µm. Occurs in wounds on conifer wood, mostly *Picea*.

Lachnellula resinaria var. *calycina* ⋉ has ± spherical spores measuring 2.3–3.5 × 2–2.8 µm. *Lachnellula subtillissima* ⋉ and *L. calyciformis* ◁ have longer spores.

Widespread, rare or overlooked; most of the year.

Capitotricha bicolor has white hairs and a bright yellow to orange-yellow hymenium. The hairs, which measure 150–250 × 3.5–4.5 µm, are sinuous, thick-walled, rough, somewhat pointed and crowned with large angular crystals that may disappear during preparation. Asci I+, 8-spored, without croziers. Paraphyses are broadly lanceolate and project far. The one-celled, somewhat fusiform spores measure 8.3–12.2 × 2.1–2.4 µm. On attached, corticated twigs of deciduous trees, mostly *Quercus*, in the same habitat as *Colpoma quercinum* (page 1480).

Capitoricha rubi ⋉ is almost identical but occurs on *Rubus* and *Epilobium*. *Capitotricha fagiseda* ⋉ grows on *Fagus* cupules. Phylogenetic studies could indicate that material from *Fagus* and *Alnus* may warrant treatment as separate species. Species of *Lachnellula* ◁ form somewhat more robust fruitbodies and do not have crowns of large crystals on the hairs.

Widespread in the nemoral–hemiboreal zones, common; mainly April–June.

Incrucipilum form small, rather long-stemmed, whitish to pale yellow apothecia with conspicuous marginal hairs that have large crystals at the tips. Decomposers of leaves and twigs.

Incrucipulum sulphurellum is a small, pale sulphur-yellow, short- to long-stemmed *Incrucipulum* with projecting hairs that measure 55–75 × 4–5 µm. The hair surface is rough along the entire length, and the hairs are somewhat thick-walled, cylindrical, septate, rounded above and bear a crown of large, angular, yellow-tinged crystals. Asci I+, 8-spored, without croziers. Paraphyses are large and wide (up to 5 µm), projecting, lanceolate and have refractive drops. Spores one-celled, hyaline, measure 7.5–11.5 × 1.7–2 µm (living spores). Occurs on twigs of *Myrica gale*.

The combination of substrate and distinctive hairs should make identification easy. A number of similar, white species occur on various leaves, including *Vaccinium*, *Quercus*, *Castanea* and *Fagus*.

Widespread; probably common in all or most *Myrica* stands – this is certainly the case in Denmark; May–October.

Belonidium ⋉ is possibly an earlier valid generic name for *Incrucipulum*.

Incrucipulum ciliare is a short- to somewhat long-stemmed *Incrucipulum* with white, shiny, projecting hairs that measure 65–80 × 5–6 µm and bear a crown of large crystals. The hymenium may be somewhat yellowish. Paraphyses are very narrowly lanceolate. The narrowly fusiform spores measure 19–26 × 2.7–3 µm (from living spores). On leaves from deciduous trees, mostly *Quercus*.

Similar whitish inoperculates on leaves have either different spore dimensions and/or hair types.

Widespread in the nemoral–hemiboreal zones, occasional to common; August–November.

Dasyscyphella nivea is a rather large, short- to rather long-stemmed, white, short-haired inoperculate cup fungus that dries to a ± orange colour. Hairs 60–110 × 2–4 µm, rough and somewhat widened but completely smooth at the tips. Asci I+. The paraphyses are narrow, lanceolate and short. The spores measure 6–9 × 1.7–2.3 µm. Occurs mostly on wood of *Quercus*.

Very easy to confuse with species of *Lachnum* ▽ but these do not have smooth hair tips and typically have wider and longer paraphyses.

Widespread in the nemoral–hemiboreal zones, occasional; mostly in cool months of the year.

Lachnum form rather small to small, elegant, mostly whitish apothecia with short to long stems. The hairs are clavate, cylindrical to tapering and usually rough along the entire length. The spores are one-celled and vary in shape from ellipsoid-fusiform to almost cylindrical. On all kinds of dead plant material. The genus is very large and specific identification can be challenging.

Lachnum virgineum is a non-staining, white, long-stemmed, *Lachnum* with dense, rather long hairs. The hairs are mostly cylindrical, rough along the entire length and measure 80–135 µm long. Asci I+, 8-spored, 45–55 × 4–5 µm, with croziers. Paraphyses are lanceolate, 3–5 µm wide, project 15–30 µm above the asci and do not have brown guttulation. The ± fusiform, one-celled spores measure 6–11 × 1.5–2.5 µm. May grow on almost any substrate, although typically on small twigs, cones and various fruit remnants.

Lachnum subvirgineum ⋉ has distinctly widened hair tips and somewhat shorter marginal hairs that measure 70–110 µm. There are many rather similar species of *Lachnum*.

Widespread, very common; mostly March–June.

Lachnum pudibundum is a whitish, short- to rather long-stemmed *Lachnum* that discolours red-brown. The thin-walled, multicelled, rough hairs measure 45–60 × 3.5–5 µm. Asci I+, 40–50 × 4.5–5.5 µm, without croziers. Paraphyses ± lanceolate, 3–6 µm wide and project 11–20 µm. The fusiform, one-celled spores measure 7–9 × 1.5–2.5 µm. Occurs on deciduous wood, mostly *Salix*.

Lachnum virgineum ◁◁ does not discolour and has longer hairs. Other similar species of *Lachnum* either have narrower paraphyses or paraphyses with brown guttules.

Mainly nemoral, apparently common; May–October.

Lachnum impudicum is a non-staining, white to cream, short- to rather long-stemmed *Lachnum*. Marginal hairs cylindrical to narrowly clavate with widened tips, 40–65 × 3–5 µm but up to 7 µm wide at the base, thin-walled, multicelled and rough; the hairs lower down are somewhat longer. Asci I+, 8-spored, without croziers. Paraphyses are lanceolate, 2–4 µm wide, project (0–)6–12 µm above the asci and multiguttulate. The one-celled, narrowly ellipsoid to fusiform spores measure 4–7 × 1.5–2 µm. Occurs on deciduous wood, mostly sapwood of *Fagus*, but also on e.g. *Fraxinus* and *Malus*.

Lachnum brevipilosum ⋉ discolours red-brown and has somewhat larger spores. *Lachnum pudibundum* △ also has somewhat larger spores and wider paraphyses. *Dasyscyphella nivea* ◁◁ has smooth hair tips.

Widespread, at least in the nemoral–hemiboreal zones, common; mainly during winter.

Lachnum rhytismatis is a very small, whitish to yellowish *Lachnum*. The hairs are hyaline, cylindrical, multicelled, crystal-covered on the tips and measure 50–60 × 3.5–4 µm (note that the prominent, apical crystals are easily lost in preparations). Asci I+, 23–32 × 3–4 µm, with croziers. Paraphyses are lanceolate, 3–3.5 µm wide, project 20–23 µm above the asci and do not have obvious drops. The spores are clavate, one-celled and measure 4–8 × 1.3–1.5 µm. Occurs on leaves of *e.g. Quercus* and Vaccinium; on leaves of *Acer* it may overgrow *Rhytisma acerinum* stromata (page 1483).

Incrucipulum ciliare ⊲⊲ also occurs on *Quercus* leaves and has similar, large crystals at the hair tips, but the spores are completely different and measure 15–22 × 2–3 µm.

Widespread, common, at least in the nemoral zones; March–August.

Lachnum nudipes is a small, reddening, short-stemmed *Lachnum* with a cream hymenium. The hairs are cylindrical, wider at the tips, thin-walled, rough, hyaline or brownish, measure 50–85 × 3–4 µm and usually have crystals at the tips. Asci I+, 45–55 × 4–5 µm. The paraphyses are lanceolate, 3–6 µm wide and project up to 20 µm above the asci. The fusiform, one-celled spores measure 9–12.5 × 1.5–2.2 µm. Occurs on dead herbaceous stems, mostly *Filipendula* and *Epilobium hirsutum*.

Lachnum morthieri ⋉ has smaller spores measuring 6–11 × 1.2–1.5 µm.

Widespread, apparently common but under-reported; May–July.

Lachnum apalum is a small, white to cream, short-stemmed, when drying out yellowish or reddish *Lachnum*. The hairs are cylindrical or tapering, thin-walled, rough and measure 50–70 × 4 µm. Asci I+, 55–65 × 6 µm, without croziers. Paraphyses lanceolate, 4–8 µm wide and project 15 µm above asci. The cylindrical, one-celled spores have many drops and measure 35–49 × 1.3–1.5 µm. Mostly on *Juncus effusus* and *J. conglomeratus*.

Similar species of *Lachnum* on *Juncus* have much shorter spores.

Widespread, at least in the nemoral zones, very common, but under-reported; most of the year.

Lachnum controversum is a large, short-stemmed, white *Lachnum* that discolours brownish-red. Hairs are 70–120 µm long, rough along entire length, cylindrical or somewhat attenuated, multicelled and thin-walled. Asci I+, 40–50 × 4–5 µm, without croziers. Paraphyses are lanceolate, 4–8 µm wide, project 20–35 µm above asci, multiguttulate. Spores ± clavate, one-celled and measure 6–10(–12) × 1.8–2.2 µm. Mainly on *Phragmites* culms

Lachnum tenuipilosum ⋉ is smaller, longer stemmed and mainly occurs on less coarse grass culms.

Mainly nemoral, common; April–September.

Lachnum imbecille is a sessile to short-stemmed *Lachnum* that, with age, acquires ± apricot-orange hymenial colours. The hairs are cylindrical, 3–7-celled, rough, hyaline, lack crystals and measure 50–80 × 3–4 µm. Asci I+, 50–70 × 6–8 µm, without croziers. The paraphyses are lanceolate, 4–5 µm wide, project 20–30 µm above the asci and have orange-red contents. The fusiform, 1–2-celled spores measure 15–26 × 2.3–3 µm. Occurs on leaves of *Eriophorum angustifolium* and perhaps other species of *Eriophorum*.

Lachnum sydowii ⋉, also on *Eriophorum,* has a white to cream hymenium and much smaller spores. There are several similar species of *Lachnum* on other host plants in the *Cyperaceae*.

Widespread, probably not rare, but rarely reported; mainly May–July.

Lachnum juncinum is a small, ± yellowish *Lachnum*. The hairs are 1–3-celled, rough, guttulate and 40–70 µm long. Asci I+, 70–103 × 8–9.5 µm, I+, without croziers. Paraphyses lanceolate, up to 5 µm wide, project 0–15 µm and have drops. The oil-filled spores measure 13–21 × 2.5–3.2 µm. Occurs on stems of *Juncus* and various *Cyperaceae*.

Lachnum diminutum ⋉ occurs on similar hosts but has narrower spores; *L. apalum* ◁ has much longer spores and is not yellowish; and *L. winteri* ⋉ has a low oil content in the spores.

Distribution and frequency poorly known, certainly nemoral; summer.

High or long-stemmed, smooth to felty inoperculate cup fungi

A form group of non-gelatinous to slightly gelatinous, distinctly long-stemmed inoperculate cup fungi or, if short-stemmed, growing to 2 mm or more above the substrate. The outer side is smooth to felty but does not have projecting hyphal hairs. A few species have teeth at the margin of the apothecia made up of aggregations of hyphae. Some species form apothecia from sclerotia.

Some species are decomposers, *e.g.* of wood or herbaceous stems, while others – members of the *Sclerotiniaceae* – parasitize plants.

Other similar fungi:
– downy to hairy inoperculate cup fungi are covered by ± projecting hairs. Many species also have lanceolate, ± projecting paraphyses (page 1372).
– small, short-stemmed to sessile, smooth to felty inoperculate cup fungi arise to less than 2 mm above the substrate (page 1426).

See also the general wheel, page 1348.

Further reading: 19, 21, 49, 59, 72, 78, 95, 101, 252, 267, 354, 362.

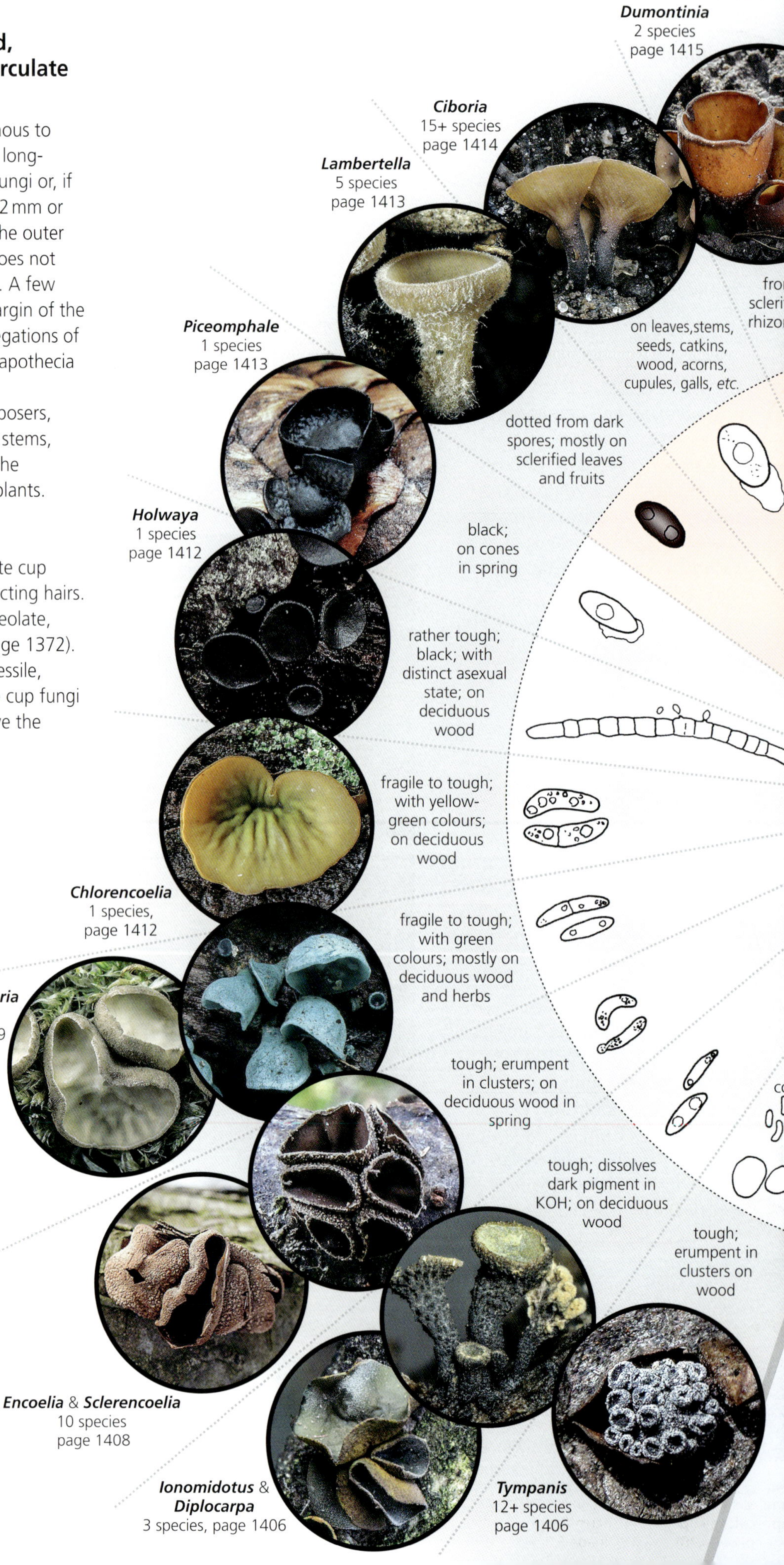

Micro-drawings: spores, followed on the inside by asci (a), conidia (con).

Approximate species number applies to temperate Europe.

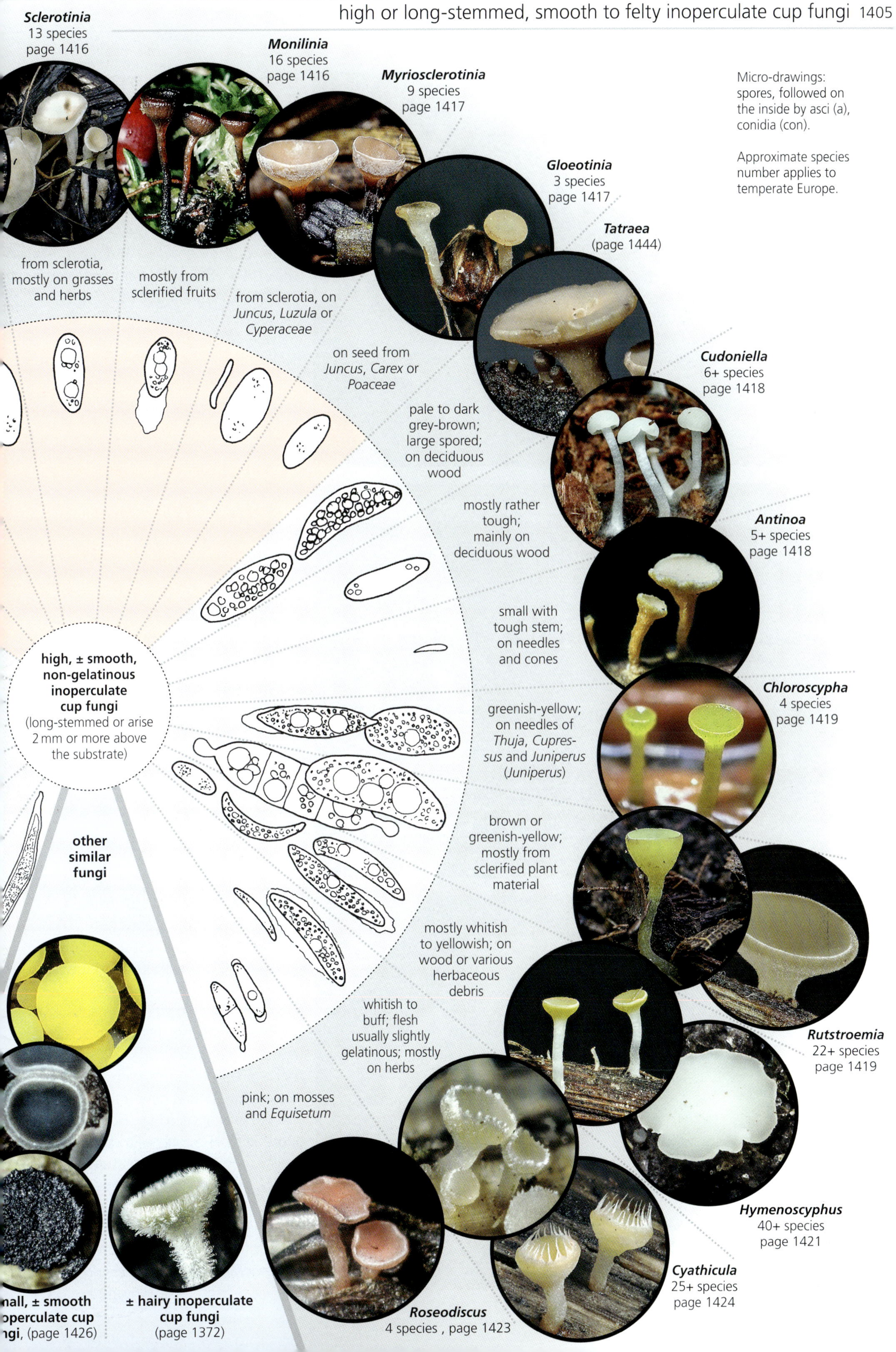

Tympanis alnea is erumpent in dense clusters. Each apothecium is black with a ± whitish granular covering and measures 0.2–1 mm wide × 0.5–3 mm high. The sexual state is often accompanied by a fleshy, asexual state (lower image – in section). The asci are I-, somewhat thick-walled and measure 135–190 × 18–22 µm. Paraphyses are ± cylindrical and imbedded in brown matter. The primary (sexual) spores are one-celled, broadly ellipsoid, hyaline and measure 5–6 × 4–5 µm, but they soon bud off numerous, tiny (≈ 2.5 × 1 µm), secondary, allantoid conidia, which fill the asci. Occurs on twigs of *Alnus*, *Malus* or other deciduous trees and possibly even on conifers.

Tympanis saligna ⋉, which occurs on *Salix*, also has one-celled primary spores but does not form dense clusters of apothecia. Various similar species of *Tympanis* occur on coniferous wood or have 2-celled primary spores.

Widespread, probably rather common; all year, but mostly during winter.

Diplocarpa bloxamii is ± clustered, has a somewhat olivaceous-buff hymenium and an almost scaly, dark, outer side. Often with its pale brown, powdery asexual state (to right on image). Margin has smooth, many-celled, ± constricted, hyaline to ± lilac hairs that measure 40–100 × 2.5–4 µm. Asci I-, 50–98 × 5–8 µm, with croziers. Paraphyses have a flame-shaped, several-celled tip. Pigment purple in KOH. Spores ellipsoid, hyaline, one-celled and measure 6–9 × 2.5–3.5 µm. Occurs on deciduous wood, mostly *Fagus*; perhaps a parasite on *Armillaria* (page 232).

A distinctive species.

Nemoral, rather rare or overlooked; April–May, September–November.

Ionomidotis form rather large apothecia with dull colours. The pigments become red-brown to purple in KOH (ionomidotic reaction). Should probably be included in *Diplocarpa* ◁.

Ionomidotis fulvotingens is sessile, ± clustered and ± erumpent. The hymenium is at first olive-green, later blackish. The flesh is rather tough and the outer side is brownish-olive to rust-brown and felty-squamulose. Has yellow lumps of amorphous pigment in the hymenium. KOH dissolves the abundant red-brown pigment from the flesh – a so-called ionomidotic reaction. Asci I-, 8-spored, 50– 67 × 5–7.5 µm, with a thickened tip and croziers. Paraphyses ± cylindrical, with septa, possibly somewhat branched, hyaline. The hyaline, bent spores contain many drops and measure (5–)6.3–8(–9.4) × 1.8–2.6 µm. On attached deciduous branches, mostly *Salix*. Possibly parasitic on other fungi.

Ionomidotis irregularis ▽ has much larger apothecia on large, dead, deciduous wood. *Diplocarpa bloxamii* ◁ has a similar ionomidotic reaction but is long-stemmed and more squamulose. *Bulgariella pulla* (page 1370) has larger, coloured spores. Similar greenish species in *Chlorencoelia* ▷▷ and *Chloroscypha* ▷▷ lack the ionomidotic reaction.

Nemoral–hemiboreal, rather rare; mainly November–April.

Ionomidotis irregularis is a large, ± veined to spatula-shaped, sessile to short-stemmed, rather thin-fleshed *Ionomidotis* with a grainy, dark outer side and an almost black hymenium that has a ± distinct purple tinge. KOH dissolves a purple pigment. Asci I-, with croziers. Paraphyses wider at the tip and septate. Spores fusiform, contain two drops and measure (6.5–) 7.5–10 × 3.5–4(–4.5) µm. Occurs on dead wood of deciduous trees in old-growth forests.

May recall species of *Pseudoplectania* (page 1296) but these have spherical spores and operculate asci.

Widespread in central and eastern Europe, rare; autumn.

Encoelia and ***Sclerencoelia*** tolerate desiccation and are ± leathery, brown, rather large, clustered inoperculate cup fungi that are erumpent through bark on deciduous trees. Spores are typically allantoid, one-celled, smooth and hyaline.

Encoelia furfuracea is a large, clustered, cinnamon-brown to greyish-brown, desiccation-tolerant inoperculate cup fungus that is bladder-like erumpent through bark at first, but later opens to reveal a floccose-serrated margin and a pale orange-brown to red-brown hymenium. The paler outer side is densely grainy-mealy. Asci I+, 8-spored and measure 90–100 × 6–7 µm. Paraphyses somewhat clavate, with large refractive drops. The allantoid, hyaline spores measure (6–)8–11.5 × 1.8–2.8 µm. Occurs on attached, corticated branches of *Corylus* and more rarely *Alnus*.

A highly distinctive species that should be straightforward to identify.

Widespread and rather common; mainly in spring but also present during mild winters.

Gilbert Moyne

Encoelia fimbriata is an erumpent, distinctly clustered, deeply cupulate inoperculate cup fungus with a pale, hairy margin and yellow- to red-brown colours on the outer side and on the hymenium. It has lanceolate marginal hairs. The asci are I- and 8-spored. The allantoid spores measure 7–11 × 2.5–3 µm. Occurs mainly on *Salix* bark, typically with *Hymenochaetopsis tabacina* (page 973) and sometimes *Hypocreopsis lichenoides* (page 1526).

A distinctive species that is phylogenetically out of place in *Encoelia*.

Nemoral, rare; most of the year.

Sclerencoelia fascicularis is a clustered, red-brown, at first bladder-like, later cupulate to expanded *Sclerencoelia* with a pale margin. The asci are I-(+), 8-spored and measure 100–130 × 8.5–12 µm. The spores are allantoid and measure 13–15.5 × 3.5–4.2 µm. Erumpent on corticated *Populus* branches, including *P. tremula*, perhaps also on *Carpinus* and *Corylus*.

Also known as *Encoelia fascicularis*. *Sclerencoelia fraxinicola* ⋉ is very similar but occurs on *Fraxinus*.

Widespread, fairly common, becoming scarcer towards the north; January–June.

Chlorociboria form small to large, ± green apothecia with a floccose to downy outer side. Spores are usually ± allantoid, 1–2-celled, smooth, hyaline. Most species are decomposers of wood but *Chlorociboria aeruginella* ▷▷ occurs on herbs and leaves. All, except *C. glauca*, stain the substrate ± green.

Chlorociboria glauca is a ± grey-green, felty-hairy to grainy, rather large, short-stemmed inoperculate cup fungus that often produces small clusters of apothecia. Asci I+, 8-spored, 38–55 × 4.7–5.6 µm, with croziers. The almost cylindrical to ± allantoid, hyaline spores measure 5.5–7.5(–8.5) × 1.5–1.7 µm. It can co-occur with its asexual state, which forms groups of pycnidia. Occurs on attached twigs of deciduous trees, e.g. *Corylus*, *Prunus spinosa*, *Rosa*, *Rubus* and *Salix*; the wood is not stained by the fungus.

This species was formerly recognized as a species of *Encoelia* ◁. *Chlorencoelia versiformis* ▷▷ is more olive-coloured and smooth, and occurs on damp wood with soil contact. The other species of *Chlorociboria* ▷▷ are a more turquoise colour and stain the substrate ± green.

Mainly nemoral Atlantic, rare; autumn.

Chlorociboria aeruginascens is a rather short-stemmed to narrowly attached, irregularly shaped, dark turquoise *Chlorociboria* with a somewhat felty outer side. The asci are I+, 8-spored and measure 70–90 × 4–4.5 µm. The paraphyses are somewhat wider at the tips. The one-celled, somewhat allantoid, hyaline spores measure 5.5–8 × 1.5–2 µm. Occurs in ± damp woodlands on deciduous wood, *e.g.* on *Corylus* and *Quercus*, staining the substrate turquoise. Sometimes found with its ± cushion-shaped asexual state (formerly known as *Dothiorina tulasnei*).

Chlorociboria aeruginosa ▷ has larger spores and typically a paler hymenium. *Chlorociboria aeruginella* ▷ is much smaller and occurs on herbaceous stems and leaves from deciduous trees. In southern Europe the green-staining *Aeruginoscyphus sericeus* ⋉ can be found on deciduous wood; it has small apothecia, hairs 225–250 µm long and 1–4-celled spores that measure, 43–57 × 3.8–4 µm. *Chlorociboria glauca* ◁◁ is a paler grey-green colour and does not stain the substrate.

Widespread, common; all year, but mostly August–November.

Chlorociboria aeruginosa is a rather large *Chlorociboria* with a turquoise, felty outer side and a whitish hymenium. The asci are I+, 8-spored, measure 75–100 × 6–7.5 µm and have croziers. The 1(–2)-celled, hyaline spores measure 9–16 × 1.5–3.3 µm. Occurs on deciduous wood, staining the substrate turquoise.

Chlorociboria aeruginascens ◁ has smaller spores and typically a greener hymenium.

Mainly hemiboreal–boreal, occasional to locally common; most of the year.

Chlorociboria aeruginella is a tiny, bright green or turquoise, somewhat hairy, very short-stemmed *Chlorociboria*. The hairs are somewhat pointed and have incrusted material on the walls. The asci are I+, 8-spored, measure 55–70 × 6–7.5 µm and have croziers. Spores 1–2-celled, somewhat allantoid, 7–13 × 2.5–3.5 µm. Occurs on herbaceous stems, grass culms and deciduous leaves in the litter layer; the stems are ± green-stained.

A very distinctive species. *Aeruginoscyphus sericeus* ⋉ occurs on deciduous wood and has much longer spores and longer hairs.

Mainly nemoral, rare, but probably overlooked; autumn.

Thomas Læssøe

Chlorencoelia versiformis is a short-stemmed, yellowish to olive-green, large inoperculate cup fungus with a central or oblique stem and an almost smooth outer side. The asci are I+ and 95–110 µm long. The paraphyses and flesh cells have greenish-yellow drops. The 1–2-celled, ± cylindrical spores measure 10–16 × 2.5–3.5 µm. Occurs on decomposed, non-stained, damp deciduous wood.

Species of *Chlorociboria* ⊲⊲ have turquoise colours, and all except *C. glauca* ⊲⊲ stain the substrate green.

Mainly hemiboreal–boreal, rare; late summer-autumn.

Holwaya mucida is an inoperculate cup fungus which has a matchstick-like asexual state that may often occur without the sexual state. The single to clustered apothecia are tough, black and short-stemmed. The asci are I-, 8-spored, measure (120–)160–220 × 9–13.5 µm and have croziers. Paraphyses are cylindrical, have widened tips and are imbedded in brown matter. The cylindrical, 12–20-celled spores measure 30–65 × 2.5–4 µm and bud off conidia. Occurs mainly on bark of *Tilia*, typically on relatively recently fallen trunks or large fallen branches. May rarely grow on *Sorbus*, *Acer platanoides* and *Salix caprea*.

A highly distinctive species, particularly due to its asexual state. This state is frequently misidentified as an asexual *Xylaria* (page 1548), but it has a rubber-like texture unlike the firm texture of *Xylaria* and the flesh is dark, not white.

Widespread in the nemoral–hemiboreal zones, rare to occasional; mainly July–March.

Piceomphale bulgarioides is a large, grey-black to black, sometimes green-tinged, ± cupulate, short-stemmed species with a tough texture. Asci I+, 8-spored, measure 110–130 × 8–9 µm, lack croziers. Paraphyses somewhat wider at the tips (up to 2 µm wide). Spores ellipsoid, one-celled, hyaline, 7–13 × 3.5–6.5 µm and have a loosening outer wall. On *Picea* cones.

No other similar species occur on *Picea* cones. Species of *Pseudoplectania* (page 1296) have globose spores and operculate asci.

Widespread, but eastern and northern and absent from the UK; rare to common; December–May.

Lambertella corni-maris is a rather short- to long-stemmed, minutely hairy inoperculate cup fungus with grey-olive colours and a hymenium that becomes sooty at maturity. The asci are I+ and 8-spored. The paraphyses are cylindrical and lack distinct contents. The asymmetrical-ellipsoid spores contain two drops, measure 8.5–11.5 × 4.3–6 µm and are hyaline at first but later dark brown and very finely rough. Occurs on mummified, black *Malus* and *Cornus* fruits in the litter, *e.g.* in damp, tall grass under *Malus* trees.

Other species of *Lambertella* occur on different substrates. *Monilinia fructigena* ⋉ and *M. laxa* ⋉ occur almost exclusively in their asexual hyphomycete stages, but may rarely produce brown, stemmed apothecia with hyaline, smooth spores.

Nemoral, very rarely recorded; late summer-early autumn.

The genera ***Ciboria*** and ***Dumontinia*** form brown, long-stemmed apothecia from mummified plant parts, mostly in spring. All species have smooth, hyaline spores.

Ciboria rufofusca is a mostly long-stemmed, yellow-brown to grey-brown *Ciboria*. Asci I+, 8-spored and measure 65–105 × 5.5–6.5 µm. The ellipsoid spores measure (5–)5.5–7.5 × 2.7–3.5 µm. Occurs on ± buried or superficial cone scales from *Abies*; usually abundant.

The distinctive substrate makes this species easy to identify.

Nemoral–hemiboreal, rare to rather common; February–June.

Ciboria amentacea is a thin- and ± long-stemmed, yellow-brown, almost smooth to downy *Ciboria*. Asci I+, 8-spored, 95–145 × 6–9 µm. The ellipsoid to ovoid, one-celled spores measure 7.5–10.5 × 4.2–6 µm and have a loosening outer wall. Occurs on male catkins of *Alnus* and *Corylus* in damp woodlands.

Ciboria caucus ⋉ occurs on *Salix* and *Populus* catkins and *C. betulicola* ⋉ occurs on *Betula* catkins; *C. viridifusca* ⋉ is associated with female *Alnus* cones; and *C. lentiformis* ⋉ occurs on *Alnus* seeds and has larger, fusiform spores.

Mainly nemoral–hemiboreal, very common; January–May.

Ciboria betulae is a small, pale grey-brown, ± long-stemmed, almost smooth *Ciboria*. Asci I+, 8-spored and measure 110–160 × 6–8 µm. The ellipsoid, one-celled spores measure 9.5–15 × 4–5.5 µm. Occurs on *Betula* seeds in the litter in damp forests or marshes.

The smaller-spored *C. viridifusca* ⋉ is associated with *Alnus* cones; *C. lentiformis* ⋉ occurs on *Alnus* seeds and has larger, fusiform spores; and *C. betulicola* ⋉ grows on male *Betula* catkins.

Mainly known in the nemoral zones but probably more widespread and perhaps very common; April–May.

Ciboria batschiana is a short- to long- and dark-stemmed, minutely downy, ± clay-buff to purple-brown, rather large *Ciboria*. Asci I+, 8-spored and measure 110–160 × 6–9 µm. Paraphyses cylindrical and 3.5 µm wide at the tips. The ellipsoid, one-celled, hyaline spores measure 6–11 × 4–6 µm. Occurs on mummified acorns (*Quercus* seeds).

This species is easy to recognize due to the distinctive substrate and its large size. *Ciboria americana* ⋉ has narrower spores and smaller apothecia.

Nemoral–hemiboreal, occasional to rather common; June–November.

Dumontinia tuberosa is a large, brown, smooth, wine glass-shaped inoperculate cup fungus with a long and dark stem. The asci are I+, 8-spored and measure 140–190 × 8–12 µm. Paraphyses are cylindrical and 3 µm wide at the tips. The spores are ellipsoid, one-celled, ± lack drops, measure 12–16.5 × 6–9 µm and have an indistinct slime coat. Parasitic on *Anemone* rhizomes, from which the apothecia develop on the blackened sclerified parts. *Ranunculus* (*Ficaria*) *verna* is also claimed to act as host. Can cause damage to cultivated garden anemones, including the autumn flowering *Anemone hupehensis*.

The somewhat smaller *D. ulmariae* ⋉ parasitizes *Filipendula ulmaria*; the spores measure 10–12 × 5–7 µm.

Widespread and rather common, becoming scarcer towards the north; mainly March–May.

Jens Maarbjerg

The genera **Monilinia**, **Sclerotinia** and **Myriosclerotinia** all form brown apothecia from sclerotia.

Monilinia johnsonii is a ± long-stemmed, clay-buff *Monilinia*. Asci I+, 8-spored, 140–170 × 8–10.5 µm. Paraphyses cylindrical. Spores are ellipsoid, one-celled and measure 10–16 × 5–8 µm. Conidia citriform to spherical, 10–21.5 µm in diameter. On mummified *Crataegus* fruits and causes a leaf blight on *Crataegus* leaves.

Similar species occur on other hosts, e.g. *M. aucupariae* ⋉ on *Sorbus* and *M. padi* ⋉ on *Prunus padus*.

Widespread and probably common; mainly March–April (apothecial state).

Thomas Læssøe

Monilinia oxycocci is a long-stemmed, dark red-brown, cupulate *Monilinia*. Asci I+, 150–190 × 8–12 µm. The ellipsoid, hyaline spores can be divided into two size groups: 9.5–14.5 × 5.5–8.5 µm and 6–10 × 4–5 µm. Conidia 14.5–25 × 9.5–15 µm, egg-shaped. Occurs on mummified *Oxycoccus* berries in mires and raised bogs.

The substrate makes this species easy to recognize. Similar species on other fruits include *M. baccarum* ⋉ on *Vaccinium myrtillus* and *M. megalospora* ⋉ on *V. uliginosum*.

Widespread at least to the north and east, occasional; May–June (apothecial state).

Sclerotinia sclerotiorum is a long-stemmed, ± apricot-orange *Sclerotinia*. Asci I+, 8-spored,110–150 × 6–10 µm. Paraphyses thread-like, up to 1.5 µm wide. The flesh is ± staining in iodine reagents. Spores narrowly ellipsoid, one-celled, uniform, 9–13.5 × 4–6 µm. The black sclerotia measure 4–25 × 2–10 mm. Parasitizes roots, stems, flowers, fruits and seeds of different herbs.

Sclerotinia trifoliorum ⋉ has spores in two size classes and produces apothecia in the autumn; *S. minor* ⋉ produces tiny sclerotia.

Widespread, probably common, but easily overlooked; apparently mainly April–June (apothecial state).

Myriosclerotinia curreyana is a short-stemmed, yellow-brown to pale vinaceous-brown *Myriosclerotinia*. Asci I+, 8-spored, 60–95 × 4–5.5 µm. Paraphyses cylindrical. Spores allantoid and measure 7.5–15 × 1–1.5 µm. Sclerotia black (inset image), internally pink, and occur in dead stems of *Juncus*, mostly *J. effusus*.

Myriosclerotinia juncida ⋉, which mainly occurs on *J. filiformis*, has smaller spores measuring 7–11 × 2–4 µm. A number of related species occur on *Cyperaceae* and *Luzula*.

Widespread, but mainly recorded in the nemoral zones, common; March–July (apothecial state).

Myriosclerotinia scirpicola is a short-stemmed, ± clay-buff, rather large *Myriosclerotinia*. Asci I+, 8-spored, 110–150 × 5.5–9 µm. Paraphyses cylindrical. The spores measure 9–16 × 4.5–7 µm. The large, 5–22 × 1–10 mm, bulbous to fusiform sclerotia form in the dead stems of *Schoenoplectus lacustris*, *Scirpus sylvaticus* and *S. maritimus*, and *Eleocharis*.

Myriosclerotinia dennisii ⋉ has cylindrical, typically very long sclerotia and is known from *Scirpus*, *Eleocharis* and *Eriophorum*.

Widespread, probably rather common; May–July (apothecial state).

Gloeotinia juncorum is a rather long-stemmed, small, yellowish *Gloeotinia*. The asci are I+, 8-spored and measure 110–150 × 7–10 µm. Paraphyses cylindrical, up to 2 µm wide. The egg-shaped spores measure 8–11 × 4–5 µm. Occurs on mummified *Juncus* fruits.

Gloeotinia aschersoniana ⋉ occurs on mummified *Carex* fruits and *G. granigena* ⋉ occurs on *Poaceae* seeds.

Mainly nemoral, rare but probably overlooked; mainly June.

Cudoniella acicularis is a long-stemmed, whitish, with age brown- or black-spotted, slightly hairy *Cudoniella*. The hymenium is convex. Asci I+, 8-spored, with croziers. The paraphyses contain many drops. The spores are hyaline, cucumber-shaped, 1–2 to rarely 4-celled, and measure 15–24 × 4–6 µm. Occurs almost exclusively on old, sometimes moss-covered, *Quercus* stumps.

May recall an immature agaric but the dark spots are a good distinguishing character for species of *Cudoniella*.

Nemoral–hemiboreal, occasional; August–March.

Antinoa form small, thin- and usually dark-stemmed apothecia, often with a convex hymenium. All have narrowly pointed spores. On cones and needles.

Antinoa pulchella is a ± long-stemmed, curry-yellow *Antinoa* with a paler, ± irregularly concave to convex hymenium. Outer side is felty and stem ± black at the base. Asci I+, about 35 µm long. Paraphyses cylindrical and almost lack refractive contents. Spores one-celled, ± fusiform, (4–)5–7.5(–8.5) × 1.4–1.6 µm. Occurs on *Pinus* and *Picea* needles.

'*Calycina*' *subtilis* ⋉ is a similar, but paler fungus.

Nemoral–hemiboreal, rare; autumn.

Thomas Kehlet

Antinoa juniperinella is a small, dark- and slender-stemmed, almost nail-shaped *Antinoa* with a white, convex hymenium. The stem is hairy, as is the asexual state, which is black and thread-like (a '*Chalara*'). Asci I+, 8-spored, 14–21 × 3.2–3.7 µm, with croziers. The ± fusiform spores measure 4–5.2 × 1.1–1.4 µm. Occurs on needles of conifers, *e.g. Picea* and *Juniperus*.

May recall a tiny *Gloiocephala* or *Marasmius* (pages 66 & 302) but has asci in the hymenium. '*Calycina*' *subtilis* ⋉ is larger, and has much longer asci and more narrowly ellipsoid spores.

Nemoral–hemiboreal, very rare or overlooked; mainly January-March.

Chloroscypha alutipes is a long- and felty-stemmed, bright greenish-yellow to curry-yellow inoperculate cup fungus. Asci I+ (blue/red), 152–170 µm long. Paraphyses pear-shaped at the tips, imbedded in amorphous matter. Spores fusiform, contain many drops, measure 22–26 × 6–6.7 µm and have a loosening outer wall. On coniferous litter from *Juniperus* and *Calocedrus decurrens*.

Chloroscypha limonicolor ⋉ seems to be a very similar species described from *Thuja* litter. Other similar species occur on litter of other conifers such as *Cryptomeria* (see page 1450).

Only recorded in southern parts of temperate Europe; mainly winter.

Michel Hairaud

Rutstroemia form short- to long-stemmed, yellowish to brown inoperculate cup fungi on darkened stems, leaves, and wood. Mostly autumn.

Rutstroemia luteovirescens is a ± greenish-yellow *Rutstroemia* with a rather long stem. Asci I+, 8-spored, 120–150 × 10–12 µm. Paraphyses cylindrical, up to 4 µm wide at the tips. Spores ellipsoid, one-celled, contain two or more drops, 15–18 × 6.5–8 µm. On dark-stained petioles of *Acer*.

The colour and substrate make this species easy to identify.

Nemoral–hemiboreal, rather common; July–November.

Rutstroemia conformata is a short- to rather long-stemmed *Rutstroemia* that arises from black, flake-like spots on petioles and leaf veins. Asci I+, 8-spored, 65–95 × 6–8 µm. Paraphyses cylindrical, 1–2 µm wide. Spores ± fusiform, hyaline and measure 6.5–11 × 4–5.5 µm. Occurs mostly on *Alnus* leaves, but also reported from *Populus* and *Betula*.

Similar species of *Rutstroemia* occur on *Quercus* and *Fagus* leaves, *e.g. R. sydowiana* ▷▷.

Widespread, very common, at least in the nemoral zones; April–July.

Rutstroemia sydowiana is a short-stemmed, finely marginally furrowed and haired, yellow-brown *Rutstroemia* with a somewhat paler hymenium. Asci I+, 8-spored, 110–130 × 9–12 µm. Paraphyses cylindrical, up to 4 µm wide at the tips. Spores curved-ellipsoid, one-celled, contain many drops and measure 11.5–13.5 × 5–6.5 µm. Mostly on *Quercus* petioles, rarely on acorns (*Quercus* seeds) or on the leaves of *Castanea*.

Rutstroemia petiolorum ⋉ has 1–4-celled spores and occurs mainly on *Fagus* petioles, although it can also be found on *Quercus*.

Nemoral–hemiboreal, fairly common; August–November.

Thomas Læssøe

Rutstroemia firma is a large, short- to long-stemmed *Rutstroemia* with a very dark stem, a cupulate to more expanded, yellow-brown hymenium and a somewhat crenulate apothecial margin. The texture is tough. The asci are I+, 8-spored and measure 125–170 × 9–12 µm. Paraphyses cylindrical, up to 3.5 µm wide at the tips. The flesh is not gelatinous and has large, parallel, oblong cells. The ± curved, 2–6-celled spores contain drops and measure 12–19 × 3.5–6.5 µm; they bud off conidia at maturity. Occurs on twigs and branches of *Quercus*, usually rather recently dead, and even on attached wood.

Rutstroemia alni ⋉ occurs on branches of *Alnus* and *Corylus* and has flesh that is more gelatinous with a more hyphoid composition.

Widespread in the nemoral–hemiboreal zones, common; mainly September–December.

Hymenoscyphus form ± smooth, usually whitish to yellowish apothecia with a short to long stem. Most have rather large, asymmetrical spores with a rounded and a tapering end and many drops. On stems, leaves and wood. A large, difficult genus.

Hymenoscyphus menthae is a straw-yellow to yellow, rather long-stemmed *Hymenoscyphus* with a pale stem and a more coloured hymenium. The stem base is mostly swollen and pale. Asci I+, measure 90–120(–130) × 8.5–10.5(–12) µm, lack croziers. Paraphyses cylindrical, with many drops. The fusiform to banana-shaped, one-celled spores contain many drops, measure (15–)17–22(–26) × 3.5–4.5 µm and have a loosening outer wall. Overmature spores may develop septa. Occurs mostly on dicot herbaceous stems in boggy habitats.

Hymenoscyphus repandus ⋉ is macroscopically identical but has smaller spores with fewer oil-drops. *Hymenoscyphus macroguttulatus* ⋉ has approximately the same spore dimensions but the spores contain very large drops and the asci have croziers. It is also generally paler, and the non-bulbous stem base can be coloured.

Nemoral–hemiboreal, common; mainly May–September.

Hymenoscyphus serotinus is a ± long-stemmed, 2–5 mm wide, yellow *Hymenoscyphus* with a pale outer side. The hyphae in the flesh lack glassy walls. Asci I+, measure110–140 × 8.5–11 µm, with croziers. Paraphyses thread-like, 2–3 µm wide. Spores curved, clavate, one-celled and measure 19–30 × 2.5–3.5 µm. Occurs on blackened twigs of *Fagus*.

Hymenoscyphus lepismoides ⋉ decomposes *Carpinus* twigs. *Hymenoscyphus virgultorum* ⋉ and *H. conscriptus* ⋉ forms a similar species complex on twigs of *Salix*.

Nemoral, common; mainly September–December.

Hymenoscyphus scutula is a rather long-stemmed, ± cream *Hymenoscyphus*. The asci are I+, 8-spored and lack croziers. Paraphyses have many drops. Flesh does not have glassy hyphal walls. The 1–2-celled spores have a rounded and a more pointed end, contain many drops, measure 18–27 × 3.5–5 µm, and usually have hair-like appendages at the ends. Occurs on stems of herbs, mostly in damp habitats.

Various similar species of *Hymenoscyphus* have shorter spores or more vivid colours.

Widespread, very common; June–November.

Hymenoscyphus fructigenus* var. *fructigenus is a straw-yellow to buff, short- to long-stemmed *Hymenoscyphus*. Asci I+, lack croziers. Paraphyses with many drops. Hyphae do not have glassy walls. Spores 1–2-celled, 13–19 × 3–4 µm, have a rounded and a tapering end, and lack hair-like appendages. Occurs on acorns (*Quercus* seeds) in the litter layer.

A number of closely related species or varieties occur on fruits of *e.g.* *Corylus*, *Carpinus*. Other species of *Hymenoscyphus* may occasionally occur on fruits, some of which have hair-like appendages on the spores.

Nemoral–hemiboreal, common; mainly August–November.

Hymenoscyphus fraxineus is a whitish, rather short-stemmed *Hymenoscyphus* that discolours ± black. Asci I+, with croziers. The spores are obliquely rounded at one end and pointed at the other, contain drops and measure 15–20 × 4–4.6 µm. Has a so-called '*Chalara*'-state. Causes ash dieback, which has killed a high number of *Fraxinus* trees throughout temperate Europe. Apothecia on petioles in the litter.

Also known as *H. pseudoalbidus*. It has ± replaced the native *H. albidus* ⋉, which has asci that lack croziers.

Widespread, very common; May–December (apothecia).

Thomas Kehlet

Roseodiscus form ± rose apothecia on liverworts, bryophytes and *Equisetum*.

Roseodiscus formosus is a short-stemmed, ± pinkish-tinged *Roseodiscus*. Asci I+ (red), 8-spored, 89–140 × 8–12 µm, tip conical, with croziers. Paraphyses mostly somewhat branched. The ± ellipsoid, 1–(2)-celled spores measure 11–20 × 3–4.5 µm and have small drops at the poles. Occurs on the bryophytes *Ceratodon* and *Polytrichum* in pioneer communities on sandy soils.

Roseodiscus subcarneus ⋉ has, in outline, ± foot-shaped spores measuring (5.2–)6–8.5(–10) × (1.8–)2–2.6(–3) µm and parasitizes liverworts.

Nemoral, locally common and probably overlooked elsewhere; spring and late autumn.

Roseodiscus rhodoleucus is typically a rather long-stemmed, grey-rose *Roseodiscus* with a minutely hairy stem. The sharply delimited margin is typically pale and somewhat downy. Hyphae in the fleshwithout glassy walls. Asci I+, 8-spored, measure 85–105 × 9–11 µm, with croziers. Paraphyses lack drops. The ellipsoid, one-celled spores do not contain drops and measure 9–12 × 5–6 µm. Occurs on stems of *Equisetum*, *often* in marshes and fens.

Roseodiscus equisetinus ⋉ also occurs on *Equisetum* but has narrower spores with a tapering end that measure 9–13 × 2.5–3.5 µm; the ends have slimy appendages.

Widespread, rather rare to occasional; mainly April–May.

Cyathicula form small, cupulate apothecia with distinct stems and usually toothed or fimbriate apothecial margins. Most have a rather tough texture.

Cyathicula coronata is a short- to long-stemmed, buff to cinnamon-buff *Cyathicula* with a cupulate fertile part that usually has up to 40 pale, pointed teeth at the margin. The hyphae in the teeth are thin-walled. Asci I+, 80–100 × 6–8 µm, with croziers. Paraphyses with drops. Flesh non-gelatinous and has large, angular crystals. Spores ± fusiform, contain many drops, measure 14–23 × 3.5–4.5 µm and have a ± grainy slime coat. Mostly on ± coarse herbaceous stems, rarely on wood.

Also known as *Crocicreas coronatum*. *Cyathicula pallida* ⋉ is very similar but is smaller and has up to 50 teeth at the margin; the spores have fewer drops and measure 11.5–18 × 3–3.8 µm. The ± pinkish-tinged *C. culmicola* ▽, on grass, has 4–8-celled spores. Species of *Pezoloma* (page 1431) have sessile apothecia and the teeth are farther apart; they are white and occur on other substrates.

Widespread, common; mainly September–November.

Cyathicula culmicola is a ± pinkish-tinged, cupulate *Cyathicula* with ± distinct teeth at the margin. The outer side may have brown, vertical stripes. Asci I+, 8-spored, 120–140 × 12–14 µm. Paraphyses clavate, 4–7 µm wide at the tips. Hyphae in the flesh have glassy walls. The ellipsoid to fusiform, 4–8-celled spores measure 20–30 × 3–5 µm and are surrounded by a colourless coat. Occurs mainly on grass culms.

Cyathicula coronata △ has longer, more widely spaced teeth and one-celled spores.

Mainly nemoral–hemiboreal, common; mostly May–September.

Cyathicula cyathoidea is a pale, ± cupulate *Cyathicula* with a ± fringed-dentate apothecial margin. The asci are I+, 8-spored, measure 45–55 × 4–5 µm and have croziers. Paraphyses with drops, narrowly clavate, 2 µm wide at the tips and hardly project above the asci. The one-celled, hyaline, fusiform to cylindrical spores measure 6.5–12.5 × 1.7–2.5 µm and contain varying amounts of drops. Occurs on coarse herbaceous stems, but also on leaves and petioles.

Also known as *Crocicreas cyathoidea*. Varieties that are more brown-coloured have been described, including from *Pteridium* stems and grass culms. There are also several yet-to-be-described closely related species that either have or lack croziers.

Widespread, very common; March–November.

Cyathicula starbaeckii is a whitish to brownish, cupulate *Cyathicula* with an erect pale collar at the apothecial margin. Asci I+, 8-spored, 38–45 × 3–4 µm, with croziers. Paraphyses with many drops, narrowly club-shaped to lanceolate, 2–2.5 µm wide. Hyphae have glassy walls. The fusiform, one-celled spores measure 7–8(–9) × 1.5–2 µm. On herbaceous stems, including *Ranunculus repens*, *Caltha palustris* and *Equisetum*.

Also known as *Crocicreas starbaeckii*. A similar, unnamed species on *Caltha palustris* lacks croziers and has larger spores.

Nemoral–hemiboreal, uncommon; mainly April–June.

Low, smooth to felty inoperculate cup fungi

This group includes sessile to very short-stemmed, less than 2 mm high, inoperculate cup fungi with an outer side and margin that is smooth to felty or rarely toothed, but which never has projecting, hyphal hairs. Large genera in the group include *Mollisia* and *Orbilia*.

OTHER SIMILAR FUNGI:
– long-stemmed or high (growing to 2 mm or more above the substrate), ± smooth inoperculate cup fungi (page 1404).
– downy to hairy inoperculate cup fungi may be sessile to long-stemmed but have ± projecting hairs. Many species also have lanceolate, ± projecting paraphyses (page 1372).
– erumpent inoperculate cup fungi are typically ± immersed in the substrate from where the apothecia emerge by splitting the substrate or by pushing a lid aside. The fruitbodies can often close again during dry spells and reopen when wetted (page 1468).

See also the general wheel, page 1348.

FURTHER READING: 20, 21, 49, 78, 79, 153, 177, 210, 211, 212, 213, 214, 216, 286, 291, 334.

Continued on next page spread . . .

Micro-drawings: spores first, followed by paraphyses (p) possibly with refractive vacuolar bodies (VBs), conidia (con) and asci (a).

Approximate species number applies to temperate Europe.

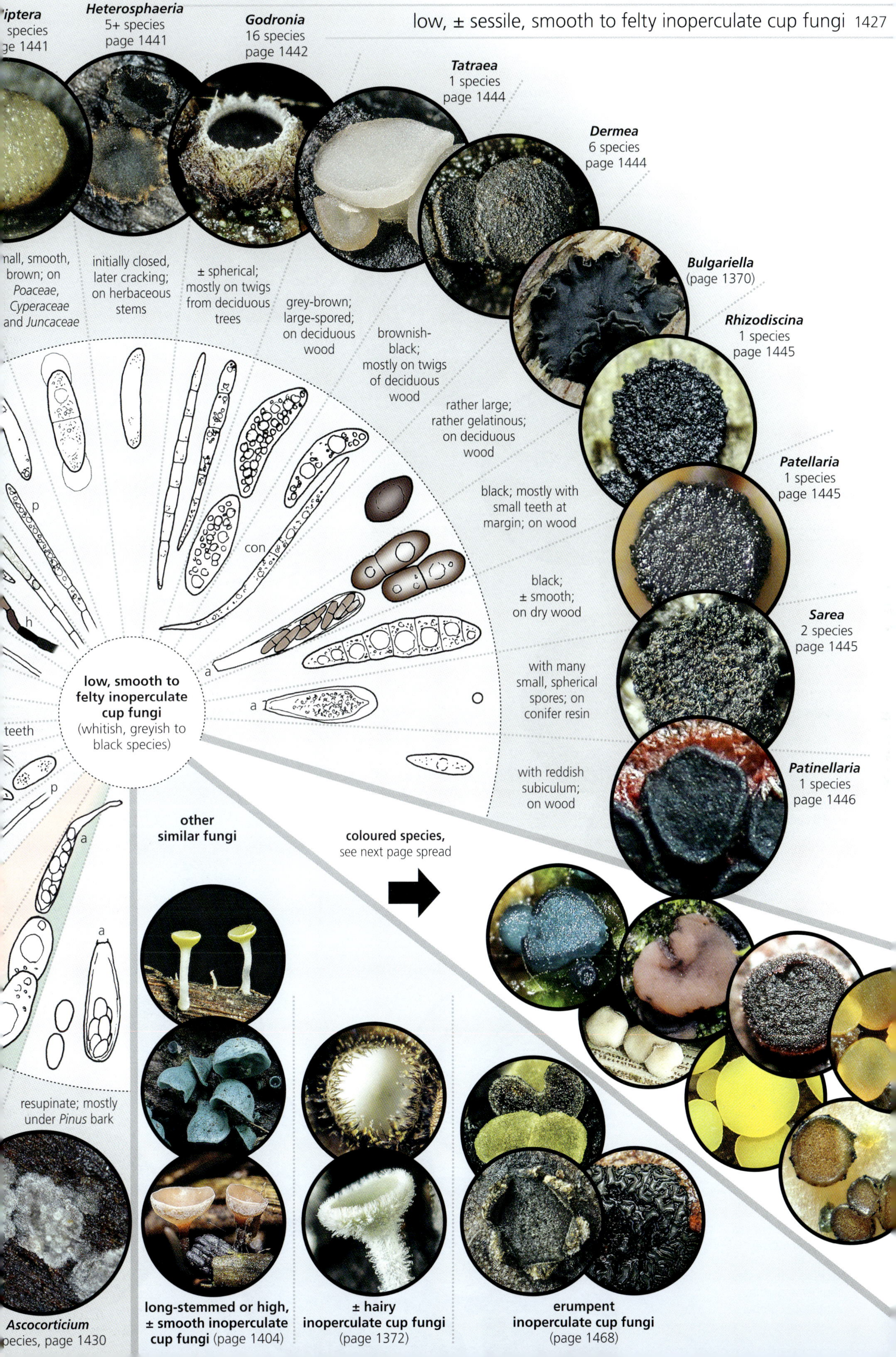

iptera
species
ge 1441
Heterosphaeria
5+ species
page 1441
Godronia
16 species
page 1442
Tatraea
1 species
page 1444
Dermea
6 species
page 1444
Bulgariella
(page 1370)
Rhizodiscina
1 species
page 1445
Patellaria
1 species
page 1445
Sarea
2 species
page 1445
Patinellaria
1 species
page 1446
mall, smooth, brown; on *Poaceae*, *Cyperaceae* and *Juncaceae*
initially closed, later cracking; on herbaceous stems
± spherical; mostly on twigs from deciduous trees
grey-brown; large-spored; on deciduous wood
brownish-black; mostly on twigs of deciduous wood
rather large; rather gelatinous; on deciduous wood
black; mostly with small teeth at margin; on wood
black; ± smooth; on dry wood
with many small, spherical spores; on conifer resin
with reddish subiculum; on wood
p
h
con
a
a
teeth
p
a
a
low, smooth to felty inoperculate cup fungi
(whitish, greyish to black species)
other similar fungi
coloured species, see next page spread
resupinate; mostly under *Pinus* bark
Ascocorticium
pecies, page 1430
long-stemmed or high, ± smooth inoperculate cup fungi (page 1404)
± hairy inoperculate cup fungi (page 1372)
erumpent inoperculate cup fungi (page 1468)

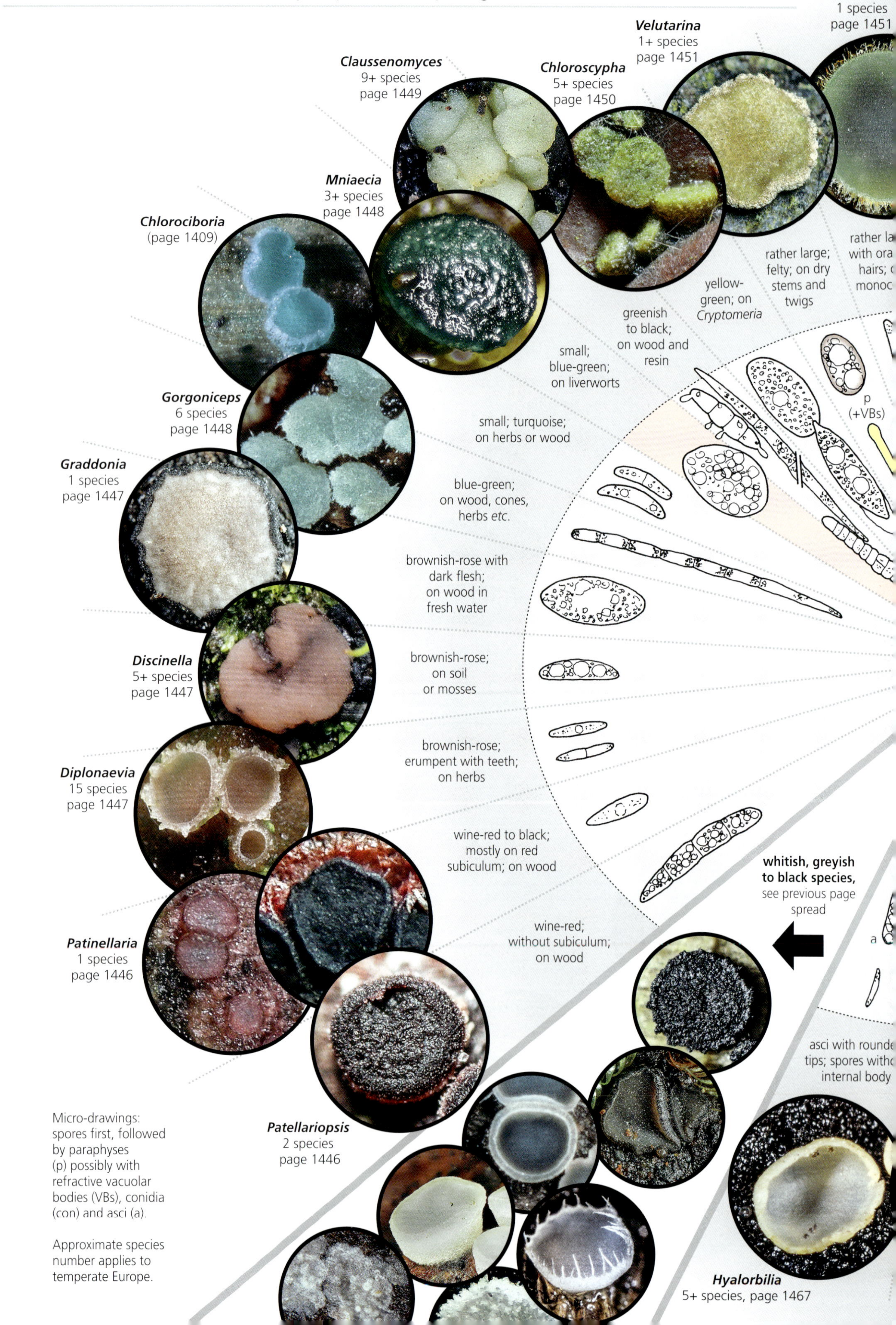

Micro-drawings: spores first, followed by paraphyses (p) possibly with refractive vacuolar bodies (VBs), conidia (con) and asci (a).

Approximate species number applies to temperate Europe.

Catinella
1 species
page 1450
Godronia
16 species
page 1442
Micropeziza
2+ species
page 1452
Pseudohelotium
3 species
page 1452
Bryoscyphus
7 species
page 1453
Vibrissea
7 species
page 1455
Podophacidium
1 species
page 1455
Cenangium
3 species
page 1454
Strossmayeria
1+ species
page 1456
Mollisia & Phaeangella
50+ species
page 1457
Bisporella & Calycina
12+ species
page 1458
Hymenoscyphus
40+ species
page 1460
Calloria
3 species
page 1461
Zytia
1 species
page 1461
Pezicula
12 species, page 1462
Orbilia
species, page 1464
rather large; felty; on large, rotten trunks of deciduous trees
± spherical; mostly on twigs of deciduous trees
very small, erumpent on Poaceae, Cyperaceae and Juncaceae
pale pinkish-buff; mostly on conifer debris
pale pinkish-buff; on mosses and liverworts
spores thread-like; somewhat gelatinous; on deciduous wood in fresh water
erumpent from brown flesh; on needle debris
erumpent; on conifers
immersed in black asexual state; on deciduous wood and Rubus
paraphyses with oblong VBs; on wet deciduous wood, Phragmites, etc.
yellowish; on deciduous wood, bamboos and reeds
apricot-yellow; on wood, leaves and soil
± cushion-shaped; mostly on stems of Urtica
with numerous small, spherical spores; on conifer resins
erumpent; mostly clusterered; on deciduous and coniferous wood
i flat-topped; spores with nternal body
low, oth to felty operculate cup fungi
oured species)
p (+VBs)
p
p
a
p
a
con

Ascocorticium anomalum is a small, flat, corticioid inoperculate cup fungus that lacks a distinct margin and has more or less confluent apothecia of variable size. The asci are I-, 8-spored, thick-walled at the top and measure 15–24 × 5–8 µm. The one-celled, hyaline spores measure 3.5–6 × 1.5–2.5 µm. It also has an asexual state ('*Rhinocladiella*') that almost resembles bunches of grapes; conidia hyaline, ellipsoid to globose, approximately 2.5 µm long. Occurs on conifer wood and bark, mostly on the inside of loosening *Pinus* bark on relatively fresh stumps.

Ascocorticium anomalum var. *juniperi* ⋉ occurs on *Juniperus* and has larger spores measuring 4.5–8 × 2–3 µm. *Ascocorticiellum vermisporum* ⋉ has allantoid, (1–2–) 4–8-celled spores that measure 20–30 × 1–1.5(–2) µm.

Widespread and common but much overlooked; most of the year.

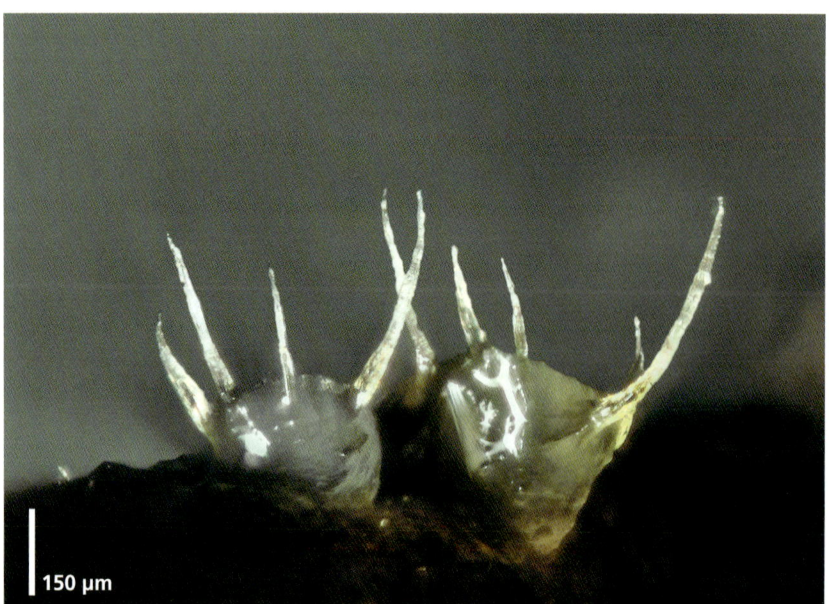

Echinula asteriadiformis is a tiny, whitish inoperculate cup fungus with a margin that has a few, very long teeth that consist of numerous, thin-walled hyphae. Asci I+, 8-spored. Paraphyses simple, branched. The hyaline, ellipsoid spores measure 10–11 × 4 µm. Occurs on dead, damp *Rubus fruticosus* leaves.

A number of undescribed, rather similar species occur on other substrates. *Pezoloma* ▷ has much larger apothecia and *Cyathicula coronata* (page 1424) is much larger and long-stemmed.

Mainly recorded in the UK but probably overlooked elsewhere; summer–autumn.

Species of ***Pezoloma*** are small, whitish somewhat gelatinous inoperculate cup fungi that often have teeth at the margin. There are many undescribed species.

Pezoloma ciliifera is a white, almost translucent inoperculate cup fungus with well-separated, pointed teeth at the margin. Asci I+ (red), 8-spored. Paraphyses simple. The hyaline, one-celled spores measure 10–13(–15) × (3.5–)4.8–5.3 µm. On very damp litter in mires and marshes, mostly with *Sphagnum*.

Pezoloma marchantiae ▽ has somewhat shorter teeth and occurs on *Marchantia*. *Cyathicula coronata* (page 1424) is larger, has a long stem and the teeth are closer together.

Widespread, rare; summer–autumn.

Pezoloma marchantiae is a ± translucent, white *Pezoloma* with scattered, rather short teeth at the apothecial margin. The asci are I+ (red) and 8-spored. The one-celled, hyaline spores measure 6.9–9.1 × 4.1–4.6 µm. Occurs on the liverwort *Marchantia*.

Pezoloma ciliifera △ is similar but has longer teeth and occurs on wet plant material, not on living *Marchantia*.

Very rare or overlooked; June–October.

Trizodia acrobia is a very small, white to pinkish-tinged, ± translucent, ± cushion-shaped inoperculate cup fungus. Asci I-, 8-spored, with croziers. Paraphyses thread-like to ± branched, do not project above asci. The smooth, ellipsoid to shoe-shaped, hyaline spores contain many drops and measure 12–20.5 × 7.5–12 µm. Occurs on the tips of living *Sphagnum*, where it is associated with cyanobacteria (*e.g. Nostoc*, see inset image), which it surrounds with hyphae. It is unclear what the precise nature of the association entails, but there is probably a true mutualistic symbiosis and fruitbodies of *Trizodia acrobia* seem only to occur where the cyanobacteria are present. The *Sphagnum* is unharmed and possibly also benefits from the association. *Trizodia* occupies a basal and very isolated position within the inoperculate cup fungi.

Discinella schimperi ▽ occurs on the same substrate but has far larger apothecia, a more delimited hymenial disc and narrower spores.

Probably widespread but rarely recorded; summer–autumn.

Discinella schimperi is a small, whitish, almost translucent inoperculate cup fungus. Asci I+, 81–95 × 9–11 µm. Spores hyaline, one-celled, 15–20 × 4.5–5.2 µm. Occurs on living *Sphagnum squarrosum*.

Possibly better placed in *Pezoloma* ◁◁. '*Pezizella*' *muscicola* ⋉ is longer-stemmed and has very small spores and inamyloid asci. *Bryoscyphus phascoides* ⋉ occurs on other mosses. Species of *Muscicola* ⋉ have verrucose spores. The lichenized *Absconditella sphagnorum* ⋉ has 2-celled spores. See also *Discinella boudieri* ▷▷.

Probably widespread and common, where the host occurs; mainly May–November.

Hymenoscyphus fagineus is a dingy white, almost sessile, occasionally almost turbinate *Hymenoscyphus* with a ± flattened hymenium. Smells markedly of perfumed chewing gum. Asci I+ (indistinctly), 8-spored, with croziers. Paraphyses have many drops. The (in outline) almost shoe-shaped spores measure (9–)10–14(–16) × (3.8–)4–5 µm. Occurs on *Fagus* cupules in the litter.

The combination of habitat and unusual smell makes identification easy. See also long-stemmed species of *Hymenoscyphus* (page 1421) and the yellowish species ▷▷.

Nemoral, common; mostly September–October.

Phaeohelotium geogenum is a short-stemmed, somewhat turbinate, white to pale cream, rather large *Phaeohelotium*. The flesh partly comprises ± spherical cells. Asci I- to somewhat bluish, 8-spored, 125–190 × 9–16 µm, without croziers. The one-celled, club-shaped to fusiform spores contain many drops and measure 25–37.5 × 4–6 µm. Occurs on damp soil or litter, possibly even very rotten wood.

Other species of *Phaeohelotium* are typically more colourful and have smaller spores ▷▷.

Mainly nemoral, occasional; June–January.

Calycina herbarum is a whitish to cream or pale yellowish, short-stemmed, somewhat downy inoperculate cup fungus. The flesh comprises spherical to angular cells. The outer surface has clavate cells. Asci I+ (red), 73–95 × 4.5–8 µm, lacks croziers. Paraphyses cylindrical, 1.5–3 µm wide and have oblong, yellowish drops. Spores cucumber-shaped, 2-celled, 14–17 × 2.8–3.3 µm (living). Occurs on herbaceous stems, mainly *Urtica*, rarely on leaf petioles.

Similar species are typically more colourful and have a different spore shape or are I+ blue.

Widespread, common, becoming scarcer towards the north; mostly September–November.

Polydesmia pruinosa is a small, 0.5–1 mm wide, white, somewhat convex inoperculate cup fungus with projecting paraphyses, which make the hymenial surface downy. It occurs either as single apothecia or in clusters. The outer side is covered by paraphysis-like hairs. The asci are I+ (blue), 8-spored and measure 95–120 × 8–10 µm. Paraphyses have branched tips. The spores are hyaline, finally 4-celled, and measure 14–22 × 3.7–5.5 µm. In living spores, each cell is filled with small drops; dead spores have large drops. Occurs on the surface of old stromatic pyrenomycetous fungi, apparently rarely on wood.

Distinctive and unlikely to be misidentified. *Polydesmia lichenis* ⋉ occurs on lichens and has even smaller apothecia.

Widespread and common; all year, but mainly in the cooler part of the year.

Coronellaria typhae is a small, whitish, sessile inoperculate cup fungus with projecting paraphyses, which makes the hymenial surface downy. The flesh consists of rather thick-walled, angular cells. Asci I+ (violet-brown), 8-spored, 60–70 × 10–14 µm. Paraphyses lanceolate, smooth. The hyaline, one-celled spores measure 18–22 × 4–5 µm. Occurs on *Typha*.

Other species of *Coronellaria* may occur on other substrates, including, for example, *C. acori* ⋉ on *Acorus* and *C. pulicaris* ⋉ on *Schoenoplectus*.

Possibly widespread, but very rarely reported; mainly summer–autumn.

Cejpia hystrix is a small, broadly attached to very short-stemmed inoperculate cup fungus with a pale grey to bluish, downy hymenium, a strigose-hairy, black outer side and a white margin. The base has radiating black setae and the flesh is pale brown with spherical to angular cells. The asci are I+ (red), 8-spored and measure 75–100 × 10–12 µm. The ellipsoid to cylindrical, one-celled spores measure 13–21 × 2.5–5 µm. Occurs on stems of *Molinia* in damp habitats.

This species should be easy to recognize due to the combination of substrate, dark hairs and downy hymenium. Various other species of inoperculate cup fungi on grasses mostly have longer, paler hairs and are usually distinctly stipitate.

Nemoral but probably more widespread, likely to be common, but rarely reported; mainly June–January.

Mollisia form small, grey to yellowish apothecia that have a smooth, felty or downy margin and outer side. Paraphyses have strongly refractive, large, oblong drops. A large, difficult genus; yellowish species on page 1457.

Mollisia cinerea is a pale grey to brown-grey *Mollisia* with a somewhat darker outer side. Usually has pale, radiating hyphae at the base. The margin is minutely downy and somewhat paler. The flesh does not stain yellow in KOH. The basal, spherical cells are brown. Asci I+, 47–73 × 5–6 µm. Paraphyses cylindrical, 2–3 µm wide, with large oblong, refractive drops. The 1(–2)-celled, cylindrical spores measure 6–12 × 1.5–2.5 µm. Mostly on fairly decayed deciduous wood.

This species belongs to a poorly understood complex: *M. melaleuca* ⋉ is apparently more contrasty with a darker outer side and paler hymenium; *M. clavata* ⋉ has conspicuous marginal cells and occurs on *Rubus*; *M. ladae* ⋉ has rather long marginal cells and narrow spores. Other species stain ± olive in ammonia or produce a yellow pigment in KOH.

Widespread and very common when defined in a broad sense; all year.

Mollisia palustris is a grey, sessile, 0.3–2 mm wide *Mollisia* with a grey-brown outer side and an almost white margin. The asci are I+ and measure 50–70 × 4–5.5 µm. Paraphyses up to 2–3 µm wide. Spores are one-celled and measure 7.5–12 × 1–1.5 µm. Occurs on *Poaceae*, *Juncus* and *Cyperaceae* in damp or wet habitats.

This species belongs to a poorly understood complex and may, in a strict sense, be confined to grasses.

Widespread, probably common, but rarely reported; mainly May–September.

Mollisia rosae is a rather cupulate *Mollisia* with crowded apothecia that have a pale, ± downy margin, a felty, brownish outer side and thereunder a thick brown subiculum. The hairs are up to 100 µm long. Abundant yellow pigment is dissolved in KOH. The asci are I+, 8-spored, measure 31–70 × 4–5.5 µm and lack croziers. Spores one-celled and measure 6–9 × 2–3 µm. Occurs on stems of *Rosa*; probably also *Cornus* and perhaps other hosts.

Mollisia prunicola ⋉ occurs on *Prunus spinosa* and the asci have croziers.

Nemoral–hemiboreal; mainly March–September.

Mollisia ligni tolerates desiccation and has dark brown, felty apothecia with a paler margin that dry to a ± triangular shape. The outer side has ± brown, projecting, 2–4-celled hairs that measure 21–60 × 6–7 µm and are wider towards the tip. Lacks yellow pigment in KOH. Asci I-, 41–52 × 5–6 µm. Paraphyses cylindrical, 2–3 µm wide. Spores clavate to cylindrical, one-celled, measure 6–10 × 1.5–3 µm, with few drops. On decorticated, firm wood of deciduous trees, mostly on standing trees.

Other species with this ecology are paler or have yellow or green pigment in KOH.

Widespread, common; all year.

Mollisia discolor is a pale yellowish, mostly clustered and ± veined/wrinkled *Mollisia* that stains ± yellow in a dry condition. The outer side is felty and brown. The asci are I+, 8-spored and measure 55–92 × 5–6.5 µm. Paraphyses cylindrical with 2–2.5 µm wide tips. The one-celled, typically somewhat allantoid spores measure 9–16(–19) × 2–3.2 µm and have drops at the poles. Occurs on deciduous wood, mostly twigs of *Quercus*.

This species belongs to a difficult, poorly understood complex.

Nemoral–hemiboreal, probably rather common; all year.

Pyrenopeziza form small to very small, grey to grey-brown, mostly erumpent apothecia that usually have a serrated/hairy margin and an outer side with ± brownish hairs. Paraphyses have somewhat refractive, spherical drops. A large, difficult genus.

Pyrenopeziza baraliana is a sturdy, clustered *Pyrenopeziza* with a distinctly pale, serrated margin and a grey hymenium. The hairs are short, 1(–2)-celled, somewhat tapering, thicker-walled and brown at the base. The asci are I+, 8-spored and 82–90 µm long. Paraphyses have rather small, refractive, spherical drops. The sinuous-fusiform spores measure 20–23(–26) × 3–4 µm (living material) and have drops at the poles. Occurs on rotten stems of *Petasites* in moist habitats.

Pyrenopeziza chailletii ⋉ (ss. Rehm), *e.g.* on *Chaerophyllum* and *Aconitum*, has blunt hair tips and smaller spores; *P. milligrana* ▷ has smaller apothecia and ± multicelled hairs; *P. carduorum* ⋉ is less cupulate and dentate, and also has blunt-tipped marginal hairs.

Nemoral, rarely reported; May–July.

Pyrenopeziza inapiculata is a greyish *Pyrenopeziza* with a pale, minutely downy margin. Flesh without KOH reaction, comprising ± spherical to angular, grey-brown cells. Lacks conspicuous marginal hairs. Asci I+, 40–65 µm long, with croziers. Paraphyses lack refractive contents, somewhat widened at the tips. The spores measure 10.5–15(–18) × 2–2.5 µm, with few drops. Occurs on *Iris* stems in boggy habitats.

Belonium nigromaculatum ⋉, also on *Iris*, has somewhat smaller spores and more projecting marginal cells on the apothecium.

Nemoral, rarely reported; mainly May–June.

Pyrenopeziza millegrana is an at first cupulate, later more expanded *Pyrenopeziza* with a grey hymenium, a red-brown to blackish outer side and a wide, pale margin. Has cylindrical, 2–3 µm wide hairs: at the margin one-celled and hyaline, lower down multicelled and brown. Asci I+, 50–56 × 7.5–9 µm. Paraphyses cylindrical, 2–3 µm wide. Spores one-celled, cylindrical to oblong fusiform, 15–18 × 3–3.5 µm. Occurs on dead stems of *Filipendula*.

Two species of *Pyrenopeziza* on *Arctium* also have large spores, *P. arctii* ⋉ and *P. inornata* ⋉, and the same applies to *P. carduorum* ⋉.

Widespread, common; May–June.

Pyrenopeziza rubi is an immersed/erumpent, cupulate *Pyrenopeziza* with long, 40–90 µm, multicelled, pale brown hairs at the margin. Asci I+, 45–55 µm long, with croziers. Paraphyses somewhat lanceolate. The clavate, 1(–2)-celled spores measure 9–12 × 2.5–3 µm, have two rather large and some smaller drops and a loosening outer wall; they bud off semiglobose conidia. Occurs on *Rubus*, both *R. fruticosus* and *R. idaeus*.

There are several similar species but they occur on other substrates, *e.g.* *P. thalictri* ⋉ on *Thalictrum*.

Widespread, rather common; mainly March–September.

Pyrenopeziza foliicola is a small, grey-brown *Pyrenopeziza* with rather long (up to 60 µm) marginal hairs. Asci I+, 8-spored, 35–50 µm long. Paraphyses somewhat clavate and have small drops. The clavate spores measure 6.5–8 × 2 µm; with few drops. Occurs on *Alnus* leaves in the litter layer.

Pyrenopeziza fuckelii ⋉ has somewhat larger spores and shorter marginal hairs. *Pyrenopeziza betulina* ⋉ has a colourless outer cell layer.

Distribution and frequency poorly known, probably common and overlooked; mainly spring.

Pyrenopeziza petiolaris is a rather large, erumpent, somewhat elongated *Pyrenopeziza* with a grey hymenium and a somewhat paler, finely floccose margin that has projecting cells up to 20 µm long. Asci I+, 8-spored, 40–60 µm long, with croziers. Paraphyses somewhat widened towards the tips and have spherical drops with refractive contents. Spores one-celled, hyaline and measure 6.5–11.5 × 2–2.5 µm. Occurs on petioles from previous season's fallen *Acer* leaves.

The combination of relatively large, oblong-shaped apothecia and the host make this species easy to identify.

Nemoral–hemiboreal, probably common; April–June.

Tremella discicola ⋉ occurs in the hymenium of various species of *Mollisia* and *Pyrenopeziza*.

Pirottaea nigrostriata is a small, dark, vertically furrowed-striate, cupulate *Pirottaea* with a paler hairy margin. Outer side has stripes of dark brown setose hairs. Setae to 50 µm long, ± widened at the tips. Asci I+, 8-spored, 50–65 µm long. Paraphyses ± cylindrical. Spores one-celled and measure 9–12 × 2–2.5 µm. Occurs on coarse *Apiaceae* stems.

Pirottaea paupercula ⋉ has shorter or absent setae; *P. veneta* ⋉, on *Helleborus*, has dark cells in dense clusters. A number of *Pirottaea* species on herbaceous stems, including *Apiaceae*, have significantly longer spores.

Mainly nemoral, probably common; January–September.

Niptera eriophori is a very small, sordid yellow-brown *Niptera* with dark appressed hairs on the outer side. Asci I+, 8-spored, up to 132 µm long. Paraphyses with swollen tips. Spores 2–(4)-celled, 15–20 × (4–)5–6 µm, with a thin slime coating that is drawn out at the ends. Occurs on stems and leaves of *Cyperaceae* and *Juncaceae*.

Niptera lacustris ⋉ has slightly larger spores. The 2-spored *N. macrospora* ⋉, on the same hosts, also has larger spores; *N. trichophoricola* on *Eriophorum vaginatum* has shorter spores. *Mollisia ellisii* var. *eriophori* ⋉ has narrower spores and larger apothecia.

Widespread, but rarely reported; mainly July–October.

Heterosphaeria patella forms erumpent, closed fruitbodies (inset image) that later open in the top; margin dentate. Outer side dark. Asci I+ (blue then red), 62–98 × 9.5–11.5 µm, with croziers. Paraphyses flame-shaped at the tips; in dead condition have yellow drops. Spores (1–)2-celled, hyaline and measure (9–)12–16(–20) × 2.5–5.3 µm. Also with an asexual state that has conidia with up to 25 µ long appendages. Occurs on dead stems, mostly from *Apiaceae*.

There are similar alpine species of *Heterosphaeria* but all lack thread-like appendages on the conidia.

Widespread, common; all year, but mainly recorded in spring–summer.

Godronia form at first cupulate, mostly somewhat marginally furrowed apothecia that may expand with age and in wet weather. The spores are long, cylindrical to almost thread-like and mostly multicelled. All species have ± conspicuous asexual states. Occur on various woody plants, rarely grasses. Species not treated include: *G. muehlenbeckii* ⋉ on *Phragmites*; *G. symphoricarpi* ⋉ on *Symphoricarpos*; *G. spiraeae* ⋉ on *Spiraea*, *G. viburni* ⋉ on *Viburnum*; *G. cassandrae* ⋉ on various species of *Ericaceae*; *G. foliicola* ⋉ and *G. urceoliformis* ⋉ on *Vaccinium*; *G. callunigera* ⋉ and *G. bicellulata* ⋉ on *Calluna*; *G. andromedae* ⋉ on *Rosmarinus*; and *G. empetri* ⋉ on *Empetrum*.

Godronia uberiformis has solitary or clustered apothecia. It is initially a closed urn-shape but opens in the top at maturity. The outer side is dark olive-brown and felty. The hymenium is paler, greyish. Asci I+, 100–165 × 7.5–9.5 µm. Spores cylindrical, 4–8-celled, measure 67–73 × 2 µm. Has an asexual pycnidial state (lower image). Occurs on branches of *Ribes nigrum*.

Godronia ribis ▷ has smaller spores and occurs on other species of *Ribes*.

Widespread, occasional; February–June (apothecia).

Godronia ribis is a ± clustered, olive-yellow *Godronia* with a brownish-black outer side and a projecting, floccose margin. The flesh consists of grey-brown, angular cells. The asci are I+ and measure 137–163 × 6–8 µm. Paraphyses cylindrical, slightly wider at the tips. The elongated, fusiform spores measure 31–40 × 2.5–4 µm, either have or lack septa, and usually have oil drops. Occurs on branches of *Ribes rubrum* and related species (but not on *R. nigrum*).

Godronia uberiformis ◁, on *Ribes nigrum*, has longer spores.

Widespread, rather common; January–August.

Godronia rosae is an urn-shaped, erumpent, ± grey-green to black, coarsely furrowed *Godronia*. The outermost flesh has thick-walled hyphae with incrusted pigment. Asci I+, 8-spored, approx. 100–160 × 8–9 µm, with croziers. Paraphyses straight, cylindrical, approx. 1.5 µm wide. Spores fusiform, 2–4-celled, 10–20.5 × 2.8–4.2 µm. Forms a black, pycnidial state that produces one-celled, hyaline conidia measuring 4.5–6.3 × 1.3–2 µm. Occurs on wood of *Rosa*.

The colours and substrate should make identification straightforward.

Very southern within temperate Europe, rare; spring–summer.

Godronia urceolus is a solitary or clustered *Godronia* with an olive-green to brown outer side and a grey hymenium. The opening is paler with irregular teeth. The asci are I+, 8-spored and measure 90–130 × 6–8 µm. The spores are cylindrical, 6–8-celled and measure 50–75 × 1.5 µm. Occurs on twigs of *Betula* and *Alnus*.

Godronia fuliginosa ⋉ is a parasite on twigs of *Salix* and has somewhat larger spores measuring 53–91 × 2–3.5 µm.

Widespread, but rarely reported; mainly May–September.

Tatraea dumbirensis is a sessile to short-stemmed, whitish, greyish to purple-brown inoperculate cup fungus with a ± pruinose to downy outer side. The outer flesh has almost spherical cells. Does not have stromatic tissue. Asci I+, 120–200 × 11–13 µm. Paraphyses cylindrical, 2–2.5 µm wide, with refractive drops. The curved-ellipsoid to almost comma-shaped spores contain many drops and measure 16–28 × 5–8.5 µm; they do not bud conidia. Occurs on deciduous wood.

The large, distinctively shaped spores are the best character.

Mainly nemoral, rare or overlooked; autumn.

Dermea prunastri is a ± clustered, erumpent, nearly sessile, discoid, dark brown to black *Dermea*. The asci are I+, 8-spored and have croziers. The 1-(2)-celled, hyaline, narrowly ellipsoid spores measure 15–20 × 5–7.5 µm. Found mostly in its asexual, black, spike-like state, which has large, strongly curved, only somewhat tapering conidia that measure 19–26(–30) × 2.8–3.2 µm. Occurs on ± living branches of *Prunus spinosa* and *P. domestica*.

The very similar *D. padi* ⋉ has very pointed, only slightly curved conidia that measure 20–35 × 3.5–4.5 µm. A number of similar species occur on other hosts, *e.g. D. cerasi* ⋉ (spores 15–25 × 4–5 µm) on *P. avium* and *P. cerasus*; *D. ariae* ⋉ (spores 12–14 × 3–4 µm) on *Sorbus*; and *D. tulasnei* ⋉ (spores 15–23 × 6–9 µm) on *Fraxinus*.

Nemoral–hemiboreal, probably common; April–July.

Rhizodiscina lignyota has sessile, black apothecia that are toothed at the margin and which have thick-walled, brown hyphae from the base. Asci 8-spored, clavate, 35–55 × 8–12 µm and double-walled (bitunicate); wall pale blue in iodine reagents. Spores 2-celled, brown, asymmetrically shoe-shaped and measure 9–14 × 3.5–5 µm. Occurs on wood, old polypores, *etc*.

Dactylospora stygia ⋉ has a thick outer amyloid layer on the asci and more symmetrical spores. *Patellaria atrata* ▽ has larger, 6–12-celled spores. Other similar cup fungi have single-walled (unitunicate) asci.

Widespread, common; all year.

Patellaria atrata is a sessile, somewhat gelatinous to tough-fleshed, black, ± flat, ± clustered *Patellaria*. Asci I-, 80–150 × 15–20 µm, double-walled (bitunicate), (4–)8-spored. Paraphyses (pseudoparaphyses) branched. Spores are clavate, 6–12-celled, thick-walled, hyaline, and measure 30–55 × 7–10 µm. Occurs on wood of deciduous trees and on coarse herbaceous stems.

The combination of large, hyaline spores and double-walled asci make this species relatively easy to identify. Species of *Durella* ⋉ that grow superficially have single-walled asci.

Nemoral–hemiboreal, common; all year.

Sarea difformis is a rather small, black *Sarea* with a somewhat convex hymenium that is filled with red-brown pigment. The asci are I+, large, thick-walled at the tips and filled with many spherical spores that measure 2–3 µm in diameter. A pycnidial state has ± spherical, pale brown conidia measuring 2–3.5 µm. Occurs on ± blackened conifer resin.

Species of *Claussenomyces* ▷▷ also occur on resin but have inamyloid ascus walls. *Tromeropsis microtheca* ⋉ has thick-walled, inamyloid asci and occurs on coniferous wood. See also the orange *Zythia resinae* ▷▷.

Widespread, probably common but easily overlooked; all year.

Patellariopsis atrovinosa is a ± sessile, discoid inoperculate cup fungus with greyish wine-red colours and a distinct, wide apothecial margin. The outermost flesh cells are purple-brown. Asci I+, 8-spored, 90–113 × 8–12 µm, with croziers. Paraphyses swollen at the tips (up to 2.5 µm) and hyaline. The spores are clavate, 4-celled, hyaline, contain many drops and measure 20–30 × 3–4.6 µm. On deciduous wood, including various shrubs.

Patinellaria sanguinea ▽ has a blood-red subiculum and much smaller spores.

Widespread, very rare or overlooked; all year.

Hans-Otto Baral

Patinellaria sanguinea is an at first pinkish, later black, sharp-margined, sessile inoperculate cup fungus on a pinkish to blood-red subiculum. The asci are I+, 8-spored, measure 40–55 × 7–9 µm and have croziers. The paraphyses are cylindrical, 1.5 µm wide, broadening to 4 µm at the tips, imbedded in brown pigment and have large internal drops. The ellipsoid-fusiform to clavate, one-celled spores measure 7–13 × 2.5–3.5 µm. Occurs on hard, decorticated deciduous wood, *e.g. Betula*.

The red subiculum and relatively small spores should make identification straightforward. *Patellariopsis atrovinosa* △ lacks a subiculum and has completely different spores.

Widespread, occasional to rare; mainly found during spring.

Jean-Pierre Dechaume

Diplonaevia bresadolae is a yellowish flesh-pink to pale rose, erumpent, at first ellipsoid, later more discoid *Diplonaevia*. The marginal cells are up to 20 × 2.5 µm. The asci are I+, 8-spored and measure 45–55 × 5–6 µm. Paraphyses cylindrical, simple, 1–2 µm wide. The spores are cylindrical to fusiform, one-celled, rarely 2-celled, and measure (8–)10–12(–16) × 2–2.5 (–3) µm. Occurs on *Urtica* stems.

Diplonaevia perpusilla ⋉ (and *D. mollisioides* ⋉, possibly a synonym) occur on stems of *Euphorbia*.

Nemoral, probably rather common; mainly May–August.

Discinella boudieri is a sessile, pale brown to purple-brown, depressed to convex *Discinella* that usually has a depression at the centre and a rooting, short stem. Asci I+ (red), 89–130 × 8–13 µm, with croziers. Paraphyses cylindrical, widened to approx. 3 µm at the tips. Spores ± ellipsoid, measure (10–)12–15 × 4–5.5 µm, contain many drops. Occurs on soil amongst bryophytes.

May recall an operculate cup fungus in the genus *Peziza* (page 1313). *Discinella purpurascens* ⋉ has larger spores and *D. limosella* ⋉ has smaller spores. See also *D. schimperi* ◁◁.

Widespread, very rare; autumn.

Graddonia coracina is a sessile, whitish, pinkish-buff to brown, ± convex inoperculate cup fungus with a darker margin and outer side. The flesh is several-layered, consisting of both hyphoid and more rounded elements. Asci I-, 8-spored, (100–) 150–185 × 12–16 µm, with croziers. Paraphyses wider at the tips, with yellow-green refractive material. Spores 1–2-celled, contain many drops, 16–24 × 7–10.5 µm. Occurs on wood, mostly in temporarily flooded forests, even on submerged wood.

Recalls *Adelphella babingtonii* (page 1322) but that species has one-celled, broader spores and operculate asci.

Widespread, occasional; mainly summer.

Gorgoniceps viridula is a pale turquoise, disc- to cushion-shaped inoperculate cup fungus with a dark brown base. The surface may appear hairy due to projecting spores. The flesh cells have green drops. Asci I+, 102–136 × 9.5–12 µm, with croziers. Spores mainly 8-celled, thread-like, taper towards one end and measure 48–75 × 2–2.5 µm. Occurs on dark-coloured *Pinus* bark and cones.

Gorgoniceps aridula ⋉ is very similar to *G. viridula* but lacks the green tinges.

Widespread, but rarely reported; mainly October–November.

Knud Knudsen

Mniaecia jungermanniae is a small, sessile, vivid blue-green, smooth inoperculate cup fungus with a flattened to convex hymenium. The asci are I-, 8-spored and measure up to 180 × 20 µm. Paraphyses green, distinctly widened to almost capitate at the tips. The spores are ellipsoid to pear-shaped, hyaline to greenish, contain many drops and measure 16–24 × 8–15 µm. Occurs on small, fine-leaved liverworts (*Jungermanniales*), mostly in spring but also during mild winters.

Chlorociboria aeruginella (page 1411) occurs on herbaceous stems and dicot leaves, and is microscopically very different. *Mniaecia nivea* ⋉ and *M. gloeocapsae* ⋉ have the same ecology as *M. jungermanniae* but are at first white, later ± buffish.

Widespread, occasional to rather common; mainly December–May.

Claussenomyces form small, greenish to almost black, discoid, mostly convex apothecia. Asci I-, rather thick-walled, and have 8 several-celled spores that at maturity may divide into many, tiny secondary spores. In living asci these secondary spores are grouped in 8 clumps. All occur on wood or resin. *Sarea* ◁◁ and *Tromeropsis* ⋉ have asci that are multispored from the start. *Claussenomyces* is a very difficult genus with many undescribed species. *Dendrostilbella smaragdina* (page 1629) may be the asexual state of an undescribed species of *Claussenomyces*.

Claussenomyces prasinulus is a sessile, gelatinous, pale grey-green *Claussenomyces*, which is typically accompanied by a pale green, matchstick-shaped conidial stage (lower image). Asci I-, 80–90 × 7–8 µm. Paraphyses cylindrical, branched at the tips. Spores 10–13 × 2.5–3.5 µm, 4-celled, do not dissolve into secondary spores. Occurs on strongly decomposed deciduous wood.

The spores of most other species of *Claussenomyces* soon dissolve into secondary spores; they also do not have a pale green, erect asexual state. The similarly pale *C. dacrymycetoideus* ⋉ has more transverse septa in the spores.

Mainly nemoral–hemiboreal, occasional; most of the year.

Claussenomyces atrovirens is a dark olive-green to black *Claussenomyces*. Asci I-, 90–120 × 10–12 µm. Paraphyses cylindrical, branched, widening to 2 µm at the tip. Spores cylindrical to fusiform, with 5–11 transverse septa, rarely with a few longitudinal septa, 15–30 × 3.5–5 µm, produce secondary spores measuring 2–3 × 1 µm that eventually fill the asci. Occurs on deciduous and coniferous wood.

Claussenomyces dacrymyceto-ideus ⋉ is paler and has smaller spores with more transverse septa. There are a number of similar species, some of which occur on resin.

Widespread and probably common; all year.

Andreas Gminder

Catinella olivacea is a rather large (up to 13 mm wide), sessile inoperculate cup fungus with a yellowish, furrowed margin and dark olive-green, somewhat concave hymenium. A red-brown pigment dissolves in KOH. The asci are I-, 8-spored and measure 90–110 × 5–6 µm. The paraphyses are cylindrical, 2–2.5 µm wide, and the tips are imbedded in mucus. The spores are ellipsoid to peanut-shaped, brown, one-celled, smooth to finely spiny and measure 9–10.5 × 3.5–4.5 µm. Mostly occurs well hidden under large, rotten trunks of deciduous trees.

A very distinctive species. Old specimens that lack green colours can easily be identified with the aid of a microscope. *Peziza apiculata* (page 1318) and other greenish operculate cup fungi on wood have large, hyaline, ornamented spores and amyloid asci with a lid.

Mainly nemoral, occasional, rare in the hemiboreal zone; June–March.

Chloroscypha seaveri is an almost sessile to short-stemmed, dark green to olive-brown inoperculate cup fungus. Asci I-, 8-spored, 120–130 × 20–30 µm. Paraphyses thread-like, wider at the tips, imbedded in mucus. The hyaline spores are broadly fusiform to clavate, sometimes constricted at the ends, (15–)23–30 × 8–12 µm. Occurs on twigs of *Cryptomeria*, *Thuja* and *Chamaecyparis*.

Chloroscypha sabinae ⋉ and the long-stemmed *C. alutipes* (page 1419) occur on *Juniperus*.

Nemoral, occasional to rare; mainly December–April.

Velutarina rufo-olivacea is a cupuluate, later ± expanded inoperculate cup fungus with a pale, felty outer side and red-brown to olive-brown hymenium. Asci typically I+ (red), 120–160 × 10–15 µm, 8-spored, without croziers. Spores one-celled, brownish, 10–15 × 6–9 µm. On bark of many deciduous hosts and *Juniperus*.

Species of *Encoelia* and *Sclerencoelia* (page 1408) have hyaline spores.

Widespread, occasional; all year, mainly May–July.

An undescribed, black, bitunicate pyrenomyceous fungus may occur in the hymenium.

Dennisiodiscus prasinus is a rather large inoperculate cup fungus with ± orange to rust-brown, projecting, curly hairs, up to approx. 180 µm long, at the margin of the quickly flattening apothecia, which has an olive-grey hymenium. The flesh consists of dark, spherical cells. The asci are I+ (reddish-brown to blue), and at maturity typically 4-spored. The spores are narrowly ellipsoid to curved-fusiform, one-celled, hyaline and measure 10–22 × 2–3 µm. Occurs on decaying monocots, *e.g. Glyceria* and *Carex*.

Typical specimens cannot be mistaken for any other inoperculate cup fungus, but the hairs are not always as well developed as on these images. *Dennisiodiscus* is related to species of *Mollisia* ◁◁ and, like that genus, belongs in the family *Dermateaceae*.

Nemoral, occasional; mainly May–July.

Micropeziza cornea is a sessile, 0.2–0.5 mm wide inoperculate cup fungus that breaks through a dark shield that remains as small, dark teeth at the margin (see lower image). The hymenium is pale yellow-brown. The flesh consists of spherical cells. The asci are I+, 8-spored, measure 53–85 × 7–11.5 µm and have croziers. Paraphyses have large oil drops, are ± branched, somewhat wider at the tips and imbedded in brown matter. The rather curved, pointed, 1–2-celled spores contain many drops and measure 12–18 × 3–3.5 µm; they have a slime coating. Occurs on *Carex*, *Juncus* and *Poaceae*.

Pseudonaevia caricina ⋉ has larger, 1–6-celled spores and paler, somewhat smaller apothecia.

Widespread, probably common; mainly May–September.

Pseudohelotium pineti is a short-stemmed, flat to convex *Pseudohelotium* with a clay-buff to red-brown outer side and a pale buff hymenium. The flesh hyphae are ± brown and lack glassy walls. Asci I+, 8-spored, 62–100 × 6–7 µm. Paraphyses cylindrical, 2–3 µm wide at the tips, with refractive drops. Spores cylindrical to somewhat pointed, ± curved, 1–4-celled, 18–37 × 1.5–2.5 µm. Occurs on *Pinus* needles, rarely on cones.

Pseudohelotium sordidulum ⋉, mainly on *Pinus* twigs and cones, has spores 17–26 × 2.5–3.5 µm.

Mainly nemoral–hemiboreal, common; mainly January–April.

Bryoscyphus parasitizes bryophytes and liverworts. The spores tend to be rather large and contain drops.

Bryoscyphus atromarginatus is a sessile to short-stemmed *Bryoscyphus* with a pale buff hymenium and a darker margin. The flesh hyphae do not have glassy walls. Asci I+, 8-spored, 100–125 × 5.5–7 µm. Paraphyses narrowly cylindrical with ± widened tips. Spores ellipsoid to clavate, one-celled, 11–13 × 3.5–4 µm. Occurs on *Marchantia*.

Bryoscyphus marchantiae ⋉, on *Reboulia*, has wider spores; the spores are also larger in *B. conocephali* ⋉, which grows on *Conocephalum*. Various other inoperculate cup fungi occur on bryophytes, sphagni and liverworts including the white species *Pezoloma marchantiae* ◁◁, *Hymenoscyphus phascioides* ◁◁ and *Trizodia acrobia* ◁◁ and the green *Mniaecia jungermanniae* ◁◁. Several genera of ± orange, operculate cup fungi also occur on such substrates, see *e.g.* page 1338.

Nemoral, rarely reported; mainly June–October.

Bryoscyphus dicrani is a short-stemmed, pale brown, smooth *Bryoscyphus*. It does not have glassy hyphal walls. Asci I-, 8-spored, 140–180 × 11.5–13 µm. Paraphyses narrowly cylindrical, project somewhat above the asci. Spores fusiform to rhombic, 1- to several-celled, 16–24 × 6.5–10 µm. Occurs on *Didymodon fallax*, *Ceratodon purpureus* and *Paraleucobryum longifolium*.

Bryoscyphus turbinatus ⋉ occurs on *Dicranum* and *Hypnum*; *Hymenoscyphus phascoides* ⋉ is very pale, downy and occurs on *Pleurozium*.

Nemoral–hemiboreal; mainly during the cooler months of the year.

Cenangium ferruginosum is an erumpent, leathery, usually clustered, grey-olive to cinnamon-buff inoperculate cup fungus with a ± dentate, mostly paler margin. The outside is at first covered by small brown scales. The flesh is gelatinous and the asci I-. Paraphyses septate, lanceolate. The pointed, egg-shaped, multiguttulate, hyaline spores measure 11–14 × 5–6.5 µm. Occurs on recently dead, attached *Pinus* branches.

Cenangium acuum ▽ is browner and occurs on dead needles of *Pinus*, *Picea* and *Juniperus*. *Therrya pini* ⋉ is more immersed in the substrate, darker and has worm-shaped, very pointed, large spores. *Crumenulopsis pinicola* ⋉ has hairy apothecia and larger, tapering spores. *Tympanis hypopodia* ⋉ has asci that gradually fill with microspores derived from the sexual spores.

Widespread and probably rather common; mainly in spring.

Cenangium acuum is a short-stemmed, 1–3 mm wide, red-brown *Cenangium* with a dark brown outer side and a dentate margin. The asci are I+ (red), measure 120–143 × 14–16.5 µm and have croziers. Paraphyses clavate, wider at the tips. Spores one-celled, narrowly ellipsoid to fusiform and measure 12–17.5 × 3.5–6 µm. Occurs on dead needles of *Picea*, *Pinus* or *Juniperus*.

Cenangium ferruginosum △ is more yellowish and is erumpent on dead twigs of *Pinus*. Other inoperculate cup fungi on conifer needles are different colours or shapes.

Widespread, occasional; mainly February–July.

Podophacidium xanthomelum is a rather large inoperculate cup fungus that is closed, brown and warty at first, but later opens and becomes dentate at the margin. Hymenium yellowish. Asci I+, 8-spored, measure 90–150 × 7–10 µm, apical apparatus tall. Paraphyses branched and curved at the tips. Spores fusiform, hyaline, contain large drops, are one-celled and measure 10–15.5 × 5–6 µm. Occurs in clusters on a dark brown subiculum on soil amongst leaf-litter and mosses in coniferous and mixed forests.

A distinctive species.

Widespread, mainly hemiboreal; summer–autumn.

Jean-Marc Moingeon

Vibrissea form small, sessile to stipitate apothecia with a somewhat yellowish hymenium and long, thread-like spores that after ejection may break into several pieces. Mostly on wet wood. See page 1360 for the stipitate-clavate species.

Vibrissea flavovirens is a ± sessile, 1–3.5 mm wide *Vibrissea* with a brown to black, above striate-dotted, outer side and a pale yellow to greenish-yellow hymenium, which at most is one-third of the total apothecial thickness. Asci I- or somewhat blue, 250–340 × 6–9 µm, with croziers; ascus stem more than 50 µm long. Paraphyses widened to 2–4 µm at the tips, sometimes branched, with very refractive contents. Spores (living) 132–195 × 1–2(–3.3) µm, straight, 16-celled, split into 4-celled parts after ejection. Occurs on deciduous and coniferous wood in clean streams.

Vibrissea decolorans ⋉ has I+ asci with shorter stems and a thicker hymenium in relation to the flesh. *Vibrissea filisporia* ▷▷ also has a thick hymenium, but I- asci.

Nemoral, uncommon, but very faithful in known sites; mainly February–August.

Vibrissea filisporia is a rather small, sessile, grey to buff *Vibrissea* with a pale to dark brown outer side and margin, and a paler yellowish hymenium. The asci are I-, measure 100–290(–465) × 6–10 µm and have croziers. Paraphyses widened to 2–5 µm at the tips, are branched or unbranched, and have highly refractive drops. The spores are straight, measure 100–305(–325) × 1–2 µm, 16-celled and split into two after ejection. Occurs on deciduous wood in very damp or wet habitats.

A number of forms of *Vibrissea filisporia* are separated on the basis of spore length: *V. filisporia* f. *gigantospora* ⋉ has spores longer than 400 µm; *V. filisporia* f. *boudieri* ⋉ has sinuous spores 210–290 µm long; *V. filisporia* f. *scella* ⋉ has spores 100–210 µm long spores and simple paraphyses; and *V. filisporia* f. *filisporia* ⋉ has spores 100–210 µm long and branched paraphyses.

Widespread, occasional; mainly spring.

Strossmayeria basitricha is a cushion- to disc-shaped, whitish to grey inoperculate cup fungus with a brown base. The outermost flesh is amyloid. Asci I-, 101–137 × 11–16 µm, with croziers. Paraphyses thread-like, refractive. Spores hyaline, fusiform to cylindrical, mostly 7–8-celled, 26–40 × 2.9–5.1 µm. Has a black asexual state, known as *Pseudospiropes simplex,* with dark, 6–12-celled conidia, 29–41 × 7.5–15 µm. Occurs on deciduous wood.

The unusual spores and the asexual state make identification straightforward.

Nemoral–hemiboreal, occasional; all year.

Mollisia retincola is a rather large, yellowish *Mollisia* with a dark outer side that is seated on a thick, dark subiculum. A yellow pigment may dissolve in KOH. The flesh contains imbedded angular crystals. The margin has indistinct hairs with a large inner drop. Asci I+, 78–126 × 5–6 µm. Paraphyses have a long, yellowish, refractive inner drop. Spores one-celled, cylindrical, hyaline and measure (18–)20–26(–28) × 2.8–3.5 µm. Occurs on standing, previous season's culms of *Phragmites*, typically just above water level, rarely on *Carex*.

Mollisia mediella ⋉ has no evident subiculum and 2–4-celled spores. See also page 1436.

Nemoral–hemiboreal, common; mainly April–July.

Phaeangella ventosa is a greyish-yellow inoperculate cup fungus with a paler margin and an almost black outer side – especially at the attachment point. The flesh consists of globose to pear-shaped cells. Asci I+, 8-spored, 110–165 × 5–7.5 µm. Paraphyses cylindrical, approximately 1 µm wide, with refractive contents. Spores cylindrical or ± curved, contain many drops, 1–4-celled and measure 13–17 × 2.5–3.5 µm. Occurs on deciduous wood in damp or wet habitats.

Mollisia olivella ⋉ and *M. olivascens* ⋉ have smaller spores; *M. ramealis* ⋉ has longer spores. Species of *Vibrissea* ◁ have thread-like spores.

Mainly nemoral, occasional; mostly May–September, but can be found year-round.

Bisporella subpallida has yellowish, ± sessile to short-stemmed, 0.5–1.5 mm wide, mostly irregularly formed apothecia. Asci somewhat I+, 50–65 × 4.5–5 µm. Paraphyses narrow, straight and contain small drops. The ellipsoid-fusiform, one-celled spores measure (5–)6–9 × (2–)2.5–3 µm. In crowded groups on deciduous wood or old stromata, and has no obvious association with a dark hyphomycete.

Bisporella pallescens ⋉ is similar but has spores measuring 9–13 × 2.5–4 µm and is clearly associated with the hyphomycete *Bispora antennata* (page 1637).

Nemoral–hemiboreal, rather common; mainly September–February.

Calycina form short-stemmed, ± smooth, mostly yellow apothecia with rather small, mostly 2-celled spores and paraphyses that have a few, large, long drops in the upper part. Many species were formerly treated in *Bisporella* △. See also the downy species on page 1389.

Calycina citrina is a bright yellow, very short-stemmed inoperculate cup fungus that lacks hairs and has a flat hymenium. The flesh is white with an outermost yellow layer that consists of very thick-walled, oblong cells. Asci I+ (greyish to reddish), 80–125 × 7–10 µm, with croziers. Paraphyses cylindrical and have large, oblong drops with yellow, grainy contents. Spores narrowly ellipsoid, contain many drops and measure 9–14 × 3–5 µm; when overmature, 2-celled with a constriction at the septum. Occurs on decorticated wood of deciduous trees, mostly on branches, particularly *Fagus*.

Bisporella confluens ⋉ (belongs in *Calycina*) apparently has somewhat larger spores that may become more than 2-celled when overmature. *Calycina claroflava* ▷ is bright lemon- to sulphur-yellow, somewhat smaller and occurs in smaller groups. *Bisporella subpallida* △ and *B. pallescens* ⋉ are paler yellow and have somewhat smaller spores.

Widespread, common to very common; mainly August–March.

Calycina claroflava is a rather small, bright lemon-yellow to sulphur-yellow, almost smooth *Calycina*. The outside has short, crowded, thick-walled, projecting cells. Asci I- or slightly I+, 120–140 × 6.5–7.3 µm when alive (smaller when dead), with or without croziers. The paraphyses are filled with bright yellow drops. The 2-celled spores measure (7.3–) 7.7–11.1(–11.8) × (2.2–)2.4–2.9 (–3.1) µm. Occurs on rotten bark of deciduous trees, mostly in connection with dead pyrenomycetous fungi and other fungi, and often accompanied by a '*Chalara*'-state (with dark phialides), which is presumed to be its asexual state.

Also known as *Bisporella sulphurina*. *Calycina citrina* ◁ has larger, slightly duller apothecia in greater aggregations on decorticated wood and has bigger spores. *Calycina scolochloae* ▽ occurs on *Phragmites* and bamboos (*Poaceae*). Species of *Hamatocanthoscypha* (page 1386) have very small, hooked hairs at the margin.

Mainly nemoral, occasional; mainly August–December.

Calycina scolochloae has small, ± smooth, vivid yellow apothecia in connection with a dark subiculum/ hyphomycete (a so-called '*Chalara*'-state). Asci I-, lack croziers. Spores 2–4-celled, contain many drops and measure 10–22 × 2.5–4 µm. Occurs on stems of *Phragmites* and bamboos.

Calycina claroflava △ has 2-celled, smaller spores. A similar fungus is known from *Vaccinium* branches. *Rodwayella citrinula* ⋉ may occur on the same substrate but has a brownish outer side and the asci are I+ and have croziers.

Nemoral, probably rather rare, but may be overlooked; mainly July–October.

Hymenoscyphus epiphyllus is a large, vivid yellow to yellow-orange, ± reddening, short-stemmed *Hymenoscyphus*. Hyphae in the flesh not glassy. Asci I+, 90–120 × 8.5–10 µm. Paraphyses cylindrical, ± branched, 2–3 µm wide, with drops. Spores 1–2-celled, ± fusiform, with many drops, 11–20(–24) × 3.5–5.5 µm. On litter, mostly from *Quercus*, but also from conifers, rarely on wood.

Hymenoscyphus caudatus ⋉ has larger spores with a tapering end. *Hymenoscyphus lutescens* ⋉, on *Pinus* cones, has smaller spores. See also pages 1421 & 1433.

Widespread and probably very common; mainly July–October.

Hymenoscyphus imberbis is a short-stemmed, whitish to yellowish, later discolouring clay-pink *Hymenoscyphus*. Flesh consists of oblong, 5–10 µm wide, non-glassy cells. Asci I+, 65–85 × 5–7.5 µm, with croziers. Paraphyses cylindrical, widening to 1.5–3 µm at the tips, with many drops. The ellipsoid to fusiform, 1–2-celled spores measure 9–12 × 3–4 µm. Occurs on deciduous wood or other woody litter in damp or wet habitats.

Most similar species have larger spores. *Hymenoscyphus epigaeus* ⋉, on soil amongst mosses, has longer asci. See also the larger species page 1421.

Widespread, common; April–November.

Phaeohelotium monticola is a rather large, ± sessile, turbinate, vivid orange, non-reddening *Phaeohelotium*. Flesh consists of ± spherical cells; surface has small clavate protuberances. The asci are I-/+, measure 90–110 × 9–12 µm and have croziers. The cylindrical paraphyses contain many drops. Spores 1–(2)-celled, contain large and small drops, ± fusiform, smooth, hyaline and measure 12–20 × 3–5 µm. Occurs on deciduous wood and woody fruits or on soil.

There are macroscopically similar inoperculate cup fungi on soil, but with completely different spores.

Mostly nemoral, apparently rare; mainly August–October.

Calloria neglecta is an orange, erumpent, sessile inoperculate cup fungus. The flesh consists of ± spherical cells. The asci are I-, 8-spored, measure 70–130 × 8–13 µm, rather thick-walled, ± clavate and have croziers. Paraphyses are unbranched or somewhat branched, and have clavate, ± curved tips. The almost allantoid, ellipsoid to shoe-shaped, 1–2(–4)-celled spores contain some small, hardly visible drops, measure 11–15 × 3–4 µm, and have a loosening outer wall. The more abundant asexual state, known as *Cylindrocolla urticae*, is more cushion-shaped and darker orange (lower image), and has ± cylindrical conidia in chains. Occurs on dead, standing stems of *Urtica*.

May superficially recall the asexual state of a species of *Dacrymyces* (page 1154), but occurs on a somewhat unusual substrate for such basidiomycotes. A few further species of *Calloria* occur on other herbaceous substrates, *e.g. C. galiorum* ⋉ on *Galium*.

Nemoral–hemiboreal, very common but becoming scarcer towards the north; mainly March–June.

Zythia resinae is an orange, ± urn-shaped, ± marginally toothed to floccose inoperculate cup fungus. The asci are clavate, have thick, I+ walls, lack an apical apparatus and are filled with small, almost spherical spores approx. 2 µm in diameter. Also found in an asexual, pycnidial state with spherical, bright yellow conidia. Occurs on resin on living conifers, mostly *Picea*, but also *Pinus* and *Larix*.

A very distinctive species. See also the black *Sarea difformis* ◁◁.

Widespread, rather common; all year.

Pezicula form grey-olive, yellowish-brown to brownish-orange apothecia, which mostly develop on an erumpent stroma. Spores large and multicellular. All occur on bark.

Pezicula ocellata has erumpent, ± solitary apothecia with a distinct, floccose, white margin and an apricot-orange to orange-brown hymenium. The asci are I-, mainly 8-spored, measure 90–150 × 19.5–36 µm, widest below the middle, and have croziers. The paraphyses are narrowly cylindrical and have yellowish contents. The ellipsoid, 1(–5)-celled spores measure 23–42 × 9.5–17.5 µm. Occurs on recently dead, corticated deciduous branches, mainly *Salix*.

Formerly known as *Ocellaria ocellata*. The substrate and the white-rimmed margin should make identification straightforward. The alpine *P. aurantiaca* ⋉, on *Alnus*, has a darker margin and less vivid hymenial colours. *Karstenia rhopaloides* (page 1470) has smaller, deeply immersed apothecia and different spores.

Widespread, occasional; probably all year.

Pezicula carpinea has erumpent stromata upon which clusters of yellow, mealy apothecia develop. The asci are I+ (red to indistinctly blue), 8-spored, measure (150–)190–260 × 22–30 µm and have croziers. The tapering, ellipsoid spores are filled with greenish oil drops and measure 20–31 × 8–12 µm; Qav. 2.1–2.4. An asexual state may occur on the stromatic tissue. Found on attached branches of *Carpinus*, rarely *Fagus*.

Other species occur either on other hosts or have a Qav. above 2.7.

Nemoral, occasional; almost all year.

Pezicula frangulae is erumpent either as solitary apothecia or in clusters. Apothecia pale grey to olive, blackening. Asci 4-spored, 65–160 × 11–18 µm, I+ (reddish to greyish). Paraphyses yellowish, cylindrical with somewhat wider tips, ± branched. The ellipsoid, hyaline to yellowish, at maturity typically 2–6-celled spores measure 17–25 × 6–10 µm. Occurs on recently dead, corticated, attached branches of *Frangula alnus*.

The combination of 4-spored asci, dull colours and the host characterize this species.

Nemoral–hemiboreal, probably common; most of the year.

Pezicula cinnamomea has mainly clustered, ± cinnamon-coloured, convex apothecia. The asci are I+ (red and KOH blue), mostly 8-spored and measure 90–135 × 15–20 µm. Spores ellipsoid, ± curved, 2, 4 or 6-celled, hyaline to yellowish and measure 16–30 × 6–10 µm; Qav. 2.7–3.4. Occurs on deciduous or coniferous wood or on cones, *e.g.* of *Pinus*, *Picea*, *Larix*, *Thuja*, *Abies*, *Quercus*, *Corylus*, *Alnus*, *Aesculus* and *Fagus*.

Pezicula corticola ⋉, on *Pyrus* and *Malus,* is very similar, but *P. cinnamomea* seemingly does not occur on these hosts.

Widespread, rather common; most of the year.

Pezicula rostrupii is a flat-topped to somewhat convex, red-brown *Pezicula* with a distinct margin and ± solitary apothecia on indistinct stromata. Asci 8-spored, 132–195 × 19–26 µm. The spores have a very variable septation, old spores may even have both transverse and longitudinal septa (muriform); they measure (28–)30.7–36(-37.5) × (8.5–)8.9-11.1(–12) µm; Qav. 3.3. Occurs on recently dead, corticated *Fraxinus* branches.

Pezicula aesculea ⋉, usually on *Acer*, is very similar but apparently forms more conspicuous stromata.

Distribution and frequency poorly known but likely mainly nemoral; probably most of the year.

Orbilia form small, semi-translucent, mostly lens-shaped apothecia. Asci always I-, usually have a truncate top, basally mostly bifurcate but lack croziers; paraphyses have swollen tips. Spores small and variable; they contain an internal refractive body at one end and the shape of this is crucial for identification. Occur on wood or less frequently on herbs or old fungi. Many species tolerate desiccation.

The intrahymenial asscomycote parasite *Helicogonium orbiliarum* ⋉ occurs in some of the species.

Orbilia xanthostigma is a bright yellow to orange, rarely white *Orbilia*. The margin is sharp or crenulate. Flesh comprises angular to ± spherical cells. Asci 8-spored, truncate, 25–45 × 2.5–4 µm. Paraphyses ± wider at tips. Spores 'C'-shaped, verrucose, 2.5–3 × 1–1.5 µm and contain a spherical spore body. The asexual state produces 'Y'-shaped, septate, hyaline conidia. Gregarious on deciduous and coniferous wood, mostly large decorticated trunks, often with *e.g. Mollisia cinerea* ◁◁, *Nemania* (page 1564) and *Chaetosphaeria* (page 1598).

Also known as *O. delicatula*. The combination of 'C'-shaped, verrucose spores and substrate make identification easy.

Nemoral–hemiboreal, common to very common; mainly May–November.

Orbilia comma is a ± apricot to vinaceous-brown *Orbilia*. The asci are truncate and measure 40–51 × 4–5 µm. Paraphyses 1–2.5 µm wide at the tips, and either have or lack a waxy layer. The comma-shaped to tadpole-shaped spores measure 7–8 × 1.5–2 µm and contain a spherical spore body. Conidia large, worm-shaped, septate. Mostly on loosening *Ulmus* bark.

Orbilia carpoboloides ▷▷ is dentate at the margin, the spores are less distinctly comma-shaped and the spore body is elongated.

Widespread, but has declined due to Dutch Elm Disease; mainly October–May.

Orbilia sarraziniana is a ± pinkish, desiccation intolerant *Orbilia* with a smooth margin. The asci measure 38–45 × 3.8–4.8 µm. Paraphyses smooth or have incrusted material at the tips. Spores (5.5–)6–9 × 1.3–1.6 (–1.8) µm, almost straight, with a blunt and a pointed end; the pointed end contains an elongated spore body with a small head. Conidia cylindrical, septate. Occurs on wood in damp habitats, *e.g.* in *Alnus* carrs.

Orbilia vinosa ⋉ has longer spores; *O. auricolor* ⋉ has curved, very narrow spores; and *O. luteorubella* ⋉ has more fusiform, somewhat larger spores.

Nemoral–hemiboreal, rather common; mainly May–October.

Orbilia rectispora is an almost colourless to yellowish *Orbilia*. Asci 29–37 × 3–3.5 µm, long-stemmed. Paraphyses have capitate tips imbedded in waxy matter. Spores almost straight, but with a tendency to be comma-shaped, and measure (5.5–)7–9.2 × 0.6–1.7 µm; contain a round spore body in the wide end. Conidia (3–)4–6(–8)-celled, 37–67 × 5.3–7.8 µm. Occurs on coarse herbaceous stems in wet habitats.

The combination of distinctive habitat and the spore shape and size should make identification easy.

Nemoral, rarely reported; mostly May–July.

Orbilia dryadum is an *Orbilia* with very pale, crowded apothecia. Asci measure 29–41(–55) × 3.2–4 µm, 8-spored, top truncate. The swollen paraphysis tips are imbedded in a thin sticky layer. The somewhat allantoid spores measure 2.7–4 × 1–1.3 µm and contain a ± globose spore body at one end. Occurs on rotten wood and dead polypores.

The very pale apothecia and small spores make this species rather distinctive. *Orbilia epipora* ⋉ is similar but has 0.8–1 µm wide spores.

Widespread, rarely reported; mainly May–October.

Orbilia carpoboloides is an erumpent, pink, reddish-grey, orange-yellow to red-brown *Orbilia* with a pale, dentate margin. The asci are 8-spored and measure 80–105 × 5–6.2 µm. The paraphyses have widened, lance-shaped tips. The somewhat comma-shaped to fusiform spores measure 10.5–16 × 2–2.7 µm; the spore body is worm-like and has a small head at the free end. Occurs on dead branches of deciduous wood, mostly *Ulmus*.

Also known as *Habrostictis rubra*. *Orbilia occulta* ⋉ has spores with a long narrow end, a broadly drop-shaped spore body and rounded paraphysis tips.

Nemoral, rare, and has probably declined due to the loss of elms as a consequence of Dutch Elm Disease; most of the year.

Orbilia aprilis is a small, ± dingy apricot-orange *Orbilia* with a smooth to ± distinctly dentate apothecial margin. The teeth consist of glassy hairs. Asci 8-spored, flat-topped. Paraphyses have rounded heads in a waxy layer. Spores fusiform, measure (6–)6.5–9.5(–10.5) × (1.5–)1.6–2 (–2.1) µm, and contain a fusiform spore body at the top up to 2.2 µm long. Asexual state has large, 3-armed, hyaline conidia with 2–3 cells in each arm. Occurs on wood, *e.g.* dead, attached deciduous and coniferous branches or in old beetle cavities.

Orbilia vinosa ⋉ has longer spores.

Nemoral, rarely reported; most of the year.

Orbilia crenatomarginata is a small, very translucent, ± rose *Orbilia* with a finely dentate margin. The asci are 8-spored and measure 28–35 × 3 µm. Paraphyses clavate, tips imbedded in a waxy layer. Spores worm-like/sinuous, taper towards one end, measure 6–9 × 0.5–0.8 µm, and contain a short, cylindrical spore body at the top. Conidia 4–8-celled, 31–50 × 6.8–12 µm. Occurs on rotten bark of deciduous trees, *e.g. Tilia* and *Salix*, in damp habitats.

No similar *Orbilia* species have such distinctive spores.

Nemoral, rarely reported; mostly during winter.

Hyalorbilia form small, semi-translucent, cupulate to lens-shaped, pale apothecia. Asci have a rounded top, I-. Paraphyses wide, ± cylindrical, with croziers. Spores egg-shaped to oblong, usually have drops but lack spore bodies.

Hyalorbilia inflatula is an almost cup-shaped, sessile to very short-stemmed, pale yellow or straw-yellow, translucent *Hyalorbilia* that usually has hyphal strings at the base. Flesh consists of angular cells. Asci 8-spored, stemless, with croziers, imbedded in waxy material. Paraphyses almost cylindrical, same width and length as the asci. Spores cylindrical, straight or somewhat curved, have some small, spherical drops at the ends, 5–7 × 0.8–1 µm. Occurs on rotten wood of deciduous trees, mostly with other fungi, *e.g.* ascomycote stromata.

Species of *Orbilia* △ ◁ have long-stemmed, flat-topped asci. *Hyalorbilia berberidis* ⋉ and *H. fusispora* ⋉ have longer spores.

Widespread, common, becoming scarcer towards the north; mainly March–November.

Erumpent inoperculate cup fungi

A form group of inoperculate cup fungi that develop fruitbodies (apothecia) within the substrate (typically in stems, leaves or wood). At maturity the apothecia breaks through the surface (erumpent). They may be disc-shaped or elongated and are often surrounded by torn parts of host tissue, which is sometimes evident as distinct teeth or 'lips'. The teeth may involve both host and fungus tissue. Many species have a black, almost carbonous outer side.

OTHER SIMILAR FUNGI:

– grades into the sessile, smooth inoperculate cup fungi (page 1426).

– some large, inoperculate cup fungi are also erumpent, but form higher fruitbodies (page 1404).

– hysteroids have a slit-like opening but are hard and the hymenium is rarely exposed (page 1614).

See also the general wheel, page 1348.

FURTHER READING: 18, 21, 49, 78, 139, 215, 297, 298.

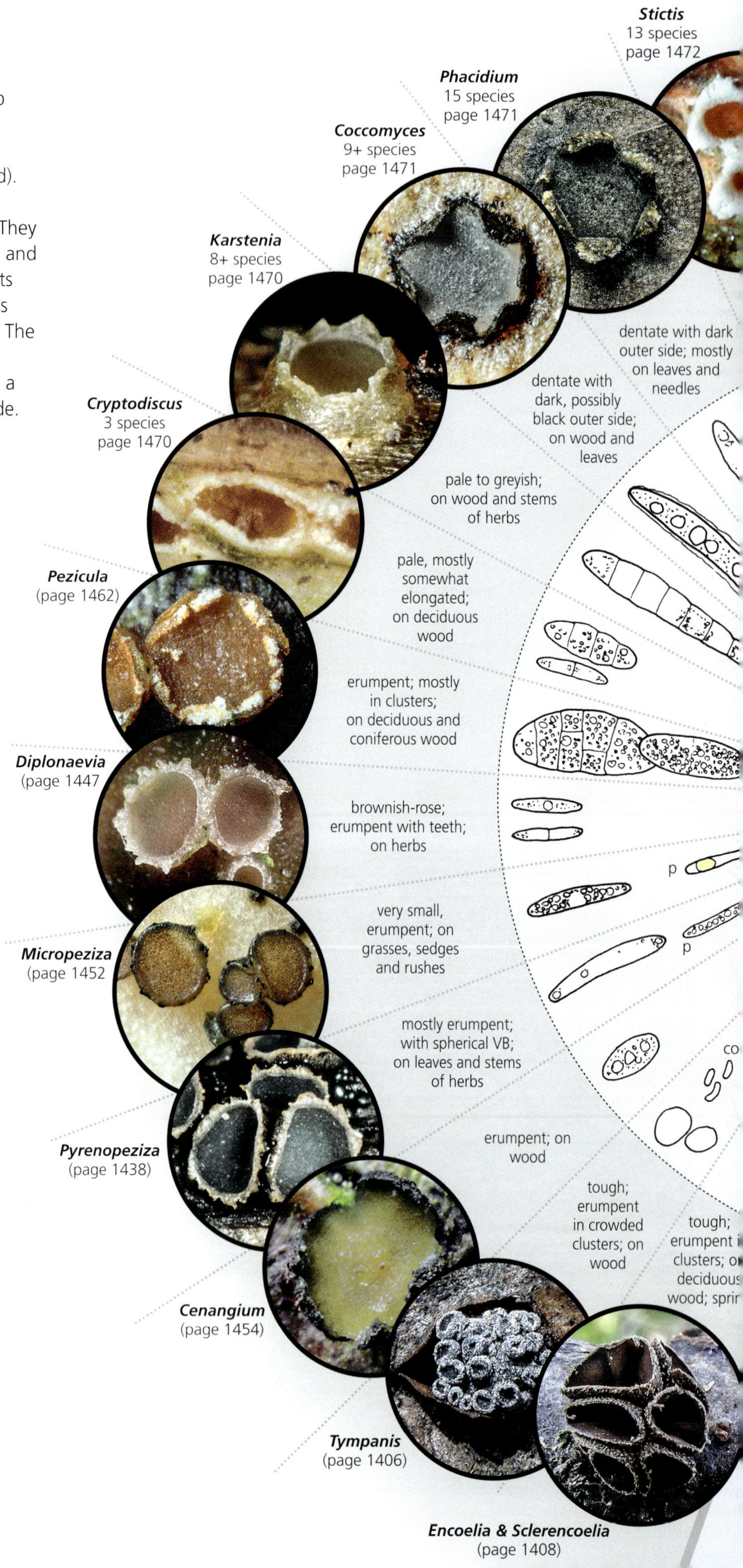

Micro-drawings: spores first, followed by conidia (con), paraphyses (p), refractive vacuolar bodies (VB), asci (a) and ascus tops (at)

Approximate species number applies to temperate Europe.

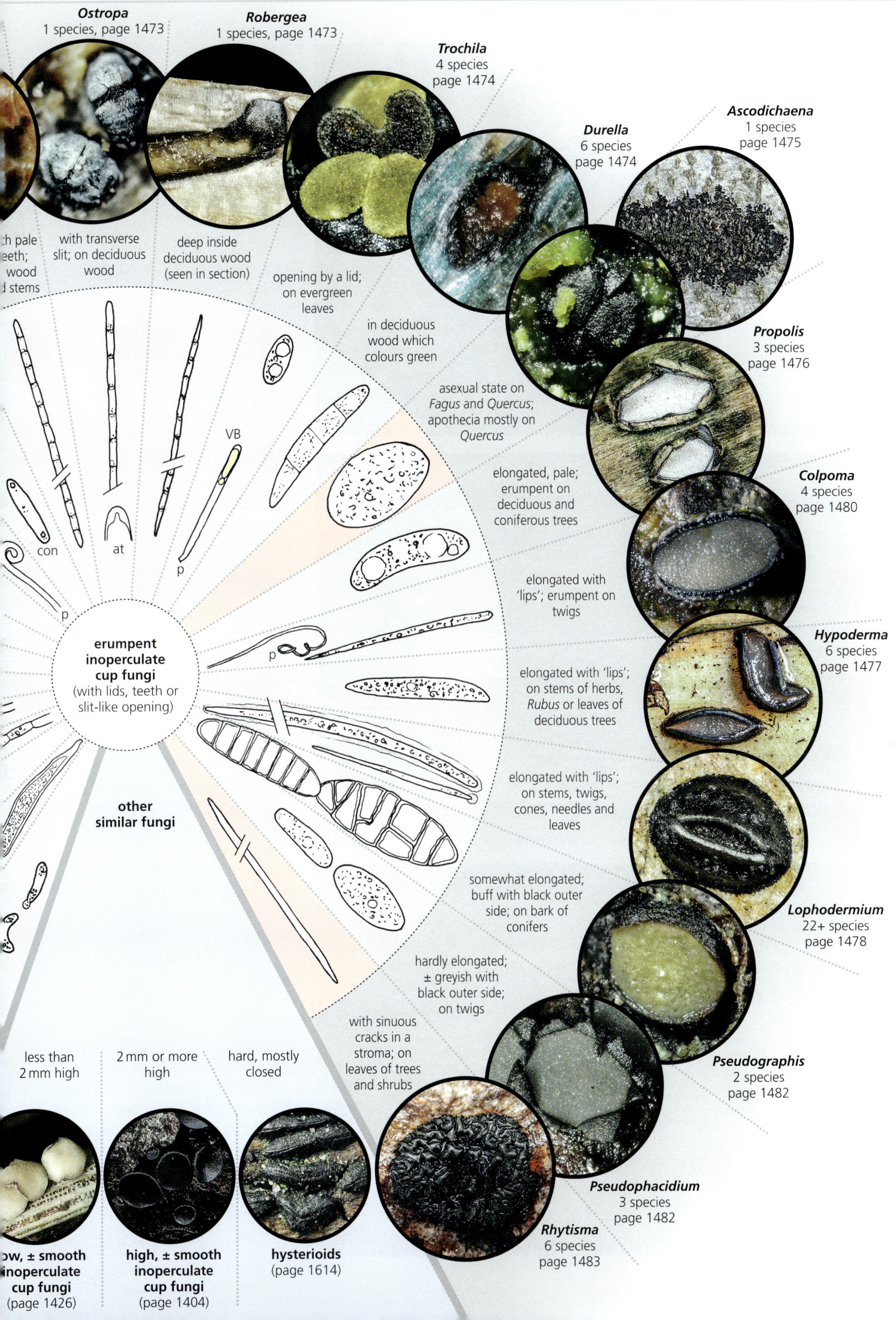

Ostropa
1 species, page 1473
Robergea
1 species, page 1473
Trochila
4 species
page 1474
Durella
6 species
page 1474
Ascodichaena
1 species
page 1475
Propolis
3 species
page 1476
Colpoma
4 species
page 1480
Hypoderma
6 species
page 1477
Lophodermium
22+ species
page 1478
Pseudographis
2 species
page 1482
Pseudophacidium
3 species
page 1482
Rhytisma
6 species
page 1483
:h pale
eeth;
wood
l stems
with transverse
slit; on deciduous
wood
deep inside
deciduous wood
(seen in section)
opening by a lid;
on evergreen
leaves
in deciduous
wood which
colours green
asexual state on
Fagus and Quercus;
apothecia mostly on
Quercus
elongated, pale;
erumpent on
deciduous and
coniferous trees
elongated with
'lips'; erumpent on
twigs
elongated with 'lips';
on stems of herbs,
Rubus or leaves of
deciduous trees
elongated with 'lips';
on stems, twigs,
cones, needles and
leaves
somewhat elongated;
buff with black outer
side; on bark of
conifers
hardly elongated;
± greyish with
black outer side;
on twigs
with sinuous
cracks in a
stroma; on
leaves of trees
and shrubs
VB
con
at
p
p
p
erumpent
inoperculate
cup fungi
(with lids, teeth or
slit-like opening)
other
similar fungi
less than
2 mm high
2 mm or more
high
hard, mostly
closed
ow, ± smooth
inoperculate
cup fungi
(page 1426)
high, ± smooth
inoperculate
cup fungi
(page 1404)
hysterioids
(page 1614)

Karstenia rhopaloides is an inoperculate cup fungus that initially is completely immersed in the substrate as a closed white stroma, but which later bursts through forming ± obvious teeth. The hymenium is cream to greyish and deeply immersed; it stains blue with iodine after KOH pretreatment. The asci are I+ and measure 70–100 × 10–15 µm. Paraphyses cylindrical to clavate at the tips. The spores are 7–10-celled, hyaline, clavate and measure 27–43 × 4–6 µm. Occurs on bark of deciduous trees and scrub, rarely in conifer bark.

Also known as *Cryptodiscus rhopaloides*. *Karstenia clematidis* ⋉ has narrower spores measuring 35–45 × 3–4 µm; *K. lonicerae* ⋉ apparently has spores that measure 19–35 × 3–4.2 µm, and *K. ideae* ⋉ has ± cylindrical, (7–)12–16-celled spores that measure (32–)48–80 × 3–5.2 µm. An undescribed species, provisionally named *K. guttulata* ⋉, can be found on *Quercus* bark on living trees; it has large, almost cylindrical, multi-guttulate, multi-septate spores. Species of *Cryptodiscus* ▽ typically produce somewhat oblong apothecia (in outline) and have smaller, 2–4-celled spores

Widespread, common; mostly during winter.

Cryptodiscus foveolaris is a deeply immersed, pale buff, in outline somewhat elongated inoperculate cup fungus with a pale margin. The asci are I- and measure 50–65 × 4–5 µm. Paraphyses unbranched, swollen at the tips. The hyaline, ellipsoid, 2-celled spores measure 6–9 × 3–6 µm. Occurs on wood of deciduous trees.

Cryptodiscus pallidus ⋉ has large, 10–17 × 3–6 µm, 4-celled spores. *Karstenia rhophaloides* △ may look very similar but the spores are 7–10-celled and measure 27–43 × 4–6 µm.

Widespread, rare or overlooked; most of the year.

Coccomyces tumidus is an inoperculate cup fungus with a brown, fleshy lid that forms teeth when it bursts open. The apothecia are 2–3 mm wide and the hymenium is cream to buff. The asci are I- and measure 160–240 × 12–16 µm. Paraphyses have bent to twisted tips. The spores measure 35–50 × 3.5–4.5 µm and contain 8–10 large drops. Occurs on dead leaves of *Betula*, *Fagus* and *Quercus*.

There has been and still is confusion over both taxonomy and nomenclature within *Coccomyces* that most likely is a polyphyletic genus. *Coccomyces coronatus* ⋉ has somewhat smaller apothecia, a more carbonous and irregularly cracking lid and longer spores measuring 60–80 × 2–2.5 µm. *Coccomyces dentatus* ⋉ is also smaller and carbonous, but has a star-like opening and spores that measure 45–65 × 2.5 µm. *Rhytisma salicinum* ▷▷ is closely related to some species of *Coccomyces* as currently defined

Widespread, occasional; April–October.

Phacidium lauri is a black, circular, inoperculate cup fungus with four to six teeth at the margin. The hymenium is grey-brown. Asci I+, 80–115 × 7–10 µm. Paraphyses cylindrical, wider at the tips (up to 3 µm). The ellipsoid to clavate, one-celled spores measure 8–11 × 3–5 µm. Asexual state recalls a volcano with a paler opening (to the right on image); conidia hyaline, ± cylindrical, 12–15 × 2–3 µm. On leaves of *Ilex*, *Prunus laurocerasus* and *Vinca*.

Also called *P. multivalve*. *Phacidium pseudophacidioides* ⋉, on *Ilex*, has smaller conidia, (11–)11.5–12.5(–13) × (2–)2.5 µm.

Widespread, rarely found in the sexual state, but common in the asexual state; most of the year.

Stictis, **Ostropa** & **Robergea** belong in a group of immersed inoperculate cup fungi that have long, thread-like spores. *Stictis* species reveal a hymenium at maturity, while *Ostropa* and *Robergea* hardly open.

Stictis radiata is a 0.4–1.2 mm wide, immersed inoperculate cup fungus with a star-shaped opening and a pale buff to brownish-orange hymenium. The mealy margin is chalk-white. The sterile cells (periphyses) on the inside of the rays are branched and measure 15–30 × 1–1.5 µm. Hymenium distinctly I+, rarely I-, and has a well-defined crystalline layer. The asci are I-, measure 120–250 × 5.5–6(–8) µm and have a distinct, high and very refractive apical apparatus. Paraphyses thread-like, somewhat branched above. Spores hyaline, multicelled, contain many drops and measure 120–250 × 2–2.5 µm. Occurs in bark, large stems of herbs, needles, and even found on old leather.

Stictis stellata ⋉ is apparently separated by having unbranched periphyses and somewhat larger apothecia. There are other similar species, and exact determination is difficult in this genus.

Probably widespread, occasional; most of the year, except in cold winter periods.

Stictis arundinacea is a small *Stictis* with a ± entire apothecial margin. The asci are I-, measure 150–200 × 5–6 µm and have a high, refractive apical apparatus. Paraphyses thread-like, somewhat wider at the tips. Spores thread-like, measuring 150–180 × 1.5 (–2) µm. Occurs on herbaceous stems, including various monocots.

Stictis graminicola ⋉ also occurs on grasses but the asci are around 130 µm long and the spores about 100 µm long. *Stictis pusilla* ⋉ has even shorter spores. The paraphyses of both species have head-like, swollen tips.

Widespread, but poorly known; probably most of the year.

Ostropa barbara is a rather aberrant inoperculate cup fungus with two thick, white to greyish 'lips' and a dark slit down to the greyish, deeply immersed hymenium (inset image shows longitudinal section). Asci I-, cylindrical, 300–500 × 6–10 µm, and have a distinct apical apparatus. Paraphyses branched at the tips. Spores thread-like, approx. 200 × 2 µm. Occurs on bark and wood of attached deciduous branches, *e.g. Salix*, *Populus* and *Fraxinus*.

Difficult to spot but easy to identify.

Apparently northern-continental within temperate Europe and not recorded in *e.g.* Denmark and the Netherlands; all year.

Robergea cubicularis is a deeply immersed, pyrenomycetous inoperculate cup fungus with an almost cylindrical fruitbody that is positioned ± in parallel with the substrate (usually bark) surface; it is only visible as a pale, 0.25 mm wide disc with a tiny, slit-like opening. The asci are I-, measure 800–2000 × 8–9(–10) µm and have a cap-like top. The paraphyses are thread-like. The spores are multicelled and thread-like, measuring 800–2,000 × 1.5(–2) µm, and are twisted within the asci. Occurs on twigs of deciduous trees, *e.g. Fraxinus*, but also found in rotten grass leaves.

A highly distinctive species but very hard to detect and therefore easily overlooked.

Widespread in the nemoral–hemiboreal zones, but very rarely reported; probably most of the year.

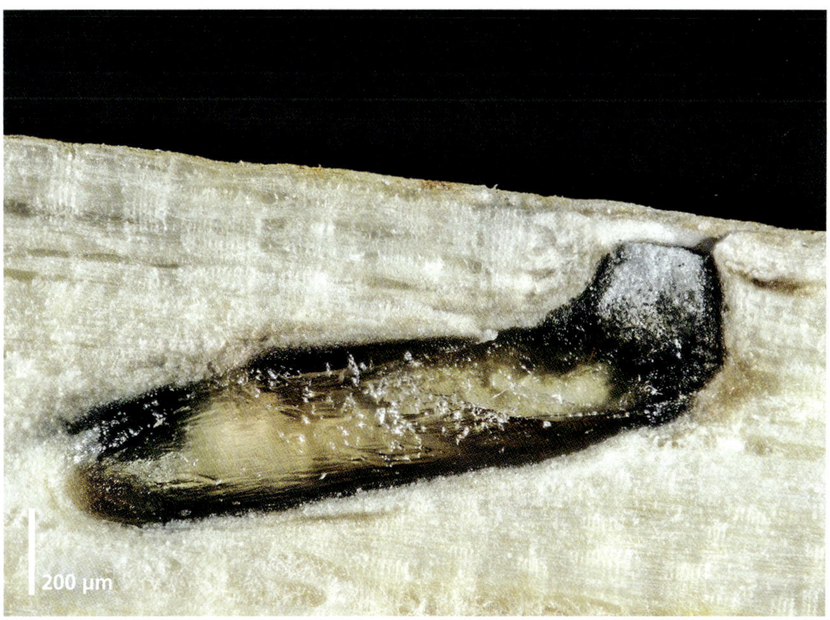

Trochila ilicina is a small inoperculate cup fungus with a greenish-yellow, cushion-shaped hymenium that is revealed when a dark lid is pushed aside at maturity. The asci are I+, somewhat clavate, measure 96–120 × 10–13.5 µm, and have croziers and a high, apical apparatus. The paraphyses have refractive greenish-yellow vacuoles in the somewhat wider tips. The ellipsoid-fusiform spores measure 11.5–13.5 × 5–5.5 µm (when living), and contain both large and small drops. Occurs on leaves of *Ilex aquifolium*.

There are no similar fungi on the leaves of this host. *Trochila laurocerasi* ⋉ occurs on leaves of *Prunus laurocerasus* and *T. craterium* ⋉ on leaves of *Hedera*.

Mainly nemoral, common; all year, but in best condition during the summer.

Durella atrocyanea is an, in outline, ellipsoid, immersed inoperculate cup fungus with a brown hymenium and a black, ± dentate margin. Asci I- and measure 78–122 × 10–13 µm. The tips of the paraphyses are imbedded in a brownish matter that dissolves in KOH. The ellipsoid, 4–6-celled spores measure 15–23 × 5.5–7 µm. Occurs on decorticated wood and woody herbaceous stems that are stained blue-green by the fungus.

Easy to identify due to the blue-green stain of the substrate. Should probably be transferred to *Xylogramma* ⋉.

Mainly nemoral–hemiboreal, probably rather common; mostly November–May.

Ascodichaena rugosa is a very distinctive species that mostly occurs as asexual black, erumpent crusts (lower image), from where large, 15–30 × 10–16 µm, hyaline, somewhat thick-walled, ellipsoid conidia are produced on simple conidiophores. During the summer the sexual state, which recalls a small, approx. 0.3–0.5 mm wide, black mouth can be found. The asci are widest at the centre and open by a slit. The ellipsoid, hyaline, one-celled spores contain many drops and measure 18–24 × 13–16 µm. Occurs on living *Fagus* bark and the bark of rather thin, living *Quercus* branches. The sexual state is most easily found on *Quercus*.

The interaction between *Ascodichaena rugosa* and the host is not clear. Electromicroscopical studies have revealed that the fungus (at least on *Fagus*) forms haustorium-like hyphae that penetrate the living bark cells walls. Whether the fungus simply acts as a biotrophic parasite or whether the relationship is of a more complex nature is unclear. Seemingly, to a certain degree, it protects *Fagus* against attack by *Cryptococcus fagisuga* (Woolly Beech Scale), an insect in the *Coccoidea* that is responsible for Beech Bark Disease, and in this way may limit the likelihood of a severe infestation.

Although it is distinctive, *A. rugosa* could initially be mistaken for a lichenized fungus.

Widespread in the nemoral–hemiboreal zones, very common; all year round, but occurs in a sexual state in the summer.

Propolis is an, in outline, oblong, erumpent inoperculate cup fungus with a mealy hymenium caused by a layer of branched paraphysis tips. Mostly occurring on decorticated, sun-exposed wood. Also known under the name *Propolomyces*.

Propolis farinosa is a ± immersed, erumpent, irregularly shaped inoperculate cup fungus that usually has flakes from the substrate at the apothecial margin. It can withstand desiccation. The hymenium is whitish and mealy. The asci are I-, 8-spored, measure 130–150 × 12–18 µm and have croziers. Paraphyses irregularly branched above and have brown incrusted material. The ± allantoid, one-celled, hyaline spores contain 2(–4) large drops, many small drops and measure 20–37 × 5–9 µm. Occurs in decorticated, dry and exposed, mostly deciduous wood, rarely in coniferous wood or *Pinus* cones.

The broader-spored *P. leonis* ⋉ occurs on *Pinus* bark. *Propolis viridis* ▽ has grey-green apothecia.

Widespread, common; all year.

Propolis viridis is a ± grey-green *Propolis*. Both hymenium and teeth are usually tinged green but occasionally only the teeth. The asci are I- and measure 75–103(–115) × 12.5–18 µm. Paraphyses branched. Spores ellipsoid to somewhat allantoid, contain 2–3 drops and measure 14–21(–26) × 6.5–8.5 µm (when living). Occurs on coniferous and deciduous wood and cones.

Propolis farinosa △ has spores that are more allantoid, and lacks green tinges on the hymenium and teeth.

Mainly occurs in the south of temperate Europe, not known in the hemiboreal–boreal zones; all year.

Hypoderma produce black, long and sinuous, mouth-like apothecia that when fresh and damp may have bluish 'lips'. Occurs on various herbs and species of *Rubus*. The spores are narrow, ± banana-shaped and produced in clavate, inamyloid asci.

Hypoderma rubi is an, in outline, elongated *Hypoderma* with black apothecia up to 10 × 0.8 mm. It opens with a rather narrow slit to reveal a ± olive-brown to greyish hymenium. The asci are I-, clavate and measure approx. 120 × 10 µm. The paraphyses have twisted tips. The ± cylindrical, hyaline, one-celled spores contain many drops and measure 20–25 × 3–4 µm. Has an asexual, shield-like '*Leptostroma*'-state. Conidia 4–6 × 1 µm. Occurs on stems of *Rubus*.

The substrate should make this species easy to recognize. Some mycologists treat *H. rubi* as a species with a much broader host range.

Mainly nemoral–hemiboreal, probably common; mostly September–October.

Hypoderma commune is a 1–2.5 mm long, black *Hypoderma* that opens with a broad slit to reveal a grey to greenish-yellow hymenium. The asci are I-, clavate and measure 65–80 × 9–12 µm. Paraphyses thread-like, twisted at the tips. Spores 1–2-celled, banana-shaped, contain both large and small drops and measure 17–26 × 3–5.3 µm. Occurs on various dicot herbaceous stems, e.g. *Solidago* and *Epilobium*.

Hypoderma rubi △, on *Rubus*, is larger and has narrower spores. Species of *Lophodermium* ▷▷ on similar substrates are smaller and have longer, narrower, 1-celled spores.

Widespread, probably occasional; mainly May–September.

Species of ***Lophodermium*** are black, slit-like to ellipsoid, ± mouth-shaped, inoperculate cup fungi, which, when living, have slimy, usually coloured apothecial 'lips'. Occur on various herbaceous and woody substrates. The spores are very narrow, ± cylindrical and usually one-celled. A genus with many, mostly host-specific, species.

Lophodermium pinastri is an ellipsoid, 0.7–1.2 mm long, black *Lophodermium* that opens by a single furrow. The 'lips' may apparently be blue or red. The asci are I- and measure 110–155 × 9.5–11.5 µm. The spores are thread-like and measure 70–110 × 2 µm; they are rounded at one end and pointed at the other. The paraphyses are clavate at the tips. Occurs on needles or cones of *Pinus*; on needles, many black, transverse zones delimit the individuals.

Lophodermium seditiosum ⋉ also occurs on *Pinus* needles but has asci 140–170 mm long and the needles do not usually have transverse black zones. It may damage young *Pinus* trees.

Widespread, very common; most of the year.

Lophodermium oxycocci is a *Lophodermium* with, in outline ellipsoid apothecia that measures up to 0.6 × 0.35–0.5 mm and have a black outer side and a pale, blue margin ('lip'). The asci are I- and measure 39–68 × 4–7 µm. The spores measure 17–25 × 1–1.5 µm. Occurs on leaves of *Vaccinium oxycoccos* in mires and raised bogs.

Other species of *Lophodermium* on *Vaccinium* have longer spores, including *L. melaleucum* ⋉ that apparently also occurs on *V. oxycoccos* and which has spores that measure 33–55 × 1.5–2.5 µm.

Widespread but rarely reported; summer–autumn.

Lophodermium foliicola is a *Lophodermium* with spherical to ellipsoid apothecia that open with a pale slit. The asci are I-, 8-spored, cylindrical to clavate and measure 80–100 × 8–10 µm. Paraphyses thread-like with straight or downcurved tips. Spores 60–70 × 0.8–1.5 µm, thread-like, lack septa. Known from leaves of *Crataegus* and *Berberis*.

Lophodermium hedericola ⋉ occurs on leaves of *Hedera*; *L. neesii* on leaves of *Ilex*; *L. aucupariae* ⋉ on leaves of *Sorbus*; and *L. petiolicola* ⋉ occurs on *Quercus* petioles.

Widespread, rarely reported, but probably overlooked; mainly late spring.

Lophodermium versicolor is a *Lophodermium* with short, black apothecia that open with a broad slit to reveal reddish 'lips'. The asci are I-, 8-spored, cylindrical to almost clavate and measure 30–75 × 5–8 µm. Paraphyses may be straight or curved and form a yellowish protective layer (an epithecium) above the asci. The spores are thread-like and measure 25–45 × 0.75–1 µm. Occurs on *Salix* leaves from the previous season.

Provided the host is obvious, this species should be easy to identify.

Widespread, mostly recorded in the boreal zone, status not well known; spring–summer.

Lophodermium conigenum is a relatively large, 0.9–2.0 mm long, ellipsoid or tapering at both ends, black *Lophodermium* that usually has green 'lips'. Opens with one or rarely several slits. Asci I-, 160–215 × 11.5–14 µm. Spores thread-like, measure 165–215 × 2 µm. Occurs on needles or cones of *Pinus*; when on needles, may form brown transverse zones.

Lophodermium seditiosum ⋉ and *L. pinastri* ◁, both mostly on *Pinus* needles, have shorter asci, and the latter forms conspicuous black transverse zones on the needles.

Widespread but rarely recorded, probably rather common, at least in the UK; probably most of the year.

Colpoma form rather large, ± elongated, slit-like, erumpent apothecia. The spores are long, somewhat curved, and multicelled; the paraphyses have twisted tips.

Colpoma quercinum is an elongated, erumpent *Colpoma* that measures 2–15 × 0.5–3 mm. The outer side is ± grey-brown. At maturity the grey to buff hymenium is exposed in wet weather. The apothecial margin has bluish 'lips'. The asci are I- and measure (130–)150–230 × 8–11 µm. The paraphyses are helically twisted. The spores are 1–multicelled, curved, pointed at the base, measure 45–95 × 1.5–2 µm and have a slime coat. It has an asexual pycnidial state '*Conostroma didymum*'. Occurs on attached twigs of *Quercus*.

No similar fungi occur on attached *Quercus* twigs. The somewhat similar *C. ledi* ⋉ occurs on *Rhododendron tomentosum* (syn. *Ledum palustre*).

Nemoral–hemiboreal, very common; most of the year, but at its best in April–June.

May be parasitized by the yellowish jelly fungus *Tremella colpomaticola* ⋉ (see tremellas on page 1193).

Colpoma crispum is an elongated to irregularly shaped *Colpoma* with dark brown to black apothecia. The hymenium and thickened margin are greenish to almost bluish. The asci are I- and measure 95–180 × 6–10 µm. The paraphyses are narrow and downcurved at the tips. The spores are somewhat curved towards the tip, pointed towards the base and measure 40–50 × 2–2.5 µm. Occurs on twigs of *Picea*.

Pseudophacidium piceae ▷▷ is not elongated and slit-like, and has smaller spores.

Hemiboreal–boreal, rarely reported; mainly during warmer part of the year.

Colpoma juniperi is a ± elongated, *Colpoma* with a ± black outer side and blue-green marginal 'lips'. It opens by a slit or more irregularly and can either have or lack recurved teeth. The hymenium is buff to yellow. The asci are I- and measure approx. 125 × 8 µm. The paraphyses are twisted at the tips and imbedded in slime. Spores measure 40–50 × 1–2 µm, one-celled, pointed at the base and have a slime coat. Occurs on twigs of *Juniperus*.

Easily identified based on host and the ± elongated apothecia.

Widespread; mainly during warmer parts of the year.

May be parasitized by *Tremella karstenii* ⋉ (see also page 1191).

Pseudographis pinicola is an immersed-erumpent, black, slightly compressed inoperculate cup fungus with a pale hymenium and a dark apothecial margin. Asci I-, 8-spored, measure 180–250 × 18–20 µm and have croziers. Paraphyses have branched upper parts. Spores are thick-walled, hyaline to yellowish, I+ (violet!), 2–8-celled and measure 2–38 × 7–9 µm. Occurs on bark of conifers.

Pseudographis elatina ⋉ has muriform spore septation. *Podophacidium xanthomelum* (page 1455) is larger and has a serrated margin and smaller, inamyloid spores.

Mainly in the boreal zone, rather common; most of the year.

Pseudophacidium is a genus of inoperculate cup fungi with ± convex hymenia; all species are erumpent from a dark, stromatic layer. Spores rather large and one-celled.

Pseudophacidium piceae is a solitary or clustered, erumpent, black, irregularly splitting inoperculate cup fungus with a grey to brown hymenium and a dentate margin. Asci are I+, 96–118 × 10.5–12 µm. Spores ellipsoid, one-celled, 11–16 × 5–9 µm. Occurs on twigs of *Picea*.

Pseudophacidium ledi ▽ has narrower spores.

Mainly boreal, rarely reported; summer–autumn.

Pseudophacidium ledi is an erumpent, black inoperculate cup fungus that opens to reveal marginal, triangular teeth. The hymenium is pale grey or yellowish to buff. Asci I-, 60–100 × 8–12 µm. Paraphyses thread-like, dissolve into slime. Spores 8–18 × 3–6 µm, narrowly ellipsoid to cylindrical, straight or somewhat curved, one-celled. Asexual state stromatic with an irregularly split opening; conidia (8–)9–11(–12) × (4–)5(–6) µm. Occurs on *Vaccinium uliginosum*, *Rhododendron* and even conifers.

Pseudophacidium piceae △ has wider spores and I+ asci.

Mainly boreal, rarely reported; mainly March–May.

Species of ***Rhytisma***, so-called tar spot fungi, produce black stromata on living leaves in the autumn, in which the fruitbodies are initially formed. The fruitbodies mature and open on dead leaves during wet periods the following spring.

Rhytisma acerinum initially produces small spots on living leaves that at maturity become black, ± wrinkled and 10–25 mm in diameter (lower image). The stromata crack open in an irregular pattern during spring and expose the greyish hymenia (upper image). The fungus attacks living leaves and continues its development after the leaf has died and fallen into the litter. Asci I-, 120–130 × 9–10 µm, clavate, 8-spored. Paraphyses thread-like with twisted tips. Spores thread-like, non-septate, 60–80 × 1.5–3 µm. An asexual state (a so-called '*Melasmia*'-state) can be seen during the summer on attached living leaves (middle image). Occurs on leaves of *Acer pseudoplatanus*, *A. platanoides*, and very rarely *A. campestre*.

Rhytisma punctatum ⋉, also on *Acer*, forms up to 15 mm wide groups of small, approx. 1 mm wide stromata; the attacked parts of the leaf stay ± green and the asci and spores are shorter.

Widespread in the nemoral–hemiboreal zones, very common; all year, but mature apothecia April–June.

Rhytisma salicinum forms 2–7 mm wide stromata that overwinter in the leaf-litter and later crack open in an irregular pattern to reveal a yellowish hymenium. The asci are I- and measure 120–150 × 10–15 µm. The paraphyses are thread-like with downturned tips. Spores are non-septate, thread-like and measure 60–90 × 1.5–3 µm. Occurs on living, later dead *Salix* leaves.

Rhytisma amphigenum ⋉ occurs on *Salix purpurea*. *Rhytisma salicinum* is related to some species of *Coccomyces* (page 1471).

Widespread, rather common; most of the year, but mature during spring.

Thomas Læssøe

Lichens

Lichenized fungi live in a mutualistic symbiosis with green algae and/or cyanobacteria that are able to perform photosynthesis. The algae/cyanobacteria (called the photobiont) provide the fungus with sugars and the fungus affords shelter for the photobiont in the form of a ± well-developed thallus (see page 19). Most species are perennial and slow growing.

Lecanoralean lichens are by far the largest group of lichens, with more than 1,500 species in temperate Europe. They are traditionally studied by lichenologists and are only briefly introduced on the following pages. When sexual, the lecanoralean lichens produce apothecia on a flattened, clavate or branched thallus. They have thick-walled asci with amyloid walls.

The calicioid lichens constitute an aberrant group that form tiny fruitbodies with a dusty spore mass on the top of a small stem (page 1494).

Pyrenolichens form perithecioid fruitbodies and belong to the bitunicate, pyrenomycetous fungi, pages 1545 and 1613.

Basidiolichens are lichenized basidiomycotes. They may be agarics, clavarioids or (in warmer climates) corticioids (pages 130 & 1114).

The most frequent photobiont is the green alga *Trebouxia*, more rarely *e.g.* the yellowish-green alga *Trentepohlia*. *Nostoc*, a genus of cyanobacteria, can also act or co-act as the photobiont, and can benefit the symbiosis by fixing nitrogen from the atmosphere.

Identification of lichens normally involves the use of a series of chemical characters but these are not covered here.

Other similar fungi:
– the inoperculate cup fungi have similar, small, discoid fruitbodies but usually much simpler asci (page 1348).
– *Trizodia* has a not fully understood symbiosis with *Sphagnum* and cyanobacteria (page 1432).
– mycocalicioids recall the calicioids but are not lichenized (page 1497)
– 'bitunicates' and hysterioids may recall crustose (crust-like) lichens (page 1602 & 1614).

Further reading: 302, 316, 355.

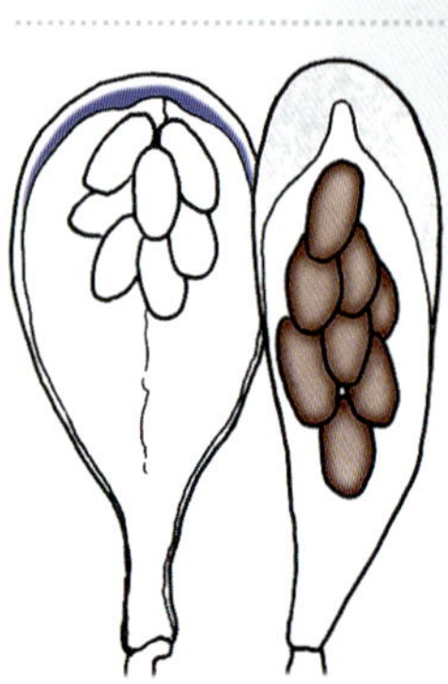

Lecanoralean asci with thick-walled tips.

crustose apothecial lichens
page 1492

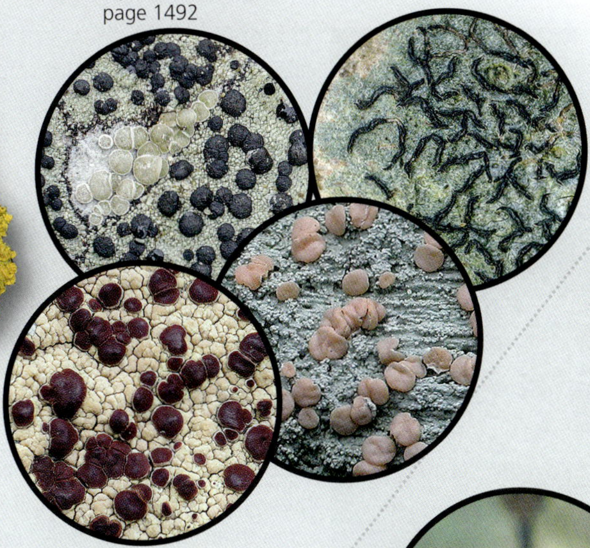

calicioid lichens
page 1494

Ascomycota with leaf-shaped thallus and open fruitbodies (apothecia) when present

Ascomycota with completely adhering thallus and open fruitbodies (apothecia)

Ascomycota with stipitate fruitbodies and passive spore dispersal

Ascomycota with ± closed (perithecioid) fruitbodies

lichens
(fungi in symbiosis with algae and/or cyanobacteria)

pyrenolichens
pages 1545 & 1613

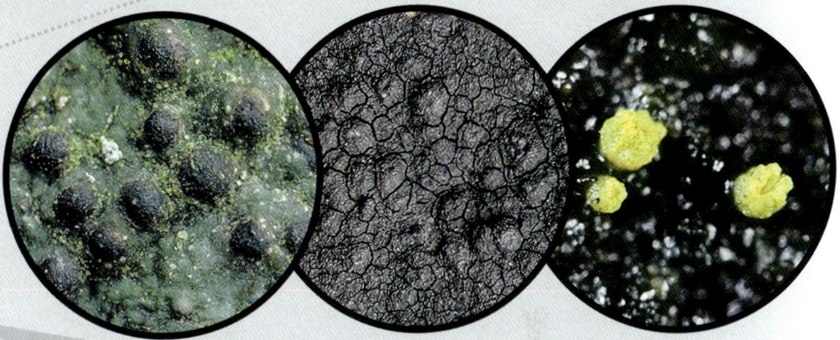

similar, non-lichenized groups

with apothecia

mycocalicioids have active spore release from fruitbodies recalling those of calicioids

with ± closed fruitbodies and active spore release

inoperculate cup fungi
(page 1348)

Trizodia
(page 1432)

mycocalicioids
(page 1497)

non-lichenized 'bitunicates'
(page 1602 & 1614)

Cladonia is a large genus of mostly grey-green, branched or trumpet-shaped secondary thalli (also termed podetia) that usually arise from smaller, scaly primary thalli. The genus is divided into two subgenera: *Cladina* (reindeer lichens) and *Cladonia* (± trumpet-shaped, may have so-called scyphi at the tips). The photobiont is a *Trebouxia*. Species of *Cladonia* are common on poor soil and rotten wood, both in open and woodland habitats.

Cladonia portentosa is a yellow-grey-tinged reindeer lichen. It forms large thalli with thick, finely felty main branches that end in open, ± erect, acute branchlets which branch again 3–5 times in all directions. The lower side of the thallus emits white light in UV-light. Fruitbodies are never produced. Occurs in heath-like habitats.

Cladonia stellaris ⋉ has several heads from the main branches and is more finely branched. The branches off the tips of other similar species have a tendency to be one-sided.

Widespread and very common; all year.

A range of fungi and slime moulds occur on thalli of *Cladonia*, including *Roselliniella cladoniae* ⋉ – a black pyrenomycetous fungus.

Cladonia rangiformis has green-dotted, highly branched podetia and no primary scales. The tips acquire a violet tinge in sunlight. May form small, brown, flask-shaped pycnidia (inset image) and brown apothecia on the tips. Occurs on sandy, mostly calcareous soils, usually coastal.

Cladonia furcata ⋉ stains brown at the tips in sunlight and is more adpressed. *Cladonia scabriuscula* ⋉ is grainy on the surfaces (sorediose).

Widespread, common; all year.

This species is often parasitized by the jelly fungus *Heterocephalacria bachmannii* (page 1196).

Cladonia floerkeana has prominent red apothecia at the tips (inset image) and persistent primary scales on the upright thalli, which do not have cups (scyphi). The surface is never finely mealy. The scales are small and have indentations. In heath-like habitats, on decomposed wood, and on sand in dunes. May dominate after burning.

Cladonia macilenta ⋉ has a finely mealy (sorediose) surface. Other *Cladonia* species with red apothecia have distinct scyphi with fruitbodies at the margin.

Widespread and rather common, but disappearing from many places due to eutrophication and other human-induced changes; all year.

Cladonia fimbriata is a typical trumpet-shaped *Cladonia* with rather wide scyphi (inset image) on top of the unbranched, finely mealy (sorediose) podetia. Margin has brownish pycnidia and, rarely, brown apothecia. Spores hyaline, one-celled. Mostly on rotten wood but has a rather wide ecology.

Cladonia chlorophaea ⋉ is not sorediose; *C. ochrochlora* ⋉ has scyphi that are only slightly wider than the stem; *C. deformis* ⋉ is more yellowish and typically has red pycnidia or apothecia.

Common and widespread; all year.

This species can be parasitized by *Tremella cladoniae* ⋉.

Cladonia gracilis does not have obvious primary scales at the base of the finely white-spotted, crowded, mostly unbranched podetia. The tips are either tapering or have a 'crown' with a serrated margin, from where new crowns may form. Margin has small, brownish apothecia (inset image) and pycnidia. Spores hyaline, one-celled. Occurs on sandy soils or on rocks covered in a thin layer of humus.

Cladonia furcata ⋉ is less erect and does not have crowns, which are also often absent in *C. gracilis*. *Cladonia crispata* ⋉ has perforated scyphi and angled branches.

Widespread, common; all year.

Walter Siegmund (Wikipedia)/Håkan Alfredsson

††***Letharia vulpina*** is a bushy (fruticose), upright to hanging, yellow-green to sulphur-yellow lichen that very rarely forms apothecia. The surface is coarsely mealy (sorediose) and has pin-like protuberances (isidia). The photobiont is a *Trebouxia*. It contains 'vulpinic acid', a poison that, amongst other things, has been used to kill wolves. Occurs on sun-exposed wood.

Very distinctive and unlikely to be misidentified.

Mainly found in the hemiboreal–boreal parts of Scandinavia, very rare or absent from the nemoral zones; all year.

Cetraria islandica is a bushy, curled, brownish-green, shiny lichen with a porous surface to the lower side of the branches; has pins (isidia) at the margin. Occasionally has brown apothecia at the branch margins. The photobiont is a *Trebouxia.* Occurs on heathlands and similar habitats.

Can be split into two subspecies: *C. islandica* ssp. *islandica* is large with wide branches; ssp. *crispiformis* is small with slender branches. *Cetraria ericetorum* ⋉ has narrower branches with pores (pseudocyphellae) at the margin but not further in.

Widespread, but has declined due to eutrophication; all year.

Jens H. Petersen/Thomas Læssøe

Usnea florida is a hanging, bushy lichen that has smooth branches which lack grains (soredia) and pins (isidia), but typically has large, pale apothecia that have thin branches from the margins. The spores measure 7–11 × 6–7 µm. The photobiont is a *Trebouxia*. Occurs on wood in ± damp woodlands.

Usnea subfloridana ⋉ is possibly just a sorediose form, which only rarely forms small apothecia.

Nowadays mainly found in western Atlantic parts of temperate Europe; all year.

Ramalina fastigiata is a rather short-branched, bushy lichen with large apothecia at the branch tips. It does not have pins and grains (isidia/soredia) or macrochemical reactions. The spores measure 10–17 × 5–7 µm. The photobiont is a *Trebouxia*. Occurs on nutrient rich bark from various deciduous trees in open habitats.

Ramalina fraxinea ⋉ has ± lax, flat and much longer branches; *R. calicaris* has inrolled branches that form a longitudinal tube.

Widespread and common; all year.

Evernia prunastri is a rather lax, flat-branched bushy lichen that has algae in the grey-green upper parts and almost white lower parts with a few spots. Has scattered grainy spots (soralia – inset image). Rarely has apothecia. Spores 7–11 × 4–6 µm. The photobiont is a *Trebouxia*. On bark and branches of mostly deciduous trees, rarely on other substrates. Fairly pollution-tolerant. It is used in the perfume industry and *Aegithalos caudatus* use the thalli as nesting material.

Pseudevernia furfuracea ⋉ has a dark lower side. Species of *Ramalina* have uniform upper and lower sides.

Widespread and very common; all year.

Hypogymnia physodes is a grey to grey-green foliose lichen that forms ± regular rosettes with ± lip-shaped grainy spots (soralia) at the tips. The lower side is black. Apothecia are rarely formed; it may form pycnidia at the tips. The photobiont is a *Trebouxia*. Mostly on branches and twigs, but may also occur on rocks and isolated stones. It is relatively pollution-tolerant.

Hypogymnia tubulosa ⋉ has spherical soralia.

Very common and widespread; all year round.

This species may be parasitized by *Tremella hypogymniae* ⋉.

Jan Vesterholt/Jens H. Petersen

Lobaria pulmonaria is a very large, when wet bright green, foliose lichen with a pitted, veined surface, hence the name 'pulmonaria'. When dry it is yellow-brown to olive-brown. It has pins and coarse grain spots (isidia and soralia) on raised ridges of the thallus. The lower side has a ± dark felt. Rarely forms orange-brown apothecia (inset image). Spores 1–3-septate, measure 18–30 × 5–9 µm. The green alga *Dictyochloropsis reticulata* is the dominant photobiont, but the cyanobacterium *Nostoc* is a co-photobiont in special so-called cephalodia inside the thallus. Occurs on bark of very mature trees in old deciduous forests, but also on rocks.

Lobaria scrobiculata ⋉ is more blue-green (photobiont is a cyanobacteria) and has spherical soralia and isidia near the lobe tips. *Lobaria virens* ⋉ and *L. amplissima* ⋉ do not have soredia or isidia; the former may have coral-like, black protuberances.

Widespread and common in the north, but very rare or absent from the nemoral zones due to air pollution and forestry practices; all year.

Peltigera hymenina is foliose with a bluish-grey, smooth upper side and scattered, pale hyphal strings below. Has ± saddle-shaped, orange-brown apothecia. The spores measure 47–70(–90) × 3–5 µm. Occurs both in forest and heathland habitats, mostly overgrowing mosses. The photobiont is *Nostoc*.

Distinguished from other non-felty *Peltigera* species by the matt surface and the pale underside; *P. neckeri* ⋉ has ± erect lobes and dark apothecia, and a darker underside.

Widespread and rather common; all year.

Many parasitic fungi occur on *Peltigera*, e.g. *Arrhenia peltigerina* (page 135).

Nephroma arcticum is a bright green, foliose lichen with very large, orange-brown, somewhat ear-shaped apothecia. The lower side is paler and matt. Spores measure 23–30 × 4–5 µm. Symbiotic with both green algae (*Coccomyxa*, dominant) and the cyanobacterium *Nostoc* (in so-called cephalodia). Occurs on soil.

Easily told by the bright green colours and the very large, dominating apothecia. Other species of *Nephroma* have only cyanobacteria as the photobiont.

A common species in the boreal and alpine zones, but absent from the nemoral–hemiboreal lowlands; all year.

Melanohalea exasperatula is an appressed, rosette-forming, greenish-brown foliose lichen with hollow pins (isidia – inset image) at the centre and rounded lobes at the margin. Apothecia are rarely formed. The spores measure 8–10 × 4–8 µm. The photobiont is a *Trebouxia*. Mostly on nutrient rich bark of deciduous trees.

Melanohalea exasperata ⋉ has a solid mass of isidia over the entire surface. There are also similar species in the genera *Melanelia* ⋉ and *Melanelixia* ⋉.

Widespread, common; all year.

Xanthoria parietina is an orange-yellow to greenish-yellow foliose lichen with conspicuous, ± yellow apothecia that have sharp, smooth margins. Does not have pins and grains (isidia/ soralea). The spores measure (10–)12–16 × (6–)7–9 µm. The photobiont is a *Trebouxia*. Occurs on brick walls, roofs, bark on tree trunks, *etc*. Thrives in areas of high ammonia deposition from intensive farming and can be the dominant lichen in heavily farmed landscapes.

Other species in the genus typically have warts, soralea or isidia on the surface of the thalli.

Widespread, very common; all year.

Lecidella elaeochroma is a thin, ± pale yellow-green crustose lichen with conspicuous black apothecia (inset image); occurs in a mosaic with a brownish-green species of *Lecanora* ⋉. It is blue-green in section. May have grains in distinctive spots (soralia) but usually not present. Hymenium 40–70 µm thick. Asci I+, 8-spored. The spores measure 8–17 × 6–10 µm. The photobiont is a *Trebouxia*. Occurs on ± smooth bark of deciduous trees, *e.g. Populus tremula*, *Sorbus*, *Malus*, *etc.*, rarely on conifer bark.

Lecidella euphorea ⋉ is very similar, but has *e.g.* a thinner hymenium. Species of *Lecidea* ⋉ have a different ascus morphology and another chemistry (composition of lichen compounds).

Widespread, very common; all year.

The *Lecanora* species on the image above forms discoid, brownish-green apothecia with a distinct margin. All *Lecanora* species are also associated with *Trebouxia*.

Thelotrema lepadinum is a pale grey crustose lichen with volcano- or barnacle-like apothecia. Spores hyaline, 10–18-septate and also longitudinally septate, and measure 60–140 × 12–28 µm. Photobiont is an orange-coloured *Trentepohlia*. Occurs on bark of old trees in ancient woodlands, mostly on *Fagus*, usually in sites with relatively high precipitation.

A fairly easy species to identify.

Local and rather rare – more common in western parts of temperate Europe; all year.

Ophioparma ventosa is a rather thick crustose lichen with spectacular, orange-brown, pale-margined apothecia on a grainy-scaly, pale grey to yellowish-grey thallus. Spores hyaline, 3–7-septate and measure (30–)40–60 × 3–4.5(–6) µm. Conidia 7–10 × 1 µm. The photobiont is a *Trebouxia*. Occurs on acidic stones and rocks.

Ophioparma lapponica ⋉ is similar but has smaller spores.

Mostly found in northern parts of temperate Europe and apparently extinct in some parts (*e.g.* Denmark); all year.

Icmadophila ericetorum is a crustose lichen with large, ± granulate, greyish, to almost blue-green thalli, and typically large, pink, up to 4 mm wide, sessile, but basally constricted, somewhat irregularly shaped, convex apothecia. Spores 2–4-celled and measure 13–27 × 4–8 µm. The photobiont is a *Coccomyxa*. Occurs on peaty soils or rarely on living mosses or rotten wood.

The large, pink apothecia make this species easy to identify. *Baeomyces carneus* ⋉ is similar but has distinctly stemmed apothecia.

Widespread and common in the hemiboreal–boreal zones but much rarer in the nemoral lowlands; all year.

Thomas Læssøe

Graphis scripta is a grey-green, thin crustose lichen with script-like, narrow, branched, black apothecia (lirellae) that do not have furrows at the margin. Asci 8-spored. Paraphyses unbranched. Spores hyaline, stain purple in iodine, 5–15-septate, 20–70 × 6–10 µm. The photobiont is a *Trentepohlia*. On bark of older deciduous trees, mostly *Fagus*.

This species can apparently be split into several closely related species. *Graphis elegans* ⋉ is similar but has furrows at the margin of the apothecia. *Opegrapha* ⋉ is also similar but has branched paraphyses, I- spores and, typically, smaller thalli.

Widespread, mainly nemoral–hemiboreal, common to occasional; all year.

Calicioids and mycocalicioids

Calicioids and mycocalicioid fungi are two groups of distantly related ascomycotes with very similar, tiny, pin-shaped fruitbodies. In some species the spores are deposited in a powdery column on the top and passively dispersed. The calicioids live in a mutualistic symbiosis with green algae such as *Trebouxia*, *Stichococcus*, *Dictyochloropsis* and *Trentepolia*; the mycocalicioid fungi are decomposers.

Many species can be found on old, sun-exposed wood, but the small fruitbodies are easily overlooked; they are best found by using the sky as a background when inspecting potential substrates.

OTHER SIMILAR FUNGI:
– *Phleogena* has larger fruitbodies; it is a basidiomycote with transversely divided basidia (page 1246).
– asexual fungi may form very similar structures (page 1626).
– many slime moulds may look similar but they are often larger and may have spores in a thread-like capillitium (page 1646).

FURTHER READING: 302, 316, 326, 355.

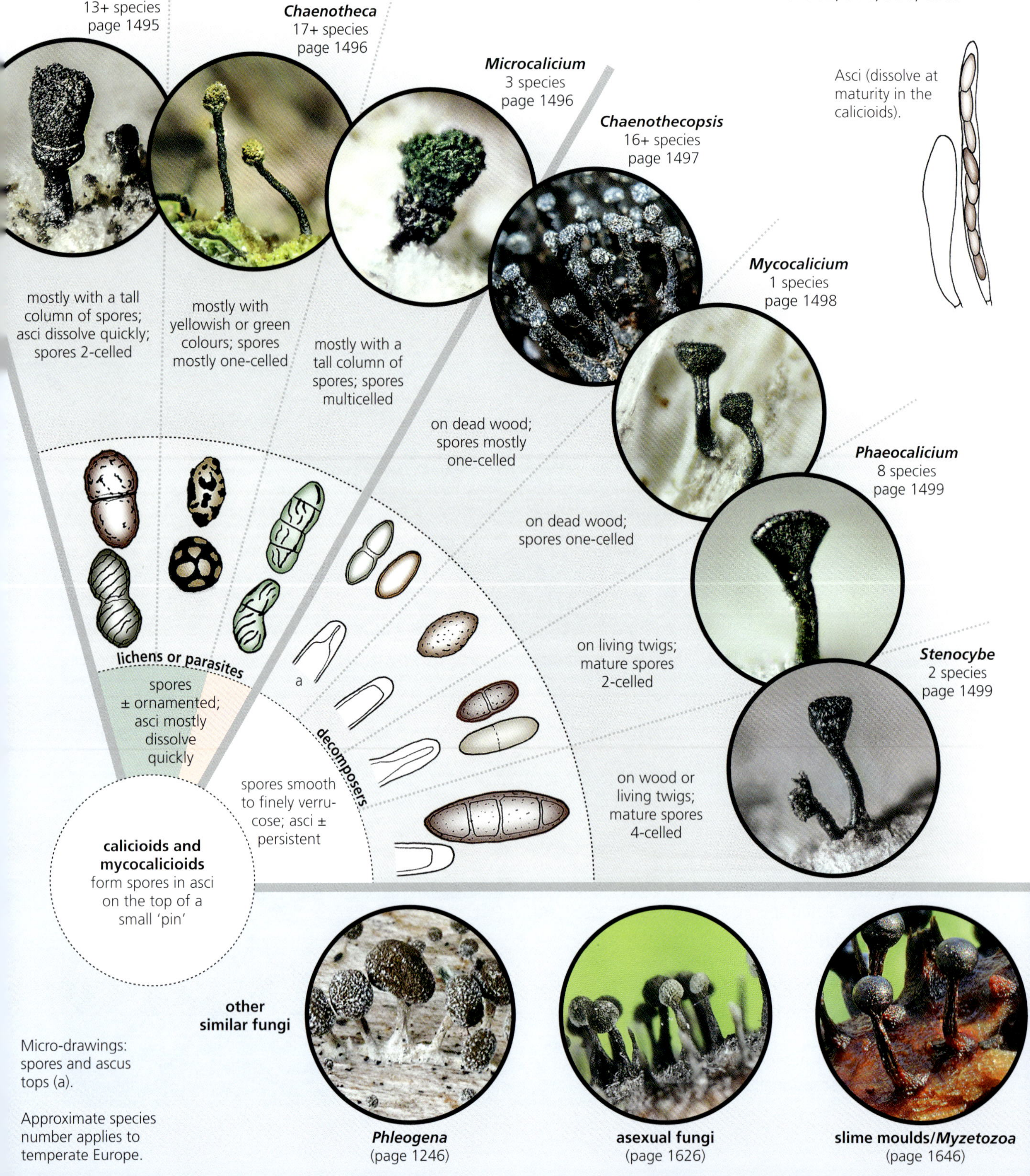

Calicium form very small, stipitate fruitbodies with a widened head, the top of which is powdery from a thick, dark spore mass. Thalli are either superficial or partly immersed in the substrate. Asci dissolve quickly; spores 2-celled, thick-walled and usually ornamented. Lichenized with green algae belonging to the genus *Trebouxia*. Typically on sun-exposed bark and wood.

Calicium viride is a calicioid with a finely grainy, yellow-green thallus. The fruitbodies have rather long, black stems and a convex, black spore mass on the top. The lower side of the head may be ± brown and finely powdery. Asci clavate. Spores dark brown, 2-celled, have a ± regular to irregularly spiralling pattern, and measure 12–14 × 6–7 µm. The algal partner (photobiont) is a *Trebouxia*. Occurs mostly on dry bark and wood, both deciduous and coniferous; mostly in a sterile version. It is rather pollution tolerant.

Calicium salicinum ▽ usually has an immersed thallus, browner fruitbodies, cylindrical asci and smaller spores; *C. glaucellum* ▷▷ has small, dark and very open fruitbodies

Widespread, common to occasional; all year.

Calicium salicinum is a calicioid with an almost invisible, ± immersed thallus, fruitbodies with a brown, almost globose head topped with a black spore mass, and stems that are black towards the base. Asci cylindrical. The spores are 2-celled, black, have a spiralling or irregularly cracking pattern and measure 8–11 × 3.5–4.5 µm. The photobiont is a *Trebouxia*. Occurs on dry bark and wood.

Calicium viride △ has a superficial, yellow-green thallus, clavate asci, larger spores and less brown powder on the outside of the head.

Widespread, common to occasional; all year.

Calicium glaucellum is a calicioid with an almost concealed thallus and rather small, mostly black fruit-bodies with an expanded head that has a black spore mass on the top. Asci cylindrical. The black to greyish-blue spores measure 9–13 × 5–7 µm and have longitudinal or diagonal furrows. The photobiont is a *Trebouxia*. Occurs mostly on dry, decorticated conifer wood.

Calicium viride ◁◁ has a distinct, yellow-green thallus and somewhat larger fruitbodies.

Widespread, rather common to occasional; all year.

Chaenotheca chrysocephala is a calicioid with a yellow-green, grainy thallus and almost tulip-shaped fruit-bodies that have ± yellow-green heads. Asci dissolve quickly. The brownish spores collect in a column on the top; they measure 6–9 × 4–5 µm; and have coarse, irregular furrows. Photobiont is a *Trebouxia*. Occurs mostly on dry bark, rarely wood, of both coniferous and deciduous trees. *Chaenothecopsis consociata* ⋉ may occur (possibly as a parasite) on the same thallus.

The vivid yellow colours on the thallus and fruitbodies should make identification straightforward.

Widespread, occasional to rare; all year.

Microcalicium disseminatum is a very small ascomycote with black, egg cup-shaped fruitbodies topped with a green, upright, columnar spore mass. The asci dissolve quickly. The greenish, ellipsoid to subcylindrical spores measure 11–13(–28!) × 3–4 µm. Occurs on wood and bark, where it parasitizes free-living algae and both the thallus and fruitbodies of various other species of calicioid; despite its parasitic nature it belongs to the calicioids.

The short stem and distinctly moss-green spore mass should make identification straightforward.

Widespread, occasional to rare, but easily overlooked; all year.

Chaenothecopsis is a genus of mycocalicioids with stems that may be branched, and a swollen, ± smooth head. The asci are ± persistent (do not dissolve at maturity) and the top has a distinct channel.

Chaenothecopsis caespitosa is a mycocalicioid that has compressed, thick mats of branched fruitbodies. The fertile heads are bluish and up to 0.35 mm wide. The asci are cylindrical. The 2-celled, brown spores measure 9–14 × 3–4.5 µm. A decomposer, mainly on *Taxus* wood.

A very distinctive species that is unlikely to be misidentified.

Poorly known, but present in *e.g.* France and the UK, apparently rare; all year.

Chaenothecopsis debilis is a ± black mycocalicioid that sometimes has a greenish tinge in the hymenium. Head up to 0.4 mm wide; stem up to 1 × 0.1 mm. Asci cylindrical, I-, 8-spored. The 2-celled, brown-walled, smooth to furrowed spores measure 7–9 × 2.5–3 µm; the septum is also brown. Probably a decomposer. Occurs mostly on wind-exposed, dry, decorticated wood.

Chaenothecopsis nigra ⋉ has shorter spores and a darker spore septum; *C. pusilla* ⋉ has shorter spores with a very pale septum. Some species have a bluish tinge in the hymenium.

Widespread, rather rare to rare, or overlooked; all year.

Mycocalicium subtile is a very small, ± black, pin-shaped mycocalicioid. The stem hyphae are ± bluish in preparations. Asci cylindrical, persistent, have uniform, thickened tips that lack a channel, I-. Has paraphysis-like, ± branched cells between and above the asci. The one-celled, brown, ± verrucose, ± flattened spores measure 5.5–9 × 2.3–4 µm. Occurs on dry, sun-exposed wood.

Similar calicioids have thalli and the asci dissolve quickly. Species of *Chaenothecopsis* △ have a channel through the ascus top.

Widespread, rather common to occasional; all year.

Phaeocalicium betulinum is a mycocalicioid with a compressed/flattened head. The outer flesh has a layer of spherical cells. Asci persistent, 8-spored, ± cylindrical, have a distinct channel, 48–71 µm long, I-. The pale brown, smooth, one-celled spores measure 10–13 × 4.5–5.5 µm. Occurs on bark of *Betula*.

Phaeocalicium compressulum ⋉, on *Alnus*, is very similar but has darker, finely verrucose spores and no spherical cells in the flesh. *Phaeocalicium praecedens* ⋉ is known from bark of *Populus tremula* and is less compressed.

Boreal and very rarely recorded; all year.

Stenocybe pullatula is a less than 0.5 mm high, black, mycocalicioid that either has one head or is branched with up to five heads. The stem has an outer gelatinous layer. Asci cylindrical, 8-spored, measure 84–89 × 4.5–6 µm, I-. The brown spores are finely verrucose, 2–4-celled and measure 11–16 × 4–5 µm. Occurs on thin branches of living *Alnus*, possibly a parasite.

Stenocybe bryophila ⋉ occurs on liverworts. The genus is most easily recognized by its non-dissolving asci and the several-celled spores.

Widespread, occasional, easily overlooked; mostly during the summer.

Mildews and cleistothecial fungi

Two form groups of ascomycotes with spherical fruitbodies are included here: the mildews (*Erysiphales*), which crack open at maturity and disperse spores actively; and the cleistothecial fungi, which do not open at maturity but decay and disperse the spores passively. In both groups the fruitbodies are very small and may act as dispersal units. To this end, many have a surface with branched or hooked appendages that may attach to passing animals, *etc.*

The mildews are biotrophic plant parasites that form whitish coverings of mycelia on stems, leaves and twigs, from which haustoria can penetrate the living plant tissues (see page 16). The mealy hyphal covering also produces asexual, wind-disseminated conidia. As leaves get older (typically in the autumn), mildews often switch to producing tiny fruitbodies (chasmothecia) that overwinter and crack open when new leaves are available for infection.

The cleistothecial fungi are decomposers of organic material. The genera *Aphanoascus*, *Arthroderma*, *Gymnoascus* and *Onygena* (*Eurotiomycetidae*) are specialists in degrading keratin (hair, horn, *etc.*). Many have prominent asexual states, but they may also produce tiny spherical fruitbodies (cleistothecia, see page 30) with ± globose asci; the spores are released passively as the ascus walls and fruitbodies dissolve/decay.

Other similar fungi:
– pyrenomycetous fungi may have an almost invisible opening (ostiole), see page 1510.
– truffles are ± spherical and decay at maturity but are much larger, see page 1256.

Further reading: 35, 47, 68, 72, 73, 78, 79, 80, 164, 188.

Micro-drawings: outermost asci (a) and sometimes spores followed by peridial appendages. Three inset images depict peridia, as seen in a compound microscope (per).

Approximate species number applies to temperate Europe.

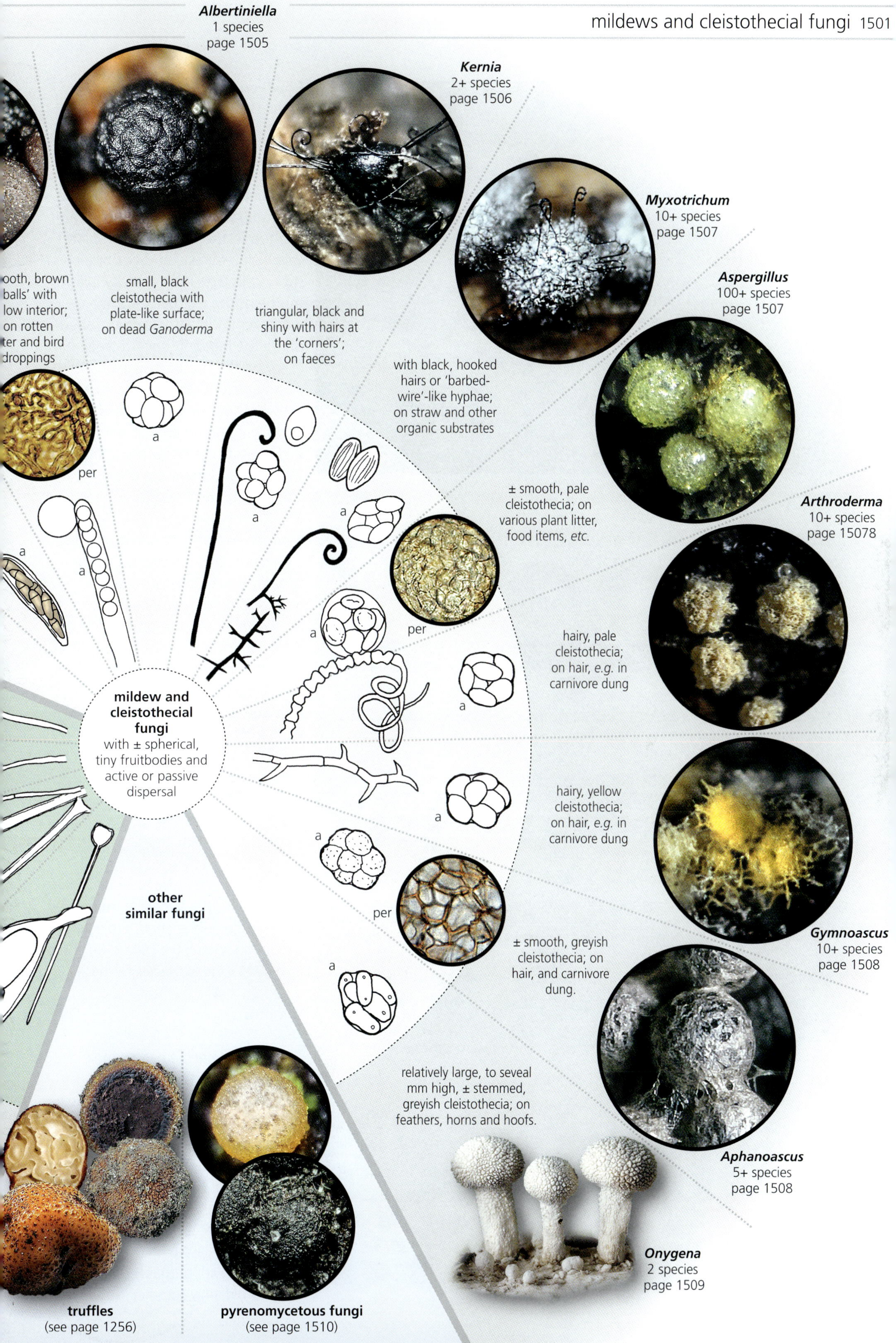
Albertiniella
1 species
page 1505
Kernia
2+ species
page 1506
Myxotrichum
10+ species
page 1507
Aspergillus
100+ species
page 1507
Arthroderma
10+ species
page 15078
Gymnoascus
10+ species
page 1508
Aphanoascus
5+ species
page 1508
Onygena
2 species
page 1509
ooth, brown balls' with low interior; on rotten ter and bird droppings
small, black cleistothecia with plate-like surface; on dead *Ganoderma*
triangular, black and shiny with hairs at the 'corners'; on faeces
with black, hooked hairs or 'barbed-wire'-like hyphae; on straw and other organic substrates
± smooth, pale cleistothecia; on various plant litter, food items, *etc.*
hairy, pale cleistothecia; on hair, *e.g.* in carnivore dung
hairy, yellow cleistothecia; on hair, *e.g.* in carnivore dung
± smooth, greyish cleistothecia; on hair, and carnivore dung.
relatively large, to seveal mm high, ± stemmed, greyish cleistothecia; on feathers, horns and hoofs.
a
per
mildew and cleistothecial fungi
with ± spherical, tiny fruitbodies and active or passive dispersal
other similar fungi
truffles
(see page 1256)
pyrenomycetous fungi
(see page 1510)

Phyllactinia guttata is a mildew, with ± lens-shaped fruitbodies that initially have a number of equatorially projecting, pointed hairs with a swollen base. Late in the autumn the hairs bend at the swelling and raise the fruitbodies free of the substrate. The top of the fruitbody is at first yellow, later black, and has branched, sticky cells that form a sticky 'pillow'; this helps the fruitbody adhere to whatever it hits after being pushed off by the bending hairs. Sometime in spring the fruitbody will open along a suture and the spores will be actively shot out; some will land on a living leaf and be able to germinate and reinfect the host. The asci are 2-spored and there are many in each fruitbody. The spores measure 25–40 × 15–25 µm. Asexual spores are formed from hyphae on the leaf surface (mealy covering on image). A parasite on *Corylus* leaves.

Many almost identical species of *Phyllactinia* occur on other hosts; *P. orbicularis* ⋉, for example, forms very conspicuous white coverings on *Fagus* leaves.

Widespread and common, becoming scarcer towards the north; all year most visible late in the autumn.

Erysiphe form ± spherical fruitbodies with simple or branched, projecting appendages. Asci typically 3–8-spored. *Erysiphe* includes species formerly accepted in *Microsphaera* and *Uncinula*.

Erysiphe ulmariae has a simple appearance with hyphae-like appendages on the fruitbodies. Asci (3–)6–8-spored, 6–16 per fruitbody. Spores 15–25 × 8–15 µm. Conidia 22–40 × 10–22 µm. Occurs on leaves and shoots of *Filipendula*.

Podosphaera filipendulae ⋉ occurs on the same host but the fruitbodies have only one ascus.

Widespread, probably rather common; mainly summer–autumn.

Erysiphe alphitoides is a mildew with a very prominent, whitish asexual state on the surface of living leaves. In the autumn yellow, later black, fruitbodies with long, 3-dimensionally branched appendages that have hooked tips are formed (inset image shows this through a compound microscope). The asci are (4–)6–8-spored, 4–16 per fruitbody. The spores measure (13–)16–26 × (7–)9–15 µm. Occurs mainly on leaves of *Quercus* and *Castanea*, but also known from *Fagus*.

This species belongs to a complex, and there are many closely related species. However, all except the following are absent from or very rare in Europe. *Erysiphe flexuosa* ⋉, on *Aesculus*, has been accidentally introduced from North America.

A widespread and very common species; all year, but mainly visible in the autumn.

Erysiphe adunca is a mildew with many long, straight appendages on the fruitbodies; the appendages have a small hook at the tip (presumably to aid dissemination). The asci are 3–6(–7)-spored, 4–15 per fruitbody. The spores measure 18–30(–33) × 9–18 µm. Conidia measure 25–40 × 11–22 µm. Occurs on leaves of *Salix* and *Populus*.

A species of *Phyllactinia* ◁, with long straight, unhooked appendages and a basal swelling, also occurs on *Salix* and *Populus* leaves

Widespread and common, at least in the nemoral–hemiboreal zones; all year but mainly observed in the autumn.

Species of ***Podosphaera*** typically have only one ascus per fruitbody.

Podosphaera epilobii is a mildew with long, simple, narrow appendages on, at first yellow, later black, small fruitbodies, which are usually immersed in a mealy, asexual surface covering. The single ascus is (6–)8-spored. The spores measure (15–)16–24(–26) × 9–16 µm. Conidia measure 19–38 × 11–120 µm. Occurs on living leaves of *Epilobium*.

On the host genus identification should be straightforward.

Widespread, but apparently rather rare; mainly observed summer–autumn.

Podosphaera leucotricha is a mildew with a dense, felty asexual covering, in which very small fruitbodies may be formed with stiff, simple, at the tips often bifurcated, appendages. The single ascus is 8-spored and the spores measure 20–32 × 12–17.5 µm. Causes damage to leaves, shoots and fruits of apples (*Malus*) and pears (*Pyrus*), especially during periods of dry weather; it may survive the winter inside living shoots.

On the hosts, identification should be straightforward.

Widespread, in the nemoral–hemiboreal zones, rather common; mostly in the autumn.

Blumeria graminis is a mainly asexual mildew, but dark fruitbodies are occasionally formed; they tend to be immersed in a white, dense mycelium and have hyphae-like appendages. Asci (4–)8-spored, 6–30 per fruitbody. Spores measure 20–24 × 10–14 µm. Conidia ellipsoid to lemon-shaped, measure (20–)24–35(–45) × (8–)12–16(–20) µm. Occurs on a range of different genera and species of *Poaceae*; it causes significant economic damage to cereal crops.

This is the only mildew on grasses and identification should be easy.

Widespread and common; summer–autumn.

Amylocarpus encephaloides is a buff to buff-yellow, semiglobose to globose cleistothecial ascomycote with a smooth to wrinkled/veined surface. Asci 8-spored, broadly clavate, 28–36 × 20–24 µm, have a stem 8–16 × 3 µm, but lack an opening apparatus. Paraphyses absent. Spores are hyaline, ± amyloid, spherical, one-celled, and have long, scattered, projecting spines 8–11(–16) µm in diameter. Occurs on the surface of driftwood on seaside beaches.

The habitat and very distinctive spores are distinguishing characters.

Widespread, rarely reported, but probably fairly common; most of the year.

Orbicula parietina is an ascomycote with ± lens-shaped, slime mould-like, cleistothecial fruitbodies that have a thin brownish outer covering and a powdery, yellow interior. Asci 8-spored, ± cylindrical, lack an opening apparatus, measure up to 136 × 14.4 µm, I-. Paraphyses simple to branched. Spores smooth, spherical, ± yellow and measure 7–11.5 µm. Occurs on excrement or other decaying organic substrates.

The smooth spores should prevent confusion with similar-looking species of *Mycetozoa* (see page 1646).

Widespread, but very rarely reported; all year.

Albertiniella polyporicola is a cleistothecial ascomycote that forms ± gregarious, tiny, shiny, rugose, spherical, closed fruitbodies on the hymenophore of rotten fruitbodies of *Ganoderma*. The fruitbody wall cracks into polygons. Asci spherical, 8-spored. Spores lens-shaped, dark, finely echinulate (a net-pattern can be seen with SEM) and measure 3–3.7 × 2.5–8 µm. Occurs on dead, mostly large specimens of *Ganoderma applanatum* (page 872).

The distinctive substrate makes identification straightforward.

Nemoral–hemiboreal, rather rarely reported, but probably not rare; mostly in the autumn, but all year.

Kernia nitida is a very distinctive cleistothecial ascomycote with ± angular, shiny, black fruitbodies that have groups of very long, shiny black hairs at the 'corners'; the hairs are twisted at the tips. The asci dissolve quickly. Spores yellowish, thick-walled, ± ellipsoid, have a germ pore at both ends and measure 3–5.7 × 3.5–4.2 µm. Mostly on herbivore excrement.

A number of other species of *Kernia* are described with spores that are more asymmetrical or fruitbodies that are less asymmetrical; they either have or lack hairs.

Very rarely reported, and frequency not known; all year.

Lasiobotrys lonicerae is a very peculiar ascomycote that is erumpent as black patches from living *Lonicera periclymenum* leaves. When the epidermis breaks, numerous black sclerotium-like bodies, 200-350 µm in diameter, become visible and below the margin of these bodies small, spherical, black fruitbodies with up to 85 µm in diameter are formed (arrows and lower picture). The margin also produces tufts of black hyphae and the entire unit is released from the spots and thereafter adhere to surrounding vegetation (lower picture). Later spores are released from double-walled asci and may re-infect *Lonicera* leaves. The spores are golden brown, 2-celled and measure 8.5–11.5 × 2.5–3.5 µm.

A highly unique fungus that should be readily recognizable.

Widespread but apparently rare; summer-autumn.

Species of ***Myxotrichum*** have black, projecting setose hairs and occur on ± cellulose-rich substrates.

Myxotrichum chartarum is a very loosely built, whitish, ± globose ascomycote with long, projecting black setae that have bishop's crook-like bent tips. The asci dissolve quickly. Spores ± brown, longitudinally striate, ± ellipsoid and measure 3.8–5.2 × 1.9–3.3 µm. Occurs on cellulose-rich substrates, including excrement; can also be isolated from soil.

Myxotrichum deflexum ▽ has setae that recall a thorny tree.

Probably widespread, rarely reported; most of the year.

Myxotrichum deflexum is a very loosely built, ± globose ascomycote that is covered by long projecting, shiny setae that recall miniature, thorny trees. The asci dissolve quickly. The spores are ± brown-tinged, ± fusiform, longitudinally furrowed and measure 4.7–5.6 × 2.4–3.8 µm. Occurs on various substrates, *e.g.* paper, excrement and animal fur.

Myxotrichum chartarum △ has setae with bishop's crook-like tips and ± ellipsoid spores.

Presumably widespread, rarely reported; most of the year.

Aspergillus glaucus is a species of *Aspergillus* with tiny, yellowish, ± globose and loosely built cleistothecial fruitbodies (inset image). Asci spherical, 8-spored, 10–12 µm in diameter, dissolve at maturity. Spores shaped like two deep plates that have been put together with a furrow at the join; 6–7 × 3.2–5.1 µm. The asexual state (a hyphomycete) is dark grey-green with an erect hypha with a swollen head with conidia producing cells. Conidia spiny, greenish, globose 4.5–7.5 µm. Occurs on dead plant material, on old jam, *etc.*

Also known as *Eurotium herbariorum*, a name based on the sexual state. See also page 1626 and 1637.

Widespread, very common; all year.

Gymnoascus reessii is a loosely built cleistothecial ascomycote with spherical, ± crowded fruitbodies that may be immersed in a subiculum. The golden to copper-coloured peridium forms a grainy reticulum that has hooked appendages in several directions. Asci 8-spored and measure 7–8.9 × 6.4–8.5 µm. Spores smooth, yellowish, flattened, measuring 3–4.7 × 1.9–2.8 µm (widest side). No asexual state is known. Occurs on excrement, remnants of dead animals and in soil.

The golden colours and distinctive peridial appendages should make identification relatively straightforward.

Widespread, occasional; all year.

Aphanoascus fulvescens is a globose, cleistothecial ascomycote with a ± shiny, felty surface. Peridium consists of angular, brownish cells that lack appendages. Asci 8-spored, spherical, 8.5–11.3 × 5.6–11.3 µm. Spores ± brown, flattened, have a reticulate ornament (best seen with SEM), measure 3.3–5 × 2.3–4.2 µm. Conidia pear-shaped, smooth, hyaline, 5.7–17 × 3.8–6.7 µm. Occurs on animal remnants, including excrement, old nests, owl pellets, *etc.*

Aphanoascus canadensis ⋉ has a finer spore ornament and a different asexual state.

Widespread, rarely reported; all year.

Arthroderma curreyi forms loose, ± spherical, whitish to pale yellowish-brown cleistothecia possibly in a pale subiculum. Peridium of spiralling, septate hyphae with grainy-verrucose swellings. Asci ± spherical, 8-spored, 5.2–6.6 × 3.8–5.6 µm, dissolve quickly. Spores smooth, hyaline, flattened, 2.8–4 × 1.9–3.3 µm (widest side). Asexual state simple with ± sessile, hyaline, one-celled, 2.4–5.6 × 1.9–2.8 µm large conidia. A decomposer of keratin; occurs on hairs from various animals.

Species of *Ctenomyces* ⋉ have comb-like protuberances from the peridium. Other species of *Arthroderma uncinatum* ⋉ and *A. quadrificum* ⋉ have large, multicelled conidia.

Widespread, common; all year.

Onygena form relatively large, up to several mm high, cleistothecia with a distinct stem. Both species occur on keratinous substrates such as hooves, feathers and hair.

Onygena corvina has a long, whitish stem and a round, grainy head that contains asci and spores. Asci 9–14 × 6.5–9 µm, dissolve at maturity. Spores 4–7.5 × 2.5–4 µm, almost hyaline, ± smooth but a reticulum is visible in SEM. On feathers, owl pellets, fur, *etc.* Often on dead roooks (*Corvus frugilegus*) in their colonies.

Onygena equina ▷ has larger spores and mainly occurs on hooves.

Widespread, occasional; mainly August–November.

Thomas Læssøe

Onygena equina is a whitish cleistothecial ascomycote with a short but distinct stem, and a grainy, round head that contains asci and spores. The asci measure 14–24 × 11–16 µm and dissolve at maturity. The nearly hyaline, almost ellipsoid spores are somewhat thick-walled and ± smooth when seen through a compound microscope but have a net-like ornament when observed in a scanning electron microscope; they measure 5.5–9.5 × 3–6 µm. Occurs mainly on hooves from various mammals, *e.g.* horses, but also on horns.

Onygena corvina ◁ is typically longer-stemmed, has smaller spores, and occurs on feathers, owl pellets, fur, *etc.*

Widespread, rather rare and has declined dramatically in parts of the nemoral zones; most of the year.

Pyrenomycetous fungi

This is a form group of fungi with small, mostly spherical, flask-shaped or sac-like fruitbodies with an opening (ostiole) from where the spores, in most cases, are shot out. The fruitbodies are here all termed 'perithecia' (see more on page 30), although a more precise terminology is often used by mycologists that reflects ontogeny and phylogeny, whereby the fruitbodies of ascomycotes with double-walled asci (right-hand side of the wheel) may be termed pseudothecia, thyriothecia and hysterothecia.

In some species the perithecia are ± imbedded in a uniting flesh, a stroma (pl. stromata), which may be flattened (applanate), cushion-shaped (pulvinate) or club-shaped (clavate).

Many species are decomposers, decaying dead wood, herbs, dung, *etc*, while others are parasites on plants, mosses, insects, *etc*. Some kill insects and spiders and some are lichenized. Many live inside living plant tissues (endophytic) and only fruit when the plant dies.

The pyrenomycetous fungi is a very large group with more than 2,500 species in temperate Europe. Parallel evolution has led to similar-looking structures in many ascomycote orders and only very pyrenomycetous fungi can be identified without microscopy. Experience and specialized literature are required for detailed studies of pyrenomycetous fungi, and there are also good internet fora available (for references, see the following wheels).

Other similar fungi:
– *Laboulbeniomycetes* has very small, peritecioid fruitbodies on living insects, millipedes, *etc*. (page 1620).
– some erumpent inoperculate cup fungi may be completely closed, but the hymenium is revealed in wet weather (page 1468).
– fungi with clavate, non-composite fruitbodies may look similar, but the hymenium is on the outside, completely exposed (pages 1092 & 1350).

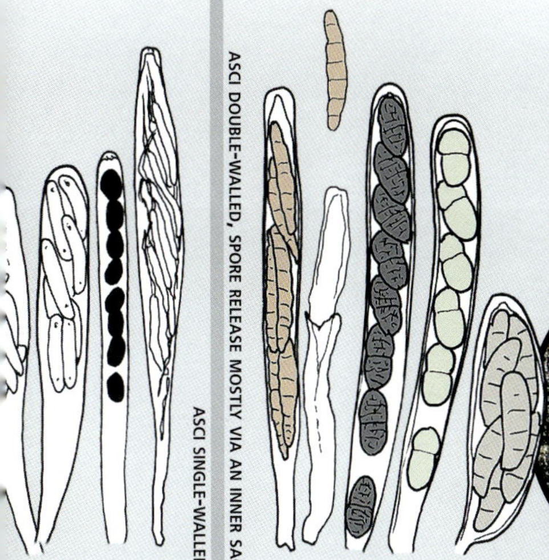

ASCI SINGLE-WALLED

ASCI DOUBLE-WALLED, SPORE RELEASE MOSTLY VIA AN INNER SAC

non-hysterioid, bitunicate pyrenomycetous fungi
page 1602

hysterioid pyrenomycetous fungi
page 1614

pyrenomycetous fungi
(forming perithecial fruitbodies)

perithecia ± black

perithecia free or in host tissue

stromatic

perithecia with ± circular opening; asci single-walled

perithecia (pseudothecia, thyrothecia) superficial or immersed with ± circular to oval opening; asci double-walled

perithecia (hysterothecia) superficial and ± compressed; opening ± slit-like; asci double-walled

erithecia and tromata white or ± brightly coloured and rather soft

other similar fungi

fruitbodies perithecioid; on living insects, millipedes, *etc.*

Laboulbeniales
(page 1620)

at first closed, later opening to reveal the hymenium

erumpent inoperculate cup fungi
(page 1468)

club-shaped, without dotted surface; hymenium exposed

clavarioids and tongue-shaped inoperculate cup fungi
(page 1092 & 1350)

Hypocrealean pyrenomycetous fungi and others

A form group of fungi with small, mostly globose to flask-shaped perithecia (page 30). The perithecia are typically relatively soft (leathery) and often brightly coloured or whitish; they may be free, sit on a soft, often brightly coloured stroma or immersed in a stroma (page 30). The dots on the surface of the stromata represent the ostioles (perithecial openings). The spores are typically hyaline, or nearly so, and vary from multicelled, thread-like (some split into part-spores) to one-celled; 2-celled spores are common and may split into two at maturity.

Many species are parasitic on plants, animals or fungi. Species of *Hypomyces* may replace parts of the living fruitbodies of larger fungi with perithecia. *Cordyceps* and *Ophiocordyceps* are predatory fungi that kill insects and spiders, later producing stromata from the carcasses. *Epichloë* and *Claviceps* live as endophytes in grasses and other monocots; they are well known due to their toxic alkaloids (ergotism is caused by *Claviceps*).

Other similar fungi:
- hard, ± black pyrenomycetous fungi are more carbonous (page 1546).
- clavarioid basidiomycotes and tongue-shaped inoperculate cup fungi, have external hymenia, and thus no dots from immersed perithecia (pages 1092 & 1350).

See also the general wheel, page 1510.

Further reading: 48, 55, 60, 96, 106, 107, 128, 129, 131, 132, 175, 248, 274, 275, 278, 279, 280, 284, 302, 311.

Micro-drawings: spores first, followed on the inside by asci (a) and ascus tops (at).

Approximate species number applies to temperate Europe.

Protocrea
1 species
page 1533
Arachnocrea
1 species
page 1533
Nectria
3 species
page 1534
Neonectria
8+ species
page 1535
Cosmospora, Dialonectria & Macroconia
16+ species
page 1536
effused,
ushion- or
b-shaped;
pores split
nto two;
on fungi,
vood, etc.
± free or immersed; spores split into two; on fungi
± free; spores split into two; on leaves and twigs
free to stromatic; on living and dead wood
± free; on living and dead wood
free; on fungi
Flammocladiella & Thyronectria
11+ species
page 1539
free to stromatic; on wood and pyrenomycetous fungi
Hydropisphaera
3 species
page 1537
splitting spores
a
free; on wood and fungi
Lasionectria
4 species
page 1538
free; on wood, stems of herbs and lichens
hypocrealean fungi
have ± soft, white or brightly coloured perithecia
free or immersed; on fungi and myzetozoa
other similar fungi
a
at
stromatic or rarely ± free; mostly on fleshy fungi
Nectriopsis
8+ species
page 1540
free; with thick ascus walls; on wood, fungi (incl. lichens) or lichenized
spores formed inside perithecia
spores formed externally on club-shaped fruitbodies
Hypomyces & Sphaerostilbella
38+ species
page 1541
Thelocarpon
12+ species
page 1545
hard, ± black pyrenomycetous fungi
see groups page 1546
tongue-shaped inoperculate cup fungi
(page 1350)
clavarioids
(page 1092)

Torrubiella albolanata has white, almost chimney-like stromata on top of the host's body. The pale yellow perithecia are ± immersed, and all are positioned vertically. The conical ostioles are clearly visible on top of the fluffy stroma. The asci measure up to 340 × 5.5 μm and have a cap-like, pierced, refractive top. The thread-like spores are ± twisted within the asci and split into part-spores that measure 8.4–9.8 × 1.4–1.5 μm. Occurs on tiny spiders (possibly exclusively members of *Linyphiidae*) in swampy habitats with tall grass vegetation, *e.g.* reedbeds.

Torrubiella aranicida ⋉ occurs on somewhat larger spiders and the perithecia are more scattered. The most commonly encountered fungi on dead spiders belong to the asexual genus *Gibellula*, particularly *G. pulchra* (page 1630); these hyphomycetes are closely related to *Torrubiella*.

Mainly nemoral, rarely reported, but probably overlooked; mostly June–October.

Cordyceps and ***Ophiocordyceps*** form club-shaped stromata from insects that they have killed and devoured. The spores are ± thread-like and split into part-spores after having been released. Both genera have asexual states with morphologies corresponding to the asexual genera *Lecanicillium* ⋉, *Hirsutella* ⋉, *Isaria* (page 1631), *Paraisaria* ⋉ and *Septofusidium* ⋉.

Cordyceps militaris is an apricot-orange to reddish-orange, club-shaped *Cordyceps* with reddish-orange immersed perithecia that have conical ostioles at the surface. Several stromata may emerge from the same cadaver. The asci are cylindrical, measure about 300 × 3–6 µm, and have a cap-like, pierced, refractive top. The thread-like spores split into part-spores that measure 4.5–5(–7) × 1.5 µm. The asexual state is a hyphomycete with *Lecanicillium* characters. Occurs on pupae and larvae of moths, both in forests and in open habitats, *e.g.* non-fertilized grasslands.

Distinctive and unlikely to be misidentified. *Cordyceps bifusispora* ▽ is yellower, and *C. memorabilis* ⋉ appears from beetle larvae cadavers.

Widespread and rather common, becoming scarcer towards the north; mainly July–December.

Cordyceps bifusispora is a sulphur-yellow to lemon-yellow, simple or forked *Cordyceps* with semi-immersed perithecia that have projecting conical ostioles on the upper part and a rather prominent, pale, sterile lower part. The asci are cylindrical, about 200–220 × 3–11.5 µm. Spores thread-like, with fusiform, swollen ends (bola-shaped): main part about 145–220 µm, end parts 3-septate and measure about 30 × 1.6 µm. The asexual state has *Septofusidium* characters. Occurs on moth pupae/larvae in *Alnus* carrs, *etc.*

A distinctive species, both macro- and microscopically.

Widespread, very rare, but possibly somewhat overlooked; mainly August–October.

Erik Arnfred Thomsen/Jens H. Petersen

Ophiocordyceps gracilis is a tall and slender *Ophiocordyceps* that has a sterile, pale stem and a ± egg-shaped, yellow-brown, fertile head with completely immersed perithecia that are only visible as ostiolar dots on the surface of the stromatal head. The asci are cylindrical, measure 300–500 × 4–5 µm, and have a cap-like, pierced, refractive top. The spores are thread-like and split into part-spores that measure 5–9 × 1.5–2 µm. The asexual state is of the *Paraisaria*-type. The conidia measure 4–6 × 1 µm. Occurs on buried larvae of moths, perhaps exclusively species of *Hepialus*.

The combination of early fruiting and the large stroma size is unique to this species. The somewhat similar *O. larvicola* ⋉ kills wood-inhabiting beetle larvae and has a ± pinkish tinge.

Widespread, occasional to rare, becoming scarcer towards the north; April–June.

Waldek Czerniawski

Ophiocordyceps ditmarii is a rather small, pale-stemmed *Ophiocordyceps* that has a ± egg-shaped, buff, fertile head with completely immersed perithecia; a small collar is often present at the base of the head. Asci cylindrical and have a cap-like, pierced, refractive top. Spores thread-like and split into part-spores that measure 8–14 × 1.5–2 µm. The asexual state is of the *Polycephalomyces*-type. Occurs on fly and wasp cadavers.

Ophiocordyceps forquinoni ⋉ has tiny stromata on flies; *O.myrmecophila* ▷ is very elegant, kills ants and has more distinct perithecial contours.

Mainly nemoral, very rare, but perhaps also somewhat overlooked; mostly July–September.

Ophiocordyceps myrmecophila is a very elegant, long-stemmed, yellowish *Ophiocordyceps* that has obliquely positioned perithecia, where the contours are just visible. Asci are cylindrical, 500–680 × 6–7 µm, and have a cap-like, pierced, refractive top. The thread-like spores (seen on top of the stroma on the image) split into part-spores and measure 7–11 × 1–2 µm. Occurs on cadavers of ants, mostly species of *Formica*.

Ophiocordyceps ditmarii ◁ does not have perithecial contours and kills flies and wasps.

Widespread, but mainly northern, rarely reported; summer–autumn.

Ophiocordyceps entomorrhiza is a grey to almost black, very long-stemmed *Ophiocordyceps* with an oval fertile head that at maturity is finely dotted-warty due to ostioles and perithecial contours; the base of the head is somewhat collar-like. The stem is greyish on the upper parts but becomes very dark at the base; it is rather stiff. The asci are cylindrical, 300–600 × 4–6 µm, and have a cap-like, pierced, refractive top. The thread-like spores split into part-spores that measure 6–8(–10) × 1.5–2µm. Often found in a parasitized state with asexual, white clubs from the parasite *Polycephalomyces ramosus* on the stem (see the lower image and page 1630). The asexual state of all *Ophiocordyceps* is of the *Hirsutella*-type. Occurs on buried larvae or more superficial imagines of carabid beetles, mainly from the genus *Carabus*.

Provided the host is known, identification should be straightforward. However, even without a host it should be possible to reach a correct identification as it is mainly detected by the presence of the white parasite.

Mainly nemoral–hemiboreal, occasional; mostly May–October.

Jens Ockelmann/Jens H. Petersen

Tolypocladium (formerly *Elaphocordyceps*) form club-shaped stromata on brown species of *Elaphomyces* (page 1276). The stromata may appear directly from the truffle itself or may be connected via thick, yellow hyphal cords; in the latter case it may be necessary to dig deep to find the truffle.

Tolypocladium ophioglossoides is a curry-yellow to olive-brown and finally almost black *Tolypocladium* that has an elongated, initially more brown fertile part with somewhat projecting ostioles, but with completely immersed perithecia (seen in section in lower image). The stem is connected to the host via thick, yellowish hyphal cords (upper image). The asci are cylindrical and have a cap-like, pierced, refractive top. The cylindrical spores split into part-spores that measure 2.5–5 × 1.5–2 µm. Occurs on brown species of *Elaphomyces* (on *E. muricatus* in upper image, see page 1276).

Also known as *Elaphocordyceps ophioglossoides*. The other *Elaphomyces*-associated species ▷ have more delimited, ± egg-shaped fertile parts.

Widespread, common, rare in the boreal zone; July–December.

Tolypocladium capitatum is a very fleshy, ± curry-yellow to clay-buff *Tolypocladium* that does not have hyphal cords at the base but sits directly on the host truffle. The ± egg-shaped fertile part is sharply delimited from the stem. The asci are cylindrical and have a cap-like, pierced, refractive top. Spores cylindrical and split into part-spores that measure 12.6–30 × 2.3–3.1 µm, av. 18 × 3 µm, not thickened at the ends. Occurs on brown species of Elaphomyces (page 1276).

Tolypocladium longisegmentatum ⋉ is very similar but on average has longer part-spores. *Tolypocladium rouxii* ▽ has more greyish-purple colours and the part-spores are somewhat shorter and broader.

Widespread, occasional to very rare; June–November.

Thomas Læssøe

Tolypocladium rouxii has very fleshy stromata with a greyish-green stem and an olivaceous-brown to brownish-black fertile head that has an incurved base. The perithecia are completely immersed. The part-spores measure (13–)15–22(–24) × 2.5–3.6 µm. Occurs directly on brown species of *Elaphomyces* (page 1276).

Tolypocladium longisegmentatum ⋉ has longer part-spores measuring 31.7–47.1 µm and brighter stromata; *T. capitatum* △ has narrower part-spores of ± the same length, a paler stem and brighter colours on the fertile part when young; the head grades into the stem.

Widespread, but rather rare; autumn.

Gerhard Koller

Claviceps parasitize grasses and *Eleocharis*. The infected flowers are transformed into black, elongated sclerotia (ergots). The sclerotia overwinter on the soil and in the spring purplish stromata appear. The ergots are very poisonous and have caused outbreaks of ergotism when grain crops have been contaminated.

††***Claviceps purpurea*** is a stromatic pyrenomycetous fungus that forms small, pinkish-buff fertile heads with immersed, purplish/lilac perithecia on a violet stem that arises from a black sclerotium. The sclerotium is formed in the previous season's grass flowers (image on right) and overwinters on the soil to produce new stromata once the grasses flower again. The stem has white hairs at the base. The asci are cylindrical and have a cap-like, pierced, refractive top, I-. Spores thread-like, septate. Parasitizes a range of different grass genera and species.

Claviceps microcephala ▽ is a closely related species on *Phragmites* that typically has smaller stromata. *Claviceps nigricans* ⋉ parasitizes *Eleocharis* (*Cyperaceae*) flowers.

Widespread, common; mainly May–July (stromata).

††***Claviceps microcephala*** is a very gracile *Claviceps* with ± purple, long-stemmed stromata that appear from small, black sclerotia in late spring. Microscopically the same as *C. purpurea* △. Typically abundant in downturned old *Phragmites* flower heads in reedbeds.

Determination relies on correct host identification. It is not currently clear whether *C. microcephala* should be treated as a separate species or is just part of the variation within *C. purpurea* △.

Probably widespread and likely to be common; May–July (stromata).

Epichloë form tubular stromata on living grass culms and is an endophyte, obtaining nutrients from the cells of the host. The fungus renders the grass poisonous, thus protecting it against herbivores. Typically, however, the ability of the grass to flower is reduced. The fly *Botanophila* (*Phorbia*) *phrenione* lays its eggs in the stroma, but also aids in transferring fungus spermatia (page 11).

Thomas Læssøe

†?*Epichloë typhina* is a stromatic pyrenomycetous fungus that has at first whitish, later yellowish-brown, tubular, 3–8 cm long stromata with crowded, ± immersed perithecia. The asci are cylindrical and have a cap-like, pierced, refractive top. The spores measure 150–210 × 1.5–2 µm and do not split into part-spores before release. Attacks a range of grasses; the host genera include e.g. *Anthoxanthum*, *Dactylis*, *Deschampsia*, *Phleum* and *Poa*.

A number of very similar, to morphologically identical species with different host preferences have been described (see below).

Widespread, common; May–September.

Thomas Læssøe

†?*Epichloë baconii* is a typical *Epichloë* with 10–17 mm long, yellowish-brown, tubular stromata. The asci are cylindrical and have a cap-like, pierced, refractive top. The spores split into two part-spores within the asci and measure 15–30 × (1.5–)1.8–2 µm. Occurs on *Agrostis* and *Calamagrostis*.

Further ± host-specific species include: Epichloë *clarkii* ⋉ on *Holcus*; *E. festucae* ⋉ on *Festuca*; *E. sylvatica* ⋉ on *Brachypodium*; *E. bromicola* ⋉ on *Bromus* and *E. elymi* ⋉ on *Elymus*.

Probably widespread and rather common; mainly July–August.

Neobarya parasitica is a non-stromatic pyrenomycetous fungus that forms olive-green to dark olive-brown, narrowly pear-shaped perithecia that are KOH inert; they may be ± solitary or in small clusters and typically have a cream subiculum. The asci are cylindrical, 120–180 µm long, have a cap-like, pierced, refractive top and are I-. The spores are thread-like (see lower image), somewhat shorter than the asci and entire when released. A whitish, asexual state has small, slimy heads, within which the conidiophores are situated; the conidia are 1–3-celled and measure (7.2–)13–18(–19) × (3.5–)4–5.2(–6.7) µm. Occurs on fruitbodies of *Bertia moriformis* (page 1598).

A very distinctive species. A range of *Neobarya* species occur on other fungi, including lichen thalli.

Widespread, but rarely reported, probably not rare; most of the year, although mostly recorded in mid-autumn.

Species of ***Acrospermum*** are very peculiar ascomycotes that form up to 3 mm high, almost clavate, rubbery, perithecioid structures with internal spore production and an opening at the top. The spores are thread-like and entire. Occur on dead herbaceous stems.

Acrospermum compressum is an up to 3 mm high, rubbery-tough, narrowly clavate, at first sessile, later ± stemmed, almost black *Acrospermum* with a small opening at the conical tip. The fruitbodies may be situated individually or in small clusters. The asci are cylindrical, very long (up to 500 × 3–4 µm), and I+. Paraphyses thread-like. The thread-like spores measure 280–440(–450) × 0.5–1 µm. It may produce several-celled, hyaline conidia on a hyphomycete state. Occurs on the previous year's herbaceous stems, mainly on *Urtica*.

Acrospermum graminum ▽ is smaller, paler and occurs on grasses; *A. pallidulum* ⋉ occurs on *Galium* stems; and the rare *A. adeanum* ⋉ occurs on mosses.

Widespread, common, becoming scarcer towards the north and rare in the boreal zone; all year, but mostly during spring.

Acrospermum graminum is an up to 1 mm high *Acrospermum* that has rather pale, grey-brown fruitbodies with a rather indistinct stem. The asci are cylindrical, 400–600 × 4–6 µm, and I+. Paraphyses thread-like. Spores thread-like, measuring up to 550 × 0.5–1 µm. Produces one-celled conidia on a hyphomycete state. Occurs on culms and leaves of various grasses.

The other three species of *Acrospermum* (see above) grow on dicot stems or on mosses.

Widespread, rarely reported, but probably rather common; April–June.

Tubeufia is a genus that has tiny, often yellow-brown, ± hairy or setose pseudothecia. Asci thick-walled. Spores slender and septate.

Tubeufia cerea is a bitunicate pyrenomycetous fungus with perithecioid, ± flat, rough to hairy, yellowish-brown to grey-olive pseudothecia that have a papilla-like ostiole. The hairs are ± cylindrical, yellow-brown and septate. Asci double-walled, 70–120 × 9–12 µm, I-. Spores hyaline, asymmetrically curved-fusiform, (5–)7–10(–13)-septate, and measure (27–)30–52 × 2.5–3.5(–4.5) µm. Occurs on old, ± dead stromata of *Eutypa*, *Diatrype* and similar stromatic pyrenomycetous fungi (pages 1569 & 1572)

The combination of substrate and distinctive spores in bitunicate asci makes identification straightforward.

Widespread, probably common; all year with most records in the warmer months.

Tubeufia helicomyces is a bitunicate pyrenomycetous fungus with barrel-shaped, yellowish to yellow-brown pseudothecia that have a circle of short, dark setae at the top. The asci are double-walled and measure up to 160 × 14 µm. Spores thread-like, measuring 66–150 × 2–3 µm, 5–7-septate and pointed at both ends. The asexual state forms pink colonies with large, helical conidia. Occurs on coarse swamp grasses, *e.g. Phragmites*, *Glyceria* and *Phalaris*.

A very distinctive species; *T. paludosa* is a synonym.

Probably widespread, rarely reported, but likely to be common; mainly June–September.

Acanthophiobolus helicosporus is a superficial, black, pear-shaped, bitunicate pyrenomycetous fungus that has up to 280 µm long, shiny, black, septate setae. Asci double-walled, measure 110–170(–200) × 6.5–8(–12) µm, I-; there are numerous paraphysis-like cells between the asci. The spores are thread-like and spun in a dense spiral inside the asci; they measure up to 170 × 1.5–2(–2.5) µm, are multicelled and contain many drops. Mostly occurs on leaves and stems of various monocots, *e.g. Typha* and *Carex*, in wet habitats.

Acanthophiobolus chaetophorus is a synonym. This species should be easy to identify. Other small, setose, bitunicate pyrenomycetous fungi can occur on monocots but they typically have shorter, sometimes branched setae.

Probably widespread and common; mainly April–September.

Hypocreopsis form large, fingered stromata that have imbedded perithecia in the inner parts but the tips are sterile. The spores are large and septate. All are parasites on the mycelium of *Hydnoporia* (pages 973 and 1033).

Hypocreopsis lichenoides is a spectacular hypocrealean fungus with dark red-brown ostioles on the inner parts of the finger-like stroma. The spores are hyaline, fusiform, 2-celled, almost smooth and measure (16–)24–30 × 8–9 µm. Occurs on and next to fruitbodies of *Hydnoporia tabacina* (see page 973), mostly on *Salix*.

The more Atlantic *H. rhododendri* ▷ is very similar but parasitizes *Hydnoporia corrugata* (page 1033) and has distinctly verrucose, 2-celled spores that may adhere to each other.

Widespread, occasional to rare; all year.

A parasite with small yellow perithecia recalling an *Ijuhya* ⋉ has been found on the stromata of *Hypocreopsis lichenoides*, which can also be accompanied by *Encoelia fimbriata* (page 1408).

Paul Cannon

Hypocreopsis rhododendri is a large, spectacular hypocrealean fungus with ± raised and projecting, stromatic, red-brown finger-like stromata from a central point. The red-brown perithecia are found on the inner parts of the fingers. Spores verrucose, 12–17(–25) × 12–13.5 (–17) µm; 2-celled, but may adhere to each other to create several-septate, *e.g.* four-spore-long, entities (see the drawing in the wheel). Parasitic on *Hydnoporia corrugata* (page 1033), *e.g.* on *Corylus* and *Salix*.

Hypocreopsis lichenoides ◁ is very similar but parasitizes *Hydnoporia tabacina* (page 973) and the spores are larger and smoother.

Only known from nemoral Atlantic parts of the UK within temperate Europe, rare; all year.

Selinia pulchra is a ± irregularly cushion-shaped, red-brown to clay-buff hypocrealean fungus with almost chimney-like, somewhat paler ostioles. It has up to 12 perithecia per stroma. The asci are normally 8-spored, measure 220–335 × 30–50 µm and lack a visible apical apparatus. The ± fusiform, smooth, somewhat golden, thick-walled spores measure 55–63 × 21.5–29(–37) µm and have a ± evident slime coat. Occurs on dung, *e.g.* from sheep, cows and deer.

This species should be very easy to identify. *Selinia intermedia* ⋉ apparently has smaller spores.

Distribution and frequency poorly known, very rarely reported; most of the year.

Thomas Læssøe

Trichoderma is a very large and difficult genus of soft-fleshed, effused, cushion- to club-shaped fungi with 2-celled spores that split into two at maturity. Formerly known as *Hypocrea*, with *Trichoderma* used for most asexual states. All species are probably parasites on other fungi.

Trichoderma leucopus has ivory, club-shaped stromata. The perithecia are only visible as tiny dots on the surface. Asci cylindrical; become 16-spored at maturity as the eight 2-celled spores split into two unequal parts; the upper, almost spherical part-spores measure (2.7–)3–3.5(–4) × (2.3–)2.8–3.2(–3.5) µm and the lower, somewhat fusiform part-spores measure 3.4–4.2(–5) × (2–)2.4–2.8(–3) µm. Occurs on soil and litter, mostly in coniferous forests, but is probably a fungus parasite and perhaps on species of *Cudonia* (page 1359) and/or *Spathularia* (page 1358).

Trichoderma seppoi ⋉ has smaller stromata and turns darker when dry; there is hardly any change in *T. leucopus*, which also has a clearer separation of the fertile part and the stem. *Trichoderma alutaceum* ▽ is more irregular in shape, somewhat darker and occurs on wood of deciduous trees.

Hemiboreal–boreal, occasional, generally rare in the nemoral zones but occasional in Denmark; July–December.

Niels V. Mogensen/Jens H. Petersen

Trichoderma alutaceum has fleshy, dark buff, club-shaped to antler-shaped stromata. The asci are cylindrical and, at maturity, have 16 unequal part-spores; the upper measure (3–)3.3–4(–4.8) × (2.8–)3–3.6(–4) µm and the lower (3.2–)4–5(–6) × (2.3–)2.7–3.1(–3.5) µm. Occurs on wood of deciduous trees, particularly *Fagus*, but probably a fungus parasite with an unknown host.

Trichoderma leucopus △ is paler, longer-stemmed and occurs on soil/litter; *T. nybergianum* ⋉ is a northern, more reddish-brown, but similarly fleshy species.

Mainly nemoral, rare; July–December.

Trichoderma pulvinatum is a vivid yellow to more cream, cushion-shaped *Trichoderma*. The perithecia (seen in longitudinal section in lower image) are only visible as tiny, dark dots on the stromatal surface. There is a darker, more intense pigmentation surrounding the ostioles. The surface is finely felty (microscopically, verrucose cells can be seen). The asci are cylindrical and, at maturity, have 16 part-spores that are all of a similar shape (monomorphic) and measure (2.5–)3.3–5(–6.3) × (2–)2.5–3.2(–3.7) µm. Occurs on old fruitbodies of *Fomitopsis betulina* (page 833) and *Fomitopsis pinicola* (page 867).

Hypocrea protopulvinata ⋉ also occurs on *Fomitopsis pinicola* but is smooth not felty, browner, and has part-spores of unequal shape (dimorphic).

Widespread, common; all year, but typically at its best in late summer.

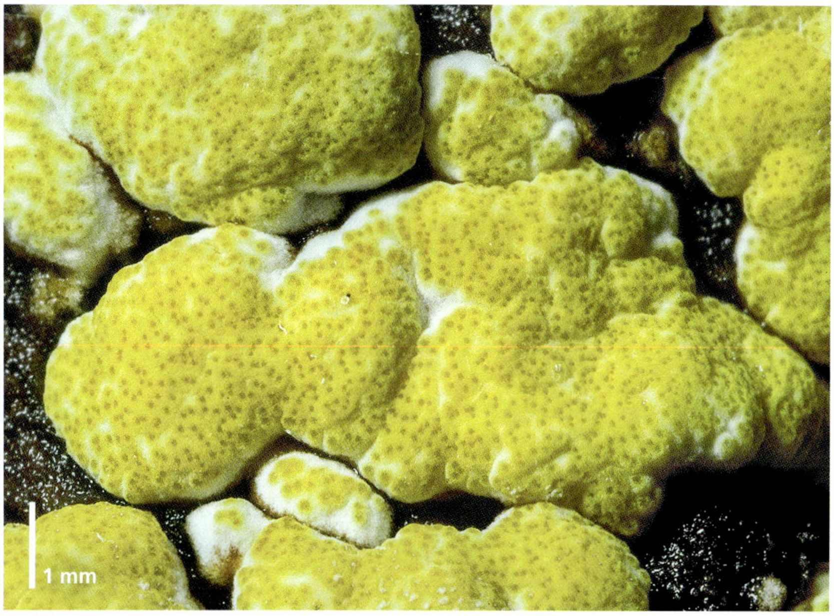

Trichoderma citrinum is a thick and effused, almost corticioid, pale yellow *Trichoderma* where the perithecia are just visible as tiny dots on the surface. The stromata may incorporate items in the litter. Asci are cylindrical and have 16 unequal part-spores that measure (3.7–)4.4–5.2(–5.8) × (3.5–)3.8–4.4(–4.8) µm and (3.6–)4.7–6(–7) × (2.8–)3.3–4(–4.3) µm. Occurs on and around deciduous and coniferous stumps and also on bare soil, but probably a fungus parasite on an unknown host.

Thicker than most corticioids, and a hand lens will reveal the ostiolar dots.

Widespread, rather common to occasional, becoming scarcer towards the north; mainly July–November.

Trichoderma europaeum is a distinctly cushion-shaped, vinaceous-brown to greyish-brown, smooth *Trichoderma* that has completely immersed perithecia. The stromata may be singular or, as in the top image, aggregated. The asci are cylindrical and have 16, unequal spores; in the upper part they are egg-shaped, measuring (3.7–)4–4.8(–6) × (3.2–)3.5–4(–5) µm, and in the lower part conical, measuring (4.2–)4.8–6(–7.2) × (2.7–)3–3.5(–4) µm. Has a turquoise-green asexual state. Conidia smooth. Mostly on rather fresh to partially decomposed deciduous wood, usually *Fagus*, rarely directly on other fungal fruitbodies or on coniferous wood.

Formerly known as *T. minutisporum,* a North American species. *Trichoderma atlanticum* ⋉ has somewhat smaller spores; *T. pachybasioides* ⋉ is very similar but normally lacks the pinkish tinges. A rather large, mostly ± cherry-red *Trichoderma* on *Quercus* branches is currently included in *T. europaeum*. *Trichoderma neorufoides* ⋉ and *T. rufa* ⋉ have often been confused with *T. europaeum*. *Trichoderma rufa* has very small, finely downy stromata, prefers coniferous wood and has verrucose conidia.

Nemoral–hemiboreal, common; mainly September–November.

Trichoderma foliicola is a cushion-shaped *Trichoderma* that forms small, at first white, later yellowish-brown, smooth or squamulose stromata with very obvious, dark ostioles. The asci are cylindrical and, at maturity, have 16 part-spores; the upper measure (4.5–)5–6.3(–7.5) × (3.2–)3.8–4.5(–4.7) µm and the lower (4.8–)5.3–6.5(–7.5) × (3–)3.3–3.8(–4.2). The asexual state is white. Occurs on leaf-litter from *Quercus*.

The combination of substrate and a white asexual state is diagnostic.

Nemoral, rarely reported, likely to occur in the hemiboreal zone but not recorded; mainly August–September.

Trichoderma gelatinosum is a small, cushion-shaped, green-spored, waxy/gelatinous, ± translucent, ± amber-yellow *Trichoderma*. When dry it is almost black. Both the greenish ostiolar dots and the outlines of the sparse perithecia are seen clearly on the surface. The asci are cylindrical and, at maturity, have 16 unequal part-spores; the upper measure (3.3–)4–4.8(–5.4) × (3.2–)3.5–4.3(–5) µm and the lower (3.4–)4.3–5.7(–6.7) × (2.5–)3–3.7 (4.2) µm. The asexual state has green conidia. Occurs on rotten wood of deciduous trees.

Trichoderma gelatinosum is the only green-spored, waxy/gelatinous *Trichoderma* in temperate Europe. The non-gelatinous *Trichoderma strictipile* ▽ is less amber-coloured and has more indistinct perithecial contours. *Trichoderma thelephoricolum* ⋉ is found in connection with *Steccherinum ochraceum* (page 1049); it is not gelatinous, but otherwise is quite similar.

Nemoral–hemiboreal, rarely reported and formerly much confused with *T. strictipile*; mostly August–November.

Trichoderma strictipile is a green-spored, matt, smooth *Trichoderma* with small, rounded to ± flattened stromata that lack very obvious perithecial contours; cream to buff with green ostioles. Part-spores unequal, (4–)4.7–5.5(–6.5) × (3.7–)4.4–5(–5.4) and (4.5–)5–6.5(–8) × (3.3–)3.7–4.3(–5) µm. On decorticated deciduous wood, mostly *Fagus*, but probably a fungus parasite.

Trichoderma aureoviride ▷▷ is somewhat larger, darker yellow and ± minutely scaly on the surface; *T. gelatinosum* △ has obvious perithecial contours on the gelatinous stroma.

Mainly nemoral, rather common; July–March.

Trichoderma spinulosum is a very small, cushion-shaped, green-spored *Trichoderma* with rather projecting dark green ostioles. The part-spores are unequal, the upper measuring (2.7–)3.5–4.8(–5.7) × (2.5–)3–4 (–4.7) µm and the lower (2.8–)3.4–5 (–6.4) × (2.2–)2.7–3.4(–4.3) µm. Has a white asexual state. Occurs on grass culms, rarely on *Juncus* or dicot herbs.

Trichoderma aerugineum ⋉ has a green asexual state; *T. danica* ⋉ is almost identical but can be identified by cultivation and sequence-analysis. There are also white-spored species of *Trichoderma* on grass culms.

Probably widespread, rarely reported; mainly July–October.

Thomas Læssøe

Trichoderma aureoviride is a rather large, lentil-shaped, initially finely-scaly, yellow *Trichoderma* with a green-dotted surface and green spores. The asci are cylindrical and have 16 part-spores. Spores mainly of uniform size, (2.8–)3.2–4(–5) × (3–)3.3–4.3(–5) µm, except the lowest positioned. Has a green asexual state. Found as groups of stromata on dry, exposed decorticated wood of deciduous trees, e.g. *Sambucus* in coastal scrub.

The combination of habitat, rather vivid colour and scaly surface should make identification easy.

Nemoral, rather common; mostly August–November.

Trichoderma atrobrunneum is a greenish to blackish, ± lentil-shaped *Trichoderma* with dark green spores. The asci soon have 16 unequal part-spores: the lower measure (3–)4–4.3 (–6) × (2.8–)3.7–4(–5) µm, and the upper (3.3–)4.2–4.5(–7) × (2.5–)3–3.7 (–4.4) µm. Occurs on rotten wood with other fungi.

A number of related species rarely form sexual states; *T. simmonsii* ⋉ is very similar and can hardly be separated on morphological characters; *T. lixii* ⋉ is a tropical species.

Nemoral, rarely reported; most of the year.

Protocrea farinosa is a ± cream, somewhat mealy-grainy, thin, stromatic hypocrealean fungus. The pale amber-coloured ostioles are visible in the mealy covering. The perithecia, including the ostioles, are KOH inert. Asci cylindrical, 75–90 × 3–4 µm; at maturity have 16 part-spores that are finely spiny and similar, measuring 3–4.5 × 2–3 µm. The asexual state is of the *Gliocladium*-type. Occurs on underside of old fruitbodies of the polypore *Skeletocutis nemoralis* (page 902).

Easily identified when the host is known. *Protocrea pallida* ▽ stains purple with KOH.

Widespread, common, scarcer towards the north; most of the year.

Protocrea pallida is a hypocrealean fungus that covers the entire pore surface of the host with a thin, pale, loose stroma that contains pale orange to flesh-pink perithecia. The ostioles typically stain purple with KOH. The asci are cylindrical and have 16 unequal part-spores that measure 2.8–3.1 × 2.2–2.6 and 2.9–3.6 × 2.2–2.4 µm. Has a '*Gliocladium*'-state. Possibly exclusively on *Postia lactea* (page 895).

Hypomyces aurantius ▷▷ has larger, more vivid orange perithecia and fusiform, non-splitting spores.

Nemoral–hemiboreal, occasional; August–April.

Arachnocrea stipata is an almost cobweb-like hypocrealean fungus that has yellowish-brown perithecia in a pale, hyphoid stroma. Perithecia KOH inert and collapse sideways when drying. The asci are cylindrical and have 16 part-spores that are finely rough, almost conical and measure 3.6–5.2 × 1.8–2.4 µm, one part spore having a more distinct acute tip than the other. Occurs on litter, including twigs, in different forest types, and apparently also on old fungi.

The spore shape is characteristic and should make identification straightforward. *Trichoderma delicatulum* ⋉ is somewhat thicker and has rounded spores.

Nemoral–hemiboreal; all year.

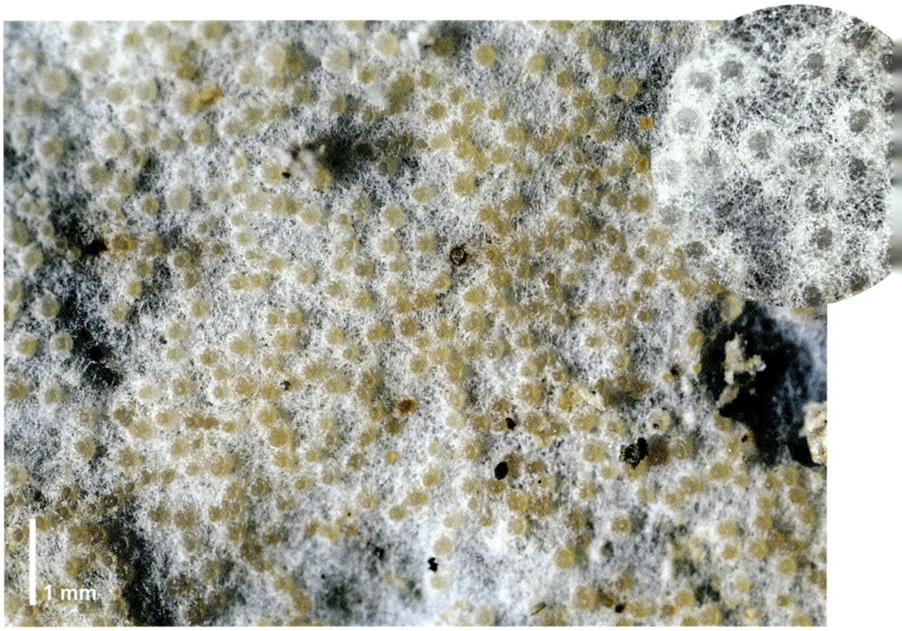

Cosmospora, ***Dialonectria***, ***Flammocladiella***, ***Macroconia***, ***Nectria***, ***Neonectria*** **and** ***Thyronectria*** all form small, red to orange perithecia that may be situated on a small stroma. All typically have 2-celled spores that remain entire. Many species are parasites, either on trees or on other fungi, and several species, *e.g.* of *Neonectria*, are considered highly harmful pathogens. All have traditionally been treated in a broadly circumscribed *Nectria*.

Nectria cinnabarina has a sexual state on raised cushions upon which many aggregated, finely verrucose perithecia are situated. The perithecial wall is rather complex when seen in section. Asci have a small apical apparatus, 8-spored, I-. The hyaline, smooth, (0–)1(–2)-septate spores measure (11.5–)14–17.5(–21.5) × (3–)4–5.5(–7) µm. The pink to orange, rarely darker, asexual 'sporodochium'-state (see also page 1632) usually has a short stem or a distinct constriction at the base. The conidia measure (4–)5.2–7(–8.5) × (1.3–)1.9–2.7 (–3.4) µm. Occurs on a number of deciduous hosts, including *Acer*, *Fagus* and *Aesculus*.

Nectria cinnabarina has recently been divided into a number of very similar species based on, for example, the appearance of the asexual cushions, whether sessile or stemmed, and whether ± blackening. The following species in the complex occur in Europe: *N. cinnabarina*, *N. nigrescens* ⋉ (with up to 3-septate spores) and *N. dematiosa s.l.* ⋉ (with sessile sporodochia).

Widespread, very common; all year.

Nectria cinnabarina is sometimes seen with the presumed parasitic *Nitschkia parasitans* (page 1597).

Neonectria ditissima is a nectrioid fungus with red, smooth perithecia, which are formed in and around cankers. The ostioles are somewhat darker and somewhat raised. The perithecia maintain their shape upon drying and stain ± purple in KOH. Asci are clavate and lack evident apical apparatus. The 2-celled, almost smooth spores measure 15–19 × 6–9 µm. The asexual, cream, cushion-shaped state produces micro- and macroconidia. Causes cankerous wounds on deciduous trees, *e.g. Fraxinus*; particularly problematic in apple orchards.

Also known as *Nectria galligena*.

Widespread, common; all year.

Neonectria coccinea is a nectrioid fungus with small, smooth, red perithecia in crowded groups, but not on obvious stromatic cushions. Spores 2-celled, brownish, somewhat verrucose, 12–15 × 5–6 µm. The asexual '*Cylindrocarpum*'-state forms two kinds of conidia; the larger are 4–8-celled, 46–80 × 6–7 µm. Occurs together with *Cryptococcus fagisuga* (Woolly Beech Scale) on *Fagus* bark, where it may produce cankers; it may, perhaps, also occur as a decomposer on other hosts.

A difficult species. Published keys use the growth rate in culture as the main distinguishing character.

Mainly nemoral, common – at least in a broad sense; all year.

Neonectria fuckeliana is a bright red, smooth nectrioid fungus that forms crowded groups of mature and immature perithecia. The ostiole is somewhat darker and raised. The asci have ± evident ring-shaped apical apparatus. The 2-celled, finely spiny spores measure (11–)12.5–15.5 (–19.7) × (4–)4.7–6(–7) µm. The asexual state forms microconidia. Forms cankers on *Picea* and *Pinus* and may cause serious damage.

The similar *N. neomacrospora* ⋉ causes problems in Christmas tree (*Abies*) cultivation.

Widespread, probably common; most of the year.

Cosmospora arxii is a nectrioid fungus that has small, bright red perithecia with indistinct, papillate ostioles. The 2-celled, finely ornamented spores measure 6.5–8.5 (–10) × 3–4 µm. Has a slimy, asexual '*Acremonium*'-state with conidia that measure 4–8.5 × 1.5–2.5 µm. Occurs on stromata of *Hypoxylon fragiforme* and *H. howeanum* (page 1556).

The similar *C. annulohypoxyli* ⋉ occurs on *Jackrogersella multiforme* and *J. cohaerens* (page 1561), and *C. deustae* ⋉ occurs on *Kretzschmaria deusta* (page 1563).

Nemoral–hemiboreal, probably common; mostly during winter.

Cosmospora coccinea is a bright red nectrioid fungus with rather distinctly delimited ostioles. The spores are 2-celled, pale brown, have large warts and measure (13.5–) 14.5–17(–18.5) × (8.5–)10–12.5 (–14.5) µm. Occurs on very decomposed fruitbodies, mainly of *Mensularia nodulosa* (page 926).

Also known as *Nectria cosmariospora*. The combination of coarsely ornamented spores and the host should make identification easy. *Cosmospora fomiticola* ⋉ occurs on *Fomes fomentarius* (page 870).

Presumably widespread, rarely reported; all year.

Dialonectria episphaeria is a small, red nectrioid fungus that, upon drying, becomes laterally compressed. Has somewhat raised ostioles. The ± yellowish, 2-celled spores measure 7–12 × 3.5–5 µm. Has a *Fusarium* asexual state. Mostly on stromata of *Diatrype* (page 1572).

Part of a species complex, with the original species described from *Diatrype* on *Crataegus*. Very similar species are found in the genus *Cosmospora* △. *Stylonectria purtonii* ⋉ has a different asexual state and occurs mainly on *Cytospora*-like pyrenomycetous fungi (page 1576).

Widespread, common; all year.

Macroconia leptosphaeriae is a small, pear-shaped, red nectrioid fungus with ± obvious pale basal hyphae and surface hairs. The 2-celled spores measure 14–18 × 5–5.5 µm. Occurs on and around fruitbodies of the genus *Leptosphaeria* (page 1607), mostly on stems of *Urtica*.

Macroconia sphaeriae is a very closely related or perhaps identical species on the same substrate; it appears to have somewhat larger spores, 19–25 × 5.8–6.5 µm. If it proves to be the same species, *M. sphaeriae* will be the correct name.

Probably widespread, rarely recorded; mainly June–October.

Flammocladiella decora forms small clusters of vivid, golden, finely warty-hairy perithecia on the ostioles of the host; the inset image shows a longitudinal section of a *Massaria* fruitbody with *Flammocladiella* perithecia in the ostiole. Before the perithecia are formed a '*Fusarium*'-state may be present. The spores are hyaline, fusiform, 4-celled and measure 20–30 × 5.5–7 µm. Parasitic on *Massaria inquinans* (page 1606) on *Acer*.

Superficially recalls an orange *Nectria cinnabarina s.l.* ◁◁.

Mainly nemoral, rare but more likely overlooked; mostly in the autumn.

Thomas Læssøe/Jens H. Petersen

Thyronectria sinopica is an erumpent, reddish, ± yellow to yellow-green scaly nectrioid fungus with darker, papillate ostioles. The perithecia are situated on a reddish, stromatic cushion. Asci measure 65–70 × 6–8 µm, I-. The 2-celled, ± smooth spores measure 9–12 × 4–6 µm. Usually has red pycnidia. Occurs on roots and branches of *Hedera*.

Also known as *Nectria sinopica* or *Pleonectria sinopica*. The combination of substrate and distinctive coloration should make identification straightforward.

Nemoral, occasional; mostly during winter.

Lasionectria vulpina has pale cinnamon-buff, very hairy perithecia; the hairs are 10–50 µm long and tinged orange. The perithecial wall has somewhat widened cells in the outer layer and more compressed cells farther in; inert to KOH. The asci are ± club-shaped, measure 37–58 × 6–8(–10) µm and have a ring-shaped apical apparatus; I-. The longitudinally striate, ± hyaline, 2-celled spores measure (7–)8–11 (–13) × 3–4 µm. Has an asexual *'Acremonium'*-state. Mostly on decorticated deciduous wood.

Lasionectria sylvana ⋉ occurs on herbaceous stems and has longer hairs and larger spores; *L. mantuana* ⋉ has finely dotted, non-striate spores; and *L. flavida* ⋉ is more vividly coloured.

Possibly widespread, rarely reported; probably most of the year.

Lasionectria lecanodes sits on an indistinct pale subiculum and has ± apricot-orange, finely white-haired perithecia with a reddish, papillate ostiole. The hyaline, 2-celled spores have small warts in longitudinal rows and measure 9–12 × 3–4 µm. Occurs on foliose lichens (in *Peltigerales*), e.g. *Peltigera* (page 1490) and *Nephroma* (page 1491).

Also known as *Nectriopsis lecanodes*. The combination of substrate and spore morphology should make identification straightforward.

Widespread, occasional; most of the year.

Hydropisphaera peziza has ± smooth, almost rubbery, orange-yellow perithecia that have a central depression when dry (inset image). It occurs mostly in large aggregations. The ostiole is seen as a small papilla. The perithecial wall is inert to KOH and has an outer layer of ± spherical cells. The asci are 8-spored, club-shaped and have a ring-shaped apical apparatus; I-. The 2-celled spores are longitudinally striate (best seen in cotton blue) and measure (9–)11–14(–17) × (3–)5–7 µm. Occurs on wood of deciduous trees, rotten polypores and even on dung.

Hydropisphaera arenula ⋉ occurs on monocot stems and has somewhat longer spores. Other species of *Hydropisphaera* have ± well-developed hairs on the perithecia. There are various other species and genera with striate spores within the hypocrealean fungi but these differ in other ways.

Mainly nemoral–hemiboreal, common; mostly August–November.

Nectriopsis form violet or whitish perithecia in a pale subiculum. Spores small and 2-celled. Occur on fungal fruitbodies and myzetozoa.

Nectriopsis oropensoides is a very pale, minutely hairy hypocrealean fungus with ± crowded or rarely more solitary perithecia on a thin, pale subiculum. The ostioles are dark olivaceous-buff and raised. The hyphae in the subiculum have a spiny surface. Asci cylindrical, 8-spored, I-. The hyaline, 2-celled, minutely spiny spores measure 4.5–5.5 × 2.8–3 µm. Occurs on fruitbodies of various lignicolous fungi *e.g. Heterobasidion* (page 856) and *Phellinopsis conchata* (page 885), and on bark of *Salix* and other deciduous trees.

Nectriopsis queletii ⋉ occurs on *Phlebia albida* ⋉ and also on species of *Peniophora* (page 1020); it has even smaller spores, 3–4.5 × 2–3 µm. *Sphaerostilbella broomeana* ▷ occurs on *Heterobasidion* but has much larger spores.

Nemoral–hemiboreal, rarely reported; mainly August–October.

Jens H. Petersen/Thomas Læssøe

Nectriopsis violacea is a violet hypocrealean fungus that forms a dense layer of perithecia on a felty subiculum on top of the host. Asci 8-spored, cylindrical, I-. The spores are 2-celled, cylindrical, finely spiny and measure (5–)7–8 × 2.5–3 µm. Has a white '*Acremonium*'-state with slimy heads and conidia that measure 6–9.5 × 2–3 µm. Occurs on aethalia of myzetozoa, *e.g.* on *Fuligo* (page 1648).

The combination of colour and host makes identification straightforward.

Widespread, probably common; May–January.

Sphaerostilbella broomeana is a hypocrealean fungus with pale brown, minutely pale-haired perithecia in a pale subiculum. The asci are cylindrical and 8-spored. The hyaline, minutely spiny, 2-celled, spores measure 12–16 × 3.5–4 µm; they are constricted at the central septum and have very short, pointed appendages at the ends. Has an asexual '*Gliocladium*'-state with hyaline conidia in slimy heads that are 0–1-septate and measure (3–)5–9 (–11) × 2–3.5 µm. Occurs on old fruitbodies of *Heterobasidion* (page 856).

Provided the host is correctly identified, identification should be straightforward.

Mainly nemoral, rarely reported, but probably overlooked; mostly August–October.

Hypomyces tubariicola is rarely seen with perithecia. When present, they are well hidden in a white subiculum and rather pale, but are somewhat brownish when dry. The 2-celled, fusiform spores are finely verrucose and measure 12–16 × 3–4 µm. The conidia measure 6–8.5 × 3.5–4.5 µm and are formed in phialides; the conidiophores are shorter than 300 µm. Occurs mostly on fruitbodies of *Tubaria furfuracea* (page 645).

Provided the host is correctly identified, identification should be easy. *Hypomyces tremellicola* ⋉, mostly on *Crepidotus* (page 90), has longer conidiophores, up to 800 µm.

Nemoral, probably rather common; mainly August–January.

T'homas Læssøe

Jan Vesterholt

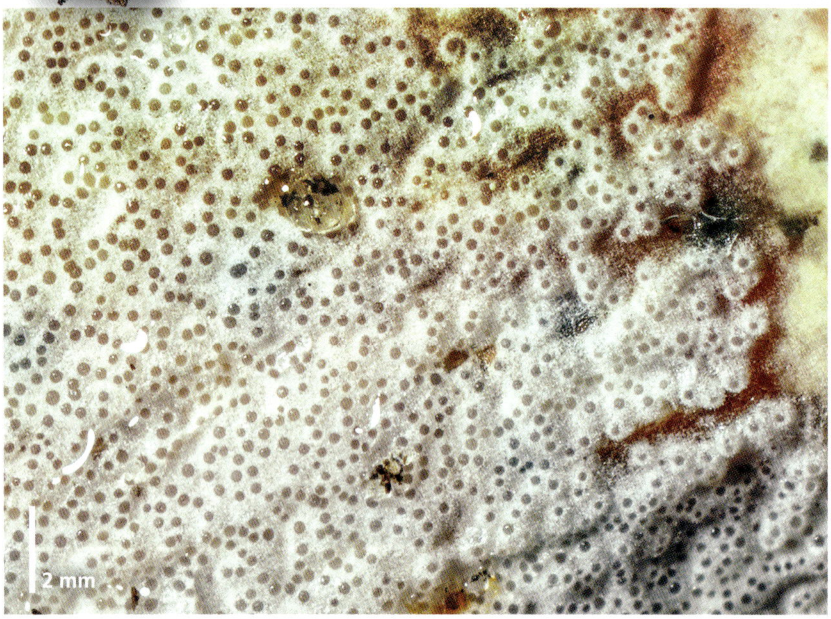

Species of ***Hypomyces*** parasitize other fungi. In some cases, the host fruitbody is allowed to stay fresh, but the *Hypomyces* takes over the entire hymenium. In other cases, the host rots away, sometimes with the flesh replaced by the asexual resting spores of the parasite. Spores mostly fusiform, 1–2-celled, and have delimited, pointed ends.

Hypomyces lateritius replaces the gills of the host with a dense layer of ± pale brown to pinkish perithecia held together by a thin stroma. The perithecia and stroma are KOH inert. The remaining host tissue stays fresh and unaltered. The asci are 8-spored, cylindrical and I-. The one-celled, ± fusiform, warty-spiny spores measure (11–)15–21(–30) × (2.5–)3.5–5(–6) µm and have pointed ends up to 4 µm long. Parasitizes species in the *Lactarius deliciosus* group (page 444).

Hypomyces spadiceus ▽ parasitizes species in the *Lactarius torminosus* group (page 418) in a similar fashion. In North and Central America a similar reddish-orange species, *H. lactifluorum* ⋉, is considered a choice edible; it is known as the 'lobster mushroom'.

Widespread, occasional; July–October.

Hypomyces spadiceus transforms the gills of the host into a brownish stroma with ± immersed, brown perithecia that stain red-brown with KOH. The asci are 8-spored, cylindrical and I-. The one-celled, spiny, fusiform spores measure (14–)17–22(–25) × (3.5–)4–5(–6) µm and have long, pointed ends. Occurs on members of the *Lactarius torminosus* group (page 418).

Hypomyces lateritius △ transforms the gills of the *L. deliciosus* group (page 444) in a similar manner.

Widespread, rather common; July–November.

Hypomyces luteovirens transforms the gills of the host and most of the stem into a thin, greenish-yellow to yellow-green stroma that has almost free perithecia on the surface. The asci are 8-spored, cylindrical and I-. The one-celled, spiny spores measure 32–35 × 4.5–5.5 µm; the ends are pointed and up to 7 µm long. Occurs on rather small fruitbodies of *Russula* (page 372), mostly in damp habitats.

The combination of host and colours should make identification straightforward.

Widespread, occasional; July–November.

Hypomyces chrysospermus is mostly seen in an initially white, later yellow asexual state, but rarely olive-green perithecia are also formed on the host. The spiny, hyaline spores have a non-central septum, measure 20–25(–30) × 4–5(–6) µm, and have long, pointed ends. The asexual '*Sepedonium*'-state is dominated by large, spherical, yellow resting spores 14–21(–25) µm in diameter. The *Boletales* act as hosts, but it is somewhat unclear how many different species and genera are involved. However, it apparently occurs on *Paxillus* (page 757), *Melanogaster* (page 1264) and various boletes, including members of *Xerocomus* (page 784) and *Boletus* (page 76).

Hypomyces microspermus ⋉ is superficially very similar, but the asci have much smaller spores, (8–)10–15 × (2.5–)3–4 µm, and the yellow, spherical conidia (resting spores) are also smaller, 9–15 µm in diameter; it occurs mostly on species of *Xerocomellus* (page 786) and similar bolete genera. A number of other *Hypomyces* /*Sepedonium* species occur on various boletes, *e.g. Sepedonium chalcipori* ⋉ on *Chalciporus* (page 812), *Hypomyces tulasneanus* ⋉ on *Suillellus luridus* (page 808) and *Sepedonium laevigatum* ⋉ on *Leccinum* (page 762).

Widespread, very common; June–November.

Thomas Læssøe/Benny Tom Olsen

Jens H. Petersen/Thomas Læssøe

Hypomyces porphyreus covers and transforms the host, so that it becomes rather similar to an *Ophiocordyceps* ◁◁. Stroma thin and brown, with perithecia on a rounded head; the stem has brownish scales. Asci 8-spored, cylindrical, I-. Spores 1–2-celled, rough, 20–30 × 3.5–5 µm; they have a central septum and somewhat pointed ends. Occurs on *Entoloma*, mainly or exclusively from the subgenus *Pouzarella* (page 476), on calcareous soils, typically in scrub.

Tolypocladium ◁◁ and *Ophiocordyceps* ◁◁ have very long, ± splitting spores.

Nemoral–hemiboreal, very rare; June–October.

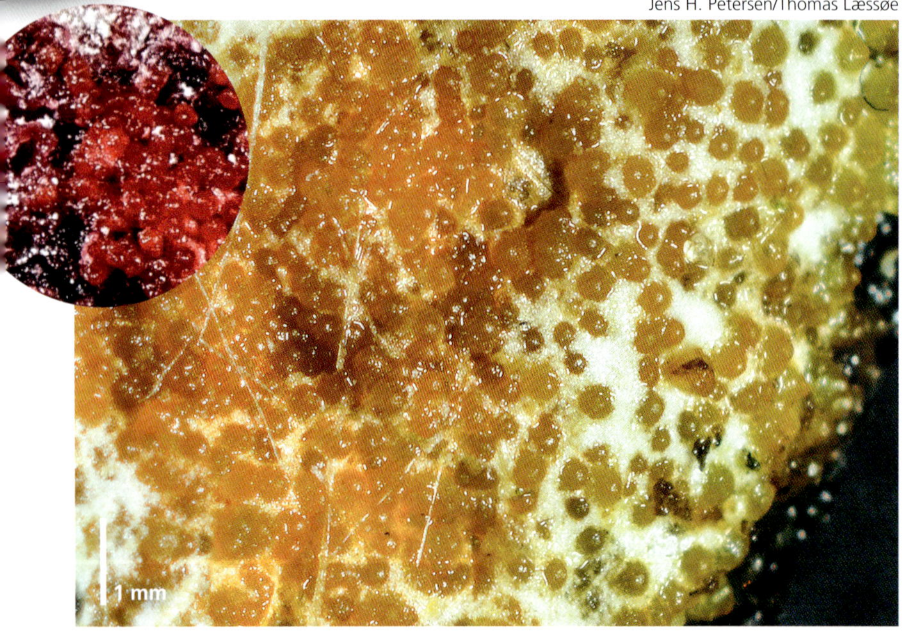

Hypomyces aurantius forms ± apricot-orange perithecia in a loose, white subiculum that at first also produces 1–2-celled, hyaline conidia measuring 8.5–21 × 5–10 µm; a '*Cladobotrym*'-state. The perithecia stain purple with KOH (inset image). The 2-celled, hyaline, spiny spores measure (15–)20–25(–27) × 4–6 (–7.5) µm; they have a septum in the middle and delimited, pointed ends up to 4.5 µm long. Occurs on polypores and other wood-degrading fungi.

Hydropisphaera peziza ◁◁ lacks an evident subiculum, is KOH inert and has rounded, striate spores.

Widespread, common; all year.

Hypomyces rosellus forms small, whitish- to pinkish-tinged membranous stromata with superficial, carmine to pale purple perithecia. Spores 2-celled, hyaline, spiny, (20–)25–30(–38) × (3–)4–5(–7.5) µm; ends mostly slightly curved, pointed and 4–6 µm long. Has a white asexual state with slender phialides and (1–2–)3–4-celled conidia that measure 16–35 × 7–13 µm. Occurs on fungi; the host is completely degraded when the perithecia are formed.

Hypomyces odoratus ⋉ is similar but the asexual state has very short, thick phialides, and the conidia are 2-celled and measure 15–35 × 7.5–12 µm.

Widespread, occasional; all year, although mainly in mid-autumn.

Thelocarpon belongs to a separate order, the *Thelocarpales*. It is a genus of both lichenized and decomposing/parasitic species. The fruitbodies are tiny, fragile and bright yellow, and thus resemble some hypocrealean pyrenomycetous fungi. The asci are generally large and thick-walled, and have numerous spores. Typically occur on rotten deciduous wood with algal films or on lichens.

Thelocarpon epibolum has tiny, perithecioid, greenish-yellow, floccose fruitbodies. The gelatinous material in the hymenium is I- or I+ (red). The asci are multispored, pointed, and may sometimes stain a weak blue at the tips with iodine. Paraphyses are simple to branched. The spores are hyaline and measure 4–6 × 1.7–2 µm. Occurs on rotten wood or on thalli of foliose lichens.

The variety *T. epibolum* var. *epithallinum* ⋉ has larger spores measuring (6–)8–10(–12) × 3 µm and typically occurs on thalli of *Peltigera* (page 1490). Similar species of *Thelocarpon* are more apothecioid or have larger spores.

Rare or overlooked; April–November.

Thelocarpon lichenicola has tiny, greenish-yellow, ± perithecioid fruitbodies with a somewhat exposed hymenium. The hymenial gel does not react with iodine. The asci are multi-spored, cylindrical and I-. Has paraphysis-like cells. The spores are hyaline, one-celled and measure (4–)5–7.5 × 1.5–2.5 µm. Occurs on rotten wood, lichen thalli, soil, rocks, *etc.*; mostly in connection with algal film.

Characterized by the cylindrical asci, presence of paraphysis-like cells and the relatively open fruitbodies.

Widespread, very rarely reported; probably most of the year.

Hard, stromatic pyrenomycetous fungi

A form group of fungi with black, mostly rather carbonized perithecia that are ± immersed in a stroma (plural stromata), and which may be club-, nail- or cushion-shaped or quite flat. The stroma may be well defined, as in most genera on the left hand side of the wheel, or composed of a mixture of tissue from the host plant and the fungus, as in many species on the right hand side of the wheel.

All species are decomposers, but some may also kill the host, *e.g. Kretzschmaria deusta* and *Eutypa spinosa*. Some species may survive as endophytes in living tissues and only become active when the host dies.

The majority of the hard, stromatic pyrenomycetous fungi are found in three families: *Xylariaceae* has brown, ± asymmetrical ellipsoid spores, mostly with a germ slit; *Diatrypaceae* has paler allantoid spores and asci with a long tail-like base; and *Diaporthaceae* has hyaline, 1–2-celled spores and loosening asci. The first two families mostly have an amyloid ascus apparatus.

The genera *Daldinia, Hypoxylon*, and *Jackrogersella* have diagnostic pigments that can be dissolved with KOH; scrape the surface onto blotting paper and add a drop of 10% KOH (corrosive) – the pigment will be drawn into the paper (see pages 1557–1560).

Other similar fungi:
– hypocrealean pyrenomycetous fungi have a softer texture and paler or brighter colours (page 1512).
– the non-stromatic pyrenomycetous fungi and the bitunicate pyrenomycetous fungi have ± free perithecia (pages 1582 & 1602).

See also the general wheel, page 1510.

Further reading: 61, 78, 82, 83, 93, 130, 179, 250, 251, 254, 255, 256, 313, 314, 353.

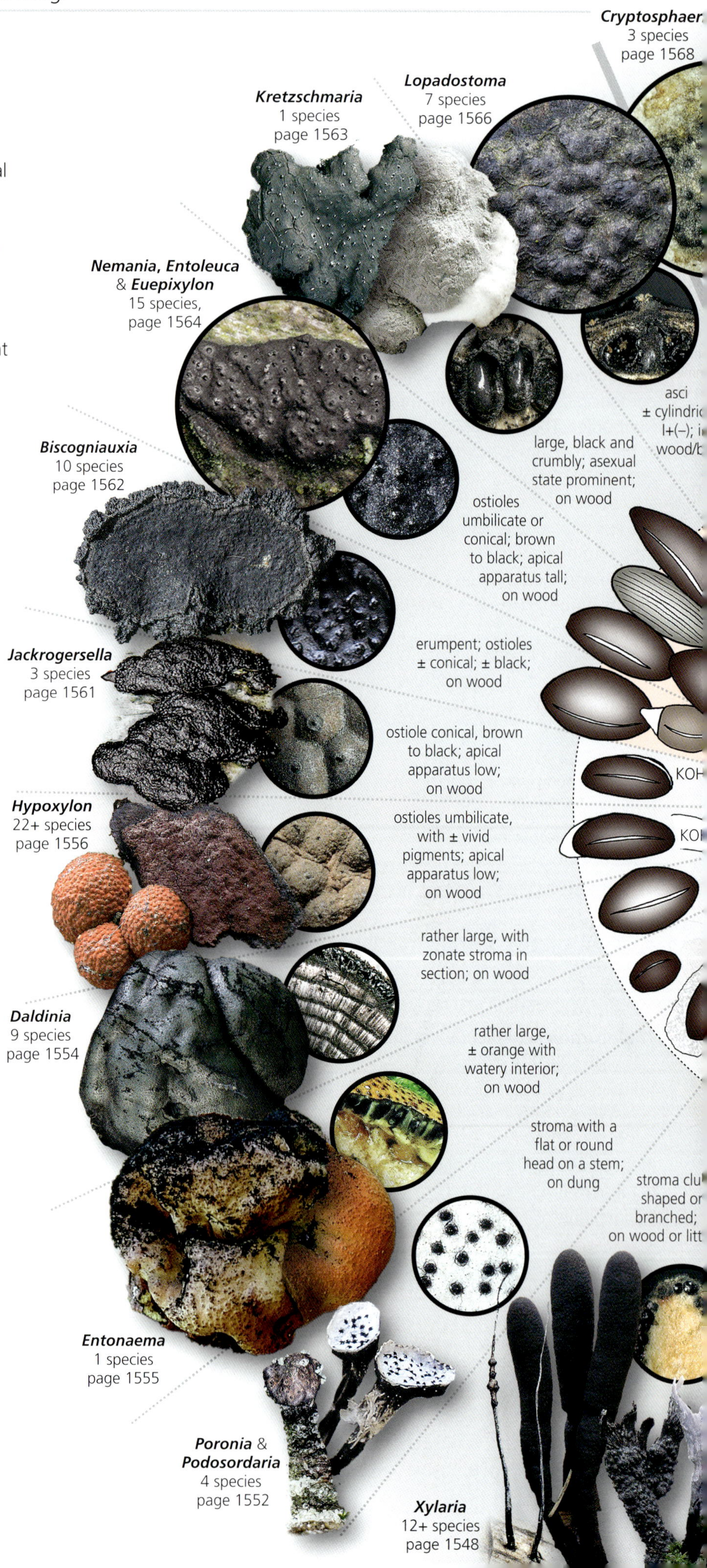

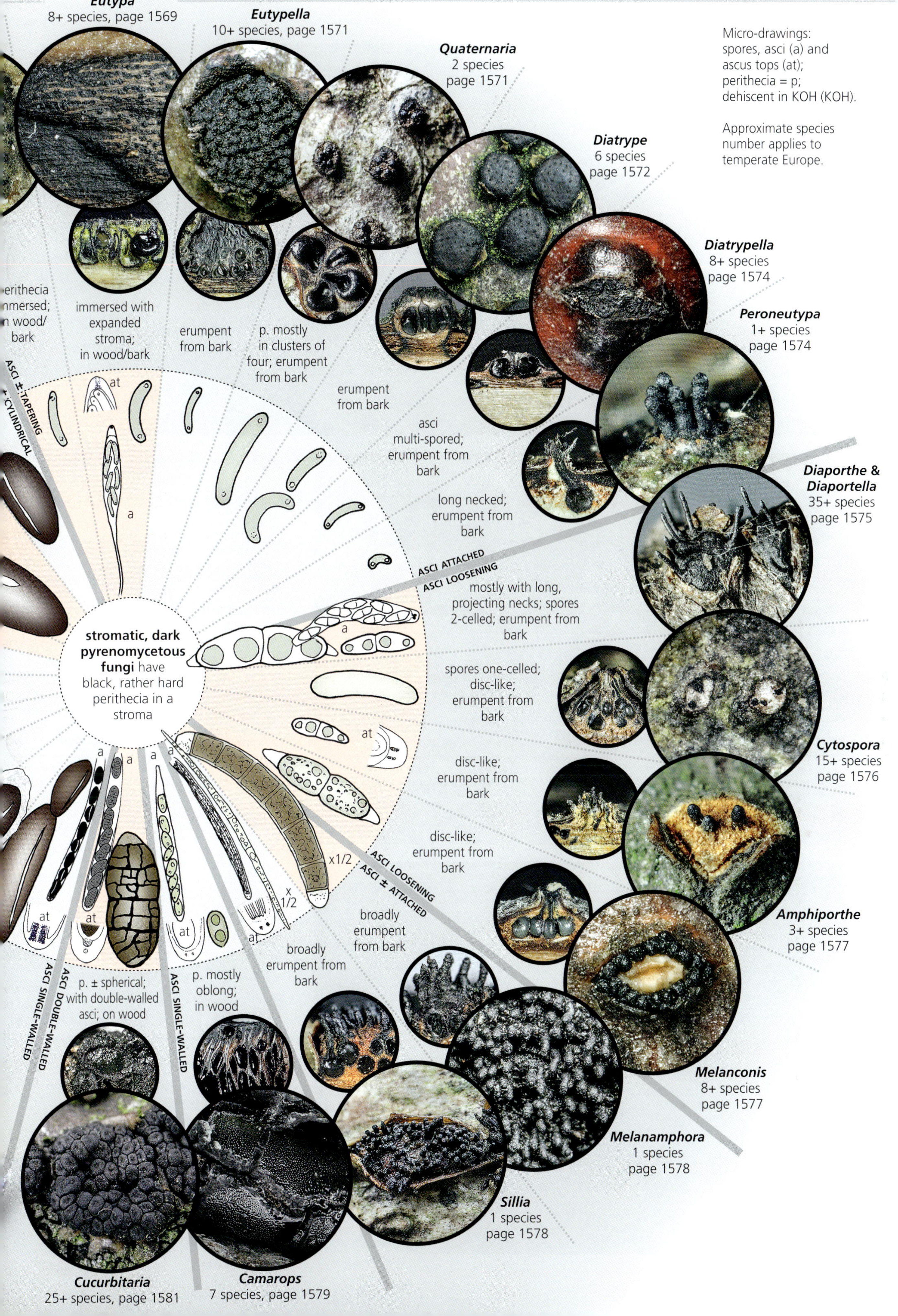
Micro-drawings: spores, asci (a) and ascus tops (at); perithecia = p; dehiscent in KOH (KOH).
Approximate species number applies to temperate Europe.
Eutypa
8+ species, page 1569
Eutypella
10+ species, page 1571
Quaternaria
2 species
page 1571
Diatrype
6 species
page 1572
Diatrypella
8+ species
page 1574
Peroneutypa
1+ species
page 1574
Diaporthe & Diaportella
35+ species
page 1575
Cytospora
15+ species
page 1576
Amphiporthe
3+ species
page 1577
Melanconis
8+ species
page 1577
Melanamphora
1 species
page 1578
Sillia
1 species
page 1578
Camarops
7 species, page 1579
Cucurbitaria
25+ species, page 1581
immersed with expanded stroma; in wood/bark
erumpent from bark
p. mostly in clusters of four; erumpent from bark
erumpent from bark
asci multi-spored; erumpent from bark
long necked; erumpent from bark
mostly with long, projecting necks; spores 2-celled; erumpent from bark
spores one-celled; disc-like; erumpent from bark
disc-like; erumpent from bark
disc-like; erumpent from bark
broadly erumpent from bark
broadly erumpent from bark
p. mostly oblong; in wood
p. ± spherical; with double-walled asci; on wood
ASCI ± TAPERING + CYLINDRICAL
ASCI ATTACHED
ASCI LOOSENING
ASCI LOOSENING
ASCI ± ATTACHED
ASCI SINGLE-WALLED
ASCI DOUBLE-WALLED
ASCI SINGLE-WALLED
stromatic, dark pyrenomycetous fungi have black, rather hard perithecia in a stroma
at
a
x1/2
x 1/2

Thomas Læssøe

Xylaria has black, ± imbedded perithecia in small to very large, thread-like, club-shaped to branched stromata that have whitish flesh. The ± black spores have a straight or spiralling germ slit. Decomposers of wood and fruits.

Xylaria polymorpha is a fleshy and variable, ± clustered, usually rooting *Xylaria* with a rough, minutely scaly, dark brownish stromatal surface. The tips are ± rounded, but may be pointed if damaged by drought. Initially covered by a greyish conidia-producing layer. Asci with large, I+ apical apparatus. Spores dark brown, with a flattened side, (19.8–)20.4–23.7(–25.6) × (6.1-)6.3–7.4 µm, with a rather short germ slit on flattened side. On deciduous trees, mostly in parks and gardens, but also in forests.

Xylaria longipes ▷, *X. digitata* ⋉ *and X. corniformis* ⋉ all have the ± same type of stromatal surface but have smaller spores and a different germ slit. *Xylaria longipes* is typically slender. more cylindrical, never rooting and with sinuous germ slit; *X. digitata* has clustered stromata from a common rooting part, the stromatal tip is always sterile, and the spores have an oblique germ slit; *X. corniformis* has a straight full length germ slit and small, often flattened stromata.

Nemoral–hemiboreal, common to rather rare; asexual from late spring, with perithecia mostly from August to early spring.

Xylaria longipes is a rather variable, but typically long and slender *Xylaria* with indistinct brown scales and a whitish flesh. Occasionally seen with flattened, small stromata. The tips may be rounded or somewhat tapering. Initially covered in grey conidiophores and conidia (lower image). Asci I+. The brown spores measure (11.4–)11.8–13.4(–13.9) × (5.2–)5.5–6.1(–6.5) µm and have a sinuous germ slit along almost their full length, with the centre on the flat side. Occurs almost exclusively on decorticated, mostly half-buried wood of *Acer pseudoplatanus* and other *Acer* species.

Xylaria corniformis ⋉ occurs on various deciduous trees, including *Carpinus* and *Fagus*; it can look similar but has smaller spores and a straight germ slit. *Xylaria polymorpha* ◁ is typically more fleshy and irregularly shaped, and has larger spores with a short, straight germ slit. *Xylaria digitata* ⋉ is clustered and has somewhat larger spores with an oblique germ slit; it typically occurs on worked timber, *e.g.* wooden gate posts, and was perhaps introduced to Europe from North America.

Nemoral–hemiboreal, common to occasional; asexual from April to June, with perithecia July–December.

Xylaria bulbosa is a rather pale, somewhat branched, at the tips sterile and flattened *Xylaria* that grows out of a rather large, black, internally firm and white sclerotium (seen in section at the bottom of the image). The brown surface layer is very thin. The brown spores measure 9.2–11.9(–13.9) × 3.4–4.6(–5.3) µm. Occurs on thick needle debris from *Picea* and *Juniperus*.

This is the only *Xylaria* that forms sclerotia in needle debris.

Nemoral, very rare and very rarely reported; summer-autumn.

Thomas Læssøe

Thomas Læssøe

Xylaria hypoxylon is a *Xylaria* with two very different morphologies. The ± branched, flattened, white, asexual state has a dark, bristly-hairy base (lower image) and is easy to detect. The sexual state is more inconspicuous, and has black, unbranched or somewhat branched stromata with rather obvious perithecial contours and a similar hairy stem base; the surface has peeling, somewhat silvery longitudinal fibres and small, discoid ostioles. The spores measure 10–12 × 4.5–5.2 µm and have a straight germ slit on the flat side that clearly does not extend along the full length. Occurs on mostly dark-stained wood of deciduous trees, mostly stumps.

At least three similar species occur, or could occur, in temperate Europe: *Xylaria karsticola* ⋉ has very constricted stromata with deep fissures and very pointed spores with longer germ slits; *X. vasconica* ⋉ has cylindrical stromata in clusters, a paler brown surface and small, conical ostioles; and *X. cinerea* ⋉ has small, black grains on a ± white surface and somewhat larger spores with a slightly oblique or sinuous germ slit. All decay deciduous wood.

Widespread, very common, but becoming scarcer towards the north; all year.

Jens H. Petersen/Thomas Læssøe

Xylaria oxyacanthae is a slender *Xylaria* that apparently only rarely develops a mature sexual state (inset image). The asexual state is pale and soft with ± distinct pink tinges. The brown spores measure (9–)10–11.5 (–12) × 4.5–5.4 µm. Occurs almost exclusively on buried *Crataegus* fruits, but there are also records from *Cotoneaster* and *Cornus*.

A distinctive species.

Nemoral, rare to occasional; June–August (asexual), September–November (sexual).

Xylaria carpophila is a very slender *Xylaria* that is mostly seen in a sterile or asexual state, but which also occurs with an almost thread-like stroma that has distinct perithecial contours and a striate, peeling surface layer. The spores measure 10–13 × 4–5 µm and have a germ slit along the full length of the flattened side. Occurs on *Fagus* cupules deep in litter, or from buried cupules.

Xylaria delitschii ⋉ occurs on *Carpinus* fruits.

Nemoral, very common; all year, but mostly sexual July–October.

Xylaria filiformis is (as interpreted here) a thread-like, smooth *Xylaria* with a long, sterile or conidiophore-bearing section above the very obvious, but very few, perithecial contours. The brown spores measure (14.2–)15.5–17.5(–18.7) × (4.6–)4.8–5.4(–5.5) µm and have an indistinct germ slit. Occurs on coarse herbaceous stems, *e.g. Aconitum* and *Epilobium*.

The taxonomy of the *Xylaria filiformis 'group'* requires further work. The original description was from material growing on *Syringa* leaves in central Europe, and the British mycologist R.W.G. Dennis described a fungus with the same ecology from the British Isles, although this is different from the fungus depicted here. There is also a 'species' with coarser stromata and larger spores that, like *X. filiformis*, occurs on herbaceous stems.

Mainly hemiboreal–boreal, occasional; summer-autumn.

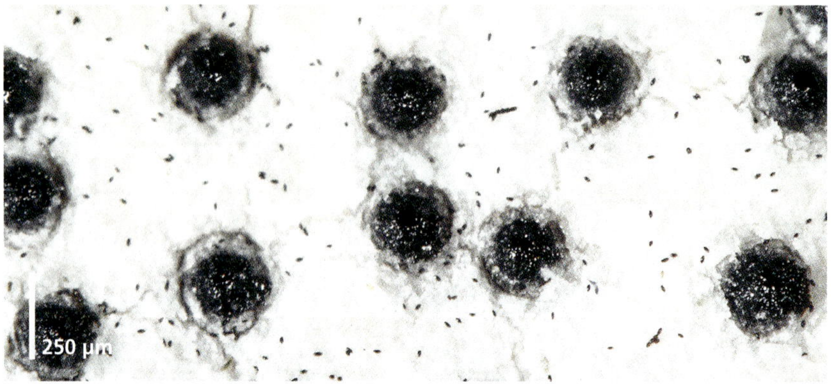

Poronia form peculiar, short- to long- stemmed stromata that have a flattened head with black ostioles in a white surface. The flesh is white and firm. The black spores are ± symmetrical and covered in a thick slime coat. Occur on herbivore dung in dry, exposed habitats.

Poronia punctata is a very distinctive, nail-shaped stromatic pyrenomycetous fungus with a white stroma surface speckled with scattered, black ostioles from immersed perithecia. The young state is white to pinkish and very powdery from conidia. The stromata have a root-like extension into the substrate. The opaque, black, one-celled spores measure 17–26 × 8.5–13 μm and have a thick slime coat. Occurs exclusively on dry, sun-exposed horse dung, mostly in dunes and heathlands, but also on alvar (Öland, Sweden).

Poronia erici ▷ has larger spores and typically smaller stromata; it is mainly recorded on hare (*Lepus*) and rabbit (*Oryctolagus*) dung, but may also occur on cow and horse dung.

Nemoral–hemiboreal, locally common; for a long period restricted to only a few sites but has increased again in recent decades; all year, but in best condition in the autumn.

Thomas Læssøe

Poronia erici is a nail-shaped, typically few millimetre wide, white, stromatic pyrenomycetous fungus with black ostioles on the white surface. The stromata are deeply rooting. The spores have a thick slime coat, a short germ slit and measure (22–)25–32(–33.5) × (12–)14–18 (–19.5) µm. Occurs on dung, mostly from hare (*Lepus*) and rabbit (*Oryctolagus*), rarely cow and horse, in dry grassland, dunes and similar places.

Poronia punctata ◁ has larger stromata and smaller spores; in temperate Europe it is only known from horse dung.

Nemoral, rare; all year.

Thomas Læssøe

Podosordaria tulasnei is a blackish, typically partly buried, usually long-stemmed *Podosordaria* with a small, pointed stromatic head that has obvious perithecial contours and an inner white flesh. Asci 8-spored, I+. The brownish-black spores measure 20–26 × 11–15 µm, have a very evident germ slit along almost their entire length and a thick slime coat. Occurs on ± buried dung of hare and rabbit, with the stromatal head just above the soil surface (insert image). Typically found in sandy habitats, such as dunes.

A distinctive species, although it is important to measure the spores. *Podosordaria pedunculata* ⋉ has larger spores that measure 40–60 × 20–30 µm.

Nemoral–hemiboreal, occasional, easily overlooked; probably mostly September–April.

Jens H. Petersen/Jan Vesterholt

Species of ***Daldinia*** have large stromata with concentric, contrasting, internal zones. Emit large amounts of black spores during the night. Occur on wood, often on fire-damaged but living deciduous trees. Most have diagnostic KOH extractable pigments; see page 1546.

Daldinia concentrica is a *Daldinia* that forms very large semiglobose, at first purple-brown, later black stromata with a finely cracked surface and indistinct ostioles. The inner stroma is distinctly zonate in shades of grey and does not have cavities. Pigment purple in KOH. The distinctly asymmetrical, somewhat pointed spores measure 13–17.5 × 6–7.5 µm and have a germ slit along the full length of the more convex side. The outer wall is dehiscent in KOH. Occurs predominantly on dead or dying *Fraxinus* (it occurs as an endophyte in healthy trees).

Daldinia petriniae ⋉ has limited amounts of pigment, smaller stromata and typically occurs on *Alnus* and other members of *Betulaceae*. *Daldinia loculata* ⋉ typically occurs on fire-damaged trees and the outer spore wall is non-dehiscent in KOH. There are other similar species of *Daldinia* just south of the area covered.

Nemoral, common in the Atlantic parts to rather rare in continental parts; mainly sporulating April–June.

Thomas Læssøe

Daldinia decipiens is a rather small, usually somewhat stemmed, purple-brown *Daldinia* that has greyish internal zones but lacks conspicuous cavities or gelatinous pockets. Pigment purple in KOH. The ± pointed spores measure 14–18 × 6.5–10 µm, have a full-length germ slit and the outer wall is dehiscent in KOH. Occurs on *Betula* and other members of *Betulaceae*.

Daldinia petriniae ⋉ is mostly larger and less stemmed, and has less pigment and smaller spores.

Nemoral–hemiboreal, rare; mostly in the autumn.

Daldinia vernicosa is a *Daldinia* that forms relatively small, mostly somewhat stemmed, ± shiny, rather brown-tinged stromata. The inner stroma has contrasting grey-black and white zones and areas of gelatinous material (inset image). Pigment purple in KOH. The somewhat asymmetrical spores measure 11–14.5 × 6–8 µm and have a non-dehiscent outer wall in KOH. Occurs on deciduous wood, mostly fire-damaged living trees.

Daldinia loculata ⋉ is not distinctly gelatinous, typically has larger stromata and is sessile.

Mainly nemoral, rather rare but possibly more common in continental parts; mainly sporulating in September–October.

Entonaema cinnabarina is a unique large, ± cushion-shaped, basally constricted, yellow-orange stromatic pyrenomycetous fungus with a surface roughened by perithecial contours and papillate ostioles. The flesh is gelatinous, centrally hollow and filled with a ± thick liquid that eventually disappears, shrinking the stroma to a irregular shape. It smells of lovage and has orange pigment in KOH. Asci I+. Spores asymmetrical, brownish-black, measure 9.5–11.5 × 4.8–6 µm and have a germ slit. Thermophilous; occurs on deciduous wood, *e.g. Fraxinus*.

A very distinctive fungus.

Nemoral, very rare and in temperate Europe only recorded in the Baltic republics – also known from Bulgaria; summer–autumn.

Thomas Læssøe/Jacques Fournier

Hypoxylon form semiglobose to flattened stromata with ± umbilicate (not raised) ostioles. Asci have a low, I+(–) apical apparatus; spores are black and have a germ slit on the more convex side. The outer, hyaline spore wall is dehiscent in KOH. Decomposers on deciduous wood, sometimes endophytic and appearing soon after host's death. KOH-dissolving pigments are useful in identification; see page 1546.

Hypoxylon fragiforme is an almost globose, brownish-orange, later black *Hypoxylon* with bumpy perithecial contours and umbilicate ostioles. Pigment orange-red in KOH. Initially has a greyish, asexual state that covers the young stroma. The spores are flat on one side, have a full-length germ slit on the convex side, and measure 10–15 × 4.8–6.8 µm, av. 12.4 × 5.7 µm. Occurs almost exclusively on bark of recently fallen *Fagus* trees, rarely on other deciduous trees.

Hypoxylon howeanum ▽ has a branched asexual state and smaller spores, av. 7.3 × 3.8 µm; it does not occur on *Fagus*.

Nemoral, very common; in best condition in late autumn, but present all year.

Sometimes covered by perithecia of *Cosmospora arxii* (page 1536).

Hypoxylon howeanum is an almost globose, brownish-orange *Hypoxylon* with a ± branched, ± pale brown, asexual state. Ostioles umbilicate. Pigment orange-red in KOH. Spores measure 6.8–8.8 × 3.4–4.8 µm, av. 7.6 × 3.8 µm. Occurs on bark of various deciduous trees, but particularly *Quercus* and *Corylus*.

Hypoxylon fragiforme △ has larger spores, av. 12.4 × 5.7 µm, and occurs almost exclusively on *Fagus*; it has a non-branched asexual state.

Mainly nemoral–hemiboreal, occasional to rather common; in best condition during the autumn.

Hypoxylon fuscum is an irregularly cushion-shaped to conical, rarely flattened *Hypoxylon* that lacks obvious perithecial contours. Pigment with KOH amber, isabelline or brownish (inset image). Asci 8-spored, I+. Spores measure 11–16 × 5–8 µm and have a somewhat 'S'-shaped germ slit on the convex side. Asexual state greyish. Occurs almost exclusively on *Corylus*.

Hypoxylon fuscoides ⋉ is very similar but has purple pigment in KOH and occurs mainly on *Alnus*; *H. porphyreum* ▽ occurs only on *Quercus*; and *H. laschii* ⋉ on *Populus*.

Widespread and common, becoming scarcer towards the north; all year.

Thomas Læssøe

Hypoxylon porphyreum is a ± irregularly cushion-shaped to conical *Hypoxylon* with lemon-yellow pigment below the purple-brown surface; olive-brown with KOH. Asci 8-spored, I+. Spores measure 11–13.5 × 4.5–5.4 µm, have a full-length germ slit on the more convex side (may be 'S'-shaped), and the outer wall is dehiscent in KOH. The asexual state is pinkish-buff. Occurs on thin *Quercus* branches.

Hypoxylon fuscum △ has orange-brown pigment just below the surface and occurs on *Corylus*.

Hemiboreal, rare; all year.

Hypoxylon perforatum is a rust-coloured, flattened *Hypoxylon* that, with age, develops white ostioles. Pigment is pale yellow-brown in KOH. Asci 8-spored, I+, ascus stem 24–51 µm long. Spores measure 9.7–11.5 × 4.7–5.3 µm and have a full-length germ slit on the convex side. Occurs on deciduous wood.

Hypoxylon rubiginosum ▷▷ does not usually develop white ostioles, the pigment is flame-red and the ascus stem is longer. *Hypoxylon fuscum* △ and *H. intermedium* ⋉ have larger spores but similar pigments.

Mainly southern within temperate Europe, uncommon; all year.

Thomas Læssøe

Hypoxylon cercidicola is an erumpent, ± disc-shaped *Hypoxylon* with a ± star-shaped thickened margin comprising cells of both host and fungus. Pigment orange with KOH. Asci 8-spored, I-. Spores measure 9.5–11.5 × 4.8–6.2 μm and have a full-length germ slit on the convex side. Mature perithecia are difficult to find. The asexual state is pale olive-brown and disc-shaped with a lobed margin. Occurs on dead branches in the crown of *Fraxinus* trees.

Hypoxylon petriniae ▽ has a sharp, dark margin and the asci are I+.

Nemoral, rare or overlooked; all year.

Thomas Læssøe

Hypoxylon petriniae is a flat, somewhat purple-tinged *Hypoxylon* that lacks obvious perithecial contours but has a sharply delimited, dark margin; may be somewhat immersed in the wood. Pigment orange in KOH (inset image). The spores measure 8.8–11.5(–13) × 4.8–6 μm, av. 10.7 × 5.1 μm and have a full-length germ slit on the convex side. Occurs almost exclusively on *Fraxinus* trunks and branches.

Hypoxylon rubiginosum ▷▷ has obvious perithecial contours and a diffuse, concolorous margin. *Hypoxylon macrocarpum* ▷▷ has dark brown pigment in KOH and a less delimited margin.

Nemoral–hemiboreal, common to occasional; all year, but in best condition late autumn.

Hypoxylon salicicola is a ± flat, ± orange-brown *Hypoxylon* with ± visible perithecial contours. The ostioles are mostly somewhat disc-shaped-umbilicate. Pigment orange-brown with KOH. Asci 8-spored, I+ (weak). The spores are brown, almost symmetrical, measure 7.2–9.6 × 3–4.2 µm and have a full-length germ slit; the outer wall is dehiscent in KOH. Occurs on attached *Salix* branches in marshes.

Hypoxylon rubiginosum ▷▷ has larger, asymmetrical spores.

Widespread, rare, apparently absent from central Europe; all year.

Thomas Læssøe

Hypoxylon vogesiacum is a ± flat, ± purple-tinged *Hypoxylon* with rather indistinct perithecial contours and a well-defined, rounded stromatal margin. Pigment purple in KOH (not old material). The spores measure 18–29 × 8–11.5 µm, av. 22.3 × 9.8 µm and have a full-length germ slit on the convex side. Occurs mostly on hard, decorticated *Ulmus* trunks.

The combination of large spores and pigment colour make this species easy to identify.

Hemiboreal–boreal, occasional; all year.

Chlorostroma vestlandicum ⋉ occurs as a parasite on the stromata.

Hypoxylon subticinense is a very vivid rust-coloured, flat *Hypoxylon* with rather small stromata, small perithecia, very indistinct perithecial contours and a somewhat felty surface, especially at the margin. Pigment vivid orange-red in KOH. The almost symmetrical spores measure 8–11.5 × 4–5.5 µm, av. 9.7 × 4.7 µm. Occurs on branches of various deciduous trees, mostly on the bark.

Hypoxylon ticinense ⋉ is very similar macroscopically but has smaller spores, av. 5.9 × 2.9 µm.

Mainly in central Europe but recently recorded as far north as Denmark, occasional to very rare; all year, in best condition in autumn.

Thomas Læssøe

Thomas Læssøe

Hypoxylon macrocarpum is a flat, ± somewhat purple-tinged *Hypoxylon* with a ± shiny surface. Pigment dark brown to grey-brown in KOH (inset image). The young stromata and the wood underneath smell strongly of vanilla. Spores brown, measure 9.5–11.5 × 4–4.8 µm, av. 10.5 × 4.4 µm, and have a full-length germ slit. Occurs on very hard wood from deciduous trees, rarely on bark (and then somewhat more cushion-shaped).

Similar species of *Hypoxylon* have ± orange pigment in KOH.

Common in the nemoral zones, rare in the hemiboreal zone; all year, but in best condition in the autumn.

Thomas Læssøe

Hypoxylon rubiginosum is a ± flat, ± orange-brown *Hypoxylon* with obvious perithecial contours (upper inset image) and a diffuse margin. Pigment vivid orange in KOH (lower inset image), and there is a shiny red layer just below the surface. Initially has a grey-brown asexual state (lower image), which almost recalls a corticioid fungus. Asci I+, ascus stem 60–98 µm long. Spores measure 8.8–11(–12) × 4–5.5 µm, av. 10.1 × 4.4 µm, are flat on one side and have a full-length germ slit on the convex side; the outer wall is hyaline and dehiscent in KOH. Occurs mostly on decorticated wood of various deciduous trees, but also on bark.

Hypoxylon perforatum ◁◁ has a more yellowish pigment in KOH. *Hypoxylon petriniae* ◁◁ has a dark margin, indistinct perithecial contours and a somewhat purplish colour. *Hypoxylon cercidicula* ◁◁ has a raised margin on smaller, circular stromata. *Hypoxylon macrocarpum* △ has dark brown pigment with KOH. *Hypoxylon ticinense* ⋉ and *H. subticinense* ◁◁ have a felty stromatal margin.

Widespread, common, becoming scarcer towards the north; in best condition in the autumn.

Thomas Læssøe

The larvae of *Mycocecis ovalis*, a gall midge, live within the perithecia and cause it to enlarge.

Jackrogersella is mainly separated from *Hypoxylon* ◁ by the projecting, papillate/conical ostioles. The species treated here were formerly placed in *Annulohypoxylon*.

Jackrogersella multiformis forms ± cushion-shaped, rarely effused, very carbonous, red-brown to black stromata with obvious perithecial contours and papillate ostioles. Pigment ± brown in KOH. Asci have small, I+ apical apparatus. Spores asymmetrical, 8.8–10.3 × 3.7–4.9 µm, and have a slightly less than full-length germ slit on the convex side; outer wall usually indehiscent in KOH but a small bump develops on the curved side. Occurs on bark and wood of deciduous trees, *e.g. Betula* and *Alnus*.

Jackrogersella minutellum ⋉ has an almost strawberry-coloured pigment in KOH; *J. cohaerens* ▽ is restricted to *Fagus*, has small, confluent stromata that are more constricted at the base, and the spores have a full-length germ slit. Species of *Hypoxylon* ◁ have ± umbilicate ostioles. Species of *Nemania* ▷▷ are less carbonous, the pigments do not dissolve and the germ slit is on the flat side of the spore.

Widespread, very common; all year.

The parasite *Immothia atrograna* ⋉, a bitunicate pyrenomycetous fungus with 2-celled, brown spores, can occur on the stromata.

Jackrogersella cohaerens has rather small, at the base somewhat constricted, brown to black stromata with obvious conical ostioles. Pigment olive to wine-red in KOH. Asci I+. The asymmetrical spores measure 8.8–12.2 × 3.4–5.5 µm and have a full-length germ slit on the convex side. Occurs on *Fagus* bark and wood.

Jackrogersella multiformis △ forms larger, less constricted stromata and has a shorter germ slit; it hardly ever occurs on *Fagus*. *Jackrogersella minutellum* ⋉ occurs on *Quercus* and has an almost strawberry-coloured pigment in KOH.

Nemoral, common; all year.

Biscogniauxia form ± erumpent, very carbonous stromata with umbilicate or conical ostioles. Asci have I+ apical apparatus. The ± brown spores are ellipsoid to almost spherical and have a germ slit. A few species have 2-celled spores with a hyaline smaller cell. Occurs on deciduous wood.

Biscogniauxia repanda forms black, erumpent, ± circular stromata that have a serrated, raised margin and a concave to flat disc with conical ostioles. The almost black, ellipsoid spores measure 10–12 × 4.5–5 µm and have a germ slit that runs all the way round the spore (double-sided). Occurs on dead, corticated or decorticated parts of standing *Sorbus*, or on fallen *Ulmus* trunks.

Biscogniauxia marginata ⋉ has almost spherical spores with a sinuous germ slit and a less raised margin. *Biscogniauxia simplicior* ⋉ has umbilicate ostioles and occurs on *Rhamnus catharticus*.

Mainly hemiboreal–boreal where rather common but very rare in the nemoral zones; all year.

Biscogniauxia nummularia has greyish-black, flat, erumpent stromata with small, conical ostioles. The brown, one-celled, symmetrical spores measure 10–13 × 7.5–8.5 µm and have a full-length germ slit. Occurs on recently dead *Fagus* branches and trunks; it is an endophyte in living *Fagus*.

Biscogniauxia granmoi ⋉ has umbilicate ostioles and occurs on *Prunus padus*. Species in the *Diatrype stigma/decorticata* group ▷▷ form thinner stromata and have allantoid, paler spores.

Nemoral, rather common to very rare in the north; all year, but in best condition during winter.

Biscogniauxia cinereolilacina forms flat, greyish-violet stromata with small, conical ostioles. The spores are brown, measure 12–14 × 5.5–7 µm, and have a double-sided, full-length germ slit. Occurs on recently dead *Tilia* branches and trunks.

Hypoxylon vogesiacum ⊲⊲ occurs mostly on *Ulmus*, has larger spores and a violet pigment in KOH. Species of *Diatrype* ⊳⊳ and *Eutypa* ⊳⊳ have allantoid spores.

Mainly hemiboreal–southern boreal, rather rare; all year.

John Bjarne Jordal

Kretzschmaria deusta forms very large, flat, bumpy stromata that are initially soft- and white-fleshed with a greyish conidial formation from the surface (lower image), but later form a sexual phase with large immersed perithecia and finally has a fragile, carbonized crust. The ostioles are conical. The asci are 8-spored and have a high, I+ apical apparatus. The dark brown, asymmetrical spores have a germ slit along half the length of the flatter side and measure 25–31 × 7–10 µm. Occurs at the base of weakened deciduous trees and at the base of stumps, particularly on *Fagus*.

Species of *Nemania* ⊳⊳ are not as thick and usually have smaller spores. Microscopically, *K. deusta* is similar to *Xylaria polymorpha* ⊲⊲ but molecular data do not place the two species very close together phylogenetically.

Nemoral–hemiboreal, common to occasional; all year, but mostly asexual March–June and sexual June–January.

The inoperculate cup fungus *Polydesmia pruinosa* (page 1434) is frequently found on the stromata of *K. deusta* and other dark stromatic pyrenomycetous fungi.

Nemania has rather thin, internally rather soft stromata, conical ostioles and ± high, I+ apical apparatus in the asci. Lacks KOH-dissolving pigments. Spores have a germ slit that, if evident, is positioned on the flat side. The outer spore wall is never dehiscent in KOH.

Nemania serpens is a red-brown-tinged to completely black *Nemania* with ± obvious perithecial contours and conical ostioles. Has a greyish hyphomycete asexual state ('*Geniculosporium*', see lower image). Asci have reddish apical apparatus in Lugol's reagent; almost invisible in Melzer's. The somewhat olive-tinged spores are asymmetrical, have a very indistinct, short germ slit and measure 10.5–14 × 4–5.5 µm. Occurs on rather decomposed wood of deciduous trees.

Nemania serpens is the only *Nemania* with a red Lugol's reaction. There are three other very similar species, but all have blue apical apparatus in Lugol's: *N. reticulata* ▽ on fruitbodies of *Fomitopsis pinicola* (page 867); *N. colliculosa* ⋉ on wood and with a very green spore-deposit; and *N. prava* ⋉ with a red-brown spore-deposit. *Nemania atropurpurea* ⋉ has slightly smaller spores with a constricted/pinched end. Other species of *Nemania* have long germ slits.

Widespread, common, becoming scarcer towards the north; all year, but in best condition in late autumn.

Nemania reticulata has rather thin, small, greyish to black stromata with conical ostioles. Asci I+ (blue). The spores measure 9.5–12 × 4.5–5 µm and have a short, usually evident, germ slit. Occurs on the top or lower side of old fruitbodies of *Fomitopsis pinicola* (page 867).

As no other species of *Nemania* has been recorded on this substrate, identification should be straightforward.

Widespread, but rarely recorded, absent from e.g. the UK; all year, but in best condition in late autumn.

Nemania chestersii is a stromatic pyrenomycetous fungus that forms rather large, effused, greyish-brown, reddish-brown to black stromata with obvious to indistinct perithecial contours and conical ostioles. Asci I+ (blue). The brown spores measure 13.5–17 × 5.5–7 µm and have longitudinal stripes (× 1,000 magnification!) but no obvious germ slit (short and indistinct). Occurs on larger, ± decorticated trunks of deciduous trees, mainly *Fagus*.

Nemania aenea ⋉ has similar spore dimensions but does not have stripes on the spores.

Nemoral, occasional; all year, but in best condition during the autumn.

Nemania confluens has only a few perithecia in a ± immersed stroma and distinct perithecial contours. The asci have an aberrant, rather low, I+ (blue) apical apparatus. The spores are very dark brown, mostly symmetrical, measure 15–18 × 8–9.5 µm, and have a full-length germ slit. On decorticated wood of deciduous trees, *e.g.* *Quercus*, *Corylus* and *Fagus*.

Euepixylon udum ▷▷ is superficially similar but has much larger and relatively narrower spores with an indistinct, oval germ pore. *Nemania maritima* ⋉ has much smaller spores.

Nemoral–hemiboreal, occasional; all year.

Rosellinia callosa has small, black, crowded, at the base somewhat constricted stromata with white flesh. Asci 8-spored; apical apparatus I+, large. Spores brown and measure 19–27(–28.5) × 6.5–7.5 µm; the germ slit is sinuous and rather short. Mostly seen in the white, cushion-shaped asexual state. Occurs on wood in winter-flooded swamp forests, mostly on *Alnus* and *Salix*.

Recalls a tiny, sessile *Xylaria* ◁◁ and is probably more closely related to some species in that genus than to typical *Rosellinia* species ▷▷ that *e.g.* lack white flesh around the perithecia.

Nemoral, rare, but also overlooked; mainly June–October.

Euepixylon udum is a pyrenomycetous fungus that forms a ± immersed, dark brown stroma with only a few perithecia, which have conspicuous, volcano-like ostioles. Asci 8-spored, apical apparatus I+, low. The rather pale brown spores have an indistinct oval germ pore and measure 23–32 × 9–12.5 µm. Occurs on decorticated, possibly still attached *Quercus* branches, rarely on other hosts.

Nemania confluens ◁◁ has smaller, darker, almost symmetrical spores with a full-length germ slit.

Nemoral–hemiboreal, occasional to rather rare; all year.

Entoleuca mammata is a pyrenomycetous fungus that forms almost circular, small stromata with very large perithecia that have clearly visible contours. Asci 8-spored, I+, large. The brown spores have full-length germ slits and measure 17.5–23 × 7.5–10 µm (from *Salix* – somewhat larger in material from *Populus*). Occurs on dead branches in the crowns of *Salix* and *Populus*. May be parasitic, at least on *Populus*.

The combination of distinctive ecology and large spores characterize this broadly circumscribed species.

Widespread, occasional to rare; all year.

Lopadostoma turgidum is an immersed/erumpent, ± cushion-shaped pyrenomycetous fungus that has numerous smaller stromata contained in a large stroma bound by a black shield. The brown spores measure 10.2–12.2 × 5.5–6 and have a one-sided, full-length germ slit. Occurs exclusively on rather fresh, corticated branches of *Fagus*.

Lopadostoma fagi ⋉ has a double-sided germ slit and the spores measure 7.7–9.8 × 3.7–4.2 µm.

Due to confusion with *L. fagi* the exact distribution and frequency is poorly known; mainly in good condition during winter.

Lopadostoma gastrinum is an immersed/erumpent, very variable fungus. The stromatal interior is very pale and surrounded by a black shield (to the right in the image), and the perithecial necks may become long and rugose. Asci 8-spored, I+. The flattened, brown spores measure 10.5–12.8 × 4.5–5.5 µm and have a double-sided germ slit on the narrow side. Occurs mostly on *Ulmus*, never on *Fagus* and *Quercus;* may occur on both corticated and decorticated substrates.

Lopadostoma lechatii ⋉ is also pale inside but occurs on *Carpinus* and has smaller spores.

Nemoral–hemiboreal, occasional; all year.

Enrique Rubio Domínguez

Lopodostoma form immersed groups of perithecia that consist of a mix of fungal and host tissue that is often bound above, and sometimes below, by a black shield. The groups may be surrounded by a larger, dark shield seen as a black line in section. The spores are brown, usually flattened and have a germ slit. Occur on deciduous trees.

Lopadostoma pouzarii is a stromatic pyrenomycetous fungus that has a brownish stroma at the level of the substrate and a black shield lower down in the wood. Ostioles are small and indistinct. Asci 8-spored, I+. The brown, ellipsoid to fusiform spores measure 12–16.5 × 5–7 µm and have a full-length germ slit. On hard, decorticated deciduous wood, mostly *Fraxinus* and *Ulmus*.

This species differs from others in this genus by having a more expanded stroma, and is more closely related to *Hypoxylon* ⋖⋖ and *Daldinia* ⋖⋖. It should be easy to identify.

Widespread, rare or more likely overlooked; probably most of the year.

Jens H. Petersen/ Lucie Zíbarová

Species of **Cryptosphaeria** are stromatic pyrenomycetous fungi that form immersed stromata with a mixture of fungal hyphae and host tissue, but without a black upper layer. The asci have long, tail-like bases, with the spores in the wider top; the spores are allantoid and contain two drops. Occur in living and recently dead wood; may cause damage in forestry.

Cryptosphaeria ligniota is an immersed stromatic pyrenomycetous fungus that has a somewhat raised stroma that is partly mixed with bark and wood, and which lacks a black upper layer. The ostioles are seen as ± star-shaped, black spots on the surface of the ± discoloured bark. Asci long-stemmed, spore part 30–60(–70) × 5–7 µm, I+/-. Spores allantoid, yellowish, 7–11.5 × (1.8–)2–2.2(–2.5) µm. The asexual state occurs in bark cavities. Conidia measure 18–32 × 1–1.2 µm and form an orange-yellow mass. Occurs on *Populus*, including *P. tremula*. May cause cankers and is considered a serious pathogen in North America.

The epithet is also seen with the spelling '*lignyota*', and this species should perhaps be accepted in *Cytospora* ▷▷ as *Cytospora populina*. *Cryptosphaeria subcutanea* ⋉, on *Salix*, has darker, larger spores; *C. eunomia* ▽ occurs on *Fraxinus*.

Widespread, rather common; mostly during winter.

Cryptosphaeria eunomia is a pyrenomycetous fungus that is revealed by the slightly raised bark of the host, when the black, simple ostioles can be seen as scattered spots. In section, a black line can be seen in the wood below the stroma, which partly consists of host tissue. Asci long-stemmed, clavate, 8-spored, I+. Spores yellowish, allantoid, one-celled and measure 11.5–16 × 2.2–3 µm. The asexual state is formed in bark cavities and has large, narrowly sickle-shaped conidia. Occurs in *Fraxinus* bark.

Cryptosphaeria eunomia var. *fraxinii* ⋉ has 2–6-celled, darker spores.

Widespread, common; mostly during winter.

Species of **Eutypa** are stromatic pyrenomycetous fungi that form immersed stromata consisting of a mixture of fungal hyphae and host tissue but with a blackish layer on the top. The asci are long-stemmed with the spores in the wider top; the spores are allantoid and contain two drops. The stromata are often parasitized by *Capronia nigerrima* (page 1612) and *Chaetosphaerella phaeostroma* (page 1596).

Eutypa flavovirens has stromatic, immersed/erumpent stromata with the grey-brown surface incorporating both fungus and host tissue. The flesh between the perithecia is yellowish-green and rather loosely built. There is hardly any stroma below the perithecia. Asci long-stemmed, spore-part measures 30–50 × 5–8 µm, I+. The allantoid, somewhat yellowish, one-celled spores measure 6–10 × (1.5–) 1.8–2 µm. The asexual state can be found at the margin of the stromata. Conidia measure 20–32 × 1–1.2 µm. Occurs on a range of deciduous trees.

The taxonomic placement of *E. flavovirens* is not clear, with recent data suggesting it belongs in its own genus. It is also often accepted in *Diatrype* ▷▷. There are no similar species in Europe.

Widespread, rather common; all year, in best condition during winter.

Eutypa maura is a flat, immersed, stromatic pyrenomycetous fungus with thin stroma that partly consist of host tissue; the surface is penetrated by star-shaped ostioles. The asci are long-stemmed and I-. Spores allantoid, one-celled, yellowish and measure 4.8–7.8 × 1.2–1.5(–1.8) µm. Occurs on branches and thin trunks of *Acer*.

Also known as *E. acharii*. Closely related to *E. lata* ⋉, which occurs on a range of hosts, including *Acer*, and has larger spores and I+ asci. *Eutypa sparsa* ⋉ occurs on *Populus*.

Widespread, common, becoming scarcer towards the north; all year.

Sirobasidium albidum (page 1168) may occur on the surface of this fungus.

Eutypa spinosa is an immersed-erumpent, typically very thick stromatic pyrenomycetous fungus that may develop very large stromata with a greyish to black, very rough and uneven surface (middle image depicts a young stage). The large perithecia (inset image) are seated deep inside the host-mixed stromatic tissue (a sharp knife is required!). Asci long-stemmed, spore part measures 25–35 × 4–6 µm, I+. The yellowish, allantoid, one-celled spores measure 5.5–9 × (1.5–)1.8–2.2 µm. Typically occurs on large trunks of *Fagus*, both standing, dying trunks and fallen trunks, and it can be present over a very long period of time. It dries out the wood so that competing species are kept out, and the degradation process proceeds at a very slow rate. Rarely found on other deciduous trees. It is rare in commercially managed *Fagus* forests.

Apparently more closely related to *Diatrype* ▷▷ than to other species of *Eutypa*. When typically developed, it is unlikely to be confused with any other stromatic species.

Nemoral, common to occasional; all year.

Eutypella sorbi is an erumpent stromatic black pyrenomycetous fungus that has long, furrowed ostioles that are gathered together in a projecting 'chimney' with more than 10 necks. Asci 8-spored, I-. Spores pale brownish-yellow, allantoid, measure 5–10 × 1.2–2 µm. Occurs on *Sorbus*.

There should be no identification problems on this host. *Biscogniauxia repanda* ⊲⊲ is a much coarser fungus and lacks furrowed, projecting ostioles. A number of similar species of *Eutypella* occur on other hosts.

Widespread, common at least in more northern parts, including the northern part of the nemoral zones; most of the year.

Quaternaria quaternata is an immersed pyrenomycetous fungus with only a few (typically four) short-necked perithecia (inset image of a tangential section), which emerge in a small cluster on the bark surface of the host. In cross section a thin, black line is seen below the perithecia down into the wood, indicating that the mycelium is held within a protective shield. Asci long-stemmed, 8-spored, I+. The smoky-grey, allantoid spores measure 12.5–19 × 3–4(–4.5) µm. The asexual state ('*Libertella faginea*' – lower image), produces long tendrils of orange-yellow conidia, each measuring 12–21 × 1–1.2 µm, that are squeezed out of cavities in the bark. Occurs on recently dead *Fagus* branches.

Sometimes accepted in the genus *Eutypella* △. Provided the microscopical features are taken into account, this species is distinctive and unlikely to be misidentified.

Nemoral, very common; most of the year.

Diatrype form small to effused, erumpent stromata with pale flesh. The stromata are surrounded by a black shield in the wood, seen as a black line in section. The asci are long-stemmed and mostly 8-spored. The spores are situated in the wider, top part of the asci, are allantoid, and contain two drops. Occur on deciduous wood.

Diatrype decorticata is an effused *Diatrype* that does not have a raised margin. The surface has deep transverse furrows. Ostioles flat or somewhat raised and star-shaped. Stromatal interior pale at the top (lower image). Asci 8-spored, I+. Spores allantoid, somewhat yellow-brown-tinged and measure 7–9(–10) × 1.2–1.5 µm. Mostly on *Fagus*, but may also occur on other deciduous trees, although probably not *Betula* and *Quercus*.

There is some confusion as to which names to apply in this group of *Diatrype*. *Diatrype undulata* ⋉, on *Betula*, has a distinct, raised margin and no conspicuous transverse furrows in the stroma; *D. stigma* ⋉, mostly on *Quercus*, has somewhat larger spores, 8–11 × 1.6–2.7 µm. *Biscogniauxia anceps* ⋉ is superficially similar, but has 2-celled spores with one dark and one pale cell.

Widespread and common, becoming scarcer towards the north; all year, in best condition during winter.

Diatrype bullata has dark, crowded, erumpent stromata that have an oblong outline, white flesh and ± star-shaped, low ostioles. Asci 8-spored, long-stemmed, I+. The yellowish, allantoid spores measure 5–7.5 × 1–1.5 µm. Occurs on *Salix*, rarely *Populus*.

Diatrype disciformis ▷ has smaller, flatter, ± circular stromata and occurs mainly on *Fagus*, never on *Salix*.

Widespread, very common; all year.

Older stromata are often red-dotted from perithecia of *Dialonectria diatrypicola* ⋉.

Diatrype disciformis has crowded, erumpent, ± circular stromata with scattered, ± star-shaped, low ostioles. The flesh is white and firm, intermixed with host tissue below the perithecia, and delimited by a black shield (seen as a line in section). Asci long-stemmed, 8-spored, I+. Spores allantoid, yellowish and measure 5.2–8.5 × 1.2–1.5 µm. Occurs almost exclusively in bark of *Fagus*, mostly on thin, recently dead branches; rarely on *Crataegus*, *Acer* and *Frangula*. It is an endophyte in *Fagus*

Diatrype bullata ◁ occurs on *Salix* and *Populus* and has larger, more elongated stromata.

Nemoral, very common; all year.

Diatrype quercina is a ± cushion-shaped, stromatic erumpent *Diatrype* with somewhat projecting, furrowed ostiolar necks and a rather pale exterior. Has a dark shield in wood and bark (seen as a dark line in section). Asci multispored, long-stemmed, I-. Spores pale yellow-brown, highly curved-allantoid (almost like half a doughnut), and measure 8–13 × 2–3 µm. Occurs on recently dead, corticated *Quercus* branches.

This species has often been placed in *Diatrypella*, principally due to its multispored asci. Distinctive and unlikely to be misidentified provided the host and microscopical details are known. Species of *Diatrypella* normally have straighter spores.

Nemoral–hemiboreal, very common; in best condition in late autumn and winter.

Old stromata may carry perithecia from *Dialonectria diatrypellicola* ⋉ and species of *Cosmospora* ◁◁.

Diatrypella favacea has immersed-erumpent stromata with ostioles that are gathered together in a brownish to black, ± elongated disc on the surface. Asci long-stemmed, multispored. Spores yellow-brown, allantoid, 5–8 × 1–2 µm. Occurs in bark of deciduous trees, *e.g. Fagus*, *Quercus*, *Corylus* and *Alnus*.

There is no consensus on how this and other species of *Diatrypella* should be delimited; some believe that only material on *Betula* should go under this name. *Diatrype quercina* ◁◁, on *Quercus*, has stromata with a rounded outline and more curved spores.

Widespread, common; mostly during winter.

Peroneutypa scoparia is a black pyrenomycetous fungus that forms perithecia deep in the host tissue, from where long, sinuous perithecial necks rise above the surface. Usually positioned in clusters of approximately five perithecia delimited by a black shield in the substrate (seen as black lines in section). Asci long-stemmed, 8-spored, distinctly I+. The almost hyaline, allantoid spores measure 3.5–6 × 1.2–1.5 µm. Also has a fingered-branched, asexual state (lower image). Occurs on wood from various deciduous trees, but is also known from bamboos (*Poaceae*), *etc.*

Also known as *Eutypella scoparia*. The combination of long perithecial necks, very small spores and distinctive asexual state makes this species easy to identify.

Nemoral–hemiboreal, occasional; all year.

Species of ***Diaporthe***, ***Diaportella*** *etc.* have asci that loosen at maturity and float freely to the ostiolar mouths, where the spores are then shot out.

Diaporthe pulla has black, deeply immersed perithecia with very long, single, projecting, smooth, black perithecial necks. Asci loosen, 8-spored, I-. Spores hyaline, 2-celled, ± non-constricted at the septum, and measure 9–13 × 2–2.5 µm. The asexual coelomycete-state ('*Phomopsis*') may also occur on the leaves, as well as the wood of the host; conidia measure 7–8 × 2–3 µm. Occurs on blackened *Hedera* wood.

Diaporthe hedera ⋉, occurs on the same host but has wider spores with a constriction at the septum.

Nemoral, occasional; mostly during winter.

Diaporthella aristata is a partly immersed, black, stromatic pyrenomycetous fungus that is erumpent with short rows of far-projecting perithecial necks. The stromatic tissue is pale and incorporates host tissue. The asci loosen, I-. The hyaline, 2-celled spores have short appendages at the ends and measure (11–)12–13.5 × 2.5–3 µm. Occurs on branches of *Betula*.

The somewhat similar *Xenotypa aterrima* ⋉ breaks through bark, as does *Vuilleminia* (page 1025), which has narrower spores.

Mainly boreal, rarely reported, but perhaps overlooked; all year.

Cytospora (formerly *Valsa*) form crowded, immersed clusters of perithecia with ostioles that are usually gathered together, forming a pale disc on the bark surface. The spores are hyaline and one-celled, and the asci loosen at maturity. Has coelomycetous asexual states (page 1628).

Cytospora intermedia is an immersed-erumpent stromatic pyrenomycetous fungus with 3–8(–25) perithecia around an asexual state at the centre. The ostioles are black and somewhat raised. Asci 8-spored, loosen, have a low, I- apical ring, which is seen as two small dots. Spores one-celled, hyaline, somewhat bent, 10–16 × 2–3 µm. The asexual state forms large, several-chambered structures with a single ostiole; conidia measure 5–6 × 1.5 µm. Occurs on corticated *Quercus* branches.

Dendrostoma leiphaemia ⋉ occurs on the same substrate but has 2-celled spores. A number of other erumpent, ± stromatic pyrenomycetous fungi also occur on *Quercus* branches but they have other spore types. One of the most common is *Diatrype quercina* ◁◁, which has multispored asci and brownish, very curved-allantoid, small spores.

Distribution not known, very rarely reported; mostly during winter.

***Cytospora abietis* s.l.** forms clusters of erumpent, dark perithecial necks. Asci 8-spored, loosen, I-. Spores hyaline, allantoid, 5–8 × 1.5–2 µm. Occurs on recently dead branches (and needles) of conifers, including *Picea*, *Abies* and *Larix*. Possibly pathogenic in *Abies* plantations; the name is also used for infections on *Pseudotsuga*.

There is continuing uncertainty as to the number of *Cytospora* species on conifer wood. Fries introduced *Valsa abietis* from *Picea abies* in Sweden and the name *Cytospora abietis* was introduced as the name for the asexual state, but now on *Abies*.

Probably widespread; mostly during winter.

Cytospora nivea is erumpent with small white discs through which black ostioles of 2–10 perithecia emerge. Asci loosen, typically 4-spored, I-. Spores hyaline, allantoid, one-celled, measure 12–16 × 3 µm. Asexual state a coelomycete (page 1628), which also appears as a white disc; conidia are approximately 7.6 × 1.9 µm. Occurs on recently dead, corticated *Populus* and *Salix* branches.

Cytospora chrysosperma ⋉, also on *Populus*, has grey discs and slightly smaller spores. *Valsa salicina* ⋉, on *Salix*, has larger spores and *Valsella salicis* ⋉ has smaller spores.

Widespread, common; mostly during winter.

Amphiporthe tiliae is a ± stromatic, erumpent black pyrenomycetous fungus with 10–20 projecting, rather long perithecial necks. Asci loosen, I-. Spores 2-celled, hyaline, measure 11–15 × 3–4 µm. Has an asexual '*Cytospora*'-state; conidia measure 4–6.5 × 1–1.5 µm. Occurs on twigs and branches of *Tilia*.

Also known as *Cryptodiaporthe hranicencis*. *Diaporthe eres* ⋉ has a completely different asexual '*Phomopsis*'-state with much longer conidia.

Nemoral–hemiboreal, occasional; all year.

Melanconis alni is an immersed black pyrenomycetous fungus where the perithecial necks break through a circle around a pale disc. Asci loosen, I-. The hyaline, 2-celled spores have short appendages at the ends and measure 20–24 × 4–7 µm. The asexual state has pale brown, thick-walled conidia that measure 10–12 × 6–8 µm. Occurs on twigs and branches of *Alnus*.

Pseudovalsella thelebola ⋉ occurs on the same host but has much larger spores with long, narrow appendages.

Widespread, probably common; all year.

Melanamphora spinifera is a black, erumpent stromatic pyrenomycetous fungus that has crowded stromata with long, black perithecial necks. Asci ± cylindrical with refractive, I- apical apparatus. Has many paraphyses. The 8-celled, brown, allantoid spores have hyaline end cells and measure 55–75 × 7–8 µm. Occurs mostly at the very base of younger, standing dead *Fagus* but also at the base of older, living trees. It may kill young trees.

A very distinctive species both in the field and through the microscope.

Nemoral, common in Denmark but apparently less common in other parts of temperate Europe; all year.

Sillia ferruginea is an immersed stromatic pyrenomycetous fungus with clustered groups of erumpent, black perithecial necks. Stroma red-brown – hence the epithet '*ferruginea*'. Asci 8-spored, non-loosening, have a small apical apparatus, I-. The hyaline, 5–6-celled spores taper at one end and measure 60–75 × 3 µm. Occurs at the base of dying *Corylus* stems.

Microscopically very distinctive, but even macroscopically rather easy to recognise.

Mainly nemoral–hemiboreal, rarely reported, but probably overlooked; mostly during winter.

Camarops form often large, erumpent stromata. The perithecia are elongated or long-necked but the spores are shot out (see middle image). Spores are one-celled, smoky-grey and have a germ pore. Wood decayers, often in forests with long continuity and plenty of dead wood.

Camarops polysperma has large, black, erumpent stromata with uniform, cylindrical perithecia that have a fine, wart-like opening. Surface shiny, recalling a black slug. When decaying it may recall an old polypore. Asci cylindrical, 8-spored, lack a distinct structure in the top, I-, but with active spore dispersal (lower image). Spores smoky-grey, one-celled, with germ pore at one end, somewhat flattened, 5–6 × 2.5–3 × 1.5–2 µm. On standing, broken, *Alnus glutinosa* trunks, but also on *e.g. Fagus*.

Camarops tubulina ▽ is more brownish and has broader perithecia at different levels in the stroma; the perithecia reach the surface via long, narrow necks. *Camarops plana* ⋉ also has perithecia at different levels but a very flat stromatal surface that lacks brown tinges; the spores measure 4.8–6.5 × 2.2–3 × 2–2.5 µm and it seems to favour *Carpinus*.

Nemoral–hemiboreal, rather common to occasional; in best condition August–November, but can be found year-round.

Camarops tubulina is a coarse, erumpent, ± brownish *Camarops* with perithecia at different levels within the stroma (inset image). The perithecia reach the surface via narrow necks. Asci 8-spored, I-. Spores flattened, 6–7.5 × 2.5–3.5 × 2 µm, and have one germ pore. On trunks and stumps of coniferous trees (*Picea* and *Abies*) and deciduous trees, mainly *Fagus*.

Camarops polysperma △ has narrow, cylindrical perithecia; *C. petersii* ⋉, on *Quercus*, is brown and initially covered by a 'veil'; and *C. plana* ⋉, on *Carpinus*, has smaller spores and a flattened stroma.

Widespread, rather rare; mostly in good condition August–November.

Thomas Læssøe/Jens H. Petersen

Thomas Læssøe

Camarops microspora is an immersed-erumpent, rather small *Camarops* with an uneven surface that remains partly covered by bark. Has short perithecial necks. Asci 8-spored, I-. The smoky-grey, somewhat flattened spores measure 4.5–6.5 × 2 × 1–1.5 µm and have one germ pore. Occurs on *Alnus*, probably with a preference for *A. incana*; the wood is not stained.

Camarops lutea ▽ stains the wood below the stromata a vivid yellow and the spores are wider.

Widespread, occasional in the nemoral zones, common farther north; most of the year.

Camarops pugillus is a small, erumpent *Camarops* with far-projecting, finger-like, black perithecial necks. Asci 8-spored, cylindrical, I-. The smoky-grey spores measure 5.5–7.5 × 3.5 × 2.5 µm and have a germ pore at one end. On deciduous wood, *e.g. Corylus* and *Alnus*; wood is not stained yellow.

Also known as *Camaropella pugillus*. *Camarops lutea* ▽ may have slightly elongated necks but stains the wood yellow. *Camarops microspora* △ has short-necked perithecia and different spores.

Nemoral–hemiboreal, very rare or overlooked; perhaps mostly during the warmer part of the year.

Jens H. Petersen/Thomas Læssøe

Camarops lutea is an erumpent, small, greyish *Camarops*. The perithecial necks are mostly immersed in the stroma but may be somewhat elongated. Asci 8-spored, I-. Spores ± flattened, (4–)5.5–6(–7) × 2–3 × 2 µm, and have a germ pore in one end. Mostly on deciduous wood, rarely on coniferous wood, including window frames; the wood is stained yellow.

Also known as *Camaropella lutea*. Very variable and probably forms a complex of closely related species; outside temperate Europe it is reported as having a yellow stroma. *Camarops pugillus* △ has long perithecial necks and does not stain wood yellow.

Mainly nemoral, rather common; most of the year.

Species of ***Cucurbitaria*** and ***Camarosporidiella*** are rather large, almost fleshy bitunicate pyrenomycetous fungi. The spores are muriform-septate.

Camarosporidiella laburni is a bitunicate pyrenomycetous fungus that forms large clusters of black, mostly smooth fruitbodies. Asci double-walled, ± cylindrical, 8-spored, I-. Spores brown, 25–35 × 10–12 µm, constricted at the centre and somewhat tapering. On leguminous shrubs, *e.g. Laburnum*.

Cucurbitaria cytisi ⋉, also on leguminous hosts, has fewer transverse septa; *C. caragana* occurs on *Caragana*.

Nemoral–hemiboreal, occasional; mainly during winter.

Thomas Læssøe

Cucurbitaria berberidis is an erumpent, bitunicate pyrenomycetous fungus with rough, at the base confluent, fruitbodies that become flat-topped with age and on drying. Each group consists of 10–30 fruitbodies. Asci double-walled, 8-spored, I-. The brown, at the ends paler, muriform-septate spores are only constricted at the centre and measure 23–42(–45) × 10–16(–17) µm. Occurs on *Berberis* and *Mahonia*.

Similar species of *Cucurbitaria* occur on other hosts and may have spores with several constrictions, smooth or hairy fruitbodies, *etc.*

Mainly nemoral–hemiboreal, rather common; mostly during winter.

Cucurbitaria coronillae is an erumpent, bitunicate pyrenomycetous fungus that flattens with age and has a somewhat roughened surface. Asci double-walled, 8-spored, I-. Spores brown, measure 19–27(–35) × 9–12 µm; muriform with 5–7 transverse septa. Occurs on *Hippocrepis* (*Coronilla*) *emerus*.

Similar species of *Cucurbitaria* on other leguminous hosts include *C. elongata* ⋉, *C. cytisi* ⋉ and *C. emeri* ⋉; all typically form fruitbodies that are more connected at the base.

Nemoral, rarely recorded; mostly found during winter months.

Non-stromatic, unitunicate, dark pyrenomycetous fungi

Species in this form group have more or less solitary, perithecia and the asci have single walls (unituncate). They are generally ± brown or black but may be covered in white or more colourful hairs. The perithecia may be seated on the surface of the substrate or immersed and almost invisible, with only the ostiolar necks appearing at or above substrate level.

The majority of the species are decomposers of leaves, herbaceous stems, wood or dung – a few are parasitic.

The non-stromatic pyrenomycetous fungi form a heterogeneous group of fungi with representatives from a range of orders, of which *Sordariales* is the most prominent. Many of these species produce large, ± brown spores.

OTHER SIMILAR FUNGI:
– the bitunicate pyrenomycetous fungi have thick, double-walled asci. The spores are mainly released when the outer wall ruptures, allowing an inner sac to stretch out (page 1602).
– hypocrealean pyrenomycetous fungi can also be non-stromatic, but the perithecia are mostly brightly coloured and rather soft (page 1512).
– the dark stromatic pyrenomycetous fungi grade into the non-stromatic. For species with immersed perithecia it can be especially difficult to determine whether or not a stroma is present (page 1546).

See also the general wheel, page 1510.

FURTHER READING: 16, 54, 58, 62, 73, 78, 185, 195, 196, 247, 257, 258, 259, 260, 261, 304, 350, 353.

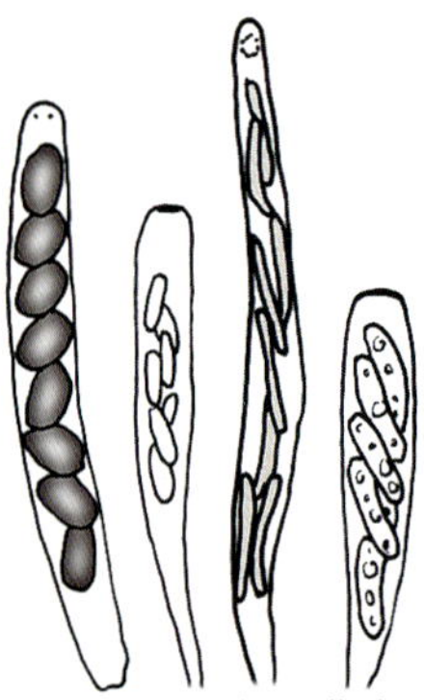

Asci with single walls (unitunicate).

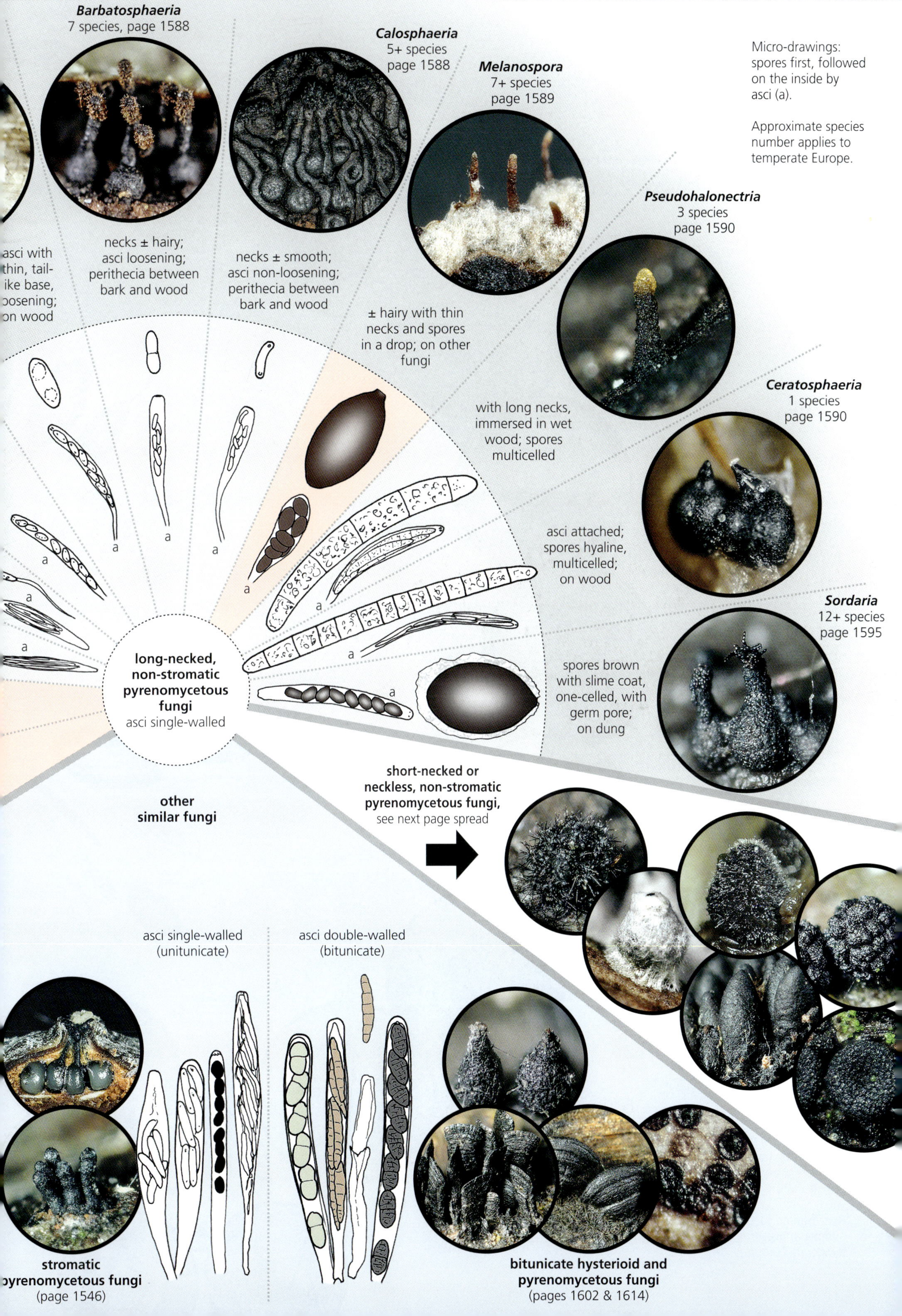
Barbatosphaeria
7 species, page 1588
Calosphaeria
5+ species
page 1588
Melanospora
7+ species
page 1589
Micro-drawings: spores first, followed on the inside by asci (a).
Approximate species number applies to temperate Europe.
Pseudohalonectria
3 species
page 1590
Ceratosphaeria
1 species
page 1590
Sordaria
12+ species
page 1595
asci with thin, tail-ike base, oosening; on wood
necks ± hairy; asci loosening; perithecia between bark and wood
necks ± smooth; asci non-loosening; perithecia between bark and wood
± hairy with thin necks and spores in a drop; on other fungi
with long necks, immersed in wet wood; spores multicelled
asci attached; spores hyaline, multicelled; on wood
spores brown with slime coat, one-celled, with germ pore; on dung
a
long-necked, non-stromatic pyrenomycetous fungi
asci single-walled
other similar fungi
short-necked or neckless, non-stromatic pyrenomycetous fungi,
see next page spread
asci single-walled (unitunicate)
asci double-walled (bitunicate)
stromatic pyrenomycetous fungi
(page 1546)
bitunicate hysterioid and pyrenomycetous fungi
(pages 1602 & 1614)

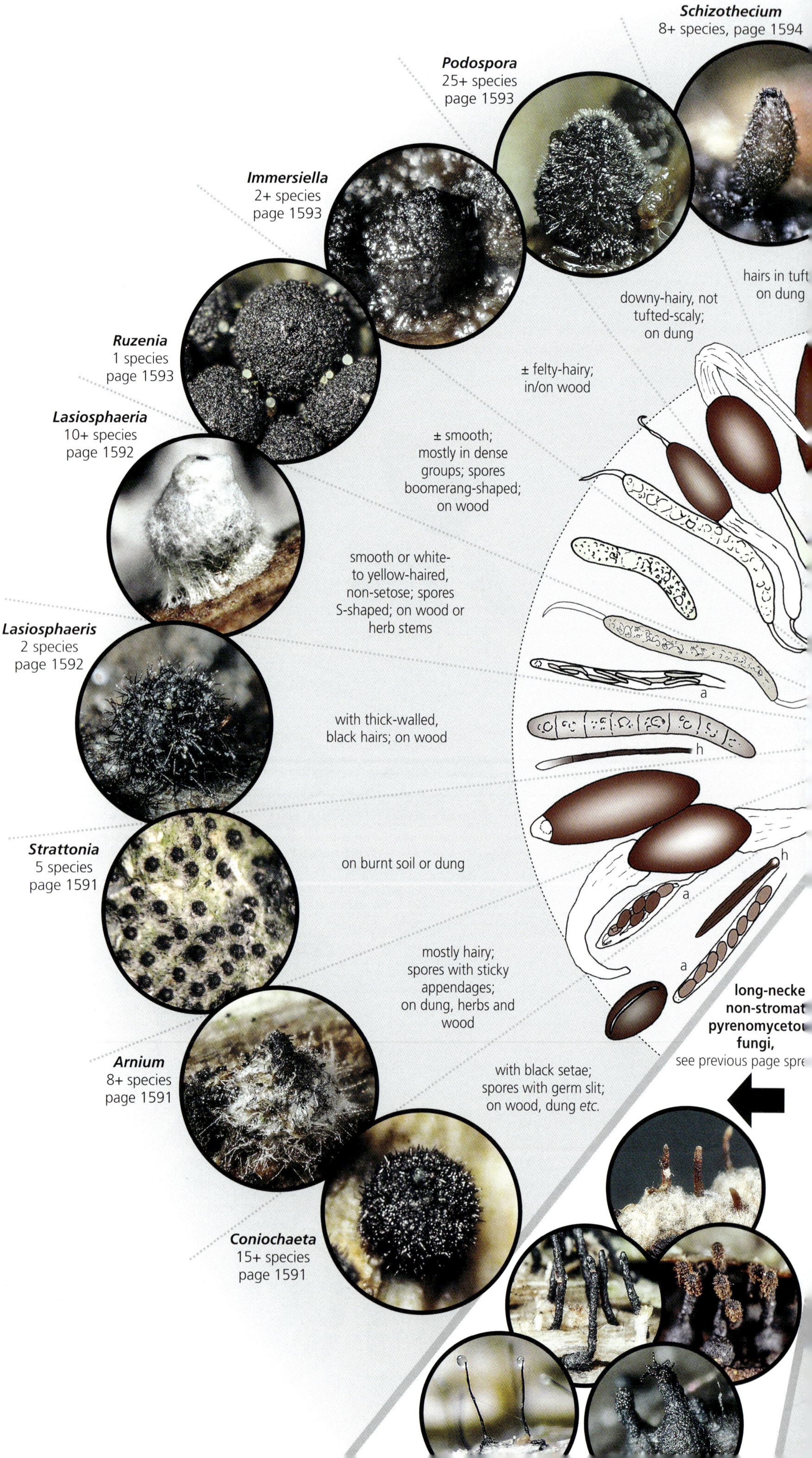

Micro-drawings: spores first, followed on the inside by asci (a), hairs (h) and conidia (con).

Approximate species number applies to temperate Europe.

short-necked
or neckless,
non-stromatic
pyrenomycetous
fungi
asci single-walled
Bombardia
1 species
page 1594
1–2 mm high, thick
and rubber-like;
on wood
Sordaria
12+ species
page 1595
smooth and pear-
shaped with a
short neck;
on dung
Echinosphaeria
4+ species
page 1595
hairs brown,
thick-walled,
septate; spores
boomerang-
shaped; on wood
Helminthosphaeria
12+ species
page 1596
with dark setae;
spores brown, with
small germ pores;
mostly on other
fungi
Chaetomium
20+ species
page 1595
spores collected in
a column; asci soon
dissolve; on litter and
dung
Chaetosphaerella
2 species
page 1596
on black hyphal mat
with production of
conidia hyphae; on
wood/fungi
Acanthonitschkea
2 species
page 1597
black setose;
± cupulate when
dry; on wood and
other fungi
Nitschkia
6+ species
page 1598
± smooth;
± cupulate when
dry; on wood and
other fungi
smooth but typically
amongst dark hair-
like asexual states;
on wood
Chaetosphaeria
15+ species
page 1598
mulberry-like;
on wood
Bertia
2 species
page 1598
confluent dark
spots; on living plant
tissue
Phyllachora
10+ species
page 1599
with shield on
top; on dung
Hypocopra
10+ species, page 1599
ithecia large; hyphal
at ± prominent, on
wood or herbs
Rosellinia
+ species, page 1600
a
h
h
h
con
h
h
a

Sporothrix polyporicola is a very long- (up to 2 mm) and thin-necked, black pyrenomycetous fungus with asci that dissolve quickly, after which the spores aggregate in a slimy drop at the tip of the neck; they are probably disseminated by insects. Spores hyaline, allantoid, 3–3.5 × 1–1.3 µm. Occurs on the pores of polypores, *e.g. Fomitopsis*, *Piptoporus* and *Tyromyces* (pages 867, 833 & 908).

Also known as *Ophiostoma polyporicola*. Species of *Melanospora* ▷ have larger perithecia and dark, much larger spores. *Sporothrix schenckii* ⋉ may infect humans (sporotrichosis).

Widespread and probably rather common; most of the year.

Syspastospora parasitica is a very long- and thin-necked, brownish pyrenomycetous fungus with asci that dissolve quickly. The spores are held in a slimy drop at the tip of the neck. The ± smoky-grey, almost cylindrical spores have a pore at each end and measure 4–10 × 2–2.5 µm. Occurs on *Cordyceps farinosa* (page 1631) and similar insect-parasitic fungi, and possibly on other hyphomycetes.

The combination of dark, strange spores and the unusual substrate make identification straightforward.

Widespread, rarely reported; mainly July–October.

Gnomonia gnomon is a black pyrenomycetous fungus with a long, central neck. Asci 8-spored, float freely at maturity, have a pointed base, measure about 36–43 × 7–9.5 µm, I-. Spores in a parallel cluster, hyaline, 2-celled, somewhat allantoid, (17–)19.5–21(–24) × (1.3–)1.6–1.8(–2.5) µm, and have 2.5–4 µm long appendages at both ends. Occurs on *Corylus* leaves, rarely *Populus*.

Ophiognomonia setacea ⋉, on *Quercus* and *Castanea* leaves, has a neck positioned obliquely or at the centre; the spores are oblique in the ascus and shorter and wider.

Widespread, rarely reported, probably common; in the cooler months.

Ophiognomonia ischnostyla has black, very long- and thin-necked perithecia with free-floating asci and a broad ring in the top, I-. Spores 2-celled, hyaline, (12.5–)13.5–15.5 (–18.5) × (1.5–)2(–2.5) µm with long, flagella-like appendages. On leaves of *Alnus*, *Betula*, *Corylus* and *Carpinus*.

Ophiognomonia intermedia ⋉ and *O. alni-viridis* ⋉ have shorter spores. All species of *Ophiognomonia* are typically permanently immersed below the host epidermis (except for the projecting perithecial neck), while species of *Gnomonia* ◁ mostly become superficial at maturity.

Widespread, rarely reported, probably common; autumn–winter.

Lentomitella conoidea is a partly immersed black, long-necked pyrenomycetous fungus; the neck is ridged and there are brown hairs on the perithecial body. Asci 8-spored, 62–70(–75) × 6.5–7.5 µm, have refractive I- apical apparatus and a short stem; non-loosening. Spores hyaline, longitudinally striate, 2–3(–4)-celled, normally lack a central septum, 8.5–9.5(–10.5) × 4– 4.5 µm (av. 9 × 4.5). In decomposed hardwood.

Very similar to *L. cirrhosa* ⋉, which has spores 8–10 × 4.5–5 µm and a central septum; *L. tenuirostris* ⋉ has somewhat longer spores.

Widespread, probably common; mostly during winter.

Natantiella ligneola is an immersed black pyrenomycetous fungus with a long, smooth neck that projects above the substrate. Asci 8-spored, ± free-floating, have a long, pointed base and a refractive ring in the top, I-. Has many septate paraphyses. Spores hyaline, contain two large drops, one-celled and measure (9–)10–12.5(–14) × 3.5–4 µm. Occurs in rotten wood of deciduous trees.

Lentomitella △ has ridged necks and septate, striate spores. *Ceratosphaeria lampadophora* ▷▷ has very long, 6–8-celled spores.

Widespread and probably common; most of the year.

Calosphaeria pulchella has very long, ± smooth-necked perithecia below loosening bark. They are arranged in a circular pattern and the necks are joined in a crack at the surface of the host (inset image). Asci clavate, flat-topped and long-stemmed, in clusters, non-loosening, 8-spored, I-. Spores one-celled, ± curved, hyaline, 4.5–5 × 1 µm. Under bark of *Prunus avium* and related species.

The peculiar pattern combined with the host make identification easy.

Nemoral–hemiboreal, rarely reported; mostly during winter.

Barbatosphaeria barbirostris has long-necked perithecia gathered in groups between wood and bark; necks brown-haired. Hairs rounded at the tip and multicelled. Asci free-floating, 8-spored, 39–47 × (5–)6–6.5 µm, I-. Paraphyses abundant. Spores narrowly ellipsoid, 2-celled, hyaline, (5–)6–7 × 1.5–2 µm. On deciduous wood, *e.g. Fagus* and *Quercus*.

Barbatosphaeria arboricola ⋉ has longer asci and spores, av. 8.6 × 2 µm; *B. dryina* ⋉ and *B. fagi* ⋉ have allantoid spores; *B. varioseptata* ⋉ has larger, up to 4-celled spores.

Nemoral–hemiboreal, rarely reported, but probably overlooked; most of the year.

Species of ***Melanospora*** have hairy, long- and thin-necked perithecia and large, brown spores with distinct germ pores at the ends. All typically occur on the fruitbodies of other fungi.

Melanospora lagenaria is a very long-necked pyrenomycetous fungus that appears in clustered groups. The perithecial body is somewhat felty and dark brown, while the long necks are paler brown and mostly smudged by the dark spores. Hyaline, rather short setae are present at the tips of the necks. Asci ± clavate, dissolve quickly, 8-spored, I-. The spores are smooth, brown, ellipsoid to lemon-shaped, have a germ pore at each end and measure 12–16(–22) × 7–10(–14) µm. Occurs on decaying polypores and stereoids, *e.g. Piptoporus* (page 833) and *Meripilus* (page 850).

The combination of substrate and long, brown perithecial necks should make identification straightforward. *Melanospora caprina* ▽ occurs on similar substrates but the perithecial body is more hairy-felty.

Widespread, occasional to rare; mostly October–December.

Melanospora caprina is a clustered, downy, white pyrenomycetous fungus with a long, brown, ± smooth neck that has easily broken setae around the opening. Asci clavate, dissolve quickly, 8-spored. The brown, ellipsoid to lemon-shaped spores have a germ pore at each end and measure 16–23 × 9–16 µm. Occurs on decaying fungi, mostly *Pseudotomentella* (page 1035) on coniferous wood.

Melanospora chionea ⋉ has flattened spores, 7.5–16 × 6–12 × 4–7 µm. *Melanospora lagenaria* △ is less woolly.

Widespread, but rarely reported; mainly in the cooler months.

Pseudohalonectria adversaria is an immersed pyrenomycetous fungus with a far-projecting, dark brown neck. Asci 8-spored, measure 120–150 × 13 –20 µm, have refractive apical apparatus, I-. Paraphyses persistent, septate. The spores are ellipsoid, 6–7-celled, measure 33.5–49 × 4.5–7 µm and are hyaline to pale orange; when in a mass, orange-yellow (seen at the tip of the perithecial neck on the images). Occurs on submerged wood of deciduous trees in unpolluted fresh water.

Pseudohalonectria lutea ⋉ also has 6-celled spores but the spores are larger, measuring 48–68 × 4.8–8.4 µm, and the end-cells are paler.

Mainly nemoral, rarely reported, but probably overlooked; mostly during the summer.

Ceratosphaeria lampadophora is a partly immersed black, rather small pyrenomycetous fungus with a brown to black neck of variable length. Asci 8-spored with a distinct refractive, I- apical apparatus. Spores hyaline, multicelled and measure 40–60 × 4–4.5 µm. Occurs on white-rotted, rather decomposed wood of deciduous trees, *e.g. Fagus*.

Platytrachelon abietis ⋉, on *Abies*, has larger spores. Other similar species in other genera have much shorter spores, *e.g. Ceratolenta caudata* ⋉, which also occurs on *Fagus*, and species of *Lentomitella* ◁◁, which all have striate spores.

Widespread, occasional; most of the year.

Coniochaeta ligniaria is a mostly superficial, black, short and bristly-hairy pyrenomycetous fungus. Setae pointed. Asci 8-spored, cylindrical, I-. The dark brown spores are somewhat flattened and measure (9–)13–18 (–20) × 8–11(–15) × 4–8 µm; have a full-length germ slit on the narrow side and possibly a slime coat. Occurs on wood of deciduous trees, but also on dung.

A rather large and difficult genus. Many species occur on dung and many have flattened spores with a low Q value. *Coniochaeta pulveracea* ⋉ has rounded hairs; *C. scatigena* ⋉ has wider spores.

Widespread, rarely reported, but probably overlooked; mainly winter.

Arnium apiculatum is a superficial, ± pear-shaped pyrenomycetous fungus with dark, pointed, 1–2-celled setae. Asci 8-spored with refractive, low apical apparatus, I-. Spores brown, 2-celled, measure 31–36 × 18–19 µm, and have a germ pore and a triangular, gelatinous appendage at both ends. Occurs on herbaceous stems, *e.g. Cirsium*.

Other similar species occur on dung.

Distribution and frequency not known, very rarely reported; perhaps most of the year.

Strattonia carbonaria is a ± immersed, dark, ± pear-shaped pyrenomycetous fungus with dark, short hairs around the furrowed ostioles. Asci cylindrical, 8-spored, have a low, refractive apical apparatus, I-. Spores 18–24 × 7–11 µm, fusiform, smooth, 2-celled, one large and dark, the other very small and colourless; immature spores may seem verrucose and contain large drops. Occurs on burnt debris and soil with some moss cover (on the image together with a species of *Anthracobia*, page 1340).

Strattonia minor ⋉ has smaller spores that appear smooth when immature.

Probably widespread, very rarely reported; most of the year.

Jürgen Marqua/Thomas Læssøe

***Lasiosphaeris hirsuta* s.l.** is a superficial, black, ± hairy pyrenomycetous fungus with rather indistinctly furrowed ostioles. Hairs vary in stiffness, septate. Asci cylindrical, 8-spored, have a refractive low ring and a star-shaped body below, I-. Spores hyaline, later pale brown and up to 7-celled, rather sharply bent close to one end and have rather long, pointed, dehiscent appendages at both ends; 60–70 × 5–7 µm. On decaying wood, also wood chips and sawdust.

The browner *Echinosphaeria canescens* ▷▷ has smaller spores that bend in the middle.

Mainly nemoral–hemiboreal, common; all year.

Lasiosphaeria species have 'S'-shaped spores with appendages; some develop a coloured end-cell at maturity.

Lasiosphaeria sulphurella has ± superficial perithecia with a sulphur-yellow, felty covering and bright yellow contents. Asci 8-spored, have a low apical apparatus and a round body below, I-. Spores 'S'-shaped with appendages at first, but later have a swollen, 16–19 × 7–8 µm end-cell and a 27–35 µm long 2–3-celled, cylindrical part. Occurs on wood.

Lasiosphaeria ovina ▽ does not have yellow hairs or a swollen end-cell.

Nemoral, very rarely reported; mainly during summer.

Lasiosphaeria ovina has superficial, white-haired perithecia with a dark ostiole. Contents greenish-yellow. Asci cylindrical, 8-spored, have a ring with a yellowish, star-shaped body below; I-. Spores hyaline to ± yellow-brown, 1–8-celled, (35–)40–46(–50) × 4–5 µm, have tapering, dehiscent appendages. Occurs on rotten deciduous wood.

Lasiosphaeria lanuginosa has longer appendages ⋉; *L. glabra* ⋉ is non-woolly; *L. sorbina* ⋉ has pinkish perithecial content. Other similar species occur in other habitats or are more dark-spored.

Widespread and rather common; most of the year.

Krieglsteinera lasiosphaeriae (page 1166) occurs on the perithecia.

Podospora fimiseda has superficial, black, pear-shaped perithecia with paler, projecting, short, but multicelled hairs. Asci clavate, 8-spored, I-. Spores have a large, dark cell 48–60 × 27–31 µm and a narrow, pale cell 36–44 × 4.5–8.5 µm; both ends have very long, longitudinally striate, pointed, slimy appendages. Occurs on dung, mostly horse and cow.

Other species of *Podospora* have asci with a different number of spores, and some of the 8-spored species have asymmetrical dark spore cells; others have dark, projecting cells on the neck or different appendages on the spores.

Widespread, rarely reported; most of the year.

Immersiella caudata has ± immersed, ± felty-hairy perithecia. The ostiole has short, rounded, dark, thick-walled hairs in clusters. Asci 8-spored, have low apical apparatus and a round body below. Spores at first snake-like, measure 42–60 × 5–6 µm, have pointed appendages and contain many drops; later 2-celled, one of which is large and dark, has a germ pore at the end and measures 16–25 × 9–12 µm. Occurs on decaying wood of deciduous trees and bark.

Also known as *Cercophora caudata*.

Widespread, rarely reported, but probably overlooked; most of the year.

Sordaria fimicola has superficial, black, ± pear-shaped, hairless perithecia. Asci 8-spored, cylindrical, have a low apical apparatus, I-. Spores one-celled, dark and measure (17–)18–24 × (9.5–)11–13 µm; they have a germ pore at the end and a very thick, completely enveloping slime coat. Occurs on all kinds of dung, but mostly known from horse, hare and rabbit.

Sordaria suberba ⋉ and *S. macrospora* ⋉ have larger spores; *S. humana* ⋉ (and others) have ± egg-shaped spores with a ± protruding germ pore.

Widespread, common; most of the year.

Schizothecium hispidulum has rather small, grey-olive, ± pear-shaped, perithecia with tufts of darker hairs, especially close to the opening. The hairs down the sides are pliable, those at the tip swollen; lacks setae. Asci 8-spored, clavate, lack apical apparatus, I-. Spores have appendages at both ends; 2-celled, one pale, one dark, dark cell measures 22–25(–27) × 11–13 µm. Occurs on dung from various herbivores, *e.g.* sheep and cow.

A number of similar species also occur on dung; *S. tetrasporum* ⋉ has 4-spored asci and is especially common on rodent dung.

Widespread, common; most of the year.

Bombardia bombarda has crowded-clustered, large (1–2 mm high), ± narrowly pear-shaped, thick-walled perithecia with a leathery texture. Asci 8-spored, ± cylindrical, I-. The spores are at first one-celled, snake-like, contain many drops and have long, pointed appendages at both ends that measure about 12 × 2 µm. Later the spores are 2-celled: one is dark brown, measures 11–15 × 7–9.5 µm and has a germ pore at the end; the other is long, colourless and measures 20–25 × 4–5 µm. Occurs on deciduous wood, mostly on the surface of rather fresh *Fagus* stumps, but also on coarse wood chips.

A distinctive species that can be recognized by its unusually large and leathery perithecia. *Ruzenia spermoides* ▷ has a similar habit, but the perithecia are less elongated, not as leathery and the spores are boomerang-shaped.

Nemoral, rather rare; in the cooler months of the year.

Chaetomium globosum has ± green-tinged, somewhat curly haired perithecia. The asci dissolve quickly and the spores are pushed up in a column on top of the fruitbody. The hairs have a grainy covering. Spores 9–11 × 7–8.5 µm, brown, contain many drops, ± lemon-shaped but flattened, and have a germ pore at one end. Occurs on decaying organic material including dung, but also on wallpaper and other cellulose-rich materials in houses.

Chaetomium is a large and difficult genus that is mainly identified through cultivation in petri dishes.

Widespread, probably common; all year.

Echinosphaeria canescens has superficial and mostly crowded, dark brown perithecia with pale brown setose hairs. Setae straight, aseptate, very pointed. Asci 8-spored, ± cylindrical, I-. Spores boomerang-shaped, hyaline to yellow-brown, 1–2-celled and measure 27–32 × 4.8–5.6 µm. Occurs on wood of deciduous trees.

Echinosphaeria heterostoma ⋉ has thick-walled, blunt hairs and 6-celled spores; *E. strigosa* ⋉ has spores up to 45 µm long. Material with shorter, more curved spores has been named *E. curvatispora*.

Widespread, common; most of the year.

Ruzenia spermoides has black, smooth, matt, almost spherical to pear-shaped perithecia in very crowded, large groups that are connected at the base. Asci 8-spored, I-. Spores almost hyaline, ± boomerang-shaped, measure (17–)20–28(–30) × (3.8–)4–4.5(–6.2) µm and contain many drops. Occurs on deciduous wood, mostly *Fraxinus*, often associated with *Armillaria* (page 232) rhizomorphs.

Bombardia bombarda ◁ has a similar habit but the perithecia are more elongated and have a leathery texture. *Lasiosphaeria glabrata* ⋉ has much larger spores that are bent near one end.

Mainly nemoral–hemiboreal, common; all year.

Helminthosphaeria form small, setose perithecia, mostly on other fungal fruitbodies.

Helminthosphaeria odontiae has partly immersed setose perithecia. Asci 8-spored, I-. Spores dark brown, 1–2 (–3)-celled, when 2-celled lack a central septum; measure (6–)8–10(–12.7) × 4–6 (7.5) µm. The dark, asexual, hyphomycete state forms large, 4–5-celled conidia. Occurs on corticioids.

Helminthosphaeria corticiorum ⋉ has longer, up to 3-celled spores; *H. hyphodermiae* ⋉ has one-celled spores.

Probably widespread, rarely reported; mostly in the autumn.

Helminthosphaeria clavariarum is a black, setose pyrenomycetous fungus that colours the host ± grey-black. Perithecia have brown setae, the basal setae measuring up to 200 × 12 µm, with a 3–5 µm thick wall. Asci 8-spored, I-. Spores brown, ± fusiform, one-celled, have a pore at each end, and measure (10–)11.5–15.6(–19) × (4.5-)5.2–7.2(–8) µm. Has large, dark, thick-walled, 2-celled conidia. Occurs on fruitbodies of *Clavulina* (page 1128).

The characteristic host make identification straightforward.

Widespread, probably common, apparently ± absent from the boreal zone; August–November.

Chaetosphaerella phaeostroma is a very crowded, black pyrenomycetous fungus, half immersed in an even darker, strigose-hairy subiculum of pointed setae. Asci 8-spored, I-. Spores somewhat curved, 4-celled, with the middle cells dark brown and the end cells hyaline; measure (20–)26–35(–42) × 6–9 µm. Occurs on deciduous wood, mostly with other fungi, *e.g.* with *Diatrype* (page 1572) and other *Diatrypaceae*.

Acanthonitschkea tristis ▷ is superficially similar, but it has one-celled spores.

Mainly nemoral–hemiboreal, rather common; all year.

Acanthonitschkea tristis is a black pyrenomycetous fungus that forms rough-surfaced perithecia that shrink to cupulate-shaped upon drying. Has a dark subiculum that partly consists of tapering setae. Asci club-shaped, 8-spored, I-. Spores one-celled, hyaline, ± allantoid and measure 6–9 × 1.5–2 µm. Occurs on wood of deciduous trees.

Chaetosphaerella phaeostroma ⊲⊲ is superficially similar but has large, multicelled, partly dark-pigmented spores. Species of *Nitschkia* ▽ do not have conspicuous setae in a subiculum.

Nemoral–hemiboreal, occasional; mostly during winter.

Nitschkia form almost spherical, rough-walled perithecia that shrink to cupulate-shaped upon drying.

Nitschkia parasitans is a clustered, black, soon ± cupulate-shaped, warty *Nitschkia* with 0.2–0.3 mm wide perithecia. Asci club-shaped, 8-spored, I-. Spores hyaline, allantoid, contain two drops, 1(–2)-celled, measure 9–16 × 2–3 µm. Overgrows cushions of *Nectria cinnabarina* (page 1534) on deciduous branches.

Nitschkia cupularis ▽ and other species of *Nitschkia* have larger perithecia and no *Nectria* association.

Widespread, occasional; mostly during winter.

Nitschkia cupularis has ± clustered sessile, black, warty perithecia that upon drying become almost cupulate-shaped. Perithecia 0.3–0.45 mm wide. Asci club-shaped, I-. Spores hyaline, allantoid, 4-guttulate, 1(–2)-celled, 12–19 x2–2.5 µm. Occurs on deciduous, typically exposed, wood where the bark has loosened, sometimes on or with other fungi.

Nitschkia parasitans △ has smaller perithecia and is found in connection with *Nectria cinnabarina* (page 1534). Other species of *Nitschkia* have shorter or wider spores. *Acanthonitschkea triste* △ has black setae in a subiculum.

Rather rarely reported; mostly during winter.

Chaetosphaeria myriocarpa is a small pyrenomycetous fungus with superficial perithecia that are mostly associated with a hair-like, asexual state. Asci 8-spored, 45–60 × 4–5 µm, I-. Spores hyaline, 2-celled, 5–7(–8) × 2.5–3 µm. The asexual state ('*Chloridium clavaeforme*') has fusiform conidia, 1.5–2.5 × 2.5–3.5 µm.

Chaetosphaeria preussii ⋉ and *C. vermicularioides* ⋉ have 2-celled spores that split into two at maturity, and ellipsoid conidia. Other species of *Chaetosphaeria* have ellipsoid conidia, different asexual states or several-celled spores.

Widespread, rarely reported, but probably common; most of the year.

Bertia moriformis is a black, superficial, clustered, mulberry-like pyrenomycetous fungus that does not have an evident ostiole. The perithecial walls are very thick except at the tip. The asci are 8-spored, club-shaped, I-. The hyaline, allantoid, guttulate spores are 1–few-celled and measure 30–50 × 4–5.5 µm. Occurs mostly on decorticated wood of deciduous trees, but also on coniferous wood, *e.g.* with *Trichaptum* (page 914).

Bertia multiseptata ⋉ is closely related but may perhaps be recognized by its more pointed, 4-celled, brown-tinged spores. A 4-spored type with very wide spores is probably also a good species.

Widespread, common; all year.

Seen occasionally with a whitish hyphomycete or with greenish perithecia from the parasite *Neobarya parasitica* (page 1522).

The genus ***Phyllachora*** is, in a European context, rather small, but it is very speciose in the tropics and subtropics. All the species are dependent upon living hosts and can be problematic pathogens, albeit rarely.

Phyllachora junci is an immersed-erumpent, black, slightly stromatic pyrenomycetous fungus. Asci 8-spored, cylindrical, have a moderate ring-like structure at the top, I-. Paraphyses abundant, simple. The hyaline, ellipsoid, biguttulate, one-celled spores are distributed in two size groups, 8.5–9.5(–10) × (3–)3.5–4 µm and 12–15 × 3–4 µm, and have an indistinct slime coat. Occurs on stems of *Juncus*, especially *J. effusus*.

Easily identified but there may be more than one species involved. A number of other species of *Phyllachora* occur on grass culms.

Mainly nemoral–hemiboreal, very common; most of the year.

Hypocopra annulata is a small, immersed pyrenomycetous fungus with a single perithecium below a small, pale stromatic ring or shield. Asci 8-spored, apical apparatus I+ (red). Spores have hyaline, small appendages and an asymmetrical dark brown part, 26–29 × 13–15 µm, a thick slime coat and a germ slit along two-thirds of the length. Occurs on herbivore dung.

Hypocopra is a rather large and difficult genus. The amount of stromatic tissue can vary, and other important characters include spore size, germ slit and ascus top appearance. The genus may be closely related to *Poronia* (page 1552).

Probably widespread, very rarely reported; mostly during winter.

Thomas Læssøe

Rosellinia form large perithecia that are initially surrounded by a hyphal mat from where conidia are produced. The perithecial walls have several layers and, technically speaking, the perithecia are one-peritheciate stromata. Spores are dark, and have a germ slit and slime and/or cellular appendages. Most species are wood decayers, but *R. desmazieri* ⋉ may attack living *Salix* and tropical species cause serious damage in *e.g.* coffee plantations. See also page 1565 for other rosellinioid species.

Rosellinia thelena is a species of *Rosellinia* with a very prominent hyphal mat, although this mostly forms deep in piles of *Picea* brush. Asci 8-spored, I+, apical apparatus 5.5–7.5 µm high. Spores, (16–)19–24 (–28) × (5.5–)–6.5–8.5(–10) µm, have full-length germ slits and long, pointed cellular appendages at both ends; young spores have a slime coat. Mostly in large piles of cut Picea branches, but also on other debris from coniferous and deciduous trees.

The long, pointed spore appendages should make this species easy to identify. *Rosellinia mycophila* ⋉, on similar substrates, often *Abies*, has somewhat smaller spores with a short, cellular appendage at one end of the spore.

Nemoral–hemiboreal, common to occasional; most of the year.

Rosellinia corticium is a large, coarse *Rosellinia* with a well-developed hyphal mat and somewhat flattened perithecia. Asci 8-spored, apical apparatus I+, 5.5–12.5 µm high. Spores measure (19–)22–26.5(–33.5) × (6.5–)8.5–10(–14) µm, have a rather oblique, ± full-length germ slit, a rounded cellular appendage in one end and the entire spore is surrounded by a thick slime coat. Occurs on deciduous wood.

Rosellinia aquila ⋉ has somewhat smaller spores with cellular appendages at both ends and the slime coat is only around the appendages.

Mainly nemoral, occasional; most of the year.

Thomas Læssøe/Jens H. Petersen

Rosellinia marcucciana is a *Rosellinia* with a very limited hyphal mat and somewhat flat, rather single-walled perithecia. Asci I+, apical apparatus 5.4–8.1 µm high. The spores measure (19.8–)21.8–25 (–27.5) × (5.2–)7–8.4(–10.4) µm, have a full-length germ slit on the flat side and lack cellular appendages.

Occurs on deciduous wood, mostly rather dry and exposed and partly decorticated.

Is better known as *R. britannicum*. *Rosellinia mammiformis* ⋉ has smaller spores and shorter ascus apical apparatus, and *R. subsimilis* ▽ has more pointed perithecia and the spores have cellular appendages.

Nemoral, rather common; most of the year, although in best condition during winter.

Rosellinia subsimilis is a *Rosellinia* with a very limited hyphal mat. Perithecia somewhat conical. Asci 8-spored, apical apparatus I+, 4–11 µm high. The spores measure (16.5–)22–27(–31) × (5–)6–8(–9.5) µm, have short cellular appendages at both ends that are surrounded by a slime coat, and a full-length germ slit that is usually somewhat oblique. Occurs on coarse herbaceous stems and wood from deciduous trees.

Similar species either do not have cellular spore appendages or have well-developed hyphal mats and coarser perithecia.

Mainly boreal, occasional; most of the year.

Non-hysterioid, bitunicate pyrenomycetous fungi

Species in this form group have thick-walled asci with several functional wall layers. Most species have small, free perithecia (also termed pseudothecia) or even smaller, circular, flattened perithecia (termed thyriothecia – page 1604).

The young asci are typically very thick-walled, especially the upper parts, where a characteristic bulge often can be seen (see arrow below). The spores are ejected from an inner sac that stretches out from the outer sac once the ascus tip ruptures – a so-called jack-in-the-box mechanism.

Most species in this form group are plant parasites, or decomposers of wood, herbs or dung.

OTHER SIMILAR FUNGI:

– the ± black, non-stromatic unitunicate pyrenomycetous fungi are similar but the asci are single-walled (page 1582).

– the hysterioid pyrenomycetous fungi have similar asci (and phylogenetically belong to the same class), but have compressed, superficial fruitbodies that open by a narrow slit (page 1614).

See the wheel page 1510.

Further reading: 61, 73, 77, 78, 84, 234, 299, 300, 302, 312, 347, 353, 357.

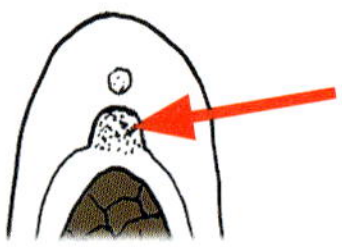

Asci in most bitunicate pyrenomycetous fungi have a characteristic thick-walled tip with a bulge, here in *Cucurbitaria*.

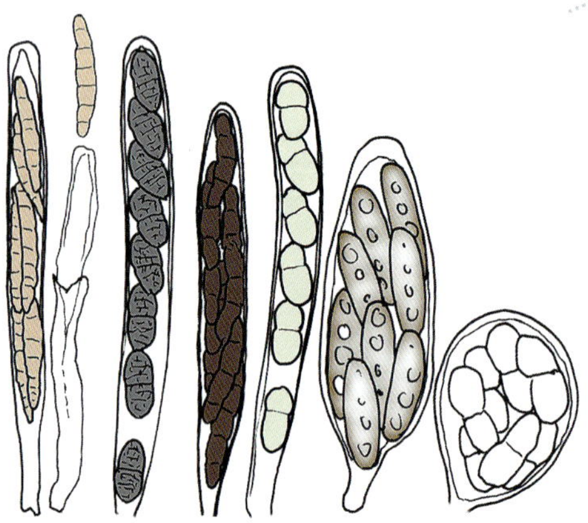

Asci with bitunicate walls – from left *Leptosphaeria* (one mature and one with the inner sac ejected), *Cucurbitaria*, *Sporormiella*, *Venturia*, *Massaria* & *Asterina*.

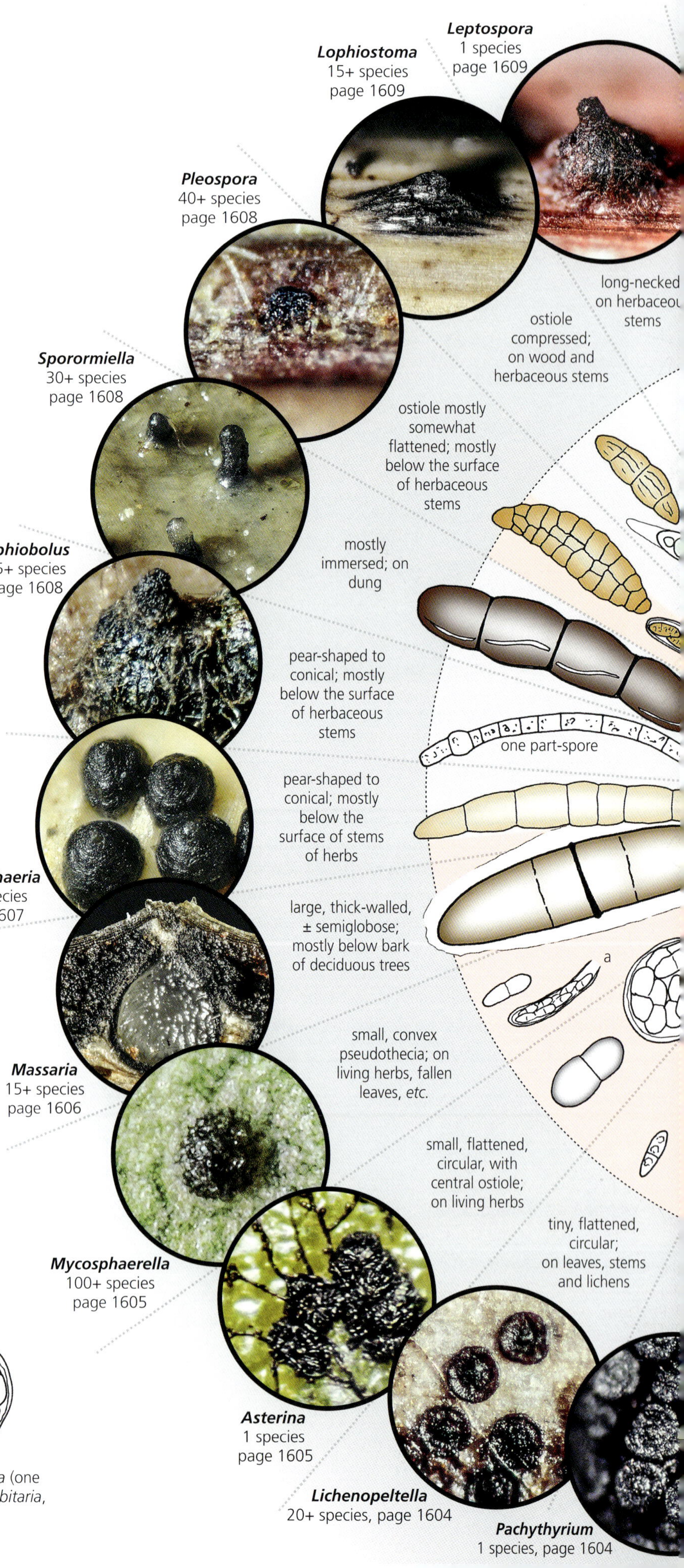

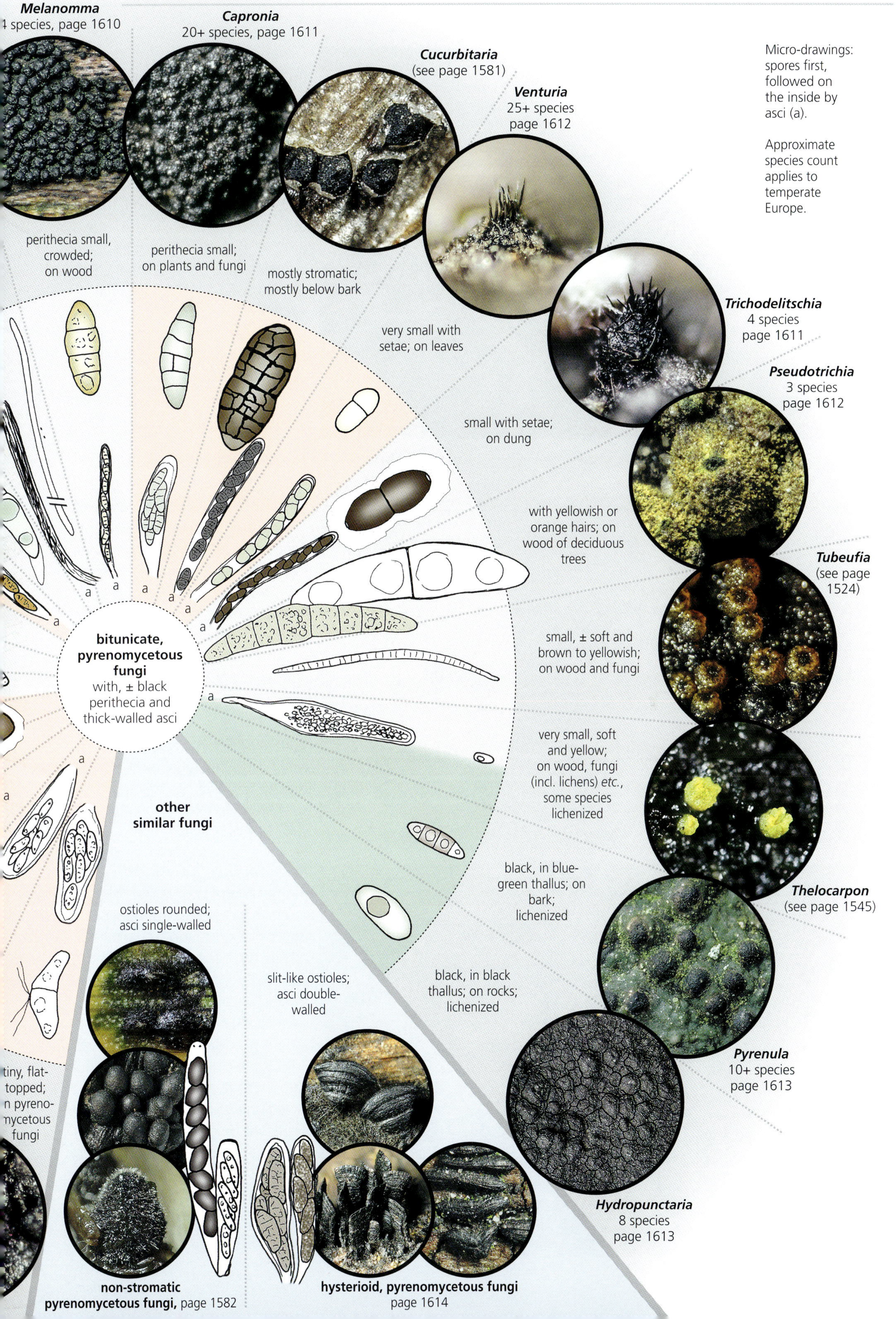
Melanomma
4 species, page 1610
Capronia
20+ species, page 1611
Cucurbitaria
(see page 1581)
Venturia
25+ species
page 1612
Micro-drawings: spores first, followed on the inside by asci (a).
Approximate species count applies to temperate Europe.
perithecia small, crowded; on wood
perithecia small; on plants and fungi
mostly stromatic; mostly below bark
very small with setae; on leaves
Trichodelitschia
4 species
page 1611
Pseudotrichia
3 species
page 1612
small with setae; on dung
with yellowish or orange hairs; on wood of deciduous trees
Tubeufia
(see page 1524)
small, ± soft and brown to yellowish; on wood and fungi
bitunicate, pyrenomycetous fungi
with, ± black perithecia and thick-walled asci
very small, soft and yellow; on wood, fungi (incl. lichens) *etc.*, some species lichenized
Thelocarpon
(see page 1545)
black, in blue-green thallus; on bark; lichenized
black, in black thallus; on rocks; lichenized
Pyrenula
10+ species
page 1613
Hydropunctaria
8 species
page 1613
other similar fungi
ostioles rounded; asci single-walled
slit-like ostioles; asci double-walled
tiny, flat-topped; n pyreno-mycetous fungi
a
non-stromatic pyrenomycetous fungi, page 1582
hysterioid, pyrenomycetous fungi
page 1614

Lichenopeltella form flat, circular fruitbodies (thyriothecia).

Lichenopeltella pinophylla has tiny, flattened, circular fruitbodies (thyriothecia) with an entire margin and a papilla-like ostiole that consists of 3–4 rings of thick-walled, dark cells. Spores hyaline, 2-celled, measure 8–11 × 1.5–3 µm. On needles of *Pinus*, *Cupressus* and perhaps other conifers.

Also known as *Trichothyrina pinophylla*. *Stomiopeltis pinastri* ⋉ has fimbriate margins. *Microthyrium pinophyllum* ⋉ does not have dark cells at the centre of the fruitbody.

Widespread and probably common; mostly during winter.

Lichenopeltella nigroannulata has tiny, circular, flat fruitbodies (thyriothecia) with a dark ostiole of 1–3 rings of dark, thick-walled cells. Spores 2-celled, smooth, somewhat asymmetrical, ± fusiform, measure 7.5–10.5 × 1.5–3 µm. Occurs mostly on *Poaceae* and *Juncaceae*, also on ferns, leaves of deciduous trees, *etc.*

Also known as *Trichothyrina nigroannulata*. *Lichenopeltella alpestris* ⋉ has holes and usually short setae around the ostiole; *L. ammophilae* ⋉ has cylindrical spores with tiny setae (setulae) from transverse septa.

Probably widespread and common; likely to be present all year.

Pachythyrium parasiticum has flat-topped but rather thick, often very crowded fruitbodies (thyriothecia/pseudothecia) with a central ostiolar papilla. Asci double-walled, 38–50 × 12–16 µm, I-. The hyaline, 2-celled spores have four, small, hair-like setae from near the slightly acentric septum, and measure 14–18 x 4–6 µm. Occurs on various carbonous pyrenomycetous fungi, including *Eutypa lata* ⋉ and *Massaria inquinans* (page 1606).

Probably widespread, but rarely reported likely to be present all year.

Asterina veronicae is a bitunicate, pyrenomycetous fungus that has flat, very small fruitbodies (thyriothecia) connected by dark, superficial hyphal cords. Asci almost spherical. Spores hyaline, 2-celled and measure 14–17 × 6–7 µm. Has a very similar asexual state with conidia that measure 18–21x 7–9 µm. Occurs on living leaves and stems of *Veronica officinalis* and perhaps other species of *Veronica*.

There are no other similar species on living *Veronica* leaves

Probably widespread, but very rarely reported, possibly overlooked; summer–autumn.

Mycosphaerella is huge genus with species that form very small fruitbodies immersed in plant tissues. Spores 2-celled.

Mycosphaerella punctiformis has tiny, immersed perithecia (pseudothecia). Asci double-walled, club-shaped, I-. Spores hyaline, 2-celled 8–14 × 2–3 µm. Occurs on fallen leaves of *Acer*, *Alnus*, *Castanea*, *Corylus*, *Fagus, Quercus*, *Salix*, *Tilia*, *Ulmus*, *etc.*

The species name is likely to change to *Ramularia enophylla*. There are numerous similar species on other hosts or with other spore dimensions.

Widespread, very common; most of the year.

Mycosphaerella podagrariae is a bitunicate pyrenomycetous fungus with small, almost superficial perithecia (pseudothecia). Asci double-walled. The hyaline, 2-celled spores have a constriction at the septum and measure 18–19 × 5 µm. Seen more often in the asexual *'Phloeospora'*-state, with 2–4-celled conidia that measure 50–80 × 2.5–4 µm. Occurs on living leaves of *Aegopodium*.

The superficially similar rust fungus *Puccinia aegopodii* ⋉ also occurs on leaves of *Aegopodium* (see *Puccinia* page 1204).

Widespread in the nemoral–hemiboreal zones, probably common; June–October.

Massaria form very large perithecia (pseudothecia) with very big asci and spores. The fruitbodies are formed in late autumn and overwinter below the bark. Eventually the bark may fall off and the empty fruitbodies, that resemble Mongolian tents with open chimneys, become visible.

Massaria inquinans has large perithecia (pseudothecia), which form raised bumps on bark. Asci double-walled, 8-spored, very large, (330–)365–415(–430) × (43–)47–55(–64) µm, I-; in a matrix of thin, branching, hyaline hyphae (trabeculae). Spores surrounded by a slime coat, brown, smooth, 4-celled with thick septa, the middle septum appearing as a pronounced waistband; measure (70–)80–89(–101) × (18–)19–22(–24) µm (visible on the substrate with a good hand lens, see upper image). Occurs on *Acer pseudoplatanus* twigs.

A range of similar species occur on other hosts, *e.g. M. aucupariae* ▽ on *Sorbus*, *M. platanoidea* ⋉ and *M. campestre* ⋉ on *Acer campestre*, and *M. vomitoria* ⋉ on *A. platanoides*.

Nemoral–hemiboreal, common; mostly during winter.

May be parasitized by *Flammocladiella decora* (page 1539) and *Pachythyrium parasiticum* ◁◁.

Massaria aucupariae has large perithecia (pseudothecia) that are seen as raised bulges on bark, and which have a black, penetrating ostiole. The pseudothecia are surrounded by a black shield within the wood. Asci double-walled, very large, I-. The 4-celled, hyaline spores turn brown after having been shot out and measure (62–)66–77(–94) × (14–)14.5–16(–17) µm; end-cells longer than mid-cells; have a slime coat. Occurs on *Sorbus* branches.

Massaria ariae ⋉ also occurs on *Sorbus* but has broader spores that are brown within the asci.

Presumably widespread, but rarely reported; mostly during winter.

Species of **Leptosphaeria** generally produce small perithecia (pseudothecia) on herbs. The spores are multicelled, pale brown and somewhat banana-shaped. Typically, a very similar looking asexual *'Phoma'*-state is produced. Species of *Phaeosphaeria* ⋉ are very similar but tend to be smaller and have different asexual states; they often occur on monocots, while *Leptosphaeria* tend to occur on dicots.

Leptosphaeria acuta is a ± smooth, erumpent, black bitunicate pyrenomycetous fungus with papilla- or chimney-like ostioles. Asci 8-spored, double-walled, narrowly club-shaped, measure 105–160 × 10–14 mm, I-. The spores are 5–9-celled, pale yellow-brown, measure (36–)45–56(–60) × 5–7(–8) µm and have constrictions, especially at the centre. Typically occurs on old stems of *Urtica*.

The very similar *L. doliolum* ▽ has ring-like structures on the fruitbody and 4-celled spores. *Leptosphaeria multiseptata* ⋉ occurs on *Lathyrus* and has longer spores.

Widespread, very common; most of the year, although mainly March–May.

Occasionally seen together with the parasitic hypocrealean *Macroconia sphaeriae* (lower image and page 1537).

Leptosphaeria doliolum has ± breast-shaped smooth fruitbodies, with two concentric, raised rings. Asci double-walled, 8-spored, I-. Spores yellow-brown, 4-celled, measure 16–23(–28) × 4–6(–7) µm. Occurs on herbaceous stems, particularly *Apiaceae* and *Urtica*.

Leptosphaeria acuta △ does not have conspicuous rings on the fruitbody and has 5–9-celled spores; *L. purpurea* ⋉ stains the substrate wine-red and occurs mostly on *Asteraceae*. There are other similar species of *Leptosphaeria*.

Widespread, very common; all year, but apparently more frequent in the autumn.

Ophiobolus acuminatus is erumpent, black and pear-shaped. Asci 8-spored, double-walled, cylindrical, 150–190 × 13–17 µm, I-. Spores yellowish to yellow-brown, 24–30-septate, 140–180 × 3–4 µm but typically split into two part-spores, each with a swollen cell that is one cell from the break point. Occurs on herbaceous stems, mostly *Asteraceae*.

Ophiobolus ophioboloides ⋉ has 16-septate spores; *O. erythrosporus* has entire spores and only one swollen cell; *O. niesslii* ⋉ has a swollen fourth cell; and *O. fruticum* ⋉, on *Ononis*, has three swellings.

Probably widespread and common, but rarely reported; all year.

Sporormiella vexans is small, dark and immersed with a ± conical, protruding top. Asci 8-spored, double-walled, narrowly club-shaped, measure 120–143 × 20–21 µm. Spores brownish-black, 7-celled, have a slime coat, measure 41–47 × 8.5–9.5 µm, and break into 7 part-spores, each with a diagonal or transverse germ slit. Occurs on dung, mostly from deer, hare and rabbit.

There are many similar species of *Sporormiella* but *S. vexans* is distinguished by its spore size and by the five middle cells being wider than long.

Widespread, rarely reported, but probably overlooked; all year.

Pleospora rubelloides has erumpent, black, pear-shaped perithecia (pseudothecia) and stains the substrate purplish. Asci 8-spored, 94–120 × 12–16 µm, I-. Spores yellow-brown, muriform-septate, ± tapering and usually constricted at the centre; measure 20–26 × 6–8 µm. Occurs on herbaceous stems, particularly grass culms.

Other species of *Pleospora* may also cause a purple discoloration: *P. rubicunda* ⋉ typically has larger fruitbodies, asci and spores; and *P. straminis* ⋉ has spores measuring 26–34 × 10–12 µm. *Leptospora rubella* ▷ has thread-like spores.

Widespread, but rarely reported; most of the year.

Leptospora rubella is erumpent to superficial, and blackish with a chimney-like ostiole and a reddish discolouration of the substrate. Asci double-walled, cylindrical, 8-spored, 170–225 × 4–5 µm, I-. Spores 15–25-septate, thread-like, 220–270 × 1–1.5 µm, twisted within the asci, lack swellings or appendages. Occurs on various herbaceous stems, and also on somewhat woody stems.

Also known as *Ophiobolus rubellus*. Species of *Pleospora* ◁ with the same discolouration of the substrate have wider, muriform-septate spores.

Widespread, but probably rather common; most of the year.

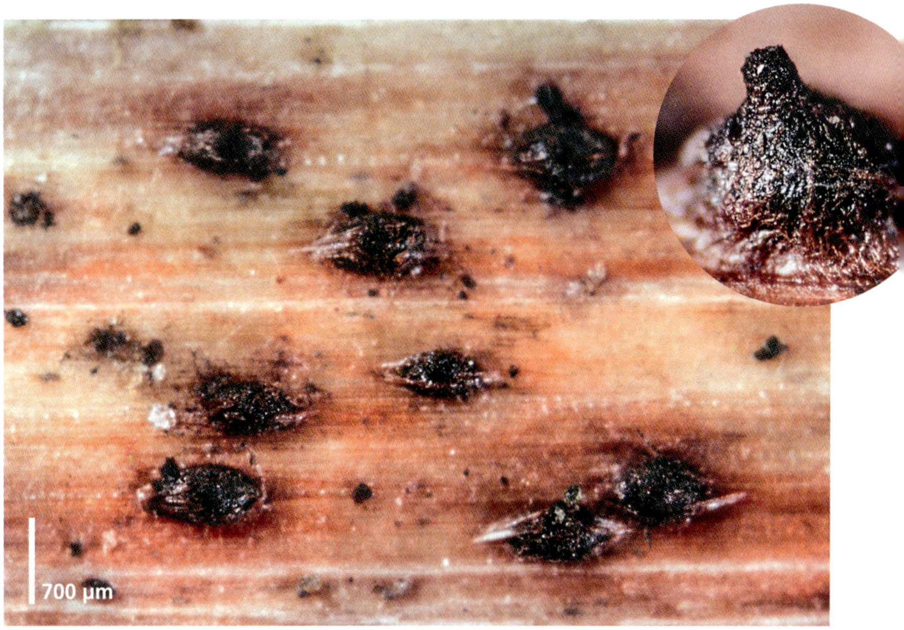

Lophiostoma form immersed perithecia (pseudothecia) that are erumpent and have an elongated crest with a ± slit-like ostiole. The spores are multicelled and may be striate, rough or smooth. Occur on herbaceous stems and wood. May recall a hysterioid pyrenomycetous fungus (page 1614), but these have superficial, more laterally compressed fruitbodies.

Lophiostoma viridarium is an immersed bitunicate pyrenomycetous fungus with a slit-like ostiole on a crest-like top. Asci double-walled. The fusiform, 4-celled, pale brown, longitudinally striate spores measure 32–46 × 8–13 µm and have a slime coat; the striation is caused by lines of warts. Occurs on turquoise-stained deciduous wood, perhaps mostly *Corylus* in calcareous woodlands, but also on *e.g. Salix* and *Cornus*.

The combination of unusual spores and characteristic discoloration makes this species easy to identify. Since the type species of *Lophiostoma* is only distantly related to *L. viridarium* a new name will probably have to be introduced.

Nemoral, rather rare and local; all year.

Lophiostoma semiliberum is an immersed-erumpent, black bitunicate pyrenomycetous fungus with a somewhat elongated and laterally compressed ostiole (crest-like with a slit). Asci double-walled, club-shaped, measure 100–125 × 12 µm, I-. The finely rough, 1–(3–5)-septate, hyaline to pale yellow-brown spores measure 32–40 × 5–6 µm, contain 3–8–drops and do not have appendages. Occurs on herbaceous stems, both monocots and dicots, but mostly on grass culms.

The similar *L. macrostomum* ▽ has appendages at the spore ends.

Widespread, common; all year.

Lophiostoma macrostomum is an immersed black bitunicate, pyrenomycetous fungus with a slit-like ostiole on a raised crest. Asci 8-spored, club-shaped, I-. Spores 1(–3)-celled, hyaline to yellow-brown, have several drops in each cell, measure 26–33(–38) × 5–7 µm, and have short appendages at the ends. Occurs both on herbaceous stems and wood of deciduous trees.

Lophiostoma semiliberum △ does not have appendages at the spore ends. *Thyridaria macrostomoides* ⋉ occurs on wood and has several cells in the darker spores.

Widespread, but rarely reported; mostly found January–July.

Melanomma pulvis-pyrius has superficial, black, irregularly angular-verrucose perithecia (pseudothecia). Asci double-walled, cylindrical, have spores in one row, I-. Spores grey-brown, 4-celled, constricted at the centre, measure 14–20 × 4–6 µm, and lack appendages. Occurs on decorticated dry wood of deciduous trees.

Melanomma fuscidula ⋉ is more immersed and has somewhat smaller spores in two rows in the shorter asci. *Melanomma* belongs to a huge family, the *Melanommataceae*, which has many genera and species.

Very common and widespread; all year.

Capronia has tiny fruitbodies with septate, grey spores, and setae or dark cells on the fruitbodies. Also make so-called black yeasts.

Capronia porothelia has solitary, ± immersed, black perithecia (pseudothecia) that lack setae but have thick-walled cells round the ostiole. Asci double-walled, club-shaped, I-. Spores ± olive-grey, have three thick septa, 10–14 × 3–4 µm. Occurs in the hymenium of *Laxitextum bicolor* (page 966).

Easily recognizable due to substrate. *Capronia spinifera* ⋉ is probably a synonym.

Widespread and probably rather common; autumn–winter.

Capronia nigerrima is a crustose or ± stromatic *Capronia*. The surface has short setae. Asci double-walled, broadly club-shaped, 8-spored, measure 40–50 × 14–17 µm, I-. Spores measure 15–22 × 5–6 µm and are olive-grey with 3–5 transverse septa and an incomplete longitudinal septum. Occurs on stromata of various members of *Diatrypaceae* (pages 1569-1574).

This is the only ± stromatic *Capronia* species – also known as *Berlesiella nigerrima*.

Widespread and common, although rarely reported, becoming scarcer towards the north; mostly during winter.

Trichodelitschia bisporula is a partly immersed bitunicate pyrenomycetous fungus with long, black setae on top of the pear-shaped fruitbody. Asci double-walled, 8-spored, cylindrical, I-. The black, 2-celled spores measure 21.5–25(–27.5) × 8–9.5(–10.5) µm and are pointed at the ends; they have germ pores and furthermore are highly constricted at the centre and surrounded by a slime coat that also has constrictions. Occurs on dung, *e.g.* from sheep, cow, hare and rabbit.

Also known as *T. minuta*. *Trichodelitschia lundqvistii* ⋉ has narrower spores.

Widespread, rarely reported, probably overlooked; all year.

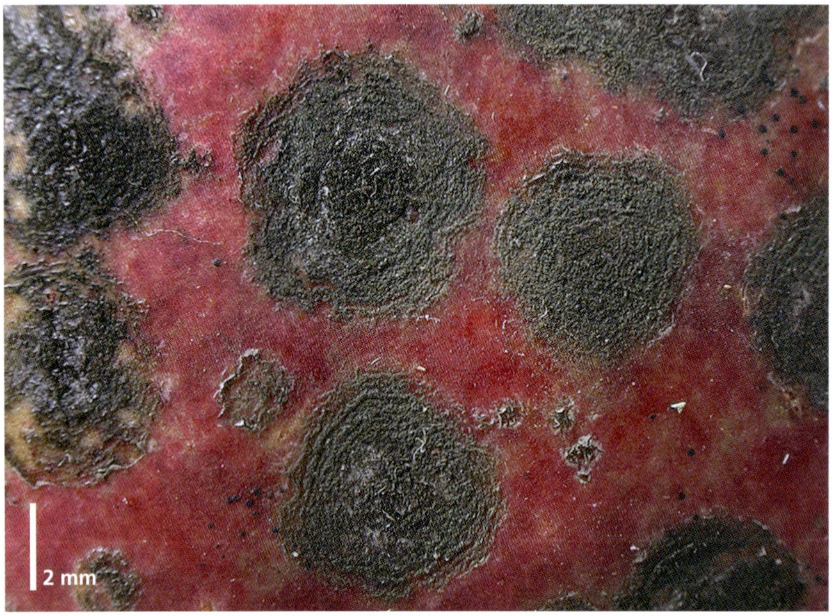

Venturia inaequalis is a bitunicate pyrenomycetous fungus with tiny, immersed fruitbodies (pseudothecia) that typically have a group of setae around the ostiole. Asci double-walled, cylindrical, 8-spored, measure 60–70 × 7–12 µm, I-. Spores olive-brown, measure 12–15 × 6–8 µm and have two uneven cells: the upper is tapering while the lower, longer cell is rounded. The asexual state ('*Spilocaea*') is formed on leaves and fruits as brownish colonies (lower image). Conidia ± pear-shaped, olive-brown, smooth, 1–2-celled, measure 12–30 × 6–10 µm. Mostly on *Malus*, but also on *Pyrus*, *Sorbus*, *Crataegus* and others; asexual reproduction on fruits and leaves and fruitbodies is on previous year's half-rotten leaves. Causes apple scab and is a huge problem in orchards, since the apples are blemished and store poorly.

The distinctive 2-celled spores makes identification straightforward, but the fruitbodies are difficult to locate – the asexual state is, however, very frequent.

Widespread and common, especially in the nemoral–hemiboreal zones; all year.

There are other serious fungus pathogens on apples, including cankers caused by *Neonectria ditissima* (page 1535).

Pseudotrichia mutabilis is a superficial bitunicate pyrenomycetous fungus with sulphur-yellow hairs and a dark ostiole. Asci double-walled, 8-spored, club-shaped, (90–)120–155 × 12–20 µm, I-. Spores 26–39 × (6–)7–9 µm, 2–4-celled, but lack longitudinal walls; hyaline to brown-tinged. Occurs on deciduous wood.

Mattirolia roseovirens ⋉ is a southern species on *Fabaceae* with yellow hairs and muriform-septate, brown spores with 3–5 transverse septa and thin-walled asci; *Mattirolia mutabilis* ⋉ has similar spores and yellow hairs, but also has 5–7 transverse septa.

Nemoral–hemiboreal, rare; most of the year.

Pyrenula nitida is a crustose lichen that is ± immersed in bark and has an olive-green to brownish-green, smooth thallus with small, whitish perforations (pseudocyphellae). It has numerous evenly distributed, black, semiglobose to conical perithecia 0.6–0.8 mm in diameter. Thallus KOH+, reddish-orange around the perithecia. Asci I-. The spores are brown, narrowly ellipsoid, 4-celled, and measure 19–24 × 6–8 µm. Associated with *Trentepohlia* algae. Occurs on bark of *Fagus*, *Carpinus*, *etc.*

Most species of *Pyrenula* are pyrenomycetous lichens, although some lack an association with algae. A large and difficult genus. The nemoral–oceanic *Pyrenula macrospora* ⋉ has larger perithecia and spores, and is KOH+ yellow. *Pyrenula nitidella* ⋉ has up to 0.35 mm wide perithecia and wider spores. *Pyrenula chlorospila* ⋉ also has small perithecia and even larger spores. *Pyrenula laevigata* ⋉ has a margin (a projecting involucrellum) at the base of the somewhat flattened perithecia, which are also small and usually have pycnidia in between.

Nemoral–hemiboreal, rather common to occasional; all year.

Erik Arnfred Thomsen

Hydropunctaria maura is a grey-black crustose lichen with a ± warty, finely cracking thallus in a polygonal pattern with immersed fruitbodies that are seen as dark dots. Asci double-walled, club-shaped, I+. The spores are hyaline, one-celled, and measure (12–)15–18 × 7–9 µm. Associated with green algae in the thallus. Occurs on sea water-sprayed rocks and stones.

Hydropunctaria rheitrophila ⋉ is associated with fresh water and is smoother. *Hydropunctaria aractina* ⋉, *H. oceanica* ⋉ and *H. orae* ⋉ are very difficult to separate from *H. maura*, but all have somewhat thinner thalli.

Widespread, coastal; all year.

Hysterioid pyrenomycetous fungi

This group of fungi includes species with mostly superficial perithecia that have distinct, ± compressed ostioles and a slit-like opening (hysterothecia). The fruitbody shape is highly variable: many are elongate, others clam- or axe-like. All have double-walled asci (bitunicate) that typically release the spores after an inner sac has stretched through the outer sac (jack-in-the box-release).

All species are decomposers of wood, cones, herbs, *etc.* They mostly occur in exposed, dry habitats, *e.g.* on bark of ± sun-exposed twigs.

OTHER SIMILAR FUNGI:
– inoperculate cup fungi with slit-like openings reveal the hymenium in wet weather. They have single-walled (unitunicate) asci (page 1468).
– some bitunicate pyrenomycetous fungi (*e.g. Lophiostoma*) have flattened ostioles, but the fruitbodies are immersed in the substrate (page 1602).

FURTHER READING: 41, 42, 368.

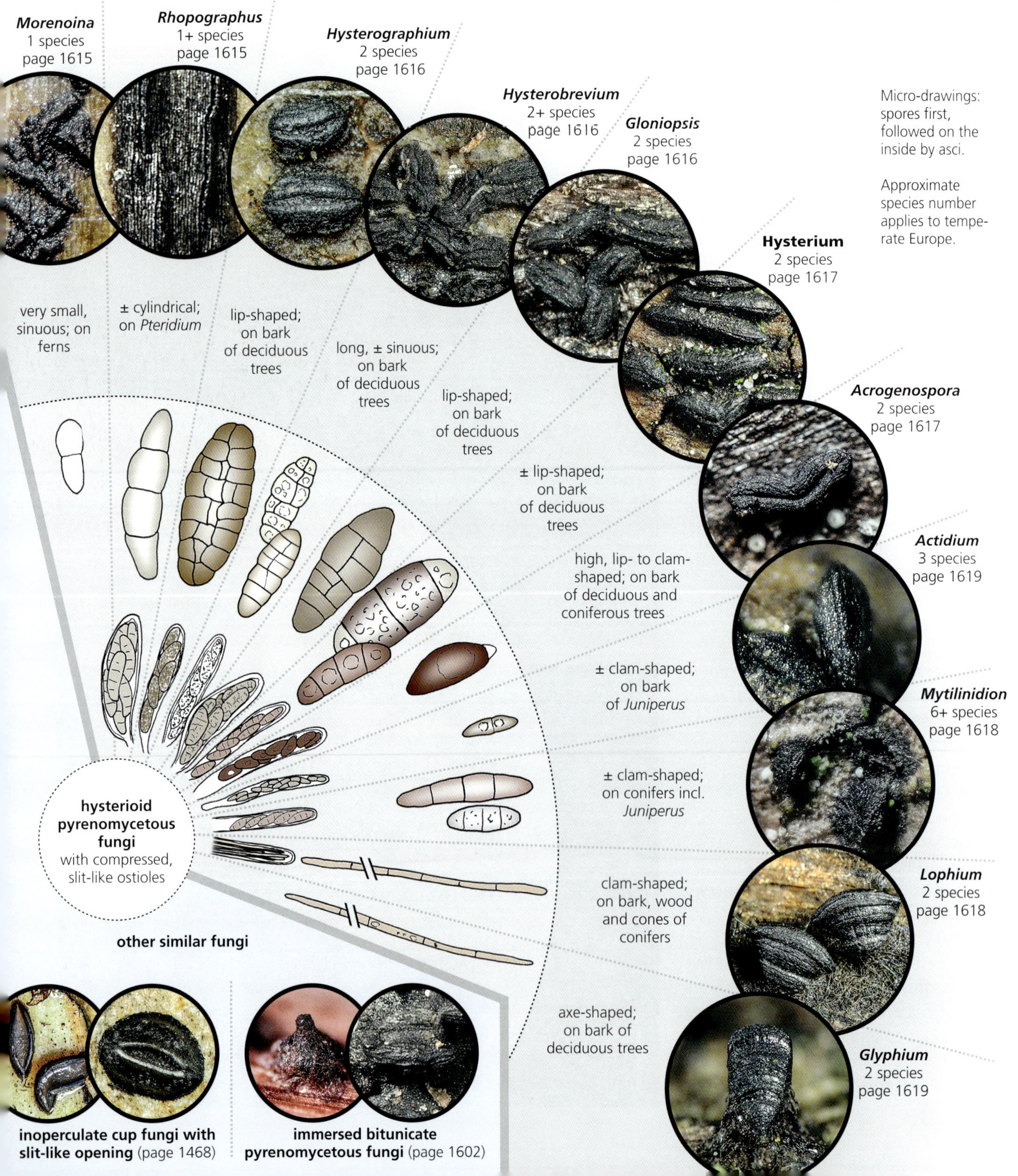

Morenoina pteridicola is a black, bitunicate pyrenomycetous fungus with oblong, somewhat branched, very small, ± flattened fruitbodies that have slit-like ostioles. The margin has brown, radiating, ± branching hyphae. Asci double-walled, measure 18–27 × 9–14 µm. The spores are 2-celled, not completely symmetrical, smooth, somewhat brownish when old, measure 9–13 × 3–4 µm. Occurs on petioles of *Dryopteris*, *Pteridium* and probably other ferns.

There are apparently no other similar species on these substrates.

Possibly widespread, but very rarely reported, most likely overlooked; most of the year.

Rhopographus filicinus is a bitunicate pyrenomycetous fungus that forms elongated-oblong, black, ± confluent fruitbodies with slit-like ostioles along the length of the hosts' 'stems'. Asci double-walled, 8-spored, measure 70–86 × 20–25 µm, I-. The somewhat bent, yellow-brown, 4–8-celled spores measure 27–35 × 7–8 µm. Occurs on petioles of *Pteridium*.

Scirrhia aspidiorum ⋉ forms grey fruitbodies/stromata and hyaline 2-celled spores.

Widespread, very common; all year.

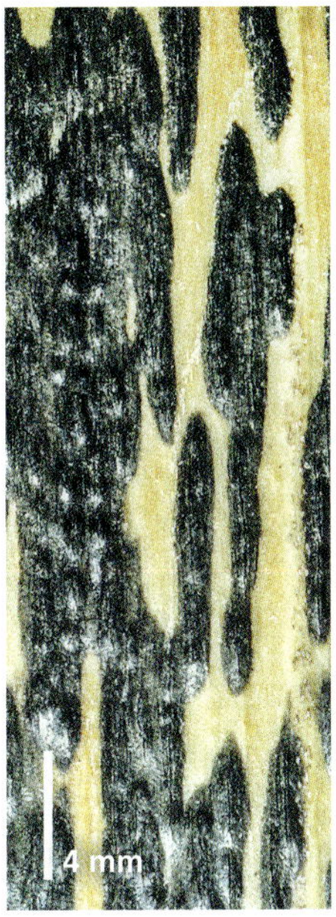

Hysterographium fraxini has elongated, black, longitudinally furrowed, half-open fruitbodies (hysterothecia) that may branch slightly. Asci double-walled, I-. Spores yellow-brown, muriform-septate with a slightly constricted middle septum and many transverse and longitudinal septa; they measure 40–48 × 16–20 µm, are mostly broadly rounded and have a slime coat. Mainly occurs on *Fraxinus* bark.

Hysterographium flexuosum ⋉ has somewhat longer, more pointed spores; *H. elongatum* ⋉ apparently has 1 or no longitudinal septa, but is perhaps just a synonym of *H. fraxini*.

Widespread, occasional; all year.

Hysterobrevium has muriform septation.

Hysterobrevium smilacis has ± erumpent, sinuous fruitbodies with a slit-like ostiole. Asci double-walled, 70–120 × 15–25 µm, I-. Spores hyaline to yellowish, (13–)15–26(–31) × (4–)5–9(–10) µm, muriform-septate with 3–5(–9) transverse and 1(–3) longitudinal septa; have a slime coat and a constriction at the centre. Occurs on wood and herbaceous stems.

Hysterobrevium mori ▽ has more pigmented spores. *Gloniopsis praelonga* ▽ has longer spores with more septa.

Nemoral, rarely reported; all year.

Hysterobrevium mori has black, oblong fruitbodies with a slit-like ostiole. Asci double-walled, I-. Spores asymmetrical, yellow-brown, muriform-septate with 3(–5–7) transverse and 1–2(–3) longitudinal septa; they measure (12–)14–22(–26) × (5–)7–10(11–) µm and are constricted at the centre. Asexual state has pycnidia; conidia measure (2–)2.5–3.5(–4) × 1–2 µm. Occurs on exposed deciduous or coniferous wood.

The very similar *H. smilacis* △ has almost hyaline spores and usually more transverse septa.

Nemoral, rarely reported, but may be overlooked; all year.

Gloniopsis praelonga has black, oblong fruitbodies with a slit-like ostiole. Asci double-walled, I-. Spores hyaline to yellowish, muriform-septate with 5–7(–10) transverse and 2–3 longitudinal septa; they measure (16–)20–32(–34) × (6–)9–12(–15) µm and have a slime coat. Occurs on wood of various deciduous trees and *Juniperus*.

Various similar species of hysterioid pyrenomycetous fungi have shorter spores; *G. subrugosa* ⋉ has pigmented spores in the asci.

Widespread, rarely reported, but probably overlooked; all year.

Species of ***Hysterium*** have low to slightly raised, ± mouth-shaped fruitbodies with elongated ostioles. The spores are brown and septate

Hysterium pulicare has black fruitbodies with a slit-like ostiole. Asci double-walled, I-. Spores 4-celled, brownish with pale end-cells, measure 20–40 × 6–12 µm. Occurs on bark and wood of a range of different deciduous trees.

Very similar to *H. acuminatum* ▽ where all four of the spore-cells are evenly pigmented. Other similar species of hysterioid pyrenomycetous fungi have different spore septations.

Widespread, common; all year.

Hysterium acuminatum has black, oblong, straight to sinuous fruitbodies with a slit-like ostiole. Asci double-walled, I-. Spores 4-celled, rounded fusiform, not constricted at the septa, grey-brown, measure (12–)14–21 (–28) × (3–)4–8(–10) µm. Occurs on a range of deciduous trees, mostly on bark of standing living trees.

Also known as *H. angustatum*. *Hysterium pulicare* △ has almost hyaline spore end-cells. Other similar species of hysterioid pyrenomycetous fungi have a different spore septation.

Widespread, common; all year.

Acrogenospora carmichaeliana has oblong, black fruitbodies (usually just a few together) with a slit-like ostiole. Asci double-walled, I-. Spores measure 19–26 × 9–12 µm, 2-celled with a large, dark brown cell and a small, colourless cell. Forms very large, brown, ± egg-shaped conidia that measure 20–40 × 15–25 µm on long, dark conidiophores. Mostly on wood of deciduous trees.

Also known as *Farlowiella carmichaeliana*. Macroscopically may look like a *Hysterium* but unmistakable microscopically.

Widespread, rarely reported, but easily overlooked; all year.

Lophium mytilinum is a bitunicate pyrenomycetous fungus with small fruitbodies that recall a vertical clam with the broader end upwards. The fruitbodies open by a narrow slit on top. Asci double-walled, (140–)180–260(–300) × 6–8(–10) µm, I-. Spores yellowish to ± brownish, thread-like, multiseptate, measure 150–170 × 2–2.5 µm. Occurs on bark and wood of *Pinus* and other conifers.

Mytilinidion mytilinellum ⋉ is very similar but the spores are short, cucumber-shaped, mostly 4-celled and measure (14–)16–22(–24) × (2.5–) 3–4(–5) µm. There are other species of *Mytilinidion* ▽ on coniferous wood with similar or shorter spores.

Widespread, probably common; all year.

Mytilinidion decipiens has small, black, clam-like fruitbodies with a slit-like ostiole. Asci double-walled, I-. Spores yellow-brown, 4-celled, slightly curved with rounded ends, do not have constrictions, measure (11–) 13–15(–21) x 3–4(–6) µm. Occurs on wood and bark from *Larix* and *Juniperus*.

Mytilinidion acicola ⋉ occurs on *Juniperus* and *Thuja* and has 4–7-celled spores measuring 14–22 (–28) x (4.5–)6–8(–10) µm; *M. tortile* ⋉ has slightly larger spores; and *M. thujarum* ⋉ has much larger spores.

Probably widespread, rarely reported; all year.

Glyphium elatum is a ± axe-shaped, shiny, black bitunicate pyrenomycetous fungus that is usually somewhat dark-felty at the base. The fruitbody opens by a slit along the 'blade'. Asci double-walled, I-. Spores thread-like, multicelled, non-fragmenting, (160–) 240–400 × 1.5–3 µm. Occurs on wood of deciduous trees, mostly attached branches, *e.g. Salix*.

Glyphium schizosporum ⋉ has fragmenting spores; *G. grisonense* ⋉ also has fragmenting, but somewhat wider spores, a somewhat different fruitbody shape and is a subarctic-alpine species.

Widespread, rather common to occasional; all year.

Actidium nitidum is a ± mussel/clam-shaped, rarely star-shaped, bitunicate pyrenomycetous fungus that opens by a narrow slit along the fruitbody margin. Asci double-walled, I-. Spores 2-celled, yellowish to brown, measure (9–)11–14(–16) × (1.5–)2–3 µm. Occurs on wood of conifers including *Juniperus*.

Actidium hysteroides ⋉ is distinctly star-shaped; *A. baccarinii* ⋉ has larger spores.

Probably widespread, rarely reported; all year.

Laboulbeniales

The *Laboulbeniales* are tiny, 0.03–2 mm high, ascomycotes that grow on living insects, millipedes and other arthropods. Each species has one or several hosts and some are always situated on specific areas of the host, *e.g.* the antenna. The entire fungus (thallus) includes a dark 'foot', a stem, a perithecium (with asci and 2-celled spores with a slime coat), antheridia, which form spermatia, and finally sterile appendages. During the sexual process spermatia are transferred from the antheridia to a trichogyne. After this the perithecium develops.

The *Laboulbeniales* do not have a true mycelium, and apparently do not enter the host cavity, but get their nutrients only from the exoskeleton. Large aggregations of fruitbodies may weaken some hosts.

The *Laboulbeniales* belong in their own class, *Laboulbeniomycetes*, with 50+ genera in temperate Europe. Only a few examples are included here – all from beetles.

OTHER SIMILAR FUNGI:
– insect parasites are also found among the hypocrealean fungi (page 1512), and within the asexual fungi (pages 1630, 1635 & 1643), but none resemble *Laboulbeniales*.

FURTHER READING: 114, 117, 187, 352.

hypocrealean pyrenomycetous fungi, (page 1512)

asexual fungi (page 1626)

other parasites on arthropods

the *Laboulbeniales* are tiny fungi on living arthropods, *e.g.* beetles and flies

Laboulbenia is the largest genus of the *Laboulbeniales*. It has a short foot cell and lateral appendages.

Laboulbenia argutoris is an approximately 275 µm long *Laboulbenia* with a dark olive-brown, perithecium about 125 µm long. The outer, simple appendages may have up to 9 cells. Spores asymmetrically 2-celled, hyaline, and have a thick slime coat that enables it to stick to the host animal. Occurs on beetles in the genus *Pterostichus*.

At least four species of *Laboulbenia* occur on this type of beetle. *Laboulbenia pseudomasei* ⋉ and *L. flagellata* ▷ have more complex outer appendages than those in *L. argutoris*. *Laboulbenia kajanensis* ⋉ is very similar to *L. argutoris* but has small differences in cell dimensions.

Distribution and frequency not well known, but probably rather common; all year.

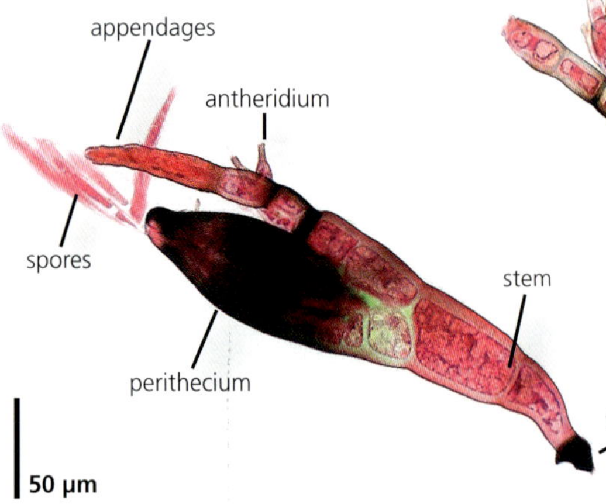

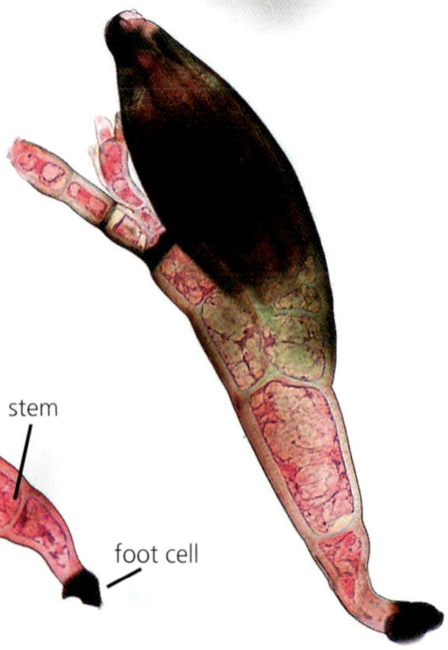

Jimmie Høier

Laboulbenia flagellata is a ± olive-green *Laboulbenia* with a thallus about 400 µm long. The perithecia appear from three large basal cells that arise from the dark attachment cell. The appendages appear from a 2-celled extension of the basal cells. Occurs on a number of carabid beetle genera and species.

A very variable and rather difficult species to identify, with identification further complicated by the long list of hosts.

Probably widespread and common; all year.

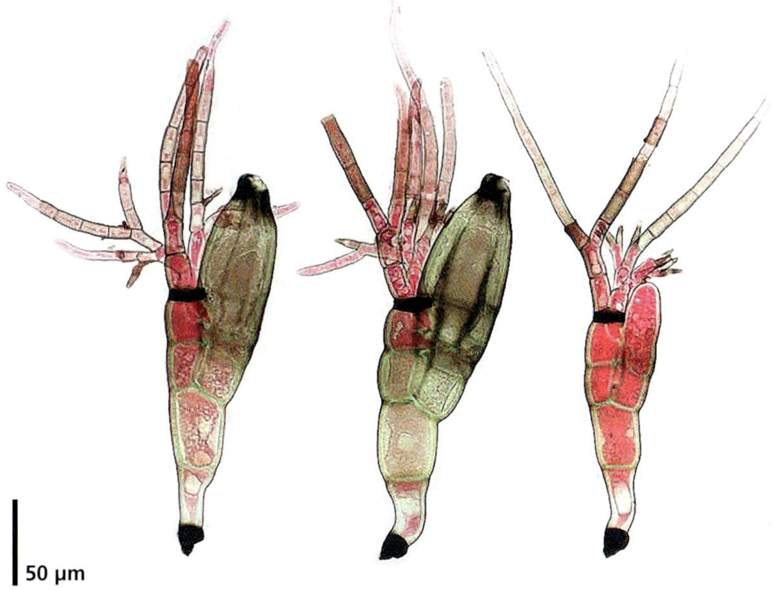

Jimmie Høier

Monoicomyces fragilis is a laboulbenialean fungus with thalli approximately 200 µm long and 140 µm wide. Each thallus divides above cell two, and one perithecium develops on each side. Two long appendages with brown basal pigmentation arise from just above the lower pigmented cell. Spores hyaline, 2-celled, and have a slime coat. Occurs on the rove beetles *Ocalea picata* and *Oxypoda opaca*.

The distinctive split thallus makes this species easy to identify.

Distribution and frequency poorly known; all year.

Jimmie Høier

Rhachomyces furcatus is a complex laboulbenialean fungus, up to 500 µm long with many dark appendages and one or two, ± centrally positioned, approximately 250 µm long perithecia. The cell size gradually increases up to the cell that supports the perithecium, and decreases again above it. The spores are 2-celled, hyaline, and have a slime coat. Occurs on rove beetles in the genus *Othius*.

A relatively easy species to recognize, provided the host animal is taken into account.

Distribution and frequency poorly known; all year.

Jimmie Høier

Taphrinales

Species in the order *Taphrinales* are all biotrophic plant parasites that never form proper fruitbodies, but mostly produce a layer of asci within the host tissue. The hosts may be galled in various ways, from small leaf blisters to huge witch's brooms in the crown of infected trees.

All belong in *Taphrinomycotina* – one of three subphyla within the phylum *Ascomycota*. The other two subphyla are the yeasts in *Saccharomycotina* and the often fruitbody-forming *Pezizomycotina* (page 12).

OTHER SIMILAR FUNGI:
- the basidiomycotes in *Exobasidium* parasitize species of *Ericaceae* (page 1214).
- *Erysiphales* are more mealy from surface production of conidia and form small spherical fruitbodies with asci (page 1500).
- some non-fungal organisms, *Albugo* for example, may look similar but do not have basidia or asci (page 1645).

FURTHER READING: 78, 80, 87, 164.

Exobasidium
(page 1214)

Erysiphales
(page 1500)

non-fungal
(page 1645)

other similar groups

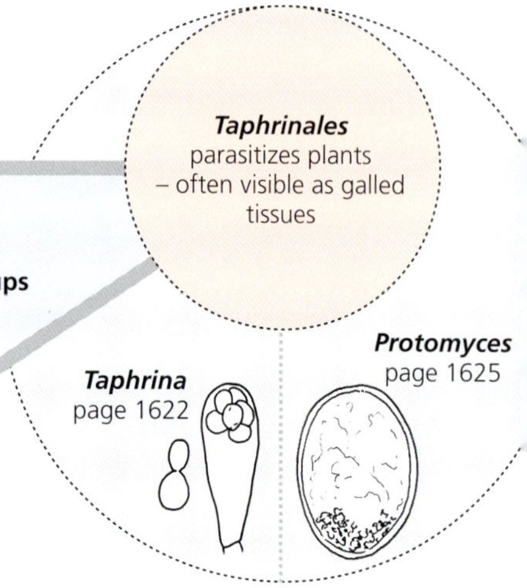

Taphrina is a large genus of plant parasites. All species deform the infected tissue in more or less spectacular ways: from curly or blistered leaves, through deforming fruits, to inducing 'witch's brooms'. The small asci are formed in a palisade directly on the surface of the host tissue. The spores mostly produce yeast stages within the asci. When sporulating the infected tissue may have a pruinose appearance.

Taphrina betulina causes the formation of Magpie nest-like witch's brooms. The asci are formed in the leaves of the host, which appear from the brooms; they measure 25–70 × 10–25 μm and are I-. Spores measure 4–6.5 × 2.5–5 μm. Occurs in the crown of living *Betula* trees.

Other organisms may also cause the formation of witch's brooms in *Betula*, so in reality reliable identification requires microscopy. *Taphrina betulae* ⋉ forms discoloured, swollen spots on the leaves of *Betula* rather than conspicuous galls.

Widespread, very common; all year (as brooms).

Taphrina wiesneri causes a very spectacular formation of drooping witch's brooms. The asci are formed on discoloured, somewhat bumpy leaves; they measure 17–35 × 15 µm and are I-. The spores measure 3.5–9 × 3–6 µm and form yeast-like cells. Occurs on both cultivated and wild species of cherry (*Prunus cerasus* and *P. avium*).

In order to ensure a correct identification, infected leaves require microscopical investigation.

Mainly nemoral–hemiboreal, rather common; all year (as brooms).

Taphrina carpini forms conspicuous witch's brooms. Asci occur on the pale leaves of the host that appear as long shoots from the brooms; they measure 20–30 × 7–15 µm. The spores measure 3.5–5 × 3–4.5 µm. Found on *Carpinus*, including in mature hedgerows.

Microscopical investigation is recommended as other organisms can induce the formation of witch's brooms in *Carpinus*.

Nemoral, rather common; all year (brooms).

Tom Smidth

Taphrina farlowii deforms leaves, shoots and fruits of the host with yellowish to reddish blisters and swollen galls. The asci are formed on the galls and measure 20–30 × 8–9 µm, the basal cell measures 8–9 × 15–25 µm. The spores are ± spherical to broadly egg-shaped and measure 4–6 × 4–5 µm. Affects *Prunus serotina*, a North American species that is widely planted and now spreading.

The very similar *T. deformans* ⋉ deforms the leaves of other *Prunus* species, including *P. amygdalus* and *P. persica*.

Nemoral, probably common; mainly June–October.

Taphrina pruni transforms young plums into oblong, flat, clavate, greenish to reddish galls that become covered by a pruinose surface (inset image). The asci form below the epicutis; they measure 30–40 × 14–17 µm and are I-. The spores measure 6–8 × 4–6 µm. Occurs on fruit of various *Prunus* species, including *P. domesticus* and *P. spinosa*.

The closely related *T. padi* ⋉ forms similar symptoms on *Prunus padus*. Other species of *Taphrina* may occur on *Prunus* leaves, e.g. *T. farlowii* ◁◁ and *T. deformans* ⋉.

Widespread, common; mainly May–July.

Taphrina johansonii causes *Populus* fruits to swell and become gall-like. The surface is yellowish. The asci are clavate and measure 67–174 × 12–32 µm; I-. The spores form secondary spores measuring 4–10 × 1.5–4 µm inside the asci. Occurs mostly on *P. tremula*.

Distinctive and unlikely to be misidentified.

Very rarely reported; mostly May–June.

Taphrina alni produces tongue-like, green to red outgrowths on the cone-like fruits of *Alnus*. The asci measure 30–50 × 10–20 µm and are at least 8-spored; the spores measure 4–5 × 3–5 µm and usually produce secondary spores within the asci.

A highly distinctive species that is unlikely to be misidentified.

Widespread and probably rather common; mostly July–October.

Protomyces has its own family within the *Taphrinales,* the *Protomycetaceae*. Large resting spores are formed in living plant tissues. These may later act as asci, within which the spores are formed. The spores may, in turn, produce yeast stages.

Protomyces macrosporus forms swollen, greenish to brown-tinged blisters on stems and leaves of the host. Asci/resting spores are yellowish, thick-walled, smooth and ± spherical, 50–65 µm in diameter. Occurs on *Aegopodium* and a number of other *Apiaceae*.

Distinctive and unlikely to be misidentified.

Widespread, very common; April–October.

Protomyces pachydermus forms reddish blisters along the midrib of living host leaves and also on the petioles. The resting spore-like cells, which later act as asci, are 30–45 µm in diameter. Occurs on *Taraxacum* leaves.

The chytridiomycote *Synchytrium taraxaci* ⨯ also forms orange to reddish galls on *Taraxacum* leaves, but the galls are much smaller and orange inside.

Widespread, rarely reported, but probably overlooked; mainly April–October.

Julia Kruse

Asexual fungi

Many fungi have asexual states where propagules are produced following mitotis rather than meiosis. These mitospores (typically termed conidia) are ± genetically identical to the mycelium that produced them. Contrary to this, meiospores (in this publication termed spores) are formed in, for example, asci, or on basidia, by a sexual process that involves nuclear fusion and subsequent meiosis, see pages 11 & 22.

Many species produce both sexual and asexual states. Where the two states occur together both states are usually included in the description of the sexual state.

Traditionally, asexual and sexual states were given separate scientific names. For example, *Aspergillus glaucus* (page 1507) was the name for an asexual state, while *Eurotium herbariorum* was the name for the sexual state of the same species. However, in 2011, the code that governs the naming of fungi was changed so that one name is used, the first described taking precedence irrespective of the state to which it was attached. Thus, *Aspergillus glaucus* became the valid name for all states of this fungus.

Asexual fungi constitute a pool of thousands of 'species'. These are not a main theme of this publication, and on the following pages only a few characteristic examples are given.

Asexual propagation is found in almost all fungal groups. However, the more spectacular examples are found mainly within the ascomycotes and zygomycotes, and less so within the basidiomycotes.

Other similar fungi:
– species of mildew form asexual spores from a mealy covering on living leaves, but later form tiny, spherical fruitbodies with internal asci (page 1500).
– rusts and smuts have a series of asexual spore states, see page 1198.
– cobweb-like corticioids may look similar to asexual moulds but form meiospores from basidia (page 986).
– *Myzetozoa* (slime moulds) may look similar to asexual fungal moulds (page 1646).

Further reading: 78, 94, 144, 145, 146, 293, 294, 321, 351, 369.

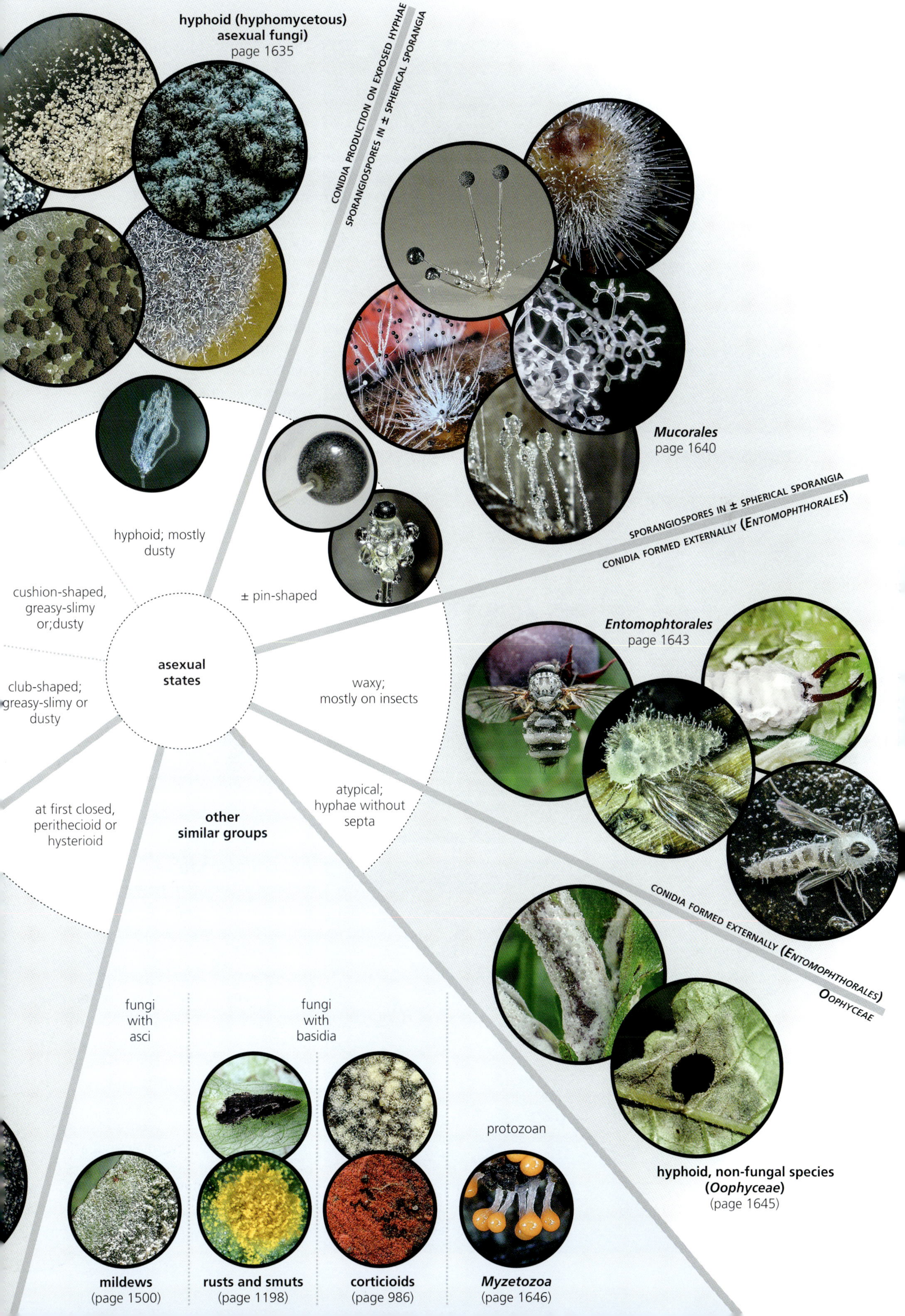
hyphoid (hyphomycetous) asexual fungi)
page 1635
CONIDIA PRODUCTION ON EXPOSED HYPHAE
SPORANGIOSPORES IN ± SPHERICAL SPORANGIA
Mucorales
page 1640
SPORANGIOSPORES IN ± SPHERICAL SPORANGIA
CONIDIA FORMED EXTERNALLY (ENTOMOPHTHORALES)
hyphoid; mostly dusty
± pin-shaped
cushion-shaped, greasy-slimy or;dusty
asexual states
club-shaped; greasy-slimy or dusty
waxy; mostly on insects
Entomophtorales
page 1643
at first closed, perithecioid or hysterioid
other similar groups
atypical; hyphae without septa
CONIDIA FORMED EXTERNALLY (ENTOMOPHTHORALES)
OOPHYCEAE
fungi with asci
fungi with basidia
protozoan
hyphoid, non-fungal species (Oophyceae)
(page 1645)
mildews
(page 1500)
rusts and smuts
(page 1198)
corticioids
(page 986)
Myzetozoa
(page 1646)

The **coelomycetous fungi** produce conidia inside a closed structure. They constitute a very form group of asexual ascomycotes.

Phacidium lauri forms both a sexual state (see page 1471) and an asexual state (*Ceuthospora*) – shown here. Texture leathery, but most noticeable character is that it recalls a volcano with a paler mouth. It has several internal cavities, within which hyaline, cylindrical conidia, 12–15 × 2–3 µm, are formed. Occurs on fallen leaves of *Ilex*.

Distinctive and unlikely to be misidentified.

Nemoral; rare as sexual, but common as asexual; most of the year.

Sphaeropsis sapinea is a coelomycetous ascomycote with black pycnidia partly immersed in host tissue. Conidia brown, 1–2-celled, measure 40–50 × 12–17 µm, finely punctate. Mostly on *Pinus* cones and other *Pinus* litter, but probably also occurs on debris from other conifers. Causes a brownish discoloration of *Pinus* needles.

As several superficially similar fungi can occur on the same substrates, microscopy is a necessity.

Widespread, very common; all year.

Diaporthopsis urticae is an ascomycote that forms ± smooth, black shields on stems of *Urtica*. The sexual state is rarely found (spores 12–18 × 4–5 µm in 4–8-spored asci) but the asexual state (also called *Apomelasmia urticae*) is very frequent; it has hyaline, one-celled conidia, 17–25 × 3.5–4.5 µm, formed on short conidiophores within the black stromatic tissue that splits along longitudinal fissures. Occurs on standing previous year's stems of *Urtica*.

Other asexual fungi may form a black covering on *Urtica* stems but this is but usually felty.

Nemoral–hemiboreal, common; mostly during spring.

The **synnematoid fungi** produce conidia from club-shaped structures (synnemata, also called coremia), which may have a powdery surface or a slimy head.

Stilbella byssiseda is a whitish to pale brown asexual ascomycote that forms rather firm, club-shaped structures (synnemata) with a round head, upon which ± branched conidiophores tipped with phialides are formed. Conidia hyaline, slender ellipsoid, measure (3.5–)4.5–6(–11) × 2–2.5 µm. Occurs on various *Myzetozoa*, particularly species of *Cribraria* (page 1655).

Polycephalomyces tomentosus ⋉ (with *Byssostilbe stilbigera* ⋉ as the supposed sexual state) has ± verrucose cells, smaller phialides and smaller conidia. It is mostly found on sporocarps of *Trichia* (page 1654) – which is not a typical host for *S. byssiseda*.

Widespread, probably rather common; mostly during summer.

Dendrostilbella smaragdina is a striking asexual fungus with small, black clubs (synnemata) topped by a blue-green, slimy head filled with conidia, which are formed in phialides. The stem base may also be ± blue-green. The black stem becomes bright green in KOH. Conidia measure 2–3.5 × 0.5–1 µm. Occurs on conifer bark and wood and possibly also on deciduous wood. May occasionally co-occur with dark apothecia from an undescribed species of *Claussenomyces* (page 1449) – the presumed sexual state.

A very distinctive species. *Claussenomyces prasinulus* (page 1449) has a similar but paler asexual state.

Mainly nemoral, occasional; most of the year.

Ole Martin

Seifertia azaleae is an asexual ascomycote that forms conspicuous black, pin-shaped synnemata on *Rhododendron* buds. Conidiophores dark with raised warts. Conidia ± lemon-shaped, thick-walled, brown, 1–(2)-celled, measure 4–12 × 4–8 µm. Occurs on dead, attached *Rhododendron* buds in gardens and parks.

Also known as *Pycnostysanus azaleae*. Identification should be straightforward due to the general appearance and the host.

Mainly nemoral, common; mostly April–August.

Polycephalomyces ramosus is an asexual ascomycote with white to yellowish club-shaped structures (synnemata). The heads have conidia gathered in slime. Has two types of conidiophores (phialides), with the conidia measuring either 2–3 × 1–2 µm or 3–4.5 × 1.5–2 µm. Occurs on the carabid beetle parasite *Ophiocordyceps entomorrhiza* (page 1517), but is apparently also known from other *Ophiocordyceps* species.

If the host is known identification should be straightforward; it is usually this striking parasite that leads to detection of the inconspicuous host.

Nemoral–hemiboreal, occasional; mainly May–August.

Gibellula pulchra is an asexual ascomycote that produces long, whitish clubs with small, ± pale lilac stems and conidiogenous heads. The conidiophores have rough walls, and the conidia are one-celled, hyaline and measure 2.5–6.5 × 1.5–2.5 µm. Attacks and kills small spiders, mostly in tall herbaceous vegetation.

Gibellula leiopus ⋉, on similar spiders, has ± sessile, more compact conidiogenous heads. It is possible that other species of *Gibellula* occur in temperate Europe; also compare with the closely related sexual *Torrubiella albolanata* (page 1514).

Nemoral–hemiboreal, common; mainly June–November.

Cordyceps farinosa is an asexual ascomycote with white-dusted, club-shaped conidiogenous structures. The sterile stems are yellow-brown. The one-celled conidia measure 2–3 × 1–1.8 µm. Kills *Lepidoptera* larvae and pupae.

Also called *Isaria farinosa. Cordyceps tenuipes* ⋉ occurs on similar hosts but apparently has larger conidia, measuring 3.5–5 × 1.8–2.4 µm, and is perhaps more lemon-yellow on the sterile parts and more branched.

Widespread, mainly nemoral–hemiboreal, rather common; June–January.

Occasionally seen with long, projecting perithecial necks from the parasite *Syspastospora parasitica* (page 1586).

Pseudonectria tilachlidii is an asexual ascomycote that is only very rarely seen in the sexual state. The asexual state (called *Tilachlidium brachiatum*) consists of very long and branched conidiogenous structures. The hyaline, smooth conidia are formed in slender phialides, measure 3–6 × 1–2.5 µm, and are gathered in slimy heads. Occurs on decaying fungal fruitbodies.

May recall a *Pterula* (page 1101), but should be relatively easy to identify, especially with microscopy.

Widespread, occasional; mainly July–November.

Penicillium vulpinum is an asexual ascomycote that forms club-shaped conidiogenous structures. The sterile stem is pinkish-tinged, while the head is blue-green. The smooth, one-celled conidia measure 4–4.5 µm × 3–3.5 µm and form chains. This species is apparently not known in a sexual state. Occurs on dung from various herbivores.

A relatively easy species to identify in an otherwise difficult genus, due to the characteristic clubs and the substrate. *Penicillium clavigerum* ⋉ has acute synnemata.

Mainly nemoral, rather common; most of the year.

Thomas Læssøe

The species on this and the next two pages produce conidia from cushion-shaped structures or spread as bulbils.

Illosporiopsis christiansenii is a shocking pink asexual ascomycote that forms dusty cushions on lichens. The conidia are ± similar to a ball of yarn and measure 17–30 × 11–20 µm. On various lichens including *Physcia* ⋉.

Erythricium aurantiacum ▽ occurs in the same habitat but has a more orange colour. *Marchandiomyces corallinus* ⋉ has pink sclerotia but does not form conidia. Both are basidiomycotes.

Widespread and probably rather common; most of the year.

Thomas Læssøe

Erythricium aurantiacum is a grainy, orange basidiomycote that normally only occurs in its asexual bulbil-forming state. The bulbils are very small, orange and measure 50–150 µm in diameter; occasionally forms basidia, but not conidia. It is an aggressive parasite on *Physcia* and adjacent thalli of other lichens.

Also known as *Marchandiobasidium aurantiacum*. *Illosporiopsis christiansenii* △ and *Marchandiomyces corallinus* ⋉ occur on the same substrate but both are shocking pink.

Widespread and rather common; most of the year.

Nectria cinnabarina forms very eye-catching asexual cushions (sporodochia) that typically occur together with the perithecia of the sexual state (see page 1534). In comparison to some closely related species the sporodochia are typically ± stemmed and do not darken. The conidia are gathered in a somewhat slimy mass on the surface and measure (4–)5.2–7(–8.5) × (1.3–)1.9–2.7(–3.4) µm.

A number of other types of asexual states are found within nectrioid (hypocrealean) fungi, and this diversity is partly the reason for the division into many genera and even families.

Widespread, very common; all year.

Nodulisporium cecidiogenes is a gall-forming asexual ascomycote that produces yellowish, bladder-like galls of various sizes and often in clusters. Has ± branching, finely serrated conidiophores. The conidia are one-celled and fall into two size groups: 8–12 × 4–6 µm and 13–24 × 4–5 µm. Occurs on fruitbodies of *Coniophora puteana* (page 1034).

An easily identified species. The generic placement in *Nodulisporium* – where most members are species of *Hypoxylon* (page 1556) – seems odd.

Nemoral, rather common; August–February.

Thomas Læssøe

Lemalis aurea is a disc- to ± cushion-shaped, vivid lemon-yellow, rarely almost white, gelatinous-waxy asexual ascomycote. The conidiophores are narrow and crowded, and conidial production is from the tips. The tiny, spherical conidia measure 1.2–1.75 µm. Occurs on litter of mostly *Pinus*, particularly on decaying cones in the litter.

There are some similar asexual fungi, and macroscopically *Lemalis aurea* may also be confused with *e.g.* species of *Orbilia* (page 1464).

Widespread, probably very common; mostly during winter.

Bactridium flavum is a cushion-shaped, rather vivid yellow to apricot-orange asexual ascomycote with short conidiophores in a palisade and very large, projecting, pale yellow (2–)3–5-celled and thick-walled conidia that measure 100–200 × 40–50 µm. Occurs on very moist, lying, decaying wood, *e.g.* in *Alnus* carrs.

This fungus can be identified in the field using a high magnification hand lens.

Apparently mainly nemoral, rarely reported, but probably rather common; mainly August–February.

Myrothecium inundatum is a ± irregularly disc- to cushion-shaped, somewhat confluent asexual ascomycote (hypocrealean) with a ± gelatinous, dark grey-green disc, within which the conidiophores are positioned. The margin is white with thick-walled hairs that measure 100–300 × 3–4 μm. Conidia cylindrical with rounded ends, 2.5–4 × 1–1.5 μm. Occurs on old, rotten fruitbodies of *e.g. Russula* and *Lactarius* (pages 372 & 414) and *Bankera* (page 1061).

The characteristic white-margined structures make it easy to identify.

Mainly nemoral, rarely reported but probably overlooked; mostly during summer.

Colletotrichum dematium is an asexual, ± disc- to cushion-shaped, mostly elongated ascomycote with large, long, black, pointed setae at the margin and on the outside that measure up to 250 × 4–8 μm. The hyaline, sickle-shaped, one-celled conidia measure 20–27 × 2–3 μm and do not have hair-like ends. Occurs on various dead herbaceous stems.

The conidia of similar-looking asexual ascomycotes have hairs at the ends.

Widespread, probably common, apparently becoming scarcer towards the north; most of the year.

Volutella ciliata is a cushion-shaped to almost cup fungus-like asexual ascomycote with white to yellowish discs that have a fringe of hyaline, distant, thick-walled hairs (setae) that measure 250–500 × (4–)8–9 μm. Conidia in a slimy mass, smooth, hyaline, one-celled, 4–9 × 1.5–2 μm. Occurs on herbaceous stems, but also on *e.g. Sarothamnus*.

The genus *Volutella* is recognized as having both asexual *Volutella* states and perithecial states (it was formerly included in *Nectria s.l.*); *V. ciliata* is part of a complex of closely related species. *Pseudonectria rousseliana* ⋉ produces a pinkish *Volutella* state on *Buxus* leaves.

Widespread, common; mainly summer.

The **hyphomycetous fungi** produce conidia from exposed, loose hyphae.

Monilinia fructigena may, on rare occasions, produce brown apothecia. The asexual state (*Monilia fructigena*) is extremely abundant on pome fruits, *e.g. Malus,* where it forms ± circular rings of buff, spore-producing hyphae on rotten brown parts of the fruit (forms a yellow monilia). The spores from the asci measure 9–11(–13) × 5–6.6 µm; the conidia measure 15–25 × 12–16 µm.

The closely related *Monilinia laxa* ⋉ mainly attacks stone fruits, *e.g.* cherries and plums (forms a grey monilia).

Widespread and common; all year.

Beauveria bassiana forms a white, felty asexual hyphal cover with conidiophores. The conidiophores have small protuberances and are up to 25 µm long; conidia are one-celled, smooth, hyaline, ± spherical and 2–3 µm in diameter. Rarely seen in the sexual *Cordyceps* state. Occurs on numerous insect hosts, *e.g.* many *Coleoptera*, and known to be a rare human pathogen.

Also known as *Cordyceps bassiana*. A rather anonymous species, which makes identification difficult. It is in use as a biological control agent against pathogenic insects.

Widespread, common; most of the year.

Botryotinia fuckeliana is an inoperculate cup fungus that is predominantly seen in the asexual, mouldy state (*Botrytis cinerea*) and as black sclerotia. Conidiophores simple or divided once, up to 25 µm wide, thick-walled. Conidia smooth, brown-tinged, egg-shaped to ellipsoid, measure 8–16 × 6–12 µm. An omnivorous rotter and plant parasite that forms grey mould on fruit. Infected grapes ('noble rot') are used in the production of highly aromatic dessert wines, *e.g.* Sauternes.

Compare with similar mildews, (page 1502).

Widespread and very common; all year as sclerotia, mainly conidial in the warmer parts of the year.

Cladosporium is a form genus of asexual ascomycotes that has a range of sexual generic names, *e.g. Mycosphaerella* (page 1605) and *Davidiella* ⋉. *Cladosporium* states have branched, brown conidiophores, where the brownish, several-celled, mostly ornamented conidia are formed in chains from small protuberances.

Species of *Cladosporium* are difficult to identify and may occur on all kinds of dead or decaying organic material. Sometimes conidia production is so high that allergies break out, and counts of airborne conidia for this genus, and for the genus *Alternaria* ⋉, are given as a warning to sensitive individuals.

Penicillium is a genus of *Ascomycota* that mostly forms conidiophores, and which has clusters of phialides on the non-swollen tips. Sexual states are rarely seen.

Some *Penicillium* species are known to be food spoilage organisms, but they are also used to enrich food with distinctive flavours, *e.g.* in cheeses (*P. camembertii* ⋉ in Camembert and *P. roqueforti* ⋉ in Roquefort). In 1929, Alexander Fleming discovered the first antibiotic, penicillin, in a culture of *P. chrysogenum* ⋉.

Penicillium italicum is an asexual ascomycote that forms greyish-blue colonies with a strong smell of lemon. The smooth-walled conidiophores may be gathered in loose structures. The spore-producing cells (the phialides) measure 8–15 × 2–5 µm. Conidia smooth, greyish-blue, measure 4–5 × 2.5–3.5 µm; formed in chains. Mostly on citrus fruits. May also be isolated from soil.

Penicillium expansum ⋉ mainly causes a rot on pome fruits. It has ellipsoid to almost spherical conidia and has a faster growth rate on standard culture media compared to *P. italicum*. *Penicillium digitatum* ⋉ may also cause citrus rot but forms looser, grey-green colonies and has somewhat larger conidia.

Common; all year.

Aspergillus is a very large ascomycote genus that mainly forms conidiophores that have swollen heads with projecting phialides. Rarely seen in cleistothecial, sexual states (page 1507).

Species of *Aspergillus* may cause serious infections on and in humans, *e.g.* in the lungs. *Aspergillus niger* ▽ may cause infections in the ears. *Aspergillus flavus* ⋉ produces a highly carcinogenic aflatoxin in fruits, nuts and grains stored in damp conditions.

Aspergillus niger forms brownish-black conidial heads on a colourless mycelium. The heads have a swollen cell upon which the 20–30 µm long conidiophores (metulae) are positioned; from these, 7–10 × 2–3 µm phialides are formed. Conidia are in chains from the phialides and are blackish-brown, verrucose and 4–5 µm in diameter. A common rotter that may cause a black mould on fruits and vegetables, *e.g.* grapes, onions and peanuts, and may produce toxins in foods and in wine. It is used in the production of citric acid and various enzymes, and can be isolated from almost anything, including soil and litter.

This species complex is characterized by the contrast between the hyaline mycelium and the dark conidia in crowded heads.

Widespread, common; all year.

Bispora antennata is a black, somewhat felty asexual ascomycote that produces stripes on exposed wood (stumps). Conidiophores short, 5–30 × 2–5 µm. Conidia blackish-brown, in chains, 2-celled with a very thick septum and similar end walls, 13–20 × 7–8 µm. Occurs on exposed deciduous wood, typically on the surface of fresh stumps. It is believed to be the asexual state of *Bisporella pallescens* ⋉ (see also page 1458).

The characteristic pattern on the wood and distinctive conidia make identification straightforward.

Mainly nemoral, very common; all year.

Botryobasidium conspersum is a corticioid that usually occurs in an asexual, ± grey-green, mould-like state. The simple conidiophores produce ellipsoid to almost spherical, pale grey conidia that measure 15–20(–22) × 9–12 µm. The hyphae do not have clamps. Sexual state has 6-spored basidia and the spores measure 7–9 × 2.5–3.5 µm. Occurs on rather decomposed, predominantly deciduous wood.

Very much recalls the asexual state of a *Nemania* (page 1564), but that has much smaller conidia.

Widespread, common, becoming scarcer towards the north; mostly July–November.

Sporophagomyces chrysostomus may produce pale perithecia with yellow ostioles, but these are rarely seen. Instead, fan-shaped asexual mycelia that trap falling spores from the host are formed. Conidia 1–4-celled, (9.5–)11.5–20(–27.5) × 2–4(–5) µm. It occurs on the underside of fruitbodies of *Ganoderma applanatum* (page 872) and devours the falling spores. The asexual state was known as *Acremonium lindtneri*.

The fan-shaped mycelium that is coloured brown by the host spores is highly distinctive.

Rarely reported, but probably overlooked; autumn–winter.

Sepedonium laevigatum is an initially greyish, later yellow, mould-like fungus. The spherical, spiny, yellow conidia measure, (13–)14–19 (–22) µm; spines cylindrical, up to 1.5 µm high. Occurs on fruitbodies of *Leccinum* (page 762), but possibly also on other boletes.

This species is best told by its host. A number of *Hypomyces* species (page 1542) occur on different groups of boletes. *Hypomyces chrysospermus* (page 1543) is very similar but has somewhat higher and more distant, rounded warts on the conidia; *H. microspermus* ⋉ has smaller conidia.

Widespread, probably rather common; summer–autumn.

Chromelosporium carneum is an asexual operculate cup fungus that forms composite, at first white, later pale pinkish-buff, dichotomous branching conidiophores that are 7–14 μm wide and form conidia towards the rounded tips. The conidia are spherical, 5–8 μm in diameter, and verrucose. Occurs on litter with soil contact.

Chromelosporium ochraceum ⋉ forms conidia on the sides of the conidiophores. Species of *Chromelosporium* are asexual states of *Peziza* species (page 1313), *e.g.* of *P. ostracoderma* ⋉.

Very rarely reported, but probably overlooked; likely to be mostly summer–autumn.

Kickxella alabastrina is a tiny ± translucent, sea anemone-like mould (*Zygomycota*, page 12) with up to 2 mm high, thick sporangiophores that are swollen and branched (bifurcate) at the top and have small, one-spored sporangia on the inner side of the branches. Spores/sporangia ellipsoid, smooth, hyaline, measure 9–10 (–15) × 3–4.5 μm. Occurs on herbivore dung.

Distinctive and unlikely to be misidentified.

Very rarely reported and distribution and frequency not known – possibly mainly nemoral; probably most of the year.

The ***Mucorales*** produce sporangiospores within spherical sporangia on often long sporangiophores.

Mucor piriformis is a species typical of the genus *Mucor*. The sporangia are 250–300 µm in diameter (inset image), and have lead-grey, ± spherical columella on long, shiny sporangiophores, mostly in clusters. It does not have basal setae or superficial hyphal cords. May form sexual zygospores that act as thick-walled resting spores. Sporangiospores almost spherical to broadly ellipsoid, 7.3–9.2 × 4.6–8.1 µm. Causes soft rot in fruits, *e.g.* apples, pears and plums.

There are a number of similar *Mucor* species. *Rhizopus* ▷ has superficial, thick hyphae and clusters of shorter sporangiophores.

Probably common, but rarely reported; all year.

Thomas Læssøe

Spinellus fusiger is a zygomycote mould with radiating, hyaline sporangiophores crowned by a dark sporangium that is 120–300 µm in diameter. The basal mycelium is branched and has pointed tips. The fusiform, hyaline, smooth sporangiospores measure 35–55 × 10–20 µm. Occurs on fruitbodies (mostly on the caps) of various *Mycena* species (page 196), rarely on other pale-spored agarics.

Distinctive and unlikely to be misidentified.

Widespread and common; mostly August–November.

Rhizopus stolonifer is a zygomycote mould with superficial, thick hyphae, from where clusters of 1–3 mm high sporangiophores develop. Sporangia spherical, grey to black; has some pale setae at the base of sporangiophores. The sporangia measure 100–300 µm and have a large columella and a basal indentation (apophysis – see inset image). The ± striate and angular sporangiospores measure 8–15(–20) × 6–8(–11) µm. Occurs on sugar-rich foods, fallen fruit, dung, *etc.*

Pilobolus ▷▷ has active dispersal of the sporangia and a swelling just below the sporangium. Species of *Mucor* ◁ and *Pilaira* ⋉ do not have thick, superficial hyphae.

Probably common, but rarely reported; all year.

Syzygites megalocarpus is a zygomycote mould with an impressive mycelium, within which branched sporangiophores are formed (inset image). The sporangia are yellow to grey and measure 50–150 µm in diameter. The ± spherical sporangiospores measure 12–30 µm in diameter. Deep within the mycelium, zygospores that act as resting spores may be formed. Occurs on various fleshy basidiomycote fruitbodies, *e.g. Leccinum* (page 762) and the amanitoids (page 352).

Spinellus ◁ occurs mostly on *Mycena* and has straight, non-branched sporangiophores.

Rarely reported; late summer–autumn.

Pilobolus crystallinus is a zygomycote mould that forms long, drop-bearing sporangiophores that are swollen like a balloon below the lens-shaped, initially yellow then black sporangium on the top. The sporangia are shot away with impressive force, and upon impact stick to the vegetation. A body within the swollen part of the sporangiophore acts as a lens and guides the direction of the 'cannon', thereby ensuring that the sporangium is shot off towards the light. The ± ellipsoid sporangiospores measure 6–12 × 4–7 µm. A coprophilous fungus that depends on gut passage in order to complete its life cycle. Occurs on herbivore dung, *e.g.* from horse.

There are some very similar species of *Pilobolus* that are hard to identify without controlled cultivation in a laboratory; they are normally not host-specific. *Pilobolus crystallinus* belongs to a group of species with small spores. Species with large spores include *e.g. P. kleinii* ⋉.

Widespread and common; mostly in the warmer months.

Pilobolus umbonatus has 3–9 mm long sporangiophores that develop from a basal, onion-shaped swelling. It has a shield-shaped sporangium (inset image) on top of the balloon-like upper part. The sporangium has a central-acute umbo and a somewhat slimy surface that secures adherence after the explosive dispersal. Sporangiospores narrowly ellipsoid, measure 5.5–6.5 × 3.5–4 µm. Occurs on herbivore dung.

Other species of *Pilobolus* △ have lens-shaped sporangia that do not have an umbo.

Distribution and frequency poorly known, but probably widespread and rather common; most of the year.

The ***Entomophthorales*** are insect-killing fungi that later produce conidia from the surface of the host. The conidia are dispersed actively. Some species have been tested for use in biological control of *Diptera*.

Entomophthora culicis forms a coating of greenish conidiophores on the host. The primary conidia are almost the shape of *Crataegus* fruits. On mosquitoes in swampy habitats.

Erynia curvispora ▽ and *Pandora dipterigena* ▷▷ are also greenish but have ± oblong primary conidia and external resting spores.

Widespread, probably common; spring–summer.

Entomophthora muscae is often seen indoors when the host is stuck to a window; outdoors, the dead flies are attached to vegetation. The carcasses have bands of conidiophores, from where ± lemon-shaped conidia surrounded by slime are actively shot out (see inset image). Has thick-walled, spherical resting spores. Best known from *Musca domestica*, but may kill flies in several dipteran families.

A range of similar species and genera can occur on *Diptera*, and identification requires specialist literature and accurate identification of the host.

Widespread, common; most of the year.

Erynia curvispora has conidiophores that measure 40–50 × 8–10 (–12) µm. The conidia are hyaline, curved and measure 30–42(–45.5) × 8–15 µm. Kills mosquitoes and flies in *Anthomyiidae* and *Simuliidae*.

The curved conidia are a particularly good separating character in comparison with other similar entomophthoralean fungi.

Probably widespread and common; during the summer.

Pandora dipterigena has conidiophores that form a greenish coating on the host. The conidia measure 18–20 × 8–15 µm. The spherical resting spores, which are produced in clusters on the outside of the infected animal, measure 20–40 µm in diameter. Occurs on *Psychodidae* mosquitoes and other small *Diptera* in moist habitats.

The combination of greenish conidiophores, external resting spores and the host is a good distinguishing feature.

Distribution and frequency poorly known, probably widespread; during the summer.

Christian Lange

Zoophthora forficulae has branched conidiophores. The primary conidia are double-walled, cylindrical-fusiform and measure 20–25 x 6–8 µm. The resting spores are spherical and dark. Kills species of *Forficula*.

There are probably no other, similar species on this host, except perhaps *Cordyceps bassiana* ⊲⊲, which produces tiny conidia.

Distribution and frequency poorly known, probably uncommon; mostly during the summer.

Albugo candida is a non-fungal mould in the *Oophyceae* (formerly *Oomycota*) within the kingdom *Chromista*. The entire plant host is infected but the most striking symptom is the white, mealy covering on the stems and fruits. The covering consists of chains of sporangia, from where flagellate zoospores are produced. A pathogen on various *Brassicaceae* crops, but also present on wild plants, *e.g. Capsella*.

Commonly known as white rust (although this is confusing since it has no relationship to rust fungi page 1198). *Peronospora parasitica* ⋉ may cause similar symptoms, but the sporangia are solitary at the tips of thin branches on 3-dimensional sporangiophores.

Widespread and common; mostly visible during summer.

Phytophthora infestans is a non-fungal *Oophyceae* (formerly *Oomycota*) that attacks both the above- and underground parts of *Solanum tuberosum* (potato). The pathogen is disseminated during wet periods by the liberation of flagellate zoospores from large sporangia on branched sporangiophores, which are formed from spherical, thick-walled, sexual oospores or asexually. Causes a very serious disease, called late blight or potato blight, which results in rotten tubers. Potato blight was one of the main causes for the mass emigration from Ireland during the Great Famine in the mid-1800s, and has also caused other famines. Other species in the *Solanaceae*, *e.g.* tomatoes, may also be attacked.

Potatoes can also sustain damage from true fungi, *e.g.* from *Rhizoctonia solani* (page 1178).

Widespread; all year in different stages.

Landbrugets rådgivningscenter

Landbrugets rådgivningscenter

Mycetozoa

Mycetozoa (slime moulds) do not belong in the fungal kingdom but within the *Protozoa*, However, their spore-producing structures may resemble those from fungi. They ingest organic particles in their mobile amoeboid stages and can be found in many environments, particularly on bark and rotten wood. The amoeboid stages range from tiny to huge. The sporulating stage mostly has internal spore production and the spores can be mixed with a hyphoid capillitium. There are approximately 48 genera in temperate Europe; a few examples are included here to show the variation.

OTHER SIMILAR FUNGI:

– asexual fungi may look similar (page 1626).

– *Phleogena* (page 1246) has dusty fruitbodies but transversely divided basidia.

– small clavarioids have basidia and are not dusty at maturity (page 1092).

FURTHER READING: 120, 217, 249.

A plasmodium of *Badhamia utricularis*.

Ceratiomyxa fruticulosa is a white to somewhat yellowish slime mould with external spore production (inset image). It forms large, dense areas of long-branched fruitbodies that ± dissolve when touched. The egg-shaped to broadly ellipsoid, mostly hyaline, smooth spores are positioned on a small stem and measure 8–15 × 6–10 µm. Occurs on rotten wood, both deciduous and coniferous.

The variety *Ceratiomyxa fruticulosa* var. *porioides* (lower image) almost looks like a resupinate polypore but is much softer and disintegrates when touched. *Mucronella* (page 1096) and *Ceratellopsis* (page 1097) are firmer when touched and have basidia. *Ceratiomyxa* is not closely related to the other groups of slime moulds.

Widespread, very common; May–October, peaking during the summer.

Thomas Læssøe

Jens Maarbjerg

Fuligo septica is a slime mould with large, ± cushion-shaped fruitbodies (aethalia) that have a loose, net-like, mostly yellow surface and brownish-black spore powder internally. The capillitium has lime in the shape of spherical, white grains. The spherical, finely ornamented spores measure 7–9 µm. Occurs on rotten, mostly mossy wood, compost, *etc.*

There are a number of varieties and closely related species. *Fuligo septica* var. *septica* (upper image) has white lime but is yellow on the surface; *F. septica* var. *candida* is entirely white; *F. septica* var. *flava* has yellow lime grains; *F. septica* var. *rufa* (lower image) is red-brown with a loose, net-like surface and often forms aethalia of an impressive size on wood chips and sawdust. *Fuligo luteonitens* ⋉ has a smoother, yellowish surface, often ovoid spores with scattered warts, and free-lying, round lime grains. *Fuligo leviderma* ⋉ is also almost smooth but has densely ornamented spores, a ± reddish-brown, thick surface and fusiform lime grains.

Widespread, common; mostly May–October.

May be covered by violet perithecia from the parasite *Nectriopsis violacea* (page 1540).

Jens H. Petersen/Jens Maarbjerg

Lycogala epidendrum is a slime mould with rather small, almost semiglobose fruitbodies (aethalia) that are initially yellow to pink (inset image) and contain fluid, but later acquire a firm, lead-grey to brownish-black surface and grey-rose to reddish-grey spore powder. Capillitium of tubes. The spherical spores have a fine, almost complete reticulum and measure 5.5–8 µm. Occurs on rotten wood.

Forms with pinkish plasmodium also go by the name *L. terrestre* ⋉. *Lycogala conicum* ⋉ forms small, but high, conical, initially redder aethalia.

Widespread, very common; mainly May–October.

Reticularia lycoperdon is a slime mould with large cushion-shaped fruitbodies (aethalia) that are initially white to yellowish (upper image), but later a shiny brown-grey with a dark brown, thread-like spore mass. The capillitium (pseudocapillitium) is plate-like but branched towards the margin and has numerous free ends. The spherical, brown spores have an incomplete reticulum and measure 7.5–10 µm. Typically appears from the bark of various living deciduous and coniferous trees, and may be situated very high up. Mature fruitbodies often have exit holes from beetles, *e.g. Anisotoma axillaris*, that spend their larval stages within the spore mass.

Reticularia intermedia ⋉ has a finer pseudocapillitium and a quickly disappearing, almost colourless peridium. There are other similar species of *Reticularia* but these have a more reticulate pseudocapillitium that lacks free ends; they are typically smaller and may have somewhat different colours.

Widespread, common to occasional towards the north; mostly April–July.

Jens Maarbjerg

Jens Maarbjerg

Mucilago crustacea is a somewhat strange-looking, loosely built slime mould that recalls a dead coral. The surface has star-shaped crystals, but these are not present in the brown spore mass. The spherical spores have spines or warts up to 1 µm high and measure 10–18 µm. Usually occurs on living herbaceous plants, which it enters from the soil.

Typical specimens are distinctive and unlikely to be confused with other slime moulds.

Widespread and common; mostly May–October.

Jens Maarbjerg

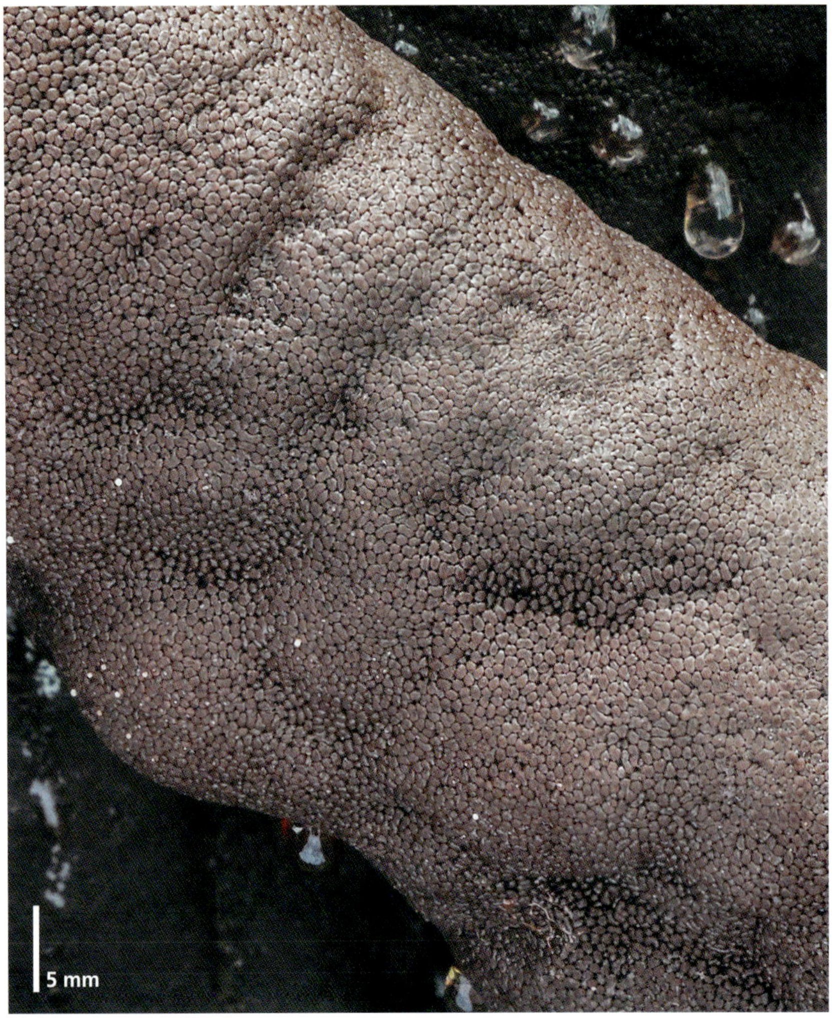

Brefeldia maxima is a huge slime mould with composite fruitbodies (pseudoaethalia) that are initially white and somewhat foam-like, but later purple-brown with a finely grainy-cellular structure on the surface (lower image). The spore mass is thread-like and dark grey-brown. The capillitium is radially expanded and bears cellular, septate swellings. The spherical, very finely spiny spores measure 9–12 µm. Mostly in deciduous forests, where it enters whatever is available and can cover many square metres.

The impressive size makes this species easy to recognize. It holds the record for being the largest spore-producing structures within the *Mycetozoa*.

Mainly nemoral–hemiboreal, occasional; mainly September–November.

Dictydiaethalium plumbeum is a slime mould with flat, well-defined, composite fruitbodies (pseudoaethalia) that have small hexagonal polygons on the surface. When immature it is pink or orange-yellow, but later the surface turns greyish and shiny. Internally it is more yellow and thread-like. The pseudoaethalium is typically surrounded by a white, film-like plate. The pseudocapillitium has long threads from small plates. The ± spherical, finely spiny, ± yellow-brown to olive-brown spores measure 9–12 µm. Occurs mostly on deciduous wood and can be some way up in a tree.

Brefeldia maxima ◁ also forms large pseudoaethalia but they are brown inside, thicker and much bigger.

Mainly nemoral–hemiboreal, rather common to occasional; mostly September–November.

Jens Maarbjerg

Tubifera ferruginosa is a slime mould with well-defined, rather thick, initially apricot-orange composite fruitbodies (pseudoaethalia) in which the individual fruitbodies can easily be made out. At maturity it is reddish-grey and powdery, but still has visible individual fruitbodies. Capillitium absent or poorly developed. The spherical spores have a reticulate ornament and measure 6–8 µm. Occurs on dead wood, mostly coniferous.

When typically developed, hard to confuse with other species.

Widespread and common; mostly July–October.

Jens Maarbjerg/Jens H. Petersen

Thomas Læssøe

Hemitrichia serpula has large, net-shaped, ± apricot-orange fruitbodies (plasmodiocarps). The spore mass is thread-like and ± curry-yellow. The capillitium has a spiralling, spiny ornament and short- tapering free ends. The spores have a coarsely reticulate ornament and measure 10–14 µm. Occurs on very rotten wood of deciduous trees and bark, mostly in swampy forests.

Typical specimens are unmistakable.

Mainly nemoral, rather rare; July–October.

Jens Maarbjerg

Diderma deplanatum is a porcelain-white slime mould that predominantly forms ring-shaped to oblong fruitbodies (plasmodiocarps) which open in an irregular fashion at maturity. The peridium is 2-layered – thick and lime-rich on the outside. The spore mass is dark brown. The finely spiny spores are ± dark brown and measure 9–10 µm. Occurs on litter and mosses in moist habitats close to soil level.

Diderma effusum ⋉ has paler, smaller spores and forms mostly small, cushion-shaped fruitbodies (sporocarps) as well as plasmodiocarps.

Nemoral–hemiboreal, rather rare; mostly September–November.

Physarum bivalve has a two-layered, pale peridium. Often has widely effused, ± net-shaped fruitbodies (plasmodiocarps), but may also form smaller sporocarps. The spore mass is almost black. The capillitium forms a distinct, lime-rich reticulum that is visible macroscopically. The spherical spores measure 8–10 µm, and are pale lilac-brown and finely warty. Occurs on litter and living plants close to soil level in damp habitats.

Physarum bitectum ⋉ has larger spores with an irregular spiny ornament and frequently forms sporocarps.

Mainly nemoral–hemiboreal, occasional; mainly June–November.

Physarum album has ± flattened, mostly long-stemmed and pendent fruitbodies (sporocarps) with white lime in the peridium. Does not have a columella/pseudocolumella. The capillitium consists of fine threads with fusiform, calcareous swellings. The spherical spores measure (7–)8–9 (–10.5) µm and have a fine ornament that appears more obvious in circular spots. Occurs on all kinds of plant litter but mostly on twigs in the litter layer.

Also known as *Physarum nutans*. *Physarum robustum* ⋉ has upright, less flattened, short-stemmed sporocarps, and typically an evident calcareous columella.

One of the most widespread and common slime moulds; mainly August–November.

Physarum psittacinum is a vividly coloured, yellow slime mould with stemmed fruitbodies (sporocarps) that have a spherical top. The peridium is shiny and covered with small, orange lime-bodies. The stem lacks lime and is bright orange and ± translucent. The spores are spherical, finely spiny, and measure (7–)8–10 µm. Occurs on various types of plant litter.

Physarum carneum ⋉ and *P. oblatum* ⋉ do not have a shiny peridium; *P. penetrale* ⋉ has a long columella; *P. viride* ⋉ has *e.g.* a different stem colour; and *P. virescens* ⋉ has sporocarps in crowded clusters.

Mainly nemoral–hemiboreal, rather rare; mainly August–October.

Trichia varia is a slime mould with sessile or very short-stemmed, initially white, but later ± yellowish-brown fruitbodies (sporocarps or plasmodiocarps). The capillitial threads have one or two spirals that are usually more prominent on one side. The spherical spores are finely spiny and measure 11–16 µm. Occurs on rotten deciduous wood, mostly large trunks, rarely on coniferous wood.

Other similar species of *Trichia* have three or several spirals on the capillitial threads. This is the case with *T. scabra* ⋉, which often occurs in the same habitats.

Widespread, common, becoming scarcer towards the north; most of the year.

Metatrichia floriformis is a slime mould with stemmed fruitbodies (sporocarps) that have a ± reddish, shiny stem and a dark peridium. The brick-red to rust-brown spore mass has capillitial threads with 4–6 smooth, spiralling ridges. The verrucose, spherical spores measure 10–13 µm. Occurs on rotten wood.

Trichia decipiens ⋉ has spores with a reticulate ornament. *Metatrichia vesparium* ⋉ opens with a lid and has spiny spirals on the capillitial threads.

Mainly nemoral–hemiboreal, occasional; most of the year.

Cribraria aurantiaca is a slime mould with yellow, stemmed fruitbodies (sporocarps) and dark spots where the peridial net joins. The spore mass is held by a basal, membranous cup that is ± smooth but has radiating furrows; lacks a columella and a capillitium (as do all *Cribraria* species). Spores ± smooth, 5–6 µm in diameter, and have a pale yellow body. Occurs mainly on coniferous wood.

The less vividly coloured *C. persoonii* ⋉ forms a cup with concentric wrinkles and has finely warty spores. *Cribraria vulgaris* ⋉ has spores with a net-like ornament.

Nemoral–hemiboreal, occasional to rare; mainly summer–autumn.

Comatricha nigra is a slime mould with small, long-stemmed fruitbodies (sporocarps) that have a ± spherical head with a quickly disappearing peridium. The main branches of the capillitium are at an oblique angle to the columella. The spore mass is dark brown. The spherical, minutely warty spores have a pale germ pore and measure 8–10(–11) µm. Occurs on rotten wood, both deciduous and coniferous.

In *C. laxa* ⋉, which may have both spherical and more elongated sporocarps, the main capillitial branches are ± perpendicular to the columella. In *C. elegans* ⋉ the columella splits into several main branches in the centre of the sporocarp (in *C. laxa* and *C. nigra* the columella is gradually pointed).

Widespread and one of the most common of all *Mycetozoa*; most of the year.

Arcyria denudata is a slime mould with carmine-red, rarely even brighter red, stemmed fruitbodies (sporocarps). The capillitium is attached to the bottom and sides of the peridium and cannot be pulled out in one piece. The capillitial threads have rings, half-rings, warts or low spines. The spherical spores are irregularly minutely warty and measure 6–8 µm. Occurs on rotten wood.

Arcyria major ⋉ has rose sporocarps, and the capillitium stretches into a long column when mature and is attached only at the bottom of the peridium (although initially also loosely attached to the sides). The capillitium is ornamented only with half-rings and rings. The brighter red *A. stipata* ⋉, *A. oerstedtii* ⋉ and *A. affinis* ⋉ also have a capillitium that is attached at the base and not at the sides; it can be drawn out with forceps.

Widespread and common; most of the year.

Arcyria obvelata is a slime mould with aggregated, almost sessile, oblong, yellowish fruitbodies (sporocarps) that, at maturity, are dominated by yellow to cinnamon-buff capillitial columns that are attached only at the bottom of the peridial cup. The capillitium has rings, half-rings and spines. The spherical spores measure 7–9 µm. Occurs on rotten wood.

Other species of *Arcyria* with a bottom-attached capillitium are redder and have shorter capillitial columns.

Widespread, common; most of the year.

Stemonitis axifera is a slime mould with cinnamon to pale reddish-brown, stemmed fruitbodies (sporocarps) that are 7–15 mm high and usually taper towards the ends. Forms drooping clusters when mature. The spherical spores measure 5–7.5 µm. Occurs on rotten wood.

Stemonitis smithii ⋉ has cinnamon-coloured, cylindrical sporocarps that are 25 mm high and have rounded ends, the spores measure 4–6 µm; *S. splendens* ⋉ and other similar species have larger spores; *S. fusca* ▽ has a reticulate ornament on the spores.

Widespread, common; mostly during the summer.

Stemonitis fusca is a slime mould with large, dark brownish-grey, stemmed fruitbodies (sporocarps) that are 6–20 mm high and lack a visible peridium at maturity. The brown spores have a reticulate ornament of spines and measure (7.5–)8–9 µm. Occurs on rotten wood.

Stemonitis trechispora ⋉ has larger spores with crests in the ornament; *S. nigrescens* ⋉ and *S. virginiensis* ⋉ has 2–6 mm high sporocarps.

Widespread, common; mostly during the summer.

Jens Maarbjerg

Stemonitopsis hyperopta is a slime mould with short-stemmed, cylindrical fruitbodies (sporocarps) that lack a peridium at maturity. The capillitium is net-like and branched, and has scattered plate-like extensions in some of the branch angles. The spores are 5–6 µm in diameter. Occurs on rotten wood and litter.

Stemonitopsis typhina ⋉ has ± distinct, silvery remnants of the peridium and verrucose spores. *Stemonitopsis microspora* ⋉ does not have a peridium and the spores are much smaller.

Mainly nemoral–hemiboreal, occasional; mainly during summer.

Jens Maarbjerg

Didymium melanospermum is a slime mould with short-stemmed to sessile fruitbodies (sporocarps) that have pale grey, ± globose heads with a mealy surface. The stem is dark and lacks crystals. Has a distinct calcareous columella. Capillitial threads dark, tapering. The spherical, spiny spores are dark brown and measure 10–12 µm. Occurs on litter and wood but favours mosses.

Didymium minus ⋉ has somewhat smaller, 8–11 µm, rather pale spores with fine warts and scattered groups of larger warts; *D. squamulosum* ⋉ has a pale, calcareous stem.

Widespread and rather common; most of the year.

Badhamia utricularis is a greyish-blue slime mould with ± spherical heads that are typically gathered at the end of long, thin, pendent, yellow threads. The young fruitbodies (sporocarps) are a vivid yellow, like the threads and the plasmodium. At maturity they become bluish grey with a black spore mass. There is no columella. The capillitium forms a reticulum of lime-filled tubes. The spores are aggregated in clumps, ± spherical, evenly spiny and measure 9–15 µm. Mostly on old polypores and other fungi in damp woodland, *e.g.* Salix carrs, but also on wood and bark.

Distinctive and unlikely to be misidentified.

Widespread and rather common, becoming scarcer towards the north; most of the year.

Diachea leucopoda is a slime mould with a conical upper, dark part that has a metallic lustre, and a thick, white stem. Lime is only present in the stem and columella. The capillitium is ± net-like and has acute tips. The globose, finely spiny spores measure 7–10 µm. Found on litter and living herbaceous plants in damp woodland and may occur in great quantities covering many square metres. It is probably a soil organism that only creeps out to sporulate.

Diachea subsessilis ⋉ has a globose spore-bearing part and a shorter stem.

Mainly nemoral, occasional; especially late summer–autumn.

Craterium minutum is a slime mould with wine glass-shaped, yellowish-brown fruitbodies (sporocarps) that are 0.7–1.5 mm high and have a well-defined lid. Lacks columella, but has lime-filled, white to pale yellow pseudocolumella. Capillitium has numerous relatively large lime bodies. Spores spherical, finely spiny, measure 7–10 µm. Occurs on litter, mostly deep in grass or other herbaceous cover.

Craterium brunneum ⋉ is a shiny brown, including the depressed lid, and has a longer stem; *C. aureonucleatum* ⋉ is only 0.5 mm high and has many branched lime bodies.

Widespread, common; most of the year.

Leocarpus fragilis is a slime mould with yellow-brown to chestnut-brown, smooth, shiny, aggregated, pear-shaped fruitbodies (sporocarps) that open by irregular cracks at the tips. It has thin, lax, pale stems, and has lime in the peridium and capillitium. The spores are globose, finely spiny, and measure (11–)12–14(–16) µm. Found mostly on litter and wood in coniferous woodland.

This is the only species of *Leocarpus* in temperate Europe, and is distinctive and unlikely to be misidentified.

Widespread, common; mainly August–October.

Jan Vesterholt

Literature

General works:
Jens H. Petersen – *The Kingdom of Fungi* (PUP).
Webster & Weber – *Introduction to Fungi* (CUP).

Important general identification resources:
British Ascomycetes (number 72, see below).
Flora Agaricina Neerlandica (number 26–28, 226–228).
Funga Nordica (number 156).
Kleine Kryptogamenflora (number 142).
Ludwig *Pilzkompendium* (number 181–184).
The website *www.ascofrance.com*.

Identification based on substrate or host:
Plant parasites of Europe (number 80).
Microfungi on land plants (number 78).
Microfungi on miscellaneous substrates (number 79).
Phytoparasitische Kleinpilze (number 164).

Articles on the latest phylogeny and naming are most easily found by googling a fungal name followed by the word "phylogeny".

Databases with scientific names and synonymy are found at *www.indexfungorum.org* and *www.mycobank.org*. These also inlcude references to the original descriptions.

Data on fungal distribution are found on *www.gbif.org*. Be aware that not all countries deliver data to GBIF, so large areas may be without coverage.

Further reading recommended on the individual identification wheels

1 Adamčik, S., Dima, B., Adamčiková, K., Harries, D., Læssøe, T., Moreau, P.-A. & Jančovicvá, S. 2018. European *Hodophilus* (Clavariaceae, Agaricales) species with yellow stipe. Mycological Progress 17(9): 1097–1111.

2 Adamčík, S., Jancoicová, S., Looney, B.P., Adamčíková, J.M., Griffith, G.Q., Læssøe, T., Moreau, P.-A. Vizzini, A. & Matheny, P.B. 2017. *Hodophilus* (Clavariaceae, Agaricales) species with dark dots on the stipe: more than one species in Europe. Mycological Progress 16: 811–821.

3 Adamčík, S., Jancoicová, S., Looney, B.P., Adamčíková, K., Birkebak, J.M., Moreau, P.-A. Vizzini, A. & Matheny, P.B. 2017. Circumscription of species in the *Hodophilus foetens* complex (Clavariaceae, Agaricales) in Europe. Mycological Progress 16(1): 47–62.

4 Adamčík, S., Ripkova, S. & Kucera, V. 2007. Re-evaluation of morphological variability of *Pseudobaeospora* group *Celluloderma* (Agaricales, Basidiomycota). Nova Hedwigia 85: 365–377.

5 Agerer, R. 1975. Flagelloscypha - Studien an cyphelloiden Basidiomyceten. Sydowia 27: 131–265.

6 Agerer, R. 1983. Typusstudien an cyphelloiden Pilzen IV - Lachnella Fr. s.l.. Mitteilungen der Botanischen Staatssammlung München 19: 163–334.

7 Agnello, C. & Kaounas, V. 2011. *Tuber asa* and *T. gennadii*. A close morphological study of two species often confused in the past with a brief historical bibliographical summary. Ascomycete.org 3(4): 65–74.

8 Alvarado, P., Moreau, P.-A., Dima, B., Vizzini, A., Consiglio, G., Moreno, G., Setti, L., Kekki, T., Huhtinen, S., Liimatainen, K. & Niskanen, T. 2018. Pseudoclitocybaceae fam. nov. (Agaricales, Tricholomatineae), a new arrangement at family, genus and species level. Fungal Diversity 90: 109–133.

9 Alvarado, P., Moreau, P.-A., Sesli, E., Youcef Khodja, L., Contu, M. & Vizzini, A. 2018. Phylogenetic studies on *Bonomyces* (Tricholomatoid clade, Agaricales) and two new combinations from *Clitocybe*. Cryptogamie, Mycologie 39(2): 149–168.

10 Alvarado, P., Moreno, G., Vizzini, A., Consiglio, G., Manjón, J.L. & Setti, L. 2015. *Atractosporocybe*, *Leucocybe* and *Rhizocybe*: three new clitocyboid genera in the tricholomatoid clade (Agaricales) with notes on *Clitocybe* and *Lepista*. Mycologia 107(1): 123–136.

11 Antonín, V. & Noordeloos, M.E. 1997. A monograph of *Marasmius*, *Collybia* and related genera in Europe. Part. 2: *Collybia*, *Gymnopus*, *Rhodocollybia*, *Crinipellis*, *Chaetocalathus*, and additions to *Marasmiellus*. Libri Botanici 17. IHW-Verlag, München.

12 Antonín, V. & Noordeloos, M.E. 2004. A monograph of the genera *Hemimycena*, *Delicatula*, *Fayodia*, *Gamundia*, *Myxomphalia*, *Resinomycena*, *Rickenella* and *Xeromphalina* (Tribus Myceneae sensu Singer, *Mycena* excluded) in Europe. IHW Verlag, Eching, 279 pp.

13 Antonín, V. & Noordeloos, M.E. 2010. A monograph of marasmioid and collybioid fungi in Europe. IHW Verlag, 480 pp.

14 Arauzo, S. & Iglesias, P. 2014. La familia Geoglossaceae ss. str. en la península Íberica y la Macaronesia. Errotari - Micología 11: 166–259.

15 Aronsen, A. & Læssøe, T. 2016. The genus *Mycena s.l.* Fungi of Northern Europe – volume 5. Svampetryk, 373 pp.

16 Asgari, B., Zare, R. & Gams, W. 2007. *Coniochaeta ershadii*, a new species from Iran, and a key to well-documented *Coniochaeta* species. Nova Hedwigia 84(1–2): 175–187.

17 Audet, S. 2010. Essai de découpage systématique du genre Scutiger (Basidiomycota): *Albatrellopsis*, *Albatrellus*, *Polyporoletus*, *Scutiger* et description de six nouveaux genres. Mycotaxon 111: 431–464.

18 Baloch, E., Döring, H. & Spooner, B.M. 2010. The genus *Cryptodiscus* in Great Britain. Field Mycology 11(1): 26–32.

19 Baral, H.-O. & Bemmann, M. 2013. *Hymenoscyphus serotinus* and *H. lipismoides* sp.nov., two lignicolous species with a high host specificity. Ascomycete.org 5(4): 109–128.

20 Baral, H.-O. & Marson, G. 2001 (2000). Monographic revision of *Gelatinopsis* and *Calloriopsis* (Calloriopsideae, Leotiales). In: Micologia 2000, A.M.B. Fondazione Centro Studi Micologici: 23–46.

21 Baral, H. O. & Weber, E. 2019. Ascomycetes taxonomy website. https://in-vivo-veritas.de.

22 Baral, H.-O. 2000-. Dichotomous key to *Lachnellula* (worldwide). Internet publication.

23 Baral, H.-O. 1993. Beiträge zur Taxonomie der Discomyceten III. Zeitschrift für Pilzkunde 59(1): 3–22.

24 Bas, C. 2002. A reconnaissance of the genus *Pseudobaeospora* in Europe I. Persoonia 18(1): 115–122.

25 Bas, C. 2003. A reconnaissance of the genus *Pseudobaeospora* in Europe II. Persoonia 18(2): 163–199.

26 Bas, C., Kuyper, T.W., Noordeloos, M.E. & Vellinga, E.C. (eds.) 1990. Flora agaricina neerlandica 2. A.A. Balkema, Rotterdam,137 pp.

27 Bas, C., Kuyper, T.W., Noordeloos, M.E. & Vellinga, E.C. (eds.) 1995. Flora agaricina neerlandica 3. A.A. Balkema, Rotterdam,183 pp.

28 Bas, C., Kuyper, T.W., Noordeloos, M.E. & Vellinga, E.C. (eds.). 1999. Flora agaricina neerlandica 4. Rotterdam 200 pp.

29 Beker, H.J., Eberhardt, U. & Vesterholt, J. 2016. Hebeloma (Fr.) P. Kumm. Fungi Europaei 14, 1218 pp. Edizioni Candusso.

30 Bellanger, J.-M., Moreau, P.-A., Corriol, G., Bidaud, A., Chalange, R., Dudova, Z. & Richard, F. 2015. Plunging hands into the mushroom jar: a phylogenetic framework for Lyophyllaceae (Agaricales, Basidiomycota). Genetica 143(2): 169–194.

31 Bendiksen, E., Bendiksen, K., Brandrud, T.E., Kytövuori, I. & Toivonen, M. 2013. Ektomykorrhizadannende korallsopper (*Ramaria p.p.*) i rik barskog i Norge. Agarica 33: 3–26.

32 Bendiksen, K., Kytövuori, I., Toivonen, M., Bendiksen, E. & Brandrud, T.E. 2015. Ectomycorrhizal *Ramaria* species in nutrient-poor Fennoscandian conifer forests including a note on the *Ramaria botrytis* complex. Agarica 36: 89–108.

33 Benkert, D. 1987. Beiträge zur Taxonomie der Gattung *Lamprospora* (Pezizales). Zeitschrift für Pilzkunde 53(2): 195–271.

34 Benkert, D. 2005. Beiträge zur Kenntniss einiger Gattungen der Pezizales (Ascomycetes): *Plectania/Pseudoplectania*, *Ramsbottomia*, *Smardaea/Greletia*, *Sowerbyella*. Zeitschrift für Mykologie 71(2): 121–164.

35 Benny, G.L. & Kimbrough, J.W. 1980. A synopsis of the orders and families of Plectomycetes with key to genera. Mycotaxon 12: 1–91.

36 Bernicchia, A. & Gorjón, S.P. 2010. Corticiaceae s.l. Fungi Europaei 12. Edizioni Candusso, 1008 pp.

37 Bernicchia, A. 2005. Polyporaceae s.l. Fungi Europaei 10, Edizioni Candusso, 807 pp.

38 Berthier, J. 1976. Monograph des Typhula Fr., *Pistillaria* Fr. et genres voisins. Bulletin mensuel de la Société Linnéenne de Lyon 45, num. spec.

39 Bezerra, J.L. & Kimbrough, J.W. 1975. The genus *Lasiobolus* (Pezizales, Ascomycetes). Canadian Journal of Botany 53: 1209–1229.

40 Bodensteiner, P. 2006. Maireina W.B. Cooke. Morphologisch-anatomische Untersuchungen an einer Gattung cyphelloider Homobasidiomyceten. Dissersation zur Erlangung des Grades eines Doktors der Naturwissenschaften der Fakultät für Biologie der Ludwig-Maximilians-Universität München.

41 Boehm, E.W.A., Marson, G., Mathiassen, G., Gardiennet, A. & Schoch, C.L. 2005. An overview of the genus *Glyphium* and its phylogenetic placement in Patellariales. Mycologia 107(3): 607–618.

42 Boehm, E.W.A., Mugambi, G.K., Miller, A.N., Huhndorf, S.M., Marincowitz, S., Spatafora, J.W. & Schoch, C.L. 2010 (2009). A molecular phylogenetic reappraisal of the Hysteriaceae, Mytilinidiaceae and Gloniaceae

(Pleosporomycetidae, Dothideomycetes) with keys to world species. In: Schoch, C.L., Spatafora, J.W., Lumbsch, T., Huhndorf, S.M., Hyde, K.D., Groenewald, J.Z. & Crous, P.W. (eds.) A phylogenetic re-evaluation of Dothideomycetes. Studies in Mycology 64: 49–83.

43 Boertmann, D. 2010. Fungi of Northern Europe 1. The genus *Hygrocybe*. 2nd ed. Danish Mycological Society, 200 pp.

44 Brandrud, T.E., Bendiksen, E., Bendiksen, K., Kytövuori, I. & Toivonen, M. 2012. Ektomykorrhizadannende korallsopper (*Ramaria p.p.*) i eikeskog og annen edellauvskog i Norge. Agarica 32: 5–19.

45 Brandrud, T.E., Bendiksen, E., Jordal, J.B., Weholt, Ø., Eidissen, S.E., Lorås, J.A., Dima, B. & Noordeloos, M.E. 2018. *Entoloma* species of the rhodopolioid clade (subgenus *Entoloma*; Tricholomatinae, Basidiomycota) in Norway. Agarica 38: 21–46.

46 Brandrud, T.E., Lindström, H., Marklund, H., Melot, J. & Muskos, S. 1989–2014. *Cortinarius*, Flora Photographica 1–5. Cortinarius HB, Matfors.

47 Braun, U. & Cook, R.T.A. 2012. Taxonomic manual of the Erysiphales (powdery mildews). CBS Biodiversity Series 11, 707 pp.

48 Brayford, D., Honda, B.M., Mantiri, F.R. & Samuels, G.J. 2004. *Neonectria* and *Cylindrocarpon*: the Nectria mammoidea group and species lacking microconidia. Mycologia 96(3): 572–597.

49 Breitenbach, J. & Kränzlin, F. 1981. Pilze der Schweiz 1/Fungi of Switzerland 1/Champignon de Suisse 1. Verlag Mykologia, Luzern, 313 pp.

50 Breitenbach, J. & Kränzlin, F. 1986. Pilze der Schweiz 2/Fungi of Switzerland 2/Champignons de Suisse 2. Verlag Mykologia, Luzern, 416 pp.

51 Bronckers, R.J.C. 2003. Een sleutel tot de europese soorten van de genera *Trichophaea*, *Trichophaeopsis* en *Paratrichophaea*. Sterbeeckia 23: 9–27.

52 Brummelen, J. van 1995. A world-monograph of the genus *Pseudombrophila* (Pezizales, Ascomycotina). Libri Botanici 14: 1–117.

53 Brummelen, J. van. 1967. A world-monograph of the genera *Ascobolus* and *Saccobolus* (Ascomycetes, Pezizales). Persoonia Suppl. 1: 1–260.

54 Candoussau, F., Fournier, J. & Magni, J.F. 2001. New and rare species of *Lasiosphaeria* in Southwestern France. Mycotaxon 80: 201–240.

55 Candoussau, F., Boqueras, M., Gómez-Bolea, A., Læssøe, T., Lowen, R., Rogers, J.D., Rossman, A.Y. & Samuels, G.J. 2007. Observations on *Neobarya*, including new species and new combinations. Sydowia 59(2): 179–215.

56 Candusso, M. & Lanzoni, G. 1990. *Lepiota s.l.* Fungi Europaei 4, Saronno, 743 pp.

57 Candusso, M. 1997. *Hygrophorus s.l.* Fungi Europaei 6. Libreria Basso, Alassio, 784 pp.

58 Cannon, P.F. & Hawksworth, D.L. 1982. A re-evaluation of *Melanospora Corda* and similar pyrenomycetes, with a revision of the British species. Botanical Journal of the Linnean Society 84(2): 115–160.

59 Carpenter, S.E. 1981. Monograph of *Crocicreas* (Ascomycetes, Helotiales, Leotiaceae). Memoirs of the New York Botanical Garden 33: 1–290.

60 Chaverri, P., Salgado, C., Hirooka, Y., Rossman, A.Y. & Samuels, G.J. 2011. Delimitation of *Neonectria* and *Cylindrocarpon* (Nectriaceae, Hypocreales, Ascomycota) and related genera with *Cylindrocarpon*-like anamorphs. Studies in Mycology 68: 57–78.

61 Checa, J. 2004. Dothideales dictiospóricos/Dictyosporic Dothideales. Flora Mycologica Iberica 6, 162 pp. J. Cramer, Madrid *etc.*

62 Checa, J., Barrasa, J.M., Moreno, G., Fort, F. & Guarro, J. 1988. The genus *Coniochaeta* (Sacc.) Cooke (Coniochaetaceae, Ascomycotina) in Spain. Cryptogamie, Mycologie 9: 1–34.

63 Christan, J. 2008. Die Gattung *Ramaria* in Deutschland. Monografie zur Gattung *Ramaria* in Deutschland, mit Bestimmungsschlüssel zu den europäischen Arten. IHW-Verlag, Eching, 352 pp.

64 Christensen, M. & Heilmann-Clausen, J. 2013. The genus *Tricholoma*. Fungi of Northern Europe 4. Danish mycological society, 228 pp.

65 Consiglio, G. & Setti, L. 2008. Il genre *Crepidotus* in Europa. Associazione Micologica Bresadola, Fondazione Centro Studi Micologici, Trento, 344 pp.

66 Consiglio, G. & Setti, L. 2018. The genera *Hohenbuehelia* and *Resupinatus* in Europe. Monografie di Pagine di Micologia, AMB, 400 pp.

67 Corfixen, P. & Parmasto, E. 2017. *Hymenochaete* and *Hymenochaetopsis* (Basidiomycota) in Europe. Karstenia 57(1–2): 49–80.

68 Currah, R.S. 1985. Taxonomy of the Onygenales: Arthrodermataceae, Gymnoascaceae, Myxotrichaceae and Onygenaceae. Mycotaxon 24: 1–216.

69 Dam, N. & Kuyper, T.W. 2011. Het geslacht *Cortinarius* in Nederland–X: Een voorlopige sleutel [The genus *Cortinarius* in the

Netherlands –X: a preliminary key]. Coolia 54(4): 189–208.

70 Dam, N. 2018. Waaszwammen [*Tulasnella*]. Coolia 61(1): 25–32.

71 Den Bakker, H.C. & Noordeloos, M.E. 2005. A revision of European species of *Leccinum* Gray and notes on extralimital species. Persoonia 18(4): 511–587.

72 Dennis, R.W.G. 1981. British Ascomycetes. Second impression, with supplement containing additional species received at the Royal Botanic Gardens, Kew, by the end of 1980. J. Cramer.

73 Doveri, F. 2004. Fungi fimicoli italici. Fondazione Centro Studi Micologici Dell´A.M.B, 1104 pp.

74 Dämmrich, F. 2004–2018. *Tomentella* und verwandte Gattungen [online keys]. www.tomentella.de.

75 Dämmrich, F. 2006. Studien der tomentelloiden Pilze in Deutschland - unter besonderer berüchsichtigung der Zeichnungen von Frau Dr. H. Maser aus den Jahren 1988–1994. Teil1: Die Gattung Tomentella. Zeitschrift für Pilzkunde 72(2): 167–212.

76 Einhellinger, A. 1985. Die Gattung *Russula* in Bayern. Hoppea 43: 5–286.

77 Ellis, J.P. 1976. British *Microthyrium* species and similar fungi. Transactions of the British Mycological Society 67(3): 381–394.

78 Ellis, M.B. & Ellis, J.P. 1985. Microfungi on land plants. Richmond Publ., 868 pp.

79 Ellis, M.B. & Ellis, J.P. 1988. Microfungi on miscellaneous substrates. An identification handbook. Croom Helm Ltd, London, 244 pp.

80 Ellis, W.N. 2018. Plant parasites of Europe, leafminers, galls and fungi. https://bladmineerders.nl.

81 Eriksson, J. & al. 1973–1988. The Corticiaceae of North Europe 1–8. Fungiflora, Oslo.

82 Fournier, J. & Magni, J.-F. 2004. Pyrenomycetes of southwestern France. http://pyrenomycetes.free.fr.

83 Fournier, J. 2014. Update on European species of *Xylaria*. Online: http://www.ascofrance.fr/uploads/xylaria/201406.pdf.

84 Friebes, G. 2012. A key to the non-lichenicolous species of the genus *Capronia* (Herpotrichiellaceae). Ascomycete.org 4(3): 55–64.

85 Ge, Z.-W., Jacobs, A., Vellinga, E.C., Sysouphanthong, P., Walt, R.v.d., Lavorato, C., An, Y.-F. & Yang, S.-L. 2018. A multi-gene phylogeny of *Chlorophyllum* (Agaricaceae, Basidiomycota): new species, new combinations and infrageneric classification. Mycokeys 32: 65–90.

86 Gerhardt, E. 1996. Taxonomische Revision der Gattungen *Panaeolus* und *Panaeolina* (Fungi, Agaricales, Coprinaceae). Bibliotheca Botanica 147: 1–149.

87 Gjærum, H.B. 1964. The genus *Taphrina* Fr. in Norway. Nytt Magasin for Botanikk 11: 5–26.

88 Gminder, A. & Krieglsteiner, G.J. 2001. Die Grosspilze Baden-Württembergs 3. Ständerpilze: Blätterpilze I. Verlag Eugen Ulmer, Stuttgart, 634 pp.

89 Gminder, A. & Krieglsteiner, G.J. 2003. Die Grosspilze Baden-Württembergs 4. Ständerpilze: Blätterpilze II. Verlag Eugen Ulmer, Stuttgart, 467 pp.

90 Gminder, A. 1994. Die trockenstielingen Arten der Gattung *Limacella* in Europa. Zeitschrift für Pilzkunde 60(2): 377–398.

91 Gminder, A. 2010. Die Grosspilze Baden-Württembergs 5. Ständerpilze: Blätterpilze III. Verlag Eugen Ulmer, Stuttgart, 671 pp.

92 Gminder, A., Krieglsteiner, G.J. & Winterhoff, W. 2000. In: Kriegelsteiner, G.J. (ed.). Die Grosspilze Baden-Württembergs. Band 2: Ständerpilze: Leisten-, Keulen-, Korallen- und Stoppelpilze, Bauchpilze, Röhrlings- und Täublingsartige. Verlag Eugen Ulmer, 620 pp.

93 Granmo, A., Læssøe, T. & Schumacher, T. 1999. The genus *Nemania s.l.* (Xylariaceae) in Norden. Sommerfeltia 27: 1–96.

94 Greaves, M.C. 2014. *Pilobolus* in Britain. Field Mycology 15(2): 62–64.

95 Groves, J.W. 1946. North American species of *Dermea*. Mycologia 38(4): 351–431.

96 Gräfenhan, T., Schroers, H.-J., Nirenberg, H.I. & Seifert, K.A. 2011. An overview of the taxonomy, phylogeny and typification of some nectriaceous fungi classifed in *Cosmospora*, *Acremonium*, *Fusarium*, *Stilbella* and *Volutella*. Studies in Mycology 68: 79–113.

97 Hahn, C. 1999. Studien zur Systematik der Paxillaceae (Boletales, Basidiomycota). Sendtnera 6: 115–133.

98 Halasz, K., Bratek, Z., Szego, D., Rudnoy, S., Racz, I., Lasztity, D. & Trappe, J. 2005. Tests of concepts of the small, white, European group of *Tuber* spp. based on morphology and rDNA ITS sequences with special reference to *Tuber rapaeodorum*. Mycological Progress 4: 281–290.

99 Hallenberg, N. & Eriksson, J. 1985. The Lachnocladiaceae and Coniophoraceae of North Europe. Fungiflora, Oslo, 96 pp.

100 Hansen, L. & Knudsen, H. (eds.) 1997. Nordic macromycetes 3. Heterobasidioid, aphyllophoroid and gastromycetoid basidiomycetes. Nordsvamp, Copenhagen, 444 pp.

101 Hansen, L. & Knudsen, H. (eds.) 2000. Nordic macromycetes 1. Ascomycetes. Nordsvamp, Copenhagen, 309 pp.

102 Hausknecht, A. 2009. *Conocybe* Fayod - *Pholiotina* Fayod in Europe. Fungi Europaei 11. Edizioni Candusso, 968 pp.

103 Hausknecht, A. 2012. Die Gattung *Simocybe* in Österreich. Österreichische Zeitschrift für Pilzkunde 21: 31–43.

104 Hausknecht, A., Krisai-Greilhuber, I. & Klofac, W. 1997. Die Gattung *Hydropus* in Österreich. Österreichische Zeitschrift für Pilzkunde 6: 181–210.

105 Heilmann-Clausen, J., Verbeken, A. & Vesterholt. J. 1998. The genus *Lactarius*. Fungi of Northern Europe 2. Danish Mycological Society, 287 pp.

106 Hirooka, Y., Rossman, A.Y. & Chaverri, P. 2011. A morphological and phylogenetic revision of the *Nectria cinnabarina* species complex. Studies in Mycology 68: 35–56.

107 Hirooka, Y., Rossman, A.Y., Samuels, G.J., Lechat, C. & Chaverri, P. 2012. A monograph of *Allantonectria*, *Nectria*, and *Pleonectria* (Nectriaceae, Hypocreales, Ascomycota) and their pycnidial, sporodochial, and synnematous anamorphs. Studies in Mycology 71: 1–210.

108 Hohmeyer, H. 1986. Ein Schlüssel zu den europäischen Arten der Gattung *Peziza* L. Zeitschrift für Pilzkunde 52(1): 161–188.

109 Hohmeyer, H. 1989. The genus *Arpinia* (Pyronemataceae, Pezizales). Mycologia Helvetica 3(2): 221–232.

110 Holec, J. & Kolank, M. 2013. *Ossicaulis lachnopus* (Agaricales, Lyophyllaceae), a species similar to *O. lignatilis*, is verified by morphological and molecular methods. Mycological Progress 12(3): 589–597.

111 Holec, J. & Kolarik, M. 2010. *Tricholomopsis flammula* (Basidiomycota, Agaricales) - molecular taxonomy, delimitation, variability and ecology. Mycological Progress 10: 93–99.

112 Holec, J. 2005. The genus *Gymnopilus* (Fungi, Agaricales) in the Czech Republic with respect to collections from other European countries. Acta Musei Nationalis Pragae, ser. B, Hist. Nat. 61(1–2): 1–52.

113 Horak, E. & Moreau, P.-A. 2004. Les "Naucoria" nouveaux de la Flore Analytique (*Alnicola*, *Flammulaster*, *Tubaria*). Bulletin Trimestriel de la Société Mycologique de France 120(1–4): 215–237.

114 Huggert, L. 2010. Laboulbeniales i Sverige. Department of Ecology and Environmental Science, Umeå University, 96 pp.

115 Huhtinen, S. 1987. Taxonomic studies in the genera *Protounguicularia*, *Arachnopeziza* and *Dematioscypha*. Mycotaxon 30: 9–28.

116 Huhtinen, S. 1990. A monograph af *Hyaloscypha* and allied genera. Karstenia 29 (2): 1–252.

117 Huldén, L. 1983. Laboulbeniales (Ascomycetes) of Finland and adjacent parts of the U.S.S.R. Karstenia 23: 31–136.

118 Häffner, J. 1986. Die apiculaten Becherlinge. Zeitschrift für Pilzkunde 52(1): 189–212.

119 Häffner, J. 1993. Die Gattung Aleuria Fuckel. Rheinland-Pfälzisches PilzJournal 3(1): 6–59.

120 Härkönen, M. & Varis, E. 2012. Suomen limasienet [The myxomycetes of Finland]. Norrlinia 25: 1–240.

121 Haan, A. de & Walleyn, R. 2002. Studies in *Galerina* - Galerinae Flandriae (1). Fungi non Delineati 23: 1- 68.

122 Haan, A. de & Walleyn, R. 2006. Studies in *Galerina* - Galerinae Flandriae (2). Fungi non Delineati 33: 1- 74.

123 Haan, A. de & Walleyn, R. 2009. Studies in *Galerina* - Galerinae Flandriae (3). Fungi non Delineati 46: 1- 84.

124 Haan, A. de, Lenaerts, L. & Volders, J. 2000. Bijdrage tot de kennis van het subgenus *Telamonia* (*Cortinarius*) in België. 6de verslag van de werkgroep Cortinarius. Sterbeeckia 19: 27–48.

125 Haan, A. de, Lenaerts, L. & Volders, J. 2001. Bijdrage tot de kennis van het subgenus *Telamonia* (*Cortinarius*) in België. 7de verslag van de werkgroep Cortinarius. Sterbeeckia 20: 21–41.

126 Haan, A. de, Volders, J. & Walleyn, R. 2003. Bijdrage tot de kennis van het subgenus *Telamonia* (*Cortinarius*) in België. 8de verslag van de werkgroep Cortinarius. Sterbeeckia 21/22: 34–63.

127 Haan, A. de, Volders, J., Gelder, J. & Walleyn, R. 2003. Bijdrage tot de kennis van het subgenus *Telamonia* (*Cortinarius*) in België. 9de verslag van de werkgroep Cortinarius. Sterbeeckia 23: 28–60.

128 Jaklitsch, W. 2009. European species of *Hypocrea*. Part I. The green-spored species. Studies in Mycology 63: 1–91.

129 Jaklitsch, W. 2011. European species of *Hypocrea* part II: species with hyaline ascospores. Fungal Diversity 48(1): 1–250.

130 Jaklitsch, W.M., Fournier, J., Rogers, J.D. & Voglmayr, H. 2014. Phylogenetic and taxonomic revision of *Lopadostoma*. Persoonia 32: 52–82.

131 Jaklitsch, W.M. & Voglmayr, H. 2014. Biodiversity of *Trichoderma* (Hypocreaceae) in Southern Europe and Macaronesia. Studies in Mycology 80:1–87.

132 Jaklitsch, W.M., Põldma, K. & Samuels, G.J. 2008. Reconsideration of *Protocrea* (Hypocreales, Hypocreaceae). Mycologia 100(6): 962–984.

133 Jargeat, P., Chaumeton, J.-P., Navaud, O., Vizzini, A. & Gryta, H. 2014. The *Paxillus involutus* (Boletales, Paxillaceae) complex in Europe: Genetic diversity and morphological description of the new species *Paxillus cuprinus*, typification of *P. involutus s.s.*, and synthesis of species boundaries. Fungal Biology 118(1): 12–31.

134 Jeppson, M. 2008. Rottryfflar och deras släktingar i Nordeuropa [The species in subordo Sclerodermatinae in North Europe]. Svensk Mykologisk Tidskrift 29(2): 46–73.

135 Jeppson, M. 2013. Jordstjärnor. En bestämningsguide [Geastrum - an identification guide]. Mykologiske publikationer 6. Sveriges Mykologiska Förening, 228 pp.

136 Jeppson, M. 2018. Puffballs of northern and central Europe. Mykologiska Publikationer 8. Sveriges Mykologiska Förening, 359 pp.

137 Jeppson, M., Altes, A., Moreno, G., Nilsson, R.H., Loarce, Y., Bustos, A. d. & Larsson, E. 2017. Unexpected high species diversity among European stalked puffballs – a contribution to the phylogeny and taxonomy of the genus Tulostoma (Agaricales). MycoKeys 21: 33–88.

138 Jeppson, M., Nilsson, R.H. & Larsson, E. 2013. European earthstars in Geastraceae (Geastrales, Phallomycetidae) – a systematic approach using morphology and molecular sequence data. Systematics and Biodiversity 11(4): 437–465.

139 Johnston, P.R. 2001. Monograph of the monocotyledon-inhabiting species of *Lophodermium*. Mycological Papers 176: 1–239.

140 Justo, A., Malysheva, E., Bulyonkova, T., Vellinga, E.C., Cobian, G., Nguyen, N. Minnis, A.M. & Hibbett, D.S. 2014. Molecular phylogeny and phylogeography of Holarctic species of *Pluteus* section *Pluteus* (Agaricales: Pluteaceae), with description of twelve new species. Phylotaxa 180(1): 1–85.

141 Justo, A., Minnis, A.M., Ghignone, S., Menolli Jr., N., Capelari, M., Rodríguez, O., Malysheva, E., Contu, M. & Vizzini, A. 2011. Species recognition in *Pluteus* and *Volvopluteus* (Pluteaceae, Agaricales): morphology, geography and phylogeny. Mycological Progress 10(4): 453–479.

142 Jülich, W. 1984. Kleine Kryptogamenflora, Band IIb/1. Die Nichtblätterpilze, Gallertpilze und Bauchpilze. Stuttgart.

143 Kautmanová, I., Tomsovsky, M., Dueñas, M. & Martín, M.P. 2012. European species of *Clavaria* (Agaricales, Agaricomycetes) with dark basidiomata - a morphological and molecular study. Persoonia 29: 133–145.

144 Keller, J. 2002. The genus *Entomophthora* (Zygomycetes, Entomophthorales) with a description of five new species. Sydowia 54(2): 157–197.

145 Keller, J. 2013. Entomophthorales (Fungi, Entomophthoromycota) attacking Coleoptera with a key for their identification. Journal of the Swiss Entomological Society 86(3–4): 261–279.

146 Keller, S. & Petrini, O. 2005. Keys to the identification of the arthropod pathogenic genera of the families Entomophthoraceae and Neozygitaceae (Zygomycetes), with descriptions of three new subfamilies and a new genus. Sydowia 57(1): 23–53.

147 Kerrigan, R.W. 2016. *Agaricus* of North America. Memoirs of the New York Botanical Garden 114, 592 pp.

148 Kibby, G. 2011. The genus *Agaricus* in Britain. 51 pp.

149 Kibby, G. 2012. The genus *Amanita* in Great Britain. 50 pp.

150 Kibby, G. 2013. British boletes: With keys to species. 77 pp.

151 Kibby, G. 2014. The genus *Russula* in Great Britain: With synoptic keys to species. 138 pp.

152 Kibby, G. 2014. British milkcaps: *Lactarius* & *Lactifluus*. 114 pp.

153 Kirk, P.M. & Spooner, B.M. 1984. An account of the fungi of Arran, Gigha and Kintyre. Kew Bulletin 38(4): 503–597.

154 Kits van Waveren, E. 1985. The Dutch, French and British species of *Psathyrella*. Persoonia Suppl. 2: 3–300.

155 Kits van Waveren, E. 1987. Additions to our monograph on Psathyrella. Thirteen

new species, some revised keys, comments on other recently described species, and corrections and additions to our monograph. Persoonia 13(3): 327–368.

156 Knudsen, H. & Vesterholt, J. (eds.) 2012. Funga nordica. Agaricoid, boletoid, clavarioid, cyphelloid and gastroid genera. Nordsvamp, 2.ed., 1083 pp.

157 Koch, R.A., Wilson, A.W., Sene, O., Henkel, T.W. & Aime, M.C. 2017. Resolved phylogeny and biogeography of the root pathogen *Armillaria* and its gasteroid relative, *Guyanagaster*. BMC Evolutionary Biology 17(33): 1–16.

158 Kokkonen, K. 2015. A survey of boreal *Entoloma* with emphasis on the subgenus *Rhodopolia*. Mycological Progress 14: 116, 52 pp.

159 Kõljalg, U. 1996. *Tomentella* (Basidiomycota) and related genera in Temperate Eurasia. Fungiflora, Oslo, 213 pp.

160 Korf, R.P. 1951. A monograph of the Arachnopezizeae. Lloydia 14(3): 129–180.

161 Korhonen, A., Seelan, J.S.S. & Miettinen, O. 2018. Cryptic species diversity in polypores: the *Skeletocutis nivea* species complex. MycoKeys 36: 45–82.

162 Kreisel, H. 1962. Die Lycoperdaceae der DDR. Bibliotheca Mycologica 36.

163 Krieglsteiner, G.J. (ed.) 2000. Die Grosspilze Baden-Württembergs Band 1: Allgemeiner Teil, Ständerpilze: Gallert-, Rinden-, Stachel- und Porenpilze. Verlag Eugen Ulmer, 629 pp.

164 Kruse, J. 2018. (Obligat) Phytoparasitische Kleinpilze Mitteleuropa mit Schwerpunkt Deutschland. http://jule.pflanzenbestimmung.de.

165 Kränzlin, F. 2005. Pilze der Schweiz/Fungi of Switzerland 6. Russulaceae. Verlag Mykologia, Luzern, 319 pp.

166 Kucera, V., Fedosova, A.G. & Arauzo, S. 2017. *Hemileucoglossum pusillum* V. Kučera, Fedosova & Arauzo, sp. nov. Fungal Planet 636. In: Crous et al. (eds.). Fungal Planet description sheets: 625–715. Persoonia 39: 304–305.

167 Kucera, V., Lizon, P. & Tomsovsky, M. 2017. Taxonomic divergence of the green naked stiped members of the genus *Microglossum* (Helotiales). Mycologia 109(1): 46–54.

168 Kucera, V., Lizon, P., Tomsovsky, M., Kucera, J. & Gaisler, J. 2014. Re-evaluation of the morphological variability of *Microglossum viride* and *M. griseoviride* sp. nov. Mycologia 106(2): 282–290.

169 Kytövuori, I. 1999. The *Stropharia semiglobata* group in NW Europe. Karstenia 39(1): 11–32.

170 Ladurner, H. & Simonini, G. 2003. *Xerocomus s.l.* Fungi Europaei 8, 528 pp, 21 col.pls. Edizioni Candusso.

171 Larsson, E. & Örstadius, L. 2008. Fourteen coprophilous species of *Psathyrella* identified in the Nordic countries using morphology and nuclear rDNA sequence data. Mycological Research 112: 1165–1185.

172 Larsson, E., Kleine, J., Jacobsson, S. & Krikorev, M. 2018. Diversity within the *Hygrophorus agathosmus* group (Basidiomycota, Agaricales) in Northern Europe. Mycological Progress 17(12): 1293–1304.

173 Larsson, E., Ryberg, M., Moreau, P.-A., Delcuse Mathiesen, Å. & Jacobsson, S. 2009. Taxonomy and evolutionary relationships within species of section *Rimosae* (*Inocybe*) based on ITS, LSU and mtSSU sequence data. Persoonia 23: 86–98.

174 Lavorato,C., Vizzini, A., Ge, Z.-W. & Contu, M. 2015. Redescription of *Clitocybe umbrinopurpurascens* (Basidiomycota, Agaricales) and revision of *Neohygrophorus* and *Pseudoomphalina*. Phytotaxa 219(1): 43–57.

175 Lechat, C. & Fournier, J. 2018 . *Flammocladiella decora*, a new combination to accommodate the hypocrealean fungus *Nectria decora*. Ascomycete.org 10(1).

176 Liu, X.Z. et al. 2015. Towards an integrated phylogenetic classification of the Tremellomycetes. Studies in Mycology 81: 85–147.

177 Lizon, P. 1992. The genus *Hymenoscyphus* (Helotiales) in Slovakia, Czechoslovakia. Mycotaxon 45: 1–59.

178 Lodge, D.J. et al. 2013. Molecular phylogeny, morphology, pigment chemistry and ecology in Hygrophoraceae (Agaricales). Fungal Diversity 64(1): 1–99.

179 Lohmeyer, T.R. & Benkert, D. 1988. *Poronia erici* - eine neue Art der Xylariales (Ascomycetes). Zeitschrift für Mykologie 54(1): 93–102.

180 Loizides, M. 2017. The morels: the story so far. Field Mycology 18(2): 42–53.

181 Ludwig, E. 2001. Pilzkompendium 1. Beschreibungen und Abbildungen. IHW-Verlag, Eching, 758 pp, 188 pls.

182 Ludwig, E. 2007. Pilzkompendium 2. Beschreibungen. Die grösseren Gattungen

der Agaricales mit farbigem Sporenpulver (ausgenommen Cortinariaceae). Fungicon-Verlag, 723 pp. & Abbildungen 209 pp.

183 Ludwig, E. 2012. Pilzkompendium 3. Beschrei-bungen. Die übrigen Gattungen der Agaricales mit weissem Sporenpulver. Fungicon-Verlag, 881 pp. & Abbildungen 298 pp.

184 Ludwig, E. 2017. Pilzkompendium 4. Beschreibungen. Cortinariaceae (*Galerina*, *Hebeloma*, *Inocybe*, *Phaeogalera*, *Cortinarius* Teil I mit den Untergattungen *Cortinarius*, *Dermocybe*, *Leprocybe*, *Phlegmacium*). Fungicon-Verlag, 793 pp. & Abbildungen 272 pp.

185 Lundqvist, N.G. 1972. Nordic Sordariaceae s.lat. Symbolae Botanicae Upsalienses 20(1): 1–374.

186 Læssøe, T., Davey, M. & Petersen, J.H. 2016. A new species of *Maireina* on *Filipendula ulmaria*. Karstenia 56(1): 39–46.

187 Majewski, T. 1994. The Laboulbeniales of Poland. Institute of Botany, Academy of Sciences, 466 pp.

188 Malloch, D. & Cain, R.F. 1971. The genus *Kernia*. Canadian Journal of Botany 49(6): 855–867.

189 Malysheva, V. & Spirin, V. 2017. Taxonomy and phylogeny of the Auriculariales (Agaricomycetes, Basidiomycota) with stereoid basidiocarps. Fungal Biology 121(8): 689–715.

190 Malysheva, V., Spirin, V., Miettinen, O., Motato-Vasquez, V., Jaya Seelan Sathiya Seelan, H. & Larsson, K-H. 2018. Revision of *Protohydnum* (Auriculariales, Basidiomycota). Mycological Progress 17(7): 805–814.

191 Marxmüller, H. 2014. Russularum Icones: Eine Täublingsikonografie. Anatis Verlag, 712 pp.

192 Miettinen, O., Spirin, V. & Niemelä, T. 2012. Notes on the genus *Aporpium* (Auriculariales, Basidiomycota), with a new species from temperate Europe. Annales Botanici Fennici 49: 359–368.

193 Miettinen, O., Vlasák, J., Rivoire, B. & Spirin, V. 2018. *Postia caesia* complex (Polyporales Basidiomycota). Fungal Systematics and Evolution 1: 101–129.

194 Miksik, M. 2017. Hribovité houby Evropy [European boletes]. Svojtka, 468 pp.

195 Miller, A.N., Huhndorf, S.M. & Fournier, J. 2014. Phylogenetic relationships of five uncommon species of *Lasiosphaeria* and three new species in the Helminthosphaeriaceae (Sordariomycetes). Mycologia 106(3): 505–524.

196 Monod, M. 1983. Monographie taxonomique des Gnomoniaceae (Ascomycètes de l'ordre des Diaporthales). I. Beihefte zur Sydowia 9: 1–315.

197 Montecchi, A. & Sarasini, M. 2000. Funghi ipogei d'Europa. AMB, Trento, 714 pp.

198 Moravec, J. 1988. A key to the species of *Sowerbyella* (Discomycetes, Pezizales). Česka Mykologie 42(4): 193–199.

199 Moravec, J. 1994. *Melastiza* (Boud.) comb. et stat.nov. - a subgenus of the genus *Aleuria* Fuck. emend. nov. (Discomycetes, Pezizales). Czech Mycology 47(4): 237–259.

200 Moravec, J. 2005. A world monograph of the genus *Cheilymenia* (Discomycetes, Pezizales, Pyronemataceae). Libri Botanici 21: 1–256.

201 Moreau, P.-A. 2005. A nomenclatural revision of the genus *Alnicola* (Cortinariaceae). Fungal Diversity 20: 121–155.

202 Moreau, P.-A., Vila, J., Aime, M.C., Antonín, V., Horak, E., Pérez-Butrón, J.L., Richard, F., Urban, A., Welti, S. & Vizzini, A. 2015. *Cibaomyces* and *Cyptotrama*, two new genera for Europe, and an emendation of *Rhizomarasmius* (Basidiomycota, Physalacriaceae). Mycological Progress 14(2–4): 1–16.

203 Moreau, P.-A., Roux, P. & Mascarell, G. 1999. Une étude du genre *Lentinellus* en Europe. Bulletin Trimestriel de la Société Mycologique de France 115: 279–373.

204 Morozova, O.V., Noordeloos, M.E. & Vila, J. 2014. *Entoloma* subgenus *Leptonia* in boreal-temperate Eurasia: towards a phylogenetic species concept. Persoonia 32: 141–169.

205 Muñoz, J.A. 2005. *Boletus s.l.* Fungi Europei vol. 2. Edizioni Candusso, 951 pp.

206 Maas Geesteranus, R.A. 1992. Mycenas of the northern hemisphere. I. Studies in Mycenas and other papers. II. Conspectus of the mycenas of the northern hemisphere. Koninklijke Nederlandse Akademie van Wetenschappen verhandelingen, afd. Natuurkunde, tweede reeks 90, 391 pp & 493pp.

207 Nagy, L.G., Desjardin, D.E., Vágvölgyi, C., Kemp, R. & Papp, T. 2013. Phylogenetic analyses of *Coprinopsis* sections Lanatuli and Atramentarii identify multiple species within morphologically defined taxa. Mycologia 105(1): 112–124.

208 Nagy, L.G., Házi, J., Vágvölgyi, C. & Papp, T. 2012. Phylogeny and species delimitation in the genus *Coprinellus* with special emphasis on the haired species. Mycologia 104(1): 254–275.

209 Nannfeldt, J.A. 1981. *Exobasidium*, a taxonomic reassessment applied to the European species. Symbolae Botanicae Upsalienses 23(2): 1–72.

210 Nauta, M.M. & Spooner, B. 1999. British Dermateaceae: 1. Introduction. Mycologist 13 (1): 3–6.

211 Nauta, M.M. & Spooner, B. 1999. British Dermateaceae: 2. Naevioideae. Mycologist 13 (2): 65–69.

212 Nauta, M.M. & Spooner, B. 1999. British Dermateaceae: 3. Peziculoideae. Mycologist 13(3): 98–101.

213 Nauta, M.M. & Spooner, B. 1999. British Dermateaceae: 4A. Dermateoideae [A key to genera of Dermateoideae recognized in Great Britian]. Mycologist 13(4): 146–148.

214 Nauta, M.M. & Spooner, B. 2000. British Dermateaceae: 4B.Dermateoideae Genera B-E. Mycologist 14(1): 21–28.

215 Nauta, M.M. & Spooner, B. 2000. British Dermateaceae: 4B. Dermateoideae Genera G-Z. Mycologist 14(2): 65–74.

216 Nauta, M.M. & Spooner, B. 2000. British Dermateaceae: 5. Index to species names. Mycologist 14(3): 121–126.

217 Neubert, H., Nowotny, W. & Baumann, K. 1993–2000. Die Myxomyceten Deutschlands und des angrenzenden Alpenraumes unter besonderer Berücksichtigung Österreichs. K. Baumann Verlag, 3 vol.

218 Nevile, P. & Poumarat, S. 2004. Amaniteae: *Amanita*, *Limacella* & *Torrendia*. Fungi Europaei 9. Edizioni Candusso.

219 Niemelä, T. 2016. Suomen käävät [Polypores of Finland]. Norrlinia 31: 1–432.

220 Niemelä, T., Kinnunen, J., Larsson, K.-H., Schigel, D.S. & Larsson, E. 2005. Genus revisions and new combinations of some North European polypores. Karstenia 45: 75–80.

221 Niskanen, T., Liimatainen, K., Nuytinck, J., Kirk, P., Olariaga Ibarguren, I., Garibay-Orijel, R., Norvell, L., Huhtinen, S., Kytövuori, I., Ruotsalainen, J., Niemelä, T., Ammirati, J.F. & Tedersoo, L. 2018. Identifying and naming the currently known diversity of the genus *Hydnum* with an emphasis on European and North American taxa. Mycologia 110(5): 890–918.

222 Nitare, J. 2014. Liten skruvbusksvamp *Tremellodendropsis helvetica* (Schild) Nitare funnen på Öland [*Tremellodendropsis helvetica* (Schild) Nitare comb. et stat.nov. found on the island of Öland, SE Sweden]. Svensk Mykologisk Tidskrift 35(2): 24–31.

223 Noordeloos, M.E. 1992. Fungi Europaei 5. *Entoloma s.l.* Libreria editrice Giovanna Biella, Saronno, 760 pp.

224 Noordeloos, M.E. 2004. Fungi Europaei 5A. *Entoloma s.l.* Supplemento. Edisioni Candusso: 761–1378.

225 Noordeloos, M.E. 2011. Strophariaceae s.l. Fungi Europaei 13. Edizioni Candusso, 648 pp.

226 Noordeloos, M.E., Kuyper, T.W. & Vellinga, E.C. (eds.) 2001. Flora agaricina neerlandica 5. A.A. Balkema Publ., Lisse. 169 pp.

227 Noordeloos, M.E., Kuyper, T.W. & Vellinga, E.C. (eds.) 2005. Flora agaricina Neerlandica 6. Taylor & Francis, 227 pp.

228 Noordeloos, M.E., Kuyper, T.W., Somhorst, I. & Vellinga, E.C. (eds). 2018. Flora agaricina Neerlandica 7. Boletales, Russulales. Candusso editrice, 632 pp.

229 Olariaga, I., Grebenc, T., Salcedo, I. & Martín, M.P. 2012. Two new species of *Hydnum* with ovoid basidiospores: *H. ovoideisporum* and *H. vesterholtii*. Mycologia 104(6): 1443–1455.

230 Olariaga, I., Jugo, B.M., García-Etxebarria, K. & Salcedo, I. 2009. Species delimitation in the European species of *Clavulina* (Cantharellales, Basidiomycota) inferred from phylogenetic analyses of ITS region and morphological data. Mycological Research 113(11): 1261–1270.

231 Olariaga, I., Moreno, G., Manjón, J.L., Salcedo, I., Hofstetter, V., Rodríguez, D. & Buyck, B. 2017. *Cantharellus* (Cantharellales, Basidiomycota) revisited in Europe through a multigene phylogeny. Fungal Diversity 83(1): 263–292.

232 Olariaga, I., Salcedo, I., Daniëls, P.P., Spooner, B. & Kautmanová, I. 2015. Taxonomy and phylogeny of yellow *Clavaria* species with clamped basidia – *Clavaria flavostellifera* sp. nov. and the typification of *C. argillacea*, *C. flavipes* and *C. sphagnicola*. Mycologia 107(1): 104–122.

233 Olariaga, I., Vooren, N. v., Carbone, M. & Hansen, K. 2015. A monograph of *Otidea* (Pyronemataceae, Pezizomycetes). Persoonia 35: 166–229.

234 Orange, A. 2012. Semi-cryptic marine species of *Hydropunctaria* (Verrucariaceae, lichenized Ascomycota) from north-west Europe. Lichenologist 44(3): 299–320.

235 Papp, V. & Dima, B. 2017. New systematic position of *Aurantiporus alborubescens* (Meruliaceae, Basidiomycota), a threatened old-growth forest polypore. Mycological Progress 17(3): 319–332.

236 Parra Sánchez, L.A. 2008. Fungi Europaei 1. Agaricus - Allopsalliota - part 1. Edizioni Candusso, Alassio, 824 pp.

237 Parra Sánchez, L.A. 2013. Fungi Europaei 1A. Agaricus s.l., part 2. Edizioni Candusso, Alassio, 824 pp.

238 Paz, A., Bellanger, J.-M., Lavoise, C., Molia, A., Ławrynowicz, M., Larsson, E., Olariaga Ibarguren, I., Jeppson, M., Læssøe, T., Sauve, M., Richard, F. & Moreau, P.-A. 2017. The genus *Elaphomyces* (Ascomycota, Eurotiales): a ribosomal DNA-based phylogeny and revised systematics of European 'deer truffles'. Persoonia 38: 197–239.

239 Paz, A., Vidal, J.M., Lavoise, C. & Moreau, P.-A. 2014. Primeros datos para una revisión del género *Octaviania* en Europa: *O. depauperata* comb. & stat. nov., *O. depauperata* var. laurarum var. nov. y *O. vacekii* sp. nov. Boletin Micológico FAMCAL 9: 77–97.

240 Pegler, D.N., Spooner, B.M. & Young, T.W.K. 1993. British truffles. A revision of Britsh hypogeous fungi. Royal Botanic Gardens, Kew, 216 pp, 26 pl.

241 Pegler, D.N., Læssøe, T. & Spooner, B.M. 1995. British puffballs, earthstars and stinkhorns. Royal Botanic Gardens, Kew, 265 pp.

242 Pegler, D.N., Roberts, P. J. & Spooner, B.M. 1997. British chanterelles and tooth-fungi. RBG, Kew, 114 pp.

243 Petersen, J.H. & Olariaga, I. 2012. Typhula Fr.: Fr. In: Læssøe, T. & Petersen, J.H. Mycokey. www.mycokey.com.

244 Petersen, J.H., Davey, M.L. & Læssøe, T. 2014. *Hirticlavula elegans*, a new clavarioid fungus from Scandinavia. Karstenia 54(1): 1–8.

245 Petersen, R.H. & Hughes, K.W. 2004. A preliminary monograph of *Lentinellus* (Russulales). Bibliotheca Mycologica 198: 1–270, pl. 1–13.

246 Petersen, R.H. & Hughes, K.W. 2010. The *Xerula/Oudemansiella* complex (Agaricales). Beihefte zur Nova Hedwigia 137: 1–625.

247 Petrini, L.E. 2013. *Rosellinia* - a world monograph. Bibliotheca Mycologica 205: 1–410.

248 Põldmaa, K. 1999. The genus *Hypomyces* and allied fungicolous fungi in Estonia. I. Species growing on aphyllophoralean basidiomycetes. Folia Cryptogamica Estonica 34: 15–31.

249 Poulain, M., Meyer, M. & Bozonnet, J. 2011. Les myxomycetès. Sarl Editions FMDS, 2 vol., 1119 pp.

250 Pouzar, Z. 1986. *Camarops* subg. *Bolinia* in Czechoslovakia. Česká Mykologie 40(4): 218–222.

251 Pouzar, Z. 1986. A key and conspectus of Central European species of *Biscogniauxia* and *Obolarina* (Pyrenomycetes). Česká Mykologie 40: 1–10.

252 Pärtel, K., Baral, H.-O., Tamm, J. & Põldmaa, K. 2016. Evidence for the polyphyly of *Encoelia* and Encoelioideae with reconsideration of respective families in Leotiomycetes. Fungal Diversity 82: 183–219.

253 Raitviir, A. 2004. Revised synopsis of the Hyaloscyphaceae. Scripta Mycologica 20. Estonian Agricultural University, Institute of Zoology and Botany, Tartu, 133 pp.

254 Rappaz, F. 1987. Taxonomie et nomenclature des Diatrypacées à asques octosporés (1). Mycologia Helvetica 2(3): 285–648.

255 Rappaz, F. 1987. Taxonomie et nomenclature des *Diatrype stigma*, *D. decorticata* et *D. undulata* (Diatrypaceae, Ascomycetes). Mycotaxon 30: 209–219.

256 Rappaz, F. 1995. *Anthostomella* and related xylariaceous fungi on hard wood from Europe and North America. Mycologia Helvetica 7: 99–168.

257 Réblová, M. & Štěpánek, V. 2009. New fungal genera, *Tectonidula* gen. nov. for Calosphaeria-like fungi with holoblastic-denticulate conidiogenesis and *Natantiella* gen. nov. for three species segregated from *Ceratostomella*. Mycological Research 113(9): 991–1002.

258 Réblová, M. 2000. The genus *Chaetosphaeria* and its anamorphs. Studies in Mycology 45: 149–168.

259 Réblová, M. 2006. Molecular systematics of *Ceratostomella sensu lato* and morphologically similar fungi. Mycologia 98(1): 68–93.

260 Réblová, M., Jaklitsch, W.M., Réblová, K. & Štěpánek, V. 2015. Phylogenetic reconstruction of the Calosphaeriales and Togniniales using five genes and predicted RNA secondary structures of ITS, and *Flabellascus tenuirostris* gen. et sp. nov.. PLOS ONE 10(12): E0144616.

261 Réblová, M., Réblová, K. & Štěpánek,V. 2015. Molecular systematics of *Barbatosphaeria* (Sordariomycetes): multigene phylogeny and secondary ITS structure. Persoonia 35: 21–38.

262 Redhead, S.A., Moncalvo, J.-M., Vilgalys, R. & Lutzoni, F. 2002. Phylogeny of agarics: partial systematics solutions for bryophilous omphalinoid agarics outside the Agaricales (euagarics). Mycotaxon 82: 151–168.

263 Redhead, S.A., Vizzini, A., Drehmel, D.C. & Contu, M. 2016. *Saproamanita*, a new name for both *Lepidella* E.-J. Gilbert and *Aspidella* E.-J. Gilbert (Amaniteae, Amanitaceae). IMA Fungus 7(1): 119–129.

264 Reid, D.A. 1965. A monograph of the stipitate stereoid fungi. Beihefte zur Nova Hedwigia 18: 1–382.

265 Reid, D.A. 1974. A monograph of the British Dacrymycetales. Transactions of the British Mycological Society 62(3): 433–494.

266 Richard, F., Bellanger J.-M., Clowez, P., Hansen, K., O'Donnell, K., Urban, A., Sauve, M., Courtecuisse, R. & Moreau, P.-A. 2014. True morels (*Morchella*, *Pezizales*) of Europe and North America: evolutionary relationship inferred from multilocus data and a unified taxonomy. Mycologia 107(2): 359–382.

267 Richter, T. & Baral, H.-O. 2008. *Coronellaria pulicaris*, *Mollisia luctuosa* und *Marasmius cornelii* - seltene Saprobionten an Cyperaceen. Boletus 31(1): 45–63.

268 Roberts, P. 1993. *Exidiopsis* species from Devon, including the new segregate genera *Ceratosebacina*, *Endoperplexa*, *Microsebacina*, and *Serendipita*. Mycological Research 97: 467–478.

269 Roberts, P. 1994. Globose and ellipsoid-spored *Tulasnella* species from Devon and Surrey, with a key to the genus in Europe. Mycological Research 98(12): 1431–1452.

270 Roberts, P. 1998. A revision of the genera *Heterochaetella*, *Myxarium*, *Protodontia*, and *Stypella* (Heterobasidiomycetes). Mycotaxon 69: 209–248.

271 Roberts, P. 1999. Rhizoctonia-forming fungi. RBG, Kew, 239 pp.

272 Robich, G. 2003. Mycena d'Europa. A.M.B. Fondazione Centro Studi Micologici, 728 pp.

273 Robich, G. 2016. Mycena d'Europa – volume 2. Associazione Micologica Bresadola; Trento, 1528 pp.

274 Rogerson, C.T. & Samuels, G.J. 1989. Boleticolous species of *Hypomyces*. Mycologia 81: 413–432.

275 Rogerson, C.T. & Samuels, G.J. 1994. Agaricolous species of *Hypomyces*. Mycologia 86: 839–866.

276 Romagnesi, H. 1985. Russules d'Europe et d'Afrique du Nord. J. Cramer, 1030 pp.

277 Roobeek, C.F. 2009. Aartongen in de duinen van Noord Kennemerland (2005 T/M 2008). RO-Rapo 09/10. Bergen NH, 45 pp.

278 Rossman, A.Y. 1983. The phragmosporous species of *Nectria* and related genera. Mycological Papers 150: 1–164.

279 Rossman, A.Y. 1987. The Tubeufiaceae and similar loculoascomycetes. Mycological Papers 157: 1–71.

280 Rossman, A.Y., Samuels, G.J., Rogerson, C.T. & Lowen R. 1999. Genera of Bionectriaceae, Hypocreaceae, and Nectriaceae (Hypocreales, Ascomycetes). Studies in Mycology 42: 1–248.

281 Rune, F. 2006. Kamskivlingar (*Amanita* sektion Vaginatae) i subarktis och skandinaviska högfjäll. Svensk Mykologisk Tidskrift 27(3): 49–65.

282 Ryvarden, L. & Gilbertson, R.L. 1993–1994. European polypores 1–2. Synopsis Fungorum 6.

283 Ryvarden, L. & Melo, I. 2014. Poroid fungi of Europe. Synopsis Fungorum 31. Fungiflora, Oslo, 455 pp.

284 Samuels, G.J. 1988. Fungicolous, lichenicolous, and myxomyceticolous species of *Hypocreopsis*, *Nectriopsis*, *Nectria*, *Peristomialis*, and *Trichonectria*. Memoirs of the New York Botanical Garden 48: 1–78.

285 Sánchez-García, M., Matheny, P.B., Palfner, G. & Lodge, D.J. 2014. Deconstructing the Tricholomataceae (Agaricales) and introduction of the new genera *Albomagister*, *Corneriella*, *Pogonoloma* and *Pseudotricholoma*. Taxon 63(5): 993–1007.

286 Sánchez, A. 1967. The sections *Apostemium* and *Microstemium* of the genus *Vibrissea* (Fungi). Journal of Agriculture of the University of Puerto Rico 51(1): 79–93.

287 Sarnari, M. 1998. Monografia illustrata del genere *Russula* in Europa. Tomo primo. A.M.B. Fondazione Centro Studi Micologici, 799 pp.

288 Sarnari, M. 2005. Monografia illustrata del genere *Russula* in Europa. Tomo secondo. A.M.B. Fondazione Centro Studi Micologici, 807–1564.

289 Scambler, R., Niskanen, T., Assyov, B., Ainsworth, A.M., Bellanger, J.-M., Loizides, M., Moreau, P.-A., Kirk, P.M. & Liimatainen, K. 2018. Diversity of *Chroogomphus* (Gomphidiaceae, Boletales) in Europe, and typification of *C. rutilus*. IMA Fungus 9(2): 271–290.

290 Schafer, D.J. 2014. The genus *Parasola* in Britain including *Parasola cuniculorum* sp. nov. Field Mycology 15(3): 77–99.

291 Schläpfer-Bernhard, E. 1969 (1968). Beitrag zur Kenntnis der Discomycetengattungen *Godronia*, *Ascocalyx*, *Neogodronia* and *Encoeliopsis*. Sydowia 22(1–4): 1–56.

292 Schumacher, T. 1990. The genus *Scutellinia* (Pyronemataceae). Opera Botanica 101: 1–107.

293 Seifert, K.A., Kendrick, B. & Murase, G. 1983. A key to hyphomycetes on dung. University of Waterloo biology series 27, 62 pp.

294 Seifert, K.A., Morgan-Jones, G., Gams, W. & Kendrick, B. 2011. The genera of Hyphomycetes. CBS Biodiversity Series 9. CBS-KNAW Fungal Biodiversity Centre, 997 pp.

295 Senn-Irlet, B. 1992. The genus *Crepidotus* in Europe. Persoonia 16(1): 1–80.

296 Senn-Irlet, B. 1995. Die Gattung *Simocybe* Karsten in Europa. Mycologia Helvetica 7(2): 27–61.

297 Sherwood, M.A. 1977. The ostropalean fungi. Mycotaxon 5: 1–277.

298 Sherwood, M.A. 1980. Taxonomic studies in the Phacidiales: The genus *Coccomyces* (Rhytismataceae). Occasional papers of the Farlow Herbarium of Cryptogamic Botany 15: 1–120.

299 Shoemaker, R.A. & Babcock, C.E. 1989. Phaeosphaeria. Canadian Journal of Botany 67: 1500–1599.

300 Sivanesan, A. 1997. The taxonomy and pathology of *Venturia* species. Bibliotheca Mycologica 59: 1–139.

301 Skrede, I., Carlsen, T. & Schumacher, T. 2017. A synopsis of the saddle fungi (*Helvella*: Ascomycota) in Europe – species delimitation, taxonomy and typification. Persoonia 39: 201–253.

302 Smith, C.W., Aptroot, A., Coppins, B.J., Fledger, A., Gilbert, O.L., James, P.V. & Wolseley, P.A. (eds.) 2009. Lichens of Great Britain and Ireland. British Lichen Society, 1046 pp.

303 Soehner, E. 1962. Die Gattung *Hymenogaster* Vitt. Eine monographische Studie mit besondere Berucksichtigung der bayerischen Arten. Beihefte zur Nova Hedwigia 2: 113 pp, 8 pl.

304 Sogonov, M.V., Castlebury, L.A., Rossman, A.Y., Mejía, L.C. & White, J.F. 2008. Leaf-inhabiting genera of the Gnomoniaceae, Diaporthales. Studies in Mycology 62(1): 1–77.

305 Soop, K. 2017. *Cortinarius* in Sweden. 181 pp.

306 Spirin, V., Malysheva, V. & Larsson, K-H. 2018. On some forgotten species of *Exidia* and *Myxarium* (Auriculariales, Basidiomycota). Nordic Journal of Botany 36(3): 11 pp.

307 Spirin, V., Malysheva, V., Yurkov, A.M., Miettinen, O. & Larsson, K.H. 2018. Studies in the *Phaeotremella foliacea* group (Tremellomycetes, Basidiomycota). Mycological Research 17(4): 451–466.

308 Spirin, V., Vlasák, J., Rivoire, B., Kotiranta, H. & Miettinen, O. 2016. Hidden diversity in the *Antrodia malicola* group (Polyporales, Basidiomycota). Mycological Progress 15: 51 (12 pp).

309 Spirin, V., Vlasák, J., Rivoire, B., Kout, J., Kotiranta, H. & Miettinen, O. 2016. Studies in the *Ceriporia purpurea* group (Polyporales, Basidiomycota), with notes on similar *Ceriporia* species. Cryptogamie, Mycologie 37(4): 421–435.

310 Spooner, B. 2001. The larger cup fungi in Britain – part 3. The genera *Peziza* and *Plicaria*. Field Mycology 2(2): 51–59.

311 Spooner, B.M. & Kemp, S.L. 2005. *Epichloë* in Britain. Mycologist 19(2): 82–87.

312 Spooner, B.M. & Kirk, P.M. 1990. Observations on some genera of Trichothyriaceae. Mycological Research 94(2): 223–230.

313 Stadler, M., Fournier, J., Granmo, A. & Beltrán-Tejera, E. 2008. The "red Hypoxylons" of the temperate and subtropical northern hemisphere. North American Fungi 3(7): 73–125.

314 Stadler, M., Læssøe, T., Fournier, J., Decock, C., Schmieschek, B., Tichy, H.-V. & Persoh, D. 2013. A polyphasic taxonomy of *Daldinia* (Xylariaceae). Studies in Mycology 77(1): 1–143.

315 Stangl, J. 1989. Die Gattung *Inocybe* in Bayern. Hoppea 46: 1–409.

316 Stenroos, S., Velmala, S., Pykälä, J. & Ahti, T. 2016. Lichens of Finland. Norrlinia 30: 1–896.

317 Stielow, B., Bratek, Z., Orczán, A.K.I., Rudnoy, S., Hensel, G., Hoffmann, P., Klenk, H.-P. & Göker, M. 2011. Species delimitation in taxonomically difficult fungi: The case of *Hymenogaster*. PLoS One 6(1): e15614.

318 Stielow, B., Bubner, B., Hensel, G., Münzenberger, B., Hoffmann, P., Klenk, H.-P. & Göker, M. 2010. The neglected hypogeous fungus *Hydnotrya bailii* Soehner (1959) is a widespread sister taxon of *Hydnotrya tulasnei* (Berk.) Berk. & Broome (1846). Mycological Progress 9(2): 195–203.

319 Suková, M. 2005. A revision of selected material of lignicolous species of *Brunnipila*, *Capitotricha*, *Dasyscyphella* and *Neodasyscypha* from the Czech Republic. Czech Mycology 57(1–2): 139–172.

320 Sunhede, S. 1989. Geastraceae (Basidiomycotina). Morphology, ecology, and systematics with special emphasis on the North European species. Synopsis Fungorum 1. Oslo, Fungiflora.

321 Sutton, B.C. 1980. The Coelomycetes. CABI Publ., 696 pp.

322 Szarkándi, J.G., Schmidt-Stohn, G., Dima, B., Hussain, S., Kocsubé, S., Papp, T., Vágvölgyi, C. & Nagy, L.G. 2017. The genus *Parasola*: phylogeny and the description of three new species. Mycologia 109(4): 620–629.

323 Saar, G., Dima, B., Schmidt-Stohn, G., Brandrud, T.-E., Bellu, F., Frøslev, T.G., Oertel, B. & Soop, K. 2014. *Cortinarius* Untergattung *Phlegmacium* Sektion *Purpurascentes* in Europa. Journal des J.E.C. 17(16): 140–161.

324 Saar, I. 2003. The genera *Cystoderma* and *Cystodermella* (Tricholomataceae) in temperate Eurasia. Mycotaxon 86: 455–473.

325 Termorshuizen, A.J. & Swertz, C.A. 2011. Roesten van Nederland/Dutch rust fungi. 423 pp.

326 Tibell, L. 1999. Calicioid lichens and fungi. In: Nordic Lichen Flora 1: 20–94.

327 Van de Putte, K., Nuytinck, J., De Crop, E. & Verbeken, A. 2015. *Lactifluus volemus* in Europe: Three species in one revealed by a multilocus genealogical approach, Bayesian species delimitation and morphology. Fungal Biology 120: 1–25.

328 Van Vooren, N. & Moreau, P.-A. 2009. Essai taxinomique sur le genre *Gyromitra* Fr. sensu lato (Pezizales). I-VII. Ascomycete.org 1.

329 Van Vooren, N., Lindemann, U. & Healy, R. 2017. Emendation of the genus *Tricharina* (Pezizales) based on phylogenetic, morphological and ecological data. Ascomycete.org 9(4).

330 Vánky, K. 1994. European smut fungi. Gustav Fischer Verlag, 570 pp.

331 Vellinga, E.C. 1986. The genus *Flammulaster* (Agaricales) in the Netherlands and adjacent regions. Persoonia 13(1): 1–26.

332 Vellinga, E.C. 2003. Type studies in Agaricaceae - the complex around *Chlorophyllum rachodes*. Mycotaxon 85: 259–270.

333 Vellinga, E.C., Blanchard, E.P., Kelly, S. & Contu, M. 2012. *Paxillus albidulus*, *P. ammoniavirescens*, and *P. validus* revisited. Mycotaxon 119: 351–359.

334 Verkley, G.J.M. 1999. A monograph of the genus *Pezicula* and its anamorphs. Studies in Mycology 44: 1–180.

335 Vila, J. & Caballero, F. 2007. Fungi non Delineati 38: *Entoloma* nuevos o interesantes de la Península Ibérica. Edizioni Candusso, 64 pp.

336 Vila, J. & Caballero, F. 2009. Fungi non Delineati 45: *Entoloma* nuevos o interessantes de la Península Ibérica (2). Edizioni Candusso, 100 pp.

337 Vila, J., Caballero, F., Carbó, J., Alvarado, P., Català, S., Higelmo, M.A. & Llimona, X. 2014. Preliminary morphologic and molecular study of the *Entoloma rusticoides* group (Agaricales - Basidiomycota). Revista Catalana de Micologia 35: 65–99.

338 Vila, J., Carbó, J. Caballero, F. Català, S., Llimona, X., Noordeloos, M.E. & Ribes, M.A. 2013. Fungi non Delineati 66: Studies on *Entoloma*. Edizioni Candusso, 150 pp.

339 Viner I., Spirin V., Zíbarová L., Larsson K.-H. 2018. Additions to the taxonomy of *Lagarobasidium* and *Xylodon* (Hymenochaetales, Basidiomycota). MycoKeys 41: 65–90.

340 Vizzini, A. & Ercole, E. 2012. *Paralepistopsis* gen. nov. and *Paralepista* (Basidiomycota, Agaricales). Mycotaxon 120: 253–267.

341 Vizzini, A., Angelini, C. & Ercole, E. 2014. Le sezioni *Velatae* e *Aporus* di *Agrocybe* sottogenere *Aporus*: rivalutazione del genere *Cyclocybe* Velen. ed un nuova specie. Bolletino dell'Associazione Micologica ed Ecologica Romana 92(2): 21–38.

342 Vizzini, A., Consiglio, G. Ercole, E. & Setti, L. 2016. *Pseudoporpoloma*, a new genus for *Agaricus pes-caprae* (Agaricales, Tricholomataceae). Phytotaxa 243(3): 271–280.

343 Vizzini, A., Consiglio, G., Setti, L. & Ercole, E. 2015. *Calycybella*, a new genus for *Rugosomyces pudicus* (Agaricales, Lyophyllaceae) and emendation of the genus *Gerhardtia*. IMA Fungus 6(1): 1–11.

344 Vizzini, A., Ercole, E. & Contu, M. 2012. A contribution to the ITS-LSU phylogeny of the genus *Leucopaxillus* (/tricholomatoid clade, Agaricales), with three new genera and notes on *Porpoloma*. Mycosphere 3(1): 79–90.

345 Vizzini, A., Para, R., Fontenla, R., Ghignone, S. & Ercole E. 2011. A preliminary ITS phylogeny of *Melanoleuca* (Agaricales), with special reference to European taxa. Mycotaxon 118: 361–381.

346 Vizzini, A., Picillo, B., Ercole, E., Voyron, S. & Contu, M. 2013. Detecting the variability of *Hydnum ovoideisporum* (Agaricomycetes,

Cantharellales) on the basis of Italian collections, and *H. magnorufescens* sp. nov. Mycosphere 4(1): 32–44.

347 Voglmayr, H. & Jaklitsch, W.M. 2011. Molecular data reveal high host specificity in the phylogenetically isolated genus *Massaria* (Ascomycota, Massariaceae). Fungal Diversity 46: 133–170.

348 Wang, P.M., Liu, X.B., Dai, Y.C., Horak, E., Steffen, K. & Yang, Z.L. 2018. Phylogeny and species delimitation of *Flammulina*: taxonomic status of winter mushroom in East Asia and a new European species identified using an integrated approach. Mycological Progress 17: 1013–1030.

349 Wang, X.H., Huhtinen, S. & Hansen, K. 2016. Multilocus phylogenetic and coalescent-based methods reveal dilemma in generic limits, cryptic species, and a prevalent intercontinental disjunct distribution in *Geopyxis* (Pyronemataceae s. l., Pezizomycetes). Mycologia 108(6): 1189–1215.

350 Wang, X.W., Lombard, L., Groenewald, J.Z., Li, J., Videira, S.I.R., Samson, R.A., Liu, X.Z. & Crous, P.W. 2016. Phylogenetic reassessment of the *Chaetomium globosum* species complex. Persoonia 36: 83–133.

351 Waterhouse, G.M. & Brady, B.L. 1982. Key to the species of *Entomophthora sensu lato*. Bulletin of the British Mycological Society 16(2): 113–143.

352 Weir, A. 1996. A preliminary host-parasite list of British Laboulbeniales (Fungi, Ascomycotina). The Entomologist 115(1): 50–58.

353 Wergen, B. 2017. Handbook of Ascomycota, Volume 1a/1b.

354 Wieschollek, D., Helleman, S., Baral, H.-O. & Richter, T. 2011. *Roseodiscus formosus* spec. nov. – ein bryophiler Pionier mit falschem Namen. Zeitschrift für Mykologie 77(2): 161–174.

355 Wirth, V. 1995. Die Flechten Baden-Württembergs 1–2. Verlag E. Ulmer, 1006 pp.

356 Wu, F., Chen, J.-.J.,Ji., X.-H., Vlasák, J. & Dai, Y.-C. 2017. Phylogeny and diversity of the morphologically similar polypore genera *Rigidoporus*, *Physisporinus*, *Oxyporus*, and *Leucophellinus*. Mycologia 109(5): 749–765.

357 Wu, H.X., Schoch, C.L., Boonmee, S., Bahkali, A.H., Chomnunti, P. & Hyde, K.H. 2011. A reappraisal of Microthyriaceae. Fungal Diversity 51(1): 189–248.

358 Yao, Y.-J. & Spooner, B.M. 1996. *Geopora seupulta* (Pezizales) in Britain, with a key to British species of the genus. Kew Bulletin 51(2): 381–383.

359 Yao, Y.-J. & Spooner, B.M. 1996. Notes on British species of *Cheilymenia*. Mycological Research 100(3): 361–367.

360 Yao, Y.-J. & Spooner, B.M. 1996. Notes on British species of *Pulvinula*, with two newly recorded species. Mycological Research 100(7): 883–884.

361 Yao, Y.-J. & Spooner, B.M. 1996. Notes on British species of *Scutellinia*. Mycological Research 100(7): 859–865.

362 Yao, Y.-J. & Spooner, B.M. 1996. Notes on British species of *Tympanis* (Leotiales) with *T. prunicola* new to Britain. Kew Bulletin 51(1): 187–191.

363 Yao, Y.-J. & Spooner, B.M. 1996. Notes on British species of *Octospora*. Mycological Research 100(2): 175–178.

364 Yao, Y.-J., Spooner, B.M. & Legon, N.W. 1998. An extraordinary species of *Anthracobia*, *A. subatra*, new to Britain with a key to British species of the genus. Mycologist 12(1): 32–34.

365 Zamora, J.C., Calonge, F.D. & Martín, M.P. 2015. Integrative taxonomy reveals an unexpected diversity in *Geastrum* section *Geastrum* (Geastrales, Basidiomycota). Persoonia 34: 130–165.

366 Zhao, C.-L., Cui, B.-K. & Steffen, K. 2013. Yuchengia, a new polypore genus segregated from *Perenniporia* (Polyporales) based on morphological and molecular evidence. Nordic Journal of Botany 31(3): 331–338.

367 Zhou, J.L., Zhu, L., Chen, H. & Cui, B.-K. 2016. Taxonomy and phylogeny of *Polyporus* group *Melanopus* (Polyporales, Basidiomycota) from China. PLoS ONE 11(8): e0159495.

368 Zogg, H. 1962. Die Hysteriaceae s. str. und Lophiaceae, unter besonderer Berücksichtigung der mitteleuropäischen Formen. Beiträge zur Kryptogamenflora der Schweiz 11: 1–190.

369 Zycha, H., Siepmann, R. & Linnemann, G. 1969. Mucorales. Verlag J. Cramer, 355 pp.

370 Örstadius, L., Ryberg, M. & Larsson, E. 2015. Molecular phylogenetics and taxonomy in Psathyrellaceae (Agaricales) with focus on psathyrelloid species: introduction of three new genera and 18 new species. Mycological Progress 14(5): 1–42.

Index

This index includes all the main genera and all species, varieties and forms included in the two volumes. For genera and species afforded full treatment, only the main entry that includes the description is indexed. Species that are only mentioned under another species are indexed wherever they occur.

B

D

E

F

G

H

J

K

P

T

U

Glossary

This glossary defines the key technical terms used throughout this publication; terms referred to in the definitions that are further defined are highlighted.

adnate gills: gills that are broadly attached to the stem, p. 29

adnexed gills: gills that are narrowly attached to the stem, p. 29

aecidium/aecidia: an asexual, discoid state in rust life-cycles, p. 1198

aecidiospore: rust **spores** (n+n) produced in **aecidia** as the result of **plasmogamy**, p. 1198

aethalium/aethalia: a sessile fruitbody of a *Mycetozoa* made by a massing of plasmodia, p. 1646

agglutinated: of hyphae that are held together by a sticky substance

allantoid (spores): curved, sausage-shaped, p. 32

amygdaliform: shaped like an almond

amyloid (I+): staining usually blue (mostly **spores** and **asci**) in iodine reagents (**Melzer's** or **Lugol's**), p. 33. I+ for **asci** refers to the reaction of the **apical apparatus** if not otherwise indicated

anastomosing: *e.g.* of gills that join to form a net-like pattern or branching **striae** on **spores**, p. 421

annulate: of stems with a ring (annulus), p. 27

antheridium: sex organ from where nuclei are transferred to the **ascogonium**, p. 11

apical apparatus: a structure within the top of an **ascus** that controls the ejection mechanism, p. 34

apiculus: small protuberance on **basidiospores** where attached to **sterigma**, p. 32

apothecium/apothecia: fruitbody of Ascomycota with exposed hymenium at maturity, p. 30

applanate: of fruitbodies or stromata that are flattened

arcuate: of decurrent gills curved upwards like a bow

ascogonium: sex organ that receive nuclei, p. 11

ascospore: spores formed within **asci**, pp. 11 & 34

ascus (pl. asci): the ± sac-like cell containing the **ascospores** in Ascomycota, pp. 11 & 34

asterosetae: branched, thick-walled, pointed setae

basidium (pl. basidia): the cell upon which the **basidiospores** are borne in Basidiomycota, p. 11 & 34

basidioles: sterile, basidium-like cells amongst **basidia**

basidiospore: **spore** formed on a **basidium**, pp. 11 & 34

binding hypha: branched flesh-**hypha** without **septa**, p. 35

biotrophy: nutrients obtained from living host or partner cells, p. 16

bifurcate(d): dividing in two, *e.g.* of gills

bitunicate: of asci with two functional walls, p. 1602

brown rot: wood rot that contains lignin – brown with transverse break points, p. 14

calciphilous: preferring calcareous habitats (soils with high pH)

calciphobic: avoiding calcareous habitats

capillitium: sterile, ± elastic cells in the **gleba** of some gasteroid fungi (p. 1228) and some *Mycetozoa* (p. 1646)

capitate: with a distinct, typically rounded head (*e.g.* of cystidia, p. 36)

caulocystidia: **cystidia** on the stem of basidiomycotes, p. 36

cellular: tissue of ± rounded cells that may crack in a random way (*e.g.* cap **cuticle**), p. 26

chasmothecia: ± round, closed fruitbodies that later break open in a controlled way and have active **ascospore** dispersal (found in the Erysiphales, p. 1500)

cheilocystidia: **cystidia** on gill edges. p. 36

chrysocystidia: **cystidia** with prominent amorphous yellow inclusions, p. 36

citriform: lemon-shaped, p. 32

clamp: short for clamp connection; a special structure that secures the **n+n state** in **hyphae** of the Basidiomycota; also at the base of many **basidia**, p. 35

clavate: club-shaped

chlamydospore: a thick-walled, asexual resting spore

cleistothecium (pl. cleistothecia): an ascomycote fruitbody that is closed at maturity and has passive **ascospore** dispersal, p. 30

coelomycete: fungal form group with production of **conidia** in an asexual structure that resembles a **perithecium**

collarium: a ring around the stem to which the gills are attached, p. 28

columella: a sterile protuberance within the **spore**-bearing tissue *e.g.* in truffles and slime moulds (*Mycetozoa*)

congo red: a red stain used to stain hyphal walls, p. 39

conidium (pl. conidia): asexual propagule, especially in *Ascomycota*

conidiogenous cell: cell from which **conidia** are produced

conidiophore: cell that produces **conidia**

coralloid: coral-shaped (highly branched)

cotton blue: a blue stain, used both for **spore ornaments** and contents of **spores** and **basidia**, p. 39

crenulate(d): cap margin that is distinctly scalloped or notched

crozier: the **clamp**-like base of many **asci**, p. 35

cuticle: outer surface layer on fruitbodies

cyanophilous: selective uptake of **cotton blue**, p. 33

cystidium (pl. cystidia): sterile, well-differentiated surface cell in *Basidiomycota*, p. 36

decurrent gills: gills continuing down the stem, p. 29

dehiscent: falling into parts, *e.g.* **spores** (including dehiscence of an outer fruitbody layer)

dendrohyphidia: thin-walled, tree-like, hyphal structures in basidiomycote **hymenia**, p. 36

dentate: with pronounced teeth (*e.g.* of **hymenophores**)

dextrinoid (I+ dex.): staining reddish-brown in **Melzer's** reagent, p. 33

dimitic: a hyphal system with generative hyphae and either skeletal or binding **hyphae**, p. 35

dimorph(ic): with two morphological expressions, *e.g.* of **spores**

distant: of gills that are well spaced, p. 28

duplex structure: a two-layered cap surface with a loose upper and a dense lower part, often separated with a darker line

echinulate: with spines, *e.g.* on **spore** surfaces

ectomycorrhiza: a **biotrophic** mutualistic relationship between a fungus and a plant, p. 17

ellipsoid: a 3-dimensional shape resembling an ellipse in optical section

emarginate gills: gills notched where they meet the stem, also termed sinuate notched gills, p. 29

ephemeral: soon disappearing, *e.g.* of fruitbodies, **veils** or **peridia**

erumpent: developed below the surface and breaking through at maturity

evanescent: soon disappearing, of **veils**

f. (= forma): a taxonomic level below **variety** – often without genetic justification.

f. sp. (= forma specialis): informal taxonomic grouping applied to a parasite adapted to a specific host

$FeSO_4$: iron sulphate; produces stains on the flesh of certain basidiomycotes, p. 39

fibrillose: of a surface with fine fibres, p. 26

floccose/flocculose: of a surface with **floccules**

floccule: very fine downy surface structures

form group: a group of fungi with similar morphological characters (but not necessarily related phylogenetically)

free gills: gills that never reach the stem, p. 29

fusiform: spindle-shaped, p. 32

generative hyphae: the **hyphae** from which **basidia** are produced, p. 35

germ pore/slit: thin, pore-shaped area or slit on a **spore** from which germination occurs, p. 32

gleba: the fertile tissue in puffballs (p. 1228), truffles (p. 1256) and the like

globose: ± perfect round shape in all dimensions

gloeocystidia: **cystidia** with oily contents, typically staining in sulphur aniline, p. 36

guttulate: with internal drops (guttules) (in hyphae, **spores**, *etc.*)

haustoria: ± branched **hyphae** within living host tissue, used for nutrient uptake, p. 16

hemiamyloid: staining ± red with iodine reagents (**Melzer's** and **Lugol's**) on the **apical apparatus** or **ascus** walls in some discomycetes and lichens, p. 39

hyaline: transparent, seemingly colourless

hygrophanous: a reversible colour change depending on humidity, often on caps that dry pale from the centre outwards

hygroscopical: fruitbodies that open in wet conditions and close in dry conditions

hyphomycete: a mould-like asexual state mostly found in the ascomycotes

hymenium (pl. hymenia): a layer of fertile cells (**basidia** or **asci**) in Basidiomycota and Ascomycota, p. 31

hymenophore: structure holding **hymenia**, *e.g.* gills, tubes or spines in basidiomycotes, p. 27

hypha (pl. hyphae): ± branching filaments that make up the **mycelium** of a fungus, p. 10

hyphal cord: a cord of many **hyphae**; if with a dark, outer rind also termed **rhizomorph** (*e.g.* in Armillaria, p. 22)

hyphal mat: a sterile or **conidia**-producing layer of **hyphae** that may surround fruitbodies (mainly Ascomycota); see also **subiculum**

hyphidium (pl. hyphidia): specialized, often branched, **hypha** or hyphal end cells forming a continuous layer at the surface of *e.g.* some jelly fungi

hyphomycete: a mould-like asexual state, especially within ascomycotes, p. 1626

I- (inamyloid): not staining in iodine reagents (**Melzer's** and **Lugol's**), p. 39

I+ (amyloid): staining, mostly blue (**amyloid**) or brown (**dextrinoid**), in iodine reagents (**Melzer's** and **Lugol's**), p. 39; in connection with **asci**, I+ normally refers to the reaction of the apical apparatus

inamyloid (I-): not staining in iodine reagents (**Melzer's** and **Lugol's**), p. 39

incrusted hyphae: an external deposition of pigment or hyaline matter on **hyphal** walls

isidium (pl. isidia): peg-like protuberances on lichen thalli that may break off and function as propagules

karyogamy: fusion of nuclei, p 11

KOH: potassium hydroxide; used to provoke staining or to revive dried material, p. 39

lageniform: flask-shaped, typically used in relation to **cystidia**, p. 36

lecythiform: shaped like a bowling pin, p. 36

Lugol's reagent: iodine stain, less toxic than **Melzer's reagent** and essential in the study of Ascomycota (see also **hemiamyloid**), p. 39

lubricous: surface as that of cold butter

meiospore: sexual spore formed after a meiosis, p. 11

Melzer's reagent: a toxic iodine stain containing chloral hydrate; much used in the study of Basidiomycota, p. 39

mitospore: a spore formed after mitosis (p.11), p. 32

moniliform: a **hypha** or **cystidium** with many transverse constrictions

monomitic: fruitbodies with only generative **hyphae** in the flesh, p. 35

mor: humus formed under acidic conditions – soil hard, often moss-covered

mull: humus formed under non-acidic conditions – soil soft, without mosses

muriform: **spores** with transverse and lengthwise divisions (**septa**), *e.g.* pp. 1602–1603

mycelium (pl. mycelia): the vegetative mass of hyphae constituting an individual fungus, p. 10

n+n state: the state following plasmogamy (fusion of cytoplasm of two parent cells) without immediate **karyogamy**, p. 11

nom. prov.: a scientific name yet to be officially validated (provisional)

ostiole: the opening of a **perithecium**, p. 30

p.p. (pro parte): part of, not in full, *e.g.* some members of a genus

pallescent: becoming paler with age

papilla: small, mostly rounded protuberance, *e.g.* in the centre of a cap

paraphysis (pl. paraphyses): sterile cells amongst **asci**; also used for cells surrounding rust **aecidia**, p. 31

pedicel: a **sterigma**-remnant attached to *e.g.* puffball **spores**, *e.g.* p. 1228

peridium (pl. peridia): the outer skin-like layer in puffballs, *etc.*

perispore: an outer, typically **hyaline** spore wall that often loosens from the inner, pigmented wall

perithecium (pl. perithecia): a ± flask-shaped, fertile structure with a narrow opening (within the Ascomycota, p. 30)

phialide: a tiny, flask-shaped cell that produces **conidia** from within – often in chains

photobiont: the photosynthesizing partner(s) in a lichen thallus

pileocystidium: **cystidium** on the cap of an agaric or bolete

pseudocapillitium: usually interpreted as remnants of the common walls of the individual **sporocarps** constituting the spore-bearing structure (within *Mycetozoa*), p. 1646

plasmodiocarp: elongated, ± sausage-shaped or net-like fruitbodies that are formed along the plasmodial veins in *Mycetozoa*, p. 1646

plasmodium (pl. plasmodia): a mobile, amoeboid stage in slime moulds containing numerous nuclei, p. 1646

plasmogamy: fusion of cells, p. 11

pleural basidia: basidia attached on the side - formed laterally from any hyphal cell

pleurocystidium: **cystidium** on the face of gills or tubes

primordium (pl. primordia): the earliest stage of fruitbody development

pruinose: surface with a slightly frosty appearance

pseudoaethalia: closely appressed, individual sporocarps (*Mycetozoa*), ± merged together with the common wall partly or completely dissolved, p. 1646

pseudothecium: a **perithecium**-like structure in ascomycotes with two functional layers in the **ascus**, p. 1602

pycnidium: a tiny, **perithecium**-like structure producing minute **spermatia**, p. 1198

Q (Q value): length:width ratio (often referring to **spores**)

Qav.: average of Q values (if range, relates to the averages from different collections)

rectocutis: of cap cuticle of radiating hyphae

reflexed (fruitbody): fruitbodies with a **resupinate** and a protruding part, p. 962

refractive vacuolar bodies (VBs): specialized cell vacuoles with a content that makes them refractive, especially useful for taxonomy in *Helotiales* and other cup fungi

repeating (spores): **basidiospores** that can duplicate themselves, p. 32

resupinate: a flattened fruitbody, appressed to substrate

reticulum: structure forming a net-like pattern

rugose: an uneven, roughened surface

rhizomorph: a hyphal cord (of many **hyphae**) with an outer protective, often black, rind

s.l. (sensu lato): scientific name applied in a broad sense

s.str. (sensu stricto): scientific name applied in a strict or narrow sense

saprotrophic fungus: a fungus that only degrades dead organic matter, p. 14

sclerotium/sclerotia: tough, dense, resting structures, normally with hardened rind

SEM: scanning electron microscopy

semiglobose: having the shape of half a sphere

septum (pl. septa): transverse walls of **hyphae** or transverse or lengthwise walls of **spores**

serrulate: like the edge of a saw, p. 28

seta/setae: pointed, thick-walled, sterile structures on or in fruitbodies, p. 36

sigmoid (shape): ± 'S'-shaped (when referring to **spores** and **germ slits**), p. 32

sinuous: having curves and turns, of hairs, ornaments, *etc.*

skeletal hyphae: non-**septate** (no transverse walls), thick-walled, rigid, unbranched flesh-**hyphae**, p. 35

soralium (pl. soralia): a ± cup-shaped structure with formation of soredia in lichenized fungi, p. 1484

soredium (pl. soredia): asexual propagules formed on lichens, sometimes within **soralia**, with fungal **hyphae** enveloping algal cells, p. 1484

spermatium (pl. spermatia): **mitospores** with the function of performing **plasmogamy** with another **spore** or a **hypha**, p. 11 & 1198

spermogonium (pl. spermogonia): flask-shaped structure producing **spermatia**, p. 1198

sporangiospore: spore formed in a **sporangium**

sporangium (pl. sporangia): an enclosed cell containing asexual **spores** (*e.g.* in Mucorales), pp. 1626–1627

spore: used (mainly) for sexual (**meiospores**) in basidiomycotes (**basidiospores**) and ascomycotes (**ascospores**). Asexual spores are *e.g.* termed **conidia** and **sporangiospores**

spore measurements: mostly given as two dimensions, length × width, or as one in globose spores; compressed spores are given with three measurements (length, and from side and front)

sporocarp: the type of spore-bearing structure in *Mycetozoa*, p. 1646

sporodochium (pl. sporodochia): a cushion-shaped structure with production of **conidia** from the surface

squamule(s): small scale(s)

squamulose: with small scales

ss. auct. (sensu auctores): according to some authors (but not in the original sense)

sterigma (pl. sterigmata): outgrowths from **basidia** from where the **spores** are produced, p. 34

striate: of lines on agaric caps caused by underlying gills and the space between them

stroma (pl. stromata): dense tissue formed from haploid cells mostly in ascomycote groups; most stromata contain several or many **perithecia**, p. 30

strigose: with stiff, projecting hair bundles (often used for stem bases in agarics)

stromatic (tissue): dense tissue, typically with ± immersed/**erumpent** fruitbodies (often **perithecia**), p. 30

subglobose: nearly **globose**

subiculum: forming a **hyphal mat** below one or more fruitbodies

sulcate: deeply radially grooved (of agaric caps)

sulfo-vanillin: a stain made from sulfuric acid (H_2SO_4) and vanillin (vanilla)

synnema (pl. synnemata): a ± erect fusion of **conidiophores**, with **conidia** production at the tips or on the sides

teliospore: rust spore produced from **telia** – often 2–several-celled within which karyogamy (fusion of nuclei) takes place, p. 1198

telium (pl. telia): structures (sori) where **teliospores** are formed, p. 1198

thallus (pl. thalli): mainly used for lichens where the entire structure containing the two or three partners is termed a thallus; it may produce fruitbodies, pp. 18–19

thermophilic: preferring warm (often south-facing) habitats

tomentose: short and densely hairy

trichoderm: cap surface with raised **hyphal** ends (agarics)

trimitic: fruitbodies with three **hyphal** types: generative, binding and skeletal, p. 35

trichogyne: outgrowth from the **ascogonium** designed to receive a nucleus from a compatible partner, p. 11

umbilicate: with a navel-like cap centre (agarics) or simple **ostioles** (pyrenomycetous fungi)

umbo: rounded protuberance at the cap centre (agarics), p. 28

uredinium (pl. uredinia): structures where **uredospores** are formed, often non-**septate**, p. 1198

uredospores: rust **spores** with two nuclei (**n+n**) produced from **uredinia** – also called summer spores, p. 1198

utriform: a fatter version of **lageniform** (flask-shaped), p. 36

var. (= variety): a taxonomic level below species and subspecies but above **forma**

veil/velum: tissue covering parts of the young fruitbody, p. 26

VBs: see **refractive vacuolar bodies**

verrucose: with a warty surface (**spores**)

volva: sac-like remnant of the universal veil at the base of the stem, p. 27

white rot: wood rot where both lignin and cellulose have been degraded – white and fibrous, p. 14

Photographers

The photographs included in this publication were taken by Jens H. Petersen, except for 1,300 images kindly provided by the fellow fungal photographers listed below. We are very grateful to all those who allowed their images to be featured.

Fungi of Temperate Europe

Fungi of Temperate Europe is translated from *Nordeuropas svampe* by Thomas Læssøe and Jens H. Petersen.

Published by Princeton University Press
41 William Street, Princeton, New Jersey 08540
99 Banbury Road, Oxford OX2 6JX

press.princeton.edu

ISBN (two-volume set): 978-0-691-18037-3.

Project management, photo work, composition and graphic design: Jens H. Petersen.
Prepress: Jens H. Petersen and Narayana Press, Gylling, Denmark.
Editor: Axel Kielland.
Print: Grafotisak, GPS Group.

Printed and bound in Bosnia & Herzegovina.

FSC www.fsc.org MIX Supporting responsible forestry FSC® C110418

10 9 8 7 6 5

Colour coding used within the identification wheels

- other similar fungi
- the group contains fungi living in mutualistic symbiosis (*e.g.* lichens and ectomycorrhizal fungi, p. 17)
- the group contains parasitic fungi (fungi that weaken or kill their host, p. 16)
- the group contains brown rotters, p 14 (white rotters are shown on a neutral background)

Edibility symbols – only important edible and seriously or deadly poisonous species are marked.

- ● acceptable edible after proper cooking
- ●● high quality edible after proper cooking
- †? presumed seriously poisonous
- † seriously poisonous
- †† deadly poisonous

References to other species

△ = found above on the same page
▽ = found below on the same page
◁ & ▷ = found on the opposite page
◁◁ = covered earlier in the chapter
▷▷ = covered later in the chapter
⋉ = not treated further